Solid Mechanics and Its Applications

Founding Editor

G. M. L. Gladwell, University of Waterloo, Waterloo, ON, Canada

Volume 272

Series Editors

J. R. Barber, Department of Mechanical Engineering, University of Michigan, Ann Arbor, MI, USA

Anders Klarbring, Mechanical Engineering, Linköping University, Linköping, Sweden

The fundamental questions arising in mechanics are: Why?, How?, and How much? The aim of this series is to provide lucid accounts written by authoritative researchers giving vision and insight in answering these questions on the subject of mechanics as it relates to solids. The scope of the series covers the entire spectrum of solid mechanics. Thus it includes the foundation of mechanics; variational formulations; computational mechanics; statics, kinematics and dynamics of rigid and elastic bodies; vibrations of solids and structures; dynamical systems and chaos; the theories of elasticity, plasticity and viscoelasticity; composite materials; rods, beams, shells and membranes; structural control and stability; soils, rocks and geomechanics; fracture; tribology; experimental mechanics; biomechanics and machine design. The median level of presentation is the first year graduate student. Some texts are monographs defining the current state of the field; others are accessible to final year undergraduates; but essentially the emphasis is on readability and clarity.

Springer and Professors Barber and Klarbring welcome book ideas from authors. Potential authors who wish to submit a book proposal should contact Dr. Mayra Castro, Senior Editor, Springer Heidelberg, Germany, email: mayra.castro@springer.com

Indexed by SCOPUS, Ei Compendex, EBSCO Discovery Service, OCLC, ProQuest Summon, Google Scholar and SpringerLink.

More information about this series at https://link.springer.com/bookseries/6557

Paul Steinmann

Spatial and Material Forces in Nonlinear Continuum Mechanics

A Dissipation-Consistent Approach

Springer

Paul Steinmann
Applied Mechanics
Friedrich-Alexander-Universität
Erlangen-Nürnberg
Erlangen, Germany

ISSN 0925-0042 ISSN 2214-7764 (electronic)
Solid Mechanics and Its Applications
ISBN 978-3-030-89072-8 ISBN 978-3-030-89070-4 (eBook)
https://doi.org/10.1007/978-3-030-89070-4

This Springer imprint is published by the registered company Springer Nature Switzerland AG
The registered company address is: Gewerbestrasse 11, 6330 Cham, Switzerland

Preface

8,027 m 28°21'08"N 85°46'47"E

This treatise represents my occupation with material forces and configurational mechanics spanning over the past two and a half decades. My first encounter with the topic, and one of its most dominant protagonists, G. A. Maugin, was in summer 1995 at a conference in Victoria, Canada. In his, as I learned to appreciate over the following years, typical scholarly manner he outlined, in a fascinating presentation, the state of affairs regarding material forces. At the time, for me this was an impressive, however yet still entirely obscure,[1] novel theoretical concept in non-linear continuum mechanics.

In the subsequent years, I delved into the intricacies of the field and attempted to find my own way through the apparent multitude of different approaches and opinions. Soon I was obsessed with the idea that the balance of material (or rather pseudo) momentum is not only a theoretical concept but also has far-reaching consequences for computational mechanics. My straightforward and somewhat naive rationale at the time was that if there is a balance in strong form, there must also be a corresponding weak form that could be discretized by finite elements; what later became the *Material Force Method* was born. However, would this approach work and render sensible computational results, especially in the realm of computational fracture and defect mechanics? These questions took some time to be explored carefully. It was thus not before late summer 1999 that the first modest but original attempt toward computational fracture mechanics based on the concept of material forces (which in hindsight turned out to be intimately related to an ingenious configurational force-driven approach toward optimization of finite element discretizations by Braun [1]) could be presented at the ECCOMAS ECCM'99 conference (Ackermann et al. [2]). The corresponding full-length two-part journal

[1]Paraphrasing a quotation from L. N. M. Carnot ... *une notion métaphysique et obscure qui est celle de force* that authorities in the field like to cite (Maugin [8]).

publication followed shortly afterwards (Steinmann et al. [3, 4]) and (as I wish to believe when surrendering to my own vanity) initiated the subsequent disruptive development of what now might be called computational material forces (see the detailed account in the Introduction to this treatise).

With an invitation for a presentation on our developments to the group of G. A. Maugin at Université Pierre et Marie Curie, Paris, in 2000, an opportunity that made me feel both honored and intimidated, a long-lasting scientific exchange on various aspects of configurational mechanics emerged. As a result, in 2003 we jointly organized the EUROMECH Colloquium 445 on the 'Mechanics of Material Forces' (Steinmann and Maugin [5]) in Kaiserslautern, an occasion that brought together some of the leading authorities of the time in the field of material forces and configurational mechanics, e.g. M. Epstein, D. Gross, M. Gurtin, G. Herrmann, R. Kienzler, P. Podio-Guidugli, M. Šilhavý, C. Trimarco, as well as some of its rising stars like R. Müller, to name only a few of the participants. G. A. Maugin would visit our group in Kaiserslautern frequently to present and share his latest insights, or to act as a reviewer and examiner, e.g. of the Habilitation of A. Menzel. I remember vividly that he always made the journey from Paris by car accompanied by his wife, also to various conferences and meetings all around Europe. In 2008, a IUTAM symposium in Erlangen shed light on the impressive achievements thus far, as summarized in the 'Progress in the Theory and Numerics of Configurational Mechanics' (Steinmann [6]); likewise, for example, a special issue on 'Configurational Mechanics' (Kienzler and Steinmann [7]) documented the expanding interest in the topic.

In the sequel, inspired by the encounter with all these great researchers and different approaches, after the first decade of the twenty-first century I felt the strong urge to consolidate my own vistas on material forces and configurational mechanics in an extended treatise. Of course, at the time I totally overestimated my energy, available time and, much to my embarrassment in hindsight, degree of maturity; as a consequence, this endeavor was my constant companion over the past decade, thereby oftentimes only lurking beneath the surface of daily routine and the necessity to attend to other, seemingly more pressing issues. Fortunately, in the second half of 2019, I had the privilege to spend a 6-month sabbatical at the lab of D. Reddy (CERECAM), University of Cape Town, South Africa, which provided the much welcomed opportunity to review my previous drafts, re-think certain concepts and ideas, expand on various aspects, and finally start combining it all. The present treatise is the result of my own long journey in the world of configurational mechanics; I dare hope that it will be of use to also others and initiate new and exciting adventures.

Erlangen, Germany Paul Steinmann
December 2021

Acknowledgements I am deeply grateful to my family who had do endure my obsessive occupation with writing this book during the past years!

References

1. Braun M (1997) Configurational forces induced by finite-element discretization. Proceedings of the Estonian Academy of Sciences, Physics and Mathematics 46:24–31
2. D. Ackermann, Barth F.J., and Steinmann P. Theoretical and computational aspects of geometrically nonlinear problems in fracture mechanics. *Proceedings (CD-ROM) of the European Conference on Computational Mechanics ECCM'99 (ECCOMAS), August 31 to September 3, Munich, Germany*, 1999
3. P. Steinmann. Application of material forces to hyperelastostatic fracture mechanics. I. Continuum mechanical setting. *International Journal of Solids and Structures*, 37:7371–7391, 2000
4. P. Steinmann, D. Ackermann, and F.J. Barth. Application of material forces to hyperelastostatic fracture mechanics. II. Computational setting. *International Journal of Solids and Structures*, 38:5509–5526, 2001
5. P. Steinmann and G.A. Maugin. Mechanics of material forces: Proceedings of the EUROMECH Colloquium held in Kaiserslautern, Germany, May 10–24, 2003, volume 11 of Advances in Mechanics and Mathematics. Springer, 2005
6. P. Steinmann. IUTAM Symposium on Progress in the Theory and Numerics of Configurational Mechanics: Proceedings of the IUTAM Symposium held in Erlangen, Germany, October 20–24, 2008, volume 17 of IUTAM Bookseries. Springer, 2009
7. R. Kienzler and P. Steinmann. Special issue: Configurational forces. *ZAMM: Zeitschrift für Angewandte Mathematik und Mechanik*, 89:609–708, 2009
8. Maugin GA (1993) Material Inhomogeneities in Elasticity. Chapman and Hall, London

Contents

Notation and Terminology

Miscellaneous:

P, B	Physical Point, Physical Body
$\mathbb{E}^3$	Three-Dimensional Euclidean Space
$\boldsymbol{E}_A, \boldsymbol{E}^A, \boldsymbol{e}_a, \boldsymbol{e}^a$	Cartesian Basis
$A, B, a, b = 1, 2, 3$	Indices in Three Dimensions
$\mathcal{M}, T\mathcal{M}, T^*\mathcal{M}$	Generic Manifold, Tangent, Cotangent Space
$\cdots, \{\cdots\}, \#\{\cdots\}$	Elements of Set, Set, Cardinality of Set
$\mathbb{R}$	Real Number Set
$\mathbb{R}^+$	Positive Real Number Set
$\mathbb{R}_0^+$	{0} + Positive Real Number Set
$I, J = 1, 2, 3$	Indices in Three Dimensions
$\alpha, \beta = 1, 2$	Indices in Two Dimensions
Θ^I	Volume (Curvilinear) Coordinates
Θ^α	Surface (Curvilinear) Coordinates
Θ	Curve (Arc-Length) Coordinate
t	Time Coordinate
$\boldsymbol{id}$	Identity Map
$\circ$	Composition of Maps
$\perp, \parallel$	Normal, Tangential
$\{\bullet\}$	Generic Vectorial/Tensorial Quantity
$\{\bullet\} \otimes \{\bullet\}$	Dyadic Product
$\{\bullet\} \cdot \{\bullet\}$	Single Contraction
$\{\bullet\} : \{\bullet\}$	Double Contraction

Kinematics in Bulk Volumes:

$\boldsymbol{\chi}, \boldsymbol{\Xi}$	Reference Placement, Position Vector
$\mathcal{B}_\mathrm{r}, \mathcal{V}_\mathrm{r}$	Reference Configuration, Manifold
$\partial\mathcal{B}_\mathrm{r}, \partial\mathcal{V}_\mathrm{r}$	Reference Boundary Surface
$\boldsymbol{\Upsilon}, \boldsymbol{\upsilon}, \boldsymbol{X}, \boldsymbol{x}$	Material/Spatial Placement, Position Vector
$\mathcal{B}_\mathrm{m}, \mathcal{V}_\mathrm{m}, \mathcal{B}_\mathrm{s}, \mathcal{V}_\mathrm{s}$	Material/Spatial Configuration, Manifold
$\partial\mathcal{B}_\mathrm{m}, \partial\mathcal{V}_\mathrm{m}, \partial\mathcal{B}_\mathrm{s}, \partial\mathcal{V}_\mathrm{s}$	Material/Spatial Boundary Surface
$\mathrm{d}V,\ \mathrm{d}A,\ \mathrm{d}v,\ \mathrm{d}a$	Material/Spatial Volume, Area Element
$\boldsymbol{N},\ \mathrm{d}\boldsymbol{A},\ \boldsymbol{n},\ \mathrm{d}\boldsymbol{a}$	Material/Spatial Surface Normal, Area Element
$\boldsymbol{G}_I, \boldsymbol{G}^I, \boldsymbol{g}_i, \boldsymbol{g}^i$	Material/Spatial Co-, Contravariant Basis
$\boldsymbol{I}, \boldsymbol{i}$	Material/Spatial Unit Tensors
$\boldsymbol{E}, \boldsymbol{e}$	Material/Spatial Permutation Tensors
$\nabla_X\{\bullet\},\ \nabla_x\{\bullet\}$	Material/Spatial Gradient Operator
$\mathrm{Div}\{\bullet\}, \mathrm{div}\{\bullet\}$	Material/Spatial Divergence Operator
$\mathrm{Curl}\{\bullet\}, \mathrm{curl}\{\bullet\}$	Material/Spatial Curl Operator
$\mathrm{cof}\{\bullet\}, \det\{\bullet\}$	Cofactor, Determinant
$\{\bullet\} \boxtimes \{\bullet\}$	Special Dyadic Product
$\{\bullet\} \times\!\!\!\times \{\bullet\}$	Tensor-Valued Tensor Double Product
$[\{\bullet\} \times\!\!\!\times \{\bullet\}] : \{\bullet\}$	Scalar-Valued Tensor Triple Product
$\{\bullet\} \times \{\bullet\}$	Vector-Valued Vector Double Product
$[\{\bullet\} \times \{\bullet\}] \cdot \{\bullet\}$	Scalar-Valued Vector Triple Product
$\boldsymbol{y}, \boldsymbol{Y}$	Spatial/Material Deformation Map
$\boldsymbol{F}, \boldsymbol{f}$	Tangent Map: Deformation Gradients
$\boldsymbol{K}, \boldsymbol{k}$	Cotangent Map: Cofactor of Tangent Maps
J, j	Measure Map: Determinant of Tangent Maps
$\lambda_{a=1,2,3}, \Lambda_{a=1,2,3}$	Spatial/Material Principal Stretches
$\dot{\{\bullet\}},\ \mathrm{D}_t\{\bullet\},\ \mathrm{d}_t\{\bullet\}$	Total/Material/Spatial Time Derivative
$\boldsymbol{W}, \boldsymbol{V}, \boldsymbol{w}, \boldsymbol{v}$	(Total) Material/Spatial Velocity
$\mathbb{P}, \mathbb{p}$	Material/Spatial Surface Projection Operator
$\mathrm{d}\boldsymbol{X},\ \mathrm{d}\boldsymbol{x}$	Material/Spatial Line Element
$\mathbf{d}S,\ \mathbf{d}s$	Material/Spatial Level Set Normal
$\boldsymbol{C}, \boldsymbol{c}$	Right Material/Spatial Cauchy-Green Strain
$\boldsymbol{B}, \boldsymbol{b}$	Left Material/Spatial Cauchy-Green Strain
$\boldsymbol{L}, \boldsymbol{l}$	Material/Spatial Velocity Gradients
$\boldsymbol{\lambda}, \boldsymbol{\Lambda}$	Material/Spatial ‘Velocity Gradients’

Kinematics on Dimensionally Reduced Smooth Manifolds (Surfaces):

$\widehat{\boldsymbol{\chi}}, \widehat{\boldsymbol{\Xi}}$	Reference Surface Placement, Position Vector
$\partial\mathcal{B}_{\mathrm{r}}, \mathcal{S}_{\mathrm{r}}$	Reference Surface Configuration, Manifold
$\partial^2\mathcal{B}_{\mathrm{r}}, \partial\mathcal{S}_{\mathrm{r}}$	Reference Boundary Curve
$\widehat{\boldsymbol{\Upsilon}}, \widehat{\boldsymbol{v}}, \widehat{\boldsymbol{X}}, \widehat{\boldsymbol{x}}$	Material/Spatial Surface Placement, Position Vector
$\partial\mathcal{B}_{\mathrm{m}}, \mathcal{S}_{\mathrm{m}}, \partial\mathcal{B}_{\mathrm{s}}, \mathcal{S}_{\mathrm{s}}$	Material/Spatial Surface Configuration, Manifold
$\partial^2\mathcal{B}_{\mathrm{m}}, \partial\mathcal{S}_{\mathrm{m}}, \partial^2\mathcal{B}_{\mathrm{s}}, \partial\mathcal{S}_{\mathrm{s}}$	Material/Spatial Boundary Curve
$\mathrm{d}A,\ \mathrm{d}L,\ \mathrm{d}a,\ \mathrm{d}l$	Material/Spatial Area, Line Element
$\widehat{\boldsymbol{N}},\ \mathrm{d}\widehat{\boldsymbol{L}},\ \widehat{\boldsymbol{n}},\ \mathrm{d}\widehat{\boldsymbol{l}}$	Material/Spatial Curve Normal, Line Element
$\widehat{\boldsymbol{G}}_\alpha, \widehat{\boldsymbol{G}}^\alpha, \boldsymbol{N}, \widehat{\boldsymbol{g}}_\alpha, \widehat{\boldsymbol{g}}^\alpha, \boldsymbol{n}$	Material/Spatial Co-, Contravariant Surface Basis
$\widehat{\boldsymbol{I}}, \widehat{\boldsymbol{i}}$	Material/Spatial Surface Unit Tensors
$\widehat{\boldsymbol{E}}, \widehat{\boldsymbol{e}}$	Material/Spatial Surface Permutation Tensors
$\widehat{\nabla}_{\widehat{X}}\{\bullet\}, \widehat{\nabla}_{\widehat{x}}\{\bullet\}$	Material/Spatial Surface Gradient Operator
$\widehat{\mathrm{Div}}\{\bullet\}, \widehat{\mathrm{div}}\{\bullet\}$	Material/Spatial Surface Divergence Operator
$\widehat{\mathrm{Curl}}\{\bullet\}, \widehat{\mathrm{curl}}\{\bullet\}$	Material/Spatial Surface Curl Operator
$\widehat{\boldsymbol{C}}, \widehat{\boldsymbol{c}}$	Material/Spatial Surface Curvature Tensors
$\widehat{C}, \widehat{c}$	Material/Spatial Surface Total Curvature
$\widehat{\mathrm{cof}}\,\{\bullet\}, \widehat{\mathrm{det}}\{\bullet\}$	Surface Cofactor, Determinant
$\{\bullet\}\,\widehat{\boxtimes}\,\{\bullet\}$	Special Surface Dyadic Product
$\widehat{\times\!\!\!\times}(\{\bullet\})$	Tensor-Valued Tensor Single 'Product'
$\{\bullet\}\,\widehat{\times\!\!\!\times}\,\{\bullet\}$	Scalar-Valued Tensor Double Product
$\widehat{\times}(\{\bullet\})$	Vector-Valued Vector Single 'Product'
$\{\bullet\}\,\widehat{\times}\,\{\bullet\}$	Scalar-Valued Vector Double Product
$\widehat{\boldsymbol{y}}, \widehat{\boldsymbol{Y}}$	Spatial/Material Surface Deformation Maps
$\widehat{\boldsymbol{F}}, \widehat{\boldsymbol{f}}$	Tangent Map: Surface Deformation Gradients
$\widehat{\boldsymbol{K}}, \widehat{\boldsymbol{k}}$	Cotangent Map: Cofactor of Surface Tangent Maps
$\widehat{J}, \widehat{j}$	Measure Map: Determinant of Surface Tangent Maps
$\widehat{\lambda}_{\widehat{a}=1,2}, \widehat{\Lambda}_{\widehat{a}=1,2}$	Spatial/Material Principal Surface Stretches
$\widehat{\boldsymbol{W}}, \widehat{\boldsymbol{V}}, \widehat{\boldsymbol{w}}, \widehat{\boldsymbol{v}}$	(Total) Material/Spatial Surface Velocity
$\widehat{\mathbb{P}}, \widehat{\mathbb{p}}$	Material/Spatial Curve Projection Operator

Kinematics on Dimensionally Reduced Smooth Manifolds (Curves):

$\widetilde{\boldsymbol{\chi}}, \widetilde{\boldsymbol{\Xi}}$	Reference Curve Placement, Position Vector
$\partial^2\mathcal{B}_r, \mathcal{C}_r$	Reference Curve Configuration, Manifold
$\partial^3\mathcal{B}_r, \partial\mathcal{C}_r$	Reference Boundary Points
$\widetilde{\boldsymbol{\Upsilon}}, \widetilde{\boldsymbol{\upsilon}}, \widetilde{\boldsymbol{X}}, \widetilde{\boldsymbol{x}}$	Material/Spatial Curve Placement, Position Vector
$\partial^2\mathcal{B}_m, \mathcal{C}_m, \partial^2\mathcal{B}_s, \mathcal{C}_s$	Material/Spatial Curve Configuration, Manifold
$\partial^3\mathcal{B}_m, \partial\mathcal{C}_m, \partial^3\mathcal{B}_s, \partial\mathcal{C}_s$	Material/Spatial Boundary Points
$\mathrm{d}L, \mathrm{d}P, \mathrm{d}l, \mathrm{d}p$	Material/Spatial Line, Point Element
$\widetilde{\boldsymbol{N}}, \mathrm{d}\widetilde{\boldsymbol{P}}, \widetilde{\boldsymbol{n}}, \mathrm{d}\widetilde{\boldsymbol{p}}$	Material/Spatial Point Normal, Point Element
$\widetilde{\boldsymbol{G}}, \widetilde{\boldsymbol{N}}_\perp, \widetilde{\boldsymbol{N}}_\times, \widetilde{\boldsymbol{g}}, \widetilde{\boldsymbol{n}}_\perp, \widetilde{\boldsymbol{n}}_\times$	Material/Spatial Curve Basis
$\widetilde{\boldsymbol{I}}, \widetilde{\boldsymbol{i}}$	Material/Spatial Curve Unit Tensors
$\widetilde{\boldsymbol{E}}, \widetilde{\boldsymbol{E}}_\perp, \widetilde{\boldsymbol{E}}_\times, \widetilde{\boldsymbol{e}}, \widetilde{\boldsymbol{e}}_\perp, \widetilde{\boldsymbol{e}}_\times$	Material/Spatial Curve ‘Permutation’ Vectors
$\widetilde{\nabla}_{\widetilde{X}}\{\bullet\}, \widetilde{\nabla}_{\widetilde{x}}\{\bullet\}$	Material/Spatial Curve Gradient Operator
$\widetilde{\mathrm{Div}}\{\bullet\}, \widetilde{\mathrm{div}}\{\bullet\}$	Material/Spatial Curve Divergence Operator
$\widetilde{\mathrm{Curl}}_{\perp,\parallel}\{\bullet\}, \widetilde{\mathrm{curl}}_{\perp,\parallel}\{\bullet\}$	Material/Spatial Curve Curl Operators
$\widetilde{\boldsymbol{C}}_\parallel, \widetilde{\boldsymbol{C}}_\perp, \widetilde{\boldsymbol{c}}_\parallel, \widetilde{\boldsymbol{c}}_\perp$	Material/Spatial Curve Curvature Tensors
$\widetilde{C}, \widetilde{T}, \widetilde{c}, \widetilde{t}$	Material/Spatial Curve Curvature, Torsion
$\widetilde{\mathrm{cof}}\{\bullet\}, \widetilde{\mathrm{det}}\{\bullet\}$	Curve Cofactor, Determinant
$\{\bullet\}\,\widetilde{\boxtimes}\,\{\bullet\}$	Special Curve Dyadic Product
$\widetilde{\divideontimes}_{\{\bullet\}}$	Tensor-Valued Tensor Null ‘Product’
$\widetilde{\divideontimes}(\{\bullet\})$	Scalar-Valued Tensor Single ‘Product’
$\widetilde{\times}_{\{\bullet\}}$	Vector-Valued Vector Null ‘Product’
$\widetilde{\times}(\{\bullet\})$	Scalar-Valued Vector Single ‘Product’
$\widetilde{\boldsymbol{y}}, \widetilde{\boldsymbol{Y}}$	Spatial/Material Curve Deformation Map
$\widetilde{\boldsymbol{F}}, \widetilde{\boldsymbol{f}}$	Tangent Map: Curve Deformation Gradients
$\widetilde{\boldsymbol{K}}, \widetilde{\boldsymbol{k}}$	Cotangent Map: Cofactor of Curve Tangent Maps
$\widetilde{J}, \widetilde{j}$	Measure Maps: Determinant of Curve Tangent Maps
$\widetilde{\lambda}, \widetilde{\Lambda}$	Spatial/Material Curve Stretch
$\widetilde{\boldsymbol{W}}, \widetilde{\boldsymbol{V}}, \widetilde{\boldsymbol{w}}, \widetilde{\boldsymbol{v}}$	(Total) Material/Spatial Curve Velocity
$\widetilde{\mathbb{P}}, \widetilde{\mathbb{p}}$	Material/Spatial Point Projection Operator

Kinematics at Singular Sets:

$\mathcal{S}_\mathrm{m}, \mathcal{S}_\mathrm{s}$	Singular Surface
$\boldsymbol{M}^\mp, \boldsymbol{M}, \boldsymbol{m}^\mp, \boldsymbol{m}$	Material/Spatial Singular Surface Normal
$\widehat{\boldsymbol{G}}_\alpha, \widehat{\boldsymbol{G}}^\alpha, \widehat{\boldsymbol{g}}_\alpha, \widehat{\boldsymbol{g}}^\alpha$	Material/Spatial Tangents
$[\![\{\bullet\}]\!], \{\{\bullet\}\}$	Jump/Average Operator
$\boldsymbol{J}, \boldsymbol{j}$	Material/Spatial Singular Surface Jump Vector
Ω, ω	Material/Spatial Singular Surface Jump Amplitude
$\boldsymbol{W}_\perp, W_-, \boldsymbol{w}_\perp, w_\perp$	Material/Spatial Singular Surface Normal Velocity
$\mathcal{C}_\mathrm{m}, \mathcal{C}_\mathrm{s}$	Singular Curve
$\boldsymbol{M}^\pm, \boldsymbol{N}^=, \boldsymbol{N}, \boldsymbol{m}^\pm, \boldsymbol{n}^\pm, \boldsymbol{n}$	Material/Spatial Singular Curve Normals
$\widehat{\boldsymbol{M}}^\pm, \widehat{\boldsymbol{N}}^\mp, \widehat{\boldsymbol{M}}, \widehat{\boldsymbol{m}}^\pm, \widehat{\boldsymbol{n}}^\mp, \widehat{\boldsymbol{m}}$	Material/Spatial Singular Curve Normals
$\widetilde{\boldsymbol{G}}, \widetilde{\boldsymbol{g}}$	Material/Spatial Tangent
$\widehat{[\![\{\bullet\}]\!]}$	Tangential Curve Jump Operator
$\boldsymbol{J}^\pm, \widehat{\boldsymbol{J}}^\pm, \boldsymbol{j}^\pm, \widehat{\boldsymbol{j}}^\pm$	Material/Spatial Singular Curve Jump Vectors
$\Omega^\pm, \widehat{\Omega}^\pm, \omega^\pm, \widehat{\omega}^\pm$	Material/Spatial Singular Curve Jump Amplitudes
$\widehat{\boldsymbol{W}}_\perp^\pm, \widehat{W}_\perp^\pm, \widehat{\boldsymbol{w}}_\perp^\pm, \widehat{w}\pm_\perp$	Material/Spatial Singular Curve Normal Velocity
$\widehat{\mathsf{C}}, \alpha_\mathrm{m}, \widehat{\mathsf{c}}, \alpha_\mathrm{s}$	Material/Spatial 'Singular Surface Curvature'
$\boldsymbol{\mathsf{M}}, \boldsymbol{\mathsf{m}}$	Material/Spatial Average Surface Normal
$\widehat{\mathbb{R}}, \widehat{\mathbb{r}}$	Material/Spatial Surface Normal Rotation Tensor
$\{\bullet\}\{\widehat{\times\!\!\times}\}\{\bullet\}$	Surface Scalar-Valued Tensor Double Product
$\mathcal{P}_\mathrm{m}, \mathcal{P}_\mathrm{s}$	Singular Point
$\widehat{\boldsymbol{M}}_\alpha^\pm, \widehat{\boldsymbol{N}}_\alpha^\pm, \widehat{\boldsymbol{N}}_\alpha, \widehat{\boldsymbol{m}}_\alpha^\pm, \widehat{\boldsymbol{n}}_\alpha^\pm, \widehat{\boldsymbol{n}}_\alpha$	Material/Spatial Singular Point Normals
$\widetilde{\boldsymbol{M}}^\pm, \widetilde{\boldsymbol{N}}^\mp, \widetilde{\boldsymbol{M}}, \widetilde{\boldsymbol{m}}^\pm, \widetilde{\boldsymbol{n}}^\mp, \widetilde{\boldsymbol{m}}$	Material/Spatial Singular Point Normals
$\widetilde{[\![\{\bullet\}]\!]}$	Tangential Point Jump Operator
$\widehat{\boldsymbol{J}}_\alpha^\pm, \widetilde{\boldsymbol{J}}^\pm, \widehat{\boldsymbol{j}}_\alpha^\pm, \widetilde{\boldsymbol{j}}^\pm$	Material/Spatial Singular Point Jump Vectors
$\widehat{\Omega}_\alpha^\pm, \widetilde{\Omega}^\pm, \widehat{\omega}_\alpha^\pm, \widetilde{\omega}^\pm$	Material/Spatial Singular Point Jump Amplitudes
$\widetilde{\boldsymbol{W}}_\perp^\pm, \widetilde{\boldsymbol{W}}_\perp^\pm, \widetilde{\boldsymbol{w}}_\perp^\pm, \widetilde{w}_\perp^\pm$	Material/Spatial Singular Point Normal Velocity
$\widetilde{\mathsf{C}}, \widehat{\alpha}_\mathrm{m}, \widetilde{\mathsf{c}}, \widehat{\alpha}_\mathrm{s}$	Material/Spatial 'Singular Curve Curvature'
$\widehat{\boldsymbol{\mathsf{M}}}, \widehat{\boldsymbol{\mathsf{m}}}$	Material/Spatial Average Curve Normal

Generic Balances:

$\mathfrak{V}_\mathrm{m}, \mathfrak{V}_\mathrm{s}$	Material/Spatial Control Volume
$\boldsymbol{\mathsf{Z}}(\mathfrak{V}_\mathrm{m}), \boldsymbol{\mathsf{Z}}(\mathfrak{V}_\mathrm{s})$	Resultant Volume Extensive Quantity
$z_\mathrm{m}, z_\mathrm{s}$	Volume Extensive Quantity
$\boldsymbol{\mathsf{A}}^\mathrm{ext}(\mathfrak{V}_\mathrm{m}), \boldsymbol{\mathsf{A}}^\mathrm{ext}(\mathfrak{V}_\mathrm{s})$	Resultant Volume Source, Extrinsic Surface Flux
$\overline{\boldsymbol{a}}_\mathrm{m}, \overline{\boldsymbol{a}}_\mathrm{s}$	Volume Source Density
$\widehat{\boldsymbol{a}}_\mathrm{m}, \widehat{\boldsymbol{a}}_\mathrm{m}^\mathrm{ext}, \widehat{\boldsymbol{a}}_\mathrm{s}, \widehat{\boldsymbol{a}}_\mathrm{s}^\mathrm{ext}$	Extrinsic Surface Flux Density

$\mathbf{Z}, \mathbf{Z}_\mathrm{D}, \mathbf{Z}_\mathrm{d}, \boldsymbol{\alpha}, \boldsymbol{\alpha}_\mathrm{D}, \boldsymbol{\alpha}_\mathrm{d}$	Intrinsic Surface Flux Density
$\mathfrak{S}_\mathrm{m}, \mathfrak{S}_\mathrm{s}$	Material/Spatial Control Surface
$\mathsf{Z}(\mathfrak{S}_\mathrm{m}), \mathsf{Z}(\mathfrak{S}_\mathrm{s})$	Resultant Surface Extensive Quantity
$\widehat{z}_\mathrm{m}, \widehat{z}_\mathrm{s}$	Surface Extensive Quantity
$\mathsf{A}^\mathrm{ext}(\mathfrak{S}_\mathrm{m}), \mathsf{A}^\mathrm{ext}(\mathfrak{S}_\mathrm{m})$	Resultant Surface Source, Extrinsic Curve Flux
$\widehat{\boldsymbol{a}}_\mathrm{m}, \widehat{\boldsymbol{a}}_\mathrm{s}$	Surface Source Density
$\widetilde{\boldsymbol{a}}_\mathrm{m}, \widetilde{\boldsymbol{a}}_\mathrm{m}^\mathrm{ext}, \widetilde{\boldsymbol{a}}_\mathrm{s}, \widetilde{\boldsymbol{a}}_\mathrm{s}^\mathrm{ext}$	Extrinsic Curve Flux Density
$\widehat{\mathbf{Z}}, \widehat{\mathbf{Z}}_\mathrm{D}, \widehat{\mathbf{Z}}_\mathrm{d}, \widehat{\boldsymbol{\alpha}}, \widehat{\boldsymbol{\alpha}}_\mathrm{D}, \widehat{\boldsymbol{\alpha}}_\mathrm{d}$	Intrinsic Curve Flux Density
$\widehat{\boldsymbol{\aleph}}_\mathrm{m}, z'_\mathrm{m}, \widehat{\boldsymbol{\aleph}}_\mathrm{s}, z'_\mathrm{s}$	Convection Corrected Surface Source Density
$\nabla_N\{\bullet\}, \nabla_n\{\bullet\}$	Material/Spatial Normal Gradient Operator
$\widehat{\nabla}_{\widehat{X}}\{\bullet\}, \widehat{\nabla}_{\widehat{x}}\{\bullet\}$	Material/Spatial Tangential Gradient Operator
$\mathfrak{C}_\mathrm{m}, \mathfrak{C}_\mathrm{s}$	Material/Spatial Control Curve
$\mathsf{Z}(\mathfrak{C}_\mathrm{m}), \mathsf{Z}(\mathfrak{C}_\mathrm{s})$	Resultant Curve Extensive Quantity
$\widetilde{z}_\mathrm{m}, \widetilde{z}_\mathrm{s}$	Curve Extensive Quantity
$\mathsf{A}^\mathrm{ext}(\mathfrak{C}_\mathrm{m}), \mathsf{A}^\mathrm{ext}(\mathfrak{C}_\mathrm{s})$	Resultant Curve Source, Extrinsic Point Flux
$\widetilde{\boldsymbol{a}}_\mathrm{m}, \widetilde{\boldsymbol{a}}_\mathrm{s},$	Curve Source Density
$\vec{\boldsymbol{a}}_\mathrm{m}, \vec{\boldsymbol{a}}_\mathrm{m}^\mathrm{ext}, \vec{\boldsymbol{a}}_\mathrm{s}, \vec{\boldsymbol{a}}_\mathrm{s}^\mathrm{ext}$	Extrinsic Point Flux ‘Density’
$\widetilde{\mathbf{Z}}, \widetilde{\mathbf{Z}}_\mathrm{D}, \widetilde{\mathbf{Z}}_\mathrm{d}, \widetilde{\boldsymbol{\alpha}}, \widetilde{\boldsymbol{\alpha}}_\mathrm{D}, \widetilde{\boldsymbol{\alpha}}_\mathrm{d}$	Intrinsic Point Flux ‘Density’
$\widetilde{\boldsymbol{\aleph}}_\mathrm{m}, \widehat{z}_\mathrm{m}^\perp, \widehat{z}_\mathrm{m}^\times, \widetilde{\boldsymbol{\aleph}}_\mathrm{s}, \widehat{z}_\mathrm{s}^\perp, \widehat{z}_\mathrm{s}^\times$	Convection Corrected Curve Source Density
$\nabla_{\widetilde{N}_\perp}\{\bullet\}, \nabla_{\widetilde{n}_\perp}\{\bullet\}$	Material/Spatial Normal Gradient Operator
$\nabla_{\widetilde{N}_\times}\{\bullet\}, \nabla_{\widetilde{n}_\times}\{\bullet\}$	Material/Spatial Bi-Normal Gradient Operator
$\widetilde{\nabla}_{\widetilde{X}}\ \{\bullet\}, \widetilde{\nabla}_{\widetilde{x}}\ \{\bullet\}$	Material/Spatial Tangential Gradient Operator

Kinematical Balances:

$\mathfrak{B}_\mathrm{m}, \mathfrak{B}_\mathrm{s}$	Material/Spatial Control Volume
$\overline{\mathsf{F}}\ \ (\mathfrak{B}_\mathrm{m}), \overline{\mathsf{F}}\ \ (\mathfrak{B}_\mathrm{s})$	Resultant Spatial Tangent Map
$\overline{\mathsf{F}}^\mathrm{ext}(\mathfrak{B}_\mathrm{m}), \overline{\mathsf{F}}^\mathrm{ext}(\mathfrak{B}_\mathrm{s})$	Resultant Extrinsic Kinematic Flux
$\widehat{\boldsymbol{F}}_\mathrm{m}, \widehat{\boldsymbol{F}}_\mathrm{m}^\mathrm{ext}, \widehat{\boldsymbol{F}}_\mathrm{s}, \widehat{\boldsymbol{F}}_\mathrm{s}^\mathrm{ext}$	Extrinsic Kinematic Flux
$\overline{\mathsf{K}}\ \ (\mathfrak{B}_\mathrm{m}), \overline{\mathsf{K}}\ \ (\mathfrak{B}_\mathrm{s})$	Resultant Spatial Cotangent Map
$\overline{\mathsf{K}}^\mathrm{ext}(\mathfrak{B}_\mathrm{m}), \overline{\mathsf{K}}^\mathrm{ext}(\mathfrak{B}_\mathrm{s})$	Resultant Extrinsic Kinematic Flux
$\widehat{\boldsymbol{K}}_\mathrm{m}, \widehat{\boldsymbol{K}}_\mathrm{m}^\mathrm{ext}, \widehat{\boldsymbol{K}}_\mathrm{s}, \widehat{\boldsymbol{K}}_\mathrm{s}^\mathrm{ext}$	Extrinsic Kinematic Flux
$\overline{\mathsf{J}}\ \ (\mathfrak{B}_\mathrm{m}), \overline{\mathsf{J}}\ \ (\mathfrak{B}_\mathrm{s})$	Resultant Spatial Measure Map
$\overline{\mathsf{J}}^\mathrm{ext}(\mathfrak{B}_\mathrm{m}), \overline{\mathsf{J}}^\mathrm{ext}(\mathfrak{B}_\mathrm{s})$	Resultant Extrinsic Kinematic Flux
$\widehat{J}_\mathrm{m}, \widehat{J}_\mathrm{m}^\mathrm{ext}, \widehat{J}_\mathrm{s}, \widehat{J}_\mathrm{s}^\mathrm{ext}$	Extrinsic Kinematic Flux
$\overline{\mathsf{f}}\ \ (\mathfrak{B}_\mathrm{s}), \overline{\mathsf{f}}\ \ (\mathfrak{B}_\mathrm{m})$	Resultant Material Tangent Map
$\overline{\mathsf{f}}^\mathrm{ext}(\mathfrak{B}_\mathrm{s}), \overline{\mathsf{f}}^\mathrm{ext}(\mathfrak{B}_\mathrm{m})$	Resultant Extrinsic Kinematic Flux
$\widehat{\boldsymbol{f}}_\mathrm{s}, \widehat{\boldsymbol{f}}_\mathrm{s}^\mathrm{ext}, \widehat{\boldsymbol{f}}_\mathrm{m}, \widehat{\boldsymbol{f}}_\mathrm{m}^\mathrm{ext}$	Extrinsic Kinematic Flux
$\overline{\mathsf{k}}\ \ (\mathfrak{B}_\mathrm{s}), \overline{\mathsf{k}}\ \ (\mathfrak{B}_\mathrm{m})$	Resultant Material Cotangent Map
$\overline{\mathsf{k}}^\mathrm{ext}(\mathfrak{B}_\mathrm{s}), \overline{\mathsf{k}}^\mathrm{ext}(\mathfrak{B}_\mathrm{m})$	Resultant Extrinsic Kinematic Flux

$\widehat{\boldsymbol{k}}_{\mathrm{s}}, \widehat{\boldsymbol{k}}_{\mathrm{s}}^{\mathrm{ext}}, \widehat{\boldsymbol{k}}_{\mathrm{m}}, \widehat{\boldsymbol{k}}_{\mathrm{m}}^{\mathrm{ext}}$	Extrinsic Kinematic Flux
$\bar{\mathsf{j}}\ (\mathfrak{B}_{\mathrm{s}}), \bar{\mathsf{j}}\ (\mathfrak{B}_{\mathrm{m}})$	Resultant Material Measure Map
$\bar{\mathsf{j}}^{\mathrm{ext}}(\mathfrak{B}_{\mathrm{s}}), \bar{\mathsf{j}}^{\mathrm{ext}}(\mathfrak{B}_{\mathrm{m}})$	Resultant Extrinsic Kinematic Flux
$\widehat{j}_{\mathrm{s}}, \widehat{j}_{\mathrm{s}}^{\mathrm{ext}}, \widehat{j}_{\mathrm{m}}, \widehat{j}_{\mathrm{m}}^{\mathrm{ext}}$	Extrinsic Kinematic Flux

Mechanical Balances:

$\mathsf{m}(\mathfrak{B}_{\mathrm{m}}), \mathsf{m}(\mathfrak{B}_{\mathrm{s}})$	Resultant Mass
$\rho_{\mathrm{m}}, \rho_{\mathrm{s}}$	Density of Mass
$\mathsf{m}^{\mathrm{ext}}(\mathfrak{B}_{\mathrm{m}}), \mathsf{m}^{\mathrm{ext}}(\mathfrak{B}_{\mathrm{s}})$	Resultant External Mass Supply
$\overline{m}_{\mathrm{m}}, \overline{m}_{\mathrm{s}}$	Mass Source
$\widehat{m}_{\mathrm{m}}, \widehat{m}_{\mathrm{m}}^{\mathrm{ext}}, \widehat{m}_{\mathrm{s}}, \widehat{m}_{\mathrm{s}}^{\mathrm{ext}}$	Extrinsic Mass Flux
$m_{\mathrm{m}}, m_{\mathrm{s}}$	Intrinsic Mass Flux
$\boldsymbol{R}, \boldsymbol{R}_{\mathrm{D}}, \boldsymbol{R}_{\mathrm{d}}, \boldsymbol{\mu}, \boldsymbol{\mu}_{\mathrm{D}}, \boldsymbol{\mu}_{\mathrm{d}}$	Spatial Piola-, Cauchy-Type Mass Flux
$\mathbf{p}(\mathfrak{B}_{\mathrm{m}}), \mathbf{p}(\mathfrak{B}_{\mathrm{s}})$	Resultant Spatial Momentum
$\boldsymbol{p}_{\mathrm{m}}, \boldsymbol{p}_{\mathrm{s}}$	Density of Spatial Momentum
$\mathbf{f}^{\mathrm{ext}}(\mathfrak{B}_{\mathrm{m}}), \mathbf{f}^{\mathrm{ext}}(\mathfrak{B}_{\mathrm{s}})$	Resultant External Spatial Force
$\overline{\boldsymbol{s}}_{\mathrm{m}}, \overline{\boldsymbol{s}}_{\mathrm{s}}$	Spatial Momentum Source
$\widehat{\boldsymbol{s}}_{\mathrm{m}}, \widehat{\boldsymbol{s}}_{\mathrm{m}}^{\mathrm{ext}}, \widehat{\boldsymbol{s}}_{\mathrm{s}}, \widehat{\boldsymbol{s}}_{\mathrm{s}}^{\mathrm{ext}}$	Spatial Momentum Extrinsic Flux
$\boldsymbol{s}_{\mathrm{m}}, \boldsymbol{s}_{\mathrm{s}}$	Spatial Momentum Intrinsic Flux
$\boldsymbol{P}, \boldsymbol{P}_{\mathrm{D}}, \boldsymbol{P}_{\mathrm{d}}, \boldsymbol{\sigma}, \boldsymbol{\sigma}_{\mathrm{D}}, \boldsymbol{\sigma}_{\mathrm{d}}$	Spatial Piola, Cauchy Stress
$\mathbf{l}(\mathfrak{B}_{\mathrm{m}}), \mathbf{l}(\mathfrak{B}_{\mathrm{s}})$	Resultant Vector Moment of Spatial Momentum
$\boldsymbol{l}_{\mathrm{m}}, \boldsymbol{l}_{\mathrm{s}}$	Density of Vector Moment of Spatial Momentum
$\mathbf{c}^{\mathrm{ext}}(\mathfrak{B}_{\mathrm{m}}), \mathbf{c}^{\mathrm{ext}}(\mathfrak{B}_{\mathrm{s}})$	Resultant External Spatial Couple
$\overline{\boldsymbol{t}}_{\mathrm{m}}, \overline{\boldsymbol{t}}_{\mathrm{m}}^{\mathrm{aut}}, \overline{\boldsymbol{t}}_{\mathrm{s}}, \overline{\boldsymbol{t}}_{\mathrm{s}}^{\mathrm{aut}}$	Vector Moment of Spatial Momentum Source
$\widehat{\boldsymbol{t}}_{\mathrm{m}}, \widehat{\boldsymbol{t}}_{\mathrm{m}}^{\mathrm{ext}}, \widehat{\boldsymbol{t}}_{\mathrm{s}}, \widehat{\boldsymbol{t}}_{\mathrm{s}}^{\mathrm{ext}}$	Vector Moment of Spatial Momentum Extrinsic Flux
$\boldsymbol{t}_{\mathrm{m}}, \boldsymbol{t}_{\mathrm{s}}$	Vector Moment of Spatial Momentum Intrinsic Flux
$\boldsymbol{L}, \boldsymbol{L}_{\mathrm{D}}, \boldsymbol{L}_{\mathrm{d}}, \boldsymbol{\theta}, \boldsymbol{\theta}_{\mathrm{D}}, \boldsymbol{\theta}_{\mathrm{d}}$	Spatial Piola-, Cauchy-Type Couple Stress
$\mathsf{o}(\mathfrak{B}_{\mathrm{m}}), \mathsf{o}(\mathfrak{B}_{\mathrm{s}})$	Resultant Scalar Moment of Spatial Momentum
$o_{\mathrm{m}}, o_{\mathrm{s}}$	Density of Scalar Moment of Spatial Momentum
$\mathsf{v}^{\mathrm{ext}}(\mathfrak{B}_{\mathrm{m}}), \mathsf{v}^{\mathrm{ext}}(\mathfrak{B}_{\mathrm{s}})$	Resultant External Spatial Virial
$\overline{p}_{\mathrm{m}}, \overline{p}_{\mathrm{m}}^{\mathrm{aut}}, \overline{p}_{\mathrm{s}}, \overline{p}_{\mathrm{s}}^{\mathrm{aut}}$	Scalar Moment of Spatial Momentum Source
$\widehat{p}_{\mathrm{m}}, \widehat{p}_{\mathrm{m}}^{\mathrm{ext}}, \widehat{p}_{\mathrm{s}}, \widehat{p}_{\mathrm{s}}^{\mathrm{ext}}$	Scalar Moment of Spatial Momentum Extrinsic Flux
$p_{\mathrm{m}}, p_{\mathrm{s}}$	Scalar Moment of Spatial Momentum Intrinsic Flux
$\boldsymbol{O}, \boldsymbol{O}_{\mathrm{D}}, \boldsymbol{O}_{\mathrm{d}}, \boldsymbol{\phi}, \boldsymbol{\phi}_{\mathrm{D}}, \boldsymbol{\phi}_{\mathrm{d}}$	Spatial Piola-, Cauchy-Type Virial Stress

Consequence of Mechanical Balances:

$\mathsf{K}(\mathfrak{B}_\mathrm{m}), \mathsf{K}(\mathfrak{B}_\mathrm{s})$	Resultant Kinetic Energy
$\kappa_\mathrm{m}, \kappa_\mathrm{s}$	Density of Kinetic Energy
$\widetilde{\boldsymbol{s}}_\mathrm{m}, \widetilde{\boldsymbol{s}}_\mathrm{s}$	Power Generating Traction
$\mathsf{P}^{\mathrm{ext/int}}(\mathfrak{B}_\mathrm{m}), \mathsf{P}^{\mathrm{ext/int}}(\mathfrak{B}_\mathrm{s})$	Resultant External/Internal Mechanical Power
$\mathbb{P}^{\mathrm{int}}_\mathrm{m}, \mathbb{P}^{\mathrm{int}}_\mathrm{s}$	Density of Internal Stress Power
$\boldsymbol{E}^\kappa, \boldsymbol{E}^\kappa_\mathrm{D}, \boldsymbol{E}^\kappa_\mathrm{d}, \boldsymbol{\varepsilon}^\kappa, \boldsymbol{\varepsilon}^\kappa_\mathrm{D}, \boldsymbol{\varepsilon}^\kappa_\mathrm{d}$	Piola-, Cauchy-Type Kinetic Energy Flux
$\overline{\varkappa}_\mathrm{m}, \widehat{\varkappa}_\mathrm{m}, \widetilde{\varkappa}_\mathrm{m}, \overline{\varkappa}_\mathrm{s}, \widehat{\varkappa}_\mathrm{s}, \widetilde{\varkappa}_\mathrm{s}$	Effective Kinetic Energy Source
$\mathbf{P}(\mathfrak{B}_\mathrm{m}), \mathbf{P}(\mathfrak{B}_\mathrm{s})$	Resultant Material Momentum
$\boldsymbol{P}_\mathrm{m}, \boldsymbol{P}_\mathrm{s}$	Density of Material Momentum
$\mathbf{F}^{\mathrm{ext}}(\mathfrak{B}_\mathrm{m}), \mathbf{F}^{\mathrm{ext}}(\mathfrak{B}_\mathrm{s})$	Resultant External Material Force
$\overline{\boldsymbol{S}}_\mathrm{m}, \overline{\boldsymbol{S}}_\mathrm{s}$	Material Momentum Source
$\widehat{\boldsymbol{S}}_\mathrm{m}, \widehat{\boldsymbol{S}}^{\mathrm{ext}}_\mathrm{m}, \widetilde{\boldsymbol{S}}_\mathrm{m}, \widehat{\boldsymbol{S}}_\mathrm{s}, \widehat{\boldsymbol{S}}^{\mathrm{ext}}_\mathrm{s}, \widetilde{\boldsymbol{S}}_\mathrm{s}$	Material Momentum Extrinsic Flux
$\boldsymbol{S}_\mathrm{m}, \boldsymbol{S}_\mathrm{s}$	Material Momentum Intrinsic Flux
$\boldsymbol{p}, \boldsymbol{p}_\mathrm{D}, \boldsymbol{p}_\mathrm{d}, \boldsymbol{\Sigma}, \boldsymbol{\Sigma}_\mathrm{D}, \boldsymbol{\Sigma}_\mathrm{d}$	Material Piola-, Cauchy-Type Stress
$\mathbf{L}(\mathfrak{B}_\mathrm{m}), \mathbf{L}(\mathfrak{B}_\mathrm{s})$	Resultant Vector Moment of Material Momentum
$\boldsymbol{L}_\mathrm{m}, \boldsymbol{L}_\mathrm{s}$	Density of Vector Moment of Material Momentum
$\mathbf{C}^{\mathrm{ext}}(\mathfrak{B}_\mathrm{m}), \mathbf{C}^{\mathrm{ext}}(\mathfrak{B}_\mathrm{s})$	Resultant External Material Couple
$\overline{\boldsymbol{T}}_\mathrm{m}, \overline{\boldsymbol{T}}^{\mathrm{aut}}_\mathrm{m}, \overline{\boldsymbol{T}}_\mathrm{s}, \overline{\boldsymbol{T}}^{\mathrm{aut}}_\mathrm{s}$	Vector Moment of Material Momentum Source
$\widehat{\boldsymbol{T}}_\mathrm{m}, \widehat{\boldsymbol{T}}^{\mathrm{ext}}_\mathrm{m}, \widetilde{\boldsymbol{T}}_\mathrm{m}, \widehat{\boldsymbol{T}}_\mathrm{s}, \widehat{\boldsymbol{T}}^{\mathrm{ext}}_\mathrm{s}, \widetilde{\boldsymbol{T}}_\mathrm{s}$	Vector Moment of Material Momentum Extrinsic Flux
$\boldsymbol{T}_\mathrm{m}, \boldsymbol{T}_\mathrm{s}$	Vector Moment of Material Momentum Intrinsic Flux
$\boldsymbol{l}, \boldsymbol{l}_\mathrm{D}, \boldsymbol{l}_\mathrm{d}, \boldsymbol{\Theta}, \boldsymbol{\Theta}_\mathrm{D}, \boldsymbol{\Theta}_\mathrm{d}$	Material Piola-, Cauchy-Type Couple Stress
$\mathsf{O}(\mathfrak{B}_\mathrm{m}), \mathsf{O}(\mathfrak{B}_\mathrm{s})$	Resultant Scalar Moment of Material Momentum
$O_\mathrm{m}, O_\mathrm{s}$	Density of Scalar Moment of Material Momentum
$\mathsf{V}^{\mathrm{ext}}(\mathfrak{B}_\mathrm{m}), \mathsf{V}^{\mathrm{ext}}(\mathfrak{B}_\mathrm{s})$	Resultant External Material Virial
$\overline{P}_\mathrm{m}, \overline{P}^{\mathrm{aut}}_\mathrm{m}, \overline{P}_\mathrm{s}, \overline{P}^{\mathrm{aut}}_\mathrm{s}$	Scalar Moment of Material Momentum Source
$\widehat{P}_\mathrm{m}, \widehat{P}^{\mathrm{ext}}_\mathrm{m}, \widetilde{P}_\mathrm{m}, \widehat{P}_\mathrm{s}, \widehat{P}^{\mathrm{ext}}_\mathrm{s}, \widetilde{P}_\mathrm{s}$	Scalar Moment of Material Momentum Extrinsic Flux
$P_\mathrm{m}, P_\mathrm{s}$	Scalar Moment of Material Momentum Intrinsic Flux
$\boldsymbol{o}, \boldsymbol{o}_\mathrm{D}, \boldsymbol{o}_\mathrm{d}, \boldsymbol{\Phi}, \boldsymbol{\Phi}_\mathrm{D}, \boldsymbol{\Phi}_\mathrm{d}$	Material Piola-, Cauchy-Type Virial Stress

Virtual Work:

$\mathsf{Z}(\mathfrak{V}_\mathrm{r}), z_\mathrm{r}, \boldsymbol{\mathcal{Z}}$	Volume Extensive Quantity, Referential Surface Flux
$\mathcal{W}^\mathrm{m}, \mathcal{W}^\mathrm{s}$	Referential PB of Total Material/Spatial Velocity
$\boldsymbol{\mathcal{M}}$	Referential Singular Surface Normal
$\mathrm{d}\mathcal{V}, \mathrm{d}\mathcal{A}, \mathrm{d}\mathcal{L}, \mathrm{d}\mathcal{P}$	Referential Volume, Area, Line, Point Element
$\boldsymbol{\mathcal{N}}, \mathrm{d}\boldsymbol{\mathcal{A}}, \widehat{\boldsymbol{\mathcal{N}}}, \mathrm{d}\widehat{\boldsymbol{\mathcal{L}}}, \widetilde{\boldsymbol{\mathcal{N}}}, \mathrm{d}\widetilde{\boldsymbol{\mathcal{P}}}$	Referential Normals, Area, Line, Point Element
$\nabla_\Xi\{\bullet\}, \widehat{\nabla}_{\widehat{\Xi}}\{\bullet\}, \widetilde{\nabla}_{\widetilde{\Xi}}\{\bullet\}$	Referential Gradient Operators

$\mathrm{DIV}\{\bullet\}, \widehat{\mathrm{DIV}}\{\bullet\}, \widetilde{\mathrm{DIV}}\{\bullet\}$	Referential Divergence Operators
$\boldsymbol{\mathcal{F}}^{\mathrm{m}}, \boldsymbol{\mathcal{F}}^{\mathrm{s}}$	Tangent Map: Placement Gradients
$\boldsymbol{\mathcal{K}}^{\mathrm{m}}, \boldsymbol{\mathcal{K}}^{\mathrm{s}}$	Cotangent Map: Cofactor of Tangent Maps
$\mathcal{J}^{\mathrm{m}}, \mathcal{J}^{\mathrm{s}}$	Measure Map: Determinant of Tangent Maps
$\widehat{\boldsymbol{\mathcal{F}}}^{\mathrm{m}}, \widehat{\boldsymbol{\mathcal{F}}}^{\mathrm{s}}$	Tangent Map: Surface Placement Gradients
$\widehat{\boldsymbol{\mathcal{K}}}^{\mathrm{m}}, \widehat{\boldsymbol{\mathcal{K}}}^{\mathrm{s}}$	Cotangent Map: Cofactor of Surface Tangent Maps
$\widehat{\mathcal{J}}^{\mathrm{m}}, \widehat{\mathcal{J}}^{\mathrm{s}}$	Measure Map: Determinant of Surface Tangent Maps
$\widetilde{\boldsymbol{\mathcal{F}}}^{\mathrm{m}}, \widetilde{\boldsymbol{\mathcal{F}}}^{\mathrm{s}}$	Tangent Map: Curve Placement Gradients
$\widetilde{\boldsymbol{\mathcal{K}}}^{\mathrm{m}}, \widetilde{\boldsymbol{\mathcal{K}}}^{\mathrm{s}}$	Cotangent Map: Cofactor of Curve Tangent Maps
$\widetilde{\mathcal{J}}^{\mathrm{m}}, \widetilde{\mathcal{J}}^{\mathrm{s}}$	Measure Map: Determinant of Curve Tangent Maps
$\delta\{\bullet\}, \mathrm{D}_\delta\{\bullet\}, \mathrm{d}_\delta\{\bullet\}$	Total/Material/Spatial Variation
$\mathsf{P}_\delta^{\mathrm{ine}}(\delta\boldsymbol{v})$	Inertial Spatial Virtual Work
$\mathsf{P}_\delta^{\mathrm{ext}}(\delta\boldsymbol{v})$	External Spatial Virtual Work
$\mathsf{P}_\delta^{\mathrm{int}}(\delta\boldsymbol{v})$	Internal Spatial Virtual Work
$\boldsymbol{P}_\bullet, \boldsymbol{\sigma}_\bullet$	Spatial Piola, Cauchy Stress
$\boldsymbol{\mathcal{P}}^{\mathrm{s}}, \boldsymbol{\mathcal{P}}^{\mathrm{s}}_\bullet, \boldsymbol{\mathcal{P}}^{\mathrm{s}}_{\mathrm{D}}, \boldsymbol{\mathcal{P}}^{\mathrm{s}}_{\mathrm{d}}$	Referential PB of Spatial Piola, Cauchy Stress
$\mathsf{p}_\delta^{\mathrm{ine}}(\delta\boldsymbol{\Upsilon})$	Inertial Material Virtual Work
$\mathsf{p}_\delta^{\mathrm{ext}}(\delta\boldsymbol{\Upsilon})$	External Material Virtual Work
$\mathsf{p}_\delta^{\mathrm{int}}(\delta\boldsymbol{\Upsilon})$	Internal Material Virtual Work
$\mathsf{p}_\delta^{\mathrm{ifc}}(\delta\boldsymbol{\Upsilon})$	Interfacial Material Virtual Work
$\boldsymbol{p}_\bullet, \boldsymbol{\Sigma}_\bullet$	Material Piola-, Cauchy-Type Stress
$\boldsymbol{\mathcal{P}}^{\mathrm{m}}, \boldsymbol{\mathcal{P}}^{\mathrm{m}}_\bullet, \boldsymbol{\mathcal{P}}^{\mathrm{m}}_{\mathrm{D}}, \boldsymbol{\mathcal{P}}^{\mathrm{m}}_{\mathrm{d}}$	Referential PB of Material Piola-, Cauchy-Type Stress

Variational Setting:

$\mathbb{Y}$	(Extended) Configuration Space
$\mathcal{R}_\delta$	Virtual Energy Release
$\overline{\boldsymbol{R}}_{\mathrm{r}}, \widehat{\boldsymbol{R}}_{\mathrm{r}}, \widetilde{\boldsymbol{R}}_{\mathrm{r}}$	Material Body Force, Boundary/Surface Traction
$\mathcal{A}$	Action Integral
$\bar{l}_{\mathrm{r}}$	Exterior Lagrange-Type Potential Energy Density
$\widehat{l}_{\mathrm{r}}$	B. Exterior Lagrange-Type Potential Energy Density
κ_{r}	Kinetic Energy Density
$\widehat{\kappa}_{\mathrm{r}}$	Boundary Kinetic Energy Density
λ_{r}	Interior Lagrange-Type Free Energy Density
ℓ_{r}	Interior Lagrange-Type Internal Energy Density
$\mathcal{U}$	Total Potential Energy Functional
$\overline{u}_{\mathrm{r}}$	Total Potential Energy Density
$\widehat{u}_{\mathrm{r}}$	Boundary Total Potential Energy Density
w_{r}	Internal Potential Energy Density

$\widehat{w}_{\mathrm{r}}$	Boundary Internal Potential Energy Density
$\overline{v}_{\mathrm{r}}$	External Potential Energy Density
$\widehat{v}_{\mathrm{r}}$	Boundary External Potential Energy Density
ψ_{r}	Interior Free Energy Density
ι_{r}	Interior Internal Energy Density
$\boldsymbol{\mathcal{P}}^{\mathrm{s}\prime}, \boldsymbol{P}', \boldsymbol{P}'_{\bullet}, \boldsymbol{\sigma}', \boldsymbol{\sigma}'_{\bullet}$	Exterior Spatial Piola, Cauchy Stress
$\widehat{\boldsymbol{\mathcal{P}}}^{\mathrm{s}\prime}, \widehat{\boldsymbol{P}}', \quad \widehat{\boldsymbol{\sigma}}'$	Exterior Spatial Piola, Cauchy Stress
$\overline{\boldsymbol{s}}'_{\mathrm{r}}$	Exterior Spatial Momentum Source Density
$\widehat{\boldsymbol{s}}^{\mathrm{ext}\prime}_{\mathrm{r}}$	Exterior Spatial Momentum Flux Density
$\boldsymbol{\mathcal{P}}^{\mathrm{m}\prime}, \boldsymbol{p}', \boldsymbol{p}'_{\bullet}, \boldsymbol{\Sigma}', \boldsymbol{\Sigma}'_{\bullet}$	Exterior Material Piola, Cauchy Stress
$\widehat{\boldsymbol{\mathcal{P}}}^{\mathrm{m}\prime}, \widehat{\boldsymbol{p}}', \quad \widehat{\boldsymbol{\Sigma}}'$	Exterior Material Piola, Cauchy Stress
$\overline{\boldsymbol{S}}'_{\mathrm{r}}$	Exterior Material Momentum Source Density
$\widehat{\boldsymbol{S}}^{\mathrm{ext}\prime}_{\mathrm{r}}$	Exterior Material Momentum Flux Density

Thermodynamical Balances:

$\mathsf{E}(\mathfrak{B}_{\mathrm{m}}), \mathsf{E}(\mathfrak{B}_{\mathrm{s}})$	Interior Total Energy
$\mathsf{E}'(\mathfrak{B}_{\mathrm{m}}), \mathsf{E}'(\mathfrak{B}_{\mathrm{s}})$	Exterior Total Energy
$\mathsf{I}(\mathfrak{B}_{\mathrm{m}}), \mathsf{I}(\mathfrak{B}_{\mathrm{s}})$	(Interior) Internal Energy
$\mathsf{V}(\mathfrak{B}_{\mathrm{m}}), \mathsf{V}(\mathfrak{B}_{\mathrm{s}})$	Potential External Energy
$\mathsf{K}(\mathfrak{B}_{\mathrm{m}}), \mathsf{K}(\mathfrak{B}_{\mathrm{s}})$	Kinetic Energy
$\varepsilon_{\mathrm{m}}, \varepsilon_{\mathrm{s}}$	Hamilton-Type Interior Internal Energy Density
$\ell_{\mathrm{m}}, \ell_{\mathrm{s}}$	Lagrange-Type Interior Internal Energy Density
$\iota_{\mathrm{m}}, \iota_{\mathrm{s}}$	Interior Internal Energy Density
$\overline{\varepsilon}_{\mathrm{m}}, \overline{\varepsilon}_{\mathrm{s}}, \varepsilon'_{\mathrm{m}}, \varepsilon'_{\mathrm{s}}$	Hamilton-Type Exterior Internal Energy Density
$\overline{\ell}_{\mathrm{m}}, \overline{\ell}_{\mathrm{s}}, \ell'_{\mathrm{m}}, \ell'_{\mathrm{s}}$	Lagrange-Type Exterior Internal Energy Density
$\overline{\iota}_{\mathrm{m}}, \overline{\iota}_{\mathrm{s}}, \iota'_{\mathrm{m}}, \iota'_{\mathrm{s}}$	Exterior Internal Energy Density
$\overline{v}_{\mathrm{m}}, \overline{v}_{\mathrm{s}}, v'_{\mathrm{m}}, v'_{\mathrm{s}}, \widehat{v}_{\mathrm{m}}, \widehat{v}_{\mathrm{s}}$	Potential External Energy Density
$\kappa_{\mathrm{m}}, \kappa_{\mathrm{s}}$	Kinetic Energy Density
$\mathsf{E}^{\mathrm{ext}}(\mathfrak{B}_{\mathrm{m}}), \mathsf{E}^{\mathrm{ext}}(\mathfrak{B}_{\mathrm{s}})$	External Power Supply
$\mathsf{E}^{\mathrm{ext}\prime}(\mathfrak{B}_{\mathrm{m}}), \mathsf{E}^{\mathrm{ext}\prime}(\mathfrak{B}_{\mathrm{s}})$	(Exterior) External Power Supply
$\mathsf{P}^{\mathrm{ext}}(\mathfrak{B}_{\mathrm{m}}), \mathsf{P}^{\mathrm{ext}}(\mathfrak{B}_{\mathrm{s}})$	Mechanical External Power Supply
$\mathsf{H}^{\mathrm{ext}}(\mathfrak{B}_{\mathrm{m}}), \mathsf{H}^{\mathrm{ext}}(\mathfrak{B}_{\mathrm{s}})$	Thermal External Power Supply
$\overline{e}_{\mathrm{m}}, \overline{e}_{\mathrm{s}}$	Total Energy Source Density
$\widehat{e}_{\mathrm{m}}, \widehat{e}_{\mathrm{s}}, \widehat{e}^{\mathrm{ext}}_{\mathrm{m}}, \widehat{e}^{\mathrm{ext}}_{\mathrm{s}}$	Extrinsic Total Energy Flux Density
$e_{\mathrm{m}}, e_{\mathrm{s}}, \widetilde{e}_{\mathrm{m}}, \widetilde{e}_{\mathrm{s}}$	Int-/Exterior Intrinsic Total Energy Flux Density
$\overline{h}_{\mathrm{m}}, \overline{h}_{\mathrm{s}}$	Heat Source Density
$\widehat{h}_{\mathrm{m}}, \widehat{h}_{\mathrm{s}}, \widehat{h}^{\mathrm{ext}}_{\mathrm{m}}, \widehat{h}^{\mathrm{ext}}_{\mathrm{s}}$	Extrinsic Heat Flux Density
$h_{\mathrm{m}}, h_{\mathrm{s}}, \widetilde{h}_{\mathrm{m}}, \widetilde{h}_{\mathrm{s}}$	Intrinsic Heat Flux Density
$\boldsymbol{E}, \boldsymbol{E}_{\mathrm{D}}, \boldsymbol{E}_{\mathrm{d}}, \boldsymbol{\varepsilon}, \boldsymbol{\varepsilon}_{\mathrm{D}}, \boldsymbol{\varepsilon}_{\mathrm{d}}$	Int-/Exterior Intrinsic Total Energy Flux

$\boldsymbol{H}, \boldsymbol{E}_{\mathrm{D}}, \boldsymbol{E}_{\mathrm{d}}, \boldsymbol{\eta}, \boldsymbol{\eta}_{\mathrm{D}}, \boldsymbol{\eta}_{\mathrm{d}}$	Intrinsic Heat Flux
$\overline{k}_{\mathrm{m}}, \widehat{k}_{\mathrm{m}}, \widetilde{k}_{\mathrm{m}}, \overline{k}_{\mathrm{s}}, \widehat{k}_{\mathrm{s}}, \widetilde{k}_{\mathrm{s}}$	Effective Internal Energy Source
$\mathsf{S}(\mathfrak{B}_{\mathrm{m}}), \mathsf{S}(\mathfrak{B}_{\mathrm{s}})$	Entropy
$\sigma_{\mathrm{m}}, \sigma_{\mathrm{s}}$	Entropy Density
$\mathsf{S}^{\mathrm{ext}}(\mathfrak{B}_{\mathrm{m}}), \mathsf{S}^{\mathrm{ext}}(\mathfrak{B}_{\mathrm{s}})$	External Entropy Supply
$\overline{n}_{\mathrm{m}}, \overline{n}_{\mathrm{s}}$	Entropy Source Density
$\widehat{n}_{\mathrm{m}}, \widehat{n}_{\mathrm{s}}, \widehat{n}^{\mathrm{ext}}_{\mathrm{m}}, \widehat{n}^{\mathrm{ext}}_{\mathrm{s}}$	Extrinsic Entropy Flux Density
$n_{\mathrm{m}}, n_{\mathrm{s}}, n^{\mathrm{ext}}_{\mathrm{m}}, n^{\mathrm{ext}}_{\mathrm{s}}$	Intrinsic Entropy Flux Density
$\mathsf{S}^{\mathrm{pro}}(\mathfrak{B}_{\mathrm{m}}), \mathsf{S}^{\mathrm{pro}}(\mathfrak{B}_{\mathrm{s}})$	Entropy Production
$\overline{\pi}_{\mathrm{m}}, \overline{\pi}_{\mathrm{s}}$	Entropy Production Source Density
$\widehat{\pi}_{\mathrm{m}}, \widehat{\pi}_{\mathrm{s}}, \widetilde{\pi}_{\mathrm{m}}, \widetilde{\pi}_{\mathrm{s}}$	Extrinsic Entropy Production Flux Density
$\boldsymbol{S}, \boldsymbol{S}_{\mathrm{D}}, \boldsymbol{S}_{\mathrm{d}}, \boldsymbol{\nu}, \boldsymbol{\nu}_{\mathrm{D}}, \boldsymbol{\nu}_{\mathrm{d}}$	Intrinsic Entropy Flux
$\overline{s}_{\mathrm{m}}, \widehat{s}_{\mathrm{m}}, \widetilde{s}_{\mathrm{m}}, \overline{s}_{\mathrm{s}}, \widehat{s}_{\mathrm{s}}, \widetilde{s}_{\mathrm{s}}$	Effective Entropy Source

Consequence of Thermodynamical Balances:

θ	Absolute Temperature
$\boldsymbol{\Gamma}$	Material Gradient of Absolute Temperature
$\overline{\delta}_{\mathrm{m}}, \overline{\delta}_{\mathrm{s}}$	Dissipation Power Source Density
$\overline{\delta}^{\mathrm{loc}}_{\mathrm{m}}, \overline{\delta}^{\mathrm{loc}}_{\mathrm{s}}$	Local Dissipation Power Source Density
$\overline{\delta}^{\mathrm{con}}_{\mathrm{m}}, \overline{\delta}^{\mathrm{con}}_{\mathrm{s}}$	Conductive Dissipation Power Source Density
$\overline{\delta}^{\mathrm{red}}_{\mathrm{m}}, \overline{\delta}^{\mathrm{red}}_{\mathrm{s}}$	Reduced Dissipation Power Source Density
$\overline{\delta}^{\mathrm{loc,red}}_{\mathrm{m}}, \overline{\delta}^{\mathrm{loc,red}}_{\mathrm{s}}$	Reduced Local Dissipation Power Source Density
$\widehat{\delta}_{\mathrm{m}}, \widehat{\delta}_{\mathrm{s}}, \widetilde{\delta}_{\mathrm{m}}, \widetilde{\delta}_{\mathrm{s}}$	Extrinsic Dissipation Power Flux Density
$\tau_{\mathrm{m}}, \tau_{\mathrm{s}}$	Entropergy Density
$\psi_{\mathrm{m}}, \psi_{\mathrm{s}}$	Interior Free Energy Density
$\lambda_{\mathrm{m}}, \lambda_{\mathrm{s}}$	Lagrange-Type Interior Free Energy Density
$\overline{\psi}_{\mathrm{m}}, \overline{\psi}_{\mathrm{s}}, \psi'_{\mathrm{m}}, \psi'_{\mathrm{s}}$	Exterior Free Energy Density
$\overline{\lambda}_{\mathrm{m}}, \overline{\lambda}_{\mathrm{s}}, \lambda'_{\mathrm{m}}, \lambda'_{\mathrm{s}}$	Lagrange-Type Exterior Free Energy Density
$\mathbb{P}^{\mathrm{int}}_{\mathrm{m}}, \mathbb{p}^{\mathrm{int}}_{\mathrm{m}}, \mathbb{P}^{\mathrm{int}}_{\mathrm{s}}, \mathbb{p}^{\mathrm{int}}_{\mathrm{s}}$	Spatial/Material Internal Power
$\mathfrak{S}_{\mathfrak{v}}, \mathfrak{S}^{\mathrm{ext}}_{\mathfrak{v}}, \mathfrak{S}^{\mathrm{int}}_{\mathfrak{v}}, \mathfrak{S}_{\mathfrak{f}}$	Set of State Variables, Functions
$A_{\mathrm{m}}, A_{\mathrm{s}}, \alpha$	Driving Force, Internal Variable
$k_{\mathrm{m}}, k_{\mathrm{s}}$	Heat Conductivity
$\overline{\boldsymbol{S}}^{\mathrm{int}}_{\mathrm{m}}, \overline{\boldsymbol{S}}^{\mathrm{mec}}_{\mathrm{m}}, \overline{\boldsymbol{S}}^{\mathrm{int}}_{\mathrm{s}}, \overline{\boldsymbol{S}}^{\mathrm{mec}}_{\mathrm{s}}$	Material Momentum Source
$\widehat{\boldsymbol{S}}^{\mathrm{int}}_{\mathrm{m}}, \widehat{\boldsymbol{S}}^{\mathrm{int}}_{\mathrm{s}}$	Interior Material Momentum Extrinsic Flux
$\widehat{\boldsymbol{S}}^{\mathrm{int}\prime}_{\mathrm{m}}, \widehat{\boldsymbol{S}}^{\mathrm{int}\prime}_{\mathrm{s}}$	Exterior Material Momentum Extrinsic Flux
$\widehat{\varsigma}_{\mathrm{m}}, \widehat{\varsigma}_{\mathrm{s}}, \widehat{\Delta}, \widehat{\Delta}^{*}$	Driving Force, Dissipation Potential: Boundary
$\widehat{\omega}, \widehat{\gamma}$	Evolution Condition, Multiplier: Boundary
$\widetilde{\varsigma}_{\mathrm{m}}, \widetilde{\varsigma}_{\mathrm{s}}, \widetilde{\Delta}, \widetilde{\Delta}^{*}$	Driving Force, Dissipation Potential: Singular Surface

$\widetilde{\omega}, \widetilde{\gamma}$	Evolution Condition, Multiplier: Singular Surface
$\boldsymbol{Q} \in \mathrm{SO}(3)$	Special Orthogonal Group
$\boldsymbol{c} \in \mathrm{T}(3)$	Translational Group

Computational Setting:

$\mathcal{B}^h_\mathrm{m}, \mathcal{B}^h_\mathrm{s}$	Material/Spatial Discretized Solution Domain
$\boldsymbol{\xi}, \square$	Isoparametric Coordinates, Domain
n_{dm}, N_{dm}	Number of Dimensions
$\mathcal{B}^e_\mathrm{m}, \mathcal{B}^e_\mathrm{s}$	Material/Spatial Finite Element Domain
$e = 1 \cdot\cdot n_{el}, n = 1 \cdot\cdot n_{en}$	Number of Elements, Element Nodes
$\boldsymbol{X}_n, \boldsymbol{x}_n$	Local Nodal Material/Spatial Position
$\boldsymbol{Y}_n, \boldsymbol{y}_n$	Local Nodal Material/Spatial Deformation Map
$\delta\boldsymbol{\Upsilon}_n, \delta\boldsymbol{v}_n$	Local Nodal Material/Spatial Virtual Displacement
$N^n(\boldsymbol{\xi}), \boldsymbol{J}_\mathrm{m}, \boldsymbol{J}_\mathrm{s}$	Local Shape Function, Material/Spatial Jacobian
$\mathcal{P}^N_\mathrm{m}, \mathcal{P}^N_\mathrm{s}$	Material/Spatial Finite Patch Domain
$N = 1 \cdot\cdot N_{np}$	Number of Node Points
$\boldsymbol{X}_N, \boldsymbol{x}_N$	Global Nodal Material/Spatial Position
$\boldsymbol{Y}_N, \boldsymbol{y}_N$	Global Nodal Material/Spatial Deformation Map
$\delta\boldsymbol{\Upsilon}_N, \delta\boldsymbol{v}_N$	Global Nodal Material/Spatial Virtual Displacement
$N^N(\boldsymbol{X}), N^N(\boldsymbol{x})$	Global Shape Function
y, y_N	Global/Global Nodal Spatial Deformation Map Matrix
$\mathsf{y}_e, \mathsf{y}_n$	Element/Local Nodal Spatial Deformation Map Matrix
$\mathsf{N}^e, \mathsf{N}^n$	Element/Local Nodal Shape Function Matrix
a_e	Boolean Assembly Matrix
$\mathsf{f}_\mathrm{int}, \mathsf{f}^N_\mathrm{int}$	Global/Global Nodal Spatial Internal Force Matrix
$\mathsf{f}_\mathrm{vol}, \mathsf{f}_\mathrm{sur}, \mathsf{f}^N_\mathrm{vol}, \mathsf{f}^N_\mathrm{sur}$	Global/Global Nodal Spatial External Force Matrix
$\mathsf{f}^e_\mathrm{int}, \mathsf{f}^n_\mathrm{int}$	Element/ Local Nodal Spa. Internal Force Matrix
$\mathsf{f}^e_\mathrm{vol}, \mathsf{f}^e_\mathrm{sur}, \mathsf{f}^n_\mathrm{vol}, \mathsf{f}^n_\mathrm{sur}$	Element/Local Nodal Spa. External Force Matrix
r	Global Spatial Residual Matrix
$\mathsf{k}, \mathsf{k}^{NM}$	Global/Global Nodal Spa. Tangent Stiffness Matrix
$\mathsf{k}^e, \mathsf{k}^{nm}$	Element/Local Nodal Spa. Tangent Stiffness Matrix
$\mathsf{F}_\mathrm{int}, \mathsf{F}^N_\mathrm{int}$	Global/Global Nodal Material Internal Force Matrix
$\mathsf{F}_\mathrm{vol}, \mathsf{F}_\mathrm{sur}, \mathsf{F}^N_\mathrm{vol}, \mathsf{F}^N_\mathrm{sur}$	Global/Global Nodal Material External Force Matrix
$\mathsf{F}_\mathrm{ifc}, \mathsf{F}^N_\mathrm{ifc}$	Global/Global Nodal Material Interface Force Matrix
$\mathsf{F}_\mathrm{mat}, \mathsf{Y}$	Global Material Force, Deformation Map Matrix

Chapter 1
Introduction

8,035 m 35°45'30"N 76°39'12"E

Abstract This chapter introduces the overarching topic of this monograph. After recalling the need for continuum mechanics, it contrasts the concepts of spatial and material forces and highlights why the latter are a necessary concept in various branches of defect mechanics. It then reviews various vistas on material forces before commenting on their computational implications.

1.1 The Need for Continuum Mechanics

Matter is essentially *discrete*, i.e. it consists of elementary particles, atoms or ions, molecules, and other elementary building blocks (Phillips [1], Tadmor and Miller [2]). These physical entities occupy discrete positions in *space* (at least with a particular probability) that vary (continuously) with *time*, carry physical attributes like mass and charge, and interact with each other and the external world.

However, on the one hand, to follow the space-time trajectories of only a single *mol* of substance (6×10^{23} entities) is a prohibitively complex endeavor, even on today's largest computers. Moreover, on the other hand, oftentimes the discreteness of matter is not of relevance for the particular application at hand. Therefore, the idealization of matter as being *continuously* distributed in bounded sub-domains of space is at the same time a tremendous reduction of complexity as well as a sufficiently accurate modeling approach.

The mechanics of continuously distributed matter is the realm of *continuum mechanics*, wherein all kinematic and kinetic quantities are considered as fields that are parameterized in space-time coordinates. Therein, a set of mechanical (mass, momentum, and moment of momentum) and thermo-dynamical (energy, entropy) balances together with kinematical and constitutive relations (and appropriate Dirichlet and Neumann boundary as well as initial conditions) dictate the space-time evolution of these fields. The resulting set of equations is typically *nonlinear*, whereby the nonlinearities are of kinematical and/or constitutive character, respectively.

P. Steinmann, *Spatial and Material Forces in Nonlinear Continuum Mechanics*,
Solid Mechanics and Its Applications 272,
https://doi.org/10.1007/978-3-030-89070-4_1

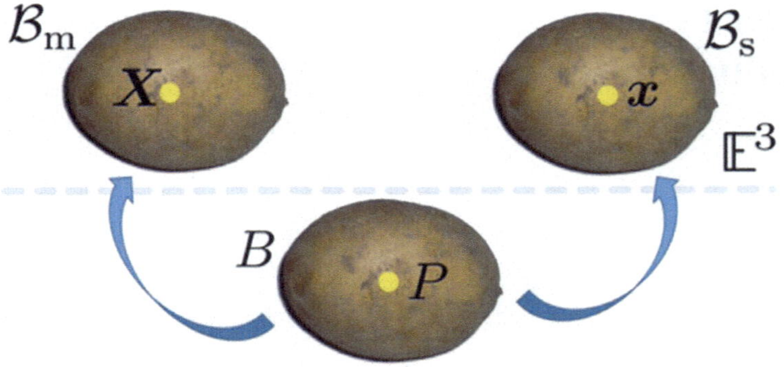

Fig. 1.1 Placement of a physical body B consisting of continuously distributed physical points P into Euclidean space $\mathbb{E}^3$ with spatial and material position vectors $\boldsymbol{x}$ and $\boldsymbol{X}$ at time zero and at any later point of time, respectively

In continuum mechanics, the physical entities or rather physical points constituting matter are assumed continuously distributed in space. Thereby, they occupy either *material* or *spatial positions* at time zero or at any later point in time, respectively, and are identified with their corresponding material and spatial position vectors in three-dimensional Euclidean space; see Fig. 1.1. As a consequence, continuum mechanics stands out from space-time parameterized field theories in that it classically allows for two different space parameterizations of all fields, i.e. parametrization in either material or spatial position vectors (material or spatial coordinates).

The *material parametrization* of all fields, in particular the continuously distributed spatial position vectors as the primary unknown field, i.e. the (spatial) deformation map parameterized in the material (space) coordinates and time, is the typical approach to *solid mechanics*. In the so-called *Lagrangian approach*, the corresponding time derivative is evaluated at fixed material space coordinates and is denoted as *material time derivative*, thus capturing temporal changes as experienced by a convected observer.

In contrast, the *spatial parametrization* of all fields is the typical approach to *fluid mechanics*. In the so-called *Eulerian approach*, the corresponding time derivative is evaluated at fixed spatial (space) coordinates and is denoted as *spatial time derivative*, thus capturing temporal changes as experienced by a space-fixed observer.

It shall be carefully noted that solid and fluid mechanics are only seemingly different subject areas of continuum mechanics since both build on the *identical* set of balance equations and kinematical relations, which are, however, only differently parameterized. Indeed, the expressions of the balances and kinematical relations characteristic for either solid or fluid mechanics may be transformed into each other by composition with the deformation map and by re-expression of the corresponding time derivative of the balanced quantities. Thus, only the constitutive relations, e.g. for a Hookean solid or a Newtonian fluid, remain as the key distinction between the two subject areas.

Continuum mechanics has been carefully elaborated in a plethora of outstanding accounts; refer, for example, to the monographs by Anand and Govindjee [3], Antmann [4], Bertram [5], Ciarlet [6], Goriely [7], Gurtin et al. [8], Haupt [9], Holzapfel [10], Malvern [11], Marsden and Hughes [12], Maugin [13], Ogden [14], Silhavy [15], Tadmor et al. [16], and Truesdell and Noll [17], among many others (e.g. Steinmann [18]).

1.2 Spatial Versus Material Forces

Way more fundamental, however, than the distinction between the material and spatial parametrization of all fields in continuum mechanics is the definition of *force*. Unlike in Newtonian mechanics, wherein force is a primary (given) object, force in the spirit of analytical, i.e. Lagrangian and Hamilton mechanics (or likewise in the spirit of D'Alembert), is a secondary (derived) object that is defined as being either *power-conjugated* or *variationally conjugated* to changes of generalized coordinates (carrying the flavor of kinematical quantities).

In the present continuum mechanics setting, the generalized coordinates of analytical mechanics are typically associated with the (spatial) deformation map, i.e. the spatial positions occupied by the continuously distributed physical points. Then temporal changes of the (spatial) deformation map correspond to its material time derivative, i.e. the classical *(spatial) velocity* field in either material or (after composition with the deformation map) spatial parametrization. In a variational context, the (spatial) velocity is substituted by the *material variation* of the (spatial) deformation map, i.e. the classical *(spatial) virtual displacement*.

Forces in their classical interpretation, that shall here be denoted as *spatial forces*, are thus either power-conjugated to the classical (spatial) velocity or variationally conjugated to the (spatial) virtual displacement, respectively; see Fig. 1.2. In this regard, traditional solid and fluid mechanics adopt a ***spatial vista*** in that they deal exclusively with spatial forces.

The material positions in continuum mechanics need, however, not be considered as temporally fixed in space, i.e. as time-independent. Indeed, if in agreement with the second law that requires positive dissipation, continuum points on the external boundary of a continuum body or at interfaces within a continuum body may well change their material position with time, thereby complying, for example, with a postulate of maximum dissipation.

A paradigmatic example is a crack tip, i.e. a singular part of the external boundary of a continuum body that is changing its material position during crack extension; see Fig. 1.2. Yet another paradigmatic example is a sharp interface between phases, i.e. a singular surface within a continuum body that is changing its material position during interface motion.

In a variational context, the *spatial variation* of the material position, i.e. its variation at fixed spatial position (resulting in *material virtual displacements*), results in a variational release of energy from the continuum body. Exchanging material

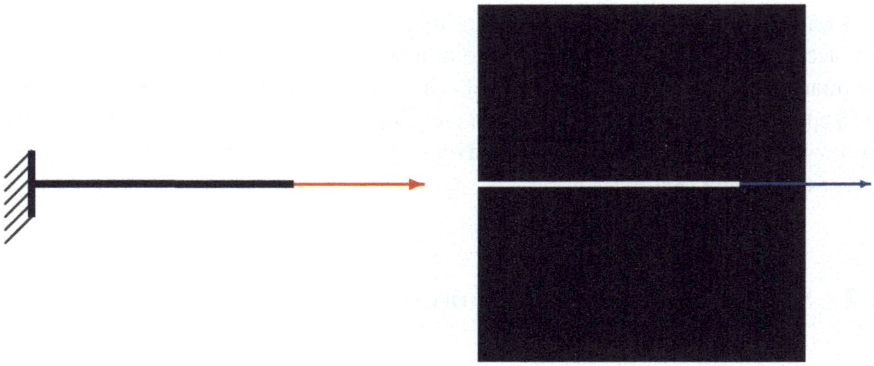

Fig. 1.2 Juxtaposition of spatial and material forces as adapted from Maugin [19]. (Left) Spatial forces are generated by spatial variations relative to the ambient space. (Right) Material forces are generated by material variations relative to the ambient material

virtual displacements for the material velocity, i.e. the spatial time derivative of the inverse or rather material deformation map, the released energy may then be re-invested into other physical processes, e.g. the creation of new crack surfaces or the accompanying phase changes on both sides of the interface, respectively.

Thus, as an alternative to the classical (spatial) velocity and (spatial) virtual displacement, we may also consider the somewhat unusual *material velocity*, i.e. the spatial time derivative of the inverse or rather material deformation map, and the likewise unusual material virtual displacement as its spatial variation. The released energy is then expressed in terms of *material forces* that are either power-conjugated or variationally conjugated to changes in material positions. As a consequence, material forces may be considered as *driving forces* for possible changes of material positions, whereby their actual evolution follows from a modeling approach that is constrained to comply with the second law requiring positive dissipation. This mindset toward changes in material positions, i.e. *material changes*, denotes the ***material vista*** on continuum mechanics.

The above concepts of spatial and material forces and the interpretation of the latter as driving force for changes of material positions are elaborated in Fig. 1.3 for the simple paradigm of an elastic spring with harmonic potential energy (Steinmann et al. [20]). The spring is supported at the left end point and loaded by a given (spatial) force at the right end point. Sub-figure **(a)** depicts the material position of the right end point at X, sub-figure **(b)** displays the spatial position of the right end point at x, sub-figure **(c)** represents the material variation $\mathrm{d}_\delta X$, sub-figure **(d)** represents the spatial variation $\mathrm{D}_\delta x$, and sub-figure **(e)** showcases the applied spatial force $\mathbf{f} := f\,\mathbf{e}$ and the resulting material force $\mathbf{F} := F\,\mathbf{e}$ (with $\mathbf{e}$ the horizontal base vector).

For unit spring stiffness, the total potential energy reads

$$E(x, X) = [x - X]^2/2 - f\,x,$$

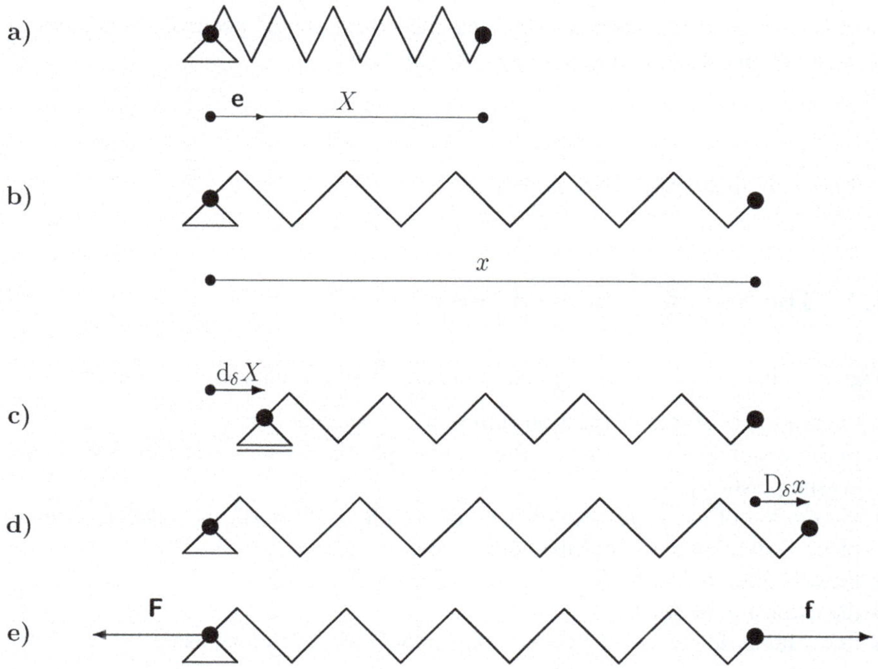

Fig. 1.3 Spatial and material forces for an elastic spring with harmonic potential energy

its spatial variation

$$D_\delta E = [x - X - f]\, D_\delta x \doteq 0 \quad \forall D_\delta x$$

results in the equilibrium of spatial forces

$$f = x - X,$$

and its material variation

$$d_\delta E = -[x - X]\, d_\delta X := F\, d_\delta X \leq 0 \quad \forall d_\delta X$$

determines the material force as

$$\mathbf{F} = -\mathbf{f},$$

i.e. here as the reaction force at the support. If we imagine the support as a frictional slider, changes in the material position of the support occur once the material force reaches a given threshold

$$|F| = F_0.$$

In a convex analysis setting, satisfying the constraint of positive dissipation as required by the second law, the temporal evolution

$$\mathrm{d}_t X \propto -\mathrm{sign}\, F$$

follows here in the positive $\mathbf{e}$ direction.

1.3 The Need for Material Forces

Material changes power-conjugated to material forces comprise, for example,

- the motion of single dislocations through a crystalline lattice,
- the homogenized, i.e. collective flow of multiple dislocations as captured by single-crystal plasticity,
- the motion of vacancies and inclusions as well as of interfaces between different phases, all relative to the embedding material,
- the dilatation of voids,
- the extension of cracks,
- the volumetric growth of living matter in the realm of biomechanics,
- and others.

Thereby, corresponding driving or rather material forces are, for example,

- the Peach–Koehler force on a single dislocation (Peach and Koehler [21]), with a delicate discussion regarding its nature as material force (Rogula [22], Ericksen [23, 24], Steinmann [25]),
- the Mandel or rather Schmid stress of single-crystal plasticity (Asaro [26]), with countless applications to computational crystal plasticity (Steinmann and Stein [27], Steinmann [28], Miehe and Schotte [29]),
- the force on an elastic singularity[1] (Eshelby [30]), a defect (Zorski [31]), or a singular surface (Abeyaratne and Knowles [32–34]),
- the J-integral and energy release of fracture mechanics (Cherepanov [35], Rice [36]), with multiple identifications as material force (Maugin [37, 38], Gurtin and Podio-Guidugli [39, 40], Agiasofitou and Kalpakides [41]),
- the M- and L-integrals of defect mechanics (Knowles and Sternberg [42], Budianski and Rice [43], Rice [44]) and their detailed analysis (Agiasofitou and Lazar [45, 46]),
- forces driving biomechanics growth (Kirchner and Lazar [47]),
- and others.

[1] A descriptive definition of material forces due to Eshelby [30] is the following: ... *the total energy of a system ... is a function of the set of parameters necessary to specify the configuration of the imperfections. The negative gradient of the total energy wrt the position of an imperfection may conveniently be called the force on it. This force, in a sense fictitious, is introduced to give a picturesque description of energy changes, and must not be confused with the ordinary surface and body forces acting on the material.*

Interestingly, all these celebrated concepts of driving or rather material forces have traditionally been developed (more or less) independently but can be demonstrated to follow from a unified approach when resorting to the material vista toward nonlinear continuum mechanics as sketched out in the above (Maugin [19, 48, 49], Gurtin [50, 51]).

For completeness, it is remarked that material forces are synonymously also denoted as *configurational forces*, likewise the changes in material positions, i.e. material changes are collectively denoted as *configurational changes*. A brief account of the historical development of configurational mechanics from its foundation in the 1950s all the way to its status at the end of the first decade of the twenty-first century is provided by Maugin [52].

1.4 Vistas on Material Forces

Since the traditional balances of continuum mechanics deal exclusively with spatial forces, the questions arise, whether there are balances that involve material forces and how these would be related to the traditional balances?

To answer these questions is the main objective of this treatise. They shall here be addressed by introducing an additional placement of physical points into absolute, time-independent *reference positions* in three-dimensional Euclidian space with respect to which both the material and spatial positions may evolve (Askes et al. [53], Kuhl et al. [54, 55], Runesson et al. [56]). The established spatial and the somewhat uncommon material vistas are then unraveled and reconciled by carefully examining the consequences of the reference embedding of the physical body when expressing the kinematical relations and formulating the mechanical and thermo-dynamical balances.

The traditional approach to material forces and configurational mechanics is based on a variational setting in the spirit of a Hamiltonian (or Dirichlet) principle in terms of a Lagrangian (or potential energy) density that is parameterized in the material positions (Rogula [22], Golebiewska-Herrmann [57–59], Maugin et al. [60, 61], Kalpakides and Maugin [62, 63], Steinmann [64, 65], Yavari et al. [66], Lazar et al. [67–69], Gupta and Markenscoff [70, 71]).

Integral to the variational setting in terms of material variations is the consideration of the energy released from a continuous body due to material (configurational) changes (Dascalu and Maugin [72], Steinmann et al. [73]).

Likewise, the variational setting is intimately connected to the notion of conservation laws as resulting from Noether's theorem (Noether [74]) and corresponding path-independent integrals (Sanders [75], Günther [76], Cherepanov [35], Rice [36], Knowles and Sternberg [42], Budianski and Rice [43], Fletcher [77], Rogula [22], Buggisch et al. [78], Francfort and Golebiewska-Herrmann [79, 80], Olver [81], Cherepanov [82], Simo and Honein [83], Honein and Herrmann [84]).

Also related to the variational setting is an approach that exchanges the material and the spatial parametrization (Shield [85], Chadwick [86], Govindjee and

Mihalic [87, 88], Kalpakides and Balassas [89]), also with extension to the analysis of ellipticity conditions (Kuhl et al. [90]) and generalized continua (Kalpakides and Agiasofitou [91], Kirchner and Steinmann [92]), thereby emphasizing the duality of flux quantities in spatial and material balances.

The variational viewpoint is admittedly very insightful and intuitive, and also easily extends to atomistic systems (Steinmann et al. [20, 93], Birang and Steinmann [94]), however, it (mostly) restricts the analysis of configurational forces to conservative systems.

Another approach to configurational mechanics resorts to a pullback of the pertinent spatial vista balances to the so-called material manifold (Maugin [19, 48, 49], Steinmann [95, 96]).

This procedure is directly extendable, for example, to thermo-mechanics (Dascalu and Maugin [97], Steinmann [96]), open system thermo-mechanics (Epstein and Maugin[98], Kuhl and Steinmann [99–102]), thermo-mechanics coupled to diffusion (Steinmann et al. [103]), poro-mechanics (Quiligotti et al. [104], Papastavrou and Steinmann [105]), electro- and magneto-mechanics (Vu and Steinmann [106–108]), plasticity (Menzel and Steinmann [109, 110], Tillberg et al. [111], Özenç et al. [112]), and other coupled and inelastic problems.

However, whereas following from more or less straightforward manipulations of the traditional spatial vista balances, it does however not explicitly exploit the implications of positive dissipation as required by the second law.

Yet another approach postulates the power expended by configurational forces and postulates and determines corresponding separate configurational balances from requiring invariance under material observer changes (Gurtin [50, 51], Cermelli and Fried [113], Mariano [114], Kalpakides and Dascalu [115], Podio-Guidugli [116, 117], Fried and Gurtin [118]).

For corresponding situations, this approach renders identical results as the variational and material manifold approaches, however at the expense of postulating separate configurational force systems and corresponding balances.

In all of the above, the tensor-valued flux quantity participating in the configurational balance of momentum or rather the *balance of material momentum* displays a so-called energy–momentum format (Eshelby [119], Chadwick [86], Hill [120], Epstein and Maugin [121, 122]), a terminology motivated by the Maxwell stress of electro-magneto-dynamics (Maugin [123], Pelteret and Steinmann [124]), and is commonly denoted as the *Eshelby stress*.

The Eshelby stress is considered the key ingredient to configurational mechanics (Kienzler and Herrmann [125, 126]). As a material vista stress measure, the Eshelby stress has also successfully been used in failure criteria for damage and fracture (Kienzler and Herrmann [127], Brünig [128], Verron et al. [129], Andriyana and Verron [130, 131], Verron [132], Previati and Kaliske [133]), partly with a view on soft materials such as rubber and elastomers.

The unifying approach advocated in this treatise attempts to reconcile all of the above viewpoints by capitalizing on the notion of an absolute, time-independent reference placement and the associated concepts of *total time derivatives* and *total variations*, whereby material (configurational) forces are responsible for *consistent* positive dissipation power upon material (configurational) changes.

1.5 Computational Material Forces

Last but not the least, a continuum formulation of configurational mechanics that unifies all aspects of defect, fracture, and failure mechanics is not only intellectually satisfying per se but also enables novel avenues in *computational mechanics* with far-reaching consequences.

A greatly simplified finite element approach to computational fracture mechanics based on the notion of material forces and their identification with the J-integral was firstly proposed within geometrically nonlinear hyperelasticity as the *Material Force Method* (Ackermann et al. [134], Steinmann et al. [95, 135]) and later extensively analyzed (Denzer et al. [136, 137]).

Computational material forces or rather the material force method prove extremely versatile and allow, for example, extension to thermo-elasticity (Kuhl et al. [138], Bargmann et al. [139]), damage (Liebe et al. [140]), visco-elasticity (Nguyen et al. [141], Näser et al. [142]), plasticity (Menzel et al. [143, 144], Kuhn et al. [145, 146]), micromorphic and gradient continua (Hirschberger et al. [147], Floros et al. [148]), elastodynamics (Kolling and Müller [149], Timmel et al. [150]), biomechanics (Kuhl and Steinmann [100–102]), electro-elasticity (Denzer and Menzel [151]), homogenization (Ricker et al. [152, 153]), atomistics (Birang and Steinmann [94]), and other topics in mechanics.

Interestingly, the concept of computational material forces also allows for the adaptivity of finite element meshes (Braun [154, 155], Müller and Maugin [156], Gross et al. [157], Kuhl et al. [54], Askes et al. [53], Heintz et al. [158], Müller et al. [159, 160], Thoutireddy and Ortiz [161], Mosler and Ortiz [162, 163], Tabarraei and Sukumar [164], Scherer et al. [165, 166], Rajagopal and Sivakumar [167]).

It is also worth noting that sensitivities in structural optimization are likewise related to computational material forces (Askes et al. [168], Materna and Barthold [169–171], Riehl and Steinmann [172]).

Last but not the least, computational material forces naturally prove beneficial in computational analyses of defect evolution (Gross et al. [173], Kolling et al. [174, 175], Timmel et al. [176]), and the propagation of cracks in computational fracture mechanics (Larsson and Fagerström [177–179], Heintz [180], Mahnken [181, 182], Miehe and Gürses [183, 184], Miehe et al. [185], Schütte [186], Brouzoulis et al. [187], Özenç et al. [188, 189], Kaczmarczyk et al. [190, 191], Bird et al. [192]).

1.6 The Nature of This Treatise

The nature of this treatise is best conveyed by commenting on its structuring into some dozen chapters spanning from continuum kinematics to computational mechanics.

Noteworthy, the present approach is characterized by the introduction of an absolute time-independent reference configuration in addition to time-dependent material and spatial configurations. It turns out that it is this key feature of the kinematic description that allows the unification of most different views on configurational mechanics.

*As a reading recommendation, the topics marked with a * may safely be ignored when first browsing through this treatise.*

After the introduction, Chap. 2 recalls the pertinent spatial and material continuum kinematics in bulk volumes, thereby focusing on the nonlinear deformation maps, their associated tangent, cotangent and measure maps, and their compatibility conditions.

Chapter 3 reviews the corresponding continuum kinematics on dimensionally reduced smooth manifolds, i.e. on boundary surfaces [and on boundary curves]*, again with an emphasis on the nonlinear deformation maps and their associated tangent, cotangent, and measure maps.

Chapter 4 revisits the relevant continuum kinematics at singular sets, i.e. at singular surfaces [and at singular curves and points]*, thereby elaborating on the jumps in the nonlinear deformation maps and their associated tangent, cotangent, and measure maps.

Chapter 5 represents the formulation of generic balances for generic volume [as well as surface and curve]* extensive quantities, thereby highlighting their global and local formats and resorting in both cases to material and spatial control volumes [as well as control surfaces and control curves]*.

[Chapter 6 applies the formats of the generic balances to the spatial and material tangent, cotangent, and measure maps to formulate what, for the sake of semantic unification, may be called kinematical 'balances']*.

Chapter 7 then details the generic balances for the case of mechanical balances of mass, spatial momentum, and its vector [and scalar]* moment, respectively, again with emphasis on their global and local formats and the distinction between material and spatial control volumes.

Chapter 8 explores the consequences of the mechanical balances by elaborating on local and global formats of the balance of kinetic energy and the balance of material momentum [and its vector and scalar moments]*, thereby again differentiating between spatial and material control volumes.

Chapter 9 capitalizes on the referential setting when introducing the notions of spatial and material virtual displacements and discussing the accompanying spatial and material virtual work principles.

Chapter 10 expands on the related variational setting in terms of extended Hamilton and Dirichlet principles for conservative elasto-dynamic and elasto-static cases, respectively, and carefully analyzes the resulting spatial and material Euler–Lagrange equations.

Chapter 11 then specifies the generic balances for the case of thermodynamical balances of energy and entropy, whereby the case of exterior energy is demarcated from the common case of (interior) energy by the formal incorporation of the external potential energy into the notion of internal energy.

Chapter 12 exploits the consequences of the thermodynamical balances and the resulting formats of the dissipation power inequalities by identifying the forces driving material (configurational) changes on the boundary and at singular surfaces as the appropriate contributions to the balance of material momentum. It is the here advocated thermodynamical derivation of material forces that qualifies the current approach as being *dissipation-consistent*.

Chapter 13 finally sketches the consequences for computational mechanics by outlining the material force method based on finite element discretization of the material virtual work principle and highlights its applicability to geometrically nonlinear fracture mechanics by some computational examples.

References

1. Phillips R (2001) Crystals, defects and microstructures: modeling across scales. Cambridge University Press, Cambridge
2. Tadmor EB, Miller RE (2011) Modeling materials: continuum, atomistic and multiscale techniques. Cambridge University Press, Cambridge
3. Anand L, Govindjee S (2020) Continuum mechanics of solids. Oxford University Press, USA
4. Antman SS (2005) Nonlinear problems of elasticity. Springer, Berlin
5. Bertram A (2012) Elasticity and plasticity of large deformations. Springer, Berlin
6. Ciarlet PG (1988) Mathematical Elasticity: Volume I: three-dimensional elasticity. North-Holland (1988)
7. Goriely A (2017) The mathematics and mechanics of biological growth. Springer, Berlin
8. Gurtin M, Fried E, Anand L (2010) The mechanics and thermodynamics of continua. Cambridge University Press, Cambridge
9. Haupt P (2013) Continuum mechanics and theory of materials. Springer, Berlin
10. Holzapfel GA (2000) Nonlinear solid mechanics: a continuum approach for engineering. Wiley, New York
11. Malvern LE (1969) Introduction to the mechanics of a continuous medium. Prentice Hall, New Jersey
12. Marsden JE, Hughes TJR (1994) Mathematical foundations of elasticity. Dover, New York
13. Maugin GA (1999) The thermomechanics of nonlinear irreversible behaviors: an introduction. World Scientific, Singapore
14. Ogden RW (1997) Non-linear elastic deformations. Dover, New York
15. Silhavy M (2013) The mechanics and thermodynamics of continuous media. Springer, Berlin
16. Tadmor EB, Miller RE, Elliott RS (2012) Continuum mechanics and thermodynamics: from fundamental concepts to governing equations. Cambridge University Press, Cambridge
17. Truesdell C, Noll W (2004) The non-linear field theories of mechanics. Springer, Berlin
18. Steinmann P (2015) Geometrical foundations of continuum mechanics. Springer, Berlin

19. Maugin GA (1993) Material inhomogeneities in elasticity. Chapman and Hall, London
20. Steinmann P, Smith A, Birang E, McBride A, Javili A (2021) Atomistic two-, three- and four-body potentials. Spatial and material settings. J Mech Phys Solids
21. Peach M, Koehler JS (1950) The forces exerted on dislocations and the stress fields produced by them. Phys Rev 80:436
22. Rogula D (1977) Forces in material space. Arch Mech 29:705–713
23. Ericksen JL (1995) Remarks concerning forces on line defects. ZAMP: Zeitschrift für Angewandte Mathematik und Physik 46:247–271
24. Ericksen JL (1998) On nonlinear elasticity theory for crystal defects. Int J Plast 14:9–24
25. Steinmann P (2002) On spatial and material settings of hyperelastostatic crystal defects. J Mech Phys Solids 50:1743–1766
26. Asaro RJ (1983) Crystal plasticity. J Appl Mech 50:921–934
27. Steinmann P, Stein E (1996) On the numerical treatment and analysis of finite deformation ductile single crystal plasticity. Comput Methods Appl Mech Eng 129:235–254
28. Steinmann P (1996) Views on multiplicative elastoplasticity and the continuum theory of dislocations. Int J Eng Sci 34:1717–1735
29. Miehe C, Schotte J (2017) Crystal plasticity and evolution of polycrystalline microstructure. In: Encyclopedia of computational mechanics, 2nd edn, pp 1–23
30. Eshelby JD (1951) The force on an elastic singularity. Philos Trans R Soc Lon. Ser A, Math Phys Sci 244:87–112
31. Zorski H (1981) Force on a defect in non-linear elastic medium. Int J Eng Sci 19:1573–1579
32. Abeyaratne R, Knowles JK (1990) On the driving traction acting on a surface of strain discontinuity in a continuum. J Mech Phys Solids 38:345–360
33. Abeyaratne R, Knowles JK (1991) Kinetic relations and the propagation of phase boundaries in solids. Arch Ration Mech Anal 114:119–154
34. Abeyaratne R, Knowles JK (2000) A note on the driving traction acting on a propagating interface: adiabatic and non-adiabatic processes of a continuum. J Appl Mech 67:829–830
35. Cherepanov GP (1967) Crack propagation in continuous media. PMM (Appl Math Mech, Translation from Russian) 31:467–488
36. Rice JR (1968) A path independent integral and the approximate analysis of strain concentration by notches and cracks. J Appl Mech 35:379–386
37. Maugin GA (1994) Eshelby stress in elastoplasticity and ductile fracture. Int J Plast 10:393–408
38. Maugin GA (1994) On the J-integral and energy-release rates in dynamical fracture. Acta Mech 105:33–47
39. Gurtin M, Podio-Guidugli P (1996) Configurational forces and the basic laws for crack propagation. J Mech Phys Solids 44:905–927
40. Gurtin M, Podio-Guidugli P (1998) Configurational forces and a constitutive theory for crack propagation that allows for kinking and curving. J Mech Phys Solids 46:1343–1378
41. Agiasofitou E, Kalpakides VK (2006) The concept of a balance law for a cracked elastic body and the configurational force and moment at the crack tip. Int J Eng Sci 44:127–139
42. Knowles JK, Sternberg E (1972) On a class of conservation laws in linearized and finite elastostatics. Arch Ration Mech Anal 44:187–211
43. Budiansky B, Rice JR (1973) Conservation laws and energy-release rates. J Appl Mech 40:201–203
44. Rice JR (1985) Conserved integrals and energetic forces. Fundamentals of deformation and fracture. Cambridge University Press, Cambridge, pp 33–56
45. Agiasofitou E, Lazar M (2017) Micromechanics of dislocations in solids: J-, M-, and L-integrals and their fundamental relations. Int J Eng Sci 114:16–40
46. Lazar M, Agiasofitou E (2018) Eshelbian dislocation mechanics: J-, m-, and l-integrals of straight dislocations. Mech Res Commun 93:89–95
47. Kirchner H, Lazar M (2008) The thermodynamic driving force for bone growth and remodelling: a hypothesis. J R Soc Interface 5:183–193

48. Maugin GA (1995) Material forces: concepts and applications. ASME Appl Mech Rev 48:213–245
49. Maugin GA (2011) Configurational forces: thermomechanics, physics, mathematics, and numerics. CRC Press, Boca Raton
50. Gurtin M (1995) The nature of configurational forces. Arch Ration Mech Anal 131:67–100
51. Gurtin M (2000) Configurational forces as basic concepts of continuum physics. Springer, New York
52. Maugin GA (2013) Sixty years of configurational mechanics (1950–2010). Mech Res Commun 50:39–49
53. Askes H, Kuhl E, Steinmann P (2004) An ALE formulation based on spatial and material settings of continuum mechanics. Part 2: Classification and applications. Comput Methods Appl Mech Eng 193:4223–4245
54. Kuhl E, Askes H, Steinmann P (2004) An ALE formulation based on spatial and material settings of continuum mechanics. Part 1: Generic hyperelastic formulation. Comput Methods Appl Mech Eng 193:4207–4222
55. Kuhl E, Steinmann P (2005) A hyperelastodynamic ALE formulation based on referential, spatial and material settings of continuum mechanics. Acta Mech 174:201–222
56. Runesson K, Larsson F, Steinmann P (2009) On energetic changes due to configurational motion of standard continua. Int J Solids Struct 46:1464–1475
57. Golebiewska-Herrmann A (1981) On conservation laws of continuum mechanics. Int J Solids Struct 17:1–9
58. Golebiewska-Herrmann A (1982) Material momentum tensor and path-independent integrals of fracture mechanics. Int J Solids Struct 18:319–326
59. Golebiewska-Herrmann A (1983) On the lagrangian formulation of continuum mechanics. Physica A 118:300–314
60. Maugin GA, Epstein M, Trimarco C (1992) Pseudomomentum and material forces in inhomogeneous materials: application to the fracture of electromagnetic materials in electromagnetoelastic fields. Int J Solids Struct 29:1889–1900
61. Maugin GA, Trimarco C (1992) Pseudomomentum and material forces in nonlinear elasticity: variational formulations and application to brittle fracture. Acta Mech 94:1–28
62. Kalpakides VK, Maugin GA (2004) Canonical formulation and conservation laws of thermoelasticity without dissipation. Rep Math Phys 53:371–391
63. Maugin GA, Kalpakides VK (2002) A Hamiltonian formulation for elasticity and thermoelasticity. J Phys A: Math Gen 35:10775
64. Steinmann P (2002) On spatial and material settings of hyperelastodynamics. Acta Mech 156:193–218
65. Steinmann P (2008) On boundary potential energies in deformational and configurational mechanics. J Mech Phys Solids 56:772–800
66. Yavari A, Marsden JE, Ortiz M (2006) On spatial and material covariant balance laws in elasticity. J Math Phys 47:042903
67. Lazar M, Kirchner H (2006) The Eshelby tensor in nonlocal elasticity and in nonlocal micropolar elasticity. J Mech Mater Struct 1:325–337
68. Lazar M, Kirchner H (2007) The Eshelby stress tensor, angular momentum tensor and dilatation flux in gradient elasticity. Int J Solids Struct 44:2477–2486
69. Lazar M, Maugin GA (2007) On microcontinuum field theories: the Eshelby stress tensor and incompatibility conditions. Phil Mag 87:3853–3870
70. Gupta A, Markenscoff X (2008) Configurational forces as dissipative mechanisms: a revisit. CR Mec 336:126–131
71. Gupta A, Markenscoff X (2012) A new interpretation of configurational forces. J Elast 108:225–228
72. Dascalu C, Maugin GA (1993) Material forces and energy-release rates in homogeneous elastic bodies with defects. Comptes Rendus de l'Académie des Sciences II(317):1135–1140
73. Steinmann P, Scherer M, Denzer R (2009) Secret and joy of configurational mechanics: from foundations in continuum mechanics to applications in computational mechanics. ZAMM - J Appl Math Mech 89:614–630

74. Noether E (1918) Invariante Variationsprobleme. Nachrichten von der Königlichen Gesellschaft der Wissenschaften zu Göttingen, Mathematisch-Physikalische Klasse 2:235–257
75. Sanders JL (1960) On the Griffith-Irwin fracture theory. J Appl Mech 27:352–353
76. Günther W (1962) Über einige Randintegrale der Elastomechanik. Abh Braunschw Wiss Ges 14:53–72
77. Fletcher DC (1976) Conservation laws in linear elastodynamics. Arch Ration Mech Anal 60:329–353
78. Buggisch H, Gross D, Krüger KH (1981) Einige Erhaltungssätze der Kontinuumsmechanik vom J-Integral-Typ. Ingenieur-Archiv 50:103–111
79. Francfort G, Golebiewska-Herrmann A (1982) Conservation laws and material momentum in thermoelasticity. J Appl Mech 49:710–714
80. Francfort G, Golebiewska-Herrmann A (1986) A contour integral and an energy release rate in thermoelasticity. Int J Solids Struct 22:759–766
81. Olver PJ (1984) Conservation laws in elasticity. Arch Ration Mech Anal 85:111–129
82. Cherepanov GP (1989) A remark on the dynamic invariant or path-independent integral. Int J Solids Struct 25:1267–1269
83. Simo JC, Honein T (1990) Variational formulation, discrete conservation laws, and path-domain independent integrals for elasto-viscoplasticity. J Appl Mech 57:488–497
84. Honein T, Herrmann G (1997) Conservation laws in nonhomogeneous plane elastostatics. J Mech Phys Solids 45:789–805
85. Shield RT (1967) Inverse deformation results in finite elasticity. ZAMP: Zeitschrift für angewandte Mathematik und Physik 18:490–500
86. Chadwick P (1975) Applications of an energy-momentum tensor in non-linear elastostatics. J Elast 5:249–258
87. Govindjee S, Mihalic PA (1996) Computational methods for inverse finite elastostatics. Comput Methods Appl Mech Eng 136:47–57
88. Govindjee S, Mihalic PA (1998) Computational methods for inverse deformations in quasi-incompressible finite elasticity. Int J Numer Meth Eng 43:821–838
89. Kalpakides VK, Balassas KG (2005) The inverse deformation mapping in the finite element method. Phil Mag 85:4257–4275
90. Kuhl E, Askes H, Steinmann P (2006) An illustration of the equivalence of the loss of ellipticity conditions in spatial and material settings of hyperelasticity. Eur J Mech-A/Solids 25:199–214
91. Kalpakides VK, Agiasofitou E (2002) On material equations in second gradient electroelasticity. J Elasticity Phys Sci Solids 67:205–227
92. Kirchner N, Steinmann P (2007) On the material setting of gradient hyperelasticity. Math Mech Solids 12:559–580
93. Steinmann P, Ricker S, Aifantis E (2011) Unconstrained and Cauchy-Born-constrained atomistic systems: deformational and configurational mechanics. Arch Appl Mech 81:669–684
94. Birang SE, Steinmann P (2021) Discrete configurational mechanics for the computational study of atomistic fracture mechanics. Forces Mech 2:100009
95. Steinmann P (2000) Application of material forces to hyperelastostatic fracture mechanics. I. Continuum mechanical setting. Int J Solids Struct 37:7371–7391
96. Steinmann P (2002) On spatial and material settings of thermo-hyperelastodynamics. J Elast 66:109–157
97. Dascalu C, Maugin GA (1995) The thermoelastic material-momentum equation. J Elast 39:201–212
98. Epstein M, Maugin GA (2000) Thermomechanics of volumetric growth in uniform bodies. Int J Plast 16:951–978
99. Kuhl E, Steinmann P (2003) On spatial and material settings of thermo-hyperelastodynamics for open systems. Acta Mech 160:179–217
100. Kuhl E, Steinmann P (2003) Theory and numerics of geometrically non-linear open system mechanics. Int J Numer Meth Eng 58:1593–1615

101. Kuhl E, Steinmann P (2004) Computational modeling of healing: an application of the material force method. Biomech Model Mechanobiol 2:187–203
102. Kuhl E, Steinmann P (2004) Material forces in open system mechanics. Comput Methods Appl Mech Eng 193:2357–2381
103. Steinmann P, McBride A, Bargmann S, Javili A (2012) A deformational and configurational framework for geometrically non-linear continuum thermomechanics coupled to diffusion. Int J Non-Linear Mech 47:215–227
104. Quiligotti S, Maugin GA, Dell'Isola F (2003) An Eshelbian approach to the nonlinear mechanics of constrained solid-fluid mixtures. Acta Mech 160:45–60
105. Papastavrou A, Steinmann P (2010) On deformational and configurational poro-mechanics: dissipative versus non-dissipative modelling of two-phase solid/fluid mixtures. Arch Appl Mech 80:969–984
106. Vu DK, Steinmann P (2007) Nonlinear electro-and magneto-elastostatics: material and spatial settings. Int J Solids Struct 44:7891–7905
107. Vu DK, Steinmann P (2010) Material and spatial motion problems in nonlinear electro-and magneto-elastostatics. Math Mech Solids 15:239–257
108. Vu DK, Steinmann P (2012) On the spatial and material motion problems in nonlinear electro-elastostatics with consideration of free space. Math Mech Solids 17:803–823
109. Menzel A, Steinmann P (2005) A note on material forces in finite inelasticity. Arch Appl Mech 74:800–807
110. Menzel A, Steinmann P (2007) On configurational forces in multiplicative elastoplasticity. Int J Solids Struct 44:4442–4471
111. Tillberg J, Larsson F, Runesson K (2010) On the role of material dissipation for the crack-driving force. Int J Plast 26:992–1012
112. Özenç K, Kaliske M, Lin G, Bhashyam G (2014) Evaluation of energy contributions in elasto-plastic fracture: a review of the configurational force approach. Eng Fract Mech 115:137–153
113. Cermelli P, Fried E (1997) The influence of inertia on configurational forces in a deformable solid. Proc R Soc Lond. Ser A: Math, Phys Eng Sci 453:1915–1927
114. Mariano PM (2000) Configurational forces in continua with microstructure. ZAMP: Zeitschrift für angewandte Mathematik und Physik 51:752–791
115. Kalpakides VK, Dascalu C (20002) On the configurational force balance in thermomechanics. Proc R Soc Lond. Ser A: Math, Phys Eng Sci 458:3023–3039
116. Podio-Guidugli P (2001) Configurational balances via variational arguments. Interfaces Free Bound 3:223–232
117. Podio-Guidugli P (2002) Configurational forces: are they needed? Mech Res Commun 29:513–519
118. Fried E, Gurtin M (2003) The role of the configurational force balance in the nonequilibrium epitaxy of films. J Mech Phys Solids 51:487–517
119. Eshelby JD (1975) The elastic energy-momentum tensor. J Elast 5:321–335
120. Hill R (1986) Energy-momentum tensors in elastostatics: some reflections on the general theory. J Mech Phys Solids 34:305–317
121. Epstein M, Maugin GA (1990) The energy-momentum tensor and material uniformity in finite elasticity. Acta Mech 83:127–133
122. Maugin GA, Epstein M (1991) The electroelastic energy–momentum tensor. Proc R Soc Lond. Ser A: Math, Phys Eng Sci 433:299–312
123. Maugin GA (2013) Continuum mechanics of electromagnetic solids. Elsevier, Amsterdam
124. Pelteret JP, Steinmann P (2019) Magneto-active polymers: fabrication, characterisation, modelling and simulation at the micro-and macro-scale. Walter de Gruyter
125. Kienzler R, Herrmann G (1997) On the properties of the Eshelby tensor. Acta Mech 125:73–91
126. Kienzler R, Herrmann G (2000) Mechanics in material space: with applications to defect and fracture mechanics. Springer, Berlin
127. Kienzler R, Herrmann G (2002) Fracture criteria based on local properties of the Eshelby tensor. Mech Res Commun 29:521–527

128. Brünig M (2004) Eshelby stress tensor in large strain anisotropic damage mechanics. Int J Mech Sci 46:1763–1782
129. Verron E, Le Cam JB, Gornet L (2006) A multiaxial criterion for crack nucleation in rubber. Mech Res Commun 33:493–498
130. Andriyana A, Verron E (2007) Prediction of fatigue life improvement in natural rubber using configurational stress. Int J Solids Struct 44:2079–2092
131. Verron E, Andriyana A (2008) Definition of a new predictor for multiaxial fatigue crack nucleation in rubber. J Mech Phys Solids 56:417–443
132. Verron E (2010) Configurational mechanics: a tool to investigate fracture and fatigue of rubber. Rubber Chem Technol 83:270–281
133. Previati G, Kaliske M (2012) Crack propagation in pneumatic tires: continuum mechanics and fracture mechanics approaches. Int J Fatigue 37:69–78
134. Ackermann D, Barth FJ, Steinmann P (1999) Theoretical and computational aspects of geometrically nonlinear problems in fracture mechanics. In: Proceedings (CD-ROM) of the European conference on computational mechanics ECCM'99 (ECCOMAS), August 31 to September 3, Munich, Germany
135. Steinmann P, Ackermann D, Barth FJ (2001) Application of material forces to hyperelastostatic fracture mechanics. II. Computational setting. Int J Solids Struct 38:5509–5526
136. Denzer R, Barth FJ, Steinmann P (2003) Studies in elastic fracture mechanics based on the material force method. Int J Numer Meth Eng 58:1817–1835
137. Denzer R, Scherer M, Steinmann P (2007) An adaptive singular finite element in nonlinear fracture mechanics. Int J Fract 147:181–190
138. Kuhl E, Denzer R, Barth FJ, Steinmann P (2004) Application of the material force method to thermo-hyperelasticity. Comput Methods Appl Mech Eng 193:3303–3325
139. Bargmann S, Denzer R, Steinmann P (2009) Material forces in non-classical thermo-hyperelasticity. J Therm Stresses 32:361–393
140. Liebe T, Denzer R, Steinmann P (2003) Application of the material force method to isotropic continuum damage. Comput Mech 30:171–184
141. Nguyen TD, Govindjee S, Klein PA, Gao H (2005) A material force method for inelastic fracture mechanics. J Mech Phys Solids 53:91–121
142. Näser B, Kaliske M, Müller R (2007) Material forces for inelastic models at large strains: application to fracture mechanics. Comput Mech 40:1005–1013
143. Menzel A, Denzer R, Steinmann P (2004) On the comparison of two approaches to compute material forces for inelastic materials. Application to single-slip crystal-plasticity. Comput Methods Appl Mech Eng 193:5411–5428
144. Menzel A, Denzer R, Steinmann P (2005) Material forces in computational single-slip crystal-plasticity. Comput Mater Sci 32:446–454
145. Kuhn C, Lohkamp R, Schneider F, Aurich J, Müller R (2015) Finite element computation of discrete configurational forces in crystal plasticity. Int J Solids Struct 56:62–77
146. Kuhn C, Müller R (2016) A discussion of fracture mechanisms in heterogeneous materials by means of configurational forces in a phase field fracture model. Comput Methods Appl Mech Eng 312:95–116
147. Hirschberger CB, Kuhl E, Steinmann P (2007) On deformational and configurational mechanics of micromorphic hyperelasticity-theory and computation. Comput Methods Appl Mech Eng 196:4027–4044
148. Floros D, Larsson F, Runesson K (2018) On configurational forces for gradient-enhanced inelasticity. Comput Mech 61:409–432
149. Kolling S, Müller R (2005) On configurational forces in short-time dynamics and their computation with an explicit solver. Comput Mech 35:392–399
150. Timmel M, Kaliske M, Kolling S, Müller R (2011) On configurational forces in hyperelastic materials under shock and impact. Comput Mech 47:93–104
151. Denzer R, Menzel A (2014) Configurational forces for quasi-incompressible large strain electro-viscoelasticity-application to fracture mechanics. Eur J Mech-A/Solids 48:3–15

152. Ricker S, Mergheim J, Steinmann P (2009) On the multiscale computation of defect driving forces. Int J Multiscale Comput Eng 7:457–474
153. Ricker S, Mergheim J, Steinmann P, Müller R (2010) A comparison of different approaches in the multi-scale computation of configurational forces. Int J Fract 166:203–214
154. Braun M (1997) Configurational forces induced by finite-element discretization. Proc Estonian Acad Sci, Phys Math 46:24–31
155. Braun M (2007) Configurational forces in discrete elastic systems. Arch Appl Mech 77:85–93
156. Müller R, Maugin GA (2002) On material forces and finite element discretizations. Comput Mech 29:52–60
157. Gross D, Kolling S, Müller R, Schmidt I (2003) Configurational forces and their application in solid mechanics. Eur J Mech-A/Solids 22:669–692
158. Heintz P, Larsson F, Hansbo P, Runesson K (2004) Adaptive strategies and error control for computing material forces in fracture mechanics. Int J Numer Meth Eng 60:1287–1299
159. Müller R, Gross D, Maugin GA (2004) Use of material forces in adaptive finite element methods. Comput Mech 33:421–434
160. Müller R, Kolling S, Gross D (2002) On configurational forces in the context of the finite element method. Int J Numer Meth Eng 53:1557–1574
161. Thoutireddy P, Ortiz M (2004) A variational r-adaption and shape-optimization method for finite-deformation elasticity. Int J Numer Meth Eng 61:1–21
162. Mosler J, Ortiz M (2006) On the numerical implementation of variational arbitrary lagrangian-eulerian (VALE) formulations. Int J Numer Meth Eng 67:1272–1289
163. Mosler J, Ortiz M (2007) Variational h-adaption in finite deformation elasticity and plasticity. Int J Numer Meth Eng 72:505–523
164. Tabarraei A, Sukumar N (2007) Adaptive computations using material forces and residual-based error estimators on quadtree meshes. Comput Methods Appl Mech Eng 196:2657–2680
165. Scherer M, Denzer R, Steinmann P (2007) Energy-based r-adaptivity: a solution strategy and applications to fracture mechanics. Int J Fract 147:117–132
166. Scherer M, Denzer R, Steinmann P (2008) On a solution strategy for energy-based mesh optimization in finite hyperelastostatics. Comput Methods Appl Mech Eng 197:609–622
167. Rajagopal A, Sivakumar SM (2007) A combined rh adaptive strategy based on material forces and error assessment for plane problems and bimaterial interfaces. Comput Mech 41:49–72
168. Askes H, Bargmann S, Kuhl E, Steinmann P (2005) Structural optimization by simultaneous equilibration of spatial and material forces. Commun Numer Methods Eng 21:433–442
169. Materna D, Barthold FJ (2007) Variational design sensitivity analysis in the context of structural optimization and configurational mechanics. Int J Fract 147:133–155
170. Materna D, Barthold FJ (2008) On variational sensitivity analysis and configurational mechanics. Comput Mech 41:661–681
171. Materna D, Barthold FJ (2009) Configurational variations for the primal and dual problem in elasticity. ZAMM: Zeitschrift für Angewandte Mathematik und Mechanik 89:666–676
172. Riehl S, Steinmann P (2014) An integrated approach to shape optimization and mesh adaptivity based on material residual forces. Comput Methods Appl Mech Eng 278:640–663
173. Gross D, Müller R, Kolling S (2002) Configurational forces-morphology evolution and finite elements. Mech Res Commun 29:529–536
174. Kolling S, Baaser H, Gross D (2002) Material forces due to crack-inclusion interaction. Int J Fract 118:229–238
175. Kolling S, Müller R, Gross D (2003) A computational concept for the kinetics of defects in anisotropic materials. Comput Mater Sci 26:87–94
176. Timmel M, Kaliske M, Kolling S (2009) Modelling of microstructural void evolution with configurational forces. ZAMM: Zeitschrift für Angewandte Mathematik und Mechanik 89:698–708
177. Fagerström M, Larsson R (2006) Theory and numerics for finite deformation fracture modelling using strong discontinuities. Int J Numer Meth Eng 66:911–948
178. Fagerström M, Larsson R (2008) Approaches to dynamic fracture modelling at finite deformations. J Mech Phys Solids 56:613–639

179. Larsson R, Fagerström M (2005) A framework for fracture modelling based on the material forces concept with xfem kinematics. Int J Numer Meth Eng 62:1763–1788
180. Heintz P (2006) On the numerical modelling of quasi-static crack growth in linear elastic fracture mechanics. Int J Numer Meth Eng 65:174–189
181. Mahnken R (2007) Material forces for crack analysis of functionally graded materials in adaptively refined fe-meshes. Int J Fract 147:269–283
182. Mahnken R (2009) Geometry update driven by material forces for simulation of brittle crack growth in functionally graded materials. Int J Numer Meth Eng 77:1753–1788
183. Gürses E, Miehe C (2009) A computational framework of three-dimensional configurational-force-driven brittle crack propagation. Comput Methods Appl Mech Eng 198:1413–1428
184. Miehe C, Gürses E (2007) A robust algorithm for configurational-force-driven brittle crack propagation with r-adaptive mesh alignment. Int J Numer Meth Eng 72:127–155
185. Miehe C, Gürses E, Birkle M (2007) A computational framework of configurational-force-driven brittle fracture based on incremental energy minimization. Int J Fract 145:245–259
186. Schütte H (2009) Curved crack propagation based on configurational forces. Comput Mater Sci 46:642–646
187. Brouzoulis J, Larsson F, Runesson K (2011) Strategies for planar crack propagation based on the concept of material forces. Comput Mech 47:295–304
188. Özenç K, Chinaryan G, Kaliske M (2016) A configurational force approach to model the branching phenomenon in dynamic brittle fracture. Eng Fract Mech 157:26–42
189. Özenç K, Kaliske M (2014) An implicit adaptive node-splitting algorithm to assess the failure mechanism of inelastic elastomeric continua. Int J Numer Meth Eng 100(9):669–688
190. Kaczmarczyk Ł, Nezhad MM, Pearce C (2014) Three-dimensional brittle fracture: configurational-force-driven crack propagation. Int J Numer Meth Eng 97:531–550
191. Kaczmarczyk Ł, Ullah Z, Pearce C (2017) Energy consistent framework for continuously evolving 3d crack propagation. Comput Methods Appl Mech Eng 324:54–73
192. Bird R, Coombs W, Giani S (2018) A quasi-static discontinuous Galerkin configurational force crack propagation method for brittle materials. Int J Numer Meth Eng 113:1061–1080

Chapter 2
Kinematics in Bulk Volumes

8,051 m 35°48’42"N 76°33’54"E

Abstract This chapter recalls the pertinent spatial and material continuum kinematics in bulk volumes, thereby focusing on the nonlinear deformation maps, their associated tangent, cotangent and measure maps, and expresses their compatibility conditions.

Nonlinear continuum kinematics captures the intricate changes in the geometry of a deforming continuum body. It is thus pivotal to the phenomenological theory of nonlinear continuum mechanics, which aims to relate continuum kinematics and corresponding continuum kinetics. Thereby, the concept of *fields*, sufficiently smooth and differentiable (almost everywhere)–allowing to apply differential operators such as *gradient*, *curl*, and *divergence*–is the underpinning cornerstone of any continuum theory. The geometry of a deforming continuum body may thus be considered a *three-dimensional differentiable manifold* and concepts from differential geometry may be applied.
Zero-, one-, two-, and three-dimensional geometrical objects of interest in the kinematics of a continuum body are *points*, *curves*, *surfaces*, and *volumes*. The continuously distributed points take the interpretation as position vectors in Euclidean space and are described in terms of their three-dimensional coordinates (representing the chart of a manifold). Nonlinear (finite) changes in the position of the continuously distributed points define the deformation of a continuum body, a field. Differential (infinitesimal) line, area, and volume elements are point-wise representatives of their finite counterparts, denoted as maps, with their changes due to the deformation entirely capturing the evolving geometry of a continuum body.

The changes of differential line, area, and volume elements follow from the space gradient of the deformation field, i.e. the deformation gradient, its cofactor, and its determinant. These shall be denoted, respectively, as the *tangent*, the *cotangent*, and the *measure map*.

P. Steinmann, *Spatial and Material Forces in Nonlinear Continuum Mechanics*,
Solid Mechanics and Its Applications 272,
https://doi.org/10.1007/978-3-030-89070-4_2

2.1 Configurations

2.1.1 Reference Configuration

Euclidian Space:
The three-dimensional Euclidean space $\mathbb{E}^3$ allows for rectilinear (Cartesian) as well as for curvilinear coordinates, moreover, in practice no distinction is necessary between vectors and covectors since $\mathbb{E}^3$ coincides with both its tangent as well as its cotangent space (see below). In an Euclidean space the multiplication of a scalar with a vector $\boldsymbol{v} = \alpha \boldsymbol{u}$, the addition of vectors $\boldsymbol{w} = \boldsymbol{u} + \boldsymbol{v}$, and the scalar product of vectors $\boldsymbol{u} \cdot \boldsymbol{v}$ are axiomatically defined. Tensor calculus as instrumental for continuum mechanics can likewise be represented in symbolic (coordinate free) and coordinate notation.

From the viewpoint of continuum mechanics a *physical body* B is a set of cardinality infinity consisting of continuously distributed *physical points* P

$$B := \{P\} \quad \text{with} \quad \#B = \infty. \tag{2.1}$$

The concept of a so defined *continuum body* is an abstraction of the discontinuous composition of matter by, e.g. individual elementary particles, atoms, or molecules. To enable a mathematical treatment, a continuum body is *embedded* into the ordinary three-dimensional ambient space that is mathematically captured by the Euclidean space $\mathbb{E}^3$. The embedding may be achieved by associating the physical points P to *reference position vectors* $\boldsymbol{\Xi} \in \mathbb{E}^3$ via the abstract one-to-one map $\chi(P)$, i.e.

$$P \mapsto \boldsymbol{\Xi} = \chi(P) \in \mathbb{E}^3. \tag{2.2}$$

Then, the embedding of the complete physical body B into $\mathbb{E}^3$ defines its *reference configuration* $\mathcal{B}_\mathrm{r} \subset \mathbb{E}^3$, i.e.

$$B \mapsto \mathcal{B}_\mathrm{r} = \chi(B) \subset \mathbb{E}^3. \tag{2.3}$$

Thereby the reference position vectors $\boldsymbol{\Xi} \in \mathbb{E}^3$ and the reference configuration $\mathcal{B}_\mathrm{r} \subset \mathbb{E}^3$ are merely mathematical representations, respectively, of the physical points P and the physical body B. The process of embedding the continuum body into Euclidean space, see Fig. 2.1, is also denoted as its *placement*. The *reference placement* $\mathcal{B}_\mathrm{r} := \{\boldsymbol{\Xi}\} \subset \mathbb{E}^3$ needs not to be actually occupied by the continuum body at any time, thus the reference placement is introduced as stationary, i.e. independent of (absolute) time. Consequently, at any generic time (instant) $t \in \mathbb{R}_0^+$ the actual, i.e. *time-dependent* placement of a continuum body into $\mathbb{E}^3$ may be captured relative to its stationary reference placement.

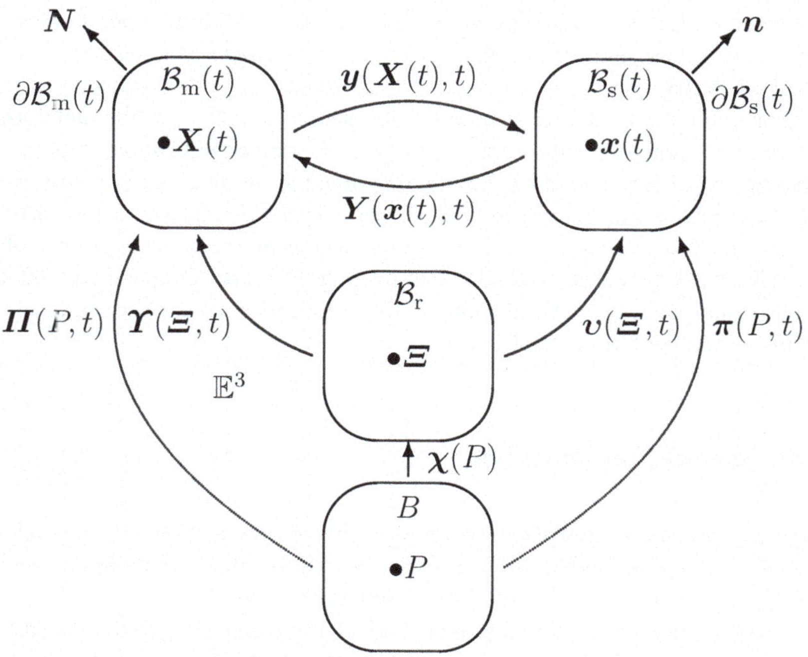

Fig. 2.1 A physical body B is embedded into Euclidean space $\mathbb{E}^3$ via its stationary placement into the reference configuration $\mathcal{B}_\mathrm{r} = \chi(B)$. The time-dependent placements into the material and the spatial configuration $\mathcal{B}_\mathrm{m}(t)$ and $\mathcal{B}_\mathrm{s}(t)$, respectively, are most conveniently captured relative to the stationary reference placement

Two distinct physically attainable configurations are considered in the sequel: the *material* and the *spatial configuration*, both being time-dependent. Note that the treatment of the material configuration as being time-dependent is specific to the here advocated approach.

2.1.2 Material Configuration

On the one hand, the continuum body may occupy the material configuration $\mathcal{B}_\mathrm{m}(t)$ consisting of all *material position vectors* $\boldsymbol{X}(t)$ whereby the *material placement* is given in terms of the map $\boldsymbol{\Upsilon} : \mathcal{B}_\mathrm{r} \times \mathbb{R}_0^+ \rightarrow \mathcal{B}_\mathrm{m}$

$$\mathcal{B}_\mathrm{m} := \{\boldsymbol{X}\} = \boldsymbol{\Upsilon}(\mathcal{B}_\mathrm{r}, t) \subset \mathbb{E}^3 \quad \text{with} \quad \boldsymbol{X} = \boldsymbol{\Upsilon}(\boldsymbol{\Xi}, t) \in \mathbb{E}^3. \tag{2.4}$$

Within the material configuration, $\mathfrak{B}_\mathrm{m} = \mathfrak{B}_\mathrm{m}(t)$ denotes a smoothly migrating time-dependent *material (bulk) control volume* such that $\mathfrak{B}_\mathrm{m} \subseteq \mathcal{B}_\mathrm{m}$. The boundary of $\mathfrak{B}_\mathrm{m}$, assumed regular (smooth) almost everywhere, is denoted by $\partial\mathfrak{B}_\mathrm{m}$ and is equipped with the outwards pointing unit normal $\boldsymbol{N}$. *Material volume elements* together with scalar- and vector-valued *material area elements* are given by $\mathrm{d}V$, $\mathrm{d}A$, and $\mathrm{d}\boldsymbol{A} := \boldsymbol{N}\,\mathrm{d}A$.

The material configuration at time $t = 0$ may be thought of as the undeformed configuration occupied by the continuum body before the application of any external loading (typically including gravity, thus making the material configuration a somewhat problematic concept in practice). For *reversible* material behavior (elasticity), the material configuration is also attained again after complete removal of the previously applied (conservative) external loading. However, as specific to the here advocated approach, the material configuration may not necessarily be stationary but may also evolve with time, see Chap. 4.

2.1.3 Spatial Configuration

In general, upon application of external loading the continuum body deforms and thus occupies a deformed configuration. Thereby the deformed configuration is denoted as the *spatial configuration* for the sake of semantic analogy.

Thus, on the other hand, the continuum body may occupy the spatial configuration $\mathcal{B}_\mathrm{s}(t)$ consisting of all *spatial position vectors* $\boldsymbol{x}(t)$ whereby the *spatial placement* is given in terms of the map $\boldsymbol{\upsilon} : \mathcal{B}_\mathrm{r} \times \mathbb{R}_0^+ \to \mathcal{B}_\mathrm{s}$

$$\mathcal{B}_\mathrm{s} := \{\boldsymbol{x}\} = \boldsymbol{\upsilon}(\mathcal{B}_\mathrm{r}, t) \subset \mathbb{E}^3 \quad \text{with} \quad \boldsymbol{x} = \boldsymbol{\upsilon}(\boldsymbol{\Xi}, t) \in \mathbb{E}^3. \tag{2.5}$$

Within the spatial configuration, $\mathfrak{B}_\mathrm{s} = \mathfrak{B}_\mathrm{s}(t)$ denotes the (convected) *spatial (bulk) control volume* such that $\mathfrak{B}_\mathrm{s} \subseteq \mathcal{B}_\mathrm{s}$. The boundary of $\mathfrak{B}_\mathrm{s}$ is denoted by $\partial\mathfrak{B}_\mathrm{s}$ and is equipped with the outwards pointing unit normal $\boldsymbol{n}$. *Spatial volume elements* together with scalar- and vector-valued *spatial area elements* are given by $\mathrm{d}v$, $\mathrm{d}a$, and $\mathrm{d}\boldsymbol{a} := \boldsymbol{n}\,\mathrm{d}a$.

The time dependence of the spatial configuration is obviously due to the application of external loading but may also indirectly be the consequence of an evolving material configuration, consider, e.g. the case of an instationary phase transformation front in a shape memory alloy and the resulting deformation due to the evolving transformation strains.

2.2 Deformation in Bulk Volumes

Geometry in Volumes:
A volume $\mathcal{V}$, i.e. a three-dimensional manifold embedded in the three-dimensional Euclidean space $\mathbb{E}^3$ with, say, coordinates $\boldsymbol{X}$ may be parameterized

by three curvilinear coordinates Θ^I with $I = 1, 2, 3$ as

$$X = X(\Theta^I). \tag{2.6}$$

The covariant (natural) basis vectors $G_I \in T\mathcal{V}$ are the tangent vectors to the curvilinear coordinate lines Θ^I

$$G_I = \partial_I X. \tag{2.7}$$

The contravariant (dual) basis vectors $G^I \in T^*\mathcal{V}$ follow from the orthonormality property $\delta^I{}_J = G^I \cdot G_J$. They are explicitly related to the covariant basis vectors G_I by the co- and contravariant metric coefficients G_{IJ} and G^{IJ}, respectively, as

$$\begin{aligned} G_I &= G_{IJ} G^J \quad \text{with} \quad G_{IJ} := G_I \cdot G_J = [G^{IJ}]^{-1}, \\ G^I &= G^{IJ} G_J \quad \text{with} \quad G^{IJ} := G^I \cdot G^J = [G_{IJ}]^{-1}. \end{aligned} \tag{2.8}$$

Consequently, the volume element $\mathrm{d}V$ follows as

$$\mathrm{d}V = G_1 \cdot [G_2 \times G_3] \, \mathrm{d}\Theta^1 \, \mathrm{d}\Theta^2 \, \mathrm{d}\Theta^3 = \sqrt{G} \, \mathrm{d}\Theta^1 \, \mathrm{d}\Theta^2 \, \mathrm{d}\Theta^3 \tag{2.9}$$

with the determinant of the covariant metric coefficients

$$G = \det[G_{IJ}]. \tag{2.10}$$

The unit tensor I of the three-dimensional, embedding Euclidean space follows as

$$I := [G^I \cdot G^J] \, G_I \otimes G_J = G_I \otimes G^I. \tag{2.11}$$

Moreover, the permutation tensor E of the three-dimensional, embedding Euclidean space follows as

$$E := \big[[G^I \times G^J] \cdot G^K\big] \, G_I \otimes G_J \otimes G_K. \tag{2.12}$$

Next, the gradient and divergence operators for vector-valued fields are defined as

$$\nabla_X(\bullet) := \partial_I(\bullet) \otimes G^I \quad \text{and} \quad \mathrm{Div}(\bullet) := \partial_I(\bullet) \cdot G^I. \tag{2.13}$$

In addition, the (vector-valued) curl operator for vector-valued fields is defined as

$$-\,\mathrm{Curl}(\bullet) := \partial_I(\bullet) \times G^I \tag{2.14}$$

The divergence and curl are alternatively expressed as the double contraction of the gradient with the unit tensor and the permutation tensor, respectively,

$$\begin{aligned} \mathrm{Div}(\bullet) &= \left[\partial_I(\bullet)\otimes \boldsymbol{G}^I\right] : \boldsymbol{I} = \nabla_X(\bullet) : \boldsymbol{I}\,, \\ -\mathrm{Curl}(\bullet) &= \left[\partial_I(\bullet)\otimes \boldsymbol{G}^I\right] : \boldsymbol{E} = \nabla_X(\bullet) : \boldsymbol{E}. \end{aligned} \tag{2.15}$$

The gradient, divergence, and curl operators for second-order tensor-valued fields $(\circ)$ follow from multiplying $(\circ)$ from the left by an arbitrary constant vector and inserting the so resulting vector-valued-field into the above definitions. Then, in $\mathcal{V}$ the product rule of differentiation $\mathrm{Div}\big((\bullet)\cdot(\circ)\big) = \nabla_X(\bullet) : (\circ) + (\bullet)\cdot \mathrm{Div}(\circ)$ holds in particular since

$$\partial_I\big[(\bullet)\cdot(\circ)\big]\cdot \boldsymbol{G}^I = [\partial_I(\bullet)\otimes \boldsymbol{G}^I] : (\circ) + (\bullet)\cdot \partial_I(\circ)\cdot \boldsymbol{G}^I. \tag{2.16}$$

Finally, the coordinate derivatives of the co- and contravariant basis vectors are expressed as

$$\partial_J \boldsymbol{G}_I = \Gamma^K{}_{IJ}\boldsymbol{G}_K \quad \text{and} \quad \partial_J \boldsymbol{G}^I = -\Gamma^I{}_{JK}\boldsymbol{G}^K. \tag{2.17}$$

Here $\Gamma^K{}_{IJ} = \partial_J \boldsymbol{G}_I \cdot \boldsymbol{G}^K = -\partial_J \boldsymbol{G}^K \cdot \boldsymbol{G}_I = \Gamma^K{}_{JI}$ denote the Christoffel symbols, symmetric in the index pair IJ.

Example: Consider a cylindrical coordinate system with Θ^I denoting the radius, the azimuth angle, and the axial length

$$\boldsymbol{X} = \Theta^1 \cos\Theta^2 \boldsymbol{E}_1 + \Theta^1 \sin\Theta^2 \boldsymbol{E}_2 + \Theta^3 \boldsymbol{E}_3. \tag{2.18}$$

Thus, the basis vectors compute with $G_{IJ} = \lceil 1, [\Theta^1]^2, 1\rfloor$ as

$$\boldsymbol{G}_1 = \cos\Theta^2 \boldsymbol{E}_1 + \sin\Theta^2 \boldsymbol{E}_2, \quad \boldsymbol{G}_2 = -\Theta^1 \sin\Theta^2 \boldsymbol{E}_1 + \Theta^1 \cos\Theta^2 \boldsymbol{E}_2 \tag{2.19}$$

and $\boldsymbol{G}_3 = \boldsymbol{G}^3 = \boldsymbol{E}_3$.

2.2.1 Nonlinear Deformation Maps

Composition of Maps The maps relating the reference position vectors to the material and spatial position vectors, respectively, are given by

$$\boldsymbol{X} = \boldsymbol{\Upsilon}(\boldsymbol{\Xi}, t) : \mathcal{B}_\mathrm{r} \times \mathbb{R}_0^+ \to \mathcal{B}_\mathrm{m} \quad \text{and} \quad \boldsymbol{x} = \boldsymbol{\upsilon}(\boldsymbol{\Xi}, t) : \mathcal{B}_\mathrm{r} \times \mathbb{R}_0^+ \to \mathcal{B}_\mathrm{s}. \tag{2.20}$$

It is noted in passing that by composition with the map $\boldsymbol{\Xi} = \boldsymbol{\chi}(P)$, i.e.

$$\boldsymbol{\Pi} := \boldsymbol{\Upsilon}(\boldsymbol{\Xi}, t) \circ \boldsymbol{\chi}(P) \quad \text{and} \quad \boldsymbol{\pi} := \boldsymbol{\upsilon}(\boldsymbol{\Xi}, t) \circ \boldsymbol{\chi}(P) \tag{2.21}$$

the material and spatial position vectors in the Euclidean space may also be given directly, albeit abstractly in terms of the physical points

$$X = \boldsymbol{\Pi}(P, t) : B \times \mathbb{R}_0^+ \to \mathcal{B}_\mathrm{m} \quad \text{and} \quad \boldsymbol{x} = \boldsymbol{\pi}(P, t) : B \times \mathbb{R}_0^+ \to \mathcal{B}_\mathrm{s}. \tag{2.22}$$

The suited composition of the maps $\boldsymbol{X} = \boldsymbol{\Upsilon}(\boldsymbol{\Xi}, t)$ or $\boldsymbol{X} = \boldsymbol{\Pi}(P, t)$ and $\boldsymbol{x} = \boldsymbol{\upsilon}(\boldsymbol{\Xi}, t)$ or $\boldsymbol{x} = \boldsymbol{\pi}(P, t)$ then eliminates either the reference position vector $\boldsymbol{\Xi} \in \mathbb{E}^3$

$$\boldsymbol{Y} := \boldsymbol{\Upsilon}(\boldsymbol{\Xi}, t) \circ \boldsymbol{\upsilon}^{-1} \quad \text{and} \quad \boldsymbol{y} := \boldsymbol{\upsilon}(\boldsymbol{\Xi}, t) \circ \boldsymbol{\Upsilon}^{-1} \tag{2.23}$$

or the physical point P, respectively,

$$\boldsymbol{Y} := \boldsymbol{\Pi}(P, t) \circ \boldsymbol{\pi}^{-1} \quad \text{and} \quad \boldsymbol{y} := \boldsymbol{\pi}(P, t) \circ \boldsymbol{\Pi}^{-1}. \tag{2.24}$$

Material and Spatial Deformation Maps

The material and spatial nonlinear *deformation maps*, respectively, thus follow as

$$\boldsymbol{X} = \boldsymbol{Y}(\boldsymbol{x}, t) : \mathcal{B}_\mathrm{s} \times \mathbb{R}_0^+ \to \mathcal{B}_\mathrm{m} \quad \text{and} \quad \boldsymbol{x} = \boldsymbol{y}(\boldsymbol{X}, t) : \mathcal{B}_\mathrm{m} \times \mathbb{R}_0^+ \to \mathcal{B}_\mathrm{s}. \tag{2.25}$$

The material and spatial deformation maps are one-to-one and in general nonlinear functions of their arguments, whereby at time $t = 0$ the deformation maps are simply the identity $\boldsymbol{Y}|_{t=0} = \boldsymbol{y}|_{t=0} = \boldsymbol{id}$, i.e. the map of a vector to itself. At any given time $t > 0$, they are inverse functions in the sense

$$\boldsymbol{Y}|_{t>0} = [\boldsymbol{y}|_{t>0}]^{-1}(\boldsymbol{x}) \quad \text{and} \quad \boldsymbol{y}|_{t>0} = [\boldsymbol{Y}|_{t>0}]^{-1}(\boldsymbol{X}). \tag{2.26}$$

The spatial deformation map uniquely relates position vectors $\boldsymbol{X}$ and $\boldsymbol{x}$ in the material and the spatial configuration. It represents the primal unknown field in conventional continuum mechanics, i.e. for given material configuration at time $t = 0$ and given external loading at time $t > 0$ typically the spatial deformation map and thus the spatial configuration are sought.

If there is no danger of confusion the explicit indication of the space-time parameterizations via the arguments $\{\boldsymbol{X}, t\}$ and $\{\boldsymbol{x}, t\}$ of any scalar-, vector- and tensor-valued function is omitted in the sequel for the sake of a more concise representation.

2.2.2 Tangent Maps: Deformation Gradients

> **Tangent and Cotangent Spaces:**
> The tangent space $T\mathcal{M}$ at a point of a generic manifold $\mathcal{M}$ is spanned by all vectors emanating from this point. Likewise, the cotangent space $T^*\mathcal{M}$ at a point of a general manifold $\mathcal{M}$ is spanned by all covectors emanating from this point. Under a change of coordinates that parameterize $\mathcal{M}$, vectors transform like coordinate differentials, whereas covectors transform like gradients. In Euclidean space (that may be parameterized by a Cartesian coordinate system with base vectors $\boldsymbol{E}_A \equiv \boldsymbol{E}^A = \boldsymbol{e}_a \equiv \boldsymbol{e}^a$), however, the corresponding tangent and the cotangent spaces are indistinguishable from a practical point of view.

Increments (total differentials) of the material and spatial position vectors (tangents) $\mathrm{d}\boldsymbol{X} \in T\mathcal{B}_\mathrm{m}$ and $\mathrm{d}\boldsymbol{x} \in T\mathcal{B}_\mathrm{s}$ are denoted as material and spatial *line elements*; they are related by spatial→material and material→spatial *tangent maps*

$$\mathrm{d}\boldsymbol{X} = \boldsymbol{f} \cdot \mathrm{d}\boldsymbol{x} \quad \text{and} \quad \mathrm{d}\boldsymbol{x} = \boldsymbol{F} \cdot \mathrm{d}\boldsymbol{X}. \tag{2.27}$$

Consequently, the spatial→material and material→spatial tangent maps follow as the spatial and material *gradients* of the material and spatial deformation maps, respectively,

$$\boldsymbol{f} := \nabla_x \boldsymbol{Y} \in T\mathcal{B}_\mathrm{m} \times T^*\mathcal{B}_\mathrm{s} \quad \text{and} \quad \boldsymbol{F} := \nabla_X \boldsymbol{y} \in T\mathcal{B}_\mathrm{s} \times T^*\mathcal{B}_\mathrm{m}. \tag{2.28}$$

Hence, $\boldsymbol{f} : T\mathcal{B}_\mathrm{s} \to T\mathcal{B}_\mathrm{m}$ and $\boldsymbol{F} : T\mathcal{B}_\mathrm{m} \to T\mathcal{B}_\mathrm{s}$ are also denoted as the material and spatial *deformation gradients*. Here, $T\mathcal{B}_\mathrm{m}$ and $T^*\mathcal{B}_\mathrm{m}$ together with $T\mathcal{B}_\mathrm{s}$ and $T^*\mathcal{B}_\mathrm{s}$ are the *tangent* and *cotangent spaces* to $\mathcal{B}_\mathrm{m}$ and $\mathcal{B}_\mathrm{s}$, respectively. Clearly, the deformation gradients are (mixed-variant) *two-point tensors* since they map vectors from different tangent spaces into each other. In curvilinear, convected coordinate representation they follow as

$$\boldsymbol{f} = \boldsymbol{G}_I \otimes \boldsymbol{g}^i \quad \text{and} \quad \boldsymbol{F} = \boldsymbol{g}_i \otimes \boldsymbol{G}^I. \tag{2.29}$$

At any given time and position (whereby the position vectors are related by the deformation maps) the material and spatial deformation gradients are algebraically *inverse* tensors

$$\boldsymbol{f}|_{t,x} = [\boldsymbol{F}|_{t,X=Y(x,t)}]]^{-1} \quad \text{and} \quad \boldsymbol{F}|_{t,X} = [\boldsymbol{f}|_{t,x=y(X,t)}]]^{-1}. \tag{2.30}$$

An indication of the space-time parametrization engraved in the above is given by $\boldsymbol{f} = \boldsymbol{F}^{-1} \circ \boldsymbol{Y}(\boldsymbol{x}, t)$ and $\boldsymbol{F} = \boldsymbol{f}^{-1} \circ \boldsymbol{y}(\boldsymbol{X}, t)$ or more explicitly

$$\boldsymbol{f} = \boldsymbol{F}^{-1}(\boldsymbol{Y}(\boldsymbol{x}, t), t) \quad \text{and} \quad \boldsymbol{F} = \boldsymbol{f}^{-1}(\boldsymbol{y}(\boldsymbol{X}, t), t). \tag{2.31}$$

Recall, however, that as a rule we will refrain from explicitly indicating the space-time parametrization as long as there is no danger of confusion.

In particular at time $t = 0$ the deformation gradients simply coincide with the second-order identity tensor $\boldsymbol{f}|_{t=0} = \boldsymbol{F}|_{t=0} = \boldsymbol{I}$ (here $\boldsymbol{I}$ is considered the appropriate two-point (mixed-variant) unit tensor).

2.2.3 Compatibility Conditions for Tangent Maps

Due to their definitions as the gradients of the deformation maps, the material and spatial tangent maps (deformation gradients) satisfy the following symmetry or rather *compatibility conditions*

> **Curl:**
> In Cartesian coordinate representation, the material curl operator $\text{Curl}(\bullet)$ of the (two-point, mixed-variant) tensor $\boldsymbol{Z} = Z^a{}_B \boldsymbol{e}_a \otimes \boldsymbol{E}^B$ expands as $\text{Curl}\boldsymbol{Z} := -Z^a{}_{B,C} E^{BC}{}_D \boldsymbol{e}_a \otimes \boldsymbol{E}^D$, thus for $Z^a{}_{B,C}$ symmetric in the index pair BC, the material curl $\text{Curl}\boldsymbol{Z}$ vanishes identically.

$$\text{curl}\boldsymbol{f} = \boldsymbol{0} \quad \text{and} \quad \text{Curl}\boldsymbol{F} = \boldsymbol{0}. \tag{2.32}$$

Observe that these compatibility conditions are consistent with the *(tangential) line theorem* equating the circuit integral of the vector-valued (tangential) line elements $\text{d}\boldsymbol{X}$ and $\text{d}\boldsymbol{x}$, respectively, over any closed circuit $\mathfrak{C}_\text{m} = \partial\mathfrak{A}_\text{m}$ and $\mathfrak{C}_\text{s} = \partial\mathfrak{A}_\text{s}$, respectively, to zero

$$\oint_{\mathfrak{C}_\text{m}} \text{d}\boldsymbol{X} = \oint_{\mathfrak{C}_\text{s}} \boldsymbol{f} \cdot \text{d}\boldsymbol{x} = \boldsymbol{0} \quad \text{and} \quad \oint_{\mathfrak{C}_\text{s}} \text{d}\boldsymbol{x} = \oint_{\mathfrak{C}_\text{m}} \boldsymbol{F} \cdot \text{d}\boldsymbol{X} = \boldsymbol{0}. \tag{2.33}$$

Applying the Stokes theorem

$$\oint_{\mathfrak{C}_\text{s}} \boldsymbol{f} \cdot \text{d}\boldsymbol{x} = \int_{\mathfrak{A}_\text{s}} \text{curl}\boldsymbol{f} \cdot \text{d}\boldsymbol{a} \quad \text{and} \quad \oint_{\mathfrak{C}_\text{m}} \boldsymbol{F} \cdot \text{d}\boldsymbol{X} = \int_{\mathfrak{A}_\text{m}} \text{Curl}\boldsymbol{F} \cdot \text{d}\boldsymbol{A}. \tag{2.34}$$

and localizing for arbitrary closed circuits establishes the result.

The compatibility conditions may be stated alternatively as

$$[\boldsymbol{A} \cdot \nabla_x \boldsymbol{f}]^{\text{skw}} = \boldsymbol{0} \quad \text{and} \quad [\boldsymbol{a} \cdot \nabla_X \boldsymbol{F}]^{\text{skw}} = \boldsymbol{0}. \tag{2.35}$$

Here, $\boldsymbol{A} \in T^*\mathcal{B}_\text{m}$ and $\boldsymbol{a} \in T^*\mathcal{B}_\text{s}$ denote arbitrary material and spatial (constant) covectors, respectively.

In Cartesian coordinate representation for the tangent maps $\boldsymbol{f} = f^A{}_b \boldsymbol{E}_A \otimes \boldsymbol{e}^b$ and $\boldsymbol{F} = F^a{}_B \boldsymbol{e}_a \otimes \boldsymbol{E}^B$, respectively, the above compatibility conditions read $f^A{}_{b,c} = f^A{}_{c,b}$ and $F^a{}_{B,C} = F^a{}_{C,B}$, respectively.

Consequently, for double contractions with (simple) second-order two-point tensors $\boldsymbol{z} := \boldsymbol{A} \otimes \boldsymbol{b}$ and $\boldsymbol{Z} := \boldsymbol{a} \otimes \boldsymbol{B}$, whereby $\boldsymbol{b} \in T\mathcal{B}_s$ and $\boldsymbol{B} \in T\mathcal{B}_m$ denote arbitrary spatial and material vectors, respectively, due to the compatibility conditions it holds that

$$\boldsymbol{z} : \nabla_x \boldsymbol{f} = \nabla_x \boldsymbol{f}^t : \boldsymbol{z} \quad \text{and} \quad \boldsymbol{Z} : \nabla_X \boldsymbol{F} = \nabla_X \boldsymbol{F}^t : \boldsymbol{Z}. \tag{2.36}$$

The extension of the above to arbitrary second-order two-point tensors $\boldsymbol{z}$ and $\boldsymbol{Z}$ is straightforward.

Clearly the compatibility conditions are rooted in the *commutation rule of partial differentiation* that leads here to the symmetry of second partial space derivatives.

2.2.4 Cotangent Maps: Cofactors

Cofactor:
The cofactor $\operatorname{cof} \boldsymbol{Z}$ of a second-order tensor $\boldsymbol{Z} \in \mathbb{E}^3 \times \mathbb{E}^3$ follows via the *tensor-valued tensor double product* (de Boer [1], Steinmann [2]) as

$$2! \operatorname{cof} \boldsymbol{Z} := \boldsymbol{Z} \times\!\!\times \boldsymbol{Z}.$$

Thereby, $\boldsymbol{Z} \times\!\!\times \boldsymbol{Z}$ is defined in Cartesian coordinates with coordinate representation for $\boldsymbol{Z} = Z^a{}_D \boldsymbol{e}_a \otimes \boldsymbol{E}^D$ as

$$\boldsymbol{Z} \times\!\!\times \boldsymbol{Z} := e_{abc} E^{DEF} Z^a{}_D Z^b{}_E \boldsymbol{e}^c \otimes \boldsymbol{E}_F.$$

Here, the spatial and material permutation symbols compute as $e_{abc} := [\boldsymbol{e}_a \times \boldsymbol{e}_b] \cdot \boldsymbol{e}_c$ and $E^{DEF} := [\boldsymbol{E}^D \times \boldsymbol{E}^E] \cdot \boldsymbol{E}^F$, respectively.

Symbolically, based on the above coordinate definition, the cofactor $\operatorname{cof} \boldsymbol{Z}$ of a second-order tensor $\boldsymbol{Z} \in \mathbb{E}^3 \times \mathbb{E}^3$ thus reads as

$$2! \operatorname{cof} \boldsymbol{Z} := [\boldsymbol{e} \boxtimes \boldsymbol{E}] :: [\boldsymbol{Z} \otimes \boldsymbol{Z}].$$

Here, $\boldsymbol{e}$ and $\boldsymbol{E}$ denote the third-order spatial and material permutation tensors with Cartesian coordinate representation $\boldsymbol{e} := e_{abc} \boldsymbol{e}^a \otimes \boldsymbol{e}^b \otimes \boldsymbol{e}^c$ and $\boldsymbol{E} := E^{DEF} \boldsymbol{E}_D \otimes \boldsymbol{E}_E \otimes \boldsymbol{E}_F$, respectively, with special dyadic product $\boldsymbol{e} \boxtimes \boldsymbol{E} := e_{abc} E^{DEF} \boldsymbol{e}^a \otimes \boldsymbol{E}_D \otimes \boldsymbol{e}^b \otimes \boldsymbol{E}_E \otimes \boldsymbol{e}^c \otimes \boldsymbol{E}_F$.

Alternatively, for linear independent and non-zero vectors $\boldsymbol{A}, \boldsymbol{B} \in \mathbb{E}^3$, the cofactor $\operatorname{cof} \boldsymbol{Z}$ of a second-order tensor $\boldsymbol{Z} \in \mathbb{E}^3 \times \mathbb{E}^3$ is defined via

$$\operatorname{cof} \boldsymbol{Z} \cdot [\boldsymbol{A} \times \boldsymbol{B}] := [\boldsymbol{Z} \cdot \boldsymbol{A}] \times [\boldsymbol{Z} \cdot \boldsymbol{B}] \quad \forall \boldsymbol{A}, \boldsymbol{B} \quad \in \mathbb{E}^3.$$

Here, the *vector-valued vector double products* $\boldsymbol{A} \times \boldsymbol{B}$ and $[\boldsymbol{Z} \cdot \boldsymbol{A}] \times [\boldsymbol{Z} \cdot \boldsymbol{B}]$ follow in Cartesian coordinates as $\boldsymbol{A} \times \boldsymbol{B} = E_{GHI} A^G B^H \boldsymbol{E}^I$ and $[\boldsymbol{Z} \cdot \boldsymbol{A}] \times [\boldsymbol{Z} \cdot \boldsymbol{B}] = e_{abc} Z^a{}_D A^D Z^b{}_E B^E \boldsymbol{e}^c$, respectively.

Finally, for an invertible second-order tensor $\boldsymbol{Z} \in \mathbb{E}^3 \times \mathbb{E}^3$, the cofactor $\operatorname{cof} \boldsymbol{Z}$ takes the closed-form expression

$$\operatorname{cof} \boldsymbol{Z} = \det \boldsymbol{Z}\, \boldsymbol{Z}^{-t} \in \mathbb{E}^3 \times \mathbb{E}^3.$$

Here, $\det \boldsymbol{Z}$ denotes the determinant of $\boldsymbol{Z}$, see below.

Material and spatial vector-valued oriented *area elements* (cotangents) $\mathrm{d}\boldsymbol{A} = \boldsymbol{N}\,\mathrm{d}A$ and $\mathrm{d}\boldsymbol{a} = \boldsymbol{n}\,\mathrm{d}a$ in the cotangent spaces $T^*\mathcal{B}_\mathrm{m}$ and $T^*\mathcal{B}_\mathrm{s}$, respectively, are related via the *cofactors* of the deformation gradients *(Nanson's formulae)*

$$\mathrm{d}\boldsymbol{A} = \operatorname{cof} \boldsymbol{f} \cdot \mathrm{d}\boldsymbol{a} \quad \text{and} \quad \mathrm{d}\boldsymbol{a} = \operatorname{cof} \boldsymbol{F} \cdot \mathrm{d}\boldsymbol{A}. \tag{2.37}$$

The cofactors of the deformation gradients are also denoted as the *area cotangent maps*

$$\boldsymbol{k} := \operatorname{cof} \boldsymbol{f} \in T^*\mathcal{B}_\mathrm{m} \times T\mathcal{B}_\mathrm{s} \quad \text{and} \quad \boldsymbol{K} := \operatorname{cof} \boldsymbol{F} \in T^*\mathcal{B}_\mathrm{s} \times T\mathcal{B}_\mathrm{m}. \tag{2.38}$$

The material and spatial cotangent maps are inverse tensors with

$$\boldsymbol{k} = \boldsymbol{K}^{-1} \quad \text{and} \quad \boldsymbol{K} = \boldsymbol{k}^{-1}. \tag{2.39}$$

2.2.5 Compatibility Conditions for Cotangent Maps

Divergence of Cofactor:
In Cartesian coordinate representation, the material divergence operator $\operatorname{Div}(\bullet)$ applied to the tensor-valued tensor double product $\boldsymbol{Z} \times\!\!\times \boldsymbol{Z}$ of the (two-point, mixed-variant) tensor $\boldsymbol{Z} = Z^a{}_B \boldsymbol{e}_a \otimes \boldsymbol{E}^B$ expands as $\operatorname{Div}(\boldsymbol{Z} \times\!\!\times \boldsymbol{Z}) := e_{abc} E^{DEF} [Z^a{}_{D,F} Z^b{}_E + Z^a{}_D Z^b{}_{E,F}] \boldsymbol{e}^c$, thus for $Z^a{}_{B,C}$ symmetric in the index pair BC, the material divergence $\operatorname{Div}(\boldsymbol{Z} \times\!\!\times \boldsymbol{Z})$ vanishes identically.

Due to their definitions as the tensor-valued tensor double products of the *compatible* deformation gradients, the material and spatial cotangent maps satisfy the following solenoidal or rather *compatibility conditions (Piola identity)*

$$\operatorname{div} \boldsymbol{k} = \boldsymbol{0} \quad \text{and} \quad \operatorname{Div} \boldsymbol{K} = \boldsymbol{0}. \tag{2.40}$$

Observe that these compatibility conditions are consistent with the *area theorem* equating the integral of the vector-valued area elements $\mathrm{d}\boldsymbol{A}$ and $\mathrm{d}\boldsymbol{a}$, respectively, over the closed boundary surface of a control volume $\mathfrak{B}_\mathrm{m} \subset \mathcal{B}_\mathrm{m}$ and $\mathfrak{B}_\mathrm{s} \subset \mathcal{B}_\mathrm{s}$, respectively, to zero

$$\int_{\partial\mathfrak{B}_\mathrm{m}} \mathrm{d}\boldsymbol{A} = \int_{\partial\mathfrak{B}_\mathrm{s}} \boldsymbol{k} \cdot \mathrm{d}\boldsymbol{a} = \boldsymbol{0} \quad \text{and} \quad \int_{\partial\mathfrak{B}_\mathrm{s}} \mathrm{d}\boldsymbol{a} = \int_{\partial\mathfrak{B}_\mathrm{m}} \boldsymbol{K} \cdot \mathrm{d}\boldsymbol{A} = \boldsymbol{0}. \tag{2.41}$$

Applying the Gauss theorem and localizing for arbitrary control volumes establishes the result.

2.2.6 Measure Maps: Determinants

Determinant:
The determinant $\det \boldsymbol{Z}$ of a second-order tensor $\boldsymbol{Z} \in \mathbb{E}^3 \times \mathbb{E}^3$ follows via the *scalar-valued tensor triple product* (de Boer [1], Steinmann [2]) as

$$3! \det \boldsymbol{Z} := [\boldsymbol{Z} \times\!\!\times \boldsymbol{Z}] : \boldsymbol{Z}.$$

Thereby, $[\boldsymbol{Z} \times\!\!\times \boldsymbol{Z}] : \boldsymbol{Z}$ is defined in Cartesian coordinates with coordinate representation for $\boldsymbol{Z} = Z^a{}_D \boldsymbol{e}_a \otimes \boldsymbol{E}^D$ as

$$[\boldsymbol{Z} \times\!\!\times \boldsymbol{Z}] : \boldsymbol{Z} := e_{abc} E^{DEF} Z^a{}_D Z^b{}_E Z^c{}_F.$$

Symbolically, based on the above coordinate definition, the determinant $\det \boldsymbol{Z}$ of a second-order tensor $\boldsymbol{Z} \in \mathbb{E}^3 \times \mathbb{E}^3$ thus reads as

$$3! \det \boldsymbol{Z} := [\boldsymbol{e} \boxtimes \boldsymbol{E}] \,\vdots\!\vdots\!\vdots\, [\boldsymbol{Z} \otimes \boldsymbol{Z} \otimes \boldsymbol{Z}].$$

Alternatively, for linear independent and non-zero vectors $\boldsymbol{A}, \boldsymbol{B}, \boldsymbol{C} \in \mathbb{E}^3$ the determinant $\det \boldsymbol{Z}$ of a second-order tensor $\boldsymbol{Z} \in \mathbb{E}^3 \times \mathbb{E}^3$ is defined via

$$\det \boldsymbol{Z}\big[\boldsymbol{A} \cdot [\boldsymbol{B} \times \boldsymbol{C}]\big] := [\boldsymbol{Z} \cdot \boldsymbol{A}] \cdot \big[[\boldsymbol{Z} \cdot \boldsymbol{B}] \times [\boldsymbol{Z} \cdot \boldsymbol{C}]\big] \quad \forall \boldsymbol{A}, \boldsymbol{B}, \boldsymbol{C} \in \mathbb{E}^3.$$

Here, the *scalar-valued vector triple products* $\boldsymbol{A} \cdot [\boldsymbol{B} \times \boldsymbol{C}]$ and $[\boldsymbol{Z} \cdot \boldsymbol{A}] \cdot \big[[\boldsymbol{Z} \cdot \boldsymbol{B}] \times [\boldsymbol{Z} \cdot \boldsymbol{C}]\big]$ follow in Cartesian coordinates as $\boldsymbol{A} \cdot [\boldsymbol{B} \times \boldsymbol{C}] = E_{GHI} A^G B^H C^I$ and $[\boldsymbol{Z} \cdot \boldsymbol{A}] \cdot \big[[\boldsymbol{Z} \cdot \boldsymbol{B}] \times [\boldsymbol{Z} \cdot \boldsymbol{C}]\big] = e_{jkl} [Z^j{}_G A^G][Z^k{}_H B^H][Z^l{}_I C^I]$, respectively.

Finally, for an invertible second-order tensor $\boldsymbol{Z} \in \mathbb{E}^3 \times \mathbb{E}^3$, the determinant $\det \boldsymbol{Z}$ takes the closed-form expression

$$3 \det \boldsymbol{Z} = \widehat{\operatorname{cof}} \boldsymbol{Z} : \boldsymbol{Z}.$$

Material and spatial scalar-valued *volume elements* (measures) $\mathrm{d}V$ and $\mathrm{d}v$ in $\mathcal{B}_\mathrm{m}$ and $\mathcal{B}_\mathrm{s}$, respectively, are related via the *determinants* of the deformation gradients

$$\mathrm{d}V = \det \boldsymbol{f}\, \mathrm{d}v \quad \text{and} \quad \mathrm{d}v = \det \boldsymbol{F}\, \mathrm{d}V. \tag{2.42}$$

The determinants of the deformation gradients are also denoted as the *Jacobians* or rather the *volume measure maps*

$$j = J^{-1} := \det \boldsymbol{f} \in \mathbb{R}^+ \quad \text{and} \quad J = j^{-1} := \det \boldsymbol{F} \in \mathbb{R}^+. \tag{2.43}$$

The Jacobians are restricted to $\mathbb{R}^+$ to ensure invertibility of the deformation maps ($j, J \neq 0$) and to avoid self-penetration ($j, J > 0$). Alternatively, the Jacobians may be expressed as

$$\frac{[\boldsymbol{F} \cdot \boldsymbol{G}_1] \cdot [[\boldsymbol{F} \cdot \boldsymbol{G}_2] \times [\boldsymbol{F} \cdot \boldsymbol{G}_3]]}{\boldsymbol{G}_1 \cdot [\boldsymbol{G}_2 \times \boldsymbol{G}_3]} \quad \text{and} \quad \frac{[\boldsymbol{f} \cdot \boldsymbol{g}_1] \cdot [[\boldsymbol{f} \cdot \boldsymbol{g}_2] \times [\boldsymbol{f} \cdot \boldsymbol{g}_3]]}{\boldsymbol{g}_1 \cdot [\boldsymbol{g}_2 \times \boldsymbol{g}_3]}. \tag{2.44}$$

Clearly, these expressions correspond to a direct computation of the volume elements via the scalar-valued vector triple product.

2.2.7 Strain Measures

The deformation gradients may conceptually serve as strain measures, however, due to reasons of objectivity (spatial isotropy) or rather invariance under rigid body motions superposed onto the spatial configuration, proper strain measures that are functions of the deformation gradients have to be used in practice. As an example, among infinitely many alternatives, the *right Cauchy-Green*-type strain measures may serve this purpose

$$\boldsymbol{c} := \boldsymbol{f}^t \cdot \boldsymbol{f} \in T^*\mathcal{B}_\mathrm{s} \times T^*\mathcal{B}_\mathrm{s} \quad \text{and} \quad \boldsymbol{C} := \boldsymbol{F}^t \cdot \boldsymbol{F} \in T^*\mathcal{B}_\mathrm{m} \times T^*\mathcal{B}_\mathrm{m}. \tag{2.45}$$

It is remarked that the right Cauchy-Green-type strain measures allow to measure the squared length of material and spatial line elements in terms of their spatial and material counterparts, respectively,

$$|\mathrm{d}\boldsymbol{X}|^2 = \mathrm{d}\boldsymbol{x} \cdot \boldsymbol{c} \cdot \mathrm{d}\boldsymbol{x} \quad \text{and} \quad |\mathrm{d}\boldsymbol{x}|^2 = \mathrm{d}\boldsymbol{X} \cdot \boldsymbol{C} \cdot \mathrm{d}\boldsymbol{X}. \tag{2.46}$$

Considering the difference in the squared length of material and spatial line elements results in the common understanding of covariant strain measures.

Additionally, the *left Cauchy-Green*-type strain measures

$$\boldsymbol{b} := \boldsymbol{F} \cdot \boldsymbol{F}^t \in T\mathcal{B}_s \times T\mathcal{B}_s \quad \text{and} \quad \boldsymbol{B} := \boldsymbol{f} \cdot \boldsymbol{f}^t \in T\mathcal{B}_m \times T\mathcal{B}_m \tag{2.47}$$

prove to be inverse to the corresponding right Cauchy-Green-type strain measures when composed with the deformation maps

$$\boldsymbol{b} = \boldsymbol{b}(\boldsymbol{X}, t) = \boldsymbol{c}^{-1} \circ \boldsymbol{y}(\boldsymbol{X}, t) \quad \text{and} \quad \boldsymbol{B} = \boldsymbol{B}(\boldsymbol{x}, t) = \boldsymbol{C}^{-1} \circ \boldsymbol{Y}(\boldsymbol{x}, t). \tag{2.48}$$

It is remarked that the left Cauchy–Green-type strain measures allow to measure the squared length of material and spatial level set normals in terms of their spatial and material counterparts, respectively,

$$|\mathbf{d}S|^2 = \mathbf{d}s \cdot \boldsymbol{b} \cdot \mathbf{d}s \quad \text{and} \quad |\mathbf{d}s|^2 = \mathbf{d}S \cdot \boldsymbol{B} \cdot \mathbf{d}S. \tag{2.49}$$

Here, $S(\boldsymbol{X}) = s(\boldsymbol{x}) \circ \boldsymbol{y}(\boldsymbol{X}) = 0$ and $s(\boldsymbol{x}) = S(\boldsymbol{X}) \circ \boldsymbol{Y}(\boldsymbol{x}) = 0$ denote a convected (zero) level set function with gradients (one-forms) $\mathbf{d}S := \nabla_X S$ and $\mathbf{d}s := \nabla_x s$, respectively. Considering the difference in the squared length of material and spatial level set normals results in the common understanding of contravariant strain measures.

2.2.8 Velocities

Time Derivatives:
The temporal rates of change of a generic time-dependent field (with space parametrization in either of the coordinates $\{\boldsymbol{\Xi}, \boldsymbol{X}, \boldsymbol{x}\}$)

$$z = z(\{\boldsymbol{\Xi}, \boldsymbol{X}, \boldsymbol{x}\}, t) \tag{2.50}$$

relative to the reference configuration $\mathcal{B}_r$, the material configuration $\mathcal{B}_m$ and the spatial configuration $\mathcal{B}_s$ are, respectively, defined by

$$\dot{z} := \partial_t z|_{\Xi} \quad \text{and} \quad \mathrm{D}_t z := \partial_t z|_X \quad \text{and} \quad \mathrm{d}_t z := \partial_t z|_x. \tag{2.51}$$

Henceforth a superposed dot denotes the *total time derivative*, whereas D_t and d_t are the *material* and *spatial time derivatives*, respectively. Finally, the total time derivative of a time-dependent quantity z may be related to its material and spatial time derivatives, respectively, via

$$\dot{z} = \mathrm{D}_t z + \nabla_X z \cdot \dot{\boldsymbol{\Upsilon}} = \mathrm{d}_t z + \nabla_x z \cdot \boldsymbol{\upsilon}. \tag{2.52}$$

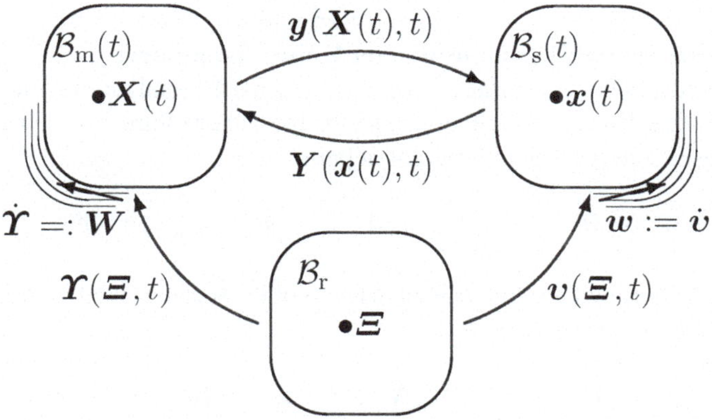

Fig. 2.2 Total material and spatial velocities

The above relations resemble the celebrated *Euler formulae* relating the spatial and material time derivatives

$$\mathrm{d}_t z = \mathrm{D}_t z + \nabla_X z \cdot \mathrm{d}_t \boldsymbol{Y} \quad \text{and} \quad \mathrm{D}_t z = \mathrm{d}_t z + \nabla_x z \cdot \mathrm{D}_t \boldsymbol{y} \tag{2.53}$$

as, for example, frequently applied in the kinematics of fluid mechanics.

The total time derivatives of the material and spatial position maps $\boldsymbol{X} = \boldsymbol{\Upsilon}(\boldsymbol{\Xi}, t)$ and $\boldsymbol{x} = \boldsymbol{\upsilon}(\boldsymbol{\Xi}, t)$ are introduced as the *total material* and *spatial velocities*, see Fig. 2.2

$$\boldsymbol{W} := \dot{\boldsymbol{\Upsilon}}(\boldsymbol{\Xi}, t) \quad \text{and} \quad \boldsymbol{w} := \dot{\boldsymbol{\upsilon}}(\boldsymbol{\Xi}, t). \tag{2.54}$$

Moreover, the spatial time derivative of the material deformation map $\boldsymbol{X} = \boldsymbol{Y}(\boldsymbol{x}, t)$ and the material time derivative of the spatial deformation map $\boldsymbol{x} = \boldsymbol{y}(\boldsymbol{X}, t)$ result in the *material* and the *spatial velocity*

$$\boldsymbol{V} := \mathrm{d}_t \boldsymbol{Y}(\boldsymbol{x}, t) \quad \text{and} \quad \boldsymbol{v} := \mathrm{D}_t \boldsymbol{y}(\boldsymbol{X}, t). \tag{2.55}$$

With these definitions at hand it is straightforward to show that the total material and spatial velocities, based on the Euler formulae for the total time derivative, are related as follows

$$\boldsymbol{W} = \boldsymbol{V} + \boldsymbol{f} \cdot \boldsymbol{w} \quad \text{and} \quad \boldsymbol{w} = \boldsymbol{v} + \boldsymbol{F} \cdot \boldsymbol{W}. \tag{2.56}$$

The relation between the material and spatial velocities via the deformation gradients is then finally a direct consequence of the above relations

$$\boldsymbol{V} = -\boldsymbol{f} \cdot \boldsymbol{v} \quad \text{and} \quad \boldsymbol{v} = -\boldsymbol{F} \cdot \boldsymbol{V}. \tag{2.57}$$

Normal-Tangential Decomposition on Volume Boundary:
On the regular boundary surface of a control volume the total material and spatial velocities that live in the tangent space to the bulk volume are conveniently decomposed into normal and tangential parts

$$\boldsymbol{W} = \boldsymbol{W}_\perp + \boldsymbol{W}_\parallel \quad \text{on} \quad \partial\mathfrak{B}_\mathrm{m} \quad \text{and} \quad \boldsymbol{w} = \boldsymbol{w}_\perp + \boldsymbol{w}_\parallel \quad \text{on} \quad \partial\mathfrak{B}_\mathrm{s}. \tag{2.58}$$

Thereby the normal parts are defined in terms of the scalar-valued *normal velocities* $W_\perp$ and $w_\perp$, respectively, as

$$\boldsymbol{W}_\perp := [\boldsymbol{W} \cdot \boldsymbol{N}]\boldsymbol{N} =: W_\perp \boldsymbol{N} \quad \text{and} \quad \boldsymbol{w}_\perp := [\boldsymbol{w} \cdot \boldsymbol{n}]\boldsymbol{n} =: w_\perp \boldsymbol{n}$$

and, consequently, the tangential parts follow as

$$\boldsymbol{W}_\parallel := \boldsymbol{W} - \boldsymbol{W}_\perp =: \mathbb{P} \cdot \boldsymbol{W} \quad \text{and} \quad \boldsymbol{w}_\parallel := \boldsymbol{w} - \boldsymbol{w}_\perp =: \mathbb{p} \cdot \boldsymbol{w}.$$

Here, $\mathbb{P}$ and $\mathbb{p}$ are the material and spatial (mixed-variant) *projection operators* onto the tangent plane to the boundary surface of the control volume

$$\mathbb{P} := \boldsymbol{I} - \boldsymbol{N} \otimes \boldsymbol{N} \quad \text{on} \quad \partial\mathfrak{B}_\mathrm{m} \quad \text{and} \quad \mathbb{p} := \boldsymbol{i} - \boldsymbol{n} \otimes \boldsymbol{n} \quad \text{on} \quad \partial\mathfrak{B}_\mathrm{s},$$

with $\boldsymbol{I}$ and $\boldsymbol{i}$ the material and spatial (mixed-variant) unit tensors, respectively.

Note that at this point no assumption is made restricting the total material velocity being exclusively normal, neither at the boundary of the control volume nor at the external boundary.

2.2.9 *Gradients of Velocities*

Material time derivatives commute with material gradients, thus the material time derivative D_t of the spatial deformation gradient $\boldsymbol{F}$ coincides with the material gradient ∇_X of the spatial velocity $\boldsymbol{v}$. Likewise, spatial time derivatives commute with spatial gradients, thus the spatial time derivative d_t of the material deformation gradient $\boldsymbol{f}$ coincides with the spatial gradient ∇_x of the material velocity $\boldsymbol{V}$. These relations read as

$$\mathrm{D}_t \boldsymbol{F} \equiv \nabla_X \boldsymbol{v} \quad \text{and} \quad \mathrm{d}_t \boldsymbol{f} \equiv \nabla_x \boldsymbol{V}. \tag{2.59}$$

Next, the spatial gradient ∇_x of the spatial velocity $\boldsymbol{v}$ and the material gradient ∇_X of the material velocity $\boldsymbol{V}$ are introduced as the spatial and the material *velocity gradient*, respectively,

$$\boldsymbol{l} := \nabla_x \boldsymbol{v} = \mathrm{D}_t \boldsymbol{F} \cdot \boldsymbol{f} \quad \text{and} \quad \boldsymbol{L} := \nabla_X \boldsymbol{V} = \mathrm{d}_t \boldsymbol{f} \cdot \boldsymbol{F}. \tag{2.60}$$

Table 2.1 Spatial velocity gradients and their pull-backs/push-forwards

	$\boldsymbol{l}$	$\mathrm{D}_t\boldsymbol{F}$	$-\mathrm{D}_t\boldsymbol{f}$	$\boldsymbol{\Lambda}$
$\boldsymbol{l}$	$\bullet$	$\mathrm{D}_t\boldsymbol{F}\cdot\boldsymbol{f}$	$-\boldsymbol{F}\cdot\mathrm{D}_t\boldsymbol{f}$	$\boldsymbol{F}\cdot\boldsymbol{\Lambda}\cdot\boldsymbol{f}$
$\mathrm{D}_t\boldsymbol{F}$	$\boldsymbol{l}\cdot\boldsymbol{F}$	$\bullet$	$-\boldsymbol{F}\cdot\mathrm{D}_t\boldsymbol{f}\cdot\boldsymbol{F}$	$\boldsymbol{F}\cdot\boldsymbol{\Lambda}$
$-\mathrm{D}_t\boldsymbol{f}$	$\boldsymbol{f}\cdot\boldsymbol{l}$	$\boldsymbol{f}\cdot\mathrm{D}_t\boldsymbol{F}\cdot\boldsymbol{f}$	$\bullet$	$\boldsymbol{\Lambda}\cdot\boldsymbol{f}$
$\boldsymbol{\Lambda}$	$\boldsymbol{f}\cdot\boldsymbol{l}\cdot\boldsymbol{F}$	$\boldsymbol{f}\cdot\mathrm{D}_t\boldsymbol{F}$	$-\quad\mathrm{D}_t\boldsymbol{f}\cdot\boldsymbol{F}$	$\bullet$

Table 2.2 Material velocity gradients and their push-forwards/pull-backs

	$\boldsymbol{L}$	$\mathrm{d}_t\boldsymbol{f}$	$-\mathrm{d}_t\boldsymbol{F}$	$\boldsymbol{\lambda}$
$\boldsymbol{L}$	$\bullet$	$\mathrm{d}_t\boldsymbol{f}\cdot\boldsymbol{F}$	$-\boldsymbol{f}\cdot\mathrm{d}_t\boldsymbol{F}$	$\boldsymbol{f}\cdot\boldsymbol{\lambda}\cdot\boldsymbol{F}$
$\mathrm{d}_t\boldsymbol{f}$	$\boldsymbol{L}\cdot\boldsymbol{f}$	$\bullet$	$-\boldsymbol{f}\cdot\mathrm{d}_t\boldsymbol{F}\cdot\boldsymbol{f}$	$\boldsymbol{f}\cdot\boldsymbol{\lambda}$
$-\mathrm{d}_t\boldsymbol{F}$	$\boldsymbol{F}\cdot\boldsymbol{L}$	$\boldsymbol{F}\cdot\mathrm{d}_t\boldsymbol{f}\cdot\boldsymbol{F}$	$\bullet$	$\boldsymbol{\lambda}\cdot\boldsymbol{F}$
$\boldsymbol{\lambda}$	$\boldsymbol{F}\cdot\boldsymbol{L}\cdot\boldsymbol{f}$	$\boldsymbol{F}\cdot\mathrm{d}_t\boldsymbol{f}$	$-\quad\mathrm{d}_t\boldsymbol{F}\cdot\boldsymbol{f}$	$\bullet$

Note that the spatial and material velocity gradients appear as right-sided push-forward and right-sided pull-back of $\mathrm{D}_t\boldsymbol{F}$ and $\mathrm{d}_t\boldsymbol{f}$, respectively. Accordingly, the left-sided pull-back of $\mathrm{D}_t\boldsymbol{F}$ and the left-sided push-forward of $\mathrm{d}_t\boldsymbol{f}$ are introduced as

$$\boldsymbol{\Lambda} := \boldsymbol{f}\cdot\mathrm{D}_t\boldsymbol{F} \quad\text{and}\quad \boldsymbol{\lambda} := \boldsymbol{F}\cdot\mathrm{d}_t\boldsymbol{f}. \tag{2.61}$$

Finally, the material time derivative D_t of the material deformation gradient $\boldsymbol{f}$ and the spatial time derivative d_t of the spatial deformation gradient $\boldsymbol{F}$ follow as

$$-\mathrm{D}_t\boldsymbol{f} = \boldsymbol{f}\cdot\mathrm{D}_t\boldsymbol{F}\cdot\boldsymbol{f} \quad\text{and}\quad -\mathrm{d}_t\boldsymbol{F} = \boldsymbol{F}\cdot\mathrm{d}_t\boldsymbol{f}\cdot\boldsymbol{F}. \tag{2.62}$$

Clearly, material and spatial time derivatives of the deformation gradients are related by corresponding Euler formulae

$$\mathrm{D}_t\boldsymbol{f} = \mathrm{d}_t\boldsymbol{f} + \nabla_x\boldsymbol{f}\cdot\boldsymbol{v} \quad\text{and}\quad \mathrm{d}_t\boldsymbol{F} = \mathrm{D}_t\boldsymbol{F} + \nabla_X\boldsymbol{F}\cdot\boldsymbol{V} \tag{2.63}$$

and

$$\mathrm{d}_t\boldsymbol{f} = \mathrm{D}_t\boldsymbol{f} + \nabla_X\boldsymbol{f}\cdot\boldsymbol{V} \quad\text{and}\quad \mathrm{D}_t\boldsymbol{F} = \mathrm{d}_t\boldsymbol{F} + \nabla_x\boldsymbol{F}\cdot\boldsymbol{v}. \tag{2.64}$$

The relations between the various types of 'velocity gradients', i.e. the gradients of the spatial and material velocity, respectively, are summarized in Tables 2.1 and 2.2.

References

1. de Boer R (1982) Vektor-und Tensorrechnung für Ingenieure. Springer, Berlin (1982)
2. Steinmann P (2015) Geometrical foundations of continuum mechanics. Springer, Berlin (2015)

Chapter 3
Kinematics on Dimensionally Reduced Smooth Manifolds

8,080 m 35°43'28"N 76°41'47"E

Abstract This chapter reviews continuum kinematics on dimensionally reduced smooth manifolds, i.e. on boundary surfaces and on boundary curves, with an emphasis on the nonlinear deformation maps and their associated tangent, cotangent, and measure maps.

In addition to the kinematics of geometrical field quantities defined previously in three-dimensional bulk volumes, the kinematics of geometrical field quantities defined on smooth two-dimensional *boundary surfaces* and smooth one-dimensional *boundary curves*, respectively, collectively denoted as *dimensionally reduced smooth manifolds*, is of relevance for a variety of the subsequent detailed discussions (Steinmann [1]).

To this end, firstly three-dimensional nonlinear deformation maps defined on two- and one-dimensional smooth manifolds are introduced. Then their corresponding manifold space gradients, i.e. the surface and curve *tangent*, *cotangent*, and *measure maps*, respectively, and their corresponding manifold time gradients, i.e. the surface and boundary *velocities*, are elaborated. Based thereon, explicit representations for the corresponding *compatibility conditions* in two- and one-dimensional smooth manifolds are considered. Purposely, the kinematics of geometrical field quantities defined on smooth two-dimensional boundary surfaces and smooth one-dimensional boundary curves, respectively, is pursued following exactly the same discussions, arguments, and formulations as in three-dimensional bulk volumes.

As a result, remarkably, the analysis of the kinematics of geometrical quantities on dimensionally reduced smooth manifolds results in respective explicit expressions for the tangent, cotangent, and measure maps and the velocities that collectively display a similar format as in the three-dimensional case.

P. Steinmann, *Spatial and Material Forces in Nonlinear Continuum Mechanics*,
Solid Mechanics and Its Applications 272,
https://doi.org/10.1007/978-3-030-89070-4_3

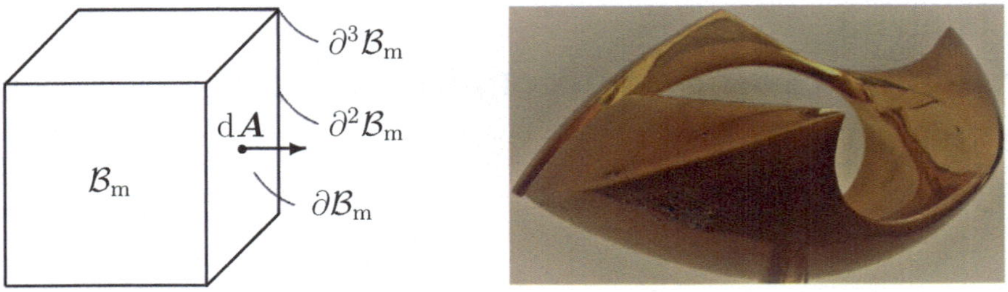

Fig. 3.1 (Left) A cube-shaped bulk volume for the material configuration $\mathcal{B}_m$ and its two-dimensional boundary surface $\partial\mathcal{B}_m$ with one- and zero-dimensional sub-sets consisting of edges and vertices. The edges (boundary curves) are denoted by $\partial^2\mathcal{B}_m \equiv \partial\partial\mathcal{B}_m$ and the vertices (boundary points) are denoted by $\partial^3\mathcal{B}_m \equiv \partial\partial\partial\mathcal{B}_m$. (Right) A sculpture artistically illustrating the concepts of a bulk volume enclosed by a boundary surface with edges and vertices (Photo ©Ella Steinmann, taken with permission of the owner)

3.1 Configurations

3.1.1 Material Configuration

The boundary (surface) of the material configuration $\mathcal{B}_m$, assumed regular (smooth) almost everywhere, is denoted by $\partial\mathcal{B}_m$ and is equipped with the outwards-pointing unit normal $\boldsymbol{N}$. Scalar- and vector-valued *material area elements* are given by $\mathrm{d}A$ and $\mathrm{d}\boldsymbol{A} := \boldsymbol{N}\,\mathrm{d}A$.

Possible one- and zero-dimensional *sub-sets* of the boundary surface $\partial\mathcal{B}_m$ are denoted by $\partial^{2-\alpha}\partial\mathcal{B}_m \equiv \partial^{3-\alpha}\mathcal{B}_m$ with $\alpha = 1, 0$. Consequently, for $\alpha = 1$ and $\alpha = 0$, one- and zero-dimensional sub-sets of the boundary $\partial\mathcal{B}_m$, i.e. boundary curves and boundary points, are explicitly given by $\partial^2\mathcal{B}_m \equiv \partial\partial\mathcal{B}_m$ and $\partial^3\mathcal{B}_m \equiv \partial\partial\partial\mathcal{B}_m$, respectively; see Fig. 3.1.

In case that the boundary (surface) $\partial\mathfrak{B}_m$ of the *material (bulk) control volume* $\mathfrak{B}_m$ introduced in Chap. 2 intersects with the external boundary (surface) $\partial\mathcal{B}_m$, the one- and zero-dimensional sub-sets $\partial^{3-\alpha}\mathfrak{B}_m$ are defined accordingly.

The boundary $\partial^2\mathcal{B}_m \equiv \partial\partial\mathcal{B}_m$ of regular (smooth) sub-parts of $\partial\mathcal{B}_m$, i.e. the set of boundary curves, is equipped with outwards-pointing unit normals $\widehat{\boldsymbol{N}}$, tangent to $\partial\mathcal{B}_m$. Scalar- and vector-valued *material line elements* are given by $\mathrm{d}L$ and $\mathrm{d}\widehat{\boldsymbol{L}} := \widehat{\boldsymbol{N}}\,\mathrm{d}L$.

Likewise, the boundary $\partial^3\mathcal{B}_m \equiv \partial\partial\partial\mathcal{B}_m$ of regular (smooth) sub-parts of $\partial^2\mathcal{B}_m \equiv \partial\partial\mathcal{B}_m$, i.e. the set of boundary points, is equipped with outwards-pointing unit normals $\widetilde{\boldsymbol{N}}$, tangent to $\partial^2\mathcal{B}_m$, i.e. the set of boundary curves. Vector-valued *material point elements* are also given by $\widetilde{\boldsymbol{N}}$ (with unit scalar-valued material point elements).

Within the regular (smooth) sub-parts of $\partial\mathcal{B}_m$, the time-dependent *material boundary control surface* $\mathfrak{S}_m = \mathfrak{S}_m(t)$ with $\mathfrak{S}_m \subseteq \partial\mathcal{B}_m$ is smoothly migrating. The boundary of $\mathfrak{S}_m$, assumed regular (smooth) almost everywhere, is denoted by $\partial\mathfrak{S}_m$ and is equipped with the outwards-pointing unit normal $\widehat{\boldsymbol{N}}$, tangent to $\mathfrak{S}_m$.

Likewise, within the regular (smooth) sub-parts of $\partial^2\mathcal{B}_\mathrm{m} \equiv \partial\partial\mathcal{B}_\mathrm{m}$, the time-dependent *material boundary control curve* $\mathfrak{C}_\mathrm{m} = \mathfrak{C}_\mathrm{m}(t)$ with $\mathfrak{C}_\mathrm{m} \subseteq \partial^2\mathcal{B}_\mathrm{m} \equiv \partial\partial\mathcal{B}_\mathrm{m}$ is smoothly migrating. The boundary of $\mathfrak{C}_\mathrm{m}$ is denoted by $\partial\mathfrak{C}_\mathrm{m}$ and is equipped with the outwards-pointing unit normal $\widetilde{\boldsymbol{N}}$, tangent to $\mathfrak{C}_\mathrm{m}$.

3.1.2 Spatial Configuration

The boundary (surface) of the spatial configuration $\mathcal{B}_\mathrm{s}$ is denoted by $\partial\mathcal{B}_\mathrm{s}$ and is equipped with the outwards-pointing unit normal $\boldsymbol{n}$. Scalar- and vector-valued *spatial area elements* are given by $\mathrm{d}a$ and $\mathrm{d}\boldsymbol{a} := \boldsymbol{n}\,\mathrm{d}a$.

Possible one- and zero-dimensional sub-sets of $\partial\mathcal{B}_\mathrm{s}$ are denoted by $\partial^{3-\alpha}\mathcal{B}_\mathrm{s}$. Accordingly, one- and zero-dimensional sub-sets contained in the intersection $\partial\mathcal{B}_\mathrm{s} \cap \partial\mathfrak{B}_\mathrm{s}$ of the boundary $\partial\mathcal{B}_\mathrm{s}$ of $\mathcal{B}_\mathrm{s}$ and the boundary $\partial\mathfrak{B}_\mathrm{s}$ of the *spatial (bulk) control volume* $\mathfrak{B}_\mathrm{s}$ are denoted as $\partial^{3-\alpha}\mathfrak{B}_\mathrm{s}$.

The boundary $\partial^2\mathcal{B}_\mathrm{s} \equiv \partial\partial\mathcal{B}_\mathrm{s}$ of regular (smooth) sub-parts of $\partial\mathcal{B}_\mathrm{s}$, i.e. the set of boundary curves, is equipped with the outwards-pointing unit normal $\widehat{\boldsymbol{n}}$, tangent to $\partial\mathcal{B}_\mathrm{s}$. Scalar- and vector-valued *spatial line elements* are given by $\mathrm{d}l$ and $\mathrm{d}\widehat{\boldsymbol{l}} := \widehat{\boldsymbol{n}}\,\mathrm{d}l$. Likewise, the boundary $\partial^3\mathcal{B}_\mathrm{s} \equiv \partial\partial\partial\mathcal{B}_\mathrm{s}$ of regular (smooth) sub-parts of $\partial^2\mathcal{B}_\mathrm{s} \equiv \partial\partial\mathcal{B}_\mathrm{s}$, i.e. the set of boundary points, is equipped with the outwards-pointing unit normal $\widetilde{\boldsymbol{n}}$, tangent to $\partial^2\mathcal{B}_\mathrm{s}$, i.e. the set of boundary curves. Vector-valued *spatial point elements* are also given by $\widetilde{\boldsymbol{n}}$ (with unit scalar-valued material point elements).

Finally, the time-dependent (convected) *spatial boundary control surface* $\mathfrak{S}_\mathrm{s} = \mathfrak{S}_\mathrm{s}(t)$ with $\mathfrak{S}_\mathrm{s} \subseteq \partial\mathcal{B}_\mathrm{s}$ is smoothly migrating within the regular (smooth) sub-parts of $\partial\mathcal{B}_\mathrm{s}$. The boundary of $\mathfrak{S}_\mathrm{s}$ is denoted by $\partial\mathfrak{S}_\mathrm{s}$ and is equipped with the outwards-pointing unit normal $\widehat{\boldsymbol{n}}$, tangent to $\mathfrak{S}_\mathrm{s}$. Likewise, the time-dependent (convected) *spatial boundary control curve* $\mathfrak{C}_\mathrm{s} = \mathfrak{C}_\mathrm{s}(t)$ with $\mathfrak{C}_\mathrm{s} \subseteq \partial^2\mathcal{B}_\mathrm{s} \equiv \partial\partial\mathcal{B}_\mathrm{s}$ is smoothly migrating within the regular (smooth) sub-parts of $\partial^2\mathcal{B}_\mathrm{s} \equiv \partial\partial\mathcal{B}_\mathrm{s}$. The boundary of $\mathfrak{C}_\mathrm{s}$ is denoted by $\partial\mathfrak{C}_\mathrm{s}$ and is equipped with the outwards-pointing unit normal $\widetilde{\boldsymbol{n}}$, tangent to $\mathfrak{C}_\mathrm{s}$.

3.2 Deformation on Boundary Surfaces

Geometry on Surfaces:
A smooth surface $\mathcal{S}$, i.e. a two-dimensional manifold embedded in the three-dimensional Euclidean space $\mathbb{E}^3$ with, say, coordinates $\widehat{\boldsymbol{X}}$ may be parameterized by two surface coordinates Θ^α with $\alpha = 1, 2$ as

$$\widehat{\boldsymbol{X}} = \widehat{\boldsymbol{X}}(\Theta^\alpha). \tag{3.1}$$

The covariant (natural) surface basis vectors $\widehat{\boldsymbol{G}}_\alpha \in T\mathcal{S}$ are the tangent vectors to the surface coordinate lines Θ^α

$$\widehat{\boldsymbol{G}}_\alpha = \partial_\alpha \widehat{\boldsymbol{X}}. \tag{3.2}$$

The contravariant (dual) surface basis vectors $\widehat{\boldsymbol{G}}^\alpha \in T^*\mathcal{S}$ follow from the orthonormality property $\delta^\alpha{}_\beta = \widehat{\boldsymbol{G}}^\alpha \cdot \widehat{\boldsymbol{G}}_\beta$. They are explicitly related to the covariant surface basis vectors $\widehat{\boldsymbol{G}}_\alpha$ by the co- and contravariant surface metric coefficients $\widehat{G}_{\alpha\beta}$ and $\widehat{G}^{\alpha\beta}$, respectively, as

$$\begin{aligned} \widehat{\boldsymbol{G}}_\alpha &= \widehat{G}_{\alpha\beta}\widehat{\boldsymbol{G}}^\beta \quad \text{with} \quad \widehat{G}_{\alpha\beta} := \widehat{\boldsymbol{G}}_\alpha \cdot \widehat{\boldsymbol{G}}_\beta = [\widehat{G}^{\alpha\beta}]^{-1}, \\ \widehat{\boldsymbol{G}}^\alpha &= \widehat{G}^{\alpha\beta}\widehat{\boldsymbol{G}}_\beta \quad \text{with} \quad \widehat{G}^{\alpha\beta} := \widehat{\boldsymbol{G}}^\alpha \cdot \widehat{\boldsymbol{G}}^\beta = [\widehat{G}_{\alpha\beta}]^{-1}. \end{aligned} \tag{3.3}$$

The $\widehat{G}_{\alpha\beta}$ are also denoted as the *first fundamental form* for the surface. Furthermore, contra- and covariant base vectors $\widehat{\boldsymbol{G}}^3$ and $\widehat{\boldsymbol{G}}_3$ are normal to $T\mathcal{S}$ with

$$\widehat{\boldsymbol{G}}^3 := \widehat{\boldsymbol{G}}_1 \times \widehat{\boldsymbol{G}}_2 \quad \text{and} \quad \widehat{\boldsymbol{G}}_3 := [\widehat{G}^{33}]^{-1}\widehat{\boldsymbol{G}}^3 \implies \widehat{\boldsymbol{G}}^3 \cdot \widehat{\boldsymbol{G}}_3 = 1. \tag{3.4}$$

Here, the contra- and covariant metric coefficients $\widehat{G}^{33}$ and $\widehat{G}_{33}$ compute alternatively via

$$\widehat{G}^{33} = |\widehat{\boldsymbol{G}}_1 \times \widehat{\boldsymbol{G}}_2|^2 = \det[\widehat{G}_{\alpha\beta}] = \big[\det[\widehat{G}^{\alpha\beta}]\big]^{-1} = [\widehat{G}_{33}]^{-1}. \tag{3.5}$$

Consequently, the surface area element $\mathrm{d}A$ and the surface unit normal $\boldsymbol{N}$ follow as

$$\mathrm{d}A = |\widehat{\boldsymbol{G}}_1 \times \widehat{\boldsymbol{G}}_2|\,\mathrm{d}\Theta^1\,\mathrm{d}\Theta^2 = \sqrt{\widehat{G}^{33}}\,\mathrm{d}\Theta^1\,\mathrm{d}\Theta^2 \tag{3.6}$$

and

$$\boldsymbol{N} = \sqrt{\widehat{G}_{33}}\,\widehat{\boldsymbol{G}}^3 = \sqrt{\widehat{G}^{33}}\,\widehat{\boldsymbol{G}}_3. \tag{3.7}$$

With the common unit tensor $\boldsymbol{I}$ of the three-dimensional, embedding Euclidean space, the surface unit tensor $\widehat{\boldsymbol{I}}$ follows as

$$\widehat{\boldsymbol{I}} := [\boldsymbol{G}^\alpha \cdot \boldsymbol{G}^\beta]\,\widehat{\boldsymbol{G}}_\alpha \otimes \widehat{\boldsymbol{G}}_\beta = \widehat{\boldsymbol{G}}_\alpha \otimes \widehat{\boldsymbol{G}}^\alpha = \boldsymbol{I} - \widehat{\boldsymbol{G}}_3 \otimes \widehat{\boldsymbol{G}}^3 = \boldsymbol{I} - \boldsymbol{N} \otimes \boldsymbol{N}. \tag{3.8}$$

Note that the surface unit tensor $\widehat{\boldsymbol{I}}$ coincides with the previously introduced (idempotent) surface projection tensor $\mathbb{P}$. Moreover, with the permutation tensor $\boldsymbol{E}$ of the three-dimensional, embedding Euclidean space, the surface permutation tensor $\widehat{\boldsymbol{E}}$ follows as

$$\widehat{\boldsymbol{E}} := \big[[\boldsymbol{G}^\alpha \times \boldsymbol{G}^\beta] \cdot \boldsymbol{N}\big]\,\boldsymbol{G}_\alpha \otimes \boldsymbol{G}_\beta := \boldsymbol{E} \cdot \boldsymbol{N}. \tag{3.9}$$

Next, the surface gradient and surface divergence operators for vector-valued (surface) fields are defined as

$$\widehat{\nabla}_{\widehat{X}}(\bullet) := \partial_\alpha(\bullet) \otimes \widehat{\boldsymbol{G}}^\alpha \quad \text{and} \quad \widehat{\text{Div}}(\bullet) := \partial_\alpha(\bullet) \cdot \widehat{\boldsymbol{G}}^\alpha. \tag{3.10}$$

Clearly, $\widehat{\nabla}_{\widehat{X}}(\bullet) \cdot \boldsymbol{N} = \boldsymbol{0}$ vanishes by definition. In addition, the (scalar-valued) curl operator for vector-valued (surface) fields is defined as

$$-\widehat{\text{Curl}}(\bullet) := \partial_\alpha(\bullet) \,\widehat{\times}\, \widehat{\boldsymbol{G}}^\alpha := [\partial_\alpha(\bullet) \times \widehat{\boldsymbol{G}}^\alpha] \cdot \boldsymbol{N}. \tag{3.11}$$

The surface divergence and surface curl are alternatively expressed as the double contraction of the surface gradient with the surface unit tensor and the surface permutation tensor, respectively

$$\begin{aligned} \widehat{\text{Div}}(\bullet) &= \big[\partial_\alpha(\bullet) \otimes \widehat{\boldsymbol{G}}^\alpha\big] : \widehat{\boldsymbol{I}} = \widehat{\nabla}_{\widehat{X}}(\bullet) : \widehat{\boldsymbol{I}}\,, \\ -\widehat{\text{Curl}}(\bullet) &= \big[\partial_\alpha(\bullet) \otimes \widehat{\boldsymbol{G}}^\alpha\big] : \widehat{\boldsymbol{E}} = \widehat{\nabla}_{\widehat{X}}(\bullet) : \widehat{\boldsymbol{E}}. \end{aligned} \tag{3.12}$$

For fields that are smooth in a three-dimensional neighborhood $\mathcal{N}$ of the surface, the surface gradient and the surface divergence operators are alternatively defined by the help of the surface unit tensor as

$$\widehat{\nabla}_{\widehat{X}}(\bullet) := \nabla_X(\bullet)|_{\mathcal{S}} \cdot \widehat{\boldsymbol{I}} \quad \text{and} \quad \widehat{\text{Div}}(\bullet) := \nabla_X(\bullet)|_{\mathcal{S}} : \widehat{\boldsymbol{I}}. \tag{3.13}$$

Likewise, in this case the surface curl operator is alternatively defined by the help of the surface permutation tensor as

$$-\widehat{\text{Curl}}(\bullet) := \nabla_X(\bullet)|_{\mathcal{S}} : \widehat{\boldsymbol{E}} = -\text{Curl}(\bullet)|_{\mathcal{S}} \cdot \boldsymbol{N}. \tag{3.14}$$

The surface gradient, surface divergence, and surface curl operators for tensor-valued fields $(\circ)$ follow from multiplying $(\circ)$ from the left by an arbitrary constant vector and inserting the so resulting vector-valued-field into the above definitions. Then, on $\mathcal{S}$ the product rule of differentiation $\widehat{\text{Div}}\big((\bullet) \cdot (\circ)\big) = \widehat{\nabla}_{\widehat{X}}(\bullet) : (\circ) + (\bullet) \cdot \widehat{\text{Div}}(\circ)$ holds in particular since

$$\partial_\alpha\big[(\bullet) \cdot (\circ)\big] \cdot \widehat{\boldsymbol{G}}^\alpha = [\partial_\alpha(\bullet) \otimes \widehat{\boldsymbol{G}}^\alpha] : (\circ) + (\bullet) \cdot \partial_\alpha(\circ) \cdot \widehat{\boldsymbol{G}}^\alpha. \tag{3.15}$$

Finally, the surface curvature tensor $\widehat{\boldsymbol{C}} = \widehat{C}_{\alpha\beta} \widehat{\boldsymbol{G}}^\alpha \otimes \widehat{\boldsymbol{G}}^\beta$ and the total surface curvature $\widehat{C} = \widehat{C}^\alpha{}_\alpha = \widehat{C}_\alpha{}^\alpha$, i.e. twice the mean surface curvature, are defined as the (negative) surface gradient

$$\widehat{\boldsymbol{C}} := -\widehat{\nabla}_{\widehat{X}} \boldsymbol{N} = -\partial_\alpha \boldsymbol{N} \otimes \widehat{\boldsymbol{G}}^\alpha \tag{3.16}$$

and (negative) surface divergence of the surface unit normal $\boldsymbol{N}$, respectively,

$$\widehat{C} := -\widehat{\mathrm{Div}}\boldsymbol{N} = -\partial_\alpha \boldsymbol{N} \cdot \widehat{\boldsymbol{G}}^\alpha. \tag{3.17}$$

Thus, the surface divergence of the surface unit tensor $\widehat{\boldsymbol{I}}$ and the surface permutation tensor result in

$$\widehat{\mathrm{Div}}\,\widehat{\boldsymbol{I}} = \widehat{C}\boldsymbol{N} \quad \text{and} \quad \widehat{\mathrm{Div}}\,\widehat{\boldsymbol{E}} = \boldsymbol{0}. \tag{3.18}$$

Note that the covariant coefficients of the surface curvature tensor are computed by $\widehat{C}_{\alpha\beta} = \widehat{\boldsymbol{G}}_\alpha \cdot \widehat{\boldsymbol{C}} \cdot \widehat{\boldsymbol{G}}_\beta = -\widehat{\boldsymbol{G}}_\alpha \cdot \partial_\beta \boldsymbol{N}$. These are also denoted as the *second fundamental form* for the surface.

The above relations are rooted in the *Gauss* formulae

$$\partial_\beta \widehat{\boldsymbol{G}}_\alpha = \widehat{\Gamma}^\gamma{}_{\alpha\beta} \widehat{\boldsymbol{G}}_\gamma + \widehat{C}_{\alpha\beta}\boldsymbol{N} \quad \text{and} \quad \partial_\beta \widehat{\boldsymbol{G}}^\alpha = -\widehat{\Gamma}^\alpha{}_{\beta\gamma} \widehat{\boldsymbol{G}}^\gamma + \widehat{C}^\alpha{}_\beta \boldsymbol{N}. \tag{3.19}$$

Here $\widehat{\Gamma}^\gamma{}_{\alpha\beta} = \partial_\beta \widehat{\boldsymbol{G}}_\alpha \cdot \widehat{\boldsymbol{G}}^\gamma = -\partial_\beta \widehat{\boldsymbol{G}}^\gamma \cdot \widehat{\boldsymbol{G}}_\alpha = \widehat{\Gamma}^\gamma{}_{\beta\alpha}$ denote the surface Christoffel symbols, symmetric in the index pair $\alpha\beta$.

Example: Consider a cylindrical surface with radius R and surface coordinates Θ^α denoting the circumferential arc-length and the axial length

$$\widehat{\boldsymbol{X}} = R\cos(\Theta^1/R)\boldsymbol{E}_1 + R\sin(\Theta^1/R)\boldsymbol{E}_2 + \Theta^2\boldsymbol{E}_3. \tag{3.20}$$

Thus, the surface basis vectors compute with $\widehat{G}_{\alpha\beta} = \delta_{\alpha\beta}$ as

$$\widehat{\boldsymbol{G}}_1 = \widehat{\boldsymbol{G}}^1 = -\sin(\Theta^1/R)\boldsymbol{E}_1 + \cos(\Theta^1/R)\boldsymbol{E}_2 \tag{3.21}$$

and $\widehat{\boldsymbol{G}}_2 = \widehat{\boldsymbol{G}}^2 = \boldsymbol{E}_3$. Thus, the *outwards*-pointing surface normal and the surface curvature tensor read as

$$\boldsymbol{N} = \cos(\Theta^1/R)\boldsymbol{E}_1 + \sin(\Theta^1/R)\boldsymbol{E}_2 \quad \text{and} \quad \widehat{\boldsymbol{C}} = -\boldsymbol{N}_{,1} \otimes \widehat{\boldsymbol{G}}^1. \tag{3.22}$$

Then, the total surface curvature computes as $\widehat{C} = -\boldsymbol{N}_{,1} \cdot \widehat{\boldsymbol{G}}^1 = -1/R$.

3.2.1 Nonlinear Deformation Maps

As preliminaries for the subsequent discussions regarding the variation of boundary surface contributions, some relevant results regarding the kinematics of the surfaces defining the regular parts of the referential, material, and spatial boundaries $\partial\mathcal{B}_\mathrm{r}$, $\partial\mathcal{B}_\mathrm{m}$, and $\partial\mathcal{B}_\mathrm{s}$, respectively, are recalled in the sequel.

The material and spatial boundary surface position vectors follow as the specialization of the material and spatial position vectors to $\partial\mathcal{B}_\mathrm{r}$, i.e.

$$\widehat{\boldsymbol{X}} = \widehat{\boldsymbol{\Upsilon}}(\widehat{\boldsymbol{\Xi}}, t) := \boldsymbol{\Upsilon}(\boldsymbol{\Xi}|_{\partial\mathcal{B}_\mathrm{r}}, t) \quad \text{and} \quad \widehat{\boldsymbol{x}} = \widehat{\boldsymbol{\upsilon}}(\widehat{\boldsymbol{\Xi}}, t) := \boldsymbol{\upsilon}(\boldsymbol{\Xi}|_{\partial\mathcal{B}_\mathrm{r}}, t). \tag{3.23}$$

Likewise, the material and spatial nonlinear boundary surface deformation maps follow as the specialization of the material and spatial deformation maps to $\partial\mathcal{B}_\mathrm{s}$ and $\partial\mathcal{B}_\mathrm{m}$, respectively, i.e.

$$\widehat{\boldsymbol{X}} = \widehat{\boldsymbol{Y}}(\widehat{\boldsymbol{x}}, t) := \boldsymbol{Y}(\boldsymbol{x}|_{\partial\mathcal{B}_\mathrm{s}}, t) \quad \text{and} \quad \widehat{\boldsymbol{x}} = \widehat{\boldsymbol{y}}(\widehat{\boldsymbol{X}}, t) := \boldsymbol{y}(\boldsymbol{X}|_{\partial\mathcal{B}_\mathrm{m}}, t). \tag{3.24}$$

We refrain in the sequel from explicitly indicating the space-time parametrization as long as there is no danger of confusion.

3.2.2 Tangent Maps: Deformation Gradients

Increments (total differentials) of the material and spatial boundary surface position vectors (tangents) $\mathrm{d}\widehat{\boldsymbol{X}} \in T\partial\mathcal{B}_\mathrm{m}$ and $\mathrm{d}\widehat{\boldsymbol{x}} \in T\partial\mathcal{B}_\mathrm{s}$ are denoted as material and spatial *boundary surface line elements*; they are related by spatial→material and material→spatial *tangent maps*

$$\mathrm{d}\widehat{\boldsymbol{X}} = \widehat{\boldsymbol{f}} \cdot \mathrm{d}\widehat{\boldsymbol{x}} \quad \text{and} \quad \mathrm{d}\widehat{\boldsymbol{x}} = \widehat{\boldsymbol{F}} \cdot \mathrm{d}\widehat{\boldsymbol{X}}. \tag{3.25}$$

Consequently, the spatial→material and material→spatial tangent maps follow as the spatial and material *boundary surface gradients* of the material and spatial boundary surface deformation maps, respectively,

$$\widehat{\boldsymbol{f}} := \widehat{\nabla}_{\widehat{x}}\widehat{\boldsymbol{Y}} \in T\partial\mathcal{B}_\mathrm{m} \times T^*\partial\mathcal{B}_\mathrm{s} \quad \text{and} \quad \widehat{\boldsymbol{F}} := \widehat{\nabla}_{\widehat{X}}\widehat{\boldsymbol{y}} \in T\partial\mathcal{B}_\mathrm{s} \times T^*\partial\mathcal{B}_\mathrm{m}. \tag{3.26}$$

Hence, $\widehat{\boldsymbol{f}} : T\partial\mathcal{B}_\mathrm{s} \to T\partial\mathcal{B}_\mathrm{m}$ and $\widehat{\boldsymbol{F}} : T\partial\mathcal{B}_\mathrm{m} \to T\partial\mathcal{B}_\mathrm{s}$ are also denoted as the material and spatial *boundary surface deformation gradients*. Here, $T\partial\mathcal{B}_\mathrm{m}$ and $T^*\partial\mathcal{B}_\mathrm{m}$ together with $T\partial\mathcal{B}_\mathrm{s}$ and $T^*\partial\mathcal{B}_\mathrm{s}$ are the *tangent* and *cotangent* planes to $\partial\mathcal{B}_\mathrm{m}$ and $\partial\mathcal{B}_\mathrm{s}$, respectively. Clearly, the boundary surface deformation gradients are (mixed-variant) *two-point tensors* since they map surface vectors from different tangent planes into each other. In surface convected coordinate representation, they follow as

$$\widehat{\boldsymbol{f}} = \widehat{\boldsymbol{G}}_\alpha \otimes \widehat{\boldsymbol{g}}^\alpha \quad \text{and} \quad \widehat{\boldsymbol{F}} = \widehat{\boldsymbol{g}}_\alpha \otimes \widehat{\boldsymbol{G}}^\alpha. \tag{3.27}$$

The material and spatial boundary surface deformation gradients are algebraically *pseudo-inverse* tensors

$$\widehat{\boldsymbol{f}} = \widehat{\boldsymbol{F}}^{\widehat{-1}} \quad \text{and} \quad \widehat{\boldsymbol{F}} = \widehat{\boldsymbol{f}}^{\widehat{-1}}. \tag{3.28}$$

In $\mathbb{E}^3 \times \mathbb{E}^3$, the boundary surface deformation gradients $\widehat{\boldsymbol{f}}$ and $\widehat{\boldsymbol{F}}$ are rank-deficient and thus non-invertible, however, being pseudo-inverse results in

$$\widehat{\boldsymbol{F}} \cdot \widehat{\boldsymbol{f}} = [\widehat{\boldsymbol{g}}_\alpha \otimes \widehat{\boldsymbol{G}}^\alpha] \cdot [\widehat{\boldsymbol{G}}_\beta \otimes \widehat{\boldsymbol{g}}^\beta] = \widehat{\boldsymbol{g}}_\alpha \otimes \widehat{\boldsymbol{g}}^\alpha =: \widehat{\boldsymbol{i}}, \qquad (3.29)$$
$$\widehat{\boldsymbol{f}} \cdot \widehat{\boldsymbol{F}} = [\widehat{\boldsymbol{G}}_\alpha \otimes \widehat{\boldsymbol{g}}^\alpha] \cdot [\widehat{\boldsymbol{g}}_\beta \otimes \widehat{\boldsymbol{G}}^\beta] = \widehat{\boldsymbol{G}}_\alpha \otimes \widehat{\boldsymbol{G}}^\alpha =: \widehat{\boldsymbol{I}}.$$

Finally, since the boundary surface deformation gradients map between the tangent planes $T\partial\mathcal{B}_{\mathrm{m}}$ and $T\partial\mathcal{B}_{\mathrm{s}}$, also the following identities hold:

$$\widehat{\boldsymbol{F}} \cdot \widehat{\boldsymbol{I}} \equiv \widehat{\boldsymbol{i}} \cdot \widehat{\boldsymbol{F}} \equiv \widehat{\boldsymbol{F}} \quad \text{and} \quad \widehat{\boldsymbol{f}} \cdot \widehat{\boldsymbol{i}} \equiv \widehat{\boldsymbol{I}} \cdot \widehat{\boldsymbol{f}} \equiv \widehat{\boldsymbol{f}}. \qquad (3.30)$$

3.2.3 Compatibility Conditions for Tangent Maps

Surface Curl:
In surface convected coordinate representation, the material surface curl operator $\widehat{\mathrm{Curl}}(\bullet)$ of the (two-point) surface tensor $\widehat{\boldsymbol{Z}} = \widehat{Z}^\alpha{}_\beta \widehat{\boldsymbol{g}}_\alpha \otimes \widehat{\boldsymbol{G}}^\beta$ expands as $\widehat{\mathrm{Curl}}\widehat{\boldsymbol{Z}} := -[\widehat{Z}^\alpha{}_{\beta,\gamma} + \widehat{\gamma}^\alpha{}_{\delta\gamma}\widehat{Z}^\delta{}_\beta]\widehat{E}^{\beta\gamma}\widehat{\boldsymbol{g}}_\alpha$ (the contribution from the material surface Christoffel symbol vanishes due to its symmetry), thus for $\widehat{Z}^\alpha{}_{\beta,\gamma} + \widehat{\gamma}^\alpha{}_{\delta\gamma}\widehat{Z}^\delta{}_\beta$ symmetric in the index pair $\beta\gamma$, the material surface curl operator $\widehat{\mathrm{Curl}}\widehat{\boldsymbol{Z}}$ vanishes identically.

Due to their definitions as the boundary surface gradients of the boundary surface deformation maps, the material and spatial tangent maps (boundary surface deformation gradients) satisfy the following symmetry or rather *compatibility conditions*

$$\widehat{\mathrm{curl}}\widehat{\boldsymbol{f}} = \boldsymbol{0} \quad \text{and} \quad \widehat{\mathrm{Curl}}\widehat{\boldsymbol{F}} = \boldsymbol{0}. \qquad (3.31)$$

Observe that these compatibility conditions are consistent with the *(tangential) line theorem* equating the circuit integral of the vector-valued (tangential) line elements $\mathrm{d}\widehat{\boldsymbol{X}}$ and $\mathrm{d}\widehat{\boldsymbol{x}}$, respectively, over any closed circuit $\mathfrak{C}_{\mathrm{m}} = \partial\mathfrak{S}_{\mathrm{m}}$ and $\mathfrak{C}_{\mathrm{s}} = \partial\mathfrak{S}_{\mathrm{s}}$, respectively, to zero

$$\oint_{\mathfrak{C}_{\mathrm{m}}} \mathrm{d}\widehat{\boldsymbol{X}} = \oint_{\mathfrak{C}_{\mathrm{s}}} \widehat{\boldsymbol{f}} \cdot \mathrm{d}\widehat{\boldsymbol{x}} = \boldsymbol{0} \quad \text{and} \quad \oint_{\mathfrak{C}_{\mathrm{s}}} \mathrm{d}\widehat{\boldsymbol{x}} = \oint_{\mathfrak{C}_{\mathrm{m}}} \widehat{\boldsymbol{F}} \cdot \mathrm{d}\widehat{\boldsymbol{X}} = \boldsymbol{0}. \qquad (3.32)$$

Applying the boundary surface Stokes theorem

$$\oint_{\mathfrak{C}_{\mathrm{s}}} \widehat{\boldsymbol{f}} \cdot \mathrm{d}\widehat{\boldsymbol{x}} = \int_{\mathfrak{S}_{\mathrm{s}}} \widehat{\mathrm{curl}}\widehat{\boldsymbol{f}}\,\mathrm{d}a \quad \text{and} \quad \oint_{\mathfrak{C}_{\mathrm{m}}} \widehat{\boldsymbol{F}} \cdot \mathrm{d}\widehat{\boldsymbol{X}} = \int_{\mathfrak{S}_{\mathrm{m}}} \widehat{\mathrm{Curl}}\widehat{\boldsymbol{F}}\,\mathrm{d}A \qquad (3.33)$$

with $\widehat{\operatorname{curl}\widehat{\boldsymbol{f}}} := \operatorname{curl}\boldsymbol{f}\cdot\boldsymbol{n}$ and $\widehat{\operatorname{curl}\widehat{\boldsymbol{f}}} := \operatorname{Curl}\boldsymbol{F}\cdot\boldsymbol{N}$, and localizing for arbitrary closed circuits establishes the result.

The compatibility conditions may be stated alternatively as

$$\big[\boldsymbol{A}\cdot\widehat{\widehat{\nabla}_{\widehat{x}}\widehat{\boldsymbol{f}}}\big]^{\text{skw}} = \boldsymbol{0} \quad\text{and}\quad \big[\boldsymbol{a}\cdot\widehat{\widehat{\nabla}_{\widehat{X}}\widehat{\boldsymbol{F}}}\big]^{\text{skw}} = \boldsymbol{0}. \tag{3.34}$$

Here $\boldsymbol{A}\in T^*\mathcal{B}_{\text{m}}$ and $\boldsymbol{a}\in T^*\mathcal{B}_{\text{s}}$ denote arbitrary material and spatial (constant) covectors, respectively. Moreover, the *tangential gradients* of the boundary surface deformation gradients are defined as

$$\widehat{\widehat{\nabla}_{\widehat{x}}\widehat{\boldsymbol{f}}} := \widehat{\nabla}_{\widehat{x}}\widehat{\boldsymbol{f}}\cdot\widehat{\boldsymbol{i}} \quad\text{and}\quad \widehat{\widehat{\nabla}_{\widehat{X}}\widehat{\boldsymbol{F}}} := \widehat{\nabla}_{\widehat{X}}\widehat{\boldsymbol{F}}\cdot\widehat{\boldsymbol{I}}. \tag{3.35}$$

In surface convected coordinate representation for the tangent maps $\widehat{\boldsymbol{f}} = \widehat{\delta}^{\alpha}{}_{\beta}\widehat{\boldsymbol{G}}_{\alpha}\otimes\widehat{\boldsymbol{g}}^{\beta}$ and $\widehat{\boldsymbol{F}} = \widehat{\delta}^{\alpha}{}_{\beta}\widehat{\boldsymbol{g}}_{\alpha}\otimes\widehat{\boldsymbol{G}}^{\beta}$, respectively, the above compatibility conditions result in the symmetry conditions for the Christoffel symbols $\widehat{\Gamma}^{\alpha}{}_{\beta\gamma} = \widehat{\Gamma}^{\alpha}{}_{\gamma\beta}$ and $\widehat{\gamma}^{\alpha}{}_{\beta\gamma} = \widehat{\gamma}^{\alpha}{}_{\gamma\beta}$, respectively.

Consequently, for double contractions with (simple) second-order two-point tensors $z := \boldsymbol{A}\otimes\boldsymbol{b}$ and $\boldsymbol{Z} := \boldsymbol{a}\otimes\boldsymbol{B}$, whereby $\boldsymbol{b}\in T\mathcal{B}_{\text{s}}$ and $\boldsymbol{B}\in T\mathcal{B}_{\text{m}}$ denote arbitrary spatial and material vectors, respectively, due to the compatibility conditions it holds that

$$z : \widehat{\widehat{\nabla}_{\widehat{x}}\widehat{\boldsymbol{f}}} = \widehat{\widehat{\nabla}_{\widehat{x}}\widehat{\boldsymbol{f}}}^{t} : z \quad\text{and}\quad \boldsymbol{Z} : \widehat{\widehat{\nabla}_{\widehat{X}}\widehat{\boldsymbol{F}}} = \widehat{\widehat{\nabla}_{\widehat{X}}\widehat{\boldsymbol{F}}}^{t} : \boldsymbol{Z}. \tag{3.36}$$

The extension of the above to arbitrary second-order two-point tensors z and $\boldsymbol{Z}$ is straightforward.

3.2.4 Cotangent Maps: Cofactors

Surface Cofactor:
The surface cofactor $\widehat{\operatorname{cof}\widehat{\mathbf{Z}}}$ of a second-order surface (tangential) tensor $\widehat{\mathbf{Z}} := \mathbf{Z}\cdot\mathbb{P}\in\mathbb{E}^3\times\mathbb{E}^3$ follows via the *tensor-valued tensor single 'product'* as

$$1!\,\widehat{\operatorname{cof}\widehat{\mathbf{Z}}} := \widehat{\times\!\!\times}(\widehat{\mathbf{Z}}).$$

Thereby, $\widehat{\times\!\!\times}(\widehat{\mathbf{Z}})$ is defined in Cartesian coordinates with coordinate representation for $\widehat{\mathbf{Z}} = \widehat{Z}^{a}{}_{C}\boldsymbol{e}_a\otimes\boldsymbol{E}^C$ as

$$\widehat{\times\!\!\times}(\widehat{\mathbf{Z}}) := \widehat{e}_{ab}\widehat{E}^{CD}\widehat{Z}^{a}{}_{C}\boldsymbol{e}^{b}\otimes\boldsymbol{E}_D.$$

Here, the spatial and material surface permutation symbols compute as $\widehat{e}_{ab} =: e_{ab}{}^{e}\,n_e$ and $\widehat{E}^{CD} =: E^{CDF}N_F$, respectively.

Symbolically, based on the above coordinate definition, the surface cofactor $\widehat{\text{cof}}\,\widehat{\boldsymbol{Z}}$ of a second-order surface (tangential) tensor $\widehat{\boldsymbol{Z}} \in \mathbb{E}^3 \times \mathbb{E}^3$ thus reads as

$$1!\,\widehat{\text{cof}}\,\widehat{\boldsymbol{Z}} := [\,\widehat{\boldsymbol{e}}\,\widehat{\boxtimes}\,\widehat{\boldsymbol{E}}\,] : \widehat{\boldsymbol{Z}}.$$

Here, $\widehat{\boldsymbol{e}}$ and $\widehat{\boldsymbol{E}}$ denote the second-order spatial and material surface permutation tensors $\widehat{\boldsymbol{e}} := \boldsymbol{e} \cdot \boldsymbol{n} = \widehat{e}_{ab}\boldsymbol{e}^a \otimes \boldsymbol{e}^b$ and $\widehat{\boldsymbol{E}} := \boldsymbol{E} \cdot \boldsymbol{N} = \widehat{E}^{CD}\boldsymbol{E}_C \otimes \boldsymbol{E}_D$, respectively, with special dyadic surface product $\widehat{\boldsymbol{e}}\,\widehat{\boxtimes}\,\widehat{\boldsymbol{E}} := \widehat{e}_{ab}\widehat{E}^{CD}\boldsymbol{e}^a \otimes \boldsymbol{E}_C \otimes \boldsymbol{e}^b \otimes \boldsymbol{E}_D$.

Alternatively, for non-zero surface vectors $\widehat{\boldsymbol{A}} \in \mathbb{E}^3$ the surface cofactor $\widehat{\text{cof}}\,\widehat{\boldsymbol{Z}}$ of a second-order surface (tangential) tensor $\widehat{\boldsymbol{Z}} \in \mathbb{E}^3 \times \mathbb{E}^3$ is defined via

$$\widehat{\text{cof}}\,\widehat{\boldsymbol{Z}} \cdot \widehat{\times}(\widehat{\boldsymbol{A}}) := \widehat{\times}(\widehat{\boldsymbol{Z}} \cdot \widehat{\boldsymbol{A}}) \quad \forall \widehat{\boldsymbol{A}} \quad \in \mathbb{E}^3.$$

Here, the *vector-valued vector single 'products'* $\widehat{\times}(\widehat{\boldsymbol{A}})$ and $\widehat{\times}(\widehat{\boldsymbol{Z}} \cdot \widehat{\boldsymbol{A}})$ follow in Cartesian coordinates as $\widehat{\times}(\widehat{\boldsymbol{A}}) := \widehat{E}_{ED}\widehat{A}^E\boldsymbol{E}^D$ and $\widehat{\times}(\widehat{\boldsymbol{Z}} \cdot \widehat{\boldsymbol{A}}) := \widehat{e}_{ab}[\widehat{Z}^a{}_C\widehat{A}^C]\boldsymbol{e}^b$, respectively.

Finally, for a pseudo-invertible second-order surface (tangential) tensor $\widehat{\boldsymbol{Z}} \in \mathbb{E}^3 \times \mathbb{E}^3$, the surface cofactor $\widehat{\text{cof}}\,\widehat{\boldsymbol{Z}}$ takes the closed-form expression

$$\widehat{\text{cof}}\,\widehat{\boldsymbol{Z}} = \widehat{\det}\widehat{\boldsymbol{Z}}\,\widehat{\boldsymbol{Z}}^{\widehat{-t}} \in \mathbb{E}^3 \times \mathbb{E}^3.$$

Here, $\widehat{\det}\widehat{\boldsymbol{Z}}$ denotes the surface determinant of $\widehat{\boldsymbol{Z}}$; see below.

Material and spatial vector-valued-oriented *line elements* (cotangents) $\mathrm{d}\widehat{\boldsymbol{L}} = \widehat{\boldsymbol{N}}\,\mathrm{d}L$ and $\mathrm{d}\widehat{\boldsymbol{l}} = \widehat{\boldsymbol{n}}\,\mathrm{d}l$ in the cotangent planes $T^*\partial\mathcal{B}_\mathrm{m}$ and $T^*\partial\mathcal{B}_\mathrm{s}$, respectively, are related via the *boundary surface cofactors* of the boundary surface deformation gradients *(boundary surface Nanson's formulae)*

$$\mathrm{d}\widehat{\boldsymbol{L}} = \widehat{\text{cof}}\,\widehat{\boldsymbol{f}} \cdot \mathrm{d}\widehat{\boldsymbol{l}} \quad \text{and} \quad \mathrm{d}\widehat{\boldsymbol{l}} = \widehat{\text{cof}}\,\widehat{\boldsymbol{F}} \cdot \mathrm{d}\widehat{\boldsymbol{L}}. \tag{3.37}$$

The boundary surface cofactors of the boundary surface deformation gradients are also denoted as the *line cotangent maps*

$$\widehat{\boldsymbol{k}} := \widehat{\text{cof}}\,\widehat{\boldsymbol{f}} \in T^*\partial\mathcal{B}_\mathrm{m} \times T^*\partial\mathcal{B}_\mathrm{s} \ \text{and}\ \widehat{\boldsymbol{K}} := \widehat{\text{cof}}\,\widehat{\boldsymbol{F}} \in T^*\partial\mathcal{B}_\mathrm{m} \times T^*\partial\mathcal{B}_\mathrm{m}. \tag{3.38}$$

The material and spatial cotangent maps are pseudo-inverse surface tensors

$$\widehat{\boldsymbol{k}} = \widehat{\boldsymbol{K}}^{\widehat{-1}} = \widehat{j}\,\widehat{\boldsymbol{F}}^t \quad \text{and} \quad \widehat{\boldsymbol{K}} = \widehat{\boldsymbol{k}}^{\widehat{-1}} = \widehat{J}\,\widehat{\boldsymbol{f}}^t. \tag{3.39}$$

3.2.5 *Compatibility Conditions for Cotangent Maps*

> **Divergence of Surface Cofactor:**
> The material surface divergence operator $\widehat{\text{Div}}(\bullet)$ applied to the tensor-valued tensor single 'product' $\widehat{\times\!\times}(\widehat{\boldsymbol{Z}}) := \widehat{\boldsymbol{e}}^t \cdot \widehat{\boldsymbol{Z}} \cdot \widehat{\boldsymbol{E}}$ of the (two-point) surface tensor $\widehat{\boldsymbol{Z}}$ expands as $\widehat{\text{Div}}\big(\widehat{\times\!\times}(\widehat{\boldsymbol{Z}})\big) = -\widehat{\boldsymbol{e}}^t \cdot \widehat{\text{Curl}\boldsymbol{Z}} - \widehat{J}\big[\boldsymbol{e} \cdot [\widehat{\boldsymbol{Z}} \cdot \widehat{\boldsymbol{f}}] \cdot \widehat{\boldsymbol{e}}\big] : \widehat{\boldsymbol{c}}$. The first term vanishes for $\widehat{\text{Curl}\boldsymbol{Z}} = \boldsymbol{0}$, i.e. for $\widehat{\nabla}_{\widehat{X}} \widehat{\boldsymbol{Z}} \cdot \widehat{\boldsymbol{I}}$ symmetric in its last two indices, and for $\widehat{\boldsymbol{Z}} = \widehat{\boldsymbol{F}}$ the last term degenerates to $\widehat{J}\,\widehat{c}\,\boldsymbol{n}$.

Due to their definitions as the tensor-valued tensor single 'products' of the *compatible* boundary surface deformation gradients, the material and spatial cotangent maps satisfy the following *compatibility conditions (boundary surface Piola identity)*

$$\widehat{\text{div}}\widehat{\boldsymbol{k}} = \widehat{j}\,\widehat{C}\,\boldsymbol{N} \quad \text{and} \quad \widehat{\text{Div}}\widehat{\boldsymbol{K}} = \widehat{J}\,\widehat{c}\,\boldsymbol{n}. \tag{3.40}$$

The boundary surface Piola identity constitutes the specialization of the Piola identity in the bulk to boundary surfaces. For flat boundary surfaces with $\widehat{c} = \widehat{C} = 0$, the similarity is particularly obvious.

Observe that these compatibility conditions are consistent with the *(normal) line theorem* equating the integral of the vector-valued line elements $\mathrm{d}\widehat{\boldsymbol{L}}$ and $\mathrm{d}\widehat{\boldsymbol{l}}$, respectively, over the closed boundary curve of a control surface $\mathfrak{S}_{\text{m}} \subseteq \partial\mathcal{B}_{\text{m}}$ and $\mathfrak{S}_{\text{s}} \subseteq \partial\mathcal{B}_{\text{s}}$, respectively, to

$$\int_{\partial\mathfrak{S}_{\text{m}}} \mathrm{d}\widehat{\boldsymbol{L}} = \int_{\partial\mathfrak{S}_{\text{s}}} \widehat{\boldsymbol{k}} \cdot \mathrm{d}\widehat{\boldsymbol{l}} = \int_{\mathfrak{S}_{\text{m}}} \widehat{C}\,\mathrm{d}\boldsymbol{A} \quad \text{and} \quad \int_{\partial\mathfrak{S}_{\text{s}}} \mathrm{d}\widehat{\boldsymbol{l}} = \int_{\partial\mathfrak{S}_{\text{m}}} \widehat{\boldsymbol{K}} \cdot \mathrm{d}\widehat{\boldsymbol{L}} = \int_{\mathfrak{S}_{\text{s}}} \widehat{c}\,\mathrm{d}\boldsymbol{a}. \tag{3.41}$$

Applying the surface Gauss theorem and localizing for arbitrary surfaces establishes the result.

3.2.6 *Measure Maps: Determinants*

Material and spatial scalar-valued boundary surface *area elements* (measures) $\mathrm{d}A$ and $\mathrm{d}a$ in $\partial\mathcal{B}_{\text{m}}$ and $\partial\mathcal{B}_{\text{s}}$, respectively, are related via the *boundary surface determinants* of the boundary surface deformation gradients

$$\mathrm{d}A = \widehat{\det}\widehat{\boldsymbol{f}}\,\mathrm{d}a \quad \text{and} \quad \mathrm{d}a = \widehat{\det}\widehat{\boldsymbol{F}}\,\mathrm{d}A. \tag{3.42}$$

Surface Determinant:
The surface determinant $\widehat{\det}\widehat{\boldsymbol{Z}}$ of a second-order surface (tangential) tensor $\widehat{\boldsymbol{Z}} := \boldsymbol{Z} \cdot \mathbb{P} \in \mathbb{E}^3 \times \mathbb{E}^3$ follows via the *scalar-valued tensor double product* as

$$2!\,\widehat{\det}\widehat{\boldsymbol{Z}} := \widehat{\boldsymbol{Z}} \widehat{\times\!\!\times} \widehat{\boldsymbol{Z}}.$$

Thereby, $\widehat{\boldsymbol{Z}} \widehat{\times\!\!\times} \widehat{\boldsymbol{Z}}$ is defined in Cartesian coordinates with coordinate representation for $\widehat{\boldsymbol{Z}} = \widehat{Z}^a{}_C \boldsymbol{e}_a \otimes \boldsymbol{E}^C$ as

$$\widehat{\boldsymbol{Z}} \widehat{\times\!\!\times} \widehat{\boldsymbol{Z}} := \widehat{e}_{ab} \widehat{E}^{CD} \widehat{Z}^a{}_C Z^b{}_D.$$

Symbolically, based on the above coordinate definition, the surface determinant $\widehat{\det}\widehat{\boldsymbol{Z}}$ of a second-order surface (tangential) tensor $\widehat{\boldsymbol{Z}} \in \mathbb{E}^3 \times \mathbb{E}^3$ thus reads as

$$2!\,\widehat{\det}\widehat{\boldsymbol{Z}} := [\widehat{\boldsymbol{e}} \boxtimes \widehat{\boldsymbol{E}}] :: [\widehat{\boldsymbol{Z}} \otimes \widehat{\boldsymbol{Z}}].$$

Alternatively, for linear independent and non-zero vectors $\widehat{\boldsymbol{A}} := \boldsymbol{A} \cdot \mathbb{P} \in \mathbb{E}^3$ and $\widehat{\boldsymbol{B}} := \boldsymbol{B} \cdot \mathbb{P} \in \mathbb{E}^3$, the surface determinant $\widehat{\det}\widehat{\boldsymbol{Z}}$ of a second-order surface (tangential) tensor $\widehat{\boldsymbol{Z}} \in \mathbb{E}^3 \times \mathbb{E}^3$ is defined via

$$\widehat{\det}\widehat{\boldsymbol{Z}}\,[\widehat{\boldsymbol{A}} \widehat{\times} \widehat{\boldsymbol{B}}] := [\widehat{\boldsymbol{Z}} \cdot \widehat{\boldsymbol{A}}] \widehat{\times} [\widehat{\boldsymbol{Z}} \cdot \widehat{\boldsymbol{B}}] \quad \forall \widehat{\boldsymbol{A}}, \widehat{\boldsymbol{B}} \quad \in \mathbb{E}^3.$$

Here, the *scalar-valued vector double products* $\widehat{\boldsymbol{A}} \widehat{\times} \widehat{\boldsymbol{B}}$ and $[\widehat{\boldsymbol{Z}} \cdot \widehat{\boldsymbol{A}}] \widehat{\times} [\widehat{\boldsymbol{Z}} \cdot \widehat{\boldsymbol{B}}]$ follow in Cartesian coordinates as $\widehat{\boldsymbol{A}} \widehat{\times} \widehat{\boldsymbol{B}} := \widehat{E}_{EF} \widehat{A}^E \widehat{B}^F$ and $[\widehat{\boldsymbol{Z}} \cdot \widehat{\boldsymbol{A}}] \widehat{\times} [\widehat{\boldsymbol{Z}} \cdot \widehat{\boldsymbol{B}}] := \widehat{e}_{gh} [\widehat{Z}^g{}_E \widehat{A}^E][\widehat{Z}^h{}_F \widehat{A}^F]$, respectively.

Finally, for a pseudo-invertible second-order surface (tangential) tensor $\widehat{\boldsymbol{Z}} \in \mathbb{E}^3 \times \mathbb{E}^3$ the surface determinant $\widehat{\det}\widehat{\boldsymbol{Z}}$ takes the closed-form expression

$$2\,\widehat{\det}\widehat{\boldsymbol{Z}} = \widehat{\operatorname{cof}}\,\widehat{\boldsymbol{Z}} : \widehat{\boldsymbol{Z}}.$$

The boundary surface determinants of the boundary surface deformation gradients are also denoted as the *boundary surface Jacobians* or rather the *area measure maps*

$$\widehat{j} = \widehat{J}^{-1} =: \widehat{\det}\widehat{\boldsymbol{f}} \in \mathbb{R}^+ \quad \text{and} \quad \widehat{J} = \widehat{j}^{-1} =: \widehat{\det}\widehat{\boldsymbol{F}} \in \mathbb{R}^+. \tag{3.43}$$

Alternatively, the boundary surface Jacobians result from Nanson's formulae in Eq. 2.37 as

$$\widehat{j} = |\operatorname{cof} \boldsymbol{f} \cdot \boldsymbol{n}| \quad \text{and} \quad \widehat{J} = |\operatorname{cof} \boldsymbol{F} \cdot \boldsymbol{N}| \tag{3.44}$$

and may thus be expressed as

$$\widehat{\det}\widehat{\boldsymbol{f}} := \frac{|[\widehat{\boldsymbol{f}} \cdot \widehat{\boldsymbol{g}}_1] \times [\widehat{\boldsymbol{f}} \cdot \widehat{\boldsymbol{g}}_2]|}{|\widehat{\boldsymbol{g}}_1 \times \widehat{\boldsymbol{g}}_2|} \quad \text{and} \quad \widehat{\det}\widehat{\boldsymbol{F}} := \frac{|[\widehat{\boldsymbol{F}} \cdot \widehat{\boldsymbol{G}}_1] \times [\widehat{\boldsymbol{F}} \cdot \widehat{\boldsymbol{G}}_2]|}{|\widehat{\boldsymbol{G}}_1 \times \widehat{\boldsymbol{G}}_2|}. \tag{3.45}$$

Note the similarity to the definition of the determinant of the deformation gradients in the bulk. The proof of the above representation follows from observing that $\widehat{\boldsymbol{f}} \cdot \widehat{\boldsymbol{g}}_\alpha = \boldsymbol{f} \cdot \widehat{\boldsymbol{g}}_\alpha$ and $\widehat{\boldsymbol{F}} \cdot \widehat{\boldsymbol{G}}_\alpha = \boldsymbol{F} \cdot \widehat{\boldsymbol{G}}_\alpha$ jointly with the Nanson's formulae in Eq. 2.37, thus

$$\frac{[\boldsymbol{f} \cdot \widehat{\boldsymbol{g}}_1] \times [\boldsymbol{f} \cdot \widehat{\boldsymbol{g}}_2]}{|\widehat{\boldsymbol{g}}_1 \times \widehat{\boldsymbol{g}}_2|} = \operatorname{cof} \boldsymbol{f} \cdot \boldsymbol{n} \quad \text{and} \quad \frac{[\boldsymbol{F} \cdot \widehat{\boldsymbol{G}}_1] \times [\boldsymbol{F} \cdot \widehat{\boldsymbol{G}}_2]}{|\widehat{\boldsymbol{G}}_1 \times \widehat{\boldsymbol{G}}_2|} = \operatorname{cof} \boldsymbol{F} \cdot \boldsymbol{N}. \tag{3.46}$$

Observe also that, for example, $\widehat{\boldsymbol{g}}_1 \mathbin{\widehat{\times}} \widehat{\boldsymbol{g}}_2 = [\widehat{\boldsymbol{g}}_1 \times \widehat{\boldsymbol{g}}_2] \cdot \boldsymbol{n} = |\widehat{\boldsymbol{g}}_1 \times \widehat{\boldsymbol{g}}_2|$ and likewise $\widehat{\boldsymbol{G}}_1 \mathbin{\widehat{\times}} \widehat{\boldsymbol{G}}_2 = [\widehat{\boldsymbol{G}}_1 \times \widehat{\boldsymbol{G}}_2] \cdot \boldsymbol{N} = |\widehat{\boldsymbol{G}}_1 \times \widehat{\boldsymbol{G}}_2|$.

3.2.7 Velocities

The total time derivatives of the material and spatial boundary surface position maps $\widehat{\boldsymbol{X}} = \widehat{\boldsymbol{\Upsilon}}(\widehat{\boldsymbol{\Xi}}, t)$ and $\widehat{\boldsymbol{x}} = \widehat{\boldsymbol{\upsilon}}(\widehat{\boldsymbol{\Xi}}, t)$, projected onto the *tangent plane* to the boundary surface, are introduced as the *total material* and *spatial boundary surface velocities*

$$\widehat{\boldsymbol{W}} := \dot{\widehat{\boldsymbol{\Upsilon}}}(\widehat{\boldsymbol{\Xi}}, t) \cdot \widehat{\boldsymbol{I}} = \boldsymbol{W}|_{\partial\mathcal{B}_r} \cdot \widehat{\boldsymbol{I}} \quad \text{and} \quad \widehat{\boldsymbol{w}} := \dot{\widehat{\boldsymbol{\upsilon}}}(\widehat{\boldsymbol{\Xi}}, t) \cdot \widehat{\boldsymbol{i}} = \boldsymbol{w}|_{\partial\mathcal{B}_r} \cdot \widehat{\boldsymbol{i}}. \tag{3.47}$$

Moreover, the spatial time derivative of the material boundary surface deformation map $\widehat{\boldsymbol{X}} = \widehat{\boldsymbol{Y}}(\widehat{\boldsymbol{x}}, t)$ and the material time derivative of the spatial boundary surface deformation map $\widehat{\boldsymbol{x}} = \widehat{\boldsymbol{y}}(\widehat{\boldsymbol{X}}, t)$, projected onto the tangent plane to the boundary surface, result in the *material* and the *spatial boundary surface velocity*

$$\widehat{\boldsymbol{V}} := \mathrm{d}_t\widehat{\boldsymbol{Y}}(\widehat{\boldsymbol{x}}, t) \cdot \widehat{\boldsymbol{I}} = \boldsymbol{V}|_{\partial\mathcal{B}_s} \cdot \widehat{\boldsymbol{I}} \quad \text{and} \quad \widehat{\boldsymbol{v}} := \mathrm{D}_t\widehat{\boldsymbol{y}}(\widehat{\boldsymbol{X}}, t) \cdot \widehat{\boldsymbol{i}} = \boldsymbol{v}|_{\partial\mathcal{B}_m} \cdot \widehat{\boldsymbol{i}}. \tag{3.48}$$

With these definitions at hand, it is straightforward to show that the total material and spatial boundary surface velocities, based on boundary surface Euler formulae for the total time derivative, are related as follows:

$$\widehat{\boldsymbol{W}} = \widehat{\boldsymbol{V}} + \widehat{\boldsymbol{f}} \cdot \widehat{\boldsymbol{w}} \quad \text{and} \quad \widehat{\boldsymbol{w}} = \widehat{\boldsymbol{v}} + \widehat{\boldsymbol{F}} \cdot \widehat{\boldsymbol{W}}. \tag{3.49}$$

The relation between the material and spatial boundary surface velocities via the boundary surface deformation gradients is then finally a direct consequence of the above relations

$$\widehat{\boldsymbol{V}} = -\widehat{\boldsymbol{f}} \cdot \widehat{\boldsymbol{v}} \quad \text{and} \quad \widehat{\boldsymbol{v}} = -\widehat{\boldsymbol{F}} \cdot \widehat{\boldsymbol{V}}. \tag{3.50}$$

Normal–Tangential Decomposition on Surface Boundary:
On the regular boundary curve of a control surface, the total material and spatial boundary surface velocities that live in the tangent plane to the boundary surface are conveniently decomposed into normal and tangential parts

$$\widehat{\boldsymbol{W}} = \widehat{\boldsymbol{W}}_\perp + \widehat{\boldsymbol{W}}_\parallel \quad \text{on} \quad \partial\mathfrak{S}_\mathrm{m} \quad \text{and} \quad \widehat{\boldsymbol{w}} = \widehat{\boldsymbol{w}}_\perp + \widehat{\boldsymbol{w}}_\parallel \quad \text{on} \quad \partial\mathfrak{S}_\mathrm{s}. \tag{3.51}$$

Thereby, the normal parts are defined in terms of the scalar-valued *normal velocities* $\widehat{W}_\perp$ and $\widehat{w}_\perp$, respectively, as

$$\widehat{\boldsymbol{W}}_\perp := [\widehat{\boldsymbol{W}} \cdot \widehat{\boldsymbol{N}}]\widehat{\boldsymbol{N}} =: \widehat{W}_\perp \widehat{\boldsymbol{N}} \quad \text{and} \quad \widehat{\boldsymbol{w}}_\perp := [\widehat{\boldsymbol{w}} \cdot \widehat{\boldsymbol{n}}]\widehat{\boldsymbol{n}} =: \widehat{w}_\perp \widehat{\boldsymbol{n}}$$

and, consequently, the tangential parts follow as

$$\widehat{\boldsymbol{W}}_\parallel := \widehat{\boldsymbol{W}} - \widehat{\boldsymbol{W}}_\perp =: \widehat{\mathbb{P}} \cdot \widehat{\boldsymbol{W}} \quad \text{and} \quad \widehat{\boldsymbol{w}}_\parallel := \widehat{\boldsymbol{w}} - \widehat{\boldsymbol{w}}_\perp =: \widehat{\mathbb{p}} \cdot \widehat{\boldsymbol{w}}.$$

Here, $\widehat{\mathbb{P}}$ and $\widehat{\mathbb{p}}$ are the material and spatial (mixed-variant) *projection operators* onto the tangent line to the boundary curve of the control surface

$$\widehat{\mathbb{P}} := \widehat{\boldsymbol{I}} - \widehat{\boldsymbol{N}} \otimes \widehat{\boldsymbol{N}} \quad \text{on} \quad \partial\mathfrak{S}_\mathrm{m} \quad \text{and} \quad \widehat{\mathbb{p}} := \widehat{\boldsymbol{i}} - \widehat{\boldsymbol{n}} \otimes \widehat{\boldsymbol{n}} \quad \text{on} \quad \partial\mathfrak{S}_\mathrm{s},$$

with $\widehat{\boldsymbol{I}}$ and $\widehat{\boldsymbol{i}}$ the material and spatial (mixed-variant) boundary surface unit tensors, respectively.

3.3 Deformation on Boundary Curves*

Geometry on Curves:
A smooth curve $\mathcal{C}$, i.e. a one-dimensional manifold embedded in the three-dimensional Euclidean space $\mathbb{E}^3$ with, say, coordinates $\widetilde{\boldsymbol{X}}$ may be parameterized by the arc-length Θ as

$$\widetilde{\boldsymbol{X}} = \widetilde{\boldsymbol{X}}(\Theta). \tag{3.52}$$

The corresponding tangent vector $\widetilde{\boldsymbol{G}} \in T\mathcal{C}$ and the (principal) normal and bi-normal vectors $\widetilde{\boldsymbol{N}}_\perp$ and $\widetilde{\boldsymbol{N}}_\times$, orthogonal to $T\mathcal{C}$, follow as

$$\widetilde{\boldsymbol{G}} = \widetilde{\boldsymbol{X}}' \quad \text{and} \quad \widetilde{\boldsymbol{N}}_\perp = \widetilde{\boldsymbol{G}}'/|\widetilde{\boldsymbol{G}}'| \quad \text{and} \quad \widetilde{\boldsymbol{N}}_\times = \widetilde{\boldsymbol{G}} \times \widetilde{\boldsymbol{N}}_\perp. \tag{3.53}$$

The tangent vector $\widetilde{\boldsymbol{G}}$ has unit length due to the parametrization of the curve in its arc-length Θ, thus, the curve line element $\mathrm{d}L$ follows as

$$\mathrm{d}L = |\widetilde{X}'|\,\mathrm{d}\Theta = |\widetilde{G}|\,\mathrm{d}\Theta = \mathrm{d}\Theta. \tag{3.54}$$

With the common (mixed-variant) unit tensor $\boldsymbol{I}$ of the three-dimensional, embedding Euclidean space, the (mixed-variant) curve unit tensor $\widetilde{\boldsymbol{I}}$ follows as

$$\widetilde{\boldsymbol{I}} := \widetilde{\boldsymbol{G}} \otimes \widetilde{\boldsymbol{G}} = \boldsymbol{I} - \widetilde{\boldsymbol{N}}_\perp \otimes \widetilde{\boldsymbol{N}}_\perp - \widetilde{\boldsymbol{N}}_\times \otimes \widetilde{\boldsymbol{N}}_\times. \tag{3.55}$$

Note that the curve unit tensor $\widetilde{\boldsymbol{I}}$ coincides with the previously introduced (idempotent) curve projection tensor $\widehat{\mathbb{P}}$. Moreover, with the permutation tensor $\boldsymbol{E}$ of the three-dimensional, embedding Euclidean space, the curve 'permutation' vector $\widetilde{\boldsymbol{E}}$ follows *formally* as

$$\widetilde{\boldsymbol{E}} := \big[[\widetilde{\boldsymbol{N}}_\perp \times \widetilde{\boldsymbol{N}}_\times] \cdot \widetilde{\boldsymbol{G}}\big]\widetilde{\boldsymbol{G}} := [\widetilde{\boldsymbol{N}}_\perp \otimes \widetilde{\boldsymbol{N}}_\times] : \boldsymbol{E} = \widetilde{\boldsymbol{N}}_\perp \times \widetilde{\boldsymbol{N}}_\times \equiv \widetilde{\boldsymbol{G}}. \tag{3.56}$$

Moreover, the curve 'permutation' tensors $\widetilde{\boldsymbol{E}}_\perp$ and $\widetilde{\boldsymbol{E}}_\times$ are introduced as

$$\widetilde{\boldsymbol{E}}_\perp := \boldsymbol{E} \cdot \widetilde{\boldsymbol{N}}_\perp \quad \text{and} \quad \widetilde{\boldsymbol{E}}_\times := \boldsymbol{E} \cdot \widetilde{\boldsymbol{N}}_\times. \tag{3.57}$$

Next, the curve gradient and curve divergence operators for vector-valued (curve) fields are defined as

$$\widetilde{\nabla}_{\widetilde{X}}(\bullet) := (\bullet)' \otimes \widetilde{\boldsymbol{G}} \quad \text{and} \quad \widetilde{\mathrm{Div}}(\bullet) := (\bullet)' \cdot \widetilde{\boldsymbol{G}}. \tag{3.58}$$

Clearly, $\widetilde{\nabla}_{\widetilde{X}}(\bullet) \cdot \widetilde{\boldsymbol{N}}_\perp = \boldsymbol{0}$ and $\widetilde{\nabla}_{\widetilde{X}}(\bullet) \cdot \widetilde{\boldsymbol{N}}_\times = \boldsymbol{0}$ vanish by definition. In addition, the (scalar-valued) curl operators for vector-valued (curve) fields are defined as

$$-\widetilde{\mathrm{Curl}}_\perp(\bullet) := [(\bullet)' \times \widetilde{\boldsymbol{G}}] \cdot \widetilde{\boldsymbol{N}}_\perp \quad \text{and} \quad -\widetilde{\mathrm{Curl}}_\times(\bullet) := [(\bullet)' \times \widetilde{\boldsymbol{G}}] \cdot \widetilde{\boldsymbol{N}}_\times. \tag{3.59}$$

The curve divergence and curve curl(s) are alternatively expressed as the double contraction of the curve gradient with the curve unit tensor and the curve permutation tensors, respectively,

$$\begin{aligned} \widetilde{\mathrm{Div}}(\bullet) &= \big[(\bullet)' \otimes \widetilde{\boldsymbol{G}}\big] : \widetilde{\boldsymbol{I}} &&= \widetilde{\nabla}_{\widetilde{X}}(\bullet) : \widetilde{\boldsymbol{I}}\,, \\ -\widetilde{\mathrm{Curl}}_\perp(\bullet) &= \big[(\bullet)' \otimes \widetilde{\boldsymbol{G}}\big] : \widetilde{\boldsymbol{E}}_\perp &&= \widetilde{\nabla}_{\widetilde{X}}(\bullet) : \widetilde{\boldsymbol{E}}_\perp, \\ -\widetilde{\mathrm{Curl}}_\times(\bullet) &= \big[(\bullet)' \otimes \widetilde{\boldsymbol{G}}\big] : \widetilde{\boldsymbol{E}}_\times &&= \widetilde{\nabla}_{\widetilde{X}}(\bullet) : \widetilde{\boldsymbol{E}}_\times. \end{aligned} \tag{3.60}$$

For fields that are smooth in a three-dimensional neighborhood $\mathcal{N}$ of the curve, the curve gradient and the curve divergence operators are alternatively defined by the help of the curve unit tensor as

$$\widetilde{\nabla}_{\widetilde{X}}(\bullet) := \nabla_X(\bullet)|_{\mathcal{C}} \cdot \widetilde{\boldsymbol{I}} \quad \text{and} \quad \widetilde{\mathrm{Div}}(\bullet) := \nabla_X(\bullet)|_{\mathcal{C}} : \widetilde{\boldsymbol{I}}. \tag{3.61}$$

Likewise, in this case the curve curl operators are alternatively defined by the help of the curve permutation tensors as

$$-\widetilde{\text{Curl}}_\times(\bullet) := \nabla_X(\bullet)|_\mathcal{C} : \widetilde{\boldsymbol{E}}_\times = -\text{Curl}(\bullet)|_\mathcal{C} \cdot \widetilde{\boldsymbol{N}}_\times, \tag{3.62}$$
$$-\widetilde{\text{Curl}}_\perp(\bullet) := \nabla_X(\bullet)|_\mathcal{C} : \widetilde{\boldsymbol{E}}_\perp = -\text{Curl}(\bullet)|_\mathcal{C} \cdot \widetilde{\boldsymbol{N}}_\perp.$$

The curve gradient, curve divergence, and curve 'curl' operators for tensor-valued fields $(\circ)$ follow from multiplying $(\circ)$ from the left by an arbitrary constant vector and inserting the so resulting vector-valued field into the above definitions. Then, on $\mathcal{C}$ the product rule of differentiation $\widetilde{\text{Div}}\big((\bullet)\cdot(\circ)\big) = \widetilde{\nabla}_{\widetilde{X}}(\bullet) : (\circ) + (\bullet)\cdot\widetilde{\text{Div}}(\circ)$ holds in particular since

$$\big[(\bullet)\cdot(\circ)\big]' \cdot \widetilde{\boldsymbol{G}} = [(\bullet)' \otimes \widetilde{\boldsymbol{G}}] : (\circ) + (\bullet)\cdot(\circ)'\cdot\widetilde{\boldsymbol{G}}. \tag{3.63}$$

Finally, the (parallel and perpendicular) curve curvature tensors $\widetilde{\boldsymbol{C}}_\parallel = \widetilde{C}\widetilde{\boldsymbol{N}}_\perp \otimes \widetilde{\boldsymbol{G}}$ and $\widetilde{\boldsymbol{C}}_\perp = \widetilde{C}\widetilde{\boldsymbol{G}} \otimes \widetilde{\boldsymbol{G}} - \widetilde{T}\widetilde{\boldsymbol{N}}_\times \otimes \widetilde{\boldsymbol{G}}$ together with the curve curvature $\widetilde{C}$ are defined as the curve gradient of the curve tangent $\widetilde{\boldsymbol{G}}$

$$\widetilde{\boldsymbol{C}}_\parallel := \widetilde{\nabla}_{\widetilde{X}}\widetilde{\boldsymbol{G}} = \widetilde{\boldsymbol{G}}' \otimes \widetilde{\boldsymbol{G}} \tag{3.64}$$

and the (negative) curve gradient

$$\widetilde{\boldsymbol{C}}_\perp := -\widetilde{\nabla}_{\widetilde{X}}\widetilde{\boldsymbol{N}}_\perp = -\widetilde{\boldsymbol{N}}'_\perp \otimes \widetilde{\boldsymbol{G}} \tag{3.65}$$

and (negative) curve divergence of the curve (principal) normal $\widetilde{\boldsymbol{N}}_\perp$, respectively,

$$\widetilde{C} := -\widetilde{\text{Div}}\widetilde{\boldsymbol{N}}_\perp = -\widetilde{\boldsymbol{N}}'_\perp \cdot \widetilde{\boldsymbol{G}}. \tag{3.66}$$

Thus, the curve divergence of the curve unit tensor $\widetilde{\boldsymbol{I}}$ and the curve 'permutation' vector result in

$$\widetilde{\text{Div}}\,\widetilde{\boldsymbol{I}} = \widetilde{C}\widetilde{\boldsymbol{N}}_\perp \quad \text{and} \quad \widetilde{\text{Div}}\,\widetilde{\boldsymbol{E}} = 0. \tag{3.67}$$

Note that the curve torsion is computed by $\widetilde{T} = \widetilde{\boldsymbol{N}}'_\perp \cdot \widetilde{\boldsymbol{N}}_\times = -\widetilde{\boldsymbol{N}}'_\times \cdot \widetilde{\boldsymbol{N}}_\perp$.

The above relations are rooted in the *Frénet–Serret* formulae

$$\widetilde{\boldsymbol{G}}' = \widetilde{C}\widetilde{\boldsymbol{N}}_\perp \quad \text{and} \quad \widetilde{\boldsymbol{N}}'_\perp = -\widetilde{C}\widetilde{\boldsymbol{G}} + \widetilde{T}\widetilde{\boldsymbol{N}}_\times \quad \text{and} \quad \widetilde{\boldsymbol{N}}'_\times = -\widetilde{T}\widetilde{\boldsymbol{N}}_\perp. \tag{3.68}$$

Alternatively, in terms of the *Darboux* (axial) vector $\widetilde{\boldsymbol{D}} := \widetilde{T}\widetilde{\boldsymbol{G}} + \widetilde{C}\widetilde{\boldsymbol{N}}_\times$, these read as

$$\widetilde{\boldsymbol{G}}' = \widetilde{\boldsymbol{D}} \times \widetilde{\boldsymbol{G}} \quad \text{and} \quad \widetilde{\boldsymbol{N}}'_\perp = \widetilde{\boldsymbol{D}} \times \widetilde{\boldsymbol{N}}_\perp \quad \text{and} \quad \widetilde{\boldsymbol{N}}'_\times = \widetilde{\boldsymbol{D}} \times \widetilde{\boldsymbol{N}}_\times. \tag{3.69}$$

Example: Consider a helical curve with radius R, pitch $2\pi H$ (thus with slope H/R) and arc-length Θ, and define $\widetilde{R}^2 := R^2 + H^2$, $\widetilde{1} := R/\widetilde{R}$ and $\widetilde{S} := H/\widetilde{R}$ together with $\widetilde{C} := \widetilde{1}/\widetilde{R}$ and $\widetilde{T} := \widetilde{S}/\widetilde{R}$. Then

$$\widetilde{\boldsymbol{X}} = R\cos(\Theta/\widetilde{R})\boldsymbol{E}_1 + R\sin(\Theta/\widetilde{R})\boldsymbol{E}_2 + \widetilde{S}\Theta\boldsymbol{E}_3. \tag{3.70}$$

Thus, the curve tangent computes as

$$\widetilde{\boldsymbol{G}} = -\widetilde{1}\sin(\Theta/\widetilde{R})\boldsymbol{E}_1 + \widetilde{1}\cos(\Theta/\widetilde{R})\boldsymbol{E}_2 + \widetilde{S}\boldsymbol{E}_3 \tag{3.71}$$

and the *inwards*-pointing (principal) normal vector reads

$$\widetilde{\boldsymbol{N}}_\perp = -\cos(\Theta/\widetilde{R})\boldsymbol{E}_1 - \sin(\Theta/\widetilde{R})\boldsymbol{E}_2, \tag{3.72}$$

while the curve bi-normal vector follows as

$$\widetilde{\boldsymbol{N}}_\times = \widetilde{S}\sin(\Theta/\widetilde{R})\boldsymbol{E}_1 - \widetilde{S}\cos(\Theta/\widetilde{R})\boldsymbol{E}_2 + \widetilde{1}\boldsymbol{E}_3. \tag{3.73}$$

Eventually, from determining $\widetilde{R}\,\widetilde{\boldsymbol{N}}'_\perp$ and $\widetilde{T}\,\widetilde{\boldsymbol{N}}'_\times$ as

$$\sin(\Theta/\widetilde{R})\boldsymbol{E}_1 - \cos(\Theta/\widetilde{R})\boldsymbol{E}_2 \quad \text{and} \quad \cos(\Theta/\widetilde{R})\boldsymbol{E}_1 + \sin(\Theta/\widetilde{R})\boldsymbol{E}_2,$$

respectively, the curve curvature and torsion are re-confirmed as $\widetilde{C} = -\widetilde{\boldsymbol{N}}'_\perp \cdot \widetilde{\boldsymbol{G}} = \widetilde{1}/\widetilde{R}$ and $\widetilde{T} = -\widetilde{\boldsymbol{N}}'_\times \cdot \widetilde{\boldsymbol{N}}_\perp = \widetilde{S}/\widetilde{R}$.

Note finally that for a circular curve with zero pitch (slope), i.e. $H = 0$, $\widetilde{R} \to R$, $\widetilde{1} \to 1$, $\widetilde{S} \to 0$ and thus $\widetilde{C} \to 1/R$ and $\widetilde{T} \to 0$ hold.

3.3.1 Nonlinear Deformation Maps

As an interesting addition, some relevant results regarding the kinematics of the curves defining the boundaries to the regular parts of the referential, material, and spatial boundaries $\partial\partial\mathcal{B}_\mathrm{r} := \partial^2\mathcal{B}_\mathrm{r}$, $\partial\partial\mathcal{B}_\mathrm{m} := \partial^2\mathcal{B}_\mathrm{m}$, and $\partial\partial\mathcal{B}_\mathrm{s} =: \partial^2\mathcal{B}_\mathrm{s}$, respectively, are recalled in the sequel.

The material and spatial boundary curve position vectors follow as the specialization of the material and spatial position vectors to $\partial^2\mathcal{B}_\mathrm{r}$, i.e.

$$\widetilde{\boldsymbol{X}} = \widetilde{\boldsymbol{\Upsilon}}(\widetilde{\boldsymbol{x}}) := \boldsymbol{\Upsilon}(\boldsymbol{x}|_{\partial^2\mathcal{B}_\mathrm{r}}, t) \quad \text{and} \quad \widetilde{\boldsymbol{x}} = \widetilde{\boldsymbol{\upsilon}}(\widetilde{\boldsymbol{X}}) := \boldsymbol{\upsilon}(\boldsymbol{X}|_{\partial^2\mathcal{B}_\mathrm{r}}, t). \tag{3.74}$$

Likewise, the material and spatial nonlinear boundary curve deformation maps follow as the specialization of the material and spatial deformation maps to $\partial^2\mathcal{B}_\mathrm{s}$ and $\partial^2\mathcal{B}_\mathrm{m}$, respectively, i.e.

$$\widetilde{X} = \widetilde{Y}(\widetilde{x}) := Y(x|_{\partial^2\mathcal{B}_s}, t) \quad \text{and} \quad \widetilde{x} = \widetilde{y}(\widetilde{X}) := y(X|_{\partial^2\mathcal{B}_m}, t). \tag{3.75}$$

We refrain in the sequel from explicitly indicating the space-time parametrization as long as there is no danger of confusion.

3.3.2 Tangent Maps: Deformation Gradients

Increments (total differentials) of the material and spatial boundary curve position vectors (tangents) $d\widetilde{X} \in T\partial^2\mathcal{B}_m$ and $d\widetilde{x} \in T\partial^2\mathcal{B}_s$ are denoted as material and spatial *boundary curve line elements*; they are related by spatial→material and material→spatial *tangent maps*

$$d\widetilde{X} = \widetilde{f} \cdot d\widetilde{x} \quad \text{and} \quad d\widetilde{x} = \widetilde{F} \cdot d\widetilde{X}. \tag{3.76}$$

Consequently, the spatial→material and material→spatial tangent maps follow as the spatial and material *boundary curve gradients* of the material and spatial boundary curve deformation maps, respectively,

$$\widetilde{f} := \widetilde{\nabla}_{\widetilde{x}}\widetilde{Y}(\widetilde{x}) \in T\partial^2\mathcal{B}_m \times T^*\partial^2\mathcal{B}_s \text{ and } \widetilde{F} := \widetilde{\nabla}_{\widetilde{X}}\widetilde{y}(\widetilde{X}) \in T\partial^2\mathcal{B}_s \times T^*\partial^2\mathcal{B}_m. \tag{3.77}$$

Hence, $\widetilde{f} : T\partial^2\mathcal{B}_s \to T\partial^2\mathcal{B}_m$ and $\widetilde{F} : T\partial^2\mathcal{B}_m \to T\partial^2\mathcal{B}_s$ are also denoted as the material and spatial *boundary curve deformation gradients*. Here, $T\partial^2\mathcal{B}_m$ and $T^*\partial^2\mathcal{B}_m$ together with $T\partial^2\mathcal{B}_s$ and $T^*\partial^2\mathcal{B}_s$ are the *tangent* and *cotangent* lines to $\partial^2\mathcal{B}_m$ and $\partial^2\mathcal{B}_s$, respectively. Clearly, the boundary curve deformation gradients are (mixed-variant) *two-point tensors* since they map curve vectors from different tangent lines into each other. In curve coordinate representation, they follow as

$$\widetilde{f} = \widetilde{\Lambda}\,\widetilde{G} \otimes \widetilde{g} \quad \text{and} \quad \widetilde{F} = \widetilde{\lambda}\,\widetilde{g} \otimes \widetilde{G}. \tag{3.78}$$

Here, $\widetilde{\lambda} = dl/dL = d\theta/d\Theta = \widetilde{\Lambda}^{-1}$ denotes the curve stretch. The material and spatial boundary curve deformation gradients are algebraically *pseudo-inverse* tensors

$$\widetilde{f} = \widetilde{F}^{\widetilde{-1}} \quad \text{and} \quad \widetilde{F} = \widetilde{f}^{\widetilde{-1}}. \tag{3.79}$$

In $\mathbb{E}^3 \times \mathbb{E}^3$, the boundary curve deformation gradients $\widetilde{f}$ and $\widetilde{F}$ are rank deficient and thus non-invertible, however, being pseudo-inverse results in

$$\begin{aligned} \widetilde{F} \cdot \widetilde{f} &= \widetilde{\lambda}\,\widetilde{\Lambda}\,[\widetilde{g} \otimes \widetilde{G}] \cdot [\widetilde{G} \otimes \widetilde{g}] = \widetilde{g} \otimes \widetilde{g} =: \widetilde{i}, \\ \widetilde{f} \cdot \widetilde{F} &= \widetilde{\Lambda}\,\widetilde{\lambda}\,[\widetilde{G} \otimes \widetilde{g}] \cdot [\widetilde{g} \otimes \widetilde{G}] = \widetilde{G} \otimes \widetilde{G} =: \widetilde{I}. \end{aligned} \tag{3.80}$$

Finally, since the boundary curve deformation gradients map between the tangent lines $T\partial^2\mathcal{B}_m$ and $T\partial^2\mathcal{B}_s$, also the following identities hold

$$\widetilde{\boldsymbol{F}} \cdot \widetilde{\boldsymbol{I}} \equiv \widetilde{\boldsymbol{i}} \cdot \widetilde{\boldsymbol{F}} \equiv \widetilde{\boldsymbol{F}} \quad \text{and} \quad \widetilde{\boldsymbol{f}} \cdot \widetilde{\boldsymbol{i}} \equiv \widetilde{\boldsymbol{I}} \cdot \widetilde{\boldsymbol{f}} \equiv \widetilde{\boldsymbol{f}}. \tag{3.81}$$

3.3.3 Compatibility Conditions for Tangent Maps

Curve Curl(s):
The material curve gradient operator $\widetilde{\nabla}_{\widetilde{X}}(\bullet)$ of the most general (two-point) curve tensor

$$\widetilde{\boldsymbol{Z}} = \widetilde{\boldsymbol{z}} \otimes \widetilde{\boldsymbol{G}} + \widetilde{\boldsymbol{z}}_\perp \otimes \widetilde{\boldsymbol{N}}_\perp + \widetilde{\boldsymbol{z}}_\times \otimes \widetilde{\boldsymbol{N}}_\times$$

computes as $\widetilde{\nabla}_{\widetilde{X}} \widetilde{\boldsymbol{Z}} = \widetilde{\boldsymbol{Z}}' \otimes \widetilde{\boldsymbol{G}}$ with curve derivative expanding as

$$\widetilde{\boldsymbol{Z}}' = \widetilde{\boldsymbol{z}}' \otimes \widetilde{\boldsymbol{G}} + \widetilde{\boldsymbol{z}}'_\perp \otimes \widetilde{\boldsymbol{N}}_\perp + \widetilde{\boldsymbol{z}}'_\times \otimes \widetilde{\boldsymbol{N}}_\times + \widetilde{\boldsymbol{z}} \otimes \widetilde{C}\widetilde{\boldsymbol{N}}_\perp + \widetilde{\boldsymbol{z}}_\perp \otimes [\widetilde{T}\widetilde{\boldsymbol{N}}_\times - \widetilde{C}\widetilde{\boldsymbol{G}}] - \widetilde{\boldsymbol{z}}_\times \otimes \widetilde{T}\widetilde{\boldsymbol{N}}_\perp.$$

Thus, $\widetilde{\mathrm{Curl}}_\perp \widetilde{\boldsymbol{Z}}$ and $\widetilde{\mathrm{Curl}}_\times \widetilde{\boldsymbol{Z}}$ result in

$$\widetilde{\mathrm{Curl}}_\perp \widetilde{\boldsymbol{Z}} = -\widetilde{\boldsymbol{z}}'_\times - \widetilde{T}\widetilde{\boldsymbol{z}}_\perp \quad \text{and} \quad \widetilde{\mathrm{Curl}}_\times \widetilde{\boldsymbol{Z}} = \widetilde{\boldsymbol{z}}'_\perp + \widetilde{C}\widetilde{\boldsymbol{z}} - \widetilde{T}\widetilde{\boldsymbol{z}}_\times.$$

Consequently, the material curve curl operators applied to $\widetilde{\boldsymbol{Z}}$ render

$$\begin{aligned} \widetilde{\mathrm{Curl}}_\perp \widetilde{\boldsymbol{Z}} = \boldsymbol{0} \quad &\text{for} \quad \widetilde{\boldsymbol{z}}'_\times = \boldsymbol{0} \quad \text{and} \quad \widetilde{\boldsymbol{z}}_\perp = \boldsymbol{0}, \\ \widetilde{\mathrm{Curl}}_\times \widetilde{\boldsymbol{Z}} = \boldsymbol{0} \quad &\text{for} \quad \widetilde{\boldsymbol{z}}'_\perp = \boldsymbol{0} \quad \text{and} \quad \widetilde{\boldsymbol{z}}_\times = \boldsymbol{0}, \quad \widetilde{\boldsymbol{z}} = \boldsymbol{0}. \end{aligned}$$

It is thus $\widetilde{\mathrm{Curl}}_\perp \widetilde{\boldsymbol{Z}} = \boldsymbol{0}$ (together with $\widetilde{\boldsymbol{Z}} \cdot \widetilde{\boldsymbol{N}}_\times = \boldsymbol{0}$) that identifies a *compatible* $\widetilde{\boldsymbol{Z}} = \widetilde{\boldsymbol{z}} \otimes \widetilde{\boldsymbol{G}}$.

Due to their definitions as the boundary curve gradients of the boundary curve deformation maps, the material and spatial tangent maps (boundary curve deformation gradients) satisfy the following *compatibility conditions*:

$$\widetilde{\mathrm{curl}}_\perp \widetilde{\boldsymbol{f}} = \boldsymbol{0} \quad \text{and} \quad \widetilde{\mathrm{Curl}}_\perp \widetilde{\boldsymbol{F}} = \boldsymbol{0}. \tag{3.82}$$

3.3.4 Cotangent Maps: Cofactors

Curve Cofactor:
The curve cofactor $\widetilde{\mathrm{cof}}\, \widetilde{\boldsymbol{Z}}$ of a second-order curve (tangential) tensor $\widetilde{\boldsymbol{Z}} := \boldsymbol{Z} \cdot \widetilde{\mathbb{P}} \in \mathbb{E}^3 \times \mathbb{E}^3$ follows via the *tensor-valued tensor null 'product'* as

$$0!\,\widetilde{\operatorname{cof}}\,\widetilde{\boldsymbol{Z}} := \widetilde{\times\!\!\times}_{\widetilde{\boldsymbol{Z}}}.$$

Thereby, $\widetilde{\times\!\!\times}_{\widetilde{\boldsymbol{Z}}}$ is defined in Cartesian coordinates (with coordinate representation for $\widetilde{\boldsymbol{Z}} = \widetilde{Z}^a{}_B \boldsymbol{e}_a \otimes \boldsymbol{E}^B$) as

$$\widetilde{\times\!\!\times}_{\widetilde{\boldsymbol{Z}}} := \widetilde{e}_a \widetilde{E}^B \boldsymbol{e}^a \otimes \boldsymbol{E}_B.$$

Here, the spatial and material curve 'permutation' symbols compute as $\widetilde{e}_a =: e_a{}^{cd}\, \widetilde{n}_{\perp c} \widetilde{n}_{\times d}$ and $\widehat{E}^B =: E^{BEF} \widetilde{N}_{\perp E} \widetilde{N}_{\times F}$, respectively.

Symbolically, based on the above coordinate definition, the curve cofactor $\widetilde{\operatorname{cof}}\,\widetilde{\boldsymbol{Z}}$ of a second-order curve (tangential) tensor $\widetilde{\boldsymbol{Z}} \in \mathbb{E}^3 \times \mathbb{E}^3$ thus reads as

$$0!\,\widetilde{\operatorname{cof}}\,\widetilde{\boldsymbol{Z}} := \widetilde{\boldsymbol{e}}\, \widetilde{\boxtimes}\, \widetilde{\boldsymbol{E}}.$$

Here, $\widetilde{\boldsymbol{e}}$ and $\widetilde{\boldsymbol{E}}$ denote the first-order spatial and material curve 'permutation' vectors $\widetilde{\boldsymbol{e}} := \boldsymbol{e} : [\widetilde{\boldsymbol{n}}_\perp \otimes \widetilde{\boldsymbol{n}}_\times] = \widetilde{e}_a \boldsymbol{e}^a$ and $\widetilde{\boldsymbol{E}} := \boldsymbol{E} : [\widetilde{\boldsymbol{N}}_\perp \otimes \widetilde{\boldsymbol{N}}_\times] = \widetilde{E}^B \boldsymbol{E}_B$, respectively, with 'special' dyadic curve product $\widetilde{\boldsymbol{e}}\, \widetilde{\boxtimes}\, \widetilde{\boldsymbol{E}} := \widetilde{e}_a \widetilde{E}^B \boldsymbol{e}^a \otimes \boldsymbol{E}_B$. Note that $\widetilde{\boxtimes}$ coincides trivially with the ordinary dyadic product $\otimes$.

Alternatively, the curve cofactor $\widetilde{\operatorname{cof}}\,\widetilde{\boldsymbol{Z}}$ of a second-order curve (tangential) tensor $\widetilde{\boldsymbol{Z}} \in \mathbb{E}^3 \times \mathbb{E}^3$ is defined via

$$\widetilde{\operatorname{cof}}\,\widetilde{\boldsymbol{Z}} \cdot \widetilde{\times}_{\widetilde{\boldsymbol{I}}} := \widetilde{\times}_{\widetilde{\boldsymbol{Z}}}.$$

Here, the *vector-valued vector null 'products'* $\widetilde{\times}_{\widetilde{\boldsymbol{I}}}$ and $\widetilde{\times}_{\widetilde{\boldsymbol{Z}}}$ follow in Cartesian coordinates as $\widetilde{\times}_{\widetilde{\boldsymbol{I}}} := \widetilde{E}_C \boldsymbol{E}^C$ and $\widetilde{\times}_{\widetilde{\boldsymbol{Z}}} := \widetilde{e}_a \boldsymbol{e}^a$, respectively.

Finally, for a pseudo-invertible second-order curve (tangential) tensor $\widetilde{\boldsymbol{Z}} \in \mathbb{E}^3 \times \mathbb{E}^3$, the curve cofactor $\widetilde{\operatorname{cof}}\,\widetilde{\boldsymbol{Z}}$ takes the closed-form expression

$$\widetilde{\operatorname{cof}}\,\widetilde{\boldsymbol{Z}} = \widetilde{\det}\,\widetilde{\boldsymbol{Z}}\, \widetilde{\boldsymbol{Z}}^{\widetilde{-}t} \in \mathbb{E}^3 \times \mathbb{E}^3.$$

Here, $\widetilde{\det}\,\widetilde{\boldsymbol{Z}}$ denotes the curve determinant of $\widetilde{\boldsymbol{Z}}$; see below.

Material and spatial vector-valued-oriented *point elements* (cotangents) $\widetilde{\boldsymbol{N}}$ and $\widetilde{\boldsymbol{n}}$ in the cotangent lines $T^*\partial^2\mathcal{B}_{\rm m}$ and $T^*\partial^2\mathcal{B}_{\rm s}$, respectively, are related via the *boundary curve cofactors* of the boundary curve deformation gradients *(boundary curve Nanson's formulae)*

$$\widetilde{\boldsymbol{n}} = \widetilde{\operatorname{cof}}\,\widetilde{\boldsymbol{F}} \cdot \widetilde{\boldsymbol{N}} \quad \text{and} \quad \widetilde{\boldsymbol{N}} = \widetilde{\operatorname{cof}}\,\widetilde{\boldsymbol{f}} \cdot \widetilde{\boldsymbol{n}}. \tag{3.83}$$

The boundary curve cofactors of the boundary curve deformation gradients are also denoted as the *point cotangent maps*

$$\widetilde{\boldsymbol{k}} := \widetilde{\text{cof}}\,\widetilde{\boldsymbol{f}} \in T^*\partial^2\mathcal{B}_\text{m} \times T^*\partial^2\mathcal{B}_\text{s} \text{ and } \widetilde{\boldsymbol{K}} := \widetilde{\text{cof}}\,\widetilde{\boldsymbol{F}} \in T^*\partial^2\mathcal{B}_\text{s} \times T^*\partial^2\mathcal{B}_\text{m}. \quad (3.84)$$

The material and spatial cotangent maps are pseudo-inverse curve tensors

$$\widetilde{\boldsymbol{k}} = \widetilde{\boldsymbol{K}}^{-1} = \widetilde{j}\,\widetilde{\boldsymbol{F}}^t \quad \text{and} \quad \widetilde{\boldsymbol{K}} = \widetilde{\boldsymbol{k}}^{-1} = \widetilde{J}\,\widetilde{\boldsymbol{f}}^t. \quad (3.85)$$

Note that $\widetilde{\boldsymbol{k}} = \widetilde{\boldsymbol{G}} \otimes \widetilde{\boldsymbol{g}}$ and $\widetilde{\boldsymbol{K}} = \widetilde{\boldsymbol{g}} \otimes \widetilde{\boldsymbol{G}}$ hold since $\widetilde{j} \equiv \widetilde{\Lambda}$ and $\widetilde{J} \equiv \widetilde{\lambda}$.

3.3.5 Compatibility Conditions for Cotangent Maps

> **Divergence of Curve Cofactor:**
> The material curve divergence operator $\widetilde{\text{Div}}(\bullet)$ applied to the tensor-valued tensor null 'product' $\widetilde{\divideontimes}_{\widetilde{\boldsymbol{Z}}} := \widetilde{\boldsymbol{e}} \otimes \widetilde{\boldsymbol{E}} \equiv \widetilde{\boldsymbol{g}} \otimes \widetilde{\boldsymbol{G}}$ of the (two-point) curve tensor $\widetilde{\boldsymbol{Z}} = Z\,\widetilde{\boldsymbol{g}} \otimes \widetilde{\boldsymbol{G}}$ expands as $\widetilde{\text{Div}}(\widetilde{\divideontimes}(\widetilde{\boldsymbol{Z}})) = \widetilde{J}\,\widetilde{c}\,\widetilde{\boldsymbol{n}}_\perp$.

Due to their definitions as the tensor-valued tensor null 'products' of the *compatible* boundary curve deformation gradients, the material and spatial cotangent maps satisfy the following *compatibility conditions (boundary curve Piola identity)*:

$$\widetilde{\text{div}}\widetilde{\boldsymbol{k}} = \widetilde{j}\,\widetilde{C}\,\widetilde{\boldsymbol{N}}_\perp \quad \text{and} \quad \widetilde{\text{Div}}\widetilde{\boldsymbol{K}} = \widetilde{J}\,\widetilde{c}\,\widetilde{\boldsymbol{n}}_\perp. \quad (3.86)$$

The boundary curve Piola identity constitutes the specialization of the Piola identity in the bulk to boundary curves. For flat boundary curves with $\widetilde{c} = \widetilde{C} = 0$, the similarity is particularly obvious.

Observe that these compatibility conditions are consistent with the *point theorem* equating the sum of the vector-valued point elements $\widetilde{\boldsymbol{N}}$ and $\widetilde{\boldsymbol{n}}$, respectively, over the boundary set of a control curve $\mathfrak{C}_\text{m} \subseteq \partial^2\mathcal{B}_\text{m}$ and $\mathfrak{C}_\text{s} \subseteq \partial^2\mathcal{B}_\text{s}$, respectively, to

$$\sum_{\partial\mathfrak{C}_\text{m}} \widetilde{\boldsymbol{N}} = \sum_{\partial\mathfrak{C}_\text{s}} \widetilde{\boldsymbol{k}} \cdot \widetilde{\boldsymbol{n}} = \int_{\mathfrak{C}_\text{m}} \widetilde{C}\,\text{d}\widetilde{\boldsymbol{L}}_\perp \quad \text{and} \quad \sum_{\partial\mathfrak{C}_\text{s}} \widetilde{\boldsymbol{n}} = \sum_{\partial\mathfrak{C}_\text{m}} \widetilde{\boldsymbol{K}} \cdot \widetilde{\boldsymbol{N}} = \int_{\mathfrak{C}_\text{s}} \widetilde{c}\,\text{d}\widetilde{\boldsymbol{l}}_\perp. \quad (3.87)$$

Applying the curve Gauss theorem and localizing for arbitrary curves establishes the result.

3.3.6 Measure Maps: Determinants

Curve Determinant:
The curve determinant $\widetilde{\det}\widetilde{\mathbf{Z}}$ of a second-order curve (tangential) tensor $\widetilde{\mathbf{Z}} := \widehat{\mathbf{Z}} \cdot \widehat{\mathbb{P}} =: \mathbf{Z} \cdot \widetilde{\mathbb{P}} \in \mathbb{E}^3 \times \mathbb{E}^3$ follows via the *scalar-valued tensor single 'product'* as

$$1!\,\widetilde{\det}\widetilde{\mathbf{Z}} := \widetilde{\times\!\!\times}(\widetilde{\mathbf{Z}}).$$

Thereby, $\widetilde{\times\!\!\times}(\widetilde{\mathbf{Z}})$ is defined in Cartesian coordinates with coordinate representation for $\widetilde{\mathbf{Z}} = \widetilde{Z}^a{}_B \boldsymbol{e}_a \otimes \boldsymbol{E}^B$ as

$$\widetilde{\times\!\!\times}(\widetilde{\mathbf{Z}}) := \widetilde{e}_a \widetilde{E}^B \widetilde{Z}^a{}_B.$$

Symbolically, based on the above coordinate definition, the curve determinant $\widetilde{\det}\widetilde{\mathbf{Z}}$ of a second-order curve (tangential) tensor $\widetilde{\mathbf{Z}} \in \mathbb{E}^3 \times \mathbb{E}^3$ thus reads as

$$1!\,\widetilde{\det}\widetilde{\mathbf{Z}} := [\widetilde{\boldsymbol{e}} \,\widetilde{\boxtimes}\, \widetilde{\boldsymbol{E}}] : \widetilde{\mathbf{Z}}.$$

Alternatively, for non-zero vectors $\widetilde{\boldsymbol{A}} := \boldsymbol{A} \cdot \widetilde{\mathbb{P}} \in \mathbb{E}^3$, the curve determinant $\widetilde{\det}\widetilde{\mathbf{Z}}$ of a second-order curve (tangential) tensor $\widetilde{\mathbf{Z}} \in \mathbb{E}^3 \times \mathbb{E}^3$ is defined via

$$\widetilde{\det}\widetilde{\mathbf{Z}}\, \widetilde{\times}(\widetilde{\boldsymbol{A}}) := \widetilde{\times}(\widetilde{\mathbf{Z}} \cdot \widetilde{\boldsymbol{A}}) \quad \forall \widetilde{\boldsymbol{A}} \quad \in \mathbb{E}^3.$$

Here, the *scalar-valued vector single 'products'* $\widetilde{\times}(\widetilde{\boldsymbol{A}})$ and $\widetilde{\times}(\widetilde{\mathbf{Z}} \cdot \widetilde{\boldsymbol{A}})$ follow in Cartesian coordinates as $\widetilde{\times}(\widetilde{\boldsymbol{A}}) := \widetilde{E}_C \widetilde{A}^C$ and $\widetilde{\times}(\widetilde{\mathbf{Z}} \cdot \widetilde{\boldsymbol{A}}) := \widetilde{e}_d[\widetilde{Z}^d{}_C \widetilde{A}^C]$, respectively.

Finally, for a pseudo-invertible second-order curve (tangential) tensor $\widetilde{\mathbf{Z}} \in \mathbb{E}^3 \times \mathbb{E}^3$, the curve determinant $\widetilde{\det}\widetilde{\mathbf{Z}}$ takes the closed-form expression

$$1\,\widetilde{\det}\widetilde{\mathbf{Z}} = \widetilde{\operatorname{cof}}\,\widetilde{\mathbf{Z}} : \widetilde{\mathbf{Z}}.$$

Material and spatial scalar-valued boundary curve *line elements* $\mathrm{d}L$ and $\mathrm{d}l$ in $\partial^2\mathcal{B}_\mathrm{m}$ and $\partial^2\mathcal{B}_\mathrm{s}$, respectively, are related via the *boundary curve determinants* of the boundary curve deformation gradients

$$\mathrm{d}L = \widetilde{\det}\widetilde{\boldsymbol{f}}\ \mathrm{d}l \quad \text{and} \quad \mathrm{d}l = \widetilde{\det}\widetilde{\boldsymbol{F}}\ \mathrm{d}L. \tag{3.88}$$

The boundary curve determinants of the boundary curve deformation gradients are also denoted as the *boundary curve Jacobians* or rather the *line measure maps*

$$\widetilde{j} = \widetilde{J}^{-1} := \widetilde{\det}\widetilde{\boldsymbol{f}} = \widetilde{\Lambda} \in \mathbb{R}^+ \quad \text{and} \quad \widetilde{J} = \widetilde{j}^{-1} := \widetilde{\det}\widetilde{\boldsymbol{F}} = \widetilde{\lambda} \in \mathbb{R}^+. \tag{3.89}$$

Alternatively, the boundary curve Jacobians result from the tangent maps in Eq. 2.27 as

$$\widetilde{j} = |\boldsymbol{f}\cdot\widetilde{\boldsymbol{g}}| \quad \text{and} \quad \widetilde{J} = |\boldsymbol{F}\cdot\widetilde{\boldsymbol{G}}| \tag{3.90}$$

and may thus be expressed as

$$\widetilde{\det}\widetilde{\boldsymbol{f}} := \frac{|\widetilde{\boldsymbol{f}}\cdot\widetilde{\boldsymbol{g}}|}{|\widetilde{\boldsymbol{g}}|} = \widetilde{\Lambda} \quad \text{and} \quad \widetilde{\det}\widetilde{\boldsymbol{F}} := \frac{|\widetilde{\boldsymbol{F}}\cdot\widetilde{\boldsymbol{G}}|}{|\widetilde{\boldsymbol{G}}|} = \widetilde{\lambda}. \tag{3.91}$$

Note the similarity to the definition of the determinant of the deformation gradients in the bulk. The proof of the above representation follows from observing that $\widetilde{\boldsymbol{f}}\cdot\widetilde{\boldsymbol{g}} = \boldsymbol{f}\cdot\widetilde{\boldsymbol{g}}$ and $\widetilde{\boldsymbol{F}}\cdot\widetilde{\boldsymbol{G}} = \boldsymbol{F}\cdot\widetilde{\boldsymbol{G}}$ jointly with the tangent maps in Eq. 2.27, thus

$$\frac{\boldsymbol{f}\cdot\widetilde{\boldsymbol{g}}}{|\widetilde{\boldsymbol{g}}|} = \boldsymbol{f}\cdot\widetilde{\boldsymbol{g}} \quad \text{and} \quad \frac{\boldsymbol{F}\cdot\widetilde{\boldsymbol{G}}}{|\widetilde{\boldsymbol{G}}|} = \boldsymbol{F}\cdot\widetilde{\boldsymbol{G}}. \tag{3.92}$$

Observe also that, for example, $\widetilde{\times}(\widetilde{\boldsymbol{g}}) = \widetilde{\boldsymbol{e}}\cdot\widetilde{\boldsymbol{g}} = |\widetilde{\boldsymbol{g}}|$ and likewise $\widetilde{\times}(\widetilde{\boldsymbol{G}}) = \widetilde{\boldsymbol{E}}\cdot\widetilde{\boldsymbol{G}} = |\widetilde{\boldsymbol{G}}|$.

3.3.7 Velocities

The total time derivatives of the material and spatial boundary curve position maps $\widetilde{\boldsymbol{X}} = \widetilde{\boldsymbol{\Upsilon}}(\widetilde{\boldsymbol{\Xi}}, t)$ and $\widetilde{\boldsymbol{x}} = \widetilde{\boldsymbol{\upsilon}}(\widetilde{\boldsymbol{\Xi}}, t)$, projected onto the *tangent line* to the boundary curve, are introduced as the *total material* and *spatial boundary curve velocities*

$$\widetilde{\boldsymbol{W}} := \dot{\widetilde{\boldsymbol{\Upsilon}}}(\widetilde{\boldsymbol{\Xi}}, t)\cdot\widetilde{\boldsymbol{I}} = \widehat{\boldsymbol{W}}|_{\partial^2\mathcal{B}_r}\cdot\widetilde{\boldsymbol{I}} \quad \text{and} \quad \widetilde{\boldsymbol{w}} := \dot{\widetilde{\boldsymbol{\upsilon}}}(\widetilde{\boldsymbol{\Xi}}, t)\cdot\widetilde{\boldsymbol{i}} = \widehat{\boldsymbol{w}}|_{\partial^2\mathcal{B}_r}\cdot\widetilde{\boldsymbol{i}}. \tag{3.93}$$

Moreover, the spatial time derivative of the material boundary curve deformation map $\widetilde{\boldsymbol{X}} = \widetilde{\boldsymbol{Y}}(\widetilde{\boldsymbol{x}}, t)$ and the material time derivative of the spatial boundary curve deformation map $\widetilde{\boldsymbol{x}} = \widetilde{\boldsymbol{y}}(\widetilde{\boldsymbol{X}}, t)$, projected onto the tangent line to the boundary curve, result in the *material* and the *spatial boundary curve velocity*

$$\widetilde{\boldsymbol{V}} := \mathrm{d}_t\widetilde{\boldsymbol{Y}}(\widetilde{\boldsymbol{x}}, t)\cdot\widetilde{\boldsymbol{I}} = \widehat{\boldsymbol{V}}|_{\partial^2\mathcal{B}_s}\cdot\widetilde{\boldsymbol{I}} \quad \text{and} \quad \widetilde{\boldsymbol{v}} := \mathrm{D}_t\widetilde{\boldsymbol{y}}(\widetilde{\boldsymbol{X}}, t)\cdot\widetilde{\boldsymbol{i}} = \widehat{\boldsymbol{v}}|_{\partial^2\mathcal{B}_m}\cdot\widetilde{\boldsymbol{i}}. \tag{3.94}$$

With these definitions at hand, it is straightforward to show that the total material and spatial boundary curve velocities, based on boundary curve Euler formulae for the total time derivative, are related as follows:

$$\widetilde{\boldsymbol{W}} = \widetilde{\boldsymbol{V}} + \widetilde{\boldsymbol{f}}\cdot\widetilde{\boldsymbol{w}} \quad \text{and} \quad \widetilde{\boldsymbol{w}} = \widetilde{\boldsymbol{v}} + \widetilde{\boldsymbol{F}}\cdot\widetilde{\boldsymbol{W}}. \tag{3.95}$$

The relation between the material and spatial boundary curve velocities via the boundary curve deformation gradients is then finally a direct consequence of the above relations

$$\widetilde{\boldsymbol{V}} = -\widetilde{\boldsymbol{f}} \cdot \widetilde{\boldsymbol{v}} \quad \text{and} \quad \widetilde{\boldsymbol{v}} = -\widetilde{\boldsymbol{F}} \cdot \widetilde{\boldsymbol{V}}. \tag{3.96}$$

Normal–Tangential Decomposition on Curve Boundary:
On the boundary points of a control curve, the total material and spatial boundary curve velocities that live in the tangent line to the boundary curve are *formally* decomposed into normal and tangential parts

$$\widetilde{\boldsymbol{W}} = \widetilde{\boldsymbol{W}}_\perp + \widetilde{\boldsymbol{W}}_\parallel \quad \text{on} \quad \partial\mathfrak{C}_\mathrm{m} \quad \text{and} \quad \widetilde{\boldsymbol{w}} = \widetilde{\boldsymbol{w}}_\perp + \widetilde{\boldsymbol{w}}_\parallel \quad \text{on} \quad \partial\mathfrak{C}_\mathrm{s}. \tag{3.97}$$

Thereby, the normal parts are defined in terms of the scalar-valued *normal velocities* $\widetilde{W}_\perp$ and $\widetilde{w}_\perp$, respectively, as

$$\widetilde{\boldsymbol{W}}_\perp := [\widetilde{\boldsymbol{W}} \cdot \widetilde{\boldsymbol{N}}]\widetilde{\boldsymbol{N}} =: \widetilde{W}_\perp \widetilde{\boldsymbol{N}} \quad \text{and} \quad \widetilde{\boldsymbol{w}}_\perp := [\widetilde{\boldsymbol{w}} \cdot \widetilde{\boldsymbol{n}}]\widetilde{\boldsymbol{n}} =: \widetilde{w}_\perp \widetilde{\boldsymbol{n}}$$

and, consequently, the tangential parts follow as

$$\widetilde{\boldsymbol{W}}_\parallel := \widetilde{\boldsymbol{W}} - \widetilde{\boldsymbol{W}}_\perp =: \widetilde{\mathbb{P}} \cdot \widetilde{\boldsymbol{W}} \quad \text{and} \quad \widetilde{\boldsymbol{w}}_\parallel := \widetilde{\boldsymbol{w}} - \widetilde{\boldsymbol{w}}_\perp =: \widetilde{\mathbb{p}} \cdot \widetilde{\boldsymbol{w}}.$$

Here, $\widetilde{\mathbb{P}}$ and $\widetilde{\mathbb{p}}$ are the material and spatial (mixed-variant) *projection operators* onto the tangent 'dot' to the boundary points of the control curve

$$\widetilde{\mathbb{P}} := \widetilde{\boldsymbol{I}} - \widetilde{\boldsymbol{N}} \otimes \widetilde{\boldsymbol{N}} \quad \text{on} \quad \partial\mathfrak{C}_\mathrm{m} \quad \text{and} \quad \widetilde{\mathbb{p}} := \widetilde{\boldsymbol{i}} - \widetilde{\boldsymbol{n}} \otimes \widetilde{\boldsymbol{n}} \quad \text{on} \quad \partial\mathfrak{C}_\mathrm{s},$$

with $\widetilde{\boldsymbol{I}}$ and $\widetilde{\boldsymbol{i}}$ the material and spatial (mixed-variant) boundary surface unit tensors, respectively. Clearly, $\widetilde{\mathbb{P}} \equiv \mathbb{O}$ and $\widetilde{\mathbb{p}} \equiv \mathbb{o}$ vanish identically.

Reference

1. Steinmann P (2008) On boundary potential energies in deformational and configurational mechanics. J Mech Phys Solids 56:772–800

Chapter 4
Kinematics at Singular Sets

8,091 m 28°35'46"N 83°49'13"E

Abstract This chapter revisits the relevant continuum kinematics at singular sets, i.e. at singular surfaces and at singular curves and points, thereby elaborating on the jumps in the nonlinear deformation maps and their associated tangent, cotangent, and measure maps.

The kinematics of geometrical field quantities that display finite jump discontinuities either across coherent singular surfaces embedded into the bulk volume or across coherent singular curves and/or points embedded into the boundary, collectively denoted as *coherent singular sets*, prove particularly intricate. The jump conditions across these two-, one-, and zero-dimensional coherent singular sets are the analogy to the corresponding compatibility conditions in bulk volumes as well as on boundary surfaces and curves, respectively. Thereby, coherence denotes the situation of no jump discontinuities of the corresponding nonlinear deformation maps, neither in bulk volumes nor on boundary surfaces and curves, respectively.

In view of the envisioned subsequent analysis of the 'forces' driving coherent singular sets, it is then of particular interest to analyze finite jump discontinuities of the corresponding space gradients of the deformation maps, i.e. of the tangent, cotangent, and measure maps, and of the corresponding time gradients of the deformation maps, i.e. the velocities. Purposely, the kinematics of geometrical field quantities at two-, one-, and zero-dimensional coherent singular sets is pursued following exactly the same discussions, arguments, and formulations.

As a result, remarkably, the kinematics of geometrical quantities at two-, one-, and zero-dimensional coherent singular sets results in a respective collection of seven formally similar jump or rather coherence conditions, i.e. explicit expressions for finite jump discontinuities at two-, one-, and zero-dimensional coherent singular sets that, respectively, display a similar format.

P. Steinmann, *Spatial and Material Forces in Nonlinear Continuum Mechanics*, Solid Mechanics and Its Applications 272,
https://doi.org/10.1007/978-3-030-89070-4_4

4.1 Configurations

Singular Sets:
At *singular sets*, the geometry and/or at least one solution field displays a *singularity*.

- The edges and vertices of arbitrary polyhedra, see, for example, Fig. 3.1, are examples for one- and zero-dimensional singular sub-sets in the *geometry* of a boundary surface.
- Singularities in a *solution field* are characterized by an r^m behavior with $m \in \mathbb{R}^-$ and $r \in \mathbb{R}_0^+$ the radial distance from the singularity.
  - A concentrated point load exerted on the boundary surface of a (linear elastic) half-space (Boussinesq's problem) with r^{-2} stress singularity is an example for a zero-dimensional singular sub-set with singularity in a solution field.
  - A (linear elastic) crack tip with $r^{-1/2}$ stress and strain singularities is an example for a one-dimensional singular sub-set with singularities in the geometry *and* at least one solution field (in fact, here two solution fields).

4.1.1 Material Configuration

In general, the total bulk volume occupied by the material configuration in $\mathbb{E}^3$ may be composed by time-dependent *sub-parts* $\mathcal{B}^+_\mathrm{m} = \mathcal{B}^+_\mathrm{m}(t)$ and $\mathcal{B}^-_\mathrm{m} = \mathcal{B}^-_\mathrm{m}(t)$ so that $\mathcal{B}_\mathrm{m} = \mathcal{B}^+_\mathrm{m} \cup \mathcal{B}^-_\mathrm{m}$ and $\mathcal{S}_\mathrm{m} := \mathcal{B}^+_\mathrm{m} \cap \mathcal{B}^-_\mathrm{m}$. Here, $\mathcal{S}_\mathrm{m} = \mathcal{S}_\mathrm{m}(t)$ denotes a time-dependent internal surface (a two-dimensional singular set or rather *singular surface*) embedded into the bulk volume, and assumed smooth for simplicity, that separates the sub-parts $\mathcal{B}^+_\mathrm{m}$ and $\mathcal{B}^-_\mathrm{m}$; see Fig. 4.1 left. At singular surfaces, at least one solution field and/or at least one material parameter exhibits a *finite jump discontinuity*. The outwards-pointing unit normals at $\mathcal{S}_\mathrm{m}$ are defined with respect to the sub-parts $\mathcal{B}^\pm_\mathrm{m}$ as $\boldsymbol{M}^\pm$ with $\boldsymbol{M} := \boldsymbol{M}^- = -\boldsymbol{M}^+$.

In the sequel, *material* (stationary) singular surfaces will be distinguished as a special case from the general case of *migrating* singular surfaces.

Furthermore, the total boundary surface $\partial\mathcal{B}_\mathrm{m}$ enclosing $\mathcal{B}_\mathrm{m}$ may be composed by time-dependent regular (smooth) *sub-parts* $\partial\mathcal{B}^+_\mathrm{m} = \partial\mathcal{B}^+_\mathrm{m}(t)$ and $\partial\mathcal{B}^-_\mathrm{m} = \partial\mathcal{B}^-_\mathrm{m}(t)$ so that $\partial\mathcal{B}_\mathrm{m} = \partial\mathcal{B}^+_\mathrm{m} \cup \partial\mathcal{B}^-_\mathrm{m}$ and $\mathcal{C}_\mathrm{m} := \partial\mathcal{B}^+_\mathrm{m} \cap \partial\mathcal{B}^-_\mathrm{m}$. Here, $\mathcal{C}_\mathrm{m} = \mathcal{C}_\mathrm{m}(t)$ denotes a set of time-dependent boundary curves (one-dimensional singular sets or rather *singular curves*) embedded into the boundary surface, and assumed piece-wise smooth, that separate the sub-parts $\partial\mathcal{B}^+_\mathrm{m}$ and $\partial\mathcal{B}^-_\mathrm{m}$. At singular curves, the geometry displays a *singularity* and/or at least one solution field and/or at least one material parameter exhibits a *finite jump discontinuity*. The outwards-pointing unit normals at $\mathcal{C}_\mathrm{m}$,

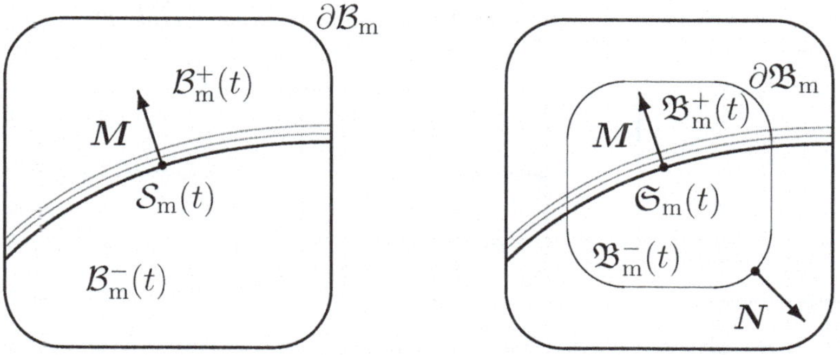

Fig. 4.1 Left: A migrating singular surface $\mathcal{S}_m(t)$ separates the total bulk volume $\mathcal{B}_m$ of the material configuration into time-dependent sub-parts $\mathcal{B}_m^\pm(t)$. For the stationary case, $\mathcal{S}_m \neq \mathcal{S}_m(t)$ is denoted as a material singular surface. Right: A migrating material control volume $\mathfrak{B}_m(t) \subseteq \mathcal{B}_m$; its intersection with the singular surface is denoted as $\mathfrak{S}_m(t)$

orthogonal to the tangent planes $T\partial\mathcal{B}_m^\pm$ as well as orthogonal to the tangent lines $T\partial^2\mathcal{B}_m^\pm$ and tangent to the tangent planes $T\partial\mathcal{B}_m^\pm$, are defined with respect to the sub-parts $\partial\mathcal{B}_m^\pm$ as $\boldsymbol{N}^\pm$ and $\widehat{\boldsymbol{N}}^\pm$, respectively.

In the sequel, *material* (stationary) singular curves will be distinguished as a special case from the general case of *migrating* singular curves.

Finally, the set of boundary curves $\partial^2\mathcal{B}_m$ enclosing the regular (smooth) subparts of $\partial\mathcal{B}_m$ may be composed by time-dependent regular (smooth) *sub-parts* $\partial^2\mathcal{B}_m^+ = \partial^2\mathcal{B}_m^+(t)$ and $\partial^2\mathcal{B}_m^- = \partial^2\mathcal{B}_m^-(t)$ so that $\partial^2\mathcal{B}_m = \partial^2\mathcal{B}_m^+ \cup \partial^2\mathcal{B}_m^-$ and $\mathcal{P}_m := \partial^2\mathcal{B}_m^+ \cap \partial^2\mathcal{B}_m^-$. Here, $\mathcal{P}_m = \mathcal{P}_m(t)$ denotes a set of time-dependent boundary points (zero-dimensional singular sets or rather *singular points*) embedded into the boundary surface that separate the sub-parts $\partial^2\mathcal{B}_m^+$ and $\partial^2\mathcal{B}_m^-$. At singular points, the geometry displays a *singularity* and/or at least one solution field exhibits a *finite jump discontinuity*. The outwards-pointing unit normals at $\mathcal{P}_m$, tangent to the tangent lines $T\partial^2\mathcal{B}_m^\pm$, are defined with respect to the sub-parts $\partial^2\mathcal{B}_m^\pm$ as $\widetilde{\boldsymbol{N}}^\pm$, respectively.

In the sequel, *material* (stationary) singular points will be distinguished as a special case from the general case of *migrating* singular points.

The (potentially empty) intersection of the singular internal surface $\mathcal{S}_m$ with a (bulk) control volume $\mathfrak{B}_m \subseteq \mathcal{B}_m$ is denoted as $\mathfrak{S}_m(t) := \mathcal{S}_m \cap \mathfrak{B}_m$; see Fig. 4.1 right. Likewise, the (potentially empty) intersection of the set of singular boundary curves $\mathcal{C}_m$ with a boundary control surface $\mathfrak{S}_m \subseteq \partial\mathcal{B}_m$ is denoted as $\mathfrak{C}_m(t) := \mathcal{C}_m \cap \mathfrak{S}_m$. Finally, the (potentially empty) intersection of the set of singular boundary points $\mathcal{P}_m$ with a boundary control curve $\mathfrak{C}_m \subseteq \partial\partial\mathcal{B}_m$ is denoted as $\mathfrak{P}_m(t) := \mathcal{P}_m \cap \mathfrak{C}_m$.

It is recalled that, as specific to the here advocated approach, the material configuration may not necessarily be stationary but may also evolve with time; as an example consider the case of interface motion in Fig. 4.1 (left) or crack propagation in Fig. 4.2.

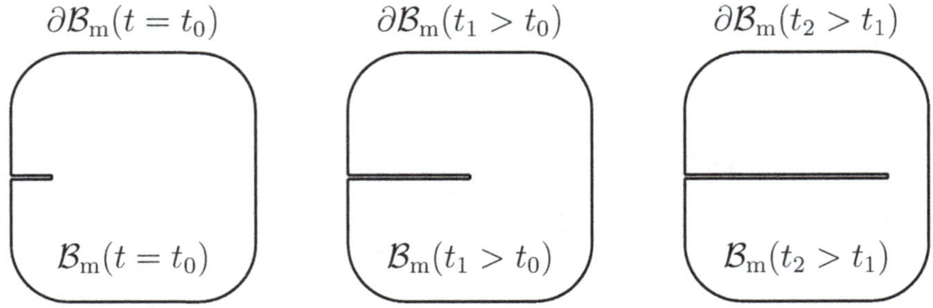

Fig. 4.2 Time evolution of material configuration and its external boundary due to a propagating crack

4.1.2 Spatial Configuration

In general, $\mathcal{B}_s = \mathcal{B}_s^+ \cup \mathcal{B}_s^-$ and $\mathcal{S}_s := \mathcal{B}_s^+ \cap \mathcal{B}_s^-$, whereby the time-dependent (convected) singular internal surface $\mathcal{S}_s = \mathcal{S}_s(t)$ separates the sub-parts $\mathcal{B}_s^+ = \mathcal{B}_s^+(t)$ and $\mathcal{B}_s^- = \mathcal{B}_s^-(t)$. The outwards-pointing unit normals at $\mathcal{S}_s$ are defined with respect to the sub-parts $\mathcal{B}_s^\pm$ as $\boldsymbol{m}^\pm$ with $\boldsymbol{m} := \boldsymbol{m}^- = -\boldsymbol{m}^+$.

Furthermore, $\partial\mathcal{B}_s = \partial\mathcal{B}_s^+ \cup \partial\mathcal{B}_s^-$ and $\mathcal{C}_s := \partial\mathcal{B}_s^+ \cap \partial\mathcal{B}_s^-$, whereby the set of time-dependent (convected) singular boundary curves $\mathcal{C}_s = \mathcal{C}_s(t)$ separates the sub-parts $\partial\mathcal{B}_s^+ = \partial\mathcal{B}_s^+(t)$ and $\partial\mathcal{B}_s^- = \partial\mathcal{B}_s^-(t)$. The outwards-pointing unit normals at $\mathcal{C}_s$, orthogonal to the tangent planes $T\partial\mathcal{B}_s^\pm$ as well as orthogonal to the tangent lines $T\partial\partial\mathcal{B}_s^\pm$ and tangent to the tangent planes $T\partial\mathcal{B}_s^\pm$, are defined with respect to the sub-parts $\partial\mathcal{B}_s^\pm$ as $\boldsymbol{n}^\pm$ and $\widehat{\boldsymbol{n}}^\pm$, respectively.

Finally, $\partial^2\mathcal{B}_s = \partial^2\mathcal{B}_s^+ \cup \partial^2\mathcal{B}_s^-$ and $\mathcal{P}_s := \partial^2\mathcal{B}_s^+ \cap \partial^2\mathcal{B}_s^-$, whereby the set of time-dependent (convected) singular boundary points $\mathcal{P}_s = \mathcal{P}_s(t)$ separates the sub-parts $\partial^2\mathcal{B}_s^+ = \partial^2\mathcal{B}_s^+(t)$ and $\partial^2\mathcal{B}_s^- = \partial^2\mathcal{B}_s^-(t)$. The outwards-pointing unit normals at $\mathcal{P}_s$, tangent to the tangent lines $T\partial^2\mathcal{B}_s^\pm$, are defined with respect to the sub-parts $\partial^2\mathcal{B}_s^\pm$ as $\widetilde{\boldsymbol{n}}^\pm$, respectively.

The intersection of the (convected) singular internal surface $\mathcal{S}_s$ with a (bulk) control volume $\mathfrak{B}_s \subseteq \mathcal{B}_s$ is denoted as $\mathfrak{S}_s(t) := \mathcal{S}_s \cap \mathfrak{B}_s$. Likewise, the intersection of the set of (convected) singular boundary curves $\mathcal{C}_s$ with a boundary control surface $\mathfrak{S}_s \subseteq \partial\mathcal{B}_s$ is denoted as $\mathfrak{C}_m(t) := \mathcal{C}_s \cap \mathfrak{S}_s$. Finally, the intersection of the set of (convected) singular boundary points $\mathcal{P}_s$ with a boundary control curve $\mathfrak{C}_s \subseteq \partial^2\mathcal{B}_s$ is denoted as $\mathfrak{P}_s(t) := \mathcal{P}_s \cap \mathfrak{C}_s$.

4.2 Coherent Singular Surfaces

In the sequel, singular surfaces are considered as embedded into the bulk volume as the intersections of its regular sub-parts. They are assumed smooth and *coherent*.

At a singular surface $\mathcal{S}_m$, the intersecting regular sub-parts $\mathcal{B}_m^{\pm}$ of the bulk volume $\mathcal{B}_m$ are characterized by their outwards-pointing normals $\boldsymbol{M}^{\pm}$, orthogonal to the tangent planes $T\partial\mathcal{B}_m^{\pm}$ to the internal boundary surfaces. Then the singular surface is endowed with its tangent vectors $\widehat{\boldsymbol{G}}_\alpha$ and its normal vector $\boldsymbol{M} := \mp\boldsymbol{M}^{\pm}$. Analogous definitions and notation hold in the spatial configuration.

Jump and Average Operators:
Let the *jump* and *average* operators as applied to arbitrary scalar-, vector-, or tensor-valued fields $\mathbf{Z}$ across a singular surface $\mathcal{S}_m$ in the material configuration be defined as

$$[\![\mathbf{Z}]\!] := \mathbf{Z}^+ - \mathbf{Z}^- \quad \text{and} \quad 2\{\mathbf{Z}\} := \mathbf{Z}^+ + \mathbf{Z}^-. \tag{4.1}$$

Thereby, it is assumed that $\mathbf{Z}$ is smooth away from the singular surface and smooth up to the singular surface from either side where it takes values $\mathbf{Z}^{\pm} = \lim_{\epsilon\to 0} \mathbf{Z}(\boldsymbol{X}|_{\mathcal{S}_m} \pm \epsilon\boldsymbol{M})$ with $\epsilon \in \mathbb{R}^+$. Jump and average operators across a singular surface $\mathcal{S}_s$ in the spatial configuration are defined accordingly. Then the two following product rules apply:

(1) Product Rule of Jumps: The jump of the generic product $\circ$ of two quantities $\mathbf{Z}_1$ and $\mathbf{Z}_2$ is related to the products of their jumps and averages as

$$[\![\mathbf{Z}_1 \circ \mathbf{Z}_2]\!] = [\![\mathbf{Z}_1]\!] \circ \{\mathbf{Z}_2\} + \{\mathbf{Z}_1\} \circ [\![\mathbf{Z}_2]\!]. \tag{4.2}$$

(2) Product Rule of Averages: The average of the generic product $\circ$ of two quantities $\mathbf{Z}_1$ and $\mathbf{Z}_2$ is related to the products of their jumps and averages as

$$\{\mathbf{Z}_1 \circ \mathbf{Z}_2\} = \{\mathbf{Z}_1\} \circ \{\mathbf{Z}_2\} + \frac{1}{4}[\![\mathbf{Z}_1]\!] \circ [\![\mathbf{Z}_2]\!]. \tag{4.3}$$

Note finally that a vanishing jump $[\![\mathbf{Z}]\!] = \mathbf{0}$ implies the average $\{\mathbf{Z}\} \equiv \mathbf{Z}$. For $[\![\mathbf{Z}_1]\!] = \mathbf{0}$, it thus follows that $[\![\mathbf{Z}_1 \circ \mathbf{Z}_2]\!] = \mathbf{Z}_1 \circ [\![\mathbf{Z}_2]\!]$ and $\{\mathbf{Z}_1 \circ \mathbf{Z}_2\} = \mathbf{Z}_1 \circ \{\mathbf{Z}_2\}$, vice versa $[\![\mathbf{Z}_2]\!] = \mathbf{0}$ results in $[\![\mathbf{Z}_1 \circ \mathbf{Z}_2]\!] = [\![\mathbf{Z}_1]\!] \circ \mathbf{Z}_2$ and $\{\mathbf{Z}_1 \circ \mathbf{Z}_2\} = \{\mathbf{Z}_1\} \circ \mathbf{Z}_2$.

4.2.1 Jump in Nonlinear Deformation Maps

Due to the coherence assumption, the material and spatial nonlinear deformation maps (and placements, respectively) are continuous across singular surfaces (coherence condition I)

$$[\![\boldsymbol{Y}]\!] = \boldsymbol{0} \quad \text{at} \quad \mathcal{S}_\mathrm{m} \quad \text{and} \quad [\![\boldsymbol{y}]\!] = \boldsymbol{0} \quad \text{at} \quad \mathcal{S}_\mathrm{s} \tag{4.4}$$

and, in addition, smooth away from singular surfaces, and smooth up to singular surfaces from either side.

4.2.2 Jump in Tangent Maps

The jump operator commutes with either the material or the spatial *tangential gradient* to singular surfaces, hence the spatial and material deformation gradients are *tangential continuous* across singular surfaces (coherence condition II)

$$[\![\boldsymbol{F}]\!] \cdot \mathbb{P} = \boldsymbol{0} \quad \text{at} \quad \mathcal{S}_\mathrm{m} \quad \text{and} \quad [\![\boldsymbol{f}]\!] \cdot \mathbb{p} = \boldsymbol{0} \quad \text{at} \quad \mathcal{S}_\mathrm{s}. \tag{4.5}$$

Here, $\mathbb{P}$ and $\mathbb{p}$ are the material and spatial (mixed-variant) projection operators onto the tangent plane to singular surfaces

$$\mathbb{P} := \boldsymbol{I} - \boldsymbol{M} \otimes \boldsymbol{M} \quad \text{at} \quad \mathcal{S}_\mathrm{m} \quad \text{and} \quad \mathbb{p} := \boldsymbol{i} - \boldsymbol{m} \otimes \boldsymbol{m} \quad \text{at} \quad \mathcal{S}_\mathrm{s}. \tag{4.6}$$

Consequently, the material and spatial projection operators $\mathbb{P}$ and $\mathbb{p}$ read specifically as

$$\mathbb{P} \equiv \widehat{\boldsymbol{I}} = \widehat{\boldsymbol{G}}_\alpha \otimes \widehat{\boldsymbol{G}}^\alpha \quad \text{at} \quad \mathcal{S}_\mathrm{m} \quad \text{and} \quad \mathbb{p} \equiv \widehat{\boldsymbol{I}} = \widehat{\boldsymbol{g}}_\alpha \otimes \widehat{\boldsymbol{g}}^\alpha \quad \text{at} \quad \mathcal{S}_\mathrm{s}. \tag{4.7}$$

It is thus obvious that the jumps in the spatial and material deformation gradients are of rank one. Observe that tangential continuity may alternatively be expressed as

$$\{\boldsymbol{F}\} \cdot \mathbb{P} = \boldsymbol{F} \cdot \mathbb{P} \quad \text{at} \quad \mathcal{S}_\mathrm{m} \quad \text{and} \quad \{\boldsymbol{f}\} \cdot \mathbb{p} = \boldsymbol{f} \cdot \mathbb{p} \quad \text{at} \quad \mathcal{S}_\mathrm{s}. \tag{4.8}$$

Thus, the action of the deformation gradients on vectors lying in the tangent planes to singular surfaces (tangent vectors) is continuous across singular surfaces.

As a consequence of the coherence condition II in Eq. 4.5, the jumps of the deformation gradients satisfy the following symmetry or rather *compatibility conditions* (coherence condition III)

$$\left[\boldsymbol{A} \cdot [\![\boldsymbol{f}]\!] \otimes \boldsymbol{m}\right]^\mathrm{skw} = \boldsymbol{0} \quad \text{and} \quad \left[\boldsymbol{a} \cdot [\![\boldsymbol{F}]\!] \otimes \boldsymbol{M}\right]^\mathrm{skw} = \boldsymbol{0}. \tag{4.9}$$

Here, $\boldsymbol{A} \in T^*\mathcal{B}_\mathrm{m}$ and $\boldsymbol{a} \in T^*\mathcal{B}_\mathrm{s}$ denote arbitrary material and spatial (constant) covectors, respectively.

In Cartesian coordinates with coordinate representation for the jumps of the tangent maps $[\![\boldsymbol{f}]\!] = [\![f^A{}_b]\!]\boldsymbol{E}_A \otimes \boldsymbol{e}^b$ and $[\![\boldsymbol{F}]\!] = [\![F^a{}_B]\!]\boldsymbol{e}_a \otimes \boldsymbol{E}^B$, respectively, the above compatibility conditions read

$$[\![f^A{}_b]\!]m_c = [\![f^A{}_c]\!]m_b \quad \text{and} \quad [\![F^a{}_B]\!]M_C = [\![F^a{}_C]\!]M_B. \tag{4.10}$$

This result is obtained by

(1) expanding the coherence condition II in Eq. 4.5 into

$$[\![\boldsymbol{f}]\!] = \boldsymbol{J} \otimes \boldsymbol{m} \quad \text{and} \quad [\![\boldsymbol{F}]\!] = \boldsymbol{j} \otimes \boldsymbol{M}$$

with jump vectors

$$\boldsymbol{J} := [\![\boldsymbol{f}]\!] \cdot \boldsymbol{m} \in T\mathcal{B}_\mathrm{m} \quad \text{and} \quad \boldsymbol{j} := [\![\boldsymbol{F}]\!] \cdot \boldsymbol{M} \in T\mathcal{B}_\mathrm{s},$$

(2) pre-multiplying with arbitrary (constant) covectors $\boldsymbol{A} \in T^*\mathcal{B}_\mathrm{m}$ and $\boldsymbol{a} \in T^*\mathcal{B}_\mathrm{s}$ to render

$$\boldsymbol{A} \cdot [\![\boldsymbol{f}]\!] = \Omega \boldsymbol{m} \quad \text{and} \quad \boldsymbol{a} \cdot [\![\boldsymbol{F}]\!] = \omega \boldsymbol{M}$$

with $\Omega := \boldsymbol{A} \cdot \boldsymbol{J}$ and $\omega := \boldsymbol{a} \cdot \boldsymbol{j}$, and finally

(3) observing the symmetry of $\boldsymbol{m} \otimes \boldsymbol{m}$ and $\boldsymbol{M} \otimes \boldsymbol{M}$ and thus of

$$\boldsymbol{A} \cdot [\![\boldsymbol{f}]\!] \otimes \boldsymbol{m} \quad \text{and} \quad \boldsymbol{a} \cdot [\![\boldsymbol{F}]\!] \otimes \boldsymbol{M},$$

respectively.

Consequently, for double contractions with (simple) second-order two-point tensors $\boldsymbol{z} := \boldsymbol{A} \otimes \boldsymbol{b}$ and $\boldsymbol{Z} := \boldsymbol{a} \otimes \boldsymbol{B}$, whereby $\boldsymbol{b} \in T\mathcal{B}_\mathrm{s}$ and $\boldsymbol{B} \in T\mathcal{B}_\mathrm{m}$ denote arbitrary spatial and material vectors, respectively, due to the compatibility conditions it holds that

$$\boldsymbol{z} : [\![\boldsymbol{f}]\!]\boldsymbol{m} = [\![\boldsymbol{f}^t]\!] \cdot \boldsymbol{z} \cdot \boldsymbol{m} \quad \text{and} \quad \boldsymbol{Z} : [\![\boldsymbol{F}]\!]\boldsymbol{M} = [\![\boldsymbol{F}^t]\!] \cdot \boldsymbol{Z} \cdot \boldsymbol{M}. \tag{4.11}$$

The extension of the above to arbitrary second-order two-point tensors $\boldsymbol{z}$ and $\boldsymbol{Z}$ is straightforward.

4.2.3 Jump in Cotangent Maps

The *area theorem* equates the integral of the vector-valued area elements $\mathrm{d}\boldsymbol{A} := \boldsymbol{N}\,\mathrm{d}A$ and $\mathrm{d}\boldsymbol{a} := \boldsymbol{n}\,\mathrm{d}a$, respectively, over the closed boundary surface of control volumes $\mathfrak{B}_\mathrm{m}$ and $\mathfrak{B}_\mathrm{s}$, respectively, to zero

$$\int_{\partial\mathfrak{B}_\mathrm{m}} \mathrm{d}\boldsymbol{A} = \boldsymbol{0} \quad \text{and} \quad \int_{\partial\mathfrak{B}_\mathrm{s}} \mathrm{d}\boldsymbol{a} = \boldsymbol{0}. \tag{4.12}$$

Note that the area theorem is in particular also valid if the control volume contains coherent singular surfaces $\mathfrak{S}_\mathrm{m}$ and $\mathfrak{S}_\mathrm{s}$, respectively.

Then upon

(1) decreasing the control volume to zero such that the surfaces $\partial\mathfrak{B}^{\pm}_\mathrm{m}$ coincide with $\mathfrak{S}_\mathrm{m}$, whereby

$$-\,\mathrm{d}\boldsymbol{A}^- = \mathrm{d}\boldsymbol{A}^+ = \mathrm{d}\boldsymbol{A} := \boldsymbol{M}\,\mathrm{d}A$$

holds, or, likewise, such that the surfaces $\partial\mathfrak{B}^{\pm}_\mathrm{s}$ coincide with $\mathfrak{S}_\mathrm{s}$, whereby

$$-\,\mathrm{d}\boldsymbol{a}^- = \mathrm{d}\boldsymbol{a}^+ = \mathrm{d}\boldsymbol{a} := \boldsymbol{m}\,\mathrm{d}a$$

holds, and

(2) localizing the result,

the area theorem degenerates to the trivial (geometric) result

$$[\![\boldsymbol{M}]\!]\,\mathrm{d}A = \boldsymbol{0} \quad \text{at} \quad \mathcal{S}_\mathrm{m} \quad \text{and} \quad [\![\boldsymbol{m}]\!]\,\mathrm{d}a = \boldsymbol{0} \quad \text{at} \quad \mathcal{S}_\mathrm{s}. \tag{4.13}$$

However, as a consequence of Nanson's formulae in Eq. 2.37, the unit normal to the singular surface lies in the null space of the jump of the cofactor of the deformation gradient (coherence condition IV)

$$[\![\boldsymbol{k}]\!] \cdot \boldsymbol{m} = \boldsymbol{0} \quad \text{at} \quad \mathcal{S}_\mathrm{m} \quad \text{and} \quad [\![\boldsymbol{K}]\!] \cdot \boldsymbol{M} = \boldsymbol{0} \quad \text{at} \quad \mathcal{S}_\mathrm{s}. \tag{4.14}$$

Observe that *normal continuity* may alternatively be expressed as

$$\{\boldsymbol{k}\} \cdot \boldsymbol{m} = \boldsymbol{k} \cdot \boldsymbol{m} \quad \text{and} \quad \{\boldsymbol{K}\} \cdot \boldsymbol{M} = \boldsymbol{K} \cdot \boldsymbol{M}. \tag{4.15}$$

Consequently, the action of the cofactor of the deformation gradients on covectors normal to the tangent plane to singular surfaces (normal vectors) is continuous across singular surfaces.

4.2.4 Jump in Measure Maps

The jump in the Jacobians $[\![J]\!]$ and $[\![j]\!]$ may straightforwardly be computed from their expressions in terms of the cofactor of the deformation gradients and the deformation gradients, i.e. from

$$3J = \boldsymbol{K} : \boldsymbol{F} \quad \text{and} \quad 3j = \boldsymbol{k} : \boldsymbol{f},$$

respectively, as (coherence condition V)

$$[\![J]\!] = \{\boldsymbol{K}\} : [\![\boldsymbol{F}]\!] \quad \text{and} \quad [\![j]\!] = \{\boldsymbol{k}\} : [\![\boldsymbol{f}]\!] \tag{4.16}$$

or, likewise, as

$$[\![J]\!] = \frac{1}{2}[\![\boldsymbol{K}]\!] : \{\boldsymbol{F}\} \quad \text{and} \quad [\![j]\!] = \frac{1}{2}[\![\boldsymbol{k}]\!] : \{\boldsymbol{f}\}. \tag{4.17}$$

This result is obtained by

(1) expanding

$$3[\![J]\!] = [\![\boldsymbol{K} : \boldsymbol{F}]\!] \quad \text{and} \quad 3[\![j]\!] = [\![\boldsymbol{k} : \boldsymbol{f}]\!]$$

into

$$\{\boldsymbol{K}\} : [\![\boldsymbol{F}]\!] + [\![\boldsymbol{K}]\!] : \{\boldsymbol{F}\} \quad \text{and} \quad \{\boldsymbol{k}\} : [\![\boldsymbol{f}]\!] + [\![\boldsymbol{k}]\!] : \{\boldsymbol{f}\},$$

(2) unfolding

$$[\![\boldsymbol{K}]\!] = \frac{1}{2}[\![\boldsymbol{F} \times\!\!\times \boldsymbol{F}]\!] \quad \text{and} \quad [\![\boldsymbol{k}]\!] = \frac{1}{2}[\![\boldsymbol{f} \times\!\!\times \boldsymbol{f}]\!]$$

into

$$[\![\boldsymbol{F}]\!] \times\!\!\times \{\boldsymbol{F}\} \quad \text{and} \quad [\![\boldsymbol{f}]\!] \times\!\!\times \{\boldsymbol{f}\},$$

(3) rearranging

$$[\![\boldsymbol{K}]\!] : \{\boldsymbol{F}\} = \big[[\![\boldsymbol{F}]\!] \times\!\!\times \{\boldsymbol{F}\}\big] : \{\boldsymbol{F}\} \quad \text{and} \quad [\![\boldsymbol{k}]\!] : \{\boldsymbol{f}\} = \big[[\![\boldsymbol{f}]\!] \times\!\!\times \{\boldsymbol{f}\}\big] : \{\boldsymbol{f}\}$$

into

$$\big[\{\boldsymbol{F}\} \times\!\!\times \{\boldsymbol{F}\}\big] : [\![\boldsymbol{F}]\!] \quad \text{and} \quad \big[\{\boldsymbol{f}\} \times\!\!\times \{\boldsymbol{f}\}\big] : [\![\boldsymbol{f}]\!],$$

(4) rewriting

$$\{\boldsymbol{F}\} \times\!\!\times \{\boldsymbol{F}\} \quad \text{and} \quad \{\boldsymbol{f}\} \times\!\!\times \{\boldsymbol{f}\}$$

with

$$[\![\boldsymbol{F}]\!] \times\!\!\times [\![\boldsymbol{F}]\!] = \boldsymbol{0} \quad \text{and} \quad [\![\boldsymbol{f}]\!] \times\!\!\times [\![\boldsymbol{f}]\!] = \boldsymbol{0}$$

into

$$\{\boldsymbol{F} \times\!\!\times \boldsymbol{F}\} = 2\{\boldsymbol{K}\} \quad \text{and} \quad \{\boldsymbol{f} \times\!\!\times \boldsymbol{f}\} = 2\{\boldsymbol{k}\},$$

(5) and finally assembling the two terms in (1) with

$$[\![\boldsymbol{K}]\!] : \{\boldsymbol{F}\} = 2\{\boldsymbol{K}\} : [\![\boldsymbol{F}]\!] \quad \text{and} \quad [\![\boldsymbol{k}]\!] : \{\boldsymbol{f}\} = 2\{\boldsymbol{k}\} : [\![\boldsymbol{f}]\!],$$

respectively.

4.2.5 Jump in Velocities

Normal–Tangential Decomposition at Singular Surfaces:
At singular surfaces, the total material and spatial velocities are conveniently decomposed into normal and tangential parts

$$\boldsymbol{W} = \boldsymbol{W}_\perp + \boldsymbol{W}_\| \quad \text{at} \quad \mathcal{S}_\text{m} \quad \text{and} \quad \boldsymbol{w} = \boldsymbol{w}_\perp + \boldsymbol{w}_\| \quad \text{at} \quad \mathcal{S}_\text{s}. \tag{4.18}$$

Thereby, the normal parts are defined in terms of the scalar-valued *normal velocities* $W_\perp$ and $w_\perp$, respectively, as

$$\boldsymbol{W}_\perp := [\boldsymbol{W} \cdot \boldsymbol{M}]\boldsymbol{M} =: W_\perp \boldsymbol{M} \quad \text{and} \quad \boldsymbol{w}_\perp := [\boldsymbol{w} \cdot \boldsymbol{m}]\boldsymbol{m} =: w_\perp \boldsymbol{m}$$

and, consequently, the tangential parts follow as

$$\boldsymbol{W}_\| := \boldsymbol{W} - \boldsymbol{W}_\perp =: \mathbb{P} \cdot \boldsymbol{W} \quad \text{and} \quad \boldsymbol{w}_\| := \boldsymbol{w} - \boldsymbol{w}_\perp =: \mathbb{p} \cdot \boldsymbol{w}.$$

The jump operator commutes with the total time derivative, hence the total material and spatial velocities are, likewise, continuous across singular surfaces (coherence condition VI)

$$[\![\boldsymbol{W}]\!] = \boldsymbol{0} \quad \text{at} \quad \mathcal{S}_\text{m} \quad \text{and} \quad [\![\boldsymbol{w}]\!] = \boldsymbol{0} \quad \text{at} \quad \mathcal{S}_\text{s}. \tag{4.19}$$

Note that the total material velocity $\boldsymbol{W} \neq \boldsymbol{0}$ for *migrating* singular surfaces $\mathcal{S}_\text{m}$, whereas $\boldsymbol{W} \equiv \boldsymbol{0}$ for *material* singular surfaces $\mathcal{S}_\text{m}$. Consequently, at material singular surfaces the total spatial velocity $\boldsymbol{w}$ coincides identically with the spatial velocity $\boldsymbol{v}$.

As a consequence of the coherence condition VI in Eq. 4.19 and the Euler formulae in Eq. 2.56, the material and spatial velocities are discontinuous across singular surfaces

$$[\![\boldsymbol{V}]\!] = -[\![\boldsymbol{f}]\!] \cdot \boldsymbol{w} \quad \text{at} \quad \mathcal{S}_\text{m} \quad \text{and} \quad [\![\boldsymbol{v}]\!] = -[\![\boldsymbol{F}]\!] \cdot \boldsymbol{W} \quad \text{at} \quad \mathcal{S}_\text{s}. \tag{4.20}$$

This result is obtained by

(1) invoking the representation for the material and spatial velocities in terms of the total velocities in Eq. 2.56, and
(2) observing the coherence condition VI in Eq. 4.19.

Furthermore, as a consequence of the coherence condition II in Eq. 4.5 (and the decomposition of the total velocities into normal and tangential parts in Eq. 4.18), the jumps in the material and spatial velocities at singular surfaces read (coherence condition VII)

$$[\![V]\!] = -[\![f]\!] \cdot \boldsymbol{w}_\perp \quad \text{at} \quad \mathcal{S}_\mathrm{m} \quad \text{and} \quad [\![\boldsymbol{v}]\!] = -[\![\boldsymbol{F}]\!] \cdot \boldsymbol{W}_\perp \quad \text{at} \quad \mathcal{S}_\mathrm{s}. \tag{4.21}$$

Observe that any potential tangential contributions to the total velocities at singular surfaces are filtered out by the coherence condition II in Eq. 4.5.

4.2.6 Summary of Coherence Conditions

The coherence conditions at coherent singular surfaces (Truesdell and Noll [1], Abeyaratne and Knowles [2]) are summarized for convenience as follows:

Summary of Coherence Conditions at Singular Surfaces:

I	:	$[\![\boldsymbol{Y}]\!]$	=	$\boldsymbol{0}$	&	$[\![\boldsymbol{y}]\!]$	=	$\boldsymbol{0}$
II	:	$[\![\boldsymbol{F}]\!] \cdot \mathbb{P}$	=	$\boldsymbol{0}$	&	$[\![\boldsymbol{f}]\!] \cdot \mathbb{p}$	=	$\boldsymbol{0}$
III	:	$[\boldsymbol{A} \cdot [\![\boldsymbol{f}]\!] \otimes \boldsymbol{m}]^{\mathrm{skw}}$	=	$\boldsymbol{0}$	&	$[\boldsymbol{a} \cdot [\![\boldsymbol{F}]\!] \otimes \boldsymbol{M}]^{\mathrm{skw}}$	=	$\boldsymbol{0}$
IV	:	$[\![\boldsymbol{k}]\!] \cdot \boldsymbol{m}$	=	$\boldsymbol{0}$	&	$[\![\boldsymbol{K}]\!] \cdot \boldsymbol{M}$	=	$\boldsymbol{0}$
V	:	$[\![j]\!] - \{\boldsymbol{k}\} : [\![\boldsymbol{f}]\!]$	=	0	&	$[\![J]\!] - \{\boldsymbol{K}\} : [\![\boldsymbol{F}]\!]$	=	0
VI	:	$[\![\boldsymbol{W}]\!]$	=	$\boldsymbol{0}$	&	$[\![\boldsymbol{w}]\!]$	=	$\boldsymbol{0}$
VII	:	$[\![\boldsymbol{V}]\!] + [\![\boldsymbol{f}]\!] \cdot \boldsymbol{w}_\perp$	=	$\boldsymbol{0}$	&	$[\![\boldsymbol{v}]\!] + [\![\boldsymbol{F}]\!] \cdot \boldsymbol{W}_\perp$	=	$\boldsymbol{0}$

4.3 Coherent Singular Curves*

In the sequel, singular curves are considered as embedded into the boundary surface as the intersections of its regular sub-parts. They are assumed piece-wise smooth and *coherent*.

At a singular curve $\mathcal{C}_\mathrm{m}$, the intersecting regular sub-parts $\partial\mathcal{B}_\mathrm{m}^\pm$ of the boundary surface $\partial\mathcal{B}_\mathrm{m}$ are characterized by their outwards-pointing normals $\boldsymbol{N}^\pm$, orthogonal to the tangent planes $T\partial\mathcal{B}_\mathrm{m}^\pm$, and by their outwards-pointing normals $\widehat{\boldsymbol{N}}^\pm$, orthogonal to the tangent lines $T\partial\partial\mathcal{B}_\mathrm{m}^\pm$ and tangential to the tangent planes $T\partial\mathcal{B}_\mathrm{m}^\pm$. Then the singular curve is endowed with its tangent vector $\widetilde{\boldsymbol{G}}$ and, for notational consistency, normal vectors $\boldsymbol{M}^\pm := \boldsymbol{N}^\pm$ and $\widehat{\boldsymbol{M}}^\pm := \mp\widehat{\boldsymbol{N}}^\pm$ are defined (with orthogonality $\boldsymbol{M}^\pm \perp \widehat{\boldsymbol{M}}^\pm$). Note that the case of a singular curve embedded into a smooth boundary surface results in $\boldsymbol{M}^\pm = \boldsymbol{N}$ and $\widehat{\boldsymbol{M}}^\pm =: \widehat{\boldsymbol{M}}$. Analogous definitions and notation hold in the spatial configuration.

Jump and Average Operators:
Jump and *average* operators as applied to arbitrary scalar-, vector-, or tensor-valued fields $\widehat{\boldsymbol{Z}}$ across a singular curve $\mathcal{C}_{\mathrm{m}}$ in the material configuration are defined in analogy to those across a singular surface in the above. Thereby, it is assumed that $\widehat{\boldsymbol{Z}}$ is smooth away from the singular curve and smooth up to the singular curve from either side where it takes values $\widehat{\boldsymbol{Z}}^{\pm} = \lim_{\epsilon \to 0} \widehat{\boldsymbol{Z}}(\widehat{\boldsymbol{\xi}}|_{\mathcal{C}_{\square}} \pm \epsilon\, \widehat{\boldsymbol{\mu}})$ with $\widehat{\boldsymbol{\xi}}$ and $\widehat{\boldsymbol{\mu}}$ the matrix arrangements of the surface coordinates θ^{α} and the normal coordinates to the singular curve in the two-dimensional surface coordinates parameter space $\partial\mathcal{B}_{\square}$, respectively, and $\epsilon \in \mathbb{R}^{+}$. In particular, the parameter curves $\widehat{\boldsymbol{X}}^{\pm}_{\epsilon} := \widehat{\boldsymbol{X}}(\widehat{\boldsymbol{\xi}}|_{\mathcal{C}_{\square}} \pm \epsilon\, \widehat{\boldsymbol{\mu}})$ render the normal vectors $\mathrm{d}_{\epsilon} \widehat{\boldsymbol{X}}^{\pm}_{\epsilon}|_{\epsilon=0} = \pm \widehat{\boldsymbol{M}}^{\pm}$, tangential to the tangent planes $T\partial\mathcal{B}^{\pm}_{\mathrm{m}}$. Jump and average operators across a singular curve $\mathcal{C}_{\mathrm{s}}$ in the spatial configuration are defined accordingly.

4.3.1 Jump in Nonlinear Deformation Maps

Due to the coherence assumption, the material and spatial nonlinear deformation maps (and placements, respectively) are continuous across singular curves (coherence condition $\widehat{\mathrm{I}}$)

$$[\![\widehat{\boldsymbol{Y}}]\!] = \boldsymbol{0} \quad \text{at} \quad \mathcal{C}_{\mathrm{m}} \quad \text{and} \quad [\![\widehat{\boldsymbol{y}}]\!] = \boldsymbol{0} \quad \text{at} \quad \mathcal{C}_{\mathrm{s}} \tag{4.22}$$

and, in addition, smooth away from singular curves, and smooth up to singular curves from either side.

4.3.2 Jump in Tangent Maps

The jump operator commutes with either the material or the spatial *tangential gradient* to singular curves, hence the spatial and material boundary surface deformation gradients are *tangential continuous* across singular curves (coherence condition $\widehat{\mathrm{II}}$)

$$[\![\widehat{\boldsymbol{F}}]\!] \cdot \widehat{\mathbb{P}} = \boldsymbol{0} \quad \text{at} \quad \mathcal{C}_{\mathrm{m}} \quad \text{and} \quad [\![\widehat{\boldsymbol{f}}]\!] \cdot \widehat{\mathbb{p}} = \boldsymbol{0} \quad \text{at} \quad \mathcal{C}_{\mathrm{s}}. \tag{4.23}$$

Here, $\widehat{\mathbb{P}}$ and $\widehat{\mathbb{p}}$ are the material and spatial (mixed-variant) projection operators onto the tangent line to singular curves

$$\widehat{\mathbb{P}} := \widehat{\boldsymbol{I}}^{\pm} - \widehat{\boldsymbol{M}}^{\pm} \otimes \widehat{\boldsymbol{M}}^{\pm} \quad \text{at} \quad \mathcal{C}_{\mathrm{m}} \quad \text{and} \quad \widehat{\mathbb{p}} := \widehat{\boldsymbol{i}}^{\pm} - \widehat{\boldsymbol{m}}^{\pm} \otimes \widehat{\boldsymbol{m}}^{\pm} \quad \text{at} \quad \mathcal{C}_{\mathrm{s}}, \tag{4.24}$$

whereby $\widehat{\boldsymbol{I}}^{\pm} := \boldsymbol{I} - \boldsymbol{M}^{\pm} \otimes \boldsymbol{M}^{\pm}$ and $\widehat{\boldsymbol{i}}^{\pm} := \boldsymbol{i} - \boldsymbol{m}^{\pm} \otimes \boldsymbol{m}^{\pm}$ denote the surface unit tensors on $\partial\mathcal{B}^{\pm}_{\mathrm{m}}$ and $\partial\mathcal{B}^{\pm}_{\mathrm{s}}$, respectively. Consequently, the material and spatial projection operators $\widehat{\mathbb{P}}$ and $\widehat{\mathbb{p}}$ read specifically as

$$\widehat{\mathbb{P}} \equiv \widetilde{\boldsymbol{I}} = \widetilde{\boldsymbol{G}} \otimes \widetilde{\boldsymbol{G}} \quad \text{at} \quad \mathcal{C}_{\mathrm{m}} \quad \text{and} \quad \widehat{\mathbb{p}} \equiv \widetilde{\boldsymbol{i}} = \widetilde{\boldsymbol{g}} \otimes \widetilde{\boldsymbol{g}} \quad \text{at} \quad \mathcal{C}_{\mathrm{s}}. \tag{4.25}$$

It is thus obvious that the jumps in the spatial and material boundary surface deformation gradients are of rank two. Observe that tangential continuity may alternatively be expressed as

$$\{\widehat{\boldsymbol{F}}\} \cdot \widehat{\mathbb{P}} = \widehat{\boldsymbol{F}} \cdot \widehat{\mathbb{P}} \quad \text{at} \quad \mathcal{C}_{\mathrm{m}} \quad \text{and} \quad \{\widehat{\boldsymbol{f}}\} \cdot \widehat{\mathbb{p}} = \widehat{\boldsymbol{f}} \cdot \widehat{\mathbb{p}} \quad \text{at} \quad \mathcal{C}_{\mathrm{s}}. \tag{4.26}$$

Thus, the action of the boundary surface deformation gradients on vectors lying in the tangent line to singular curves (tangent vectors) is continuous across singular curves.

As a consequence of the coherence condition $\widehat{\mathrm{II}}$ in Eq. 4.23, the jumps of the boundary surface deformation gradients satisfy the following symmetry or rather *compatibility conditions* (coherence condition $\widehat{\mathrm{III}}$)

$$\left[\boldsymbol{A} \cdot \widehat{[\![\widehat{\boldsymbol{f}}]\!]}^{\pm} \otimes \widehat{\boldsymbol{m}}^{\pm}\right]^{\mathrm{skw}} = \boldsymbol{0} \quad \text{and} \quad \left[\boldsymbol{a} \cdot \widehat{[\![\widehat{\boldsymbol{F}}]\!]}^{\pm} \otimes \widehat{\boldsymbol{M}}^{\pm}\right]^{\mathrm{skw}} = \boldsymbol{0}. \tag{4.27}$$

Here, $\boldsymbol{A} \in T^*\mathcal{B}_{\mathrm{m}}$ and $\boldsymbol{a} \in T^*\mathcal{B}_{\mathrm{s}}$ denote arbitrary material and spatial (constant) covectors, respectively. Moreover, the *tangential jumps* of the surface deformation gradients are defined as

$$\widehat{[\![\widehat{\boldsymbol{f}}]\!]}^{\pm} := [\![\widehat{\boldsymbol{f}}]\!] \cdot \widehat{\boldsymbol{i}}^{\pm} \quad \text{and} \quad \widehat{[\![\widehat{\boldsymbol{F}}]\!]}^{\pm} := [\![\widehat{\boldsymbol{F}}]\!] \cdot \widehat{\boldsymbol{I}}^{\pm}. \tag{4.28}$$

This result is obtained by

(1) expanding the coherence condition $\widehat{\mathrm{II}}$ in Eq. 4.23 into

$$[\![\widehat{\boldsymbol{f}}]\!] = \boldsymbol{J}^{\pm} \otimes \boldsymbol{m}^{\pm} + \widehat{\boldsymbol{J}}^{\pm} \otimes \widehat{\boldsymbol{m}}^{\pm} \quad \text{and} \quad [\![\widehat{\boldsymbol{F}}]\!] = \boldsymbol{j}^{\pm} \otimes \boldsymbol{M}^{\pm} + \widehat{\boldsymbol{j}}^{\pm} \otimes \widehat{\boldsymbol{M}}^{\pm}$$

with jump vectors

$$\boldsymbol{J}^{\pm} := [\![\widehat{\boldsymbol{f}}]\!] \cdot \boldsymbol{m}^{\pm} \in T\mathcal{B}_{\mathrm{m}} \quad \text{and} \quad \boldsymbol{j}^{\pm} := [\![\widehat{\boldsymbol{F}}]\!] \cdot \boldsymbol{M}^{\pm} \in T\mathcal{B}_{\mathrm{s}}$$

and

$$\widehat{\boldsymbol{J}}^{\pm} := [\![\widehat{\boldsymbol{f}}]\!] \cdot \widehat{\boldsymbol{m}}^{\pm} \in T\mathcal{B}_{\mathrm{m}} \quad \text{and} \quad \widehat{\boldsymbol{j}}^{\pm} := [\![\widehat{\boldsymbol{F}}]\!] \cdot \widehat{\boldsymbol{M}}^{\pm} \in T\mathcal{B}_{\mathrm{s}}$$

such that the tangential jumps of the surface deformation gradients result in

$$\widehat{[\![\widehat{\boldsymbol{f}}]\!]}^{\pm} = \widehat{\boldsymbol{J}}^{\pm} \otimes \widehat{\boldsymbol{m}}^{\pm} \quad \text{and} \quad \widehat{[\![\widehat{\boldsymbol{F}}]\!]}^{\pm} = \widehat{\boldsymbol{j}}^{\pm} \otimes \widehat{\boldsymbol{M}}^{\pm},$$

(2) pre-multiplying with arbitrary (constant) covectors $\boldsymbol{A} \in T^*\mathcal{B}_{\mathrm{m}}$ and $\boldsymbol{a} \in T^*\mathcal{B}_{\mathrm{s}}$ to render

$$\boldsymbol{A} \cdot [\![\widehat{\boldsymbol{f}}]\!] = \Omega^{\pm}\boldsymbol{m}^{\pm} + \widehat{\Omega}^{\pm}\widehat{\boldsymbol{m}}^{\pm} \quad \text{and} \quad \boldsymbol{a} \cdot [\![\widehat{\boldsymbol{F}}]\!] = \omega^{\pm}\boldsymbol{M}^{\pm} + \widehat{\omega}^{\pm}\widehat{\boldsymbol{M}}^{\pm}$$

with $\Omega^{\pm} := \boldsymbol{A} \cdot \boldsymbol{J}^{\pm}, \widehat{\Omega}^{\pm} := \boldsymbol{A} \cdot \widehat{\boldsymbol{J}}^{\pm}$ and $\omega^{\pm} := \boldsymbol{a} \cdot \boldsymbol{j}^{\pm}, \widehat{\omega}^{\pm} := \boldsymbol{a} \cdot \widehat{\boldsymbol{j}}^{\pm}$, thus

$$\boldsymbol{A} \cdot \widehat{[\![\widehat{\boldsymbol{f}}]\!]}^{\pm} = \widehat{\Omega}^{\pm}\widehat{\boldsymbol{m}}^{\pm} \quad \text{and} \quad \boldsymbol{a} \cdot \widehat{[\![\widehat{\boldsymbol{F}}]\!]}^{\pm} = \widehat{\omega}^{\pm}\widehat{\boldsymbol{M}}^{\pm},$$

and finally

(3) observing the symmetry of $\widehat{\boldsymbol{m}}^{\pm} \otimes \widehat{\boldsymbol{m}}^{\pm}$ and $\widehat{\boldsymbol{M}}^{\pm} \otimes \widehat{\boldsymbol{M}}^{\pm}$ and thus of

$$\boldsymbol{A} \cdot \widehat{[\![\widehat{\boldsymbol{f}}]\!]}^{\pm} \otimes \widehat{\boldsymbol{m}}^{\pm} \quad \text{and} \quad \boldsymbol{a} \cdot \widehat{[\![\widehat{\boldsymbol{F}}]\!]}^{\pm} \otimes \widehat{\boldsymbol{M}}^{\pm},$$

respectively.

Consequently, for double contractions with (simple) tangential second-order two-point tensors $\boldsymbol{z} := \boldsymbol{A} \otimes \boldsymbol{b}$ and $\boldsymbol{Z} := \boldsymbol{a} \otimes \boldsymbol{B}$, whereby $\boldsymbol{b} \in T\mathcal{B}_s$ and $\boldsymbol{B} \in T\mathcal{B}_m$ denote arbitrary spatial and material vectors, respectively, due to the compatibility conditions it holds that

$$\boldsymbol{z} : \widehat{[\![\widehat{\boldsymbol{f}}]\!]}^{\pm}\, \widehat{\boldsymbol{m}}^{\pm} = \widehat{[\![\widehat{\boldsymbol{f}}^{t}]\!]}^{\pm} \cdot \boldsymbol{z} \cdot \widehat{\boldsymbol{m}}^{\pm} \text{ and } \boldsymbol{Z} : \widehat{[\![\widehat{\boldsymbol{F}}]\!]}^{\pm}\, \widehat{\boldsymbol{M}}^{\pm} = \widehat{[\![\widehat{\boldsymbol{F}}^{t}]\!]}^{\pm} \cdot \boldsymbol{Z} \cdot \widehat{\boldsymbol{M}}^{\pm}. \tag{4.29}$$

The extension of the above to arbitrary second-order two-point tensors $\boldsymbol{z}$ and $\boldsymbol{Z}$ is straightforward.

4.3.3 Jump in Cotangent Maps

The *(normal) line theorem* equates the integral of the vector-valued (normal) line elements $\mathrm{d}\widehat{\boldsymbol{L}} := \widehat{\boldsymbol{N}}\,\mathrm{d}L$ and $\mathrm{d}\widehat{\boldsymbol{l}} := \widehat{\boldsymbol{n}}\,\mathrm{d}l$, respectively, over the closed boundary curve of control surfaces $\mathfrak{S}_m \subseteq \partial\mathcal{B}_m$ and $\mathfrak{S}_s \subseteq \partial\mathcal{B}_s$, respectively, to the resultant vector-valued control surface curvature (whereby $\mathrm{d}\boldsymbol{A} := \boldsymbol{M}\,\mathrm{d}A$ and $\mathrm{d}\boldsymbol{a} := \boldsymbol{m}\,\mathrm{d}a$)

$$\int_{\partial\mathfrak{S}_m} \mathrm{d}\widehat{\boldsymbol{L}} = \int_{\mathfrak{S}_m} \widehat{C}\,\mathrm{d}\boldsymbol{A} \quad \text{and} \quad \int_{\partial\mathfrak{S}_s} \mathrm{d}\widehat{\boldsymbol{l}} = \int_{\mathfrak{S}_s} \widehat{c}\,\mathrm{d}\boldsymbol{a}. \tag{4.30}$$

Note that the (normal) line theorem is in particular also valid if the control surface contains coherent singular curves $\mathfrak{C}_m$ and $\mathfrak{C}_s$, respectively (as long as the Dirac-delta-type singularity of the surface curvature across the singular curves is accounted for).

Then upon

(1) decreasing the control surface to zero such that the curves $\partial\mathfrak{S}_m^{\pm}$ coincide with $\mathfrak{C}_m$, whereby

$$-\,\mathrm{d}\widehat{\boldsymbol{L}}^{-} := \widehat{\boldsymbol{M}}^{-}\,\mathrm{d}L \quad \text{and} \quad \mathrm{d}\widehat{\boldsymbol{L}}^{+} := \widehat{\boldsymbol{M}}^{+}\,\mathrm{d}L$$

hold, or, likewise, such that the curves $\partial\mathfrak{S}_s^{\pm}$ coincide with $\mathfrak{C}_s$, whereby

$$-\,\mathrm{d}\widehat{\boldsymbol{l}}^- := \widehat{\boldsymbol{m}}^-\,\mathrm{d}l \quad\text{and}\quad \mathrm{d}\widehat{\boldsymbol{l}}^+ := \widehat{\boldsymbol{m}}^+\,\mathrm{d}l$$

hold, and

(2) localizing the result, whereby

$$\int_{\mathfrak{S}_m\to\mathfrak{C}_m} \widehat{C}\,\mathrm{d}A \to \int_{\mathfrak{C}_m} \widehat{\mathsf{C}}\,\mathbf{M}\,\mathrm{d}L \quad\text{and}\quad \int_{\mathfrak{S}_s\to\mathfrak{C}_s} \widehat{c}\,\mathrm{d}\boldsymbol{a} \to \int_{\mathfrak{C}_s} \widehat{\mathsf{c}}\,\mathbf{m}\,\mathrm{d}l \tag{4.31}$$

with 'singular surface curvature' and average surface normal

$$\widehat{\mathsf{C}} := -2\tan\alpha_m,\ \mathbf{M} := \{\boldsymbol{M}\} \quad\text{and}\quad \widehat{\mathsf{c}} := -2\tan\alpha_s,\ \mathbf{m} := \{\boldsymbol{m}\} \tag{4.32}$$

and $2\alpha_m := \angle(\boldsymbol{M}^-, \boldsymbol{M}^+)$, $2\alpha_s := \angle(\boldsymbol{m}^-, \boldsymbol{m}^+)$ is taken into account,

the (normal) line theorem degenerates to the trivial (geometric) result

$$[\![\widehat{\boldsymbol{M}}]\!]\,\mathrm{d}L = \widehat{\mathsf{C}}\,\mathbf{M}\,\mathrm{d}L \quad\text{at}\quad \mathcal{C}_m \quad\text{and}\quad [\![\widehat{\boldsymbol{m}}]\!]\,\mathrm{d}l = \widehat{\mathsf{c}}\,\mathbf{m}\,\mathrm{d}l \quad\text{at}\quad \mathcal{C}_s. \tag{4.33}$$

However, as a consequence, the boundary surface Nanson's formulae in Eq. 3.37 render (coherence condition $\widehat{\mathrm{IV}}$)

$$[\![\widehat{\boldsymbol{k}}\cdot\widehat{\boldsymbol{m}}]\!] = \widetilde{\Lambda}\widehat{\mathsf{C}}\,\mathbf{M} \text{ at } \mathcal{C}_m \quad\text{and}\quad [\![\widehat{\boldsymbol{K}}\cdot\widehat{\boldsymbol{M}}]\!] = \widetilde{\lambda}\widehat{\mathsf{c}}\,\mathbf{m} \text{ at } \mathcal{C}_s. \tag{4.34}$$

Note that, for example, for the spatially *smooth case* with $\alpha_s = 0$ and the two special cases with, for example, $\alpha_s = \pi/4$ and $\alpha_s \to \pi/2$, respectively, the coherence condition $\widehat{\mathrm{IV}}$ reads as

$$[\![\widehat{\boldsymbol{K}}]\!]\cdot\widehat{\boldsymbol{M}} \overset{\text{smooth}}{=} \mathbf{0},\ [\![\widehat{\boldsymbol{K}}\cdot\widehat{\boldsymbol{M}}]\!] = -2\widetilde{\lambda}\,\mathbf{m} \quad\text{and}\quad [\![\widehat{\boldsymbol{K}}\cdot\widehat{\boldsymbol{M}}]\!] \to -\infty\,\mathbf{m}. \tag{4.35}$$

Thus, only for a *smooth* boundary surface with $\widehat{\boldsymbol{M}}^\pm = \widehat{\boldsymbol{M}}$ and $\boldsymbol{M}^\pm = \boldsymbol{M}$ or $\widehat{\boldsymbol{m}}^\pm = \widehat{\boldsymbol{m}}$ and $\boldsymbol{m}^\pm = \boldsymbol{m}$, respectively, the right-hand sides in Eq. 4.34 vanish. In this case, the unit normal to the singular curve in the tangent plane to the smooth boundary surface lies in the null space of the jump of the boundary surface cofactor of the boundary surface deformation gradient. Observe that in this case, *normal continuity* in the tangent plane to the smooth boundary surface may alternatively be expressed as

$$\{\widehat{\boldsymbol{k}}\}\cdot\widehat{\boldsymbol{m}} \overset{\text{smooth}}{=} \widehat{\boldsymbol{k}}\cdot\widehat{\boldsymbol{m}} \quad\text{and}\quad \{\widehat{\boldsymbol{K}}\}\cdot\widehat{\boldsymbol{M}} \overset{\text{smooth}}{=} \widehat{\boldsymbol{K}}\cdot\widehat{\boldsymbol{M}}. \tag{4.36}$$

Consequently, the action of the boundary surface cofactor of the boundary surface deformation gradients on covectors normal to singular curves and tangent to the embedding smooth boundary surface (normal vectors) is continuous across singular curves.

4.3.4 Jump in Measure Maps

The jump in the boundary surface Jacobians $[\![\widehat{J}]\!]$ and $[\![\widehat{j}]\!]$ may straightforwardly be computed from their expressions in terms of the boundary surface deformation gradients, i.e. from

$$2\widehat{J} = \widehat{\boldsymbol{F}}\widehat{\times\!\!\times}\widehat{\boldsymbol{F}} = \widehat{\mathbb{R}} : [\widehat{\boldsymbol{F}}\times\!\!\times\widehat{\boldsymbol{F}}] \quad \text{and} \quad 2\widehat{j} = \widehat{\boldsymbol{f}}\widehat{\times\!\!\times}\widehat{\boldsymbol{f}} = \widehat{\mathbb{r}} : [\widehat{\boldsymbol{f}}\times\!\!\times\widehat{\boldsymbol{f}}],$$

respectively, as (coherence condition $\widehat{\mathrm{V}}$)

$$[\![\widehat{J}]\!] = \{\widehat{\boldsymbol{F}}\}\{\widehat{\times\!\!\times}\}[\![\widehat{\boldsymbol{F}}]\!] + [\![\widehat{J}]\!]^{\mathrm{ns}} \quad \text{and} \quad [\![\widehat{j}]\!] = \{\widehat{\boldsymbol{f}}\}\{\widehat{\times\!\!\times}\}[\![\widehat{\boldsymbol{f}}]\!] + [\![\widehat{j}]\!]^{\mathrm{ns}}. \tag{4.37}$$

Here, $\widehat{\mathbb{R}}$ and $\widehat{\mathbb{r}}$ denote the boundary surface normal rotation tensors

$$\widehat{\mathbb{R}} := \boldsymbol{n} \otimes \boldsymbol{N} \quad \text{and} \quad \widehat{\mathbb{r}} := \boldsymbol{N} \otimes \boldsymbol{n}. \tag{4.38}$$

These satisfy the SO(3)-like relations

$$\begin{aligned} \widehat{\mathbb{R}}^t &= \widehat{\mathbb{r}} \ \text{ with } \widehat{\mathbb{R}} \cdot \widehat{\mathbb{R}}^t = \widehat{\mathbb{R}} \cdot \widehat{\mathbb{r}} = \boldsymbol{n} \otimes \boldsymbol{n} \,, \\ \widehat{\mathbb{r}}^t &= \widehat{\mathbb{R}} \text{ with } \widehat{\mathbb{r}} \cdot \widehat{\mathbb{r}}^t = \widehat{\mathbb{r}} \cdot \widehat{\mathbb{R}} = \boldsymbol{N} \otimes \boldsymbol{N}. \end{aligned} \tag{4.39}$$

Furthermore, $[\![\widehat{J}]\!]^{\mathrm{ns}}$ and $[\![\widehat{j}]\!]^{\mathrm{ns}}$ abbreviate the contributions due to the non-smoothness of the boundary surface

$$[\![\widehat{J}]\!]^{\mathrm{ns}} := \frac{1}{2}[\![\widehat{\mathbb{R}}]\!] : \{\widehat{\boldsymbol{F}}\times\!\!\times\widehat{\boldsymbol{F}}\} \quad \text{and} \quad [\![\widehat{j}]\!]^{\mathrm{ns}} := \frac{1}{2}[\![\widehat{\mathbb{r}}]\!] : \{\widehat{\boldsymbol{f}}\times\!\!\times\widehat{\boldsymbol{f}}\}. \tag{4.40}$$

Finally, the following abbreviations are used

$$\begin{aligned} \{\widehat{\boldsymbol{F}}\}\{\widehat{\times\!\!\times}\}[\![\widehat{\boldsymbol{F}}]\!] &:= \{\widehat{\mathbb{R}}\} : \big[\{\widehat{\boldsymbol{F}}\}\times\!\!\times[\![\widehat{\boldsymbol{F}}]\!]\big], \\ \{\widehat{\boldsymbol{f}}\}\{\widehat{\times\!\!\times}\}[\![\widehat{\boldsymbol{f}}]\!] &:= \{\widehat{\mathbb{r}}\} : \big[\{\widehat{\boldsymbol{f}}\}\times\!\!\times[\![\widehat{\boldsymbol{f}}]\!]\big]. \end{aligned} \tag{4.41}$$

This result is obtained by

(1) expanding

$$2[\![\widehat{J}]\!] = \big[\!\big[\widehat{\mathbb{R}} : [\widehat{\boldsymbol{F}}\times\!\!\times\widehat{\boldsymbol{F}}]\big]\!\big] \quad \text{and} \quad 2[\![\widehat{j}]\!] = \big[\!\big[\widehat{\mathbb{r}} : [\widehat{\boldsymbol{f}}\times\!\!\times\widehat{\boldsymbol{f}}]\big]\!\big]$$

into

$$\{\widehat{\mathbb{R}}\} : [\![\widehat{\boldsymbol{F}}\times\!\!\times\widehat{\boldsymbol{F}}]\!] + [\![\widehat{\mathbb{R}}]\!] : \{\widehat{\boldsymbol{F}}\times\!\!\times\widehat{\boldsymbol{F}}\} \quad \text{and} \quad \{\widehat{\mathbb{r}}\} : [\![\widehat{\boldsymbol{f}}\times\!\!\times\widehat{\boldsymbol{f}}]\!] + [\![\widehat{\mathbb{r}}]\!] : \{\widehat{\boldsymbol{f}}\times\!\!\times\widehat{\boldsymbol{f}}\},$$

(2) unfolding

$$[\![\widehat{\boldsymbol{F}}\times\!\!\times\widehat{\boldsymbol{F}}]\!] \quad \text{and} \quad [\![\widehat{\boldsymbol{f}}\times\!\!\times\widehat{\boldsymbol{f}}]\!]$$

into

$$2\{\widehat{\boldsymbol{F}}\} \times\!\!\times [\![\widehat{\boldsymbol{F}}]\!] \quad \text{and} \quad 2\{\widehat{\boldsymbol{f}}\} \times\!\!\times [\![\widehat{\boldsymbol{f}}]\!],$$

(3) and finally assembling the various terms using the above abbreviations.

For a *smooth* boundary surface, the terms $[\![\widehat{J}]\!]^{\text{ns}}$, $[\![\widehat{j}]\!]^{\text{ns}}$ and $[\![\widehat{\mathbb{R}}]\!]$, $[\![\widehat{\mathbb{r}}]\!]$ vanish identically. Consequently, coherence condition $\widehat{\text{V}}$ degenerates to

$$[\![\widehat{J}]\!] \overset{\text{smooth}}{=} \{\widehat{\boldsymbol{K}}\} : [\![\widehat{\boldsymbol{F}}]\!] \quad \text{and} \quad [\![\widehat{j}]\!] \overset{\text{smooth}}{=} \{\widehat{\boldsymbol{k}}\} : [\![\widehat{\boldsymbol{f}}]\!]. \tag{4.42}$$

Note that for the smooth case coherence condition $\widehat{\text{V}}$ for the jumps in the boundary surface Jacobians resembles coherence condition V for the jump in the Jacobians.

4.3.5 Jump in Velocities

Normal–Tangential Decomposition at Singular Curves:
At singular curves, the total material and spatial boundary surface velocities are conveniently decomposed into normal and tangential parts

$$\widehat{\boldsymbol{W}}^{\pm} = \widehat{\boldsymbol{W}}^{\pm}_{\perp} + \widehat{\boldsymbol{W}}_{\parallel} \quad \text{at} \quad \mathcal{C}_{\text{m}} \quad \text{and} \quad \widehat{\boldsymbol{w}}^{\pm} = \widehat{\boldsymbol{w}}^{\pm}_{\perp} + \widehat{\boldsymbol{w}}_{\parallel} \quad \text{at} \quad \mathcal{C}_{\text{s}}. \tag{4.43}$$

Thereby, the normal parts are defined in terms of the scalar-valued *normal boundary surface velocities* $\widehat{W}^{\pm}_{\perp}$ and $\widehat{w}^{\pm}_{\perp}$, respectively, as

$$\widehat{\boldsymbol{W}}^{\pm}_{\perp} := [\widehat{\boldsymbol{W}} \cdot \widehat{\boldsymbol{M}}]^{\pm} \widehat{\boldsymbol{M}}^{\pm} =: \widehat{W}^{\pm}_{\perp} \widehat{\boldsymbol{M}}^{\pm} \quad \text{and} \quad \widehat{\boldsymbol{w}}^{\pm}_{\perp} := [\widehat{\boldsymbol{w}} \cdot \widehat{\boldsymbol{m}}]^{\pm} \widehat{\boldsymbol{m}}^{\pm} =: \widehat{w}^{\pm}_{\perp} \widehat{\boldsymbol{m}}^{\pm}$$

and, consequently, the tangential parts follow as

$$\widehat{\boldsymbol{W}}_{\parallel} := \widehat{\boldsymbol{W}}^{\pm} - \widehat{\boldsymbol{W}}^{\pm}_{\perp} =: \widehat{\mathbb{P}} \cdot \widehat{\boldsymbol{W}}^{\pm} \quad \text{and} \quad \widehat{\boldsymbol{w}}_{\parallel} := \widehat{\boldsymbol{w}}^{\pm} - \widehat{\boldsymbol{w}}^{\pm}_{\perp} =: \widehat{\mathbb{p}} \cdot \widehat{\boldsymbol{w}}^{\pm}.$$

The jump operator commutes with the total time derivative, hence the total material and spatial boundary surface velocities are, likewise, tangential continuous across singular curves (coherence condition $\widehat{\text{VI}}$)

$$\widehat{[\![\widehat{\boldsymbol{W}}]\!]}^{\pm} = \boldsymbol{0} \quad \text{at} \quad \mathcal{C}_{\text{m}} \quad \text{and} \quad \widehat{[\![\widehat{\boldsymbol{w}}]\!]}^{\pm} = \boldsymbol{0} \quad \text{at} \quad \mathcal{C}_{\text{s}}. \tag{4.44}$$

Note that the total material boundary surface velocity $\widehat{\boldsymbol{W}}^{\pm} \neq \boldsymbol{0}$ for *migrating* singular curves $\mathcal{C}_{\text{m}}$, whereas $\widehat{\boldsymbol{W}}^{\pm} \equiv \boldsymbol{0}$ for *material* singular curves $\mathcal{C}_{\text{m}}$. Consequently, at material singular curves the total spatial boundary surface velocity $\widehat{\boldsymbol{w}}^{\pm}$ coincides identically with the spatial velocity $\widehat{\boldsymbol{v}}^{\pm}$.

As a consequence of the coherence condition $\widehat{\text{VI}}$ in Eq. 4.44 and the Euler formulae in Eq. 3.49, the material and spatial boundary surface velocities are tangential discontinuous across singular curves

$$\widehat{[\![\widehat{\boldsymbol{V}}]\!]}^{\pm} = -\widehat{[\![\widehat{\boldsymbol{f}}]\!]}^{\pm} \cdot \widehat{\boldsymbol{w}}^{\pm} \quad \text{at} \quad \mathcal{C}_{\mathrm{m}} \quad \text{and} \quad \widehat{[\![\widehat{\boldsymbol{v}}]\!]}^{\pm} = -\widehat{[\![\widehat{\boldsymbol{F}}]\!]}^{\pm} \cdot \widehat{\boldsymbol{W}}^{\pm} \quad \text{at} \quad \mathcal{C}_{\mathrm{s}}. \tag{4.45}$$

This result is obtained by

(1) invoking the representation for the material and spatial boundary surface velocities in terms of the total boundary surface velocities in Eq. 3.49, and
(2) observing the coherence condition $\widehat{\text{VI}}$ in Eq. 4.44.

Furthermore, as a consequence of the coherence condition $\widehat{\text{II}}$ in Eq. 4.23, (and the decomposition of the total boundary surface velocities into normal and tangential parts in Eq. 4.43), the jumps in the material and spatial boundary surface velocities at singular curves read (coherence condition $\widehat{\text{VII}}$)

$$\widehat{[\![\widehat{\boldsymbol{V}}]\!]}^{\pm} = -\widehat{[\![\widehat{\boldsymbol{f}}]\!]}^{\pm} \cdot \widehat{\boldsymbol{w}}^{\pm}_{\perp} \quad \text{at} \quad \mathcal{C}_{\mathrm{m}} \quad \text{and} \quad \widehat{[\![\widehat{\boldsymbol{v}}]\!]}^{\pm} = -\widehat{[\![\widehat{\boldsymbol{F}}]\!]}^{\pm} \cdot \widehat{\boldsymbol{W}}^{\pm}_{\perp} \quad \text{at} \quad \mathcal{C}_{\mathrm{s}}. \tag{4.46}$$

Observe that any potential tangential contributions to the total boundary surface velocities at singular curves are filtered out by the coherence condition $\widehat{\text{II}}$ in Eq. 4.23.

4.3.6 Summary of Coherence Conditions

The coherence conditions at coherent singular curves are summarized for convenience as follows:

Summary of Coherence Conditions at Singular Curves:

$$\begin{array}{rclclcl}
\widehat{\mathbf{I}} & : & [\![\widehat{\boldsymbol{Y}}]\!] & = & \mathbf{0} & \& & [\![\widehat{\boldsymbol{y}}]\!] = \mathbf{0} \\
\widehat{\mathbf{II}} & : & [\![\widehat{\boldsymbol{F}}]\!] \cdot \widehat{\mathbb{P}} & = & \mathbf{0} & \& & [\![\widehat{\boldsymbol{f}}]\!] \cdot \widehat{\mathbb{p}} = \mathbf{0} \\
\widehat{\mathbf{III}} & : & \left[\boldsymbol{A} \cdot \widehat{[\![\widehat{\boldsymbol{f}}]\!]}^{\pm} \otimes \widehat{\boldsymbol{m}}^{\pm}\right]^{\mathrm{skw}} & = & \mathbf{0} & \& & \left[\boldsymbol{a} \cdot \widehat{[\![\widehat{\boldsymbol{F}}]\!]}^{\pm} \otimes \widehat{\boldsymbol{M}}^{\pm}\right]^{\mathrm{skw}} = \mathbf{0} \\
\widehat{\mathbf{IV}} & : & [\![\widehat{\boldsymbol{k}}]\!] \cdot \widehat{\boldsymbol{m}} & \overset{\text{smooth}}{=} & \mathbf{0} & \& & [\![\widehat{\boldsymbol{K}}]\!] \cdot \widehat{\boldsymbol{M}} \overset{\text{smooth}}{=} \mathbf{0} \\
\widehat{\mathbf{V}} & : & [\![\widehat{j}]\!] - \{\widehat{\boldsymbol{k}}\} : [\![\widehat{\boldsymbol{f}}]\!] & \overset{\text{smooth}}{=} & 0 & \& & [\![\widehat{J}]\!] - \{\widehat{\boldsymbol{K}}\} : [\![\widehat{\boldsymbol{F}}]\!] \overset{\text{smooth}}{=} 0 \\
\widehat{\mathbf{VI}} & : & \widehat{[\![\widehat{\boldsymbol{W}}]\!]}^{\pm} & = & \mathbf{0} & \& & \widehat{[\![\widehat{\boldsymbol{w}}]\!]}^{\pm} = \mathbf{0} \\
\widehat{\mathbf{VII}} & : & \widehat{[\![\widehat{\boldsymbol{V}}]\!]}^{\pm} + \widehat{[\![\widehat{\boldsymbol{f}}]\!]}^{\pm} \cdot \widehat{\boldsymbol{w}}^{\pm}_{\perp} & = & \mathbf{0} & \& & \widehat{[\![\widehat{\boldsymbol{v}}]\!]}^{\pm} + \widehat{[\![\widehat{\boldsymbol{F}}]\!]}^{\pm} \cdot \widehat{\boldsymbol{W}}^{\pm}_{\perp} = \mathbf{0}
\end{array}$$

4.4 Coherent Singular Points*

In the sequel, singular points are considered as embedded into the set of boundary curves as the intersections of their regular sub-parts (we restrict ourselves to the intersection of only two boundary curves, thus, triple points and generalizations thereof are here excluded for the sake of simplicity). They are assumed *coherent*.

Due to the ultimate dimensional reduction, the following results are oftentimes trivial, however, they demonstrate the consistent applicability of the approach pursued here to arbitrary dimensions.

At a singular point $\mathcal{P}_\mathrm{m}$, the intersecting regular sub-parts $\partial\partial\mathcal{B}^\pm_\mathrm{m}$ of the boundary curve $\partial\partial\mathcal{B}_\mathrm{m}$ are characterized by their mutually orthogonal and outwards-pointing normals $\widehat{\boldsymbol{N}}^\pm_\alpha$ (with $\widehat{\boldsymbol{N}}^\pm_1 := \widetilde{\boldsymbol{N}}^\pm_\perp$ and $\widehat{\boldsymbol{N}}^\pm_2 := \widetilde{\boldsymbol{N}}^\pm_\times$), orthogonal to the tangent lines $T\partial\partial\mathcal{B}^\pm_\mathrm{m}$, and by their outwards-pointing normals $\widetilde{\boldsymbol{N}}^\pm$, (orthogonal to the tangent 'dots' $T\partial\partial\partial\mathcal{B}^\pm_\mathrm{m}$) and tangential to the tangent lines $T\partial\partial\mathcal{B}^\pm_\mathrm{m}$. Then, for notational consistency, normal vectors $\widehat{\boldsymbol{M}}^\pm_\alpha := \widehat{\boldsymbol{N}}^\pm_\alpha$ and $\widetilde{\boldsymbol{M}}^\pm := \mp\widetilde{\boldsymbol{N}}^\pm$ are defined (with orthogonality $\widehat{\boldsymbol{M}}^\pm_\alpha \perp \widetilde{\boldsymbol{M}}^\pm$). Note that the case of a singular point embedded into a smooth boundary curve results in $\widehat{\boldsymbol{M}}^\pm_\alpha = \widehat{\boldsymbol{N}}_\alpha$ and $\widetilde{\boldsymbol{M}}^\pm =: \widetilde{\boldsymbol{M}}$. Analogous definitions and notation hold in the spatial configuration.

Jump and Average Operators:
Jump and *average* operators as applied to arbitrary scalar-, vector-, or tensor-valued fields $\widetilde{\boldsymbol{Z}}$ across a singular point $\mathcal{P}_\mathrm{m}$ in the material configuration are defined in analogy to those across a singular surface or curve in the above. Thereby, it is assumed that $\widetilde{\boldsymbol{Z}}$ is smooth away from the singular point and smooth up to the singular point from either side where it takes values $\widetilde{\boldsymbol{Z}}^\pm = \lim_{\epsilon\to 0} \widetilde{\boldsymbol{Z}}(\theta|_{\mathcal{P}_\square} \pm \epsilon)$ with the curve coordinate θ in the one-dimensional curve coordinate parameter space $\partial\partial\mathcal{B}_\square$ and $\epsilon \in \mathbb{R}^+$. In particular, the parameter curves $\widetilde{\boldsymbol{X}}^\pm_\epsilon := \widetilde{\boldsymbol{X}}(\theta|_{\mathcal{P}_\square} \pm \epsilon)$ render the normal vectors $\mathrm{d}_\epsilon \widetilde{\boldsymbol{X}}^\pm_\epsilon|_{\epsilon=0} = \pm\widetilde{\boldsymbol{M}}^\pm$, tangential to the tangent lines $T\partial\partial\mathcal{B}^\pm_\mathrm{m}$. Jump and average operators across a singular point $\mathcal{P}_\mathrm{s}$ in the spatial configuration are defined accordingly.

4.4.1 Jump in Nonlinear Deformation Maps

Due to the coherence assumption, the material and spatial nonlinear deformation maps (and placements, respectively) are continuous across singular points (coherence condition I)

$$[\![\widetilde{\boldsymbol{Y}}]\!] = \boldsymbol{0} \quad \text{at} \quad \mathcal{P}_\mathrm{m} \quad \text{and} \quad [\![\widetilde{\boldsymbol{y}}]\!] = \boldsymbol{0} \quad \text{at} \quad \mathcal{P}_\mathrm{s} \tag{4.47}$$

and, in addition, smooth away from singular points, and smooth up to singular points from either side.

4.4.2 Jump in Tangent Maps

The jump operator commutes with either the material or the spatial *tangential gradient* to singular points, hence the spatial and material boundary curve deformation gradients are *tangential continuous* across singular points (coherence condition $\widetilde{\text{II}}$)

$$[\![\widetilde{\boldsymbol{F}}]\!] \cdot \widetilde{\mathbb{P}} = \boldsymbol{0} \quad \text{at} \quad \mathcal{P}_\text{m} \quad \text{and} \quad [\![\widetilde{\boldsymbol{f}}]\!] \cdot \widetilde{\mathbb{p}} = \boldsymbol{0} \quad \text{at} \quad \mathcal{P}_\text{s}. \tag{4.48}$$

Here, $\widetilde{\mathbb{P}}$ and $\widetilde{\mathbb{p}}$ are the material and spatial (mixed-variant) projection operators onto the tangent 'dot' to singular points

$$\widetilde{\mathbb{P}} := \widetilde{\boldsymbol{I}}^\pm - \widetilde{\boldsymbol{M}}^\pm \otimes \widetilde{\boldsymbol{M}}^\pm \quad \text{at} \quad \mathcal{P}_\text{m} \quad \text{and} \quad \widetilde{\mathbb{p}} := \widetilde{\boldsymbol{i}}^\pm - \widetilde{\boldsymbol{m}}^\pm \otimes \widetilde{\boldsymbol{m}}^\pm \quad \text{at} \quad \mathcal{P}_\text{s}, \tag{4.49}$$

whereby $\widetilde{\boldsymbol{I}}^\pm := \widetilde{\boldsymbol{M}}^\pm \otimes \widetilde{\boldsymbol{M}}^\pm$ and $\widetilde{\boldsymbol{i}}^\pm := \widetilde{\boldsymbol{m}}^\pm \otimes \widetilde{\boldsymbol{m}}^\pm$ denote the curve unit tensors on $\partial\partial\mathcal{B}_\text{m}^\pm$ and $\partial\partial\mathcal{B}_\text{s}^\pm$, respectively. Consequently, the material and spatial projection operators $\widetilde{\mathbb{P}}$ and $\widetilde{\mathbb{p}}$ read specifically as

$$\widetilde{\mathbb{P}} \equiv \mathbb{O} \quad \text{at} \quad \mathcal{P}_\text{m} \quad \text{and} \quad \widetilde{\mathbb{p}} \equiv \mathbb{o} \quad \text{at} \quad \mathcal{P}_\text{s}. \tag{4.50}$$

It is thus obvious that the jumps in the spatial and material boundary curve deformation gradients are of rank three. Observe that tangential continuity may alternatively be expressed as

$$\{\widetilde{\boldsymbol{F}}\} \cdot \widetilde{\mathbb{P}} = \widetilde{\boldsymbol{F}} \cdot \widetilde{\mathbb{P}} \quad \text{at} \quad \mathcal{P}_\text{m} \quad \text{and} \quad \{\widetilde{\boldsymbol{f}}\} \cdot \widetilde{\mathbb{p}} = \widetilde{\boldsymbol{f}} \cdot \widetilde{\mathbb{p}} \quad \text{at} \quad \mathcal{P}_\text{s}. \tag{4.51}$$

Thus, the action of the boundary curve deformation gradients on vectors lying in the tangent 'dot' to singular points (tangent vectors) is continuous across singular points.

As a consequence of the coherence condition $\widetilde{\text{II}}$ in Eq. 4.48, the jumps of the boundary curve deformation gradients satisfy the following symmetry or rather *compatibility conditions* (coherence condition $\widetilde{\text{III}}$)

$$\left[\boldsymbol{A} \cdot \widetilde{[\![\widetilde{\boldsymbol{f}}]\!]}^\pm \otimes \widetilde{\boldsymbol{m}}^\pm\right]^\text{skw} = \boldsymbol{0} \quad \text{and} \quad \left[\boldsymbol{a} \cdot \widetilde{[\![\widetilde{\boldsymbol{F}}]\!]}^\pm \otimes \widetilde{\boldsymbol{M}}^\pm\right]^\text{skw} = \boldsymbol{0}. \tag{4.52}$$

Here, $\boldsymbol{A} \in T^*\mathcal{B}_\text{m}$ and $\boldsymbol{a} \in T^*\mathcal{B}_\text{s}$ denote arbitrary material and spatial (constant) covectors, respectively. Moreover, the *tangential jumps* of the curve deformation gradients are defined as

$$\widetilde{[\![\widetilde{\boldsymbol{f}}]\!]}^\pm := [\![\widetilde{\boldsymbol{f}}]\!] \cdot \widetilde{\boldsymbol{i}}^\pm \quad \text{and} \quad \widetilde{[\![\widetilde{\boldsymbol{F}}]\!]}^\pm := [\![\widetilde{\boldsymbol{F}}]\!] \cdot \widetilde{\boldsymbol{I}}^\pm. \tag{4.53}$$

This result is obtained by

(1) expanding the coherence condition $\widetilde{\text{II}}$ in Eq. 4.48 into

$$[\![\widetilde{\boldsymbol{f}}]\!] = \widehat{\boldsymbol{J}}^{\pm}_{\alpha} \otimes \widehat{\boldsymbol{m}}^{\pm}_{\alpha} + \widetilde{\boldsymbol{J}}^{\pm} \otimes \widetilde{\boldsymbol{m}}^{\pm} \quad \text{and} \quad [\![\widetilde{\boldsymbol{F}}]\!] = \widehat{\boldsymbol{j}}^{\pm}_{\alpha} \otimes \widehat{\boldsymbol{M}}^{\pm}_{\alpha} + \widetilde{\boldsymbol{j}}^{\pm} \otimes \widetilde{\boldsymbol{M}}^{\pm}$$

with jump vectors

$$\widehat{\boldsymbol{J}}^{\pm}_{\alpha} := [\![\widetilde{\boldsymbol{f}}]\!] \cdot \widehat{\boldsymbol{m}}^{\pm}_{\alpha} \in T\mathcal{B}_{\text{m}} \quad \text{and} \quad \widehat{\boldsymbol{j}}^{\pm}_{\alpha} := [\![\widetilde{\boldsymbol{F}}]\!] \cdot \widehat{\boldsymbol{M}}^{\pm}_{\alpha} \in T\mathcal{B}_{\text{s}}$$

and

$$\widetilde{\boldsymbol{J}}^{\pm} := [\![\widetilde{\boldsymbol{f}}]\!] \cdot \widetilde{\boldsymbol{m}}^{\pm} \in T\mathcal{B}_{\text{m}} \quad \text{and} \quad \widetilde{\boldsymbol{j}}^{\pm} := [\![\widetilde{\boldsymbol{F}}]\!] \cdot \widetilde{\boldsymbol{M}}^{\pm} \in T\mathcal{B}_{\text{s}}$$

such that the tangential jumps of the curve deformation gradients result in

$$\widetilde{[\![\widetilde{\boldsymbol{f}}]\!]}^{\pm} = \widetilde{\boldsymbol{J}}^{\pm} \otimes \widetilde{\boldsymbol{m}}^{\pm} \quad \text{and} \quad \widetilde{[\![\widetilde{\boldsymbol{F}}]\!]}^{\pm} = \widetilde{\boldsymbol{j}}^{\pm} \otimes \widetilde{\boldsymbol{M}}^{\pm},$$

(2) pre-multiplying with arbitrary (constant) covectors $\boldsymbol{A} \in T^*\mathcal{B}_{\text{m}}$ and $\boldsymbol{a} \in T^*\mathcal{B}_{\text{s}}$ to render

$$\boldsymbol{A} \cdot [\![\widetilde{\boldsymbol{f}}]\!] = \widehat{\Omega}^{\pm}_{\alpha} \widehat{\boldsymbol{m}}^{\pm} + \widetilde{\Omega}^{\pm} \widetilde{\boldsymbol{m}}^{\pm} \quad \text{and} \quad \boldsymbol{a} \cdot [\![\widetilde{\boldsymbol{F}}]\!] = \widehat{\omega}^{\pm}_{\alpha} \widehat{\boldsymbol{M}}^{\pm} + \widetilde{\omega}^{\pm} \widetilde{\boldsymbol{M}}^{\pm}$$

with $\widehat{\Omega}^{\pm}_{\alpha} := \boldsymbol{A} \cdot \widehat{\boldsymbol{J}}^{\pm}_{\alpha}$, $\widetilde{\Omega}^{\pm} := \boldsymbol{A} \cdot \widetilde{\boldsymbol{J}}^{\pm}$ and $\widehat{\omega}^{\pm}_{\alpha} := \boldsymbol{a} \cdot \widehat{\boldsymbol{j}}^{\pm}_{\alpha}$, $\widetilde{\omega}^{\pm} := \boldsymbol{a} \cdot \widetilde{\boldsymbol{j}}^{\pm}$, thus

$$\boldsymbol{A} \cdot \widetilde{[\![\widetilde{\boldsymbol{f}}]\!]}^{\pm} = \widetilde{\Omega}^{\pm} \widetilde{\boldsymbol{m}}^{\pm} \quad \text{and} \quad \boldsymbol{a} \cdot \widetilde{[\![\widetilde{\boldsymbol{F}}]\!]}^{\pm} = \widetilde{\omega}^{\pm} \widetilde{\boldsymbol{M}}^{\pm},$$

and finally

(3) observing the symmetry of $\widetilde{\boldsymbol{m}}^{\pm} \otimes \widetilde{\boldsymbol{m}}^{\pm}$ and $\widetilde{\boldsymbol{M}}^{\pm} \otimes \widetilde{\boldsymbol{M}}^{\pm}$ and thus of

$$\boldsymbol{A} \cdot \widetilde{[\![\widetilde{\boldsymbol{f}}]\!]}^{\pm} \otimes \widetilde{\boldsymbol{m}}^{\pm} \quad \text{and} \quad \boldsymbol{a} \cdot \widetilde{[\![\widetilde{\boldsymbol{F}}]\!]}^{\pm} \otimes \widetilde{\boldsymbol{M}}^{\pm},$$

respectively.

Consequently, for double contractions with (simple) tangential second-order two-point tensors $\boldsymbol{z} := \boldsymbol{A} \otimes \boldsymbol{b}$ and $\boldsymbol{Z} := \boldsymbol{a} \otimes \boldsymbol{B}$, whereby $\boldsymbol{b} \in T\mathcal{B}_{\text{s}}$ and $\boldsymbol{B} \in T\mathcal{B}_{\text{m}}$ denote arbitrary spatial and material vectors, respectively, due to the compatibility conditions it holds that

$$\boldsymbol{z} : \widetilde{[\![\widetilde{\boldsymbol{f}}]\!]}^{\pm}\, \widetilde{\boldsymbol{m}}^{\pm} = \widetilde{[\![\widetilde{\boldsymbol{f}}^{t}]\!]}^{\pm} \cdot \boldsymbol{z} \cdot \widetilde{\boldsymbol{m}}^{\pm} \text{ and } \boldsymbol{Z} : \widetilde{[\![\widetilde{\boldsymbol{F}}]\!]}^{\pm}\, \widetilde{\boldsymbol{M}}^{\pm} = \widetilde{[\![\widetilde{\boldsymbol{F}}^{t}]\!]}^{\pm} \cdot \boldsymbol{Z} \cdot \widetilde{\boldsymbol{M}}^{\pm}. \tag{4.54}$$

The extension of the above to arbitrary second-order two-point tensors $\boldsymbol{z}$ and $\boldsymbol{Z}$ is straightforward.

4.4.3 Jump in Cotangent Maps

The *point theorem* equates the sum of the vector-valued point elements $\widetilde{\boldsymbol{N}}$ and $\widetilde{\boldsymbol{n}}$, respectively, over the boundary points of (open) control curves $\mathfrak{C}_\mathrm{m} \subseteq \partial\partial\mathcal{B}_\mathrm{m}$ and $\mathfrak{C}_\mathrm{s} \subseteq \partial\partial\mathcal{B}_\mathrm{s}$, respectively, to the resultant vector-valued control curve curvature (whereby $\mathrm{d}\widehat{\boldsymbol{L}}_1 := \widehat{\boldsymbol{M}}_1\,\mathrm{d}L$ and $\mathrm{d}\widehat{\boldsymbol{l}}_1 := \widehat{\boldsymbol{m}}_1\,\mathrm{d}l$)

$$\sum_{\partial\mathfrak{C}_\mathrm{m}} \widetilde{\boldsymbol{N}} = \int_{\mathfrak{C}_\mathrm{m}} \widetilde{C}\,\mathrm{d}\widehat{\boldsymbol{L}}_1 \quad \text{and} \quad \sum_{\partial\mathfrak{C}_\mathrm{s}} \widetilde{\boldsymbol{n}} = \int_{\mathfrak{C}_\mathrm{s}} \widetilde{c}\,\mathrm{d}\widehat{\boldsymbol{l}}_1. \tag{4.55}$$

Note that the point theorem is in particular also valid if the control curve contains coherent singular points $\mathfrak{P}_\mathrm{m}$ and $\mathfrak{P}_\mathrm{s}$, respectively (as long as the Dirac-delta-type singularity of the curve curvature across the singular points is accounted for).

Then upon

(1) decreasing the control curve to zero such that the points $\partial\mathfrak{C}^\pm_\mathrm{m}$ coincide with $\mathfrak{P}_\mathrm{m}$, whereby

$$-\widetilde{\boldsymbol{N}}^- := \widetilde{\boldsymbol{M}}^- \quad \text{and} \quad \widetilde{\boldsymbol{N}}^+ := \widetilde{\boldsymbol{M}}^+$$

hold, or, likewise, such that the points $\partial\mathfrak{C}^\pm_\mathrm{s}$ coincide with $\mathfrak{P}_\mathrm{s}$, whereby

$$-\widetilde{\boldsymbol{n}}^- := \widetilde{\boldsymbol{m}}^- \quad \text{and} \quad \widetilde{\boldsymbol{n}}^+ := \widetilde{\boldsymbol{m}}^+$$

hold, and

(2) localizing the result, whereby

$$\int_{\mathfrak{C}_\mathrm{m}\to\mathfrak{P}_\mathrm{m}} \widetilde{C}\,\mathrm{d}\widehat{\boldsymbol{L}}_1 \to \widetilde{\mathsf{C}}\,\widehat{\mathbf{M}} \quad \text{and} \quad \int_{\mathfrak{C}_\mathrm{s}\to\mathfrak{P}_\mathrm{s}} \widetilde{c}\,\mathrm{d}\widehat{\boldsymbol{l}}_1 \to \widetilde{\mathsf{c}}\,\widehat{\mathbf{m}} \tag{4.56}$$

with 'singular curve curvature' and average curve normal

$$\widetilde{\mathsf{C}} := -2\tan\widehat{\alpha}_\mathrm{m},\ \widehat{\mathbf{M}} := \{\widehat{\boldsymbol{M}}_1\} \quad \text{and} \quad \widetilde{\mathsf{c}} := -2\tan\widehat{\alpha}_\mathrm{s},\ \widehat{\mathbf{m}} := \{\widehat{\boldsymbol{m}}_1\} \tag{4.57}$$

and $2\widehat{\alpha}_\mathrm{m} := \angle(\widehat{\boldsymbol{M}}^-_1, \widehat{\boldsymbol{M}}^+_1)$, $2\widehat{\alpha}_\mathrm{s} := \angle(\widehat{\boldsymbol{m}}^-_1, \widehat{\boldsymbol{m}}^+_1)$ is taken into account,

the point theorem degenerates to the trivial (geometric) result

$$[\![\widetilde{\boldsymbol{M}}]\!] = \widetilde{\mathsf{C}}\,\widehat{\mathbf{M}} \quad \text{at} \quad \mathcal{P}_\mathrm{m} \quad \text{and} \quad [\![\widetilde{\boldsymbol{m}}]\!] = \widetilde{\mathsf{c}}\,\widehat{\mathbf{m}} \quad \text{at} \quad \mathcal{P}_\mathrm{s}. \tag{4.58}$$

However, as a consequence, the boundary curve Nanson's formulae in Eq. 3.83 render (coherence condition $\widetilde{\mathrm{IV}}$)

$$[\![\widetilde{\boldsymbol{k}}\cdot\widetilde{\boldsymbol{m}}]\!] = \widetilde{\mathsf{C}}\,\widehat{\mathbf{M}} \quad \text{at} \quad \mathcal{P}_\mathrm{m} \quad \text{and} \quad [\![\widetilde{\boldsymbol{K}}\cdot\widetilde{\boldsymbol{M}}]\!] = \widetilde{\mathsf{c}}\,\widehat{\mathbf{m}} \quad \text{at} \quad \mathcal{P}_\mathrm{s}. \tag{4.59}$$

Note that, for example, for the spatially *smooth case* with $\widehat{\alpha}_s = 0$ and the two special cases with, for example, $\widehat{\alpha}_s = \pi/4$ and $\widehat{\alpha}_s \to \pi/2$, respectively, the coherence condition $\widetilde{\text{IV}}$ reads as

$$[\![\widetilde{\boldsymbol{K}}]\!] \cdot \widetilde{\boldsymbol{M}} \overset{\text{smooth}}{=} \boldsymbol{0}, \quad [\![\widetilde{\boldsymbol{K}} \cdot \widetilde{\boldsymbol{M}}]\!] = -2\,\widehat{\mathbf{m}} \quad \text{and} \quad [\![\widetilde{\boldsymbol{K}} \cdot \widetilde{\boldsymbol{M}}]\!] \to -\infty\,\widehat{\mathbf{m}}. \tag{4.60}$$

Thus, only for a *smooth* boundary curve with $\widetilde{\boldsymbol{M}}^{\pm} = \widetilde{\boldsymbol{M}}$ and $\widehat{\boldsymbol{M}}^{\pm}_{\alpha} = \widehat{\boldsymbol{M}}_{\alpha}$ or $\widetilde{\boldsymbol{m}}^{\pm} = \widetilde{\boldsymbol{m}}$ and $\widehat{\boldsymbol{m}}^{\pm}_{\alpha} = \widehat{\boldsymbol{m}}_{\alpha}$, respectively, the right-hand sides in Eq. 4.59 vanish. In this case, the unit normal to the singular point in the tangent line to the smooth boundary curve lies in the null space of the jump of the boundary curve cofactor of the boundary curve deformation gradient. Observe that in this case, *normal continuity* in the tangent line to the smooth boundary curve may alternatively be expressed as

$$\{\widetilde{\boldsymbol{k}}\} \cdot \widetilde{\boldsymbol{m}} \overset{\text{smooth}}{=} \widetilde{\boldsymbol{k}} \cdot \widetilde{\boldsymbol{m}} \quad \text{and} \quad \{\widetilde{\boldsymbol{K}}\} \cdot \widetilde{\boldsymbol{M}} \overset{\text{smooth}}{=} \widetilde{\boldsymbol{K}} \cdot \widetilde{\boldsymbol{M}}. \tag{4.61}$$

Consequently, the action of the boundary curve cofactor of the boundary curve deformation gradients on covectors normal to singular points and tangent to the embedding smooth boundary curve (normal vectors) is continuous across singular points.

4.4.4 Jump in Measure Maps

The jump in the boundary curve Jacobians $[\![\widetilde{J}]\!]$ and $[\![\widetilde{j}]\!]$ may straightforwardly be computed from their expressions in terms of the boundary curve deformation gradients, i.e. from

$$\widetilde{J} \equiv \widetilde{\lambda} = \widetilde{\boldsymbol{K}} : \widetilde{\boldsymbol{F}} \quad \text{and} \quad \widetilde{j} \equiv \widetilde{\Lambda} = \widetilde{\boldsymbol{k}} : \widetilde{\boldsymbol{f}},$$

respectively, as (coherence condition $\widetilde{\text{V}}$)

$$[\![\widetilde{J}]\!] = \{\widetilde{\boldsymbol{K}}\} : [\![\widetilde{\boldsymbol{F}}]\!] + [\![\widetilde{J}]\!]^{\text{ns}} \quad \text{and} \quad [\![\widetilde{j}]\!] = \{\widetilde{\boldsymbol{k}}\} : [\![\widetilde{\boldsymbol{f}}]\!] + [\![\widetilde{j}]\!]^{\text{ns}}. \tag{4.62}$$

It is recalled that $\widetilde{\boldsymbol{K}}$ and $\widetilde{\boldsymbol{k}}$ denote the boundary curve cofactors

$$\widetilde{\boldsymbol{K}} = \widetilde{\boldsymbol{n}} \otimes \widetilde{\boldsymbol{N}} \quad \text{and} \quad \widetilde{\boldsymbol{k}} = \widetilde{\boldsymbol{N}} \otimes \widetilde{\boldsymbol{n}}. \tag{4.63}$$

Furthermore, $[\![\widetilde{J}]\!]^{\text{ns}}$ and $[\![\widetilde{j}]\!]^{\text{ns}}$ abbreviate the contributions due to the non-smoothness of the boundary curve

$$[\![\widetilde{J}]\!]^{\text{ns}} := [\![\widetilde{\boldsymbol{K}}]\!] : \{\widetilde{\boldsymbol{F}}\} \quad \text{and} \quad [\![\widetilde{j}]\!]^{\text{ns}} := [\![\widetilde{\boldsymbol{k}}]\!] : \{\widetilde{\boldsymbol{f}}\}. \tag{4.64}$$

This result is obtained by

(1) expanding

$$[\![\widetilde{J}]\!] = [\![\widetilde{\boldsymbol{K}} : \widetilde{\boldsymbol{F}}]\!] \quad \text{and} \quad [\![\widetilde{j}]\!] = [\![\widetilde{\boldsymbol{k}} : \widetilde{\boldsymbol{f}}]\!]$$

into

$$\{\widetilde{\boldsymbol{K}}\} : [\![\widetilde{\boldsymbol{F}}]\!] + [\![\widetilde{\boldsymbol{K}}]\!] : \{\widetilde{\boldsymbol{F}}\} \quad \text{and} \quad \{\widetilde{\boldsymbol{k}}\} : [\![\widetilde{\boldsymbol{f}}]\!] + [\![\widetilde{\boldsymbol{k}}]\!] : \{\widetilde{\boldsymbol{f}}\},$$

(2) and assembling the various terms using the above abbreviations.

For a *smooth* boundary curve, the terms $[\![\widetilde{J}]\!]^{\text{ns}}$, $[\![\widetilde{j}]\!]^{\text{ns}}$ or rather $[\![\widetilde{\boldsymbol{K}}]\!]$, $[\![\widetilde{\boldsymbol{k}}]\!]$ vanish identically. Consequently, coherence condition $\widetilde{\text{V}}$ degenerates to

$$[\![\widetilde{J}]\!] \equiv [\![\widetilde{\lambda}]\!] \overset{\text{smooth}}{=} \{\widetilde{\boldsymbol{K}}\} : [\![\widetilde{\boldsymbol{F}}]\!] \quad \text{and} \quad [\![\widetilde{j}]\!] \equiv [\![\widetilde{\Lambda}]\!] \overset{\text{smooth}}{=} \{\widetilde{\boldsymbol{k}}\} : [\![\widetilde{\boldsymbol{f}}]\!]. \tag{4.65}$$

Note that for the smooth case coherence condition $\widetilde{\text{V}}$ for the jumps in the boundary curve Jacobians resembles coherence conditions V and $\widehat{\text{V}}$ for the jump in the Jacobians and the boundary surface Jacobians.

4.4.5 Jump in Velocities

Normal–Tangential Decomposition at Singular Points:
At singular points, the total material and spatial boundary curve velocities are conveniently decomposed into normal and tangential parts

$$\widetilde{\boldsymbol{W}}^{\pm} = \widetilde{\boldsymbol{W}}_{\perp}^{\pm} + \widetilde{\boldsymbol{W}}_{\parallel} \quad \text{at} \quad \mathcal{P}_{\text{m}} \quad \text{and} \quad \widetilde{\boldsymbol{w}}^{\pm} = \widetilde{\boldsymbol{w}}_{\perp}^{\pm} + \widetilde{\boldsymbol{w}}_{\parallel} \quad \text{at} \quad \mathcal{P}_{\text{s}}. \tag{4.66}$$

Thereby, the normal parts are defined in terms of the scalar-valued *normal boundary curve velocities* $\widetilde{W}_{\perp}^{\pm}$ and $\widetilde{w}_{\perp}^{\pm}$, respectively, as

$$\widetilde{\boldsymbol{W}}_{\perp}^{\pm} := [\widetilde{\boldsymbol{W}} \cdot \widetilde{\boldsymbol{M}}]^{\pm} \widetilde{\boldsymbol{M}}^{\pm} =: \widetilde{W}_{\perp}^{\pm} \widetilde{\boldsymbol{M}}^{\pm} \quad \text{and} \quad \widetilde{\boldsymbol{w}}_{\perp}^{\pm} := [\widetilde{\boldsymbol{w}} \cdot \widetilde{\boldsymbol{m}}]^{\pm} \widetilde{\boldsymbol{m}}^{\pm} =: \widetilde{w}_{\perp}^{\pm} \widetilde{\boldsymbol{m}}^{\pm}$$

and, consequently, the tangential parts follow as

$$\widetilde{\boldsymbol{W}}_{\parallel} := \widetilde{\boldsymbol{W}}^{\pm} - \widetilde{\boldsymbol{W}}_{\perp}^{\pm} =: \widetilde{\mathbb{P}} \cdot \widetilde{\boldsymbol{W}}^{\pm} \equiv \boldsymbol{0} \quad \text{and} \quad \widetilde{\boldsymbol{w}}_{\parallel} := \widetilde{\boldsymbol{w}}^{\pm} - \widetilde{\boldsymbol{w}}_{\perp}^{\pm} =: \widetilde{\mathbb{p}} \cdot \widetilde{\boldsymbol{w}}^{\pm} \equiv \boldsymbol{0}.$$

The jump operator commutes with the total time derivative, hence the total material and spatial boundary curve velocities are, likewise, tangential continuous across singular points (coherence condition $\widetilde{\text{VI}}$)

$$\widetilde{[\![\widetilde{\boldsymbol{W}}]\!]}^{\pm} = \boldsymbol{0} \quad \text{at} \quad \mathcal{P}_{\text{m}} \quad \text{and} \quad \widetilde{[\![\widetilde{\boldsymbol{w}}]\!]}^{\pm} = \boldsymbol{0} \quad \text{at} \quad \mathcal{P}_{\text{s}}. \tag{4.67}$$

Note that the total material boundary curve velocity $\widetilde{\boldsymbol{W}}^{\pm} \neq \mathbf{0}$ for *migrating* singular points $\mathcal{P}_{\mathrm{m}}$, whereas $\widetilde{\boldsymbol{W}}^{\pm} \equiv \mathbf{0}$ for *material* singular points $\mathcal{P}_{\mathrm{m}}$. Consequently, at material singular points the total spatial boundary curve velocity $\widetilde{\boldsymbol{w}}^{\pm}$ coincides identically with the spatial velocity $\widetilde{\boldsymbol{v}}^{\pm}$.

As a consequence of the coherence condition $\widetilde{\text{VI}}$ in Eq. 4.67 and the Euler formulae in Eq. 3.95, the material and spatial boundary curve velocities are tangential discontinuous across singular points

$$\widetilde{[\![\boldsymbol{V}]\!]^{\pm}} = -\widetilde{[\![\boldsymbol{f}]\!]^{\pm}} \cdot \widetilde{\boldsymbol{w}}^{\pm} \quad \text{at} \quad \mathcal{P}_{\mathrm{m}} \quad \text{and} \quad \widetilde{[\![\boldsymbol{v}]\!]^{\pm}} = -\widetilde{[\![\boldsymbol{F}]\!]^{\pm}} \cdot \widetilde{\boldsymbol{W}}^{\pm} \quad \text{at} \quad \mathcal{P}_{\mathrm{s}}. \tag{4.68}$$

This result is obtained by

(1) invoking the representation for the material and spatial boundary curve velocities in terms of the total boundary curve velocities in Eq. 3.95, and
(2) observing the coherence condition $\widetilde{\text{VI}}$ in Eq. 4.67.

Furthermore, as a consequence of the coherence condition $\widetilde{\text{II}}$ in Eq. 4.48, (and the decomposition of the total boundary curve velocities into normal and tangential parts in Eq. 4.66), the jumps in the material and spatial boundary curve velocities at singular points read (coherence condition $\widetilde{\text{VII}}$)

$$\widetilde{[\![\widetilde{\boldsymbol{V}}]\!]^{\pm}} = -\widetilde{[\![\widetilde{\boldsymbol{f}}]\!]^{\pm}} \cdot \widetilde{\boldsymbol{w}}_{\perp}^{\pm} \quad \text{at} \quad \mathcal{P}_{\mathrm{m}} \quad \text{and} \quad \widetilde{[\![\widetilde{\boldsymbol{v}}]\!]^{\pm}} = -\widetilde{[\![\widetilde{\boldsymbol{F}}]\!]^{\pm}} \cdot \widetilde{\boldsymbol{W}}_{\perp}^{\pm} \quad \text{at} \quad \mathcal{P}_{\mathrm{s}}. \tag{4.69}$$

Observe that any potential tangential contributions to the total boundary curve velocities at singular points are filtered out by the coherence condition $\widetilde{\text{II}}$ in Eq. 4.48.

4.4.6 Summary of Coherence Conditions

The coherence conditions at coherent singular points are summarized for convenience as follows:

Summary of Coherence Conditions at Singular Points:

$$\begin{array}{llllllll}
\widetilde{\text{I}} & : & [\![\widetilde{\boldsymbol{Y}}]\!] & = & \mathbf{0} & \& & [\![\widetilde{\boldsymbol{y}}]\!] & = \mathbf{0} \\
\widetilde{\text{II}} & : & [\![\widetilde{\boldsymbol{F}}]\!] \cdot \widetilde{\mathbb{P}} & = & \mathbf{0} & \& & [\![\widetilde{\boldsymbol{f}}]\!] \cdot \widetilde{\mathbb{p}} & = \mathbf{0} \\
\widetilde{\text{III}} & & \left[\boldsymbol{A} \cdot \widetilde{[\![\widetilde{\boldsymbol{f}}]\!]^{\pm}} \otimes \widetilde{\boldsymbol{m}}^{\pm} \right]^{\text{skw}} & = & \mathbf{0} & \& & \left[\boldsymbol{a} \cdot \widetilde{[\![\widetilde{\boldsymbol{F}}]\!]^{\pm}} \otimes \widetilde{\boldsymbol{M}}^{\pm} \right]^{\text{skw}} & = \mathbf{0} \\
\widetilde{\text{IV}} & : & [\![\widetilde{\boldsymbol{k}}]\!] \cdot \widetilde{\boldsymbol{m}} & \overset{\text{smooth}}{=} & \mathbf{0} & \& & [\![\widetilde{\boldsymbol{K}}]\!] \cdot \widetilde{\boldsymbol{M}} & \overset{\text{smooth}}{=} \mathbf{0} \\
\widetilde{\text{V}} & : & [\![\widetilde{j}]\!] - \{ \widetilde{\boldsymbol{k}} \} : [\![\widetilde{\boldsymbol{f}}]\!] & \overset{\text{smooth}}{=} & 0 & \& & [\![\widetilde{J}]\!] - \{ \widetilde{\boldsymbol{K}} \} : [\![\widetilde{\boldsymbol{F}}]\!] & \overset{\text{smooth}}{=} 0 \\
\widetilde{\text{VI}} & : & \widetilde{[\![\widetilde{\boldsymbol{W}}]\!]^{\pm}} & = & \mathbf{0} & \& & \widetilde{[\![\widetilde{\boldsymbol{w}}]\!]^{\pm}} & = \mathbf{0} \\
\widetilde{\text{VII}} & : & \widetilde{[\![\widetilde{\boldsymbol{V}}]\!]^{\pm}} + \widetilde{[\![\widetilde{\boldsymbol{f}}]\!]^{\pm}} \cdot \widetilde{\boldsymbol{w}}_{\perp}^{\pm} & = & \mathbf{0} & \& & \widetilde{[\![\widetilde{\boldsymbol{v}}]\!]^{\pm}} + \widetilde{[\![\widetilde{\boldsymbol{F}}]\!]^{\pm}} \cdot \widetilde{\boldsymbol{W}}_{\perp}^{\pm} & = \mathbf{0}
\end{array}$$

References

1. Truesdell C, Noll W (2004) The non-linear field theories of mechanics. Springer, Berlin
2. Abeyaratne R, Knowles JK (1990) On the driving traction acting on a surface of strain discontinuity in a continuum. J Mech Phys Solids 38:345–360

Chapter 5
Generic Balances

8,126 m 35°14'15"N 74°35'21"E

Abstract This chapter represents the formulation of generic balances for generic volume as well as surface and curve extensive quantities, thereby highlighting their global and local formats and resorting in both cases to material and spatial control volumes as well as control surfaces and control curves.

Generically, balance equations relate the temporal rate of change of a balanced quantity to a corresponding external input in terms of sources and fluxes (here tacitly also including possible production terms for the ease of presentation). Thereby, globally, the balanced quantity together with its sources and fluxes may typically be represented as appropriate integrals of their densities either in the material or the spatial (or the reference) configuration. Then the balance is established as a global statement with the *total time derivative* capturing the temporal rate of change of the balanced quantity, a characteristic feature specific for the here advocated approach. Consequently, when deducing the corresponding local versions of the balance in terms of either the material or the spatial time derivative of the balanced quantity, additional terms stemming from the *transport theorem* arise. In the sequel, balances are formulated for generic *volume*, *surface*, and *curve extensive* balanced quantities. The approach taken as well as the formulations obtained are by and large analogous for each case, however, with rather subtle differences between the various dimensionally reduced manifolds. The terminology discerning between these cases is assembled in Table 5.1.

P. Steinmann, *Spatial and Material Forces in Nonlinear Continuum Mechanics*,
Solid Mechanics and Its Applications 272,
https://doi.org/10.1007/978-3-030-89070-4_5

Table 5.1 Terminology used for three- to zero-dimensional manifolds

Manifold		Tangent		Integral	Measure		Element	
Volume	$\mathcal{V}_\mathrm{m}$	Space	$T\mathcal{V}_\mathrm{m}$	Volume	Volume	V	Volume	$\mathrm{d}V$
Surface	$\mathcal{S}_\mathrm{m}$	Plane	$T\mathcal{S}_\mathrm{m}$	Surface	Area	A	Area	$\mathrm{d}A$
Curve	$\mathcal{C}_\mathrm{m}$	Line	$T\mathcal{C}_\mathrm{m}$	Curve	Length	L	Line	$\mathrm{d}L$
Point	$\mathcal{P}_\mathrm{m}$	Dot	$T\mathcal{P}_\mathrm{m}$	Point	–	–	–	–

5.1 Generic Volume Extensive Quantity

5.1.1 Preliminaries

(Volume) Gauss Theorem (Material Control Volume):
For smooth material control volumes, $\mathfrak{V}_\mathrm{m}$ integrals extending over $\mathfrak{V}_\mathrm{m}$ are related to integrals extending over $\partial\mathfrak{V}_\mathrm{m}$ by *Gauss' theorem*, whereby $\boldsymbol{Z} = \boldsymbol{Z}(\boldsymbol{X})$ is any (scalar-,) vector-, or tensor-valued field defined over the material configuration $\mathcal{V}_\mathrm{m}$

$$\int_{\mathfrak{V}_\mathrm{m}} \mathrm{Div}\boldsymbol{Z}\,\mathrm{d}V = \int_{\partial\mathfrak{V}_\mathrm{m}} \boldsymbol{Z}\cdot\boldsymbol{N}\,\mathrm{d}A. \tag{5.1}$$

Here, $\mathrm{Div}\boldsymbol{Z}$ denotes the properly defined *material divergence* of $\boldsymbol{Z}$, whereby the *material gradient operator* contracts with $\boldsymbol{Z}$ from the right, i.e. $\mathrm{Div}\boldsymbol{Z} := \boldsymbol{Z}\cdot\nabla_X$. Extending the Gauss theorem to cases with a singular surface $\mathfrak{S}_\mathrm{m}$ contained in the material control volume $\mathfrak{V}_\mathrm{m}$ is straightforward and renders

$$\int_{\mathfrak{V}_\mathrm{m}} \mathrm{Div}\boldsymbol{Z}\,\mathrm{d}V = \int_{\partial\mathfrak{V}_\mathrm{m}} \boldsymbol{Z}\cdot\boldsymbol{N}\,\mathrm{d}A - \int_{\mathfrak{S}_\mathrm{m}} [\![\boldsymbol{Z}]\!]\cdot\boldsymbol{M}\,\mathrm{d}A. \tag{5.2}$$

Here, $[\![\boldsymbol{Z}]\!] := \boldsymbol{Z}^+ - \boldsymbol{Z}^-$ denotes the finite jump discontinuity of $\boldsymbol{Z}$ across the singular surface. Thereby, it is assumed that $\boldsymbol{Z}$ is smooth away from the singular surface and smooth up to the singular surface from either side where it takes values $\boldsymbol{Z}^\pm = \lim_{\epsilon\to 0}\boldsymbol{Z}(\boldsymbol{X}|_{\mathcal{S}_\mathrm{m}} \pm \epsilon\boldsymbol{M})$ with $\epsilon \in \mathbb{R}^+$. Considering finally the case that the boundary $\partial\mathfrak{V}_\mathrm{m}$ of the material control volume intersects with the external boundary $\partial\mathcal{V}_\mathrm{m}$ with one- or zero-dimensional singular sub-sets contained in the intersection, i.e. $\partial^{3-\alpha}\mathfrak{V}_\mathrm{m} \in \partial\mathcal{V}_\mathrm{m} \cap \partial\mathfrak{V}_\mathrm{m}$, Gauss' theorem finally expands as

$$\int_{\mathfrak{V}_\mathrm{m}} \mathrm{Div}\boldsymbol{Z}\,\mathrm{d}V = \int_{\partial\mathfrak{V}_\mathrm{m}\backslash\partial^{3-\alpha}\mathfrak{V}_\mathrm{m}} \boldsymbol{Z}\cdot\boldsymbol{N}\,\mathrm{d}A - \int_{\mathfrak{S}_\mathrm{m}} [\![\boldsymbol{Z}]\!]\cdot\boldsymbol{M}\,\mathrm{d}A + \boldsymbol{J}_Z. \tag{5.3}$$

Here, the singular contribution $\boldsymbol{J}_Z$ follows as the limiting integral

$$\boldsymbol{J}_Z := \lim_{r \to 0} \int_{\partial \mathscr{B}_\mathrm{m}(r)} \boldsymbol{Z} \cdot \boldsymbol{N} \, \mathrm{d}A, \tag{5.4}$$

whereby it is assumed that the limit exists. $\mathscr{B}_\mathrm{m}(r)$ denotes the (semi) cylinder or ball, respectively, with radius r that encloses the one- or zero-dimensional singular sub-set $\partial^{3-\alpha}\mathfrak{V}_\mathrm{m}$ and that serves to cut out the singularity.

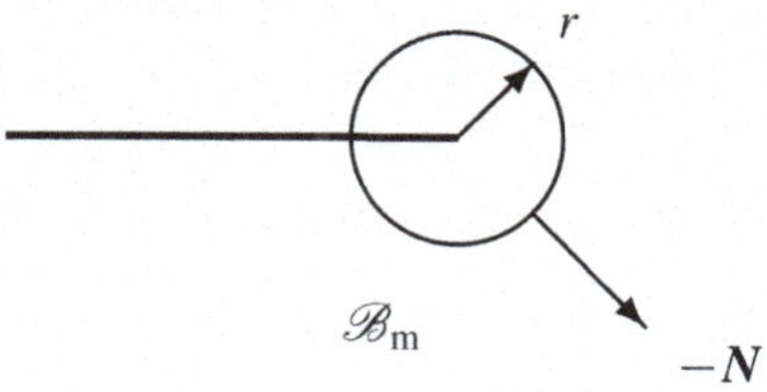

In passing, it is mentioned that applying the (volume) Gauss theorem to the (volume) unit tensor $\boldsymbol{I}$ renders the area theorem

$$\int_{\partial \mathfrak{V}_\mathrm{m}} \boldsymbol{N} \, \mathrm{d}A = \boldsymbol{0}. \tag{5.5}$$

The area theorem equates the integrals of the vectorial area element over a closed surface to zero and is instrumental, for example, in the tetrahedra argument when deriving the Cauchy theorem.

Global (Volume) Transport Theorem (Material Control Volume):
The (volume) resultant $\mathbf{Z}$ of any time-dependent scalar-, vector-, or tensor-valued field $z_\mathrm{m} = z_\mathrm{m}(\boldsymbol{X}, t)$, a density per unit volume in the material configuration, defined over the (smooth) material control volume $\mathfrak{V}_\mathrm{m}$ is given as

$$\mathbf{Z} := \int_{\mathfrak{V}_\mathrm{m}} z_\mathrm{m} \, \mathrm{d}V. \tag{5.6}$$

Then the (total) time derivative of the resultant $\mathbf{Z}$ is expressed by the help of the *transport theorem*

$$\dot{\mathbf{Z}} = \int_{\mathfrak{V}_\mathrm{m}} \mathrm{D}_t z_\mathrm{m} \, \mathrm{d}V + \int_{\partial \mathfrak{V}_\mathrm{m}} [z_\mathrm{m} \otimes \boldsymbol{W}] \cdot \boldsymbol{N} \, \mathrm{d}A. \tag{5.7}$$

The proof follows by

(1) re-expressing $\mathbf{Z}$ as an integral over the referential control volume $\mathfrak{V}_\mathrm{r}$ in the reference domain $\mathcal{V}_\mathrm{r}$,

(2) applying the chain rule to the resultant integrand $z_m \det(\nabla_\Xi \boldsymbol{\Upsilon})$,
(3) invoking the Euler formula $\dot{z}_m = D_t z_m + \nabla_X z_m \cdot \boldsymbol{W}$,
(4) combining the terms $\nabla_X z_m \cdot \boldsymbol{W}$ and $z_m \mathrm{Div} \boldsymbol{W}$ (that results from the total time derivative of $\det(\nabla_\Xi \boldsymbol{\Upsilon})$) into $\mathrm{Div}(z_m \otimes \boldsymbol{W})$, and
(5) using finally Gauss' theorem.

Extending the transport theorem to cases with a singular surface $\mathfrak{S}_m$ contained in the material control volume $\mathfrak{V}_m$ is straightforward and renders

$$\dot{\mathbf{Z}} = \int_{\mathfrak{V}_m} D_t z_m \, dV + \int_{\partial\mathfrak{V}_m} [z_m \otimes \boldsymbol{W}] \cdot \boldsymbol{N} \, dA - \int_{\mathfrak{S}_m} [\![z_m \otimes \boldsymbol{W}]\!] \cdot \boldsymbol{M} \, dA. \tag{5.8}$$

Finally, any singular contribution $\mathbf{J}_{z\otimes W}$ at one- or zero-dimensional singular sub-sets $\partial^{3-\alpha}\mathfrak{V}_m \in \partial\mathcal{V}_m \cap \partial\mathfrak{V}_m$ follows as the limiting integral

$$\mathbf{J}_{z\otimes W} := \lim_{r\to 0} \int_{\partial\mathscr{B}_m(r)} [z_m \otimes \boldsymbol{W}] \cdot \boldsymbol{N} \, dA, \tag{5.9}$$

whereby it is assumed that the limit exists and $\mathscr{B}_m(r)$ is defined as before.

The transport theorem is instrumental when balancing (total) time derivatives of global quantities that are defined as *volume* integrals with corresponding external fluxes and sources. From such global statements, local balance equations are obtained by the process of *localization*.

Material↔Spatial (Volume) Piola Transformation:
For any vector- or tensor-valued field $\boldsymbol{Z} \in \cdots \times T\mathcal{V}_m$, extending over the material configuration $\mathcal{V}_m$ and mapping from the material cotangent space $T^*\mathcal{V}_m$, the corresponding *Piola transformation* $z \in \cdots \times T\mathcal{V}_s$, extending over the spatial configuration $\mathcal{V}_s$ and mapping from the spatial cotangent space $T^*\mathcal{V}_s$, is defined as

$$z := \boldsymbol{Z} \cdot \mathrm{cof}\, \boldsymbol{f}. \tag{5.10}$$

Then the relation between $\mathrm{Div}\boldsymbol{Z}$ and the properly defined *spatial divergence* of z, whereby the *spatial gradient operator* contracts with z from the right, i.e. $\mathrm{div} z := z \cdot \nabla_x$, follows as

$$\mathrm{Div}\boldsymbol{Z} = J \mathrm{div} z. \tag{5.11}$$

The proof follows by

(1) applying the chain rule to

$$\text{Div}(z \cdot \text{cof}\, \boldsymbol{F}) = [z \otimes \nabla_X] : \text{cof}\, \boldsymbol{F} + z \cdot \text{Div}(\text{cof}\, \boldsymbol{F}),$$

whereby ∇_X is here understood as the material gradient operator,

(2) exploiting the *(volume) Piola identity* $\text{Div}(\text{cof}\, \boldsymbol{F}) = \boldsymbol{0}$ (that in turn directly follows from the area theorem), and

(3) finally re-expressing the term

$$[z \otimes \nabla_X] : \text{cof}\, \boldsymbol{F} = J[z \otimes \nabla_X] : \boldsymbol{f}^t = J[z \otimes \nabla_x] : \boldsymbol{i} = Jz \cdot \nabla_x$$

as $J \text{div} z$.

Moreover, the relation between $\boldsymbol{Z} \cdot \boldsymbol{N}$ and $z \cdot \boldsymbol{n}$, whereby $\boldsymbol{N}$ and $\boldsymbol{n}$ denote the outwards-pointing unit normals to the boundary surfaces $\partial\mathcal{V}_m$ and $\partial\mathcal{V}_s$ of the material and spatial configurations $\mathcal{V}_m$ and $\mathcal{V}_s$, respectively, follows as

$$\boldsymbol{Z} \cdot \boldsymbol{N} = \widehat{J}\, z \cdot \boldsymbol{n}. \tag{5.12}$$

Likewise, the relation between $[\![\boldsymbol{Z}]\!] \cdot \boldsymbol{M}$ and $[\![z]\!] \cdot \boldsymbol{m}$, whereby $\boldsymbol{M}$ and $\boldsymbol{m}$ denote the unit normals to the singular surfaces $\mathcal{S}_m$ and $\mathcal{S}_s$ within the material and spatial configuration $\mathcal{V}_m$ and $\mathcal{V}_s$, respectively, follows as

$$[\![\boldsymbol{Z}]\!] \cdot \boldsymbol{M} = \widehat{J}\, [\![z]\!] \cdot \boldsymbol{m}. \tag{5.13}$$

The proof follows in both cases by

(1) recalling the (volume) Piola transformation $\boldsymbol{Z} \cdot \text{cof}\, \boldsymbol{f} = z$,

(2) multiplying with the vectorial area element $\text{d}\boldsymbol{a}$ from the right,

(3) involving the Nanson's formula $\text{d}\boldsymbol{a} = \text{cof}\, \boldsymbol{F} \cdot \text{d}\boldsymbol{A}$ in terms of the cofactor of the deformation gradient as well as the surface elements $\text{d}A$ and $\text{d}a$ (together with the coherence condition IV in the case of a singular surface), and

(4) defining the *surface Jacobian* as $\widehat{J} := \text{d}a/\text{d}A = J|\boldsymbol{N} \cdot \boldsymbol{f}|$.

Material↔Spatial Local (Volume) Transport Theorems:
For any time-dependent scalar-, vector-, or tensor-valued field $z_m = z_m(\boldsymbol{X}, t)$, a density per unit volume in the material configuration, and its pendant $z_s = jz_m \circ \boldsymbol{Y}(\boldsymbol{x}, t) = z_s(\boldsymbol{x}, t)$, a density per unit volume in the spatial configuration, the local version of the transport theorem reads

$$\begin{aligned} j\, \text{D}_t z_m &= \text{d}_t z_s + \text{div}(z_s \otimes \boldsymbol{v}), \\ J\, \text{d}_t z_s &= \text{D}_t z_m + \text{Div}(z_m \otimes \boldsymbol{V}) \end{aligned} \tag{5.14}$$

point-wise in smooth parts of any control volume. The proof, for example, for Eq. 5.14.1, follows by

(1) applying the product rule $\mathrm{D}_t z_\mathrm{m} = \mathrm{D}_t[J\, z_\mathrm{s}] = J\,\mathrm{D}_t z_\mathrm{s} + z_\mathrm{s}\,\mathrm{D}_t J$,
(2) invoking the Euler formula $\mathrm{D}_t z_\mathrm{s} = \mathrm{d}_t z_\mathrm{s} + \nabla_x z_\mathrm{s} \cdot \boldsymbol{v}$,
(3) invoking $\mathrm{D}_t J = J\,\mathrm{div}\boldsymbol{v}$,
(4) combining the terms $\nabla_x \widehat{z}_\mathrm{s} \cdot \boldsymbol{v}$ and $z_\mathrm{s}\,\mathrm{div}\boldsymbol{v}$ into $\mathrm{div}(\widehat{z}_\mathrm{s} \otimes \boldsymbol{v})$.

Moreover, for regular points on the boundary of the control volume, the (volume) Piola transformation (PT) (together with $\boldsymbol{W} \cdot \boldsymbol{F}^t = [\boldsymbol{w} - \boldsymbol{v}]$ and $\boldsymbol{w} \cdot \boldsymbol{f}^t = [\boldsymbol{W} - \boldsymbol{V}]$) and the (volume) transport theorem (TT) read

$$\widehat{j}\,[z_\mathrm{m} \otimes \boldsymbol{W}] \cdot \boldsymbol{N} \overset{\mathrm{PT}}{=} \big[z_\mathrm{s} \otimes [\boldsymbol{w} - \boldsymbol{v}]\big] \cdot \boldsymbol{n} \overset{\mathrm{TT}}{=} [z_\mathrm{s} \otimes \boldsymbol{w} - z_\mathrm{s} \otimes \boldsymbol{v}] \cdot \boldsymbol{n} \tag{5.15}$$

and

$$\widehat{J}\,[z_\mathrm{s} \otimes \boldsymbol{w}] \cdot \boldsymbol{n} \overset{\mathrm{PT}}{=} \big[z_\mathrm{m} \otimes [\boldsymbol{W} - \boldsymbol{V}]\big] \cdot \boldsymbol{N} \overset{\mathrm{TT}}{=} [z_\mathrm{m} \otimes \boldsymbol{W} - z_\mathrm{m} \otimes \boldsymbol{V}] \cdot \boldsymbol{N}.$$

Likewise, for regular points at singular surfaces within the control volume, the (volume) Piola transformation and the (volume) transport theorem read

$$\widehat{j}\,[\![z_\mathrm{m} \otimes \boldsymbol{W}]\!] \cdot \boldsymbol{M} \overset{\mathrm{PT}}{=} [\![z_\mathrm{s} \otimes [\boldsymbol{w} - \boldsymbol{v}]]\!] \cdot \boldsymbol{m} \overset{\mathrm{TT}}{=} \big[[\![z_\mathrm{s} \otimes \boldsymbol{w}]\!] - [\![z_\mathrm{s} \otimes \boldsymbol{v}]\!]\big] \cdot \boldsymbol{m} \tag{5.16}$$

and

$$\widehat{J}\,[\![z_\mathrm{s} \otimes \boldsymbol{w}]\!] \cdot \boldsymbol{m} \overset{\mathrm{PT}}{=} [\![z_\mathrm{m} \otimes [\boldsymbol{W} - \boldsymbol{V}]]\!] \cdot \boldsymbol{M} \overset{\mathrm{TT}}{=} \big[[\![z_\mathrm{m} \otimes \boldsymbol{W}]\!] - [\![z_\mathrm{m} \otimes \boldsymbol{V}]\!] \cdot \boldsymbol{M}.$$

The proof follows from noting, for example, that $\boldsymbol{W} \cdot \boldsymbol{N} = \widehat{J}\,j[\boldsymbol{w} - \boldsymbol{v}] \cdot \boldsymbol{n}$ and $\boldsymbol{w} \cdot \boldsymbol{n} = \widehat{j}\,J[\boldsymbol{W} - \boldsymbol{V}] \cdot \boldsymbol{N}$, respectively. Here, the (volume) transport theorem is merely a trivial application of the dyadic product distributivity.

Global (Volume) Transport Theorem (Spatial Control Volume):
The (volume) resultant Z of any time-dependent scalar-, vector-, or tensor-valued field $z_\mathrm{s} = z_\mathrm{s}(\boldsymbol{x}, t)$, a density per unit volume in the spatial configuration, defined over the spatial control volume $\mathfrak{V}_\mathrm{s}$ is expressed as

$$\mathsf{Z} := \int_{\mathfrak{V}_\mathrm{s}} z_\mathrm{s}\,\mathrm{d}v. \tag{5.17}$$

Then for cases with a singular surface $\mathfrak{S}_s$ contained in the spatial control volume $\mathfrak{V}_s$, on the one hand, the transport theorem reads when proceeding in analogy to the proof of Eq. 5.8

$$\dot{\mathsf{Z}} = \int_{\mathfrak{V}_s} \mathrm{d}_t z_s \,\mathrm{d}v + \int_{\partial\mathfrak{V}_s} [z_s \otimes \boldsymbol{w}] \cdot \boldsymbol{n} \,\mathrm{d}a - \int_{\mathfrak{S}_s} [\![z_s \otimes \boldsymbol{w}]\!] \cdot \boldsymbol{m} \,\mathrm{d}a. \tag{5.18}$$

On the other hand, when simply using the relations between the material and spatial volume and vectorial area elements, i.e. $\mathrm{d}V = j\,\mathrm{d}v$ and $\mathrm{d}\boldsymbol{A} = \operatorname{cof} \boldsymbol{f} \cdot \mathrm{d}\boldsymbol{a}$, and changing the integration domains in Eq. 5.8, the transport theorem is, likewise, expressed as

$$\dot{\mathsf{Z}} = \int_{\mathfrak{V}_s} j\,\mathrm{D}_t z_m \,\mathrm{d}v + \int_{\partial\mathfrak{V}_s} \big[z_s \otimes [\boldsymbol{w} - \boldsymbol{v}]\big] \cdot \boldsymbol{n} \,\mathrm{d}a - \int_{\mathfrak{S}_s} [\![z_s \otimes [\boldsymbol{w} - \boldsymbol{v}]]\!] \cdot \boldsymbol{m} \,\mathrm{d}a. \tag{5.19}$$

The two different expansions of the global transport theorem in Eqs. 5.18 and 5.19 prove identical when the local transport theorem in Eq. 5.14 is invoked.

Finally, using again the relations between the spatial and material volume and vectorial area elements, i.e. $\mathrm{d}v = J\,\mathrm{d}V$ and $\mathrm{d}\boldsymbol{a} = \operatorname{cof} \boldsymbol{F} \cdot \mathrm{d}\boldsymbol{A}$, and changing the integration domains in Eq. 5.18, the transport theorem reads equivalently

$$\begin{aligned}\dot{\mathsf{Z}} = \int_{\mathfrak{V}_m} J\,\mathrm{d}_t z_s \,\mathrm{d}V &+ \int_{\partial\mathfrak{V}_m} \big[z_m \otimes [\boldsymbol{W} - \boldsymbol{V}]\big] \cdot \boldsymbol{N} \,\mathrm{d}A \\ &- \int_{\mathfrak{S}_m} [\![z_m \otimes [\boldsymbol{W} - \boldsymbol{V}]]\!] \cdot \boldsymbol{M} \,\mathrm{d}A.\end{aligned} \tag{5.20}$$

Again the two different expansions of the global transport theorem in Eqs. 5.8 and 5.20 prove identical when the local transport theorem in Eq. 5.14 is invoked.

Any singular contributions to the transport theorem are here neglected in the presentation for the sake of conciseness.

5.1.2 Global Format (Material Control Volume)

The balance equation for a generic *volume extensive* global quantity Z may be stated as follows.

Global Balance of Generic Volume Extensive Quantity (Material Control Volume):

$$\dot{\mathbf{Z}}(\mathfrak{V}_{\mathrm{m}}) = \mathbf{A}^{\mathrm{ext}}(\mathfrak{V}_{\mathrm{m}}) \quad \forall \mathfrak{V}_{\mathrm{m}} \subseteq \mathcal{V}_{\mathrm{m}}. \tag{5.21}$$

Here, $\mathbf{Z}$ is expressed as the resultant over the material control volume $\mathfrak{V}_{\mathrm{m}}$ of the generic volume extensive local quantity z_{m}, a density per unit volume in the material configuration, i.e.

$$\mathbf{Z}(\mathfrak{V}_{\mathrm{m}}) := \int_{\mathfrak{V}_{\mathrm{m}}} z_{\mathrm{m}} \, \mathrm{d}V \tag{5.22}$$

and $\mathbf{A}^{\mathrm{ext}}$ consists of volume and boundary surface integrals together with singular contributions (in case the material control volume boundary intersects with the singular part of the external boundary)

$$\mathbf{A}^{\mathrm{ext}}(\mathfrak{V}_{\mathrm{m}}) := \int_{\mathfrak{V}_{\mathrm{m}}} \overline{\boldsymbol{a}}_{\mathrm{m}} \, \mathrm{d}V + \int_{\partial\mathfrak{V}_{\mathrm{m}} \setminus \partial^{3-\alpha}\mathcal{V}_{\mathrm{m}}} \widehat{\boldsymbol{a}}_{\mathrm{m}} \, \mathrm{d}A + \mathbf{J}_{\hat{a}} \tag{5.23}$$

in terms of the generic (volume) source density $\overline{\boldsymbol{a}}_{\mathrm{m}}$ per unit volume in the material configuration (distant interaction) and the generic *extrinsic* (surface) flux density $\widehat{\boldsymbol{a}}_{\mathrm{m}}$ per unit area in the material configuration (contact interaction).

At the intersection of the material control volume boundary with the regular part of the external boundary, a Neumann-type boundary condition

$$\widehat{\boldsymbol{a}}_{\mathrm{m}} \equiv \widehat{\boldsymbol{a}}_{\mathrm{m}}^{\mathrm{ext}} \quad \text{on} \quad \partial\mathcal{V}_{\mathrm{m}} \setminus \partial^{3-\alpha}\mathcal{V}_{\mathrm{m}} \tag{5.24}$$

holds for the extrinsic (surface) flux density $\widehat{\boldsymbol{a}}_{\mathrm{m}}$ with $\widehat{\boldsymbol{a}}_{\mathrm{m}}^{\mathrm{ext}}$ externally *prescribed.*

The extrinsic (surface) flux density along with the convective term stemming from the (volume) transport theorem is *postulated*[1] to satisfy a *Cauchy theorem* for the *intrinsic* (surface) flux density $\boldsymbol{Z} \cdot \boldsymbol{N}$ (that is expressed in terms of the *intrinsic (surface) flux* $\boldsymbol{Z}$) both on the regular part

$$\boldsymbol{Z} \cdot \boldsymbol{N} := \widehat{\boldsymbol{a}}_{\mathrm{m}} - [z_{\mathrm{m}} \otimes \boldsymbol{W}] \cdot \boldsymbol{N} \quad \text{on} \quad \partial\mathfrak{V}_{\mathrm{m}} \setminus \partial^{3-\alpha}\mathcal{V}_{\mathrm{m}} \tag{5.25}$$

as well as on the singular part of the boundary (surface)

[1] The intrinsic flux $\boldsymbol{Z}$ is related to the material configuration and the material time derivative of the balanced quantity z_{m}. The *dynamic* flux $\boldsymbol{Z} + z_{\mathrm{m}} \otimes \boldsymbol{W}$ may be considered as a (right-sided referential→material) Piola transformation of a corresponding flux $\boldsymbol{\mathcal{Z}}$ as related to the reference configuration and the total time derivative of the balanced quantity z_{r}. The latter denotes a density per unit volume in the reference configuration. The referential perspective on continuum mechanics is elaborated in much more detail within Chaps. 9 and 10 on Virtual Work and the Variational Setting, respectively.

$$\boldsymbol{J}_Z := \mathbf{J}_{\hat{a}} - \mathbf{J}_{z\otimes W} \quad \text{on} \quad \partial^{3-\alpha}\mathcal{V}_\mathrm{m}. \tag{5.26}$$

Here, the singular contributions $\boldsymbol{J}_Z$ and $\mathbf{J}_{z\otimes W}$ are defined as in the above.

Consequently, the global format of the balance equation for a generic *volume extensive* quantity expands as

$$\int_{\mathfrak{V}_\mathrm{m}} \mathrm{D}_t z_\mathrm{m} \,\mathrm{d}V = \tag{5.27}$$

$$\int_{\mathfrak{V}_\mathrm{m}} \overline{\boldsymbol{a}}_\mathrm{m} \,\mathrm{d}V + \int_{\partial\mathfrak{V}_\mathrm{m}\backslash\partial^{3-\alpha}\mathcal{V}_\mathrm{m}} \boldsymbol{Z}\cdot\boldsymbol{N}\,\mathrm{d}A + \boldsymbol{J}_Z + \int_{\mathfrak{S}_\mathrm{m}} [\![z_\mathrm{m}\otimes \boldsymbol{W}]\!]\cdot\boldsymbol{M}\,\mathrm{d}A.$$

5.1.3 Local Format (Material Control Volume)

For the process of localization, four different scenarios may be considered:

i. Consider a smooth material control volume $\mathfrak{V}_\mathrm{m}$, completely contained within $\mathcal{V}_\mathrm{m}$. Using the (volume) Gauss theorem to transform the surface integral in Eq. 5.27 to a volume integral renders

$$\int_{\mathfrak{V}_\mathrm{m}} \mathrm{D}_t z_\mathrm{m} \,\mathrm{d}V = \int_{\mathfrak{V}_\mathrm{m}} [\overline{\boldsymbol{a}}_\mathrm{m} + \mathrm{Div}\boldsymbol{Z}]\,\mathrm{d}V. \tag{5.28}$$

Localization, i.e. requiring this equation to hold for arbitrary material control volumes $\mathfrak{V}_\mathrm{m}$ thus renders the point-wise generic balance equation at regular points in the (volume) domain $\mathcal{V}_\mathrm{m}$.

ii. Consider a smooth material control volume $\mathfrak{V}_\mathrm{m}$ with boundary $\partial\mathfrak{V}_\mathrm{m}$ intersecting with the regular part of the external boundary surface $\partial\mathcal{V}_\mathrm{m}$. Then the Cauchy theorem renders the point-wise generic balance equation at regular points on the boundary surface with the Neumann-type boundary condition $\widehat{\boldsymbol{a}}_\mathrm{m} \equiv \widehat{\boldsymbol{a}}_\mathrm{m}^\mathrm{ext}$.

iii. Consider a material control volume $\mathfrak{V}_\mathrm{m}$, completely contained within $\mathcal{V}_\mathrm{m}$, that includes an internal coherent singular surface $\mathfrak{S}_\mathrm{m}$. Then the material control volume $\mathfrak{V}_\mathrm{m}$ is decreased to zero such that the surfaces $\partial\mathfrak{V}_\mathrm{m}^\pm$ coincide with $\mathfrak{S}_\mathrm{m}$ and $-\boldsymbol{N}^- = \boldsymbol{N}^+ = \boldsymbol{M}$. Thus, all terms in Eq. 5.27 involving volume integrals vanish to render

$$\boldsymbol{0} = \int_{\mathfrak{S}_\mathrm{m}} [\![\boldsymbol{Z} + z_\mathrm{m}\otimes\boldsymbol{W}]\!]\cdot\boldsymbol{M}\,\mathrm{d}A. \tag{5.29}$$

Localization, i.e. requiring this equation to hold for arbitrary material control volumes $\mathfrak{V}_\mathrm{m}$ and thus arbitrary $\mathfrak{S}_\mathrm{m}$ thus renders the point-wise generic balance equation at singular surfaces $\mathcal{S}_\mathrm{m}$.

iv. Consider a smooth material control volume $\mathfrak{V}_{\rm m}$ with boundary $\partial\mathfrak{V}_{\rm m}$ intersecting with the singular part of the external boundary $\partial\mathcal{V}_{\rm m}$. Then the Cauchy theorem renders the point-wise generic balance equation at the singular part of the external boundary $\partial^{3-\alpha}\mathcal{V}_{\rm m}$.

The results of localizing the generic balance equation are summarized as follows.

Local Balance of Generic Volume Extensive Quantity (Material Control Volume):

i. Regular Points in the (Volume) Domain

$$\mathrm{D}_t z_{\rm m} - \mathrm{Div}\boldsymbol{Z} = \overline{\boldsymbol{a}}_{\rm m} \quad \text{in} \quad \mathcal{V}_{\rm m}. \tag{5.30}$$

ii. Regular Points on the Boundary (Surface)

$$\left[z_{\rm m} \otimes \boldsymbol{W} + \boldsymbol{Z}\right] \cdot \boldsymbol{N} = \widehat{\boldsymbol{a}}^{\rm ext}_{\rm m} \quad \text{on} \quad \partial\mathcal{V}_{\rm m} \setminus \partial^{3-\alpha}\mathcal{V}_{\rm m}. \tag{5.31}$$

iii. Regular Points at Singular Surfaces

$$[\![z_{\rm m} \otimes \boldsymbol{W} + \boldsymbol{Z}]\!] \cdot \boldsymbol{M} = \boldsymbol{0} \quad \text{at} \quad \mathcal{S}_{\rm m}. \tag{5.32}$$

iv. Singular Points on the Boundary (Surface)

$$\mathbf{J}_{z\otimes W} + \boldsymbol{J}_Z = \mathbf{J}_{\hat{a}} \quad \text{on} \quad \partial^{3-\alpha}\mathcal{V}_{\rm m}. \tag{5.33}$$

When stating the local generic balance equations in Eqs. 5.30–5.33, the three occurring terms in each statement have been ordered by purpose: the first set of terms $\mathrm{D}_t z_{\rm m}$, $z_{\rm m} W_\perp$, $[\![z_{\rm m}]\!] W_\perp$, and $\mathbf{J}_{z\otimes W}$ capture *temporal sensitivities* (contrast the operators $\mathrm{D}_t\{\bullet\}$ and $[\![\{\bullet\}]\!] W_\perp$ in the domain and across the singular surface); the second set of terms $\mathrm{Div}\boldsymbol{Z}$, $\boldsymbol{Z}\cdot\boldsymbol{N}$, $[\![\boldsymbol{Z}]\!]\cdot\boldsymbol{M}$, and $\boldsymbol{J}_Z$ capture the *spatial sensitivities* (contrast the operators $\mathrm{Div}\{\bullet\}$ and $[\![\{\bullet\}]\!]\cdot\boldsymbol{M}$ in the domain and across the singular surface); and the third set of terms $\overline{\boldsymbol{a}}_{\rm m}$, $\widehat{\boldsymbol{a}}^{\rm ext}_{\rm m}$, and $\mathbf{J}_{\hat{a}}$ on the right-hand side represent the *external data*. This logic will help to better understand the structure of the specific balance equations in the sequel.

Cancelation of Extrinsic Flux Density

It is noted that the generic balance equation at a singular surface $\mathcal{S}_{\rm m}$ implies the *cancelation of the extrinsic flux density*

$$\widehat{\boldsymbol{a}}^{+}_{\rm m} + \widehat{\boldsymbol{a}}^{-}_{\rm m} = \boldsymbol{0} \quad \text{on} \quad \mathcal{S}_{\rm m}, \tag{5.34}$$

i.e. the generic extrinsic flux densities at either side of the singular surface cancel one another. This can be deduced by restricting the generic balance equation at regular points on the boundary to the singular surface, considered two-sided, with the sign convention $\boldsymbol{M}^- = -\boldsymbol{M}^+ = \boldsymbol{M}$ to render

$$-\,[\![z_\mathrm{m} \otimes \boldsymbol{W} + \boldsymbol{Z}]\!] \cdot \boldsymbol{M} = \widehat{\boldsymbol{a}}_\mathrm{m}^+ + \widehat{\boldsymbol{a}}_\mathrm{m}^-. \tag{5.35}$$

Thereby, the left-hand side is zero by virtue of the generic balance equation at a singular surface, thus the result is established.

5.1.4 Global Format (Spatial Control Volume)

The balance equation for a global generic *volume extensive* quantity Z may be stated as follows.

Global Balance of Generic Volume Extensive Quantity (Spatial Control Volume):

$$\dot{\mathsf{Z}}(\mathfrak{V}_\mathrm{s}) = \mathsf{A}^\mathrm{ext}(\mathfrak{V}_\mathrm{s}) \quad \forall \mathfrak{V}_\mathrm{s} \subseteq \mathcal{V}_\mathrm{s}. \tag{5.36}$$

Here, Z is expressed as the resultant over the spatial control volume $\mathfrak{V}_\mathrm{s}$ of the local generic volume extensive quantity z_s, a density per unit volume in the spatial configuration, i.e.

$$\mathsf{Z}(\mathfrak{V}_\mathrm{s}) := \int_{\mathfrak{V}_\mathrm{s}} z_\mathrm{s}\,\mathrm{d}v \tag{5.37}$$

and A^ext consists of volume and boundary surface integrals (any singular contributions are here ignored for the sake of presentation)

$$\mathsf{A}^\mathrm{ext}(\mathfrak{V}_\mathrm{s}) := \int_{\mathfrak{V}_\mathrm{s}} \overline{\boldsymbol{a}}_\mathrm{s}\,\mathrm{d}v + \int_{\partial\mathfrak{V}_\mathrm{s}} \widehat{\boldsymbol{a}}_\mathrm{s}\,\mathrm{d}a \tag{5.38}$$

in terms of the generic (volume) source density $\overline{\boldsymbol{a}}_\mathrm{s} = j\,\overline{\boldsymbol{a}}_\mathrm{m}$ per unit volume in the spatial configuration and the generic extrinsic (surface) flux density $\widehat{\boldsymbol{a}}_\mathrm{s} = \widehat{j}\,\widehat{\boldsymbol{a}}_\mathrm{m}$ per unit area in the spatial configuration (with $\widehat{j} := \mathrm{d}A/\,\mathrm{d}a = j|\boldsymbol{n} \cdot \boldsymbol{F}|$ denoting the *surface Jacobian*). The intrinsic (surface) flux density is defined as the (surface) Piola transformation $\boldsymbol{\alpha} := \boldsymbol{Z} \cdot \operatorname{cof} \boldsymbol{f}$, thus, the corresponding Cauchy theorem for $\boldsymbol{\alpha} \cdot \boldsymbol{n}$ reads in terms of the extrinsic (surface) flux density along with the convective term stemming from the (volume) transport theorem

$$\boldsymbol{\alpha} \cdot \boldsymbol{n} := \widehat{\boldsymbol{a}}_\mathrm{s} - \left[z_\mathrm{s} \otimes [\boldsymbol{w} - \boldsymbol{v}]\right] \cdot \boldsymbol{n} \quad \text{on} \quad \partial\mathfrak{V}_\mathrm{s}. \tag{5.39}$$

Consequently, the global format of the generic balance equation for a generic *volume extensive* quantity expands as

$$\int_{\mathfrak{V}_s} \mathrm{d}_t z_s \,\mathrm{d}v = \int_{\mathfrak{V}_s} \overline{a}_s \,\mathrm{d}v + \int_{\partial\mathfrak{V}_s} [\boldsymbol{\alpha} - z_s \otimes \boldsymbol{v}] \cdot \boldsymbol{n} \,\mathrm{d}a + \int_{\mathfrak{S}_s} [\![z_s \otimes \boldsymbol{w}]\!] \cdot \boldsymbol{m} \,\mathrm{d}a. \quad (5.40)$$

The process of localizing the above global format to the corresponding local generic balance equations then follows the same rationale as already outlined earlier.

5.1.5 Local Format (Spatial Control Volume)

The local balances for the generic *volume extensive* local quantity z_s, a density per unit volume in the spatial configuration, expressed in terms of the Piola-transformed (surface) flux $\boldsymbol{\alpha}$, the generic (volume) source density $\overline{\boldsymbol{a}}_s$ per unit volume in the spatial configuration, and the generic extrinsic (surface) flux density $\widehat{\boldsymbol{a}}_s^{\text{ext}}$ per unit surface in the spatial configuration, are summarized as follows.

Local Balance of Generic Volume Extensive Quantity (Spatial Control Volume):

i. Regular Points in the (Volume) Domain

$$\mathrm{d}_t z_s - \operatorname{div}(\boldsymbol{\alpha} - z_s \otimes \boldsymbol{v}) = \overline{\boldsymbol{a}}_s \quad \text{in} \quad \mathcal{V}_s. \quad (5.41)$$

ii. Regular Points on the Boundary (Surface)

$$\left[z_s \otimes \boldsymbol{w} + [\boldsymbol{\alpha} - z_s \otimes \boldsymbol{v}]\right] \cdot \boldsymbol{n} = \widehat{\boldsymbol{a}}_s^{\text{ext}} \quad \text{on} \quad \partial\mathcal{V}_s. \quad (5.42)$$

iii. Regular Points at Singular Surfaces

$$[\![z_s \otimes \boldsymbol{w} + [\boldsymbol{\alpha} - z_s \otimes \boldsymbol{v}]]\!] \cdot \boldsymbol{m} = \boldsymbol{0} \quad \text{at} \quad \mathcal{S}_s. \quad (5.43)$$

Note that when stating the local generic balance equations in Eqs. 5.41–5.43, the four occurring terms in each statement have been ordered again by purpose: the first set of terms $\mathrm{d}_t z_s$, $z_s w_\perp$, and $[\![z_s]\!] w_\perp$ capture *temporal sensitivities* (contrast the operators $\mathrm{d}_t\{\bullet\}$ and $[\![\{\bullet\}]\!] w_\perp$ in the domain and across the singular surface); the second set of terms $\operatorname{div} z$, $z \cdot \boldsymbol{n}$, and $[\![z]\!] \cdot \boldsymbol{m}$ capture the *spatial sensitivities* (contrast the operators $\operatorname{div}\{\bullet\}$ and $[\![\{\bullet\}]\!] \cdot \boldsymbol{m}$ in the domain and across the singular surface); the third set of terms $\operatorname{div}(z_s \otimes \boldsymbol{v})$, $z_s v_\perp$, and $[\![z_s v_\perp]\!]$ capture the *convective sensitivities*; and the fourth set of terms $\overline{\boldsymbol{a}}_s$, and $\widehat{\boldsymbol{a}}_s^{\text{ext}}$ on the right-hand side represent the *external data*.

5.1.6 Balance Tetragon

By exploiting the version of the (volume) transport theorem in Eq. 5.19, the global format of the generic balance equation is stated alternatively as

$$\int_{\mathfrak{V}_s} j\, \mathrm{D}_t z_{\mathrm{m}}\, \mathrm{d}v = \int_{\mathfrak{V}_s} \overline{\boldsymbol{a}}_s\, \mathrm{d}v + \int_{\partial\mathfrak{V}_s} \boldsymbol{\alpha} \cdot \boldsymbol{n}\, \mathrm{d}a + \int_{\mathfrak{S}_s} [\![z_s \otimes [\boldsymbol{w} - \boldsymbol{v}]]\!] \cdot \boldsymbol{m}\, \mathrm{d}a. \tag{5.44}$$

Likewise, by invoking the version of the (volume) transport theorem in Eq. 5.20 the global format of the generic balance equation reads equivalently

$$\int_{\mathfrak{V}_\mathrm{m}} J\, \mathrm{d}_t z_s\, \mathrm{d}v = \tag{5.45}$$

$$\int_{\mathfrak{V}_\mathrm{m}} \overline{\boldsymbol{a}}_\mathrm{m}\, \mathrm{d}V + \int_{\partial\mathfrak{V}_\mathrm{m}} [\boldsymbol{Z} + z_\mathrm{m} \otimes \boldsymbol{V}] \cdot \boldsymbol{N}\, \mathrm{d}A + \int_{\mathfrak{S}_\mathrm{m}} [\![z_\mathrm{m} \otimes [\boldsymbol{W} - \boldsymbol{V}]]\!] \cdot \boldsymbol{M}\, \mathrm{d}A.$$

Thus, upon localizing the various versions of the global format of the generic balance equation, the local generic balance equation *at regular points in the (volume) domain* may be expressed in *four* different but fully equivalent versions; see the arrangement in a tetragon in Table 5.2.

Moreover, in accordance with Eqs. 5.31, 5.42 and 5.32, 5.43, local generic balance equations *at regular points on the boundary (surface)* and *at regular points at singular surfaces*, respectively, may, likewise, be expressed in *four* different, but fully equivalent versions; see the arrangement in the corresponding tetragons in Table 5.2.

Here, *dynamic versions* of the intrinsic (surface) flux densities have been introduced as $\boldsymbol{\alpha}_\mathrm{D} = \boldsymbol{Z}_\mathrm{D} \cdot \operatorname{cof} \boldsymbol{f}$ and $\boldsymbol{\alpha}_\mathrm{d} = \boldsymbol{Z}_\mathrm{d} \cdot \operatorname{cof} \boldsymbol{f}$, respectively, whereby the subscripts D and d indicate that the corresponding intrinsic (surface) flux is consistent with either the material or the spatial time derivative D_t or d_t, respectively, of the balanced generic local quantity, a volume extensive density.

Table 5.2 Tetragons of fully equivalent local generic balance equations for a volume extensive quantity at regular points in the (volume) domain, on the boundary (surface) and at singular surfaces. The various versions are related by (volume) Piola transformation (PT) and/or the (volume) transport theorem (TT), respectively

Regular Points in the (Volume) Domain

$\mathrm{D}_t \boldsymbol{z}_\mathrm{m} = \mathrm{Div}\boldsymbol{Z}_\mathrm{D} + \overline{\boldsymbol{a}}_\mathrm{m}$	←PT→	$j\,\mathrm{D}_t \boldsymbol{z}_\mathrm{m} = \mathrm{div}\boldsymbol{\alpha}_\mathrm{D} + \overline{\boldsymbol{a}}_\mathrm{s}$
↕ TT		↕ TT
$J\,\mathrm{d}_t \boldsymbol{z}_\mathrm{s} = \mathrm{Div}\boldsymbol{Z}_\mathrm{d} + \overline{\boldsymbol{a}}_\mathrm{m}$	←PT→	$\mathrm{d}_t \boldsymbol{z}_\mathrm{s} = \mathrm{div}\boldsymbol{\alpha}_\mathrm{d} + \overline{\boldsymbol{a}}_\mathrm{s}$

Regular Points on the Boundary (Surface)

$\left[\boldsymbol{z}_\mathrm{m} \otimes \boldsymbol{W} + \boldsymbol{Z}_\mathrm{D}\right] \cdot \boldsymbol{N} = \widehat{\boldsymbol{a}}_\mathrm{m}^\mathrm{ext}$	←PT→	$\left[\boldsymbol{z}_\mathrm{s} \otimes [\boldsymbol{w} - \boldsymbol{v}] + \boldsymbol{\alpha}_\mathrm{D}\right] \cdot \boldsymbol{n} = \widehat{\boldsymbol{a}}_\mathrm{s}^\mathrm{ext}$
↕ TT		↕ TT
$\left[\boldsymbol{z}_\mathrm{m} \otimes [\boldsymbol{W} - \boldsymbol{V}] + \boldsymbol{Z}_\mathrm{d}\right] \cdot \boldsymbol{N} = \widehat{\boldsymbol{a}}_\mathrm{m}^\mathrm{ext}$	←PT→	$\left[\boldsymbol{z}_\mathrm{s} \otimes \boldsymbol{w} + \boldsymbol{\alpha}_\mathrm{d}\right] \cdot \boldsymbol{n} = \widehat{\boldsymbol{a}}_\mathrm{s}^\mathrm{ext}$

Regular Points at Singular Surfaces

$[\![\boldsymbol{z}_\mathrm{m} \otimes \boldsymbol{W} + \boldsymbol{Z}_\mathrm{D}]\!] \cdot \boldsymbol{M} = \boldsymbol{0}$	←PT→	$[\![\boldsymbol{z}_\mathrm{s} \otimes [\boldsymbol{w} - \boldsymbol{v}] + \boldsymbol{\alpha}_\mathrm{D}]\!] \cdot \boldsymbol{m} = \boldsymbol{0}$
↕ TT		↕ TT
$[\![\boldsymbol{z}_\mathrm{m} \otimes [\boldsymbol{W} - \boldsymbol{V}] + \boldsymbol{Z}_\mathrm{d}]\!] \cdot \boldsymbol{M} = \boldsymbol{0}$	←PT→	$[\![\boldsymbol{z}_\mathrm{s} \otimes \boldsymbol{w} + \boldsymbol{\alpha}_\mathrm{d}]\!] \cdot \boldsymbol{m} = \boldsymbol{0}$

with *Dynamic Versions* of Intrinsic (Surface) Flux Densities

$\boldsymbol{Z}_\mathrm{D} := \boldsymbol{Z}$	←PT→	$\boldsymbol{\alpha}_\mathrm{D} := \boldsymbol{\alpha}$
↕ TT		↕ TT
$\boldsymbol{Z}_\mathrm{d} := \boldsymbol{Z} + \boldsymbol{z}_\mathrm{m} \otimes \boldsymbol{V}$	←PT→	$\boldsymbol{\alpha}_\mathrm{d} := \boldsymbol{\alpha} - \boldsymbol{z}_\mathrm{s} \otimes \boldsymbol{v}$

5.2 Generic Surface Extensive Quantity*

5.2.1 Preliminaries

Surface Gauss Theorem (Material Control Surface):
Let $\mathfrak{S}_\mathrm{m}$ denote a smooth material control surface, i.e. a sub-surface of a material surface $\mathcal{S}_\mathrm{m}$. The outwards-pointing unit normal to the boundary curve $\partial\mathfrak{S}_\mathrm{m}$ of the sub-surface $\mathfrak{S}_\mathrm{m}$ that lives in the cotangent space $T^*\mathfrak{S}_\mathrm{m}$ is denoted by $\widehat{\boldsymbol{N}}$. Then the surface Gauss theorem for vector-valued fields $\widehat{\mathbf{Z}} = \widehat{\mathbf{Z}}(\widehat{\boldsymbol{X}})$ reads

$$\int_{\mathfrak{S}_\mathrm{m}} \widehat{\mathrm{Div}}\,\widehat{\mathbf{Z}}\,\mathrm{d}A = \int_{\partial\mathfrak{S}_\mathrm{m}} \widehat{\mathbf{Z}}\cdot\widehat{\boldsymbol{N}}\,\mathrm{d}L - \int_{\mathfrak{S}_\mathrm{m}} \widehat{C}\,\widehat{\mathbf{Z}}\cdot\boldsymbol{N}\,\mathrm{d}A. \tag{5.46}$$

Here, $\widehat{\mathrm{Div}}\,\widehat{\mathbf{Z}}$ denotes the properly defined *material surface divergence* of $\widehat{\mathbf{Z}}$, whereby the *material surface gradient operator* contracts with $\widehat{\mathbf{Z}}$ from the right, i.e. $\widehat{\mathrm{Div}}\,\widehat{\mathbf{Z}} := \widehat{\mathbf{Z}}\cdot\nabla_{\widehat{X}}$. Extending the surface Gauss theorem to cases with a singular curve $\mathfrak{C}_\mathrm{m}$ contained in the material control surface $\mathfrak{S}_\mathrm{m}$ is straightforward and renders

$$\int_{\mathfrak{S}_\mathrm{m}} \widehat{\mathrm{Div}}\,\widehat{\mathbf{Z}}\,\mathrm{d}A = \int_{\partial\mathfrak{S}_\mathrm{m}} \widehat{\mathbf{Z}}\cdot\widehat{\boldsymbol{N}}\,\mathrm{d}L - \int_{\mathfrak{S}_\mathrm{m}} \widehat{C}\,\widehat{\mathbf{Z}}\cdot\boldsymbol{N}\,\mathrm{d}A - \int_{\mathfrak{C}_\mathrm{m}} [\![\widehat{\mathbf{Z}}]\!]\cdot\widehat{\boldsymbol{M}}\,\mathrm{d}L. \tag{5.47}$$

Here, $[\![\widehat{\mathbf{Z}}]\!] := \widehat{\mathbf{Z}}^+ - \widehat{\mathbf{Z}}^-$ denotes the finite jump discontinuity of $\widehat{\mathbf{Z}}$ across the singular curve. Thereby, it is assumed that $\widehat{\mathbf{Z}}$ is smooth away from the singular curve and smooth up to the singular curve from either side where it takes values $\widehat{\mathbf{Z}}^\pm = \lim_{\epsilon\to 0}\widehat{\mathbf{Z}}(\widehat{\boldsymbol{\xi}}|_{\mathcal{C}_\square} \pm \epsilon\,\widehat{\boldsymbol{\mu}})$ with $\widehat{\boldsymbol{\xi}}$ and $\widehat{\boldsymbol{\mu}}$ the matrix arrangements of the surface coordinates θ^α and the normal coordinates to the singular curve in the two-dimensional surface coordinates parameter space $\mathcal{S}_\square$, respectively, and $\epsilon \in \mathbb{R}^+$. Any singular contributions are here ignored for the sake of presentation.

The corresponding surface Gauss theorem for tensor-valued fields $\widehat{\mathbf{Z}}$ follows from multiplying $\widehat{\mathbf{Z}}$ from the left by an arbitrary constant vector and inserting the so resulting vector-valued-field into the above relation.

The surface Gauss theorem is the specialization of the ordinary Gauss theorem to surfaces, which is particularly obvious if the vector- or tensor-valued field $\widehat{\mathbf{Z}}$ is tangent in the sense that $\widehat{\mathbf{Z}}\cdot\boldsymbol{N} = \mathbf{0}$.

Applying the surface Gauss theorem to the surface unit tensor $\widehat{\boldsymbol{I}}$ renders eventually the (normal) line theorem

$$\int_{\mathfrak{S}_m} \widehat{C}\boldsymbol{N}\,\mathrm{d}A = \int_{\partial\mathfrak{S}_m} \widehat{\boldsymbol{N}}\,\mathrm{d}L. \tag{5.48}$$

The (normal) line theorem is the specialization of the ordinary area theorem to surfaces, which is particularly obvious if the surface is flat with $\widehat{C} = 0$. The (normal) line theorem is also valid if the control surface contains a singular curve $\mathfrak{C}_m$ as long as the Dirac-delta-type singularity of the surface curvature across the singular curve is accounted for.

Global Surface Transport Theorem (Material Control Surface):
The (surface) resultant Z of any time-dependent scalar-, vector-, or tensor-valued surface field $\widehat{\mathfrak{z}}_m = \widehat{\mathfrak{z}}_m(\widehat{\boldsymbol{X}}, t)$, a density per unit area in the material configuration, defined over the smooth material control surface $\mathfrak{S}_m$ is given as

$$\mathsf{Z} := \int_{\mathfrak{S}_m} \widehat{\mathfrak{z}}_m\,\mathrm{d}A. \tag{5.49}$$

Then the (total) time derivative of the resultant Z is expressed by the help of the *surface transport theorem* (compare Petryk and Mroz [1])

$$\begin{aligned}\dot{\mathsf{Z}} = &\int_{\mathfrak{S}_m} \Big[\mathrm{D}_t\widehat{\mathfrak{z}}_m + \big[[\nabla_N\widehat{\mathfrak{z}}_m - \widehat{\mathfrak{z}}_m\widehat{C}] \otimes \boldsymbol{W}\big]\cdot\boldsymbol{N}\Big]\,\mathrm{d}A \\ &+ \int_{\partial\mathfrak{S}_m} [\widehat{\mathfrak{z}}_m \otimes \widehat{\boldsymbol{W}}]\cdot\widehat{\boldsymbol{N}}\,\mathrm{d}L.\end{aligned} \tag{5.50}$$

The proof follows by

(1) re-expressing Z as an integral over the referential control surface $\mathfrak{S}_r$ in the reference surface $\mathcal{S}_r$,
(2) applying the chain rule to the resultant integrand $\widehat{\mathfrak{z}}_m \det(\widehat{\nabla}_{\widehat{\Xi}}\widehat{\boldsymbol{\Upsilon}})$,
(3) invoking the Euler formula $\dot{\widehat{\mathfrak{z}}}_m = \mathrm{D}_t\widehat{\mathfrak{z}}_m + \nabla_{\widehat{X}}\widehat{\mathfrak{z}}_m \cdot \boldsymbol{W}$,
(4) expanding $\nabla_{\widehat{X}}\widehat{\mathfrak{z}}_m$ into its normal and tangential contributions, i.e. $\nabla_{\widehat{X}}\widehat{\mathfrak{z}}_m = \nabla_N\widehat{\mathfrak{z}}_m \otimes \boldsymbol{N} + \widehat{\nabla}_{\widehat{X}}\widehat{\mathfrak{z}}_m$,
(5) combining the terms $\widehat{\nabla}_{\widehat{X}}\widehat{\mathfrak{z}}_m \cdot \boldsymbol{W}$ and $\widehat{\mathfrak{z}}_m \widehat{\mathrm{Div}}\boldsymbol{W}$ (that results from the total time derivative of $\det(\nabla_{\widehat{\Xi}}\widehat{\boldsymbol{\Upsilon}})$) into $\widehat{\mathrm{Div}}(\widehat{\mathfrak{z}}_m \otimes \boldsymbol{W})$,
(6) and finally using the surface Gauss theorem with $\widehat{\boldsymbol{W}} := \boldsymbol{W} \cdot \widehat{\boldsymbol{I}}$.

Extending the surface transport theorem to cases with a singular curve $\mathfrak{C}_m$ contained in the material control surface $\mathfrak{S}_m$ is straightforward and renders

$$\dot{\mathsf{Z}} = \int_{\mathfrak{S}_\mathrm{m}} \Big[\mathrm{D}_t \widehat{z}_\mathrm{m} + \big[[\nabla_N \widehat{z}_\mathrm{m} - \widehat{z}_\mathrm{m} \widehat{C}] \otimes \boldsymbol{W}\big] \cdot \boldsymbol{N} \Big] \,\mathrm{d}A \tag{5.51}$$
$$+ \int_{\partial \mathfrak{S}_\mathrm{m}} [\widehat{z}_\mathrm{m} \otimes \widehat{\boldsymbol{W}}] \cdot \widehat{\boldsymbol{N}} \,\mathrm{d}L - \int_{\mathfrak{C}_\mathrm{m}} [\![\widehat{z}_\mathrm{m} \otimes \widehat{\boldsymbol{W}}]\!] \cdot \widehat{\boldsymbol{M}} \,\mathrm{d}L .$$

Furthermore, any singular contributions are here ignored for the sake of presentation.

The surface transport theorem is instrumental when balancing (total) time derivatives of global quantities that are defined as *surface* integrals with corresponding external fluxes and sources.

Material↔Spatial Surface Piola Transformation:
For any vector- or tensor-valued surface (*superficial*) field $\widehat{\boldsymbol{Z}}_\parallel \in \cdots \times T\mathcal{S}_\mathrm{m}$ extending over the smooth surface $\mathcal{S}_\mathrm{m}$ in the material configuration and mapping from the material surface cotangent space $T^*\mathcal{S}_\mathrm{m}$, the corresponding *surface Piola transformation* $\widehat{z}_\parallel \in \cdots \times T\mathcal{S}_\mathrm{s}$ extending over the smooth surface $\mathcal{S}_\mathrm{s}$ in the spatial configuration and mapping from the spatial surface cotangent space $T^*\mathcal{S}_\mathrm{s}$ is defined as

$$\widehat{z}_\parallel := \widehat{\boldsymbol{Z}}_\parallel \cdot \widehat{\mathrm{cof}}\, \widehat{\boldsymbol{f}} . \tag{5.52}$$

Then, the surface Piola identity results in

$$\widehat{\mathrm{Div}} \widehat{\boldsymbol{Z}}_\parallel = \widehat{J}\, \widehat{\mathrm{div}} \widehat{z}_\parallel . \tag{5.53}$$

The proof follows by

(1) applying the chain rule to

$$\widehat{\mathrm{Div}}(\widehat{z}_\parallel \cdot \widehat{\mathrm{cof}}\, \widehat{\boldsymbol{F}}) = [\widehat{z}_\parallel \otimes \widehat{\nabla}_{\widehat{X}}] : \widehat{\mathrm{cof}}\, \widehat{\boldsymbol{F}} + \widehat{z}_\parallel \cdot \widehat{\mathrm{Div}}(\widehat{\mathrm{cof}}\, \widehat{\boldsymbol{F}}),$$

whereby $\widehat{\nabla}_{\widehat{X}}$ is here understood as the material surface gradient operator,

(2) exploiting the *surface Piola identity* $\widehat{\mathrm{Div}}(\widehat{\mathrm{cof}}\, \widehat{\boldsymbol{F}}) = \widehat{J}\, \widehat{c}\, \boldsymbol{n}$ (that in turn directly follows from the line theorem) together with $\widehat{z}_\parallel \cdot \boldsymbol{n} = 0$, and

(3) finally re-expressing the term

$$[\widehat{z}_\parallel \otimes \widehat{\nabla}_{\widehat{X}}] : \widehat{\mathrm{cof}}\, \widehat{\boldsymbol{F}} = \widehat{J}\, [\widehat{z}_\parallel \otimes \widehat{\nabla}_{\widehat{X}}] : \widehat{\boldsymbol{f}}^t = \widehat{J}\, [\widehat{z}_\parallel \otimes \widehat{\nabla}_{\widehat{x}}] : \widehat{\boldsymbol{i}} = \widehat{J}\, \widehat{z}_\parallel \cdot \widehat{\nabla}_{\widehat{x}}$$

as $\widehat{J}\, \widehat{\mathrm{div}} \widehat{z}_\parallel$.

Moreover, the relation between $\widehat{\boldsymbol{Z}}_{\|} \cdot \widehat{\boldsymbol{N}}$ and $\widehat{\boldsymbol{z}}_{\|} \cdot \widehat{\boldsymbol{n}}$, whereby $\widehat{\boldsymbol{N}}$ and $\widehat{\boldsymbol{n}}$ denote the outwards-pointing unit normals to the boundary curves $\partial\mathcal{S}_{\mathrm{m}}$ and $\partial\mathcal{S}_{\mathrm{s}}$, tangential to the material and spatial manifolds $\mathcal{S}_{\mathrm{m}}$ and $\mathcal{S}_{\mathrm{s}}$, respectively, follows as

$$\widehat{\boldsymbol{Z}}_{\|} \cdot \widehat{\boldsymbol{N}} = \widetilde{J}\,\widehat{\boldsymbol{z}}_{\|} \cdot \widehat{\boldsymbol{n}}. \tag{5.54}$$

Likewise, the relation between $[\![\widehat{\boldsymbol{Z}}_{\|}]\!] \cdot \widehat{\boldsymbol{M}}$ and $[\![\widehat{\boldsymbol{z}}_{\|}]\!] \cdot \widehat{\boldsymbol{m}}$, whereby $\widehat{\boldsymbol{M}}$ and $\widehat{\boldsymbol{m}}$ denote the unit normals to the singular curves $\mathcal{C}_{\mathrm{m}}$ and $\mathcal{C}_{\mathrm{s}}$, tangential to the material and spatial manifolds $\mathcal{S}_{\mathrm{m}}$ and $\mathcal{S}_{\mathrm{s}}$, respectively, follows as

$$[\![\widehat{\boldsymbol{Z}}]\!] \cdot \widehat{\boldsymbol{M}} = \widetilde{J}\,[\![\widehat{\boldsymbol{z}}]\!] \cdot \widehat{\boldsymbol{m}}. \tag{5.55}$$

The proof follows in both cases by

(1) recalling the surface Piola transformation $\widehat{\boldsymbol{Z}}_{\|} \cdot \widehat{\operatorname{cof} \boldsymbol{f}} = \widehat{\boldsymbol{z}}_{\|}$,
(2) multiplying with the vectorial curve element $\mathrm{d}\widehat{\boldsymbol{l}}$ from the right,
(3) involving the surface Nanson formula $\mathrm{d}\widehat{\boldsymbol{l}} = \widehat{\operatorname{cof} \boldsymbol{F}} \cdot \mathrm{d}\widehat{\boldsymbol{L}}$ in terms of the surface cofactor of the surface deformation gradient as well as the surface elements $\mathrm{d}L$ and $\mathrm{d}l$ (together with the coherence condition $\widehat{\mathrm{IV}}$ in the case of a singular curve), and
(4) defining the *curve Jacobian* as $\widetilde{J} := \mathrm{d}l/\mathrm{d}L = |\boldsymbol{F} \cdot \boldsymbol{T}|$.

The surface Piola transformation is a specialization of the Piola transformation in the bulk to surfaces.

Material↔Spatial Local Surface Transport Theorems:
For any time-dependent scalar-, vector-, or tensor-valued field $\widehat{z}_{\mathrm{m}} = \widehat{z}_{\mathrm{m}}(\widehat{\boldsymbol{X}}, t)$, a density per unit area in the material configuration, and its pendant $\widehat{z}_{\mathrm{s}} = \widehat{j}\widehat{z}_{\mathrm{m}} \circ \widehat{\boldsymbol{Y}}(\widehat{\boldsymbol{x}}, t) = \widehat{z}_{\mathrm{s}}(\widehat{\boldsymbol{x}}, t)$, a density per unit area in the spatial configuration, the local version of the surface transport theorem reads

$$\begin{aligned} \widehat{j}\,\mathrm{D}_t\widehat{z}_{\mathrm{m}} &= \mathrm{d}_t\widehat{z}_{\mathrm{s}} + [\nabla_n \widehat{z}_{\mathrm{s}} \otimes \boldsymbol{v}] \cdot \boldsymbol{n} + \widehat{\operatorname{div}}(\widehat{z}_{\mathrm{s}} \otimes \boldsymbol{v}), \\ \widehat{J}\,\mathrm{d}_t\widehat{z}_{\mathrm{s}} &= \mathrm{D}_t\widehat{z}_{\mathrm{m}} + [\nabla_N\widehat{z}_{\mathrm{m}} \otimes \boldsymbol{V}] \cdot \boldsymbol{N} + \widehat{\operatorname{Div}}(\widehat{z}_{\mathrm{m}} \otimes \boldsymbol{V}) \end{aligned} \tag{5.56}$$

point-wise in smooth parts of any control surface. The proof, for example, for Eq. 5.56.1, follows by

(1) applying the product rule $\mathrm{D}_t\widehat{z}_{\mathrm{m}} = \mathrm{D}_t[\widehat{J}\,\widehat{z}_{\mathrm{s}}] = \widehat{J}\,\mathrm{D}_t\widehat{z}_{\mathrm{s}} + \widehat{z}_{\mathrm{s}}\,\mathrm{D}_t\widehat{J}$,
(2) invoking the Euler formula $\mathrm{D}_t\widehat{z}_{\mathrm{s}} = \mathrm{d}_t\widehat{z}_{\mathrm{s}} + \nabla_x\widehat{z}_{\mathrm{s}} \cdot \boldsymbol{v}$,
(3) expanding $\nabla_x\widehat{z}_{\mathrm{s}} = \nabla_n\widehat{z}_{\mathrm{s}} \otimes \boldsymbol{n} + \widehat{\nabla}_{\widehat{x}}\widehat{z}_{\mathrm{s}}$ into normal and tangential contributions,
(4) invoking $\mathrm{D}_t\widehat{J} = \widehat{J}\,\widehat{\operatorname{div}}\boldsymbol{v}$,
(5) and combining the terms $\widehat{\nabla}_{\widehat{x}}\widehat{z}_{\mathrm{s}} \cdot \boldsymbol{v}$ and $\widehat{z}_{\mathrm{s}}\,\widehat{\operatorname{div}}\boldsymbol{v}$ into $\widehat{\operatorname{div}}(\widehat{z}_{\mathrm{s}} \otimes \boldsymbol{v})$.

With $\boldsymbol{v} = \boldsymbol{v} \cdot [\boldsymbol{n} \otimes \boldsymbol{n}] + \widehat{\boldsymbol{v}}$ where $\widehat{\boldsymbol{v}} := \boldsymbol{v} \cdot \widehat{\boldsymbol{i}}$ and $\boldsymbol{V} = \boldsymbol{V} \cdot [\boldsymbol{N} \otimes \boldsymbol{N}] + \widehat{\boldsymbol{V}}$ where $\widehat{\boldsymbol{V}} := \boldsymbol{V} \cdot \widehat{\boldsymbol{I}}$, the surface divergence terms may be re-expressed as

$$\begin{aligned} \widehat{\operatorname{div}}(\widehat{z}_s \otimes \boldsymbol{v}) &= \widehat{\operatorname{div}}(\widehat{z}_s \otimes \widehat{\boldsymbol{v}}) - \widehat{c}[\widehat{z}_s \otimes \boldsymbol{v}] \cdot \boldsymbol{n} \\ \widehat{\operatorname{Div}}(\widehat{z}_m \otimes \boldsymbol{V}) &= \widehat{\operatorname{Div}}(\widehat{z}_m \otimes \widehat{\boldsymbol{V}}) - \widehat{C}[\widehat{z}_m \otimes \boldsymbol{V}] \cdot \boldsymbol{N}. \end{aligned} \tag{5.57}$$

Thus taken together, combining Eqs. 5.56 and 5.57 alongside with the Piola transformation results in the identity

$$z_s' := \nabla_n \widehat{z}_s - \widehat{z}_s \widehat{c} = j[\nabla_N \widehat{z}_m - \widehat{z}_m \widehat{C}] =: j z_m'. \tag{5.58}$$

This identity is crucial when verifying the global surface transport theorem within a spatial control surface in the following.

As a consequence of the above, the local version of the surface transport theorem expands alternatively as

$$\begin{aligned} \widehat{j}\, \mathrm{D}_t \widehat{z}_m &= \mathrm{d}_t \widehat{z}_s + \widehat{\operatorname{div}}(\widehat{z}_s \otimes \widehat{\boldsymbol{v}}) + [z_s' \otimes \boldsymbol{v}] \cdot \boldsymbol{n}, \\ \widehat{J}\, \mathrm{d}_t \widehat{z}_s &= \mathrm{D}_t \widehat{z}_m + \widehat{\operatorname{Div}}(\widehat{z}_m \otimes \widehat{\boldsymbol{V}}) + [z_m' \otimes \boldsymbol{V}] \cdot \boldsymbol{N}. \end{aligned} \tag{5.59}$$

Moreover, for regular points on the boundary of the control surface the surface Piola transformation (PT) (together with $\widehat{\boldsymbol{W}} \cdot \widehat{\boldsymbol{F}}^t = \widehat{\boldsymbol{w}} - \widehat{\boldsymbol{v}}$ and $\widehat{\boldsymbol{w}} \cdot \widehat{\boldsymbol{f}}^t = \widehat{\boldsymbol{W}} - \widehat{\boldsymbol{V}}$) and the surface transport theorem (TT) read

$$\widetilde{j}\, [\widehat{z}_m \otimes \widehat{\boldsymbol{W}}] \cdot \widehat{\boldsymbol{N}} \overset{\text{PT}}{=} \big[\widehat{z}_s \otimes [\widehat{\boldsymbol{w}} - \widehat{\boldsymbol{v}}]\big] \cdot \widehat{\boldsymbol{n}} \overset{\text{TT}}{=} [\widehat{z}_s \otimes \widehat{\boldsymbol{w}} - \widehat{z}_s \otimes \widehat{\boldsymbol{v}}] \cdot \widehat{\boldsymbol{n}} \tag{5.60}$$

and

$$\widetilde{J}\, [\widehat{z}_s \otimes \widehat{\boldsymbol{w}}] \cdot \widehat{\boldsymbol{n}} \overset{\text{PT}}{=} \big[\widehat{z}_m \otimes [\widehat{\boldsymbol{W}} - \widehat{\boldsymbol{V}}]\big] \cdot \widehat{\boldsymbol{N}} \overset{\text{TT}}{=} [\widehat{z}_m \otimes \widehat{\boldsymbol{W}} - \widehat{z}_m \otimes \widehat{\boldsymbol{V}}] \cdot \widehat{\boldsymbol{N}}.$$

Likewise, for regular points at singular curves within the control surface, the surface Piola transformation and the surface transport theorem read

$$\widetilde{j}\, [\![\widehat{z}_m \otimes \widehat{\boldsymbol{W}}]\!] \cdot \widehat{\boldsymbol{M}} \overset{\text{PT}}{=} [\![\widehat{z}_s \otimes [\widehat{\boldsymbol{w}} - \widehat{\boldsymbol{v}}]]\!] \cdot \widehat{\boldsymbol{m}} \overset{\text{TT}}{=} \big[[\![\widehat{z}_s \otimes \widehat{\boldsymbol{w}}]\!] - [\![\widehat{z}_s \otimes \widehat{\boldsymbol{v}}]\!]\big] \cdot \widehat{\boldsymbol{m}} \tag{5.61}$$

and

$$\widetilde{J}\, [\![\widehat{z}_s \otimes \widehat{\boldsymbol{w}}]\!] \cdot \widehat{\boldsymbol{m}} \overset{\text{PT}}{=} [\![\widehat{z}_m \otimes [\widehat{\boldsymbol{W}} - \widehat{\boldsymbol{V}}]]\!] \cdot \widehat{\boldsymbol{M}} \overset{\text{TT}}{=} \big[[\![\widehat{z}_m \otimes \widehat{\boldsymbol{W}}]\!] - [\![\widehat{z}_m \otimes \widehat{\boldsymbol{V}}]\!] \cdot \widehat{\boldsymbol{M}}.$$

The proof follows from noting, for example, that $\widehat{\boldsymbol{W}} \cdot \widehat{\boldsymbol{N}} = \widetilde{J}\,\widehat{j}[\widehat{\boldsymbol{w}} - \widehat{\boldsymbol{v}}] \cdot \widehat{\boldsymbol{n}}$ and $\widehat{\boldsymbol{w}} \cdot \widehat{\boldsymbol{n}} = \widetilde{j}\,\widehat{J}[\widehat{\boldsymbol{W}} - \widehat{\boldsymbol{V}}] \cdot \widehat{\boldsymbol{N}}$, respectively. Here, the surface transport theorem is merely a trivial application of the dyadic product distributivity.

Global Surface Transport Theorem (Spatial Control Surface):
The (surface) resultant Z of any time-dependent scalar-, vector-, or tensor-valued surface field $\widehat{z}_s = \widehat{z}_s(\widehat{\boldsymbol{x}}, t)$, a density per unit area in the spatial configuration, defined over the spatial control surface $\mathfrak{S}_s$ is given as

$$\mathsf{Z} := \int_{\mathfrak{S}_s} \widehat{z}_s \, \mathrm{d}a. \tag{5.62}$$

Then for cases with a singular curve $\mathfrak{C}_s$ contained in the spatial control surface $\mathfrak{S}_s$, on the one hand, the surface transport theorem reads when proceeding in analogy to the proof of Eq. 5.50

$$\begin{aligned}\dot{\mathsf{Z}} = &\int_{\mathfrak{S}_s} \Big[\mathrm{d}_t \widehat{z}_s + [[\nabla_n \widehat{z}_s - \widehat{z}_s \widehat{c}] \otimes \boldsymbol{w}] \cdot \boldsymbol{n} \Big] \mathrm{d}a \\ &+ \int_{\partial\mathfrak{S}_s} [\widehat{z}_s \otimes \widehat{\boldsymbol{w}}] \cdot \widehat{\boldsymbol{n}} \, \mathrm{d}l - \int_{\mathfrak{C}_s} [\![\widehat{z}_s \otimes \widehat{\boldsymbol{w}}]\!] \cdot \widehat{\boldsymbol{m}} \, \mathrm{d}l.\end{aligned} \tag{5.63}$$

On the other hand, when simply using the relations between the material and the spatial scalar and vectorial area elements and vectorial line elements, i.e. $\mathrm{d}A = \widehat{j}\,\mathrm{d}a$ and $\mathrm{d}\boldsymbol{A} = \operatorname{cof} \boldsymbol{f} \cdot \mathrm{d}\boldsymbol{a}$, and $\mathrm{d}\widehat{\boldsymbol{L}} = \widehat{\operatorname{cof} \widehat{\boldsymbol{f}}} \cdot \mathrm{d}\boldsymbol{l}$ and changing the integration domain in Eq. 5.50, the surface transport theorem is, likewise, expressed as

$$\begin{aligned}\dot{\mathsf{Z}} = &\int_{\mathfrak{S}_s} \Big[\widehat{j}\, \mathrm{D}_t \widehat{z}_m + [j[\nabla_N \widehat{z}_m - \widehat{z}_m \widehat{C}] \otimes [\boldsymbol{w} - \boldsymbol{v}]] \cdot \boldsymbol{n} \Big] \mathrm{d}a \\ &+ \int_{\partial\mathfrak{S}_s} [\widehat{z}_s \otimes [\widehat{\boldsymbol{w}} - \widehat{\boldsymbol{v}}]] \cdot \widehat{\boldsymbol{n}} \, \mathrm{d}l - \int_{\mathfrak{C}_s} [\![\widehat{z}_s \otimes [\widehat{\boldsymbol{w}} - \widehat{\boldsymbol{v}}]]\!] \cdot \widehat{\boldsymbol{m}} \, \mathrm{d}l.\end{aligned} \tag{5.64}$$

The two different expansions of the global surface transport theorem in Eqs. 5.63 and 5.64 prove identical when the local surface transport theorem in Eq. 5.56 and the identity in Eq. 5.58 are invoked.

Finally, using again the relations between the spatial and the material scalar and vectorial area elements and vectorial line elements, i.e. $\mathrm{d}a = \widehat{J}\,\mathrm{d}A$ and $\mathrm{d}\boldsymbol{a} = \operatorname{cof} \boldsymbol{F} \cdot \mathrm{d}\boldsymbol{A}$, and $\mathrm{d}\widehat{\boldsymbol{l}} = \widehat{\operatorname{cof} \widehat{\boldsymbol{F}}} \cdot \mathrm{d}\widehat{\boldsymbol{L}}$ and changing the integration domain in

Eq. 5.63, the surface transport theorem reads equivalently

$$\dot{\mathsf{Z}} = \int\limits_{\mathfrak{S}_m} \Big[\widehat{J}\,\mathrm{d}_t\widehat{z}_s + \big[J[\nabla_n\widehat{z}_s - \widehat{z}_s\widehat{C}] \otimes [\boldsymbol{W} - \boldsymbol{V}]\big] \cdot \boldsymbol{N}\Big]\,\mathrm{d}A \tag{5.65}$$
$$+ \int\limits_{\partial\mathfrak{S}_m} [\widehat{z}_m \otimes [\widehat{\boldsymbol{W}} - \widehat{\boldsymbol{V}}]] \cdot \widehat{\boldsymbol{N}}\,\mathrm{d}L - \int\limits_{\mathfrak{C}_m} [\![\widehat{z}_m \otimes [\widehat{\boldsymbol{W}} - \widehat{\boldsymbol{V}}]]\!] \cdot \widehat{\boldsymbol{M}}\,\mathrm{d}L.$$

Again the two different expansions of the global surface transport theorem in Eqs. 5.50 and 5.65 prove identical when the local surface transport theorem in Eq. 5.56 and the identity in Eq. 5.58 are invoked.

Any singular contributions to the surface transport theorem are here neglected in the presentation for the sake of conciseness.

5.2.2 Global Format (Material Control Surface)

The balance equation for a global generic *surface extensive* quantity Z may be stated as follows.

Global Balance of Generic Surface Extensive Quantity (Material Control Surface):

$$\dot{\mathsf{Z}}(\mathfrak{S}_m) = \mathsf{A}^{ext}(\mathfrak{S}_m) \quad \forall \mathfrak{S}_m \subseteq \mathcal{S}_m. \tag{5.66}$$

Here, Z is expressed as the resultant over the material control surface $\mathfrak{S}_m$ of the local generic surface extensive quantity $\widehat{z}_m$, a density per unit area in the material configuration, i.e.

$$\mathsf{Z}(\mathfrak{S}_m) := \int\limits_{\mathfrak{S}_m} \widehat{z}_m\,\mathrm{d}A \tag{5.67}$$

and A^{ext} consists of surface and boundary curve integrals (any singular contributions are here ignored for the sake of presentation)

$$\mathsf{A}^{ext}(\mathfrak{S}_m) := \int\limits_{\mathfrak{S}_m} \widehat{a}_m\,\mathrm{d}A + \int\limits_{\partial\mathfrak{S}_m} \widetilde{a}_m\,\mathrm{d}L \tag{5.68}$$

in terms of the generic surface source density $\widehat{\boldsymbol{a}}_m$ per unit area in the material configuration (distant interaction) and the generic *extrinsic* curve flux density $\widetilde{\boldsymbol{a}}_m$ per unit length in the material configuration (contact interaction).

At the intersection of the material control surface boundary with the external boundary, a Neumann-type boundary condition

$$\widetilde{\boldsymbol{a}}_\mathrm{m} \equiv \widetilde{\boldsymbol{a}}_\mathrm{m}^\mathrm{ext} \quad \text{on} \quad \partial\mathcal{S}_\mathrm{m} \tag{5.69}$$

holds for the extrinsic curve flux density $\widetilde{\boldsymbol{a}}_\mathrm{m}$ with $\widetilde{\boldsymbol{a}}_\mathrm{m}^\mathrm{ext}$ externally *prescribed*.

The extrinsic curve flux density along with the convective term stemming from the surface transport theorem is *postulated* to satisfy a *Cauchy theorem* for the *intrinsic* curve flux density $\widehat{\mathbf{Z}} \cdot \widehat{\boldsymbol{N}}$ (that is expressed in terms of the *intrinsic curve flux* $\widehat{\mathbf{Z}}$, a surface (*superficial*) tensor field with $\widehat{\mathbf{Z}} \cdot \boldsymbol{N} = \mathbf{0}$)

$$\widehat{\mathbf{Z}} \cdot \widehat{\boldsymbol{N}} := \widetilde{\boldsymbol{a}}_\mathrm{m} - [\widehat{\mathfrak{z}}_\mathrm{m} \otimes \widehat{\boldsymbol{W}}] \cdot \widehat{\boldsymbol{N}} \quad \text{on} \quad \partial\mathfrak{S}_\mathrm{m}. \tag{5.70}$$

Moreover, the surface source density $\widehat{\boldsymbol{a}}_\mathrm{m}$ combines with the convective term stemming from the surface transport theorem into

$$\widehat{\aleph}_\mathrm{m} := \widehat{\boldsymbol{a}}_\mathrm{m} - [z'_\mathrm{m} \otimes \boldsymbol{W}] \cdot \boldsymbol{N} \quad \text{in} \quad \mathfrak{S}_\mathrm{m} \tag{5.71}$$

with the abbreviation $z'_\mathrm{m} := \nabla_N \widehat{\mathfrak{z}}_\mathrm{m} - \widehat{\mathfrak{z}}_\mathrm{m}\widehat{C}$.

Consequently, the global format of the balance equation for a generic *surface extensive* global quantity expands as

$$\int_{\mathfrak{S}_\mathrm{m}} \mathrm{D}_t \widehat{\mathfrak{z}}_\mathrm{m} \,\mathrm{d}A = \tag{5.72}$$
$$\int_{\mathfrak{S}_\mathrm{m}} \widehat{\aleph}_\mathrm{m} \,\mathrm{d}A + \int_{\partial\mathfrak{S}_\mathrm{m}} \widehat{\mathbf{Z}} \cdot \widehat{\boldsymbol{N}} \,\mathrm{d}L + \int_{\mathfrak{C}_\mathrm{m}} [\![\widehat{\mathfrak{z}}_\mathrm{m} \otimes \widehat{\boldsymbol{W}}]\!] \cdot \widehat{\boldsymbol{M}} \,\mathrm{d}L.$$

5.2.3 *Local Format (Material Control Surface)*

For the process of localization, three different scenarios may be considered:

i. Consider a smooth material control surface $\mathfrak{S}_\mathrm{m}$, completely contained within $\mathcal{S}_\mathrm{m}$. Using the surface Gauss theorem to transform the curve integral in Eq. 5.72 to a surface integral renders

$$\int_{\mathfrak{S}_\mathrm{m}} \mathrm{D}_t \widehat{\mathfrak{z}}_\mathrm{m} \,\mathrm{d}A = \int_{\mathfrak{S}_\mathrm{m}} [\widehat{\aleph}_\mathrm{m} + \widehat{\mathrm{Div}}\widehat{\mathbf{Z}}] \,\mathrm{d}A. \tag{5.73}$$

Localization thus renders the point-wise generic balance equation in the surface domain $\mathcal{S}_\mathrm{m}$ with $\widehat{\aleph}_\mathrm{m} = \widehat{\boldsymbol{a}}_\mathrm{m} - [z'_\mathrm{m} \otimes \boldsymbol{W}] \cdot \boldsymbol{N}$.

ii. Consider a smooth material control surface $\mathfrak{S}_m$ with boundary curve $\partial\mathfrak{S}_m$ intersecting with the regular part of the external boundary $\partial\mathcal{S}_m$. Then the Cauchy theorem renders the point-wise generic balance equation on the boundary curve with the Neumann-type boundary condition $\widetilde{\boldsymbol{a}}_m \equiv \widetilde{\boldsymbol{a}}_m^{ext}$.

iii. Consider a material control surface $\mathfrak{S}_m$, completely contained within $\mathcal{S}_m$, that includes a coherent singular curve $\mathfrak{C}_m$. Then the material control surface $\mathfrak{S}_m$ is decreased to zero such that the curves $\partial\mathfrak{S}_m^{\pm}$ coincide with $\mathfrak{C}_m$ and $-\widehat{\boldsymbol{N}}^- = \widehat{\boldsymbol{N}}^+ = \widehat{\boldsymbol{M}}$. Thus, all terms in Eq. 5.72 involving surface integrals vanish to render

$$\mathbf{0} = \int_{\mathfrak{C}_m} [\![\widehat{\mathbf{Z}} + \widehat{\mathbf{z}}_m \otimes \widehat{\boldsymbol{W}}]\!] \cdot \widehat{\boldsymbol{M}}\, \mathrm{d}L. \tag{5.74}$$

Localization thus renders the point-wise generic balance equation at singular curves $\mathcal{C}_m$.

The results of localizing the generic balance equation are summarized as follows.

Local Balance of Generic Surface Extensive Quantity (Material Control Surface):

i. Regular Points in the Surface Domain

$$\mathrm{D}_t\widehat{\mathbf{z}}_m - \widehat{\mathrm{Div}\widehat{\mathbf{Z}}} = \widehat{\mathbf{\aleph}}_m \quad \text{in} \quad \mathcal{S}_m. \tag{5.75}$$

ii. Regular Points on the Boundary Curve

$$\left[\widehat{\mathbf{z}}_m \otimes \widehat{\boldsymbol{W}} + \widehat{\mathbf{Z}}\right] \cdot \widehat{\boldsymbol{N}} = \widetilde{\boldsymbol{a}}_m^{ext} \quad \text{on} \quad \partial\mathcal{S}_m. \tag{5.76}$$

iii. Regular Points at Singular Curves

$$[\![\widehat{\mathbf{z}}_m \otimes \widehat{\boldsymbol{W}} + \widehat{\mathbf{Z}}]\!] \cdot \widehat{\boldsymbol{M}} = \mathbf{0} \quad \text{at} \quad \mathcal{C}_m. \tag{5.77}$$

5.2.4 *Global Format (Spatial Control Surface)*

The balance equation for a global generic *surface extensive* quantity $\mathbf{Z}$ may be stated as follows.

Global Balance of Generic Surface Extensive Quantity (Spatial Control Surface):

$$\dot{\mathsf{Z}}(\mathfrak{S}_s) = \mathsf{A}^{\text{ext}}(\mathfrak{S}_s) \quad \forall \mathfrak{S}_s \subseteq \mathcal{S}_s. \tag{5.78}$$

Here, Z is expressed as the resultant over the spatial control surface $\mathfrak{S}_s$ of the local generic surface extensive quantity $\widehat{z}_s$, a density per unit area in the spatial configuration, i.e.

$$\mathsf{Z}(\mathfrak{S}_s) := \int_{\mathfrak{S}_s} \widehat{z}_s \, \mathrm{d}a \tag{5.79}$$

and A^{ext} consists of surface and boundary curve integrals (any singular contributions are here ignored for the sake of presentation)

$$\mathsf{A}^{\text{ext}}(\mathfrak{S}_s) := \int_{\mathfrak{S}_s} \widehat{\boldsymbol{a}}_s \, \mathrm{d}a + \int_{\partial\mathfrak{S}_s} \widetilde{\boldsymbol{a}}_s \, \mathrm{d}l \tag{5.80}$$

in terms of the generic surface source density $\widehat{\boldsymbol{a}}_s = \widehat{j}\,\widehat{\boldsymbol{a}}_m$ per unit area in the spatial configuration and the generic extrinsic curve flux density $\widetilde{\boldsymbol{a}}_s = \widetilde{j}\,\widetilde{\boldsymbol{a}}_m$ per unit length in the spatial configuration (with $\widetilde{j} := \mathrm{d}L/\,\mathrm{d}l = |\boldsymbol{f} \cdot \boldsymbol{t}|$, where $\boldsymbol{t}$ is the curve tangent, denoting the *curve Jacobian*). The intrinsic curve flux density is defined as the surface Piola transformation $\widehat{\boldsymbol{\alpha}} := \widehat{\boldsymbol{Z}} \cdot \widehat{\operatorname{cof}}\,\widehat{\boldsymbol{f}}$ (a surface (*superficial*) tensor field with $\widehat{\boldsymbol{\alpha}} \cdot \boldsymbol{n} = \boldsymbol{0}$), thus the corresponding Cauchy theorem for $\widehat{\boldsymbol{\alpha}} \cdot \widehat{\boldsymbol{n}}$ reads in terms of the extrinsic curve flux density along with the convective term stemming from the surface transport theorem

$$\widehat{\boldsymbol{\alpha}} \cdot \widehat{\boldsymbol{n}} := \widetilde{\boldsymbol{a}}_s - \left[\widehat{z}_s \otimes [\widehat{\boldsymbol{w}} - \widehat{\boldsymbol{v}}]\right] \cdot \widehat{\boldsymbol{n}} \quad \text{on} \quad \partial\mathfrak{S}_s. \tag{5.81}$$

Moreover, the surface source density $\widehat{\boldsymbol{a}}_s$ combines with the convective term stemming from the surface transport theorem into

$$\widehat{\boldsymbol{\aleph}}_s := \widehat{\boldsymbol{a}}_s - \left[z'_s \otimes [\boldsymbol{w} - \boldsymbol{v}]\right] \cdot \boldsymbol{n} \quad \text{in} \quad \mathfrak{S}_s \tag{5.82}$$

with the abbreviation $z'_s := j z'_m := \nabla_n \widehat{z}_s - \widehat{z}_s \widehat{c}$.

Consequently, the global format of the generic balance equation for a generic *surface extensive* quantity expands as

$$\int_{\mathfrak{S}_s} \mathrm{d}_t \widehat{z}_s \, \mathrm{d}a = \tag{5.83}$$

$$\int_{\mathfrak{S}_s} \left[\widehat{\boldsymbol{\aleph}}_s - [z'_s \otimes \boldsymbol{v}] \cdot \boldsymbol{n}\right] \mathrm{d}a + \int_{\partial\mathfrak{S}_s} [\widehat{\boldsymbol{\alpha}} - \widehat{z}_s \otimes \widehat{\boldsymbol{v}}] \cdot \widehat{\boldsymbol{n}} \, \mathrm{d}l + \int_{\mathfrak{C}_s} [\![\widehat{z}_s \otimes \widehat{\boldsymbol{w}}]\!] \cdot \widehat{\boldsymbol{m}} \, \mathrm{d}l.$$

The process of localizing the above global format to the corresponding local generic balance equations then follows the same rationale as already outlined earlier.

5.2.5 *Local Format (Spatial Control Surface)*

The local balances for the generic *surface extensive* local quantity $\widehat{z}_s$ (a density per unit area in the spatial configuration), expressed in terms of the Piola-transformed curve flux $\widehat{\boldsymbol{\alpha}}$, the generic surface source density $\widehat{\boldsymbol{a}}_s$ per unit area in the spatial configuration, and the generic extrinsic curve flux density $\widetilde{\boldsymbol{a}}_s^{\text{ext}}$ per unit length in the spatial configuration, are summarized as follows.

Local Balance of Generic Surface Extensive Quantity (Spatial Control Surface):

i. Regular Points in the Surface Domain

$$\mathrm{d}_t\widehat{z}_s - \widehat{\mathrm{div}}(\widehat{\boldsymbol{\alpha}} - \widehat{z}_s \otimes \widehat{\boldsymbol{v}}) = \widehat{\aleph}_s - \left[z_s' \otimes \boldsymbol{v}\right] \cdot \boldsymbol{n} \quad \text{in} \quad \mathcal{S}_s. \tag{5.84}$$

ii. Regular Points on the Boundary Curve

$$\left[\widehat{z}_s \otimes \widehat{\boldsymbol{w}} + [\widehat{\boldsymbol{\alpha}} - \widehat{z}_s \otimes \widehat{\boldsymbol{v}}]\right] \cdot \widehat{\boldsymbol{n}} = \widetilde{\boldsymbol{a}}_s^{\text{ext}} \quad \text{on} \quad \partial\mathcal{S}_s. \tag{5.85}$$

iii. Regular Points at Singular Curves

$$[\![\widehat{z}_s \otimes \widehat{\boldsymbol{w}} + [\widehat{\boldsymbol{\alpha}} - \widehat{z}_s \otimes \widehat{\boldsymbol{v}}]]\!] \cdot \widehat{\boldsymbol{m}} = \boldsymbol{0} \quad \text{at} \quad \mathcal{C}_s. \tag{5.86}$$

5.2.6 *Balance Tetragon*

By exploiting the version of the surface transport theorem in Eq. 5.64, the global format of the generic balance equation is stated alternatively as

$$\int_{\mathfrak{S}_s} \widehat{j}\,\mathrm{D}_t\widehat{z}_m \,\mathrm{d}a = \int_{\mathfrak{S}_s} \widehat{\aleph}_s \,\mathrm{d}a + \int_{\partial\mathfrak{S}_s} \widehat{\boldsymbol{\alpha}} \cdot \widehat{\boldsymbol{n}} \,\mathrm{d}l + \int_{\mathfrak{C}_s} [\![\widehat{z}_s \otimes [\widehat{\boldsymbol{w}} - \widehat{\boldsymbol{v}}]]\!] \cdot \widehat{\boldsymbol{m}} \,\mathrm{d}l. \tag{5.87}$$

Likewise, by invoking the version of the transport theorem in Eq. 5.65, the global format of the generic balance equation reads equivalently

Table 5.3 Dynamic versions of effective surface source densities

$\widehat{\aleph}_{\mathrm{mD}} := \widehat{\aleph}_{\mathrm{m}}$	⟷ PT	$\widehat{\aleph}_{\mathrm{sD}} := \widehat{\aleph}_{\mathrm{s}}$
↕ TT		↕ TT
$\widehat{\aleph}_{\mathrm{md}} := \widehat{\aleph}_{\mathrm{m}} + [\boldsymbol{z}'_{\mathrm{m}} \otimes \boldsymbol{V}] \cdot \boldsymbol{N}$	⟷ PT	$\widehat{\aleph}_{\mathrm{sd}} := \widehat{\aleph}_{\mathrm{s}} - [\boldsymbol{z}'_{\mathrm{s}} \otimes \boldsymbol{v}] \cdot \boldsymbol{n}$

$$\int_{\mathfrak{S}_{\mathrm{m}}} \widehat{J}\, \mathrm{d}_t \widehat{z}_{\mathrm{s}}\, \mathrm{d}a = \int_{\mathfrak{S}_{\mathrm{m}}} \left[\widehat{\aleph}_{\mathrm{m}} + [z'_{\mathrm{m}} \otimes V] \cdot N\right] \mathrm{d}A + \tag{5.88}$$

$$\int_{\partial \mathfrak{S}_{\mathrm{m}}} [\widehat{Z} + \widehat{z}_{\mathrm{m}} \otimes \widehat{V}] \cdot \widehat{N}\, \mathrm{d}L + \int_{\mathfrak{C}_{\mathrm{m}}} [\![\widehat{z}_{\mathrm{m}} \otimes [\widehat{W} - \widehat{V}]]\!] \cdot \widehat{M}\, \mathrm{d}L.$$

Thus, upon localizing the various versions of the global format of the generic balance equation, the local generic balance equation *at regular points in the surface domain* may be expressed in *four* different but fully equivalent versions; see the arrangement in a tetragon in Table 5.4. Thereby, *dynamic versions* of the effective surface source densities have been introduced; see Table 5.3.

Moreover, in accordance with Eqs. 5.76, 5.85 and 5.77, 5.86, local generic balance equations *at regular points on the boundary curve* and *at regular points at singular curves*, respectively, may, likewise, be expressed in *four* different, but fully equivalent versions; see the arrangement in the corresponding tetragons in Table 5.4.

5.3 Generic Curve Extensive Quantity*

5.3.1 Preliminaries

Curve Gauss Theorem (Material Control Curve):
Let $\mathfrak{C}_{\mathrm{m}}$ denote a smooth material control curve, i.e. a sub-curve of a material curve $\mathcal{C}_{\mathrm{m}}$. The outwards-pointing unit normal to the boundary points $\partial \mathfrak{C}_{\mathrm{m}}$ of the sub-curve $\mathfrak{C}_{\mathrm{m}}$ that lives in the cotangent space $T^*\mathfrak{C}_{\mathrm{m}}$ is denoted by $\widetilde{N}$. Then the curve Gauss theorem for vector-valued fields $\widetilde{Z} = \widetilde{Z}(\widetilde{X})$ reads

$$\int_{\mathfrak{C}_{\mathrm{m}}} \widetilde{\mathrm{Div}}\, \widetilde{Z}\, \mathrm{d}L = \sum_{\partial \mathfrak{C}_{\mathrm{m}}} \widetilde{Z} \cdot \widetilde{N} - \int_{\mathfrak{C}_{\mathrm{m}}} \widetilde{C}\, \widetilde{Z} \cdot \widetilde{N}_{\perp}\, \mathrm{d}L. \tag{5.89}$$

Table 5.4 Tetragons of fully equivalent local generic balance equations for a surface extensive quantity at regular points in the surface domain, on the boundary curve and at singular curves. The various versions are related by surface Piola transformation (PT) and/or the surface transport theorem (TT), respectively

Regular Points in the Surface Domain

$\mathrm{D}_t\widehat{\boldsymbol{z}}_\mathrm{m} = \widehat{\mathrm{Div}}\widehat{\boldsymbol{Z}}_\mathrm{D} + \widehat{\aleph}_\mathrm{mD}$	⟷ PT	$\widehat{j}\,\mathrm{D}_t\widehat{\boldsymbol{z}}_\mathrm{m} = \widehat{\mathrm{div}}\widehat{\boldsymbol{\alpha}}_\mathrm{D} + \widehat{\aleph}_\mathrm{sD}$
↕ TT		↕ TT
$\widehat{J}\,\mathrm{d}_t\widehat{\boldsymbol{z}}_\mathrm{s} = \widehat{\mathrm{Div}}\widehat{\boldsymbol{Z}}_\mathrm{d} + \widehat{\aleph}_\mathrm{md}$	⟷ PT	$\mathrm{d}_t\widehat{\boldsymbol{z}}_\mathrm{s} = \widehat{\mathrm{div}}\widehat{\boldsymbol{\alpha}}_\mathrm{d} + \widehat{\aleph}_\mathrm{sd}$

Regular Points on the Boundary Curves

$[\widehat{\boldsymbol{z}}_\mathrm{m} \otimes \widehat{\boldsymbol{W}} + \widehat{\boldsymbol{Z}}_\mathrm{D}] \cdot \widehat{\boldsymbol{N}} = \widetilde{\boldsymbol{a}}^\mathrm{ext}_\mathrm{m}$	⟷ PT	$[\widehat{\boldsymbol{z}}_\mathrm{s} \otimes [\widehat{\boldsymbol{w}} - \widehat{\boldsymbol{v}}] + \widehat{\boldsymbol{\alpha}}_\mathrm{D}] \cdot \widehat{\boldsymbol{n}} = \widetilde{\boldsymbol{a}}^\mathrm{ext}_\mathrm{s}$
↕ TT		↕ TT
$[\widehat{\boldsymbol{z}}_\mathrm{m} \otimes [\widehat{\boldsymbol{W}} - \widehat{\boldsymbol{V}}] + \widehat{\boldsymbol{Z}}_\mathrm{d}] \cdot \widehat{\boldsymbol{N}} = \widetilde{\boldsymbol{a}}^\mathrm{ext}_\mathrm{m}$	⟷ PT	$[\widehat{\boldsymbol{z}}_\mathrm{s} \otimes \widehat{\boldsymbol{w}} + \widehat{\boldsymbol{\alpha}}_\mathrm{d}] \cdot \widehat{\boldsymbol{n}} = \widetilde{\boldsymbol{a}}^\mathrm{ext}_\mathrm{s}$

Regular Points at Singular Curves

$[\![\widehat{\boldsymbol{z}}_\mathrm{m} \otimes \widehat{\boldsymbol{W}} + \widehat{\boldsymbol{Z}}_\mathrm{D}]\!] \cdot \widehat{\boldsymbol{M}} = \boldsymbol{0}$	⟷ PT	$[\![\widehat{\boldsymbol{z}}_\mathrm{s} \otimes [\widehat{\boldsymbol{w}} - \widehat{\boldsymbol{v}}] + \widehat{\boldsymbol{\alpha}}_\mathrm{D}]\!] \cdot \widehat{\boldsymbol{m}} = \boldsymbol{0}$
↕ TT		↕ TT
$[\![\widehat{\boldsymbol{z}}_\mathrm{m} \otimes [\widehat{\boldsymbol{W}} - \widehat{\boldsymbol{V}}] + \widehat{\boldsymbol{Z}}_\mathrm{d}]\!] \cdot \widehat{\boldsymbol{M}} = \boldsymbol{0}$	⟷ PT	$[\![\widehat{\boldsymbol{z}}_\mathrm{s} \otimes \widehat{\boldsymbol{w}} + \widehat{\boldsymbol{\alpha}}_\mathrm{d}]\!] \cdot \widehat{\boldsymbol{m}} = \boldsymbol{0}$

with *Dynamic Versions* of Intrinsic Curve Flux Densities

$\widehat{\boldsymbol{Z}}_\mathrm{D} := \widehat{\boldsymbol{Z}}$	⟷ PT	$\widehat{\boldsymbol{\alpha}}_\mathrm{D} := \widehat{\boldsymbol{\alpha}}$
↕ TT		↕ TT
$\widehat{\boldsymbol{Z}}_\mathrm{d} := \widehat{\boldsymbol{Z}} + \widehat{\boldsymbol{z}}_\mathrm{m} \otimes \widehat{\boldsymbol{V}}$	⟷ PT	$\widehat{\boldsymbol{\alpha}}_\mathrm{d} := \widehat{\boldsymbol{\alpha}} - \widehat{\boldsymbol{z}}_\mathrm{s} \otimes \widehat{\boldsymbol{v}}$

Here, $\widetilde{\mathrm{Div}}\,\widetilde{\mathbf{Z}}$ denotes the properly defined *material curve divergence* of $\widetilde{\mathbf{Z}}$, whereby the *material curve gradient operator* contracts with $\widetilde{\mathbf{Z}}$ from the right, i.e. $\widetilde{\mathrm{Div}}\,\widetilde{\mathbf{Z}} := \widetilde{\mathbf{Z}} \cdot \widetilde{\nabla}_{\widetilde{X}}$. Extending the curve Gauss theorem to cases with a singular point $\mathfrak{P}_{\mathrm{m}}$ contained in the material control curve $\mathfrak{C}_{\mathrm{m}}$ is straightforward and renders

$$\int_{\mathfrak{C}_{\mathrm{m}}} \widetilde{\mathrm{Div}}\,\widetilde{\mathbf{Z}}\,\mathrm{d}L = \sum_{\partial\mathfrak{C}_{\mathrm{m}}} \widetilde{\mathbf{Z}} \cdot \widetilde{\boldsymbol{N}} - \int_{\mathfrak{C}_{\mathrm{m}}} \widetilde{C}\,\widetilde{\mathbf{Z}} \cdot \widetilde{\boldsymbol{N}}_{\perp}\,\mathrm{d}L - \big[[\![\widetilde{\mathbf{Z}}]\!] \cdot \widetilde{\boldsymbol{M}}\big]_{\mathfrak{P}_{\mathrm{m}}}. \tag{5.90}$$

Here, $[\![\widetilde{\mathbf{Z}}]\!] := \widetilde{\mathbf{Z}}^{+} - \widetilde{\mathbf{Z}}^{-}$ denotes the finite jump discontinuity of $\widetilde{\mathbf{Z}}$ across the singular point. Thereby, it is assumed that $\widetilde{\mathbf{Z}}$ is smooth away from the singular point and smooth up to the singular point from either side where it takes values $\widetilde{\mathbf{Z}}^{\pm} = \lim_{\epsilon\to 0} \widetilde{\mathbf{Z}}(\theta|_{\mathcal{P}_{\square}} \pm \epsilon)$ with the curve coordinate θ in the one-dimensional curve coordinate parameter space $\mathcal{C}_{\square}$ and $\epsilon \in \mathbb{R}^{+}$. Any singular contributions are here ignored for the sake of presentation.

The corresponding curve Gauss theorem for tensor-valued fields $\widetilde{\mathbf{Z}}$ follows from multiplying $\widetilde{\mathbf{Z}}$ from the left by an arbitrary constant vector and inserting the so resulting vector-valued field into the above relation.

The curve Gauss theorem is the specialization of the fundamental theorem of calculus to curves, which is particularly obvious if the vector- or tensor-valued field $\widetilde{\mathbf{Z}}$ is tangent in the sense that $\widetilde{\mathbf{Z}} \cdot \widetilde{\boldsymbol{N}}_{\perp} = \mathbf{0}$.

Applying the curve Gauss theorem to the curve unit tensor $\widetilde{\boldsymbol{I}}$ renders eventually the point theorem

$$\int_{\mathfrak{C}_{\mathrm{m}}} \widetilde{C}\widetilde{\boldsymbol{N}}_{\perp}\,\mathrm{d}L = \sum_{\partial\mathfrak{C}_{\mathrm{m}}} \widetilde{\boldsymbol{N}}. \tag{5.91}$$

The point theorem is the specialization of the (normal) line theorem to curves, which is particularly obvious if the curve is flat with $\widetilde{C} = 0$. The point theorem is also valid if the control curve contains a singular point $\mathfrak{P}_{\mathrm{m}}$ as long as the Dirac-delta-type singularity of the curve curvature across the singular point is accounted for.

Global Curve Transport Theorem (Material Control Curve):
The (curve) resultant $\mathbf{Z}$ of any time-dependent scalar-, vector-, or tensor-valued curve field $\widetilde{\mathrm{z}}_{\mathrm{m}} = \widetilde{\mathrm{z}}_{\mathrm{m}}(\widetilde{\boldsymbol{X}}, t)$, a density per unit length in the material configuration, defined over the smooth material control curve $\mathfrak{C}_{\mathrm{m}}$ is given as

$$\mathbf{Z} := \int_{\mathfrak{C}_m} \widetilde{\mathfrak{z}}_m \, dL. \tag{5.92}$$

Then the (total) time derivative of the resultant $\mathbf{Z}$ is expressed by the help of the *curve transport theorem* (compare Petryk and Mroz [1])

$$\begin{aligned} \dot{\mathbf{Z}} = & \int_{\mathfrak{C}_m} \Big[\mathrm{D}_t \widetilde{\mathfrak{z}}_m + \big[[\nabla_{\widetilde{N}_\perp} \widetilde{\mathfrak{z}}_m - \widetilde{\mathfrak{z}}_m \widetilde{C}] \otimes \boldsymbol{W}\big] \cdot \widetilde{\boldsymbol{N}}_\perp \Big] \, dL \\ & + \int_{\mathfrak{C}_m} \big[\nabla_{\widetilde{N}_\times} \widetilde{\mathfrak{z}}_m \otimes \boldsymbol{W} \big] \cdot \widetilde{\boldsymbol{N}}_\times \, dL \\ & + \sum_{\partial \mathfrak{C}_m} [\widetilde{\mathfrak{z}}_m \otimes \widetilde{\boldsymbol{W}}] \cdot \widetilde{\boldsymbol{N}}. \end{aligned} \tag{5.93}$$

The proof follows by

(1) re-expressing $\mathbf{Z}$ as an integral over the referential control curve $\mathfrak{C}_r$ in the reference curve $\mathcal{C}_r$,
(2) applying the chain rule to the resultant integrand $\widetilde{\mathfrak{z}}_m \det(\widetilde{\nabla}_{\widetilde{\Xi}} \widetilde{\boldsymbol{\Upsilon}})$,
(3) invoking the Euler formula $\dot{\widetilde{\mathfrak{z}}}_m = \mathrm{D}_t \widetilde{\mathfrak{z}}_m + \nabla_{\widetilde{X}} \widetilde{\mathfrak{z}}_m \cdot \boldsymbol{W}$,
(4) expanding $\nabla_{\widetilde{X}} \widetilde{\mathfrak{z}}_m$ into its normal, bi-normal, and tangential contributions, i.e. $\nabla_{\widetilde{X}} \widetilde{\mathfrak{z}}_m = \nabla_{\widetilde{N}_\perp} \widetilde{\mathfrak{z}}_m \otimes \widetilde{\boldsymbol{N}}_\perp + \nabla_{\widetilde{N}_\times} \widetilde{\mathfrak{z}}_m \otimes \widetilde{\boldsymbol{N}}_\times + \widetilde{\nabla}_{\widetilde{X}} \widetilde{\mathfrak{z}}_m$,
(5) combining the terms $\widetilde{\nabla}_{\widetilde{X}} \widetilde{\mathfrak{z}}_m \cdot \boldsymbol{W}$ and $\widetilde{\mathfrak{z}}_m \widetilde{\mathrm{Div}} \boldsymbol{W}$ (that results from the total time derivative of $\det(\nabla_{\widetilde{\Xi}} \widetilde{\boldsymbol{\Upsilon}})$) into $\widetilde{\mathrm{Div}}(\widetilde{\mathfrak{z}}_m \otimes \boldsymbol{W})$,
(6) and finally using the curve Gauss theorem with $\widetilde{\boldsymbol{W}} := \boldsymbol{W} \cdot \widetilde{\boldsymbol{I}}$.

Extending the curve transport theorem to cases with a singular point $\mathfrak{P}_m$ contained in the material control curve $\mathfrak{C}_m$ is straightforward and renders

$$\begin{aligned} \dot{\mathbf{Z}} = & \int_{\mathfrak{C}_m} \Big[\mathrm{D}_t \widetilde{\mathfrak{z}}_m + \big[[\nabla_{\widetilde{N}_\perp} \widetilde{\mathfrak{z}}_m - \widetilde{\mathfrak{z}}_m \widetilde{C}] \otimes \boldsymbol{W}\big] \cdot \widetilde{\boldsymbol{N}}_\perp \Big] \, dL \\ & + \int_{\mathfrak{C}_m} \big[\nabla_{\widetilde{N}_\times} \widetilde{\mathfrak{z}}_m \otimes \boldsymbol{W} \big] \cdot \widetilde{\boldsymbol{N}}_\times \, dL \end{aligned} \tag{5.94}$$

$$+ \sum_{\partial \mathfrak{C}_m} [\widetilde{\mathfrak{z}}_m \otimes \widetilde{\boldsymbol{W}}] \cdot \widetilde{\boldsymbol{N}} - \Big[[\![\widetilde{\mathfrak{z}}_m \otimes \widetilde{\boldsymbol{W}}]\!] \cdot \widetilde{\boldsymbol{M}} \Big]_{\mathfrak{P}_m}. \tag{5.95}$$

Furthermore, any singular contributions are here ignored for the sake of presentation.

The curve transport theorem is instrumental when balancing (total) time derivatives of global quantities that are defined as *curve* integrals with corresponding external fluxes and sources.

Material↔Spatial Curve Piola Transformation:
For any vector- or tensor-valued curve field $\widetilde{\boldsymbol{Z}}_\| \in \cdots \times T\mathcal{C}_\mathrm{m}$ extending over the smooth curve $\mathcal{C}_\mathrm{m}$ in the material configuration and mapping from the material curve cotangent space $T^*\mathcal{C}_\mathrm{m}$, the corresponding *curve Piola transformation* $\widetilde{z}_\| \in \cdots \times T\mathcal{C}_\mathrm{s}$ extending over the smooth curve $\mathcal{C}_\mathrm{s}$ in the spatial configuration and mapping from the spatial curve cotangent space $T^*\mathcal{C}_\mathrm{s}$ is defined as

$$\widetilde{z}_\| := \widetilde{\boldsymbol{Z}}_\| \cdot \widetilde{\mathrm{cof}}\,\widetilde{\boldsymbol{f}}. \tag{5.96}$$

Then, the curve Piola identity results in

$$\widetilde{\mathrm{Div}}\widetilde{\boldsymbol{Z}}_\| = \widetilde{J}\,\widetilde{\mathrm{div}}\widetilde{z}_\|. \tag{5.97}$$

The proof follows by

(1) applying the chain rule to

$$\widetilde{\mathrm{Div}}(\widetilde{z}_\| \cdot \widetilde{\mathrm{cof}}\,\widetilde{\boldsymbol{F}}) = [\widetilde{z}_\| \otimes \widetilde{\nabla}_{\widetilde{X}}] : \widetilde{\mathrm{cof}}\,\widetilde{\boldsymbol{F}} + \widetilde{z}_\| \cdot \widetilde{\mathrm{Div}}(\widetilde{\mathrm{cof}}\,\widetilde{\boldsymbol{F}}),$$

whereby $\widetilde{\nabla}_{\widetilde{X}}$ is here understood as the material curve gradient operator,

(2) exploiting the *curve Piola identity* $\widetilde{\mathrm{Div}}(\widetilde{\mathrm{cof}}\,\widetilde{\boldsymbol{F}}) = \widetilde{J}\,\widetilde{c}\,\widetilde{\boldsymbol{n}}_\perp$ (that in turn directly follows from the point theorem) together with $\widetilde{z}_\| \cdot \widetilde{\boldsymbol{n}}_\perp = 0$, and

(3) finally re-expressing the term

$$[\widetilde{z}_\| \otimes \widetilde{\nabla}_{\widetilde{X}}] : \widetilde{\mathrm{cof}}\,\widetilde{\boldsymbol{F}} = \widetilde{J}\,[\widetilde{z}_\| \otimes \widetilde{\nabla}_{\widetilde{X}}] : \widetilde{\boldsymbol{f}}^t = \widetilde{J}\,[\widetilde{z}_\| \otimes \widetilde{\nabla}_{\widetilde{x}}] : \widetilde{\boldsymbol{i}} = \widetilde{J}\,\widetilde{z}_\| \cdot \widetilde{\nabla}_{\widetilde{x}}$$

as $\widetilde{J}\,\widetilde{\mathrm{div}}\widetilde{z}_\|$.

Moreover, the relation between $\widetilde{\boldsymbol{Z}}_\| \cdot \widetilde{\boldsymbol{N}}$ and $\widetilde{z}_\| \cdot \widetilde{\boldsymbol{n}}$, whereby $\widetilde{\boldsymbol{N}}$ and $\widetilde{\boldsymbol{n}}$ denote the outwards-pointing unit normals to the boundary points $\partial\mathcal{C}_\mathrm{m}$ and $\partial\mathcal{C}_\mathrm{s}$, tangential to the material and spatial manifolds $\mathcal{C}_\mathrm{m}$ and $\mathcal{C}_\mathrm{s}$, respectively, follows as

$$\widetilde{\boldsymbol{Z}}_\| \cdot \widetilde{\boldsymbol{N}} = \widetilde{z}_\| \cdot \widetilde{\boldsymbol{n}}. \tag{5.98}$$

Likewise, the relation between $[\![\widetilde{\boldsymbol{Z}}_\|]\!] \cdot \widetilde{\boldsymbol{M}}$ and $[\![\widetilde{z}_\|]\!] \cdot \widetilde{\boldsymbol{m}}$, whereby $\widetilde{\boldsymbol{M}}$ and $\widetilde{\boldsymbol{m}}$ denote the unit normals to the singular points $\mathcal{P}_\mathrm{m}$ and $\mathcal{P}_\mathrm{s}$, tangential to the material and spatial manifolds $\mathcal{C}_\mathrm{m}$ and $\mathcal{C}_\mathrm{s}$, respectively, follows as

$$[\![\widetilde{\boldsymbol{Z}}]\!] \cdot \widetilde{\boldsymbol{M}} = [\![\widetilde{z}]\!] \cdot \widetilde{\boldsymbol{m}}. \tag{5.99}$$

The proof follows in both cases by

(1) recalling the curve Piola transformation $\widetilde{\boldsymbol{Z}}_{\|} \cdot \widehat{\operatorname{cof}} \widehat{\boldsymbol{f}} = \widehat{z}_{\|}$,
(2) multiplying with the vectorial point element $\widetilde{\boldsymbol{n}}$ from the right,
(3) involving the curve Nanson formula $\widetilde{\boldsymbol{n}} = \widetilde{\operatorname{cof}} \widetilde{\boldsymbol{F}} \cdot \widetilde{\boldsymbol{N}}$ in terms of the curve cofactor of the curve deformation gradient (together with the coherence condition $\widetilde{\text{IV}}$ in the case of a singular point).

The curve Piola transformation is a specialization of the Piola transformation in the bulk and on surfaces to curves.

Material↔Spatial Local Curve Transport Theorems:
For any time-dependent scalar-, vector-, or tensor-valued field $\widetilde{z}_{\mathrm{m}} = \widetilde{z}_{\mathrm{m}}(\widetilde{\boldsymbol{X}}, t)$, a density per unit length in the material configuration, and its pendant $\widetilde{z}_{\mathrm{s}} = \widetilde{j}\widetilde{z}_{\mathrm{m}} \circ \widetilde{\boldsymbol{Y}}(\widetilde{\boldsymbol{x}}, t) = \widetilde{z}_{\mathrm{s}}(\widetilde{\boldsymbol{x}}, t)$, a density per unit length in the spatial configuration, the local version of the curve transport theorem reads

$$
\begin{aligned}
\widetilde{j}\, \mathrm{D}_t \widetilde{z}_{\mathrm{m}} &= \mathrm{d}_t \widetilde{z}_{\mathrm{s}} + [\nabla_{\widetilde{n}_\perp} \widetilde{z}_{\mathrm{s}} \otimes \boldsymbol{v}\,] \cdot \widetilde{\boldsymbol{n}}_\perp + \widetilde{\operatorname{div}}(\widetilde{z}_{\mathrm{s}} \otimes \boldsymbol{v}\,) \\
&+ \qquad [\nabla_{\widetilde{n}_\times} \widetilde{z}_{\mathrm{s}} \otimes \boldsymbol{v}\,] \cdot \widetilde{\boldsymbol{n}}_\times, \\
\widetilde{J}\, \mathrm{d}_t \widetilde{z}_{\mathrm{s}} &= \mathrm{D}_t \widetilde{z}_{\mathrm{m}} + [\nabla_{\widetilde{N}_\perp} \widetilde{z}_{\mathrm{m}} \otimes \boldsymbol{V}] \cdot \widetilde{\boldsymbol{N}}_\perp + \widetilde{\operatorname{Div}}(\widetilde{z}_{\mathrm{m}} \otimes \boldsymbol{V}) \\
&+ \qquad [\nabla_{\widetilde{N}_\times} \widetilde{z}_{\mathrm{m}} \otimes \boldsymbol{V}] \cdot \widetilde{\boldsymbol{N}}_\times
\end{aligned}
\tag{5.100}
$$

point-wise in smooth parts of any control curve. The proof, for example, for Eq. 5.100.1, follows by

(1) applying the product rule $\mathrm{D}_t \widetilde{z}_{\mathrm{m}} = \mathrm{D}_t[\widetilde{J}\, \widetilde{z}_{\mathrm{s}}] = \widetilde{J}\, \mathrm{D}_t \widetilde{z}_{\mathrm{s}} + \widetilde{z}_{\mathrm{s}}\, \mathrm{D}_t \widetilde{J}$,
(2) invoking the Euler formula $\mathrm{D}_t \widetilde{z}_{\mathrm{s}} = \mathrm{d}_t \widetilde{z}_{\mathrm{s}} + \nabla_x \widetilde{z}_{\mathrm{s}} \cdot \boldsymbol{v}$,
(3) expanding $\nabla_x \widetilde{z}_{\mathrm{s}} = \nabla_{\widetilde{n}_\perp} \widetilde{z}_{\mathrm{s}} \otimes \widetilde{\boldsymbol{n}}_\perp + \nabla_{\widetilde{n}_\times} \widetilde{z}_{\mathrm{s}} \otimes \widetilde{\boldsymbol{n}}_\times + \widetilde{\nabla}_{\widetilde{x}} \widetilde{z}_{\mathrm{s}}$ into normal, bi-normal, and tangential contributions,
(4) invoking $\mathrm{D}_t \widetilde{J} = \widetilde{J}\, \widetilde{\operatorname{div}} \boldsymbol{v}$,
(5) combining the terms $\widetilde{\nabla}_{\widetilde{x}} \widetilde{z}_{\mathrm{s}} \cdot \boldsymbol{v}$ and $\widetilde{z}_{\mathrm{s}}\, \widetilde{\operatorname{div}} \boldsymbol{v}$ into $\widetilde{\operatorname{div}}(\widetilde{z}_{\mathrm{s}} \otimes \boldsymbol{v})$.

With $\boldsymbol{v} = \boldsymbol{v} \cdot [\boldsymbol{i} - \widetilde{\boldsymbol{i}}] + \widetilde{\boldsymbol{v}}$ where $\widetilde{\boldsymbol{v}} := \boldsymbol{v} \cdot \widetilde{\boldsymbol{i}}$ and $\boldsymbol{V} = \boldsymbol{V} \cdot [\boldsymbol{I} - \widetilde{\boldsymbol{I}}] + \widetilde{\boldsymbol{V}}$ where $\widetilde{\boldsymbol{V}} := \boldsymbol{V} \cdot \widetilde{\boldsymbol{I}}$, the curve divergence terms may be re-expressed as

$$
\begin{aligned}
\widetilde{\operatorname{div}}(\widetilde{z}_{\mathrm{s}} \otimes \boldsymbol{v}\,) &= \widetilde{\operatorname{div}}(\widetilde{z}_{\mathrm{s}} \otimes \widetilde{\boldsymbol{v}}\,) - \widetilde{c}\,[\widetilde{z}_{\mathrm{s}} \otimes \boldsymbol{v}\,] \cdot \widetilde{\boldsymbol{n}}_\perp \\
\widetilde{\operatorname{Div}}(\widetilde{z}_{\mathrm{m}} \otimes \boldsymbol{V}) &= \widetilde{\operatorname{Div}}(\widetilde{z}_{\mathrm{m}} \otimes \widetilde{\boldsymbol{V}}) - \widetilde{C}[\widetilde{z}_{\mathrm{m}} \otimes \boldsymbol{V}] \cdot \widetilde{\boldsymbol{N}}_\perp.
\end{aligned}
\tag{5.101}
$$

Thus taken together, combining Eqs. 5.100 and 5.101 alongside with the Piola transformation results in the identities

$$
\begin{aligned}
\widehat{z}_{\mathrm{s}}^{\perp} &:= \nabla_{\widetilde{n}_\perp} \widetilde{z}_{\mathrm{s}} - \widetilde{z}_{\mathrm{s}} \widetilde{c} = \widehat{j}\,[\nabla_{\widetilde{N}_\perp} \widetilde{z}_{\mathrm{m}} - \widetilde{z}_{\mathrm{m}} \widetilde{C}] =: \widehat{j}\, \widehat{z}_{\mathrm{m}}^{\perp} \\
\widehat{z}_{\mathrm{s}}^{\times} &:= \nabla_{\widetilde{n}_\times} \widetilde{z}_{\mathrm{s}} \qquad\;\; = \widehat{j}\;\; \nabla_{\widetilde{N}_\times} \widetilde{z}_{\mathrm{m}} \qquad\quad =: \widehat{j}\, \widehat{z}_{\mathrm{m}}^{\times}.
\end{aligned}
\tag{5.102}
$$

These identities are crucial when verifying the global curve transport theorem within a spatial control curve in the following.

As a consequence of the above, the local version of the curve transport theorem expands alternatively as

$$\begin{aligned} \widetilde{j}\, \mathrm{D}_t \widetilde{z}_{\mathrm{m}} &= \mathrm{d}_t \widetilde{z}_{\mathrm{s}} + \widetilde{\mathrm{div}}(\widetilde{z}_{\mathrm{s}} \otimes \widetilde{\boldsymbol{v}}) + [\widehat{z}_{\mathrm{s}}^{\perp} \otimes \boldsymbol{v}] \cdot \widetilde{\boldsymbol{n}}_{\perp} \\ &+ [\widehat{z}_{\mathrm{s}}^{\times} \otimes \boldsymbol{v}] \cdot \widetilde{\boldsymbol{n}}_{\times}, \\ \widetilde{J}\, \mathrm{d}_t \widetilde{z}_{\mathrm{s}} &= \mathrm{D}_t \widetilde{z}_{\mathrm{m}} + \widetilde{\mathrm{Div}}(\widetilde{z}_{\mathrm{m}} \otimes \widetilde{\boldsymbol{V}}) + [\widehat{z}_{\mathrm{m}}^{\perp} \otimes \boldsymbol{V}] \cdot \widetilde{\boldsymbol{N}}_{\perp} \\ &+ [\widehat{z}_{\mathrm{m}}^{\times} \otimes \boldsymbol{V}] \cdot \widetilde{\boldsymbol{N}}_{\times}. \end{aligned} \tag{5.103}$$

Moreover, for the boundary points of the control curve, the curve Piola transformation (PT) (together with $\widetilde{\boldsymbol{W}} \cdot \widetilde{\boldsymbol{F}}^t = \widetilde{\boldsymbol{w}} - \widetilde{\boldsymbol{v}}$ and $\widetilde{\boldsymbol{w}} \cdot \widetilde{\boldsymbol{f}}^t = \widetilde{\boldsymbol{W}} - \widetilde{\boldsymbol{V}}$) and the curve transport theorem (TT) read

$$[\widetilde{z}_{\mathrm{m}} \otimes \widetilde{\boldsymbol{W}}] \cdot \widetilde{\boldsymbol{N}} \stackrel{\mathrm{PT}}{=} \big[\widetilde{z}_{\mathrm{s}} \otimes [\widetilde{\boldsymbol{w}} - \widetilde{\boldsymbol{v}}]\big] \cdot \widetilde{\boldsymbol{n}} \stackrel{\mathrm{TT}}{=} [\widetilde{z}_{\mathrm{s}} \otimes \widetilde{\boldsymbol{w}} - \widetilde{z}_{\mathrm{s}} \otimes \widetilde{\boldsymbol{v}}] \cdot \widetilde{\boldsymbol{n}} \tag{5.104}$$

and

$$[\widetilde{z}_{\mathrm{s}} \otimes \widetilde{\boldsymbol{w}}] \cdot \widetilde{\boldsymbol{n}} \stackrel{\mathrm{PT}}{=} \big[\widetilde{z}_{\mathrm{m}} \otimes [\widetilde{\boldsymbol{W}} - \widetilde{\boldsymbol{V}}]\big] \cdot \widetilde{\boldsymbol{N}} \stackrel{\mathrm{TT}}{=} [\widetilde{z}_{\mathrm{m}} \otimes \widetilde{\boldsymbol{W}} - \widetilde{z}_{\mathrm{m}} \otimes \widetilde{\boldsymbol{V}}] \cdot \widetilde{\boldsymbol{N}}.$$

Likewise, for singular points within the control curve, the curve Piola transformation and the curve transport theorem read

$$[\![\widetilde{z}_{\mathrm{m}} \otimes \widetilde{\boldsymbol{W}}]\!] \cdot \widetilde{\boldsymbol{M}} \stackrel{\mathrm{PT}}{=} [\![\widetilde{z}_{\mathrm{s}} \otimes [\widetilde{\boldsymbol{w}} - \widetilde{\boldsymbol{v}}]]\!] \cdot \widetilde{\boldsymbol{m}} \stackrel{\mathrm{TT}}{=} \big[[\![\widetilde{z}_{\mathrm{s}} \otimes \widetilde{\boldsymbol{w}}]\!] - [\![\widetilde{z}_{\mathrm{s}} \otimes \widetilde{\boldsymbol{v}}]\!]\big] \cdot \widetilde{\boldsymbol{m}} \tag{5.105}$$

and

$$[\![\widetilde{z}_{\mathrm{s}} \otimes \widetilde{\boldsymbol{w}}]\!] \cdot \widetilde{\boldsymbol{m}} \stackrel{\mathrm{PT}}{=} [\![\widetilde{z}_{\mathrm{m}} \otimes [\widetilde{\boldsymbol{W}} - \widetilde{\boldsymbol{V}}]]\!] \cdot \widetilde{\boldsymbol{M}} \stackrel{\mathrm{TT}}{=} \big[[\![\widetilde{z}_{\mathrm{m}} \otimes \widetilde{\boldsymbol{W}}]\!] - [\![\widetilde{z}_{\mathrm{m}} \otimes \widetilde{\boldsymbol{V}}]\!] \cdot \widetilde{\boldsymbol{M}}.$$

The proof follows from noting, for example, that $\widetilde{\boldsymbol{W}} \cdot \widetilde{\boldsymbol{N}} = \widetilde{j}[\widetilde{\boldsymbol{w}} - \widetilde{\boldsymbol{v}}] \cdot \widetilde{\boldsymbol{n}}$ and $\widetilde{\boldsymbol{w}} \cdot \widetilde{\boldsymbol{n}} = \widetilde{J}[\widetilde{\boldsymbol{W}} - \widetilde{\boldsymbol{V}}] \cdot \widetilde{\boldsymbol{N}}$, respectively. Here, the curve transport theorem is merely a trivial application of the dyadic product distributivity.

Global Curve Transport Theorem (Spatial Control Curve):
The (curve) resultant Z of any time-dependent scalar-, vector-, or tensor-valued curve field $\widetilde{z}_{\mathrm{s}} = \widetilde{z}_{\mathrm{s}}(\widetilde{\boldsymbol{x}}, t)$, a density per unit length in the spatial configuration, defined over the spatial control curve $\mathfrak{C}_{\mathrm{s}}$ is given as

$$\mathsf{Z} := \int\limits_{\mathfrak{C}_s} \widetilde{z}_s \, \mathrm{d}l. \tag{5.106}$$

Then for cases with a singular point $\mathfrak{P}_s$ contained in the spatial control curve $\mathfrak{C}_s$, on the one hand, the curve transport theorem reads when proceeding in analogy to the proof of Eq. 5.93

$$\begin{aligned}\dot{\mathsf{Z}} = &\int\limits_{\mathfrak{C}_s} \Big[\mathrm{d}_t \widetilde{z}_s + [[\nabla_{\widetilde{n}_\perp} \widetilde{z}_s - \widetilde{z}_s \widetilde{c}] \otimes \boldsymbol{w}] \cdot \widetilde{\boldsymbol{n}}_\perp \Big] \mathrm{d}l \\ + &\int\limits_{\mathfrak{C}_s} \quad\; [\; \nabla_{\widetilde{n}_\times} \widetilde{z}_s \qquad\quad \otimes \boldsymbol{w}] \cdot \widetilde{\boldsymbol{n}}_\times \;\; \mathrm{d}l \\ + &\sum_{\partial \mathfrak{C}_s} [\widetilde{z}_s \otimes \widetilde{\boldsymbol{w}}] \cdot \widetilde{\boldsymbol{n}} - [[\![\widetilde{z}_s \otimes \widetilde{\boldsymbol{w}}]\!] \cdot \widetilde{\boldsymbol{m}}]_{\mathfrak{P}_s}.\end{aligned} \tag{5.107}$$

On the other hand, when simply using the relations between the material and the spatial scalar and vectorial line elements, i.e. $\mathrm{d}L = \widetilde{j}\,\mathrm{d}l$ and $\mathrm{d}\widehat{\boldsymbol{L}} = \widehat{\mathrm{cof}\,\boldsymbol{f}} \cdot \mathrm{d}\widehat{\boldsymbol{l}}$ and changing the integration domain in Eq. 5.93, the curve transport theorem is, likewise, expressed as

$$\begin{aligned}\dot{\mathsf{Z}} = &\int\limits_{\mathfrak{C}_s} \Big[\widetilde{j}\, \mathrm{D}_t \widetilde{z}_m + [\widehat{j}\, [\nabla_{\widetilde{N}_\perp} \widetilde{z}_m - \widetilde{z}_m \widetilde{C}] \otimes [\boldsymbol{w} - \boldsymbol{v}]] \cdot \widetilde{\boldsymbol{n}}_\perp \Big] \mathrm{d}l \\ + &\int\limits_{\mathfrak{C}_s} \qquad\quad \widehat{j}\; \nabla_{\widetilde{N}_\times} \widetilde{z}_m \qquad\quad \otimes [\boldsymbol{w} - \boldsymbol{v}]] \cdot \widetilde{\boldsymbol{n}}_\times \;\; \mathrm{d}l \\ + &\sum_{\partial \mathfrak{C}_s} [\widetilde{z}_s \otimes [\widetilde{\boldsymbol{w}} - \widetilde{\boldsymbol{v}}]] \cdot \widetilde{\boldsymbol{n}} - \Big[[\![\widetilde{z}_s \otimes [\widetilde{\boldsymbol{w}} - \widetilde{\boldsymbol{v}}]]\!] \cdot \widetilde{\boldsymbol{m}} \Big]_{\mathfrak{P}_s}.\end{aligned} \tag{5.108}$$

The two different expansions of the global curve transport theorem in Eqs. 5.107 and 5.108 prove identical when the local curve transport theorem in Eq. 5.100 and the identities in Eq. 5.102 are invoked.

Finally, using again the relations between the spatial and the material scalar and vectorial line elements, i.e. $\mathrm{d}l = \widetilde{J}\,\mathrm{d}L$ and $\mathrm{d}\widehat{\boldsymbol{l}} = \widehat{\mathrm{cof}\,\boldsymbol{F}} \cdot \mathrm{d}\widehat{\boldsymbol{L}}$ and changing the integration domain in Eq. 5.107, the curve transport theorem reads equivalently

$$\dot{\mathbf{Z}} = \int\limits_{\mathfrak{C}_m} \Big[\widetilde{J}\,\mathrm{d}_t\widetilde{z}_s + \big[\widehat{J}[\nabla_{\widetilde{n}_\perp}\widetilde{z}_s - \widetilde{z}_s\widetilde{c}] \otimes [\boldsymbol{W} - \boldsymbol{V}]\big] \cdot \widetilde{\boldsymbol{N}}_\perp\Big]\,\mathrm{d}L \tag{5.109}$$
$$+ \int\limits_{\mathfrak{C}_m} \big[\widehat{J}\,\nabla_{\widetilde{n}_\times}\widetilde{z}_s \otimes [\boldsymbol{W} - \boldsymbol{V}]\big] \cdot \widetilde{\boldsymbol{N}}_\times\,\mathrm{d}L$$
$$+ \sum_{\partial\mathfrak{C}_m} [\widetilde{z}_m \otimes [\widetilde{\boldsymbol{W}} - \widetilde{\boldsymbol{V}}]] \cdot \widetilde{\boldsymbol{N}} - \Big[[\![\widetilde{z}_m \otimes [\widetilde{\boldsymbol{W}} - \widetilde{\boldsymbol{V}}]]\!] \cdot \widetilde{\boldsymbol{M}}\Big]_{\mathfrak{P}_m}.$$

Again the two different expansions of the global curve transport theorem in Eqs. 5.93 and 5.109 prove identical when the local surface transport theorem in Eq. 5.100 and the identity in Eq. 5.102 are invoked.

Any singular contributions to the curve transport theorem are here neglected in the presentation for the sake of conciseness.

5.3.2 *Global Format (Material Control Curve)*

The balance equation for a global generic *curve extensive* quantity $\mathbf{Z}$ may be stated as follows.

Global Balance of Generic Curve Extensive Quantity (Material Control Curve):

$$\dot{\mathbf{Z}}(\mathfrak{C}_m) = \mathbf{A}^{\mathrm{ext}}(\mathfrak{C}_m) \quad \forall \mathfrak{C}_m \subseteq \mathcal{C}_m. \tag{5.110}$$

Here, $\mathbf{Z}$ is expressed as the resultant over the material control curve $\mathfrak{C}_m$ of the local generic curve extensive quantity $\widetilde{z}_m$, a density per unit length in the material configuration, i.e.

$$\mathbf{Z}(\mathfrak{C}_m) := \int\limits_{\mathfrak{C}_m} \widetilde{z}_m\,\mathrm{d}L \tag{5.111}$$

and $\mathbf{A}^{\mathrm{ext}}$ consists of a curve integral and a boundary points sum (any singular contributions are here ignored for the sake of presentation)

$$\mathbf{A}^{\mathrm{ext}}(\mathfrak{C}_m) := \int\limits_{\mathfrak{C}_m} \widetilde{\boldsymbol{a}}_m\,\mathrm{d}L + \sum_{\partial\mathfrak{C}_m} \vec{\boldsymbol{a}}_m \tag{5.112}$$

in terms of the generic curve source density $\widetilde{\boldsymbol{a}}_m$ per unit length in the material configuration (distant interaction) and the generic *extrinsic* point flux $\vec{\boldsymbol{a}}_m$ (contact interaction).

At the intersection of the material control curve boundary points with the external boundary, a Neumann-type boundary condition

$$\vec{\boldsymbol{a}}_{\mathrm{m}} \equiv \vec{\boldsymbol{a}}_{\mathrm{m}}^{\mathrm{ext}} \quad \text{on} \quad \partial \mathcal{C}_{\mathrm{m}} \tag{5.113}$$

holds for the extrinsic point flux $\vec{\boldsymbol{a}}_{\mathrm{m}}$ with $\vec{\boldsymbol{a}}_{\mathrm{m}}^{\mathrm{ext}}$ externally *prescribed.*

The extrinsic point flux along with the convective term stemming from the curve transport theorem is *postulated* to satisfy a *Cauchy theorem* for the *intrinsic* point flux $\widetilde{\mathbf{Z}} \cdot \widetilde{\boldsymbol{N}}$ (that is expressed in terms of the *intrinsic point flux* $\widetilde{\mathbf{Z}}$, a curve tensor field with $\widetilde{\mathbf{Z}} \cdot \widetilde{\boldsymbol{N}}_{\perp} = \widetilde{\mathbf{Z}} \cdot \widetilde{\boldsymbol{N}}_{\times} = \mathbf{0}$)

$$\widetilde{\mathbf{Z}} \cdot \widetilde{\boldsymbol{N}} := \vec{\boldsymbol{a}}_{\mathrm{m}} - [\widetilde{\mathfrak{z}}_{\mathrm{m}} \otimes \widetilde{\boldsymbol{W}}] \cdot \widetilde{\boldsymbol{N}} \quad \text{on} \quad \partial \mathfrak{C}_{\mathrm{m}}. \tag{5.114}$$

Moreover, the curve source density $\widetilde{\boldsymbol{a}}_{\mathrm{m}}$ combines with the convective term stemming from the curve transport theorem into

$$\widetilde{\aleph}_{\mathrm{m}} := \widetilde{\boldsymbol{a}}_{\mathrm{m}} - [\widehat{\mathfrak{z}}_{\mathrm{m}}^{\perp} \otimes \boldsymbol{W}] \cdot \widetilde{\boldsymbol{N}}_{\perp} - [\widehat{\mathfrak{z}}_{\mathrm{m}}^{\times} \otimes \boldsymbol{W}] \cdot \widetilde{\boldsymbol{N}}_{\times} \quad \text{in} \quad \mathfrak{C}_{\mathrm{m}} \tag{5.115}$$

with the abbreviations $\widehat{\mathfrak{z}}_{\mathrm{m}}^{\perp} := \nabla_{\widetilde{N}_{\perp}} \widetilde{\mathfrak{z}}_{\mathrm{m}} - \widetilde{\mathfrak{z}}_{\mathrm{m}} \widetilde{C}$ and $\widehat{\mathfrak{z}}_{\mathrm{m}}^{\times} := \nabla_{\widetilde{N}_{\times}} \widetilde{\mathfrak{z}}_{\mathrm{m}}$.

Consequently, the global format of the balance equation for a generic *curve extensive* global quantity expands as

$$\int_{\mathfrak{C}_{\mathrm{m}}} \mathrm{D}_t \widetilde{\mathfrak{z}}_{\mathrm{m}} \, \mathrm{d}L = \int_{\mathfrak{C}_{\mathrm{m}}} \widetilde{\aleph}_{\mathrm{m}} \, \mathrm{d}L + \sum_{\partial \mathfrak{C}_{\mathrm{m}}} \widetilde{\mathbf{Z}} \cdot \widetilde{\boldsymbol{N}} + \left[[\![\widetilde{\mathfrak{z}}_{\mathrm{m}} \otimes \widetilde{\boldsymbol{W}}]\!] \cdot \widetilde{\boldsymbol{M}} \right]_{\mathfrak{P}_{\mathrm{m}}}. \tag{5.116}$$

5.3.3 *Local Format (Material Control Curve)*

For the process of localization, three different scenarios may be considered:

i. Consider a smooth material control curve $\mathfrak{C}_{\mathrm{m}}$, completely contained within $\mathcal{C}_{\mathrm{m}}$. Using the curve Gauss theorem to transform the point sum in Eq. 5.116 to a curve integral renders

$$\int_{\mathfrak{C}_{\mathrm{m}}} \mathrm{D}_t \widetilde{\mathfrak{z}}_{\mathrm{m}} \, \mathrm{d}L = \int_{\mathfrak{C}_{\mathrm{m}}} [\widetilde{\aleph}_{\mathrm{m}} + \widetilde{\mathrm{Div}} \widetilde{\mathbf{Z}}] \, \mathrm{d}L. \tag{5.117}$$

Localization thus renders the point-wise generic balance equation in the curve domain $\mathcal{C}_{\mathrm{m}}$ with $\widetilde{\aleph}_{\mathrm{m}} = \widetilde{\boldsymbol{a}}_{\mathrm{m}} - [\widehat{\mathfrak{z}}_{\mathrm{m}}^{\perp} \otimes \boldsymbol{W}] \cdot \widetilde{\boldsymbol{N}}_{\perp} - [\widehat{\mathfrak{z}}_{\mathrm{m}}^{\times} \otimes \boldsymbol{W}] \cdot \widetilde{\boldsymbol{N}}_{\times}$.

ii. Consider a smooth material control curve $\mathfrak{C}_{\mathrm{m}}$ with boundary points $\partial \mathfrak{C}_{\mathrm{m}}$ intersecting with the regular part of the external boundary $\partial \mathcal{C}_{\mathrm{m}}$. Then the Cauchy theorem renders the point-wise generic balance equation on the boundary points with the Neumann-type boundary condition $\vec{a}_m \equiv \vec{a}_m^{ext}$.

iii. Consider a material control curve $\mathfrak{C}_m$, completely contained within $\mathcal{C}_m$, that includes a coherent singular point $\mathfrak{P}_m$. Then the material control curve $\mathfrak{C}_m$ is decreased to zero such that the points $\partial\mathfrak{C}_m^{\pm}$ coincide with $\mathfrak{P}_m$ and $-\widetilde{\boldsymbol{N}}^- = \widetilde{\boldsymbol{N}}^+ = \widetilde{\boldsymbol{M}}$. Thus, all terms in Eq. 5.116 involving curve integrals vanish to render

$$\mathbf{0} = \left[[\![\widetilde{\boldsymbol{Z}} + \widetilde{z}_m \otimes \widetilde{\boldsymbol{W}}]\!] \cdot \widetilde{\boldsymbol{M}} \right]_{\mathfrak{P}_m}. \tag{5.118}$$

Localization thus renders the point-wise generic balance equation at singular points $\mathcal{P}_m$.

The results of localizing the generic balance equation are summarized as follows.

Local Balance of Generic Curve Extensive Quantity (Material Control Curve):

i. Regular Points in the Curve Domain

$$\mathrm{D}_t \widetilde{z}_m - \widetilde{\mathrm{Div}} \widetilde{\boldsymbol{Z}} = \widetilde{\mathfrak{s}}_m \quad \text{in} \quad \mathcal{C}_m. \tag{5.119}$$

ii. Boundary Points

$$\left[\widetilde{z}_m \otimes \widetilde{\boldsymbol{W}} + \widetilde{\boldsymbol{Z}} \right] \cdot \widetilde{\boldsymbol{N}} = \vec{a}_m^{ext} \quad \text{on} \quad \partial\mathcal{C}_m. \tag{5.120}$$

iii. Singular Points

$$[\![\widetilde{z}_m \otimes \widetilde{\boldsymbol{W}} + \widetilde{\boldsymbol{Z}}]\!] \cdot \widetilde{\boldsymbol{M}} = \mathbf{0} \quad \text{at} \quad \mathcal{P}_m. \tag{5.121}$$

5.3.4 Global Format (Spatial Control Curve)

The balance equation for a global generic *curve extensive* quantity Z may be stated as follows.

Global Balance of Generic Curve Extensive Quantity (Spatial Control Curve):

$$\dot{\mathsf{Z}}(\mathfrak{C}_s) = \mathsf{A}^{ext}(\mathfrak{C}_s) \quad \forall \mathfrak{C}_s \subseteq \mathcal{C}_s. \tag{5.122}$$

Here, Z is expressed as the resultant over the spatial control curve $\mathfrak{C}_s$ of the local generic curve extensive quantity $\widetilde{z}_s$, a density per unit length in the spatial configuration, i.e.

$$\mathsf{Z}(\mathfrak{C}_s) := \int_{\mathfrak{C}_s} \widetilde{z}_s \, \mathrm{d}l \tag{5.123}$$

and $\mathsf{A}^{\mathrm{ext}}$ consists of curve integral and a boundary points sum (any singular contributions are here ignored for the sake of presentation)

$$\mathsf{A}^{\mathrm{ext}}(\mathfrak{C}_s) := \int_{\mathfrak{C}_s} \widetilde{\boldsymbol{a}}_s \, \mathrm{d}l + \sum_{\partial\mathfrak{C}_s} \vec{\boldsymbol{a}}_s \tag{5.124}$$

in terms of the generic curve source density $\widetilde{\boldsymbol{a}}_s = \widetilde{j}\,\widetilde{\boldsymbol{a}}_m$ per unit length in the spatial configuration and the generic extrinsic point flux $\vec{\boldsymbol{a}}_s \equiv \vec{\boldsymbol{a}}_m$. The intrinsic point flux is defined as the curve Piola transformation $\widetilde{\boldsymbol{\alpha}} := \widetilde{\boldsymbol{Z}} \cdot \widetilde{\operatorname{cof}} \widetilde{\boldsymbol{f}}$ (a curve tensor field with $\widetilde{\boldsymbol{\alpha}} \cdot \widetilde{\boldsymbol{n}}_\perp = \widetilde{\boldsymbol{\alpha}} \cdot \widetilde{\boldsymbol{n}}_\times = \boldsymbol{0}$), thus the corresponding Cauchy theorem for $\widetilde{\boldsymbol{\alpha}} \cdot \widetilde{\boldsymbol{n}}$ reads in terms of the extrinsic point flux along with the convective term stemming from the curve transport theorem

$$\widetilde{\boldsymbol{\alpha}} \cdot \widetilde{\boldsymbol{n}} := \vec{\boldsymbol{a}}_s - \left[\widetilde{z}_s \otimes [\widetilde{\boldsymbol{w}} - \widetilde{\boldsymbol{v}}]\right] \cdot \widetilde{\boldsymbol{n}} \quad \text{on} \quad \partial\mathfrak{C}_s. \tag{5.125}$$

Moreover, the curve source density $\widetilde{\boldsymbol{a}}_s$ combines with the convective term stemming from the curve transport theorem into

$$\widetilde{\boldsymbol{\aleph}}_s := \widetilde{\boldsymbol{a}}_s - \left[\widehat{z}_s^{\perp} \otimes [\boldsymbol{w} - \boldsymbol{v}]\right] \cdot \widetilde{\boldsymbol{n}}_\perp - \left[\widehat{z}_s^{\times} \otimes [\boldsymbol{w} - \boldsymbol{v}]\right] \cdot \widetilde{\boldsymbol{n}}_\times \quad \text{in} \quad \mathfrak{C}_s \tag{5.126}$$

with the abbreviations $\widehat{z}_s^{\perp} := \widehat{j\,\widetilde{z}_m^{\perp}} := \nabla_{\widetilde{n}_\perp} \widetilde{z}_s - \widetilde{z}_s \widetilde{c}$ and $\widehat{z}_s^{\times} := \widehat{j\,\widetilde{z}_m^{\times}} := \nabla_{\widetilde{n}_\times} \widetilde{z}_s$.

Consequently, the global format of the generic balance equation for a generic *curve extensive* quantity expands as

$$\begin{aligned} \int_{\mathfrak{C}_s} \mathrm{d}_t \widetilde{z}_s \, \mathrm{d}l = & \int_{\mathfrak{C}_s} \left[\widetilde{\boldsymbol{\aleph}}_s - [\widehat{z}_s^{\perp} \otimes \boldsymbol{v}] \cdot \widetilde{\boldsymbol{n}}_\perp - [\widehat{z}_s^{\times} \otimes \boldsymbol{v}] \cdot \widetilde{\boldsymbol{n}}_\times\right] \mathrm{d}l \\ & + \sum_{\partial\mathfrak{C}_s} [\widetilde{\boldsymbol{\alpha}} - \widetilde{z}_s \otimes \widetilde{\boldsymbol{v}}] \cdot \widetilde{\boldsymbol{n}} + \left[[\![\widetilde{z}_s \otimes \widetilde{\boldsymbol{w}}]\!] \cdot \widetilde{\boldsymbol{m}} \right]_{\mathfrak{P}_s}. \end{aligned} \tag{5.127}$$

The process of localizing the above global format to the corresponding local generic balance equations then follows the same rationale as already outlined earlier.

5.3.5 *Local Format (Spatial Control Curve)*

The local balances for the generic *curve extensive* local quantity $\widetilde{z}_s$ (a density per unit length in the spatial configuration), expressed in terms of the Piola-transformed point flux $\widetilde{\boldsymbol{\alpha}}$, the generic curve source density $\widetilde{\boldsymbol{a}}_s$ per unit length in the spatial configuration, and the generic extrinsic point flux $\vec{\boldsymbol{a}}_s^{\text{ext}}$, are summarized as follows.

Local Balance of Generic Curve Extensive Quantity (Spatial Control Curve):

i. Regular Points in the Curve Domain

$$\mathrm{d}_t \widetilde{z}_s - \widetilde{\mathrm{div}}(\widetilde{\boldsymbol{\alpha}} - \widetilde{z}_s \otimes \widetilde{\boldsymbol{v}}) = \widetilde{\aleph}_s - \left[\widehat{z}_s^{\perp} \otimes \boldsymbol{v}\right] \cdot \widetilde{\boldsymbol{n}}_{\perp} - \left[\widehat{z}_s^{\times} \otimes \boldsymbol{v}\right] \cdot \widetilde{\boldsymbol{n}}_{\times} \quad \text{in} \quad \mathcal{C}_s. \tag{5.128}$$

ii. Boundary Points

$$\left[\widetilde{z}_s \otimes \widetilde{\boldsymbol{w}} + [\widetilde{\boldsymbol{\alpha}} - \widetilde{z}_s \otimes \widetilde{\boldsymbol{v}}]\right] \cdot \widetilde{\boldsymbol{n}} = \vec{\boldsymbol{a}}_s^{\text{ext}} \quad \text{on} \quad \partial\mathcal{C}_s. \tag{5.129}$$

iii. Singular Points

$$[\![\widetilde{z}_s \otimes \widetilde{\boldsymbol{w}} + [\widetilde{\boldsymbol{\alpha}} - \widetilde{z}_s \otimes \widetilde{\boldsymbol{v}}]]\!] \cdot \widetilde{\boldsymbol{m}} = \boldsymbol{0} \quad \text{at} \quad \mathcal{P}_s. \tag{5.130}$$

Table 5.5 Dynamic versions of effective curve source densities

$\widetilde{\aleph}_{\mathrm{mD}} := \widetilde{\aleph}_{\mathrm{m}}$	↔ PT	$\widetilde{\aleph}_{\mathrm{sD}} := \widetilde{\aleph}_{\mathrm{s}}$
↕ TT		↕ TT
$\widetilde{\aleph}_{\mathrm{md}} := \widetilde{\aleph}_{\mathrm{m}} + [\widehat{\boldsymbol{z}}_{\mathrm{m}}^{\perp} \otimes \boldsymbol{V}] \cdot \widetilde{\boldsymbol{N}}_{\perp} + [\widehat{\boldsymbol{z}}_{\mathrm{m}}^{\times} \otimes \boldsymbol{V}] \cdot \widetilde{\boldsymbol{N}}_{\times}$	↔ PT	$\widetilde{\aleph}_{\mathrm{sd}} := \widetilde{\aleph}_{\mathrm{s}} - [\widehat{\boldsymbol{z}}_{\mathrm{s}}^{\perp} \otimes \boldsymbol{v}] \cdot \widetilde{\boldsymbol{n}}_{\perp} - [\widehat{\boldsymbol{z}}_{\mathrm{s}}^{\times} \otimes \boldsymbol{v}] \cdot \widetilde{\boldsymbol{n}}_{\times}$

Table 5.6 Tetragons of fully equivalent local generic balance equations for a curve extensive quantity at regular points in the curve domain, on the boundary points and at singular points. The various versions are related by curve Piola transformation (PT) and/or the curve transport theorem (TT), respectively

Regular Points in the Curve Domain

$\mathrm{D}_t\widetilde{\boldsymbol{z}}_{\mathrm{m}} = \widetilde{\mathrm{Div}}\widetilde{\boldsymbol{Z}}_{\mathrm{D}} + \widetilde{\boldsymbol{\aleph}}_{\mathrm{mD}}$	↔ PT	$\widetilde{j}\,\mathrm{D}_t\widetilde{\boldsymbol{z}}_{\mathrm{m}} = \widetilde{\mathrm{div}}\widetilde{\boldsymbol{\alpha}}_{\mathrm{D}} + \widetilde{\boldsymbol{\aleph}}_{\mathrm{sD}}$
↕ TT		TT ↕
$\widetilde{J}\,\mathrm{d}_t\widetilde{\boldsymbol{z}}_{\mathrm{s}} = \widetilde{\mathrm{Div}}\widetilde{\boldsymbol{Z}}_{\mathrm{d}} + \widetilde{\boldsymbol{\aleph}}_{\mathrm{md}}$	↔ PT	$\mathrm{d}_t\widetilde{\boldsymbol{z}}_{\mathrm{s}} = \widetilde{\mathrm{div}}\widetilde{\boldsymbol{\alpha}}_{\mathrm{d}} + \widetilde{\boldsymbol{\aleph}}_{\mathrm{sd}}$

Boundary Points

$\left[\widetilde{\boldsymbol{z}}_{\mathrm{m}} \otimes \widetilde{\boldsymbol{W}} + \widetilde{\boldsymbol{Z}}_{\mathrm{D}}\right] \cdot \widetilde{\boldsymbol{N}} = \vec{\boldsymbol{a}}_{\mathrm{m}}^{\mathrm{ext}}$	↔ PT	$\left[\widetilde{\boldsymbol{z}}_{\mathrm{s}} \otimes [\widetilde{\boldsymbol{w}} - \widetilde{\boldsymbol{v}}] + \widetilde{\boldsymbol{\alpha}}_{\mathrm{D}}\right] \cdot \widetilde{\boldsymbol{n}} = \vec{\boldsymbol{a}}_{\mathrm{s}}^{\mathrm{ext}}$
↕ TT		TT ↕
$\left[\widetilde{\boldsymbol{z}}_{\mathrm{m}} \otimes [\widetilde{\boldsymbol{W}} - \widetilde{\boldsymbol{V}}] + \widetilde{\boldsymbol{Z}}_{\mathrm{d}}\right] \cdot \widetilde{\boldsymbol{N}} = \vec{\boldsymbol{a}}_{\mathrm{m}}^{\mathrm{ext}}$	↔ PT	$\left[\widetilde{\boldsymbol{z}}_{\mathrm{s}} \otimes \widetilde{\boldsymbol{w}} + \widetilde{\boldsymbol{\alpha}}_{\mathrm{d}}\right] \cdot \widetilde{\boldsymbol{n}} = \vec{\boldsymbol{a}}_{\mathrm{s}}^{\mathrm{ext}}$

Singular Points

$\llbracket \widetilde{\boldsymbol{z}}_{\mathrm{m}} \otimes \widetilde{\boldsymbol{W}} + \widetilde{\boldsymbol{Z}}_{\mathrm{D}} \rrbracket \cdot \widetilde{\boldsymbol{M}} = \boldsymbol{0}$	↔ PT	$\llbracket \widetilde{\boldsymbol{z}}_{\mathrm{s}} \otimes [\widetilde{\boldsymbol{w}} - \widetilde{\boldsymbol{v}}] + \widetilde{\boldsymbol{\alpha}}_{\mathrm{D}} \rrbracket \cdot \widetilde{\boldsymbol{m}} = \boldsymbol{0}$
↕ TT		TT ↕
$\llbracket \widetilde{\boldsymbol{z}}_{\mathrm{m}} \otimes [\widetilde{\boldsymbol{W}} - \widetilde{\boldsymbol{V}}] + \widetilde{\boldsymbol{Z}}_{\mathrm{d}} \rrbracket \cdot \widetilde{\boldsymbol{M}} = \boldsymbol{0}$	↔ PT	$\llbracket \widetilde{\boldsymbol{z}}_{\mathrm{s}} \otimes \widetilde{\boldsymbol{w}} + \widetilde{\boldsymbol{\alpha}}_{\mathrm{d}} \rrbracket \cdot \widetilde{\boldsymbol{m}} = \boldsymbol{0}$

with *Dynamic Versions* of Intrinsic Point Flux

$\widetilde{\boldsymbol{Z}}_{\mathrm{D}} := \widetilde{\boldsymbol{Z}}$	↔ PT	$\widetilde{\boldsymbol{\alpha}}_{\mathrm{D}} := \widetilde{\boldsymbol{\alpha}}$
↕ TT		TT ↕
$\widetilde{\boldsymbol{Z}}_{\mathrm{d}} := \widetilde{\boldsymbol{Z}} + \widetilde{\boldsymbol{z}}_{\mathrm{m}} \otimes \widetilde{\boldsymbol{V}}$	↔ PT	$\widetilde{\boldsymbol{\alpha}}_{\mathrm{d}} := \widetilde{\boldsymbol{\alpha}} - \widetilde{\boldsymbol{z}}_{\mathrm{s}} \otimes \widetilde{\boldsymbol{v}}$

5.3.6 Balance Tetragon

By exploiting the version of the curve transport theorem in Eq. 5.108, the global format of the generic balance equation is stated alternatively as

$$\int_{\mathfrak{C}_s} \widetilde{j}\, \mathrm{D}_t \widetilde{\mathfrak{z}}_{\mathrm{m}}\, \mathrm{d}l = \int_{\mathfrak{c}_s} \widetilde{\aleph}_{\mathrm{s}}\, \mathrm{d}l + \sum_{\partial \mathfrak{C}_s} \widetilde{\boldsymbol{\alpha}} \cdot \widetilde{\boldsymbol{n}} + \left[[\![\widetilde{\mathfrak{z}}_{\mathrm{s}} \otimes [\widetilde{\boldsymbol{w}} - \widetilde{\boldsymbol{v}}]]\!] \cdot \widetilde{\boldsymbol{m}} \right]_{\mathfrak{P}_s} . \tag{5.131}$$

Likewise, by invoking the version of the transport theorem in Eq. 5.109, the global format of the generic balance equation reads equivalently

$$\int_{\mathfrak{C}_m} \widetilde{J}\, \mathrm{d}_t \widetilde{\mathfrak{z}}_{\mathrm{s}}\, \mathrm{d}l = \int_{\mathfrak{C}_m} \left[\widetilde{\aleph}_{\mathrm{m}} + [\widehat{\mathfrak{z}}_{\mathrm{m}}^{\perp} \otimes \boldsymbol{V}] \cdot \widetilde{\boldsymbol{N}}_{\perp} + [\widehat{\mathfrak{z}}_{\mathrm{m}}^{\times} \otimes \boldsymbol{V}] \cdot \widetilde{\boldsymbol{N}}_{\times} \right] \mathrm{d}L \tag{5.132}$$

$$+ \sum_{\partial \mathfrak{C}_m} [\widetilde{\boldsymbol{Z}} + \widetilde{\mathfrak{z}}_{\mathrm{m}} \otimes \widetilde{\boldsymbol{V}}] \cdot \widetilde{\boldsymbol{N}} + \left[[\![\widetilde{\mathfrak{z}}_{\mathrm{m}} \otimes [\widetilde{\boldsymbol{W}} - \widetilde{\boldsymbol{V}}]]\!] \cdot \widetilde{\boldsymbol{M}} \right]_{\mathfrak{P}_m} .$$

Thus, upon localizing the various versions of the global format of the generic balance equation, the local generic balance equation *at regular points in the curve domain* may be expressed in *four* different but fully equivalent versions; see the arrangement in a tetragon in Table 5.6. Thereby, *dynamic versions* of the effective curve source densities have been introduced; see Table 5.5.

Moreover, in accordance with Eqs. 5.120, 5.129 and 5.121, 5.130, local generic balance equations *at boundary points* and *at singular points*, respectively, may, likewise, be expressed in *four* different, but fully equivalent versions; see the arrangement in the corresponding tetragons in Table 5.6.

Reference

1. Petryk H, Mroz Z (1986) Time derivatives of integrals and functionals defined on varying volume and surface domains. Arch Mech 38:697–724

Chapter 6
Kinematical 'Balances'*

8,163 m 28°32'58"N 84°33'43"E

Abstract This chapter applies the formats of the generic balances to the spatial and material tangent, cotangent, and measure maps to formulate what, for the sake of semantic unification, may be called kinematical 'balances'.

The spatial and material tangent, cotangent, and measure maps, i.e. the deformation gradients, their cofactors, and their determinants, are obviously no balanced quantities in the ordinary sense and thus there is basically no need to formulate corresponding kinematical 'balance' equations. However, motivated by concepts from homogenization, these maps may be considered as the *densities* of their (weighted) *volume-averaged*, i.e. homogenized, counterparts.

In doing so, it is then indeed illuminating to demonstrate that the formulation of global kinematical 'balances' of either spatial or material tangent, cotangent, and measure maps and their localized versions are—when following the standard procedure outlined for generic balances as formulated either in material or spatial control volumes—entirely *consistent* with the obvious rate-type kinematic compatibility conditions in the bulk and critical coherence conditions at singular surfaces.

Specifically, the proper definition of *extrinsic kinematic flux densities* per unit area on the external boundary together with the incorporation of critical coherence conditions at singular surfaces are crucial to *trivially* satisfy the emerging kinematical 'balances' (jump conditions) at singular coherent surfaces.

Summarizing, on the one hand, kinematical 'balances' are equivalent to rate-type compatibility conditions in the bulk and previously established critical coherence conditions for singular surfaces and are thus *seemingly* redundant. However, on the other hand, kinematical 'balances' are of major importance when formulating problems of, for example, shock propagation and the like.

P. Steinmann, *Spatial and Material Forces in Nonlinear Continuum Mechanics*,
Solid Mechanics and Its Applications 272,
https://doi.org/10.1007/978-3-030-89070-4_6

6.1 Spatial Tangent Map

The spatial tangent map computes as the material gradient of the spatial deformation map, aka as the spatial deformation gradient, and maps material into spatial vector-valued (differential) line elements.

6.1.1 Global Format (Material Control Volume)

The *'balance' of spatial tangent map* is *postulated* in global format as follows.

Global 'Balance' of Spatial Tangent Map (Material Control Volume):

$$\dot{\overline{\mathsf{F}}}(\mathfrak{B}_\mathrm{m}) = \overline{\mathsf{F}}^{\mathrm{ext}}(\mathfrak{B}_\mathrm{m}) \quad \forall \mathfrak{B}_\mathrm{m} \subseteq \mathcal{B}_\mathrm{m}. \tag{6.1}$$

Thereby, $\overline{\mathsf{F}}$ is expressed as the resultant over the material control volume $\mathfrak{B}_\mathrm{m}$ of the spatial tangent map $\boldsymbol{F}$, i.e. as its *weighted volume average*

$$\overline{\mathsf{F}}(\mathfrak{B}_\mathrm{m}) := \int_{\mathfrak{B}_\mathrm{m}} \boldsymbol{F} \,\mathrm{d}V. \tag{6.2}$$

Observe that the notation for the weighted volume average $\overline{\mathsf{F}}$ is a reminiscence of the notation used for macro-scale quantities in computational homogenization. Furthermore, $\overline{\mathsf{F}}^{\mathrm{ext}}$ consists solely of a boundary integral (any singular contributions are here ignored for the sake of presentation)

$$\overline{\mathsf{F}}^{\mathrm{ext}}(\mathfrak{B}_\mathrm{m}) := \int_{\partial\mathfrak{B}_\mathrm{m}} \widehat{\boldsymbol{F}}_\mathrm{m} \,\mathrm{d}A \tag{6.3}$$

in terms of a yet unspecified *extrinsic kinematic flux density* $\widehat{\boldsymbol{F}}_\mathrm{m}$ per unit area in the material configuration. The extrinsic kinematic flux density $\widehat{\boldsymbol{F}}_\mathrm{m}$ on the external boundary $\partial\mathcal{B}_\mathrm{m}$ will be determined in the sequel so as to obtain meaningful results for the postulated 'balance' of the spatial tangent map.

The extrinsic kinematic flux density along with the convective term stemming from the transport theorem is postulated to satisfy a Cauchy theorem for the *intrinsic kinematic flux density*

$$[\boldsymbol{v} \otimes \boldsymbol{I}] \cdot \boldsymbol{N} = \boldsymbol{v} \otimes \boldsymbol{N} := \widehat{\boldsymbol{F}}_\mathrm{m} - [\boldsymbol{F} \otimes \boldsymbol{W}] \cdot \boldsymbol{N} \quad \text{on} \quad \partial\mathfrak{B}_\mathrm{m}. \tag{6.4}$$

Note the postulated dyadic format of the intrinsic kinematic flux density.

As a consequence of the above definitions, the global format of the ‘balance’ of spatial tangent map expands as

$$\int_{\mathfrak{B}_m} \mathrm{D}_t \boldsymbol{F} \mathrm{d}V = \int_{\partial\mathfrak{B}_m} \boldsymbol{v} \otimes \boldsymbol{N} \mathrm{d}A + \int_{\mathfrak{S}_m} [\![\boldsymbol{F} \otimes \boldsymbol{W}]\!] \cdot \boldsymbol{M} \mathrm{d}A. \tag{6.5}$$

6.1.2 Local Format (Material Control Volume)

The results of localizing the global ‘balance’ of spatial tangent map are summarized as follows.

Local ‘Balance’ of Spatial Tangent Map (Material Control Volume):

i. Regular Points in the Domain

$$\mathrm{D}_t \boldsymbol{F} - \nabla_X \boldsymbol{v} = \boldsymbol{0} \quad \text{in} \quad \mathcal{B}_m. \tag{6.6}$$

ii. Regular Points on the Boundary

$$[\boldsymbol{F} \otimes \boldsymbol{W}] \cdot \boldsymbol{N} + \boldsymbol{v} \otimes \boldsymbol{N} = \widehat{\boldsymbol{F}}_m^{\text{ext}} \quad \text{on} \quad \partial\mathcal{B}_m. \tag{6.7}$$

iii. Regular Points at Singular Surfaces

$$[\![\boldsymbol{F} \otimes \boldsymbol{W}]\!] \cdot \boldsymbol{M} + [\![\boldsymbol{v}]\!] \otimes \boldsymbol{M} = \boldsymbol{0} \quad \text{on} \quad \mathcal{S}_m. \tag{6.8}$$

The local ‘balance’ of spatial tangent map allows the following conclusions:

i. In a smooth domain $\mathcal{B}_m$, the identity $\mathrm{Div}(\boldsymbol{v} \otimes \boldsymbol{I}) = \nabla_X \boldsymbol{v}$ is utilized for the process of localization. As a result, the rate-type kinematic compatibility condition $\mathrm{D}_t \boldsymbol{F} = \nabla_X \boldsymbol{v}$ between the material time derivative of the spatial tangent map and the material gradient of the spatial velocity is retrieved.
ii. On (the regular part of) the external boundary $\partial\mathcal{B}_m$, the postulated Cauchy theorem renders a meaningful result if the extrinsic kinematic flux density $\widehat{\boldsymbol{F}}_m^{\text{ext}}$ is indeed *defined* as

$$\widehat{\boldsymbol{F}}_m^{\text{ext}} := W_\perp \boldsymbol{F} + \boldsymbol{v} \otimes \boldsymbol{N}. \tag{6.9}$$

Thus, the boundary condition for regular points on $\partial\mathcal{B}_m$ degenerates to a trivial identity. However, remarkably, when restricted to singular surfaces, it results in the *cancelation* of extrinsic kinematic flux density

$$\widehat{\boldsymbol{F}}_m^{+} + \widehat{\boldsymbol{F}}_m^{-} = \boldsymbol{0} \quad \text{at} \quad \mathcal{S}_m \tag{6.10}$$

and translates into the jump condition across singular surfaces

$$[\![\boldsymbol{F}]\!] W_{\perp} + [\![\boldsymbol{v}]\!] \otimes \boldsymbol{M} = \boldsymbol{0} \quad \text{at} \quad \mathcal{S}_{\mathrm{m}}. \tag{6.11}$$

iii. At singular surfaces $\mathcal{S}_{\mathrm{m}}$, the jump condition may be rewritten with coherence condition VII $[\![\boldsymbol{v}]\!] = -[\![\boldsymbol{F}]\!] \cdot \boldsymbol{W}_{\perp}$ for a coherent singular surface as

$$W_{\perp} [\![\boldsymbol{F}]\!] \cdot \mathbb{P} = \boldsymbol{0}. \tag{6.12}$$

Clearly, this is coherence condition II $[\![\boldsymbol{F}]\!] \cdot \mathbb{P} = \boldsymbol{0}$ for a coherent singular surface, thus, the jump condition at $\mathcal{S}_{\mathrm{m}}$ is trivially satisfied.

6.1.3 Global Format (Spatial Control Volume)

Alternatively, the 'balance' of spatial tangent map is postulated in a global format as follows.

Global 'Balance' of Spatial Tangent Map (Spatial Control Volume):

$$\dot{\overline{\mathsf{F}}}(\mathfrak{B}_{\mathrm{s}}) = \overline{\mathsf{F}}^{\mathrm{ext}}(\mathfrak{B}_{\mathrm{s}}) \quad \forall \mathfrak{B}_{\mathrm{s}} \subseteq \mathcal{B}_{\mathrm{s}}. \tag{6.13}$$

Here, $\overline{\mathsf{F}}$ is expressed as the resultant, or rather the integral over the spatial control volume $\mathfrak{B}_{\mathrm{s}}$ of the transposed spatial cotangent map $j\boldsymbol{F} \equiv \boldsymbol{k}^{t}$, i.e.

$$\overline{\mathsf{F}}(\mathfrak{B}_{\mathrm{s}}) := \int_{\mathfrak{B}_{\mathrm{s}}} \boldsymbol{k}^{t} \mathrm{d}v. \tag{6.14}$$

Furthermore, $\overline{\mathsf{F}}^{\mathrm{ext}}$ consists solely of a boundary integral (any singular contributions are here ignored for the sake of presentation)

$$\overline{\mathsf{F}}^{\mathrm{ext}}(\mathfrak{B}_{\mathrm{s}}) := \int_{\partial \mathfrak{B}_{\mathrm{s}}} \widehat{\boldsymbol{F}}_{\mathrm{s}} \mathrm{d}a \tag{6.15}$$

in terms of the *extrinsic kinematic flux density* $\widehat{\boldsymbol{F}}_{\mathrm{s}} := \widehat{j} \widehat{\boldsymbol{F}}_{\mathrm{m}}$ per unit area in the spatial configuration.

The intrinsic kinematic flux density is defined as the Piola transformation $\boldsymbol{v} \otimes \boldsymbol{k} := [\boldsymbol{v} \otimes \boldsymbol{I}] \cdot \operatorname{cof} \boldsymbol{f}$, thus the corresponding Cauchy theorem for $[\boldsymbol{v} \otimes \boldsymbol{k}] \cdot \boldsymbol{n}$ reads in terms of the extrinsic kinematic flux density along with the convective term stemming from the transport theorem

$$[\boldsymbol{v} \otimes \boldsymbol{k}] \cdot \boldsymbol{n} := \widehat{\boldsymbol{F}}_{\mathrm{s}} - \big[\boldsymbol{k}^t \otimes [\boldsymbol{w} - \boldsymbol{v}]\big] \cdot \boldsymbol{n} \quad \text{on} \quad \partial\mathfrak{B}_{\mathrm{s}}. \tag{6.16}$$

Note the Piola-transformed format for the intrinsic kinematic flux density.

As a consequence of the above definitions, the global format of the 'balance' of spatial tangent map expands as

$$\int_{\mathfrak{B}_{\mathrm{s}}} \mathrm{d}_t \boldsymbol{k}^t \mathrm{d}v = \int_{\partial\mathfrak{B}_{\mathrm{s}}} [\boldsymbol{v} \otimes \boldsymbol{k} - \boldsymbol{k}^t \otimes \boldsymbol{v}] \cdot \boldsymbol{n} \mathrm{d}a + \int_{\mathfrak{S}_{\mathrm{s}}} [\![\boldsymbol{k}^t \otimes \boldsymbol{w}]\!] \cdot \boldsymbol{m} \, \mathrm{d}a. \tag{6.17}$$

6.1.4 Local Format (Spatial Control Volume)

The local 'balance' of spatial tangent map (per unit volume in the spatial configuration), expressed in terms of the Cauchy-type quantity $\boldsymbol{v} \otimes \boldsymbol{k} := [\boldsymbol{v} \otimes \boldsymbol{I}] \cdot \operatorname{cof} \boldsymbol{f}$ and the extrinsic kinematic flux density $\widehat{\boldsymbol{F}}_{\mathrm{s}}^{\mathrm{ext}} := \widehat{j}\,\widehat{\boldsymbol{F}}_{\mathrm{m}}^{\mathrm{ext}}$ per unit area in the spatial configuration, is summarized as follows.

Local 'Balance' of Spatial Tangent Map (Spatial Control Volume):

i. Regular Points in the Domain

$$\mathrm{d}_t \boldsymbol{k}^t - \operatorname{div}(\boldsymbol{v} \otimes \boldsymbol{k} - \boldsymbol{k}^t \otimes \boldsymbol{v}) = \boldsymbol{0} \quad \text{in} \quad \mathcal{B}_{\mathrm{s}}. \tag{6.18}$$

ii. Regular Points on the Boundary

$$\big[\boldsymbol{k}^t \otimes \boldsymbol{w} + [\boldsymbol{v} \otimes \boldsymbol{k} - \boldsymbol{k}^t \otimes \boldsymbol{v}]\big] \cdot \boldsymbol{n} = \widehat{\boldsymbol{F}}_{\mathrm{s}}^{\mathrm{ext}} \quad \text{on} \quad \partial\mathcal{B}_{\mathrm{s}}. \tag{6.19}$$

iii. Regular Points at Singular Surfaces

$$\big[\!\big[\boldsymbol{k}^t \otimes \boldsymbol{w} + [\boldsymbol{v} \otimes \boldsymbol{k} - \boldsymbol{k}^t \otimes \boldsymbol{v}]\big]\!\big] \cdot \boldsymbol{m} = \boldsymbol{0} \quad \text{at} \quad \mathcal{S}_{\mathrm{s}}. \tag{6.20}$$

The local 'balance' of spatial tangent map allows the following conclusions:

i. In a smooth domain $\mathcal{B}_{\mathrm{s}}$, the local transport theorem $j\,\mathrm{D}_t \boldsymbol{F} = \mathrm{d}_t \boldsymbol{k}^t + \operatorname{div}(\boldsymbol{k}^t \otimes \boldsymbol{v})$ and the Piola transformation $j\,\mathrm{Div}(\boldsymbol{v} \otimes \boldsymbol{I}) = \operatorname{div}(\boldsymbol{v} \otimes \boldsymbol{k})$ allow to retrieve the 'balance' $\mathrm{D}_t \boldsymbol{F} = \mathrm{Div}(\boldsymbol{v} \otimes \boldsymbol{I}) \equiv \nabla_X \boldsymbol{v}$.

ii. On (the regular part of) the external boundary $\partial\mathcal{B}_{\mathrm{m}}$, the identities $\boldsymbol{w} = \boldsymbol{v} + \boldsymbol{W} \cdot \boldsymbol{F}^t$ and $\boldsymbol{k}^t = j\,\boldsymbol{F}$ allow to rewrite the corresponding 'balance' as $[\boldsymbol{F} \otimes \boldsymbol{W} + \boldsymbol{v} \otimes \boldsymbol{I}] \cdot \boldsymbol{k} \cdot \boldsymbol{n} = \widehat{\boldsymbol{F}}_{\mathrm{s}}^{\mathrm{ext}}$. Then, with $\boldsymbol{k} \cdot \boldsymbol{n} = \widehat{j}\,\boldsymbol{N}$, the 'balance' $W_{\perp} \boldsymbol{F} + \boldsymbol{v} \otimes \boldsymbol{N} = \widehat{\boldsymbol{F}}_{\mathrm{m}}^{\mathrm{ext}}$ is retrieved.

iii. At singular surfaces $\mathcal{S}_s$, the identities $\boldsymbol{w} = \boldsymbol{v} + \boldsymbol{W} \cdot \boldsymbol{F}^t$ and $\boldsymbol{k}^t = j\,\boldsymbol{F}$ allow to rewrite the corresponding 'balance' as $[\![\boldsymbol{F} \otimes \boldsymbol{W} + \boldsymbol{v} \otimes \boldsymbol{I}]\!] \cdot \boldsymbol{k} \cdot \boldsymbol{m} = \boldsymbol{0}$. Then, with $\boldsymbol{k} \cdot \boldsymbol{m} = \widehat{j}\,\boldsymbol{M}$, the 'balance' $W_\perp [\![\boldsymbol{F}]\!] + [\![\boldsymbol{v}]\!] \otimes \boldsymbol{M} = \boldsymbol{0}$ is retrieved.

6.2 Spatial Cotangent Map

The spatial cotangent map computes as the cofactor of the spatial deformation gradient and maps material into spatial vector-valued-oriented (differential) area elements.

6.2.1 Global Formulation (Material Control Volume)

The *'balance' of spatial cotangent map* is *postulated* in global format as follows.

Global 'Balance' of Spatial Cotangent Map (Material Control Volume):

$$\dot{\overline{\mathbf{K}}}(\mathfrak{B}_m) = \overline{\mathbf{K}}^{\text{ext}}(\mathfrak{B}_m) \quad \forall \mathfrak{B}_m \subseteq \mathcal{B}_m. \tag{6.21}$$

Thereby, $\overline{\mathbf{K}}$ is expressed as the resultant, or rather the integral over the material control volume $\mathfrak{B}_m$ of the spatial cotangent map $\boldsymbol{K}$, i.e. its *weighted volume average*

$$\overline{\mathbf{K}}(\mathfrak{B}_m) := \int_{\mathfrak{B}_m} \boldsymbol{K} \,\mathrm{d}V. \tag{6.22}$$

Furthermore, $\overline{\mathbf{K}}^{\text{ext}}$ consists solely of a boundary integral (any singular contributions are here ignored for the sake of presentation)

$$\overline{\mathbf{K}}^{\text{ext}}(\mathfrak{B}_m) := \int_{\partial\mathfrak{B}_m} \widehat{\boldsymbol{K}}_m \,\mathrm{d}A \tag{6.23}$$

in terms of a yet unspecified *extrinsic kinematic flux density* $\widehat{\boldsymbol{K}}_m$ per unit area in the material configuration. The extrinsic kinematic flux density $\widehat{\boldsymbol{K}}_m$ on the external boundary $\partial\mathcal{B}_m$ will be determined in the sequel so as to obtain meaningful results for the postulated 'balance' of spatial cotangent map.

The extrinsic kinematic flux density along with the convective term stemming from the transport theorem are postulated to satisfy a Cauchy theorem for the *intrinsic kinematic flux density*

$$\big[\boldsymbol{F} \times\!\!\times [\boldsymbol{v} \otimes \boldsymbol{I}]\big] \cdot \boldsymbol{N} = \boldsymbol{F} \times\!\!\times [\boldsymbol{v} \otimes \boldsymbol{N}] := \widehat{\boldsymbol{K}}_{\mathrm{m}} - [\boldsymbol{K} \otimes \boldsymbol{W}] \cdot \boldsymbol{N} \quad \text{on} \quad \partial\mathfrak{B}_{\mathrm{m}}. \tag{6.24}$$

Note the postulated tensor cross-product format of the intrinsic kinematic flux density.

As a consequence of the above definitions, the global format of the 'balance' of spatial cotangent map expands as

$$\int_{\mathfrak{B}_{\mathrm{m}}} \mathrm{D}_t \boldsymbol{K} \mathrm{d}V = \int_{\partial\mathfrak{B}_{\mathrm{m}}} \boldsymbol{F} \times\!\!\times [\boldsymbol{v} \otimes \boldsymbol{N}] \mathrm{d}A + \int_{\mathfrak{S}_{\mathrm{m}}} [\![\boldsymbol{K} \otimes \boldsymbol{W}]\!] \cdot \boldsymbol{M} \mathrm{d}A. \tag{6.25}$$

6.2.2 *Local Format (Material Control Volume)*

The results of localizing the global 'balance' of spatial cotangent map are summarized as follows.

Local 'Balance' of Spatial Cotangent Map (Material Control Volume):

i. Regular Points in the Domain

$$\mathrm{D}_t \boldsymbol{K} - \boldsymbol{F} \times\!\!\times \nabla_X \boldsymbol{v} = \boldsymbol{0} \quad \text{in} \quad \mathcal{B}_{\mathrm{m}}. \tag{6.26}$$

ii. Regular Points on the Boundary

$$[\boldsymbol{K} \otimes \boldsymbol{W}] \cdot \boldsymbol{N} + \boldsymbol{F} \times\!\!\times [\boldsymbol{v} \otimes \boldsymbol{N}] = \widehat{\boldsymbol{K}}_{\mathrm{m}}^{\mathrm{ext}} \quad \text{on} \quad \partial\mathcal{B}_{\mathrm{m}}. \tag{6.27}$$

iii. Regular Points at Singular Surfaces

$$[\![\boldsymbol{K} \otimes \boldsymbol{W}]\!] \cdot \boldsymbol{M} + \{\boldsymbol{F}\} \times\!\!\times \big[[\![\boldsymbol{v}]\!] \otimes \boldsymbol{M}\big] = \boldsymbol{0} \quad \text{on} \quad \mathcal{S}_{\mathrm{m}}. \tag{6.28}$$

The local 'balance' of spatial cotangent map allows the following conclusions:

i. In a smooth domain $\mathcal{B}_{\mathrm{m}}$, the identity $\mathrm{Div}(\boldsymbol{F} \times\!\!\times [\boldsymbol{v} \otimes \boldsymbol{I}]) = \boldsymbol{F} \times\!\!\times \nabla_X \boldsymbol{v}$ is utilized for the process of localization. As a result, the rate-type kinematic compatibility condition $\mathrm{D}_t \boldsymbol{K} = \boldsymbol{F} \times\!\!\times \nabla_X \boldsymbol{v}$ between the material time derivative of the spatial cotangent map and the material gradient of the spatial velocity is retrieved.
ii. On (the regular part of) the external boundary $\partial\mathcal{B}_{\mathrm{m}}$, the postulated Cauchy theorem renders a meaningful result if the extrinsic kinematic flux density $\widehat{\boldsymbol{K}}_{\mathrm{m}}^{\mathrm{ext}}$ is indeed *defined* as

$$\widehat{\boldsymbol{K}}_{\mathrm{m}}^{\mathrm{ext}} := W_{\perp} \boldsymbol{K} + \boldsymbol{F} \times\!\!\times [\boldsymbol{v} \otimes \boldsymbol{N}]. \tag{6.29}$$

Thus, the boundary condition for regular points on $\partial\mathcal{B}_{\mathrm{m}}$ degenerates to a trivial identity. However, remarkably, when restricted to singular surfaces, it results in the *cancelation* of extrinsic kinematic flux density

$$\widehat{\boldsymbol{K}}_{\mathrm{m}}^{+}+\widehat{\boldsymbol{K}}_{\mathrm{m}}^{-}=\boldsymbol{0} \quad \text{at} \quad \mathcal{S}_{\mathrm{m}} \tag{6.30}$$

and translates into the jump condition across singular surfaces

$$[\![\boldsymbol{K}]\!]W_{\perp}+\{\boldsymbol{F}\} \times\!\!\times \big[[\![\boldsymbol{v}]\!]\otimes\boldsymbol{M}\big]=\boldsymbol{0} \quad \text{on} \quad \mathcal{S}_{\mathrm{m}}. \tag{6.31}$$

iii. At singular surfaces $\mathcal{S}_{\mathrm{m}}$, the jump condition may be rewritten with coherence condition VII $[\![\boldsymbol{v}]\!]=-[\![\boldsymbol{F}]\!]\cdot\boldsymbol{W}_{\perp}$ for a coherent singular surface as

$$W_{\perp}[\![\boldsymbol{K}]\!]-\{\boldsymbol{F}\} \times\!\!\times \big[[\![\boldsymbol{F}]\!]\cdot\boldsymbol{W}_{\perp}\otimes\boldsymbol{M}\big]=\boldsymbol{0}. \tag{6.32}$$

Expanding furthermore coherence condition II $[\![\boldsymbol{F}]\!]\cdot\mathbb{P}=\boldsymbol{0}$ for a coherent singular surface into $[\![\boldsymbol{F}]\!]=\boldsymbol{j}\otimes\boldsymbol{M}$ renders $[\![\boldsymbol{F}]\!]\cdot\boldsymbol{W}_{\perp}\otimes\boldsymbol{M}=W_{\perp}[\![\boldsymbol{F}]\!]$ and, together with $\{\boldsymbol{F}\} \times\!\!\times [\![\boldsymbol{F}]\!]=[\![\boldsymbol{K}]\!]$, the jump condition across $\mathcal{S}_{\mathrm{m}}$ is trivially satisfied.

6.2.3 Global Format (Spatial Control Volume)

Alternatively, the 'balance' of spatial cotangent map is postulated in a global format as follows.

Global 'Balance' of Spatial Cotangent Map (Spatial Control Volume):

$$\dot{\overline{\mathbf{K}}}(\mathfrak{B}_{\mathrm{s}})=\overline{\mathbf{K}}^{\mathrm{ext}}(\mathfrak{B}_{\mathrm{s}}) \quad \forall\mathfrak{B}_{\mathrm{s}}\subseteq\mathcal{B}_{\mathrm{s}}. \tag{6.33}$$

Here, $\overline{\mathbf{K}}$ is expressed as the resultant, or rather the integral over the spatial control volume $\mathfrak{B}_{\mathrm{s}}$ of the transposed spatial tangent map $j\boldsymbol{K}\equiv\boldsymbol{f}^{t}$, i.e.

$$\overline{\mathbf{K}}(\mathfrak{B}_{\mathrm{s}}):=\int_{\mathfrak{B}_{\mathrm{s}}}\boldsymbol{f}^{t}\,\mathrm{d}v. \tag{6.34}$$

Furthermore, $\overline{\mathbf{K}}^{\mathrm{ext}}$ consists solely of a boundary integral (any singular contributions are here ignored for the sake of presentation)

$$\overline{\mathbf{K}}^{\mathrm{ext}}(\mathfrak{B}_{\mathrm{s}}):=\int_{\partial\mathfrak{B}_{\mathrm{s}}}\widehat{\boldsymbol{K}}_{\mathrm{s}}\,\mathrm{d}a \tag{6.35}$$

in terms of the *extrinsic kinematic flux density* $\widehat{\boldsymbol{K}}_{\mathrm{s}} := \widehat{j\boldsymbol{K}}_{\mathrm{m}}$ per unit area in the spatial configuration.

The intrinsic kinematic flux density is defined as the Piola transformation $\boldsymbol{F} \times \times[\boldsymbol{v} \otimes \boldsymbol{k}] := \left[\boldsymbol{F} \times\!\!\times[\boldsymbol{v} \otimes \boldsymbol{I}]\right] \cdot \operatorname{cof} \boldsymbol{f}$, thus, the corresponding Cauchy theorem for $\left[\boldsymbol{F} \times\!\!\times[\boldsymbol{v} \otimes \boldsymbol{k}]\right] \cdot \boldsymbol{n}$ reads in terms of the extrinsic kinematic flux density along with the convective term stemming from the transport theorem

$$\left[\boldsymbol{F} \times\!\!\times[\boldsymbol{v} \otimes \boldsymbol{k}]\right] \cdot \boldsymbol{n} := \widehat{\boldsymbol{K}}_{\mathrm{s}} - \left[\boldsymbol{f}^t \otimes [\boldsymbol{w} - \boldsymbol{v}]\right] \cdot \boldsymbol{n} \quad \text{on} \quad \partial\mathfrak{B}_{\mathrm{s}}. \tag{6.36}$$

Note the Piola-transformed format for the intrinsic kinematic flux density.

As a consequence of the above definitions, the global format of the 'balance' of spatial cotangent map expands as

$$\int_{\mathfrak{B}_{\mathrm{s}}} \mathrm{d}_t \boldsymbol{f}^t \mathrm{d}v = \int_{\partial\mathfrak{B}_{\mathrm{s}}} \left[\boldsymbol{F} \times\!\!\times[\boldsymbol{v} \otimes \boldsymbol{k}] - \boldsymbol{f}^t \otimes \boldsymbol{v}\right] \cdot \boldsymbol{n} \mathrm{d}a + \int_{\mathfrak{S}_{\mathrm{s}}} [\![\boldsymbol{f}^t \otimes \boldsymbol{w}]\!] \cdot \boldsymbol{m}\, \mathrm{d}a. \tag{6.37}$$

6.2.4 Local Format (Spatial Control Volume)

The local 'balance' of spatial cotangent map (per unit volume in the spatial configuration), expressed in terms of the Cauchy-type quantity $\boldsymbol{F} \times\!\!\times[\boldsymbol{v} \otimes \boldsymbol{k}] := \boldsymbol{F} \times\!\!\times[\boldsymbol{v} \otimes \boldsymbol{I}] \cdot \operatorname{cof} \boldsymbol{f}$ and the extrinsic kinematic flux density $\widehat{\boldsymbol{K}}_{\mathrm{s}}^{\mathrm{ext}} := \widehat{j\boldsymbol{K}}_{\mathrm{m}}^{\mathrm{ext}}$ per unit area in the spatial configuration, is summarized as follows.

Local 'Balance' of Spatial Cotangent Map (Spatial Control Volume):

i. Regular Points in the Domain

$$\mathrm{d}_t \boldsymbol{f}^t - \operatorname{div}(\boldsymbol{F} \times\!\!\times[\boldsymbol{v} \otimes \boldsymbol{k}] - \boldsymbol{f}^t \otimes \boldsymbol{v}) = \boldsymbol{0} \quad \text{in} \quad \mathcal{B}_{\mathrm{s}}. \tag{6.38}$$

ii. Regular Points on the Boundary

$$\left[\boldsymbol{f}^t \otimes \boldsymbol{w} + \left[\boldsymbol{F} \times\!\!\times[\boldsymbol{v} \otimes \boldsymbol{k}] - \boldsymbol{f}^t \otimes \boldsymbol{v}\right]\right] \cdot \boldsymbol{n} = \widehat{\boldsymbol{K}}_{\mathrm{s}}^{\mathrm{ext}} \quad \text{on} \quad \partial\mathcal{B}_{\mathrm{s}}. \tag{6.39}$$

iii. Regular Points at Singular Surfaces

$$\left[\!\!\left[\boldsymbol{f}^t \otimes \boldsymbol{w} + \left[\boldsymbol{F} \times\!\!\times[\boldsymbol{v} \otimes \boldsymbol{k}] - \boldsymbol{f}^t \otimes \boldsymbol{v}\right]\right]\!\!\right] \cdot \boldsymbol{m} = \boldsymbol{0} \quad \text{at} \quad \mathcal{S}_{\mathrm{s}}. \tag{6.40}$$

The local 'balance' of spatial cotangent map allows the following conclusions:

i. In a smooth domain $\mathcal{B}_s$, the local transport theorem $j\,\mathrm{D}_t\boldsymbol{K} = \mathrm{d}_t\boldsymbol{f}^t + \mathrm{div}(\boldsymbol{f}^t \otimes \boldsymbol{v})$ and the Piola transformation $j\,\mathrm{Div}(\boldsymbol{F} \times\!\!\times [\boldsymbol{v} \otimes \boldsymbol{I}]) = \mathrm{div}(\boldsymbol{F} \times\!\!\times [\boldsymbol{v} \otimes \boldsymbol{k}])$ allow to retrieve the 'balance' $\mathrm{D}_t\boldsymbol{K} = \mathrm{Div}(\boldsymbol{F} \times\!\!\times [\boldsymbol{v} \otimes \boldsymbol{I}]) \equiv \boldsymbol{F} \times\!\!\times \nabla_X \boldsymbol{v}$.
ii. On (the regular part of) the external boundary $\partial\mathcal{B}_m$, the identities $\boldsymbol{w} = \boldsymbol{v} + \boldsymbol{W} \cdot \boldsymbol{F}^t$ and $\boldsymbol{f}^t = j\,\boldsymbol{K}$ allow to rewrite the corresponding 'balance' as $\big[\boldsymbol{K} \otimes \boldsymbol{W} + \boldsymbol{F} \times\!\!\times [\boldsymbol{v} \otimes \boldsymbol{I}]\big] \cdot \boldsymbol{k} \cdot \boldsymbol{n} = \widehat{\boldsymbol{K}}^{\mathrm{ext}}_{\mathrm{s}}$. Then, with $\boldsymbol{k} \cdot \boldsymbol{n} = \widehat{j}\,\boldsymbol{N}$, the 'balance' $W_\perp \boldsymbol{K} + \boldsymbol{F} \times\!\!\times [\boldsymbol{v} \otimes \boldsymbol{N}] = \widehat{\boldsymbol{K}}^{\mathrm{ext}}_{\mathrm{m}}$ is retrieved.
iii. At singular surfaces $\mathcal{S}_s$, the identities $\boldsymbol{w} = \boldsymbol{v} + \boldsymbol{W} \cdot \boldsymbol{F}^t$ and $\boldsymbol{f}^t = j\,\boldsymbol{K}$ allow to rewrite the corresponding 'balance' as $[\![\boldsymbol{K} \otimes \boldsymbol{W} + \boldsymbol{F} \times\!\!\times [\boldsymbol{v} \otimes \boldsymbol{I}]]\!] \cdot \boldsymbol{k} \cdot \boldsymbol{m} = \boldsymbol{0}$. Then, with $\boldsymbol{k} \cdot \boldsymbol{m} = \widehat{j}\,\boldsymbol{M}$, the 'balance' $W_\perp [\![\boldsymbol{K}]\!] + \{\boldsymbol{F}\} \times\!\!\times \big[[\![\boldsymbol{v}]\!] \otimes \boldsymbol{M}\big] = \boldsymbol{0}$ is retrieved, whereby $[\![\boldsymbol{F}]\!] \times\!\!\times [\{\boldsymbol{v}\} \otimes \boldsymbol{M}] = [\boldsymbol{j} \otimes \boldsymbol{M}] \times\!\!\times [\{\boldsymbol{v}\} \otimes \boldsymbol{M}] = \boldsymbol{0}$ has been used.

6.3 Spatial Measure Map

The spatial measure map computes as the determinant of the spatial deformation gradient and maps material into spatial scalar-valued (differential) volume elements.

6.3.1 Global Format (Material Control Volume)

The *'balance' of spatial measure map* is *postulated* in global format as follows.

Global 'Balance' of Spatial Measure Map (Material Control Volume):

$$\dot{\overline{\mathsf{J}}}(\mathfrak{B}_m) = \overline{\mathsf{J}}^{\mathrm{ext}}(\mathfrak{B}_m) \quad \forall \mathfrak{B}_m \subseteq \mathcal{B}_m. \tag{6.41}$$

Thereby, $\overline{\mathsf{J}}$ is expressed as the resultant, or rather the integral over the material control volume $\mathfrak{B}_m$ of the spatial measure map J, i.e. its *weighted volume average* or rather the total volume contained in the corresponding spatial control volume $\mathfrak{B}_s$

$$\overline{\mathsf{J}}(\mathfrak{B}_m) := \int_{\mathfrak{B}_m} J\,\mathrm{d}V = \int_{\mathfrak{B}_s} \mathrm{d}v. \tag{6.42}$$

Furthermore, $\overline{\mathsf{J}}^{\mathrm{ext}}$ consists solely of a boundary integral (any singular contributions are here ignored for the sake of presentation)

$$\overline{\mathrm{J}}^{\text{ext}}(\mathfrak{B}_{\text{m}}) := \int_{\partial\mathfrak{B}_{\text{m}}} \widehat{J}_{\text{m}} \mathrm{d}A \tag{6.43}$$

in terms of a yet unspecified *extrinsic kinematic flux density* $\widehat{J}_{\text{m}}$ per unit area in the material configuration. The extrinsic kinematic flux density $\widehat{J}_{\text{m}}$ at the external boundary $\partial\mathcal{B}_{\text{m}}$ will be determined in the sequel so as to obtain meaningful results for the postulated 'balance' of spatial measure map.

The extrinsic kinematic flux density along with the convective term stemming from the transport theorem are postulated to satisfy a Cauchy theorem for the *intrinsic kinematic flux density*

$$\big[\boldsymbol{K} : [\boldsymbol{v} \otimes \boldsymbol{I}]\big] \cdot \boldsymbol{N} = \boldsymbol{K} : [\boldsymbol{v} \otimes \boldsymbol{N}] := \widehat{J}_{\text{m}} - [J\boldsymbol{W}] \cdot \boldsymbol{N} \quad \text{on} \quad \partial\mathfrak{B}_{\text{m}}. \tag{6.44}$$

Note the postulated double contraction format of the intrinsic kinematic flux density.

As a consequence of the above definitions, the global format of the 'balance' of spatial measure map expands as

$$\int_{\mathfrak{B}_{\text{m}}} \mathrm{D}_t J \mathrm{d}V = \int_{\partial\mathfrak{B}_{\text{m}}} \boldsymbol{K} : [\boldsymbol{v} \otimes \boldsymbol{N}] \mathrm{d}A + \int_{\mathfrak{S}_{\text{m}}} [\![J\boldsymbol{W}]\!] \cdot \boldsymbol{M} \mathrm{d}A. \tag{6.45}$$

6.3.2 *Local Format (Material Control Volume)*

The results of localizing the global 'balance' of spatial measure map are summarized as follows.

Local 'Balance' of Spatial Measure Map (Material Control Volume):

i. Regular Points in the Domain

$$\mathrm{D}_t J - \boldsymbol{K} : \nabla_X \boldsymbol{v} = \boldsymbol{0} \quad \text{in} \quad \mathcal{B}_{\text{m}}. \tag{6.46}$$

ii. Regular Points on the Boundary

$$[J\boldsymbol{W}] \cdot \boldsymbol{N} + \boldsymbol{K} : [\boldsymbol{v} \otimes \boldsymbol{N}] = \widehat{J}_{\text{m}}^{\text{ext}} \quad \text{on} \quad \partial\mathcal{B}_{\text{m}}. \tag{6.47}$$

iii. Regular Points at Singular Surfaces

$$[\![J\boldsymbol{W}]\!] \cdot \boldsymbol{M} + \{\boldsymbol{K}\} : \big[[\![\boldsymbol{v}]\!] \otimes \boldsymbol{M}\big] = \boldsymbol{0} \quad \text{on} \quad \mathcal{S}_{\text{m}}. \tag{6.48}$$

The local ‘balance’ of spatial measure map allows the following conclusions:

i. In a smooth domain $\mathcal{B}_{\text{m}}$, the identity $\text{Div}(\boldsymbol{K} : [\boldsymbol{v} \otimes \boldsymbol{I}]) = \boldsymbol{K} : \nabla_X \boldsymbol{v}$ is utilized for the process of localization. As a result, the rate-type kinematic compatibility condition $\text{D}_t J = \boldsymbol{K} : \nabla_X \boldsymbol{v}$ between the material time derivative of the spatial measure map and the material gradient of the spatial velocity is retrieved.
ii. On (the regular part of) the external boundary $\partial\mathcal{B}_{\text{m}}$, the postulated Cauchy theorem renders a meaningful result if the extrinsic kinematic flux density $\widehat{J}_{\text{m}}^{\text{ext}}$ is indeed *defined* as

$$\widehat{J}_{\text{m}}^{\text{ext}} := W_\perp J + \boldsymbol{K} : [\boldsymbol{v} \otimes \boldsymbol{N}]. \tag{6.49}$$

Thus, the boundary condition for regular points on $\partial\mathcal{B}_{\text{m}}$ degenerates to a trivial identity. However, remarkably, when restricted to singular surfaces, it results in the *cancelation* of extrinsic kinematic flux density

$$\widehat{J}_{\text{m}}^{+} + \widehat{J}_{\text{m}}^{-} = 0 \quad \text{at} \quad \mathcal{S}_{\text{m}} \tag{6.50}$$

and translates into the jump condition across singular surfaces

$$[\![J]\!] W_\perp + \{\boldsymbol{K}\} : \big[[\![\boldsymbol{v}]\!] \otimes \boldsymbol{M}\big] = \boldsymbol{0} \quad \text{at} \quad \mathcal{S}_{\text{m}}. \tag{6.51}$$

iii. At singular surfaces $\mathcal{S}_{\text{m}}$, the jump condition may be rewritten with coherence condition VII $[\![\boldsymbol{v}]\!] = -[\![\boldsymbol{F}]\!] \cdot \boldsymbol{W}_\perp$ for a coherent singular surface as

$$W_\perp [\![J]\!] - \{\boldsymbol{K}\} : \big[[\![\boldsymbol{F}]\!] \cdot \boldsymbol{W}_\perp \otimes \boldsymbol{M}\big] = \boldsymbol{0}. \tag{6.52}$$

Expanding furthermore the coherence condition II $[\![\boldsymbol{F}]\!] \cdot \mathbb{P} = \boldsymbol{0}$ for a coherent singular surface into $[\![\boldsymbol{F}]\!] = \boldsymbol{j} \otimes \boldsymbol{M}$ renders $[\![\boldsymbol{F}]\!] \cdot \boldsymbol{W}_\perp \otimes \boldsymbol{M} = W_\perp [\![\boldsymbol{F}]\!]$ and, together with coherence condition V $\{\boldsymbol{K}\} : [\![\boldsymbol{F}]\!] = [\![J]\!]$ for a coherent singular surface, the jump condition across $\mathcal{S}_{\text{m}}$ is trivially satisfied.

6.3.3 Global Format (Spatial Control Volume)

Alternatively, the ‘balance’ of spatial measure map is postulated in a global format as follows.

Global ‘Balance’ of Spatial Measure Map (Spatial Control Volume):

$$\dot{\mathsf{J}}(\mathfrak{B}_{\text{s}}) = \overline{\mathsf{J}}^{\text{ext}}(\mathfrak{B}_{\text{s}}) \quad \forall \mathfrak{B}_{\text{s}} \subseteq \mathcal{B}_{\text{s}}. \tag{6.53}$$

Here, $\overline{\mathsf{J}}$ is expressed as the resultant, or rather the integral over the spatial control volume $\mathfrak{B}_\mathrm{s}$ of the spatial scalar-valued unit map $j\,J \equiv 1$, i.e.

$$\overline{\mathsf{J}}(\mathfrak{B}_\mathrm{s}) := \int_{\mathfrak{B}_\mathrm{s}} 1 \mathrm{d}v. \tag{6.54}$$

Furthermore, $\overline{\mathsf{J}}^{\mathrm{ext}}$ consists solely of a boundary integral (any singular contributions are here ignored for the sake of presentation)

$$\overline{\mathsf{J}}^{\mathrm{ext}}(\mathfrak{B}_\mathrm{s}) := \int_{\partial\mathfrak{B}_\mathrm{s}} \widehat{J}_\mathrm{s} \mathrm{d}a \tag{6.55}$$

in terms of the *extrinsic kinematic flux density* $\widehat{J}_\mathrm{s} := \widehat{j}\,\widehat{J}_\mathrm{m}$ per unit area in the spatial configuration.

The intrinsic kinematic flux density is defined as the Piola transformation $\boldsymbol{v} \equiv \boldsymbol{K} : [\boldsymbol{v} \otimes \boldsymbol{k}] := \left[\boldsymbol{K} : [\boldsymbol{v} \otimes \boldsymbol{I}]\right] \cdot \operatorname{cof} \boldsymbol{f}$, thus, the corresponding Cauchy theorem for $\boldsymbol{v} \cdot \boldsymbol{n} \equiv \left[\boldsymbol{K} : [\boldsymbol{v} \otimes \boldsymbol{k}]\right] \cdot \boldsymbol{n}$ reads in terms of the extrinsic kinematic flux density along with the convective term stemming from the transport theorem

$$\boldsymbol{v} \cdot \boldsymbol{n} \equiv \left[\boldsymbol{K} : [\boldsymbol{v} \otimes \boldsymbol{k}]\right] \cdot \boldsymbol{n} := \widehat{J}_\mathrm{s} - [\boldsymbol{w} - \boldsymbol{v}] \cdot \boldsymbol{n} \quad \text{on} \quad \partial\mathfrak{B}_\mathrm{s}. \tag{6.56}$$

Note the Piola-transformed format for the intrinsic kinematic flux density.

As a consequence of the above definitions, the global format of the 'balance' of spatial measure map expands as the trivial identity

$$\int_{\mathfrak{B}_\mathrm{s}} \mathrm{d}_t 1 \mathrm{d}v = \int_{\partial\mathfrak{B}_\mathrm{s}} \left[\boldsymbol{K} : [\boldsymbol{v} \otimes \boldsymbol{k}] - \boldsymbol{v}\right] \cdot \boldsymbol{n} \mathrm{d}a + \int_{\mathfrak{S}_\mathrm{s}} [\![\boldsymbol{w}]\!] \cdot \boldsymbol{m} \mathrm{d}a. \tag{6.57}$$

6.3.4 Local Format (Spatial Control Volume)

The local 'balance' of spatial measure map (per unit volume in the spatial configuration), expressed in terms of the Cauchy-type quantity $\boldsymbol{v} \equiv \boldsymbol{K} : [\boldsymbol{v} \otimes \boldsymbol{k}] := \boldsymbol{K} : [\boldsymbol{v} \otimes \boldsymbol{I}] \cdot \operatorname{cof} \boldsymbol{f}$ and the extrinsic kinematic flux density $\widehat{J}_\mathrm{s}^{\mathrm{ext}} := \widehat{j}\,\widehat{J}_\mathrm{m}^{\mathrm{ext}}$ per unit area in the spatial configuration, is summarized as follows.

Local 'Balance' of Spatial Measure Map (Spatial Control Volume):

i. Regular Points in the Domain

$$\mathrm{d}_t 1 - \operatorname{div}(\boldsymbol{K} : [\boldsymbol{v} \otimes \boldsymbol{k}] - \boldsymbol{v}) = 0 \quad \text{in} \quad \mathcal{B}_\mathrm{s}. \tag{6.58}$$

ii. Regular Points on the Boundary

$$\left[\boldsymbol{w} + \left[\boldsymbol{K} : [\boldsymbol{v} \otimes \boldsymbol{k}] - \boldsymbol{v}\right]\right] \cdot \boldsymbol{n} = \widehat{J}_{\mathrm{s}}^{\mathrm{ext}} \quad \text{on} \quad \partial\mathcal{B}_{\mathrm{s}}. \tag{6.59}$$

iii. Regular Points at Singular Surfaces

$$\left[\!\left[\boldsymbol{w} + \left[\boldsymbol{K} : [\boldsymbol{v} \otimes \boldsymbol{k}] - \boldsymbol{v}\right]\right]\!\right] \cdot \boldsymbol{m} = 0 \quad \text{at} \quad \mathcal{S}_{\mathrm{s}}. \tag{6.60}$$

The local 'balance' of spatial measure map allows the following conclusions:

i. In a smooth domain $\mathcal{B}_{\mathrm{s}}$, the local transport theorem $j\,\mathrm{D}_t J = \mathrm{d}_t 1 + \mathrm{div}(\boldsymbol{v})$ and the Piola transformation $j\,\mathrm{Div}(\boldsymbol{K} : [\boldsymbol{v} \otimes \boldsymbol{I}]) = \mathrm{div}(\boldsymbol{K} : [\boldsymbol{v} \otimes \boldsymbol{k}])$ allow to retrieve the 'balance' $\mathrm{D}_t J = \mathrm{Div}(\boldsymbol{K} : [\boldsymbol{v} \otimes \boldsymbol{I}]) \equiv \boldsymbol{K} : \nabla_X \boldsymbol{v}$.
ii. On (the regular part of) the external boundary $\partial\mathcal{B}_{\mathrm{m}}$, the identities $\boldsymbol{w} = \boldsymbol{v} + \boldsymbol{W} \cdot \boldsymbol{F}^t$ and $1 = j\,J$ allow to rewrite the corresponding 'balance' as $\left[J\boldsymbol{W} + \boldsymbol{K} : [\boldsymbol{v} \otimes \boldsymbol{I}]\right] \cdot \boldsymbol{k} \cdot \boldsymbol{n} = \widehat{J}_{\mathrm{s}}^{\mathrm{ext}}$. Then, with $\boldsymbol{k} \cdot \boldsymbol{n} = \widehat{j}\,\boldsymbol{N}$, the 'balance' $W_{\perp} J + \boldsymbol{K} : [\boldsymbol{v} \otimes \boldsymbol{N}] = \widehat{J}_{\mathrm{m}}^{\mathrm{ext}}$ is retrieved.
iii. At singular surfaces $\mathcal{S}_{\mathrm{s}}$, the identities $\boldsymbol{w} = \boldsymbol{v} + \boldsymbol{W} \cdot \boldsymbol{F}^t$ and $1 = j\,J$ allow to rewrite the corresponding 'balance' as $[\![J\boldsymbol{W} + \boldsymbol{K} : [\boldsymbol{v} \otimes \boldsymbol{I}]]\!] \cdot \boldsymbol{k} \cdot \boldsymbol{m} = \boldsymbol{0}$. Then, with $\boldsymbol{k} \cdot \boldsymbol{m} = \widehat{j}\,\boldsymbol{M}$, the 'balance' $W_{\perp}[\![J]\!] + \{\boldsymbol{K}\} : \left[[\![\boldsymbol{v}]\!] \otimes \boldsymbol{M}\right] = \boldsymbol{0}$ is retrieved, whereby $[\![\boldsymbol{K}]\!] : [\{\boldsymbol{v}\} \otimes \boldsymbol{M}] = \boldsymbol{0}$ has been used.

6.4 Summary: 'Balances' of Spatial Maps

The above 'balances' of the spatial deformation measures, i.e. the spatial tangent, cotangent, and measure maps, display an illuminating structure and natural sequence, both in material and spatial control volumes, when recalling that $\boldsymbol{I} = \frac{1}{0!}\boldsymbol{F}^0$, $\boldsymbol{F} = \frac{1}{1!}\boldsymbol{F}$, $\boldsymbol{K} = \frac{1}{2!}\boldsymbol{F} \times\!\!\times \boldsymbol{F}$, and $J = \frac{1}{3!}[\boldsymbol{F} \times\!\!\times \boldsymbol{F}] : \boldsymbol{F}$ are, respectively, constant, linear, quadratic, and cubic in $\boldsymbol{F}$. To bring out these correlations more clearly and for convenience of comparison, Tables 6.1 and 6.2 re-express and juxtapose the local 'balances' for the spatial deformation measures. These kinematical 'balances' proved, for example, beneficial in novel computational approaches to nonlinear solid dynamics (Bonet et al. [1–3] and references therein).

Table 6.1 Juxtaposition of local kinematic 'balance' laws for spatial deformation measures and material control volume

'Balances' of Spatial Tangent, Cotangent and Measure Map (Material Control Volume):

i. Regular Points in the Domain $\mathcal{B}_{\mathrm{m}}$

$$
\begin{aligned}
\mathrm{D}_t\left[\frac{1}{3!}[\boldsymbol{F}\times\!\!\!\times\boldsymbol{F}]:\boldsymbol{F}\right] &= \left[\frac{1}{2!}\boldsymbol{F}\times\!\!\!\times\boldsymbol{F}\right]:\nabla_X\boldsymbol{v} = \mathrm{Div}\left(\left[\frac{1}{2!}\boldsymbol{F}\times\!\!\!\times\boldsymbol{F}\right]:[\boldsymbol{v}\otimes\boldsymbol{I}]\right)\\
\mathrm{D}_t\left[\frac{1}{2!}\boldsymbol{F}\times\!\!\!\times\boldsymbol{F}\right] &= \left[\frac{1}{1!}\boldsymbol{F}\right]\times\!\!\!\times\nabla_X\boldsymbol{v} = \mathrm{Div}\left(\left[\frac{1}{1!}\boldsymbol{F}\right]\times\!\!\!\times[\boldsymbol{v}\otimes\boldsymbol{I}]\right)\\
\mathrm{D}_t\left[\frac{1}{1!}\boldsymbol{F}\right] &= \left[\frac{1}{0!}\boldsymbol{I}\right]\cdot\nabla_X\boldsymbol{v} = \mathrm{Div}\left(\left[\frac{1}{0!}\boldsymbol{I}\right]\cdot[\boldsymbol{v}\otimes\boldsymbol{I}]\right)
\end{aligned}
$$

ii. Regular Points on the Boundary $\partial\mathcal{B}_{\mathrm{m}}$

$$
\begin{aligned}
\left[\frac{1}{3!}[\boldsymbol{F}\times\!\!\!\times\boldsymbol{F}]:\boldsymbol{F}\right]W_\perp + \left[\frac{1}{2!}\boldsymbol{F}\times\!\!\!\times\boldsymbol{F}\right]:[\boldsymbol{v}\otimes\boldsymbol{N}] &= \widehat{J}_{\mathrm{m}}^{\mathrm{ext}}\\
\left[\frac{1}{2!}\boldsymbol{F}\times\!\!\!\times\boldsymbol{F}\right]W_\perp + \left[\frac{1}{1!}\boldsymbol{F}\right]\times\!\!\!\times[\boldsymbol{v}\otimes\boldsymbol{N}] &= \widehat{\boldsymbol{K}}_{\mathrm{m}}^{\mathrm{ext}}\\
\left[\frac{1}{1!}\boldsymbol{F}\right]W_\perp + \left[\frac{1}{0!}\boldsymbol{I}\right]\cdot[\boldsymbol{v}\otimes\boldsymbol{N}] &= \widehat{\boldsymbol{F}}_{\mathrm{m}}^{\mathrm{ext}}
\end{aligned}
$$

iii. Regular Points at Singular Surfaces $\mathcal{S}_{\mathrm{m}}$

$$
\begin{aligned}
\left[\!\!\left[\frac{1}{3!}[\boldsymbol{F}\times\!\!\!\times\boldsymbol{F}]:\boldsymbol{F}\right]\!\!\right]W_\perp + \left\{\frac{1}{2!}\boldsymbol{F}\times\!\!\!\times\boldsymbol{F}\right\}:[[\![\boldsymbol{v}]\!]\otimes\boldsymbol{M}] &= 0\\
\left[\!\!\left[\frac{1}{2!}\boldsymbol{F}\times\!\!\!\times\boldsymbol{F}\right]\!\!\right]W_\perp + \left\{\frac{1}{1!}\boldsymbol{F}\right\}\times\!\!\!\times[[\![\boldsymbol{v}]\!]\otimes\boldsymbol{M}] &= \boldsymbol{0}\\
\left[\!\!\left[\frac{1}{1!}\boldsymbol{F}\right]\!\!\right]W_\perp + \left\{\frac{1}{0!}\boldsymbol{I}\right\}\cdot[[\![\boldsymbol{v}]\!]\otimes\boldsymbol{M}] &= \boldsymbol{0}
\end{aligned}
$$

Table 6.2 Juxtaposition of local kinematic 'balance' laws for spatial deformation measures and spatial control volume

'Balances' of Spatial Tangent, Cotangent and Measure Map (Spatial Control Volume):

i. Regular Points in the Domain $\mathcal{B}_s$

$$\mathrm{d}_t\left[\frac{j}{3!}[\boldsymbol{F}\times\!\!\!\times\boldsymbol{F}]:\boldsymbol{F}\right]=\operatorname{div}\left(\left[\frac{1}{2!}\boldsymbol{F}\times\!\!\!\times\boldsymbol{F}\right]:[\boldsymbol{v}\otimes\boldsymbol{k}]-\left[\frac{j}{3!}[\boldsymbol{F}\times\!\!\!\times\boldsymbol{F}]:\boldsymbol{F}\right]\otimes\boldsymbol{v}\right)$$

$$\mathrm{d}_t\left[\frac{j}{2!}\boldsymbol{F}\times\!\!\!\times\boldsymbol{F}\right]=\operatorname{div}\left(\left[\frac{1}{1!}\boldsymbol{F}\right]\times\!\!\!\times[\boldsymbol{v}\otimes\boldsymbol{k}]-\left[\frac{j}{2!}\boldsymbol{F}\times\!\!\!\times\boldsymbol{F}\right]\otimes\boldsymbol{v}\right)$$

$$\mathrm{d}_t\left[\frac{j}{1!}\boldsymbol{F}\right]=\operatorname{div}\left(\left[\frac{1}{0!}\boldsymbol{I}\right]\cdot[\boldsymbol{v}\otimes\boldsymbol{k}]-\left[\frac{j}{1!}\boldsymbol{F}\right]\otimes\boldsymbol{v}\right)$$

ii. Regular Points on the Boundary $\partial\mathcal{B}_s$

$$\left[\left[\frac{j}{3!}[\boldsymbol{F}\times\!\!\!\times\boldsymbol{F}]:\boldsymbol{F}\right][\boldsymbol{w}-\boldsymbol{v}]+\left[\frac{1}{2!}\boldsymbol{F}\times\!\!\!\times\boldsymbol{F}\right]:[\boldsymbol{v}\otimes\boldsymbol{k}]\right]\cdot\boldsymbol{n}=\widehat{J}_s^{\mathrm{ext}}$$

$$\left[\left[\frac{j}{2!}\boldsymbol{F}\times\!\!\!\times\boldsymbol{F}\right][\boldsymbol{w}-\boldsymbol{v}]+\left[\frac{1}{1!}\boldsymbol{F}\right]\times\!\!\!\times[\boldsymbol{v}\otimes\boldsymbol{k}]\right]\cdot\boldsymbol{n}=\widehat{\boldsymbol{K}}_s^{\mathrm{ext}}$$

$$\left[\left[\frac{j}{1!}\boldsymbol{F}\right][\boldsymbol{w}-\boldsymbol{v}]+\left[\frac{1}{0!}\boldsymbol{I}\right]\cdot[\boldsymbol{v}\otimes\boldsymbol{k}]\right]\cdot\boldsymbol{n}=\widehat{\boldsymbol{F}}_s^{\mathrm{ext}}$$

ii. Regular Points at Singular Surfaces $\mathcal{S}_s$

$$\left[\!\left[\left[\frac{j}{3!}[\boldsymbol{F}\times\!\!\!\times\boldsymbol{F}]:\boldsymbol{F}\right][\boldsymbol{w}-\boldsymbol{v}]+\left[\frac{1}{2!}\boldsymbol{F}\times\!\!\!\times\boldsymbol{F}\right]:[\boldsymbol{v}\otimes\boldsymbol{k}]\right]\!\right]\cdot\boldsymbol{m}=0$$

$$\left[\!\left[\left[\frac{j}{2!}\boldsymbol{F}\times\!\!\!\times\boldsymbol{F}\right][\boldsymbol{w}-\boldsymbol{v}]+\left[\frac{1}{1!}\boldsymbol{F}\right]\times\!\!\!\times[\boldsymbol{v}\otimes\boldsymbol{k}]\right]\!\right]\cdot\boldsymbol{m}=\boldsymbol{0}$$

$$\left[\!\left[\left[\frac{j}{1!}\boldsymbol{F}\right][\boldsymbol{w}-\boldsymbol{v}]+\left[\frac{1}{0!}\boldsymbol{I}\right]\cdot[\boldsymbol{v}\otimes\boldsymbol{k}]\right]\!\right]\cdot\boldsymbol{m}=\boldsymbol{0}$$

6.5 Material Tangent Map

The material tangent map computes as the spatial gradient of the material deformation map, aka as the material deformation gradient, and maps spatial into material vector-valued (differential) line elements.

6.5.1 Global Format (Spatial Control Volume)

The *'balance' of material tangent map* is *postulated* in global format as follows.

Global 'Balance' of Material Tangent Map (Spatial Control Volume):

$$\dot{\bar{\mathbf{f}}}(\mathfrak{B}_s) = \bar{\mathbf{f}}^{\text{ext}}(\mathfrak{B}_s) \quad \forall \mathfrak{B}_s \subseteq \mathcal{B}_s. \tag{6.61}$$

Thereby, $\bar{\mathbf{f}}$ is expressed as the resultant, or rather the integral over the spatial control volume $\mathfrak{B}_s$ of the material tangent map $\boldsymbol{f}$, i.e. its *weighted volume average*

$$\bar{\mathbf{f}}(\mathfrak{B}_s) := \int_{\mathfrak{B}_s} \boldsymbol{f} \, \mathrm{d}v. \tag{6.62}$$

Furthermore, $\bar{\mathbf{f}}^{\text{ext}}$ consists solely of a boundary integral (any singular contributions are here ignored for the sake of presentation)

$$\bar{\mathbf{f}}^{\text{ext}}(\mathfrak{B}_s) := \int_{\partial\mathfrak{B}_s} \widehat{\boldsymbol{f}}_s \, \mathrm{d}a \tag{6.63}$$

in terms of the *extrinsic kinematic flux density* $\widehat{\boldsymbol{f}}_s$ per unit area in the spatial configuration. The extrinsic kinematic flux density $\widehat{\boldsymbol{f}}_s$ on the external boundary $\partial\mathcal{B}_s$ is determined so as to obtain meaningful results for the postulated 'balance' of material tangent map.

The extrinsic kinematic flux density along with the convective term stemming from the transport theorem is postulated to satisfy a Cauchy theorem for the *intrinsic kinematic flux density*

$$[\boldsymbol{V} \otimes \boldsymbol{i}] \cdot \boldsymbol{n} = \boldsymbol{V} \otimes \boldsymbol{n} := \widehat{\boldsymbol{f}}_s - [\boldsymbol{f} \otimes \boldsymbol{w}] \cdot \boldsymbol{n} \quad \text{on} \quad \partial\mathfrak{B}_s. \tag{6.64}$$

Note the postulated dyadic format of the intrinsic kinematic flux density.

As a consequence of the above definitions, the global format of the 'balance' of material tangent map expands as

$$\int_{\mathfrak{B}_s} \mathrm{d}_t \boldsymbol{f} \mathrm{d}v = \int_{\partial\mathfrak{B}_s} \boldsymbol{V} \otimes \boldsymbol{n} \mathrm{d}a + \int_{\mathfrak{S}_s} [\![\boldsymbol{f} \otimes \boldsymbol{w}]\!] \cdot \boldsymbol{m} \mathrm{d}a. \tag{6.65}$$

6.5.2 Local Format (Spatial Control Volume)

Following the same procedure as for the spatial tangent map that resulted in Eqs. 6.6–6.8, the results of localizing the corresponding global 'balance' of material tangent map are summarized as follows.

Local 'Balance' of Material Tangent Map (Spatial Control Volume):

i. Regular Points in the Domain

$$\mathrm{d}_t \boldsymbol{f} - \nabla_x \boldsymbol{V} = \boldsymbol{0} \quad \text{in} \quad \mathcal{B}_s. \tag{6.66}$$

ii. Regular Points on the Boundary

$$[\boldsymbol{f} \otimes \boldsymbol{w}] \cdot \boldsymbol{n} + \boldsymbol{V} \otimes \boldsymbol{n} = \widehat{\boldsymbol{f}}_s^{\mathrm{ext}} \quad \text{on} \quad \partial\mathcal{B}_s. \tag{6.67}$$

iii. Regular Points at Singular Surfaces

$$[\![\boldsymbol{f} \otimes \boldsymbol{w}]\!] \cdot \boldsymbol{m} + [\![\boldsymbol{V}]\!] \otimes \boldsymbol{m} = \boldsymbol{0} \quad \text{on} \quad \mathcal{S}_s. \tag{6.68}$$

The local 'balance' of material tangent map allows analogous conclusions as those following Eqs. 6.6–6.8.

6.5.3 Global Format (Material Control Volume)

Alternatively, the 'balance' of material tangent map is postulated in a global format as follows.

Global 'Balance' of Material Tangent Map (Material Control Volume):

$$\dot{\overline{\mathbf{f}}}(\mathfrak{B}_\mathrm{m}) = \overline{\mathbf{f}}^{\mathrm{ext}}(\mathfrak{B}_\mathrm{m}) \quad \forall \mathfrak{B}_\mathrm{m} \subseteq \mathcal{B}_\mathrm{m}. \tag{6.69}$$

Here, $\overline{\mathbf{f}}$ is expressed as the resultant, or rather the integral over the material control volume $\mathfrak{B}_\mathrm{m}$ of the transposed material cotangent map $J\boldsymbol{f} \equiv \boldsymbol{K}^t$, i.e.

$$\bar{\mathsf{f}}(\mathfrak{B}_\mathrm{m}) := \int_{\mathfrak{B}_\mathrm{m}} \boldsymbol{K}^t \mathrm{d}V. \tag{6.70}$$

Furthermore, $\bar{\mathsf{f}}^{\mathrm{ext}}$ consists solely of a boundary integral (any singular contributions are here ignored for the sake of presentation)

$$\bar{\mathsf{f}}^{\mathrm{ext}}(\mathfrak{B}_\mathrm{m}) := \int_{\partial\mathfrak{B}_\mathrm{m}} \widehat{\boldsymbol{f}}_\mathrm{m} \mathrm{d}A \tag{6.71}$$

in terms of the *extrinsic kinematic flux density* $\widehat{\boldsymbol{f}}_\mathrm{m} := \widehat{J\boldsymbol{f}}_\mathrm{s}$ per unit area in the material configuration.

The intrinsic kinematic flux density is defined as the Piola transformation $\boldsymbol{V} \otimes \boldsymbol{K} := [\boldsymbol{V} \otimes \boldsymbol{i}] \cdot \operatorname{cof} \boldsymbol{F}$, thus, the corresponding Cauchy theorem for $[\boldsymbol{V} \otimes \boldsymbol{K}] \cdot \boldsymbol{N}$ reads in terms of the extrinsic kinematic flux density along with the convective term stemming from the transport theorem

$$[\boldsymbol{V} \otimes \boldsymbol{K}] \cdot \boldsymbol{N} := \widehat{\boldsymbol{f}}_\mathrm{m} - \big[\boldsymbol{K}^t \otimes [\boldsymbol{W} - \boldsymbol{V}]\big] \cdot \boldsymbol{N} \quad \text{on} \quad \partial\mathfrak{B}_\mathrm{m}. \tag{6.72}$$

Note the Piola-transformed format for the intrinsic kinematic flux density.

As a consequence of the above definitions, the global format of the 'balance' of material tangent map expands as

$$\int_{\mathfrak{B}_\mathrm{m}} \mathrm{D}_t \boldsymbol{K}^t \mathrm{d}V = \int_{\partial\mathfrak{B}_\mathrm{m}} [\boldsymbol{V} \otimes \boldsymbol{K} - \boldsymbol{K}^t \otimes \boldsymbol{V}] \cdot \boldsymbol{N} \mathrm{d}A + \int_{\mathfrak{S}_\mathrm{m}} [\![\boldsymbol{K}^t \otimes \boldsymbol{W}]\!] \cdot \boldsymbol{M} \mathrm{d}A. \tag{6.73}$$

6.5.4 *Local Format (Material Control Volume)*

The local 'balance' of material tangent map (per unit volume in the material configuration), expressed in terms of the Cauchy-type quantity $\boldsymbol{V} \otimes \boldsymbol{K} := [\boldsymbol{V} \otimes \boldsymbol{i}] \cdot \operatorname{cof} \boldsymbol{F}$ and the extrinsic kinematic flux density $\widehat{\boldsymbol{f}}^{\mathrm{ext}}_\mathrm{m} := \widehat{J\boldsymbol{f}}^{\mathrm{ext}}_\mathrm{s}$ per unit area in the material configuration, is summarized as follows.

Local 'Balance' of Material Tangent Map (Material Control Volume):

i. Regular Points in the Domain

$$\mathrm{D}_t \boldsymbol{K}^t - \operatorname{Div}(\boldsymbol{V} \otimes \boldsymbol{K} - \boldsymbol{K}^t \otimes \boldsymbol{V}) = \boldsymbol{0} \quad \text{in} \quad \mathcal{B}_\mathrm{m}. \tag{6.74}$$

ii. Regular Points on the Boundary

$$\big[\boldsymbol{K}^t \otimes \boldsymbol{W} + [\boldsymbol{V} \otimes \boldsymbol{K} - \boldsymbol{K}^t \otimes \boldsymbol{V}]\big] \cdot \boldsymbol{N} = \widehat{\boldsymbol{f}}^{\mathrm{ext}}_\mathrm{m} \quad \text{on} \quad \partial\mathcal{B}_\mathrm{m}. \tag{6.75}$$

iii. Regular Points at Singular Surfaces

$$[\![\boldsymbol{K}^t \otimes \boldsymbol{W} + [\boldsymbol{V} \otimes \boldsymbol{K} - \boldsymbol{K}^t \otimes \boldsymbol{V}]]\!] \cdot \boldsymbol{M} = \boldsymbol{0} \quad \text{at} \quad \mathcal{S}_\mathrm{m}. \tag{6.76}$$

The local 'balance' of material tangent map allows analogous conclusions as those following Eqs. 6.18–6.20.

6.6 Material Cotangent Map

The material cotangent map computes as the cofactor of the material deformation gradient and maps spatial into material vector-valued-oriented (differential) area elements.

6.6.1 *Global Format (Spatial Control Volume)*

The *'balance' of material cotangent map* is *postulated* in global format as follows.

Global 'Balance' of Material Cotangent Map (Spatial Control Volume):

$$\dot{\overline{\mathbf{k}}}(\mathfrak{B}_\mathrm{s}) = \overline{\mathbf{k}}^{\mathrm{ext}}(\mathfrak{B}_\mathrm{s}) \quad \forall \mathfrak{B}_\mathrm{s} \subseteq \mathcal{B}_\mathrm{s}. \tag{6.77}$$

Thereby, $\overline{\mathbf{k}}$ is expressed as the resultant, or rather the integral over the spatial control volume $\mathfrak{B}_\mathrm{s}$ of the material area map $\boldsymbol{k}$, i.e. its *weighted volume average*

$$\overline{\mathbf{k}}(\mathfrak{B}_\mathrm{s}) := \int_{\mathfrak{B}_\mathrm{s}} \boldsymbol{k} \mathrm{d}v. \tag{6.78}$$

Furthermore, $\overline{\mathbf{k}}^{\mathrm{ext}}$ consists solely of a boundary integral (any singular contributions are here ignored for the sake of presentation)

$$\overline{\mathbf{k}}^{\mathrm{ext}}(\mathfrak{B}_\mathrm{s}) := \int_{\partial \mathfrak{B}_\mathrm{s}} \widehat{\boldsymbol{k}}_\mathrm{s} \mathrm{d}a \tag{6.79}$$

in terms of the *extrinsic kinematic flux density* $\widehat{\boldsymbol{k}}_{\text{s}}$ per unit area in the spatial configuration. The extrinsic kinematic flux density $\widehat{\boldsymbol{k}}_{\text{s}}$ on the external boundary $\partial\mathcal{B}_{\text{s}}$ is determined so as to obtain meaningful results for the postulated 'balance' of material cotangent map.

The extrinsic kinematic flux density along with the convective term stemming from the transport theorem is postulated to satisfy a Cauchy theorem for the *intrinsic kinematic flux density*

$$\left[\boldsymbol{f} \times\!\!\times [\boldsymbol{V} \otimes \boldsymbol{i}] \right] \cdot \boldsymbol{n} = \boldsymbol{f} \times\!\!\times [\boldsymbol{V} \otimes \boldsymbol{n}] := \widehat{\boldsymbol{k}}_{\text{s}} - [\boldsymbol{k} \otimes \boldsymbol{w}] \cdot \boldsymbol{n} \quad \text{on} \quad \partial\mathfrak{B}_{\text{s}}. \tag{6.80}$$

Note the postulated tensor cross-product format of the intrinsic kinematic flux density.

As a consequence of the above definitions, the global format of the 'balance' of material cotangent map expands as

$$\int_{\mathfrak{B}_{\text{s}}} \mathrm{d}_t \boldsymbol{k} \mathrm{d}v = \int_{\partial\mathfrak{B}_{\text{s}}} \boldsymbol{f} \times\!\!\times [\boldsymbol{V} \otimes \boldsymbol{n}] \mathrm{d}a + \int_{\mathfrak{S}_{\text{s}}} [\![\boldsymbol{k} \otimes \boldsymbol{w}]\!] \cdot \boldsymbol{m} \mathrm{d}a. \tag{6.81}$$

6.6.2 *Local Format (Spatial Control Volume)*

Following the same procedure as for the spatial cotangent map that resulted in Eqs. 6.26–6.28, the results of localizing the corresponding global 'balance' of material cotangent map are summarized as follows.

Local 'Balance' of Material Cotangent Map (Spatial Control Volume):

i. Regular Points in the Domain

$$\mathrm{d}_t \boldsymbol{k} - \boldsymbol{f} \times\!\!\times \nabla_x \boldsymbol{V} = \boldsymbol{0} \quad \text{in} \quad \mathcal{B}_{\text{s}}. \tag{6.82}$$

ii. Regular Points on the Boundary

$$[\boldsymbol{k} \otimes \boldsymbol{w}] \cdot \boldsymbol{n} + \boldsymbol{f} \times\!\!\times [\boldsymbol{V} \otimes \boldsymbol{n}] = \widehat{\boldsymbol{k}}_{\text{s}}^{\text{ext}} \quad \text{on} \quad \partial\mathcal{B}_{\text{s}}. \tag{6.83}$$

iii. Regular Points at Singular Surfaces

$$[\![\boldsymbol{k} \otimes \boldsymbol{w}]\!] \cdot \boldsymbol{m} + \{\boldsymbol{f}\} \times\!\!\times \left[[\![\boldsymbol{V}]\!] \otimes \boldsymbol{m} \right] = \boldsymbol{0} \quad \text{on} \quad \mathcal{S}_{\text{s}}. \tag{6.84}$$

The local 'balance' of material cotangent map allows analogous conclusions as those following Eqs. 6.26–6.28.

6.6.3 Global Format (Material Control Volume)

Alternatively, the 'balance' of material cotangent map is postulated in a global format as follows.

Global 'Balance' of Material Cotangent Map (Material Control Volume):

$$\dot{\overline{\mathbf{k}}}(\mathfrak{B}_\mathrm{m}) = \overline{\mathbf{k}}^{\mathrm{ext}}(\mathfrak{B}_\mathrm{m}) \quad \forall \mathfrak{B}_\mathrm{m} \subseteq \mathcal{B}_\mathrm{m}. \tag{6.85}$$

Here, $\overline{\mathbf{k}}$ is expressed as the resultant, or rather the integral over the material control volume $\mathfrak{B}_\mathrm{m}$ of the transposed material tangent map $J\boldsymbol{k} \equiv \boldsymbol{F}^t$, i.e.

$$\overline{\mathbf{k}}(\mathfrak{B}_\mathrm{m}) := \int_{\mathfrak{B}_\mathrm{m}} \boldsymbol{F}^t \mathrm{d}V. \tag{6.86}$$

Furthermore, $\overline{\mathbf{k}}^{\mathrm{ext}}$ consists solely of a boundary integral (any singular contributions are here ignored for the sake of presentation)

$$\overline{\mathbf{k}}^{\mathrm{ext}}(\mathfrak{B}_\mathrm{m}) := \int_{\partial\mathfrak{B}_\mathrm{m}} \widehat{\boldsymbol{k}}_\mathrm{m} \mathrm{d}A \tag{6.87}$$

in terms of the *extrinsic kinematic flux density* $\widehat{\boldsymbol{k}}_\mathrm{m} := \widehat{J\widehat{\boldsymbol{k}}_\mathrm{s}}$ per unit area in the material configuration.

The intrinsic kinematic flux density is defined as the Piola transformation $\boldsymbol{f} \times\!\!\times [\boldsymbol{V} \otimes \boldsymbol{K}] := \left[\boldsymbol{f} \times\!\!\times [\boldsymbol{V} \otimes \boldsymbol{i}]\right] \cdot \operatorname{cof} \boldsymbol{F}$, thus, the corresponding Cauchy theorem for $\left[\boldsymbol{f} \times\!\!\times [\boldsymbol{V} \otimes \boldsymbol{K}]\right] \cdot \boldsymbol{N}$ reads in terms of the extrinsic kinematic flux density along with the convective term stemming from the transport theorem

$$\left[\boldsymbol{f} \times\!\!\times [\boldsymbol{V} \otimes \boldsymbol{K}]\right] \cdot \boldsymbol{N} := \widehat{\boldsymbol{k}}_\mathrm{m} - \left[\boldsymbol{F}^t \otimes [\boldsymbol{W} - \boldsymbol{V}]\right] \cdot \boldsymbol{N} \quad \text{on} \quad \partial\mathfrak{B}_\mathrm{m}. \tag{6.88}$$

Note the Piola-transformed format for the intrinsic kinematic flux density.

As a consequence of the above definitions, the global format of the 'balance' of material cotangent map expands as

$$\int_{\mathfrak{B}_\mathrm{m}} \mathrm{D}_t \boldsymbol{F}^t \mathrm{d}V = \int_{\partial\mathfrak{B}_\mathrm{m}} \left[\boldsymbol{f} \times\!\!\times [\boldsymbol{V} \otimes \boldsymbol{K}] - \boldsymbol{F}^t \otimes \boldsymbol{V}\right] \cdot \boldsymbol{N} \mathrm{d}A \tag{6.89}$$
$$+ \int_{\mathfrak{S}_\mathrm{m}} [\![\boldsymbol{F}^t \otimes \boldsymbol{W}]\!] \cdot \boldsymbol{M} \mathrm{d}A.$$

6.6.4 *Local Format (Material Control Volume)*

The local 'balance' of material cotangent map (per unit volume in the material configuration), expressed in terms of the Cauchy-type quantity $\boldsymbol{f} \times\!\!\times [\boldsymbol{V} \otimes \boldsymbol{K}] := \boldsymbol{f} \times\!\!\times [\boldsymbol{V} \otimes \boldsymbol{i}] \cdot \operatorname{cof} \boldsymbol{F}$ and the extrinsic kinematic flux density $\widehat{\boldsymbol{k}}_{\mathrm{m}}^{\mathrm{ext}} := \widehat{J \boldsymbol{k}}_{\mathrm{s}}^{\mathrm{ext}}$ per unit area in the material configuration, is summarized as follows.

Local 'Balance' of Material Cotangent Map (Material Control Volume):

i. Regular Points in the Domain

$$\mathrm{D}_t \boldsymbol{F}^t - \mathrm{Div}(\boldsymbol{f} \times\!\!\times [\boldsymbol{V} \otimes \boldsymbol{K}] - \boldsymbol{F}^t \otimes \boldsymbol{V}) = \boldsymbol{0} \quad \text{in} \quad \mathcal{B}_{\mathrm{m}}. \tag{6.90}$$

ii. Regular Points on the Boundary

$$\left[\boldsymbol{F}^t \otimes \boldsymbol{W} + \left[\boldsymbol{f} \times\!\!\times [\boldsymbol{V} \otimes \boldsymbol{K}] - \boldsymbol{F}^t \otimes \boldsymbol{V}\right]\right] \cdot \boldsymbol{N} = \widehat{\boldsymbol{k}}_{\mathrm{m}}^{\mathrm{ext}} \quad \text{on} \quad \partial\mathcal{B}_{\mathrm{m}}. \tag{6.91}$$

iii. Regular Points at Singular Surfaces

$$\left[\!\left[\boldsymbol{F}^t \otimes \boldsymbol{W} + \left[\boldsymbol{f} \times\!\!\times [\boldsymbol{V} \otimes \boldsymbol{K}] - \boldsymbol{F}^t \otimes \boldsymbol{V}\right]\right]\!\right] \cdot \boldsymbol{M} = \boldsymbol{0} \quad \text{at} \quad \mathcal{S}_{\mathrm{m}}. \tag{6.92}$$

The local 'balance' of spatial cotangent map allows analogous conclusions as those following Eqs. 6.38–6.40.

6.7 Material Measure Map

The material measure map computes as the determinant of the material deformation gradient and maps spatial into material scalar-valued (differential) volume elements.

6.7.1 *Global Format (Spatial Control Volume)*

The *'balance' of material measure map* is *postulated* in global format as follows.

Global 'Balance' of Material Measure Map (Spatial Control Volume):

$$\dot{\mathrm{j}}(\mathfrak{B}_{\mathrm{s}}) = \mathrm{j}^{\mathrm{ext}}(\mathfrak{B}_{\mathrm{s}}) \quad \forall \mathfrak{B}_{\mathrm{s}} \subseteq \mathcal{B}_{\mathrm{s}}. \tag{6.93}$$

Thereby, $\bar{\mathrm{j}}$ is expressed as the resultant, or rather the integral over the spatial control volume $\mathfrak{B}_s$ of the material volume map j, i.e. its *weighted volume average* or rather the total volume contained in the corresponding material control volume $\mathfrak{B}_m$

$$\bar{\mathrm{j}}(\mathfrak{B}_s) := \int_{\mathfrak{B}_s} j\mathrm{d}v = \int_{\mathfrak{B}_m} \mathrm{d}V. \tag{6.94}$$

Furthermore, $\bar{\mathrm{j}}^{\mathrm{ext}}$ consists solely of a boundary integral (any singular contributions are here ignored for the sake of presentation)

$$\bar{\mathrm{j}}^{\mathrm{ext}}(\mathfrak{B}_s) := \int_{\partial\mathfrak{B}_s} \widehat{j}_s \mathrm{d}a \tag{6.95}$$

in terms of the *extrinsic kinematic flux density* $\widehat{j}_s$ per unit area in the material configuration. The extrinsic kinematic flux density $\widehat{j}_s$ at the external boundary $\partial\mathcal{B}_s$ is determined so as to obtain meaningful results for the postulated 'balance' of material measure map.

The extrinsic kinematic flux density along with the convective term stemming from the transport theorem is postulated to satisfy a Cauchy theorem for the *intrinsic kinematic flux density*

$$\big[\boldsymbol{k} : [\boldsymbol{V} \otimes \boldsymbol{i}]\big] \cdot \boldsymbol{n} = \boldsymbol{k} : [\boldsymbol{V} \otimes \boldsymbol{n}] := \widehat{j}_s - [j\boldsymbol{w}] \cdot \boldsymbol{n} \quad \text{on} \quad \partial\mathfrak{B}_s. \tag{6.96}$$

Note the postulated double contraction format of the intrinsic kinematic flux density.

As a consequence, the global format of the 'balance' of material measure map expands as

$$\int_{\mathfrak{B}_s} \mathrm{d}_t j \mathrm{d}v = \int_{\partial\mathfrak{B}_s} \boldsymbol{k} : [\boldsymbol{V} \otimes \boldsymbol{n}]\mathrm{d}a + \int_{\mathfrak{S}_s} [\![j\boldsymbol{w}]\!] \cdot \boldsymbol{m}\mathrm{d}a. \tag{6.97}$$

6.7.2 Local Format (Spatial Control Volume)

Following the same procedure as for the spatial measure map that resulted in Eqs. 6.46–6.48, the results of localizing the corresponding global 'balance' of material measure map are summarized as follows.

Local 'Balance' of Material Measure Map (Spatial Control Volume):

i. Regular Points in the Domain

$$\mathrm{d}_t j - \boldsymbol{k} : \nabla_x \boldsymbol{V} = \boldsymbol{0} \quad \text{in} \quad \mathcal{B}_\mathrm{s}. \tag{6.98}$$

ii. Regular Points on the Boundary

$$[j\boldsymbol{w}] \cdot \boldsymbol{n} + \boldsymbol{k} : [\boldsymbol{V} \otimes \boldsymbol{n}] = \widehat{j}_\mathrm{s}^\mathrm{ext} \quad \text{on} \quad \partial\mathcal{B}_\mathrm{s}. \tag{6.99}$$

iii. Regular Points at Singular Surfaces

$$[\![j\boldsymbol{w}]\!] \cdot \boldsymbol{m} + \{\boldsymbol{k}\} : \left[[\![\boldsymbol{V}]\!] \otimes \boldsymbol{m}\right] = \boldsymbol{0} \quad \text{on} \quad \mathcal{S}_\mathrm{s}. \tag{6.100}$$

The local 'balance' of material measure map allows analogous conclusions as those following Eqs. 6.46–6.48.

6.7.3 Global Format (Material Control Volume)

Alternatively, the 'balance' of material measure map is postulated in a global format as follows.

Global 'Balance' of Material Measure Map (Material Control Volume):

$$\dot{\bar{\mathrm{j}}}(\mathfrak{B}_\mathrm{m}) = \bar{\mathrm{j}}^\mathrm{ext}(\mathfrak{B}_\mathrm{m}) \quad \forall \mathfrak{B}_\mathrm{m} \subseteq \mathcal{B}_\mathrm{m}. \tag{6.101}$$

Here, $\bar{\mathrm{j}}$ is expressed as the resultant, or rather the integral over the material control volume $\mathfrak{B}_\mathrm{m}$ of the material scalar-valued unit map $J\,j \equiv 1$, i.e.

$$\bar{\mathrm{j}}(\mathfrak{B}_\mathrm{m}) := \int_{\mathfrak{B}_\mathrm{m}} 1 \mathrm{d}V. \tag{6.102}$$

Furthermore, $\bar{\mathrm{j}}^\mathrm{ext}$ consists solely of a boundary integral (any singular contributions are here ignored for the sake of presentation)

$$\bar{\mathrm{j}}^\mathrm{ext}(\mathfrak{B}_\mathrm{m}) := \int_{\partial\mathfrak{B}_\mathrm{m}} \widehat{j}_\mathrm{m} \mathrm{d}A \tag{6.103}$$

in terms of the *extrinsic kinematic flux density* $\widehat{j}_{\mathrm{m}} := \widehat{J}\,\widehat{j}_{\mathrm{s}}$ per unit area in the material configuration.

The intrinsic kinematic flux density is defined as the Piola transformation $\boldsymbol{V} \equiv \boldsymbol{k} : [\boldsymbol{V} \otimes \boldsymbol{K}] := \big[\boldsymbol{k} : [\boldsymbol{V} \otimes \boldsymbol{i}]\big] \cdot \operatorname{cof} \boldsymbol{F}$, thus, the corresponding Cauchy theorem for $\boldsymbol{V} \cdot \boldsymbol{N} \equiv \big[\boldsymbol{k} : [\boldsymbol{V} \otimes \boldsymbol{K}]\big] \cdot \boldsymbol{N}$ reads in terms of the extrinsic kinematic flux density along with the convective term stemming from the transport theorem

$$\boldsymbol{V} \cdot \boldsymbol{N} \equiv \big[\boldsymbol{k} : [\boldsymbol{V} \otimes \boldsymbol{K}]\big] \cdot \boldsymbol{N} := \widehat{j}_{\mathrm{m}} - [\boldsymbol{W} - \boldsymbol{V}] \cdot \boldsymbol{N} \quad \text{on} \quad \partial\mathfrak{B}_{\mathrm{m}}. \tag{6.104}$$

Note the Piola-transformed format for the intrinsic kinematic flux density.

As a consequence of the above definitions, the global format of the 'balance' of material measure map expands as the trivial identity

$$\int_{\mathfrak{B}_{\mathrm{m}}} \mathrm{D}_t 1 \mathrm{d}V = \int_{\partial\mathfrak{B}_{\mathrm{m}}} \big[\boldsymbol{k} : [\boldsymbol{V} \otimes \boldsymbol{k}] - \boldsymbol{V}\big] \cdot \boldsymbol{N} \mathrm{d}A + \int_{\mathfrak{S}_{\mathrm{m}}} [\![\boldsymbol{W}]\!] \cdot \boldsymbol{M} \mathrm{d}A. \tag{6.105}$$

6.7.4 Local Format (Material Control Volume)

The local 'balance' of material measure map (per unit volume in the material configuration), expressed in terms of the Cauchy-type quantity $\boldsymbol{V} \equiv \boldsymbol{k} : [\boldsymbol{V} \otimes \boldsymbol{K}] := \boldsymbol{k} : [\boldsymbol{V} \otimes \boldsymbol{i}] \cdot \operatorname{cof} \boldsymbol{F}$ and the extrinsic kinematic flux density $\widehat{j}^{\mathrm{ext}}_{\mathrm{m}} := \widehat{J}\,\widehat{j}^{\mathrm{ext}}_{\mathrm{s}}$ per unit area in the material configuration, is summarized as follows.

Local 'Balance' of Material Measure Map (Material Control Volume):

i. Regular Points in the Domain

$$\mathrm{D}_t 1 - \mathrm{Div}(\boldsymbol{k} : [\boldsymbol{V} \otimes \boldsymbol{K}] - \boldsymbol{V}) = 0 \quad \text{in} \quad \mathcal{B}_{\mathrm{m}}. \tag{6.106}$$

ii. Regular Points on the Boundary

$$\Big[\boldsymbol{W} + \big[\boldsymbol{k} : [\boldsymbol{V} \otimes \boldsymbol{K}] - \boldsymbol{V}\big]\Big] \cdot \boldsymbol{N} = \widehat{j}^{\mathrm{ext}}_{\mathrm{m}} \quad \text{on} \quad \partial\mathcal{B}_{\mathrm{m}}. \tag{6.107}$$

iii. Regular Points at Singular Surfaces

$$\Big[\!\Big[\boldsymbol{W} + \big[\boldsymbol{k} : [\boldsymbol{V} \otimes \boldsymbol{K}] - \boldsymbol{V}\big]\Big]\!\Big] \cdot \boldsymbol{M} = 0 \quad \text{at} \quad \mathcal{S}_{\mathrm{m}}. \tag{6.108}$$

The local 'balance' of material measure map allows analogous conclusions as those following Eqs. 6.58–6.60.

Table 6.3 Juxtaposition of local kinematic 'balance' laws for material deformation measures and spatial control volume

'Balances' of Material Tangent, Cotangent and Measure Map (Spatial Control Volume):

i. Regular Points in the Domain $\mathcal{B}_s$

$$
\begin{aligned}
\mathrm{d}_t\left[\frac{1}{3!}[\boldsymbol{f}\times\!\!\!\!\times\boldsymbol{f}] : \boldsymbol{f}\right] &= \left[\frac{1}{2!}\boldsymbol{f}\times\!\!\!\!\times\boldsymbol{f}\right] : \nabla_x \boldsymbol{V} = \operatorname{div}\left(\left[\frac{1}{2!}\boldsymbol{f}\times\!\!\!\!\times\boldsymbol{f}\right] : [\boldsymbol{V}\otimes\boldsymbol{i}]\right)\\
\mathrm{d}_t\left[\frac{1}{2!}\boldsymbol{f}\times\!\!\!\!\times\boldsymbol{f}\right] &= \left[\frac{1}{1!}\boldsymbol{f}\right] \times\!\!\!\!\times \nabla_x \boldsymbol{V} = \operatorname{div}\left(\left[\frac{1}{1!}\boldsymbol{f}\right] \times\!\!\!\!\times [\boldsymbol{V}\otimes\boldsymbol{i}]\right)\\
\mathrm{d}_t\left[\frac{1}{1!}\boldsymbol{f}\right] &= \left[\frac{1}{0!}\boldsymbol{i}\right] \cdot \nabla_x \boldsymbol{V} = \operatorname{div}\left(\left[\frac{1}{0!}\boldsymbol{i}\right] \cdot [\boldsymbol{V}\otimes\boldsymbol{i}]\right)
\end{aligned}
$$

ii. Regular Points on the Boundary $\partial\mathcal{B}_s$

$$
\begin{aligned}
\left[\frac{1}{3!}[\boldsymbol{f}\times\!\!\!\!\times\boldsymbol{f}] : \boldsymbol{f}\right] w_\perp + \left[\frac{1}{2!}\boldsymbol{f}\times\!\!\!\!\times\boldsymbol{f}\right] : [\boldsymbol{V}\otimes\boldsymbol{n}] &= \widehat{j}_s^{\text{ext}}\\
\left[\frac{1}{2!}\boldsymbol{f}\times\!\!\!\!\times\boldsymbol{f}\right] w_\perp + \left[\frac{1}{1!}\boldsymbol{f}\right] \times\!\!\!\!\times [\boldsymbol{V}\otimes\boldsymbol{n}] &= \widehat{\boldsymbol{k}}_s^{\text{ext}}\\
\left[\frac{1}{1!}\boldsymbol{f}\right] w_\perp + \left[\frac{1}{0!}\boldsymbol{i}\right] \cdot [\boldsymbol{V}\otimes\boldsymbol{n}] &= \widehat{\boldsymbol{f}}_s^{\text{ext}}
\end{aligned}
$$

iii. Regular Points at Singular Surfaces $\mathcal{S}_s$

$$
\begin{aligned}
\left[\!\!\left[\frac{1}{3!}[\boldsymbol{f}\times\!\!\!\!\times\boldsymbol{f}] : \boldsymbol{f}\right]\!\!\right] w_\perp + \left\{\frac{1}{2!}\boldsymbol{f}\times\!\!\!\!\times\boldsymbol{f}\right\} : [[\![\boldsymbol{V}]\!]\otimes\boldsymbol{m}] &= 0\\
\left[\!\!\left[\frac{1}{2!}\boldsymbol{f}\times\!\!\!\!\times\boldsymbol{f}\right]\!\!\right] w_\perp + \left\{\frac{1}{1!}\boldsymbol{f}\right\} \times\!\!\!\!\times [[\![\boldsymbol{V}]\!]\otimes\boldsymbol{m}] &= \boldsymbol{0}\\
\left[\!\!\left[\frac{1}{1!}\boldsymbol{f}\right]\!\!\right] w_\perp + \left\{\frac{1}{0!}\boldsymbol{i}\right\} \cdot [[\![\boldsymbol{V}]\!]\otimes\boldsymbol{m}] &= \boldsymbol{0}
\end{aligned}
$$

Table 6.4 Juxtaposition of local kinematic 'balance' laws for material deformation measures and material control volume

'Balances' of Material Tangent, Cotangent and Measure Map (Material Control Volume):

i. Regular Points in the Domain $\mathcal{B}_{\mathrm{m}}$

$$\mathrm{D}_t\left[\frac{J}{3!}[\boldsymbol{f} \mathbin{⨳} \boldsymbol{f}] : \boldsymbol{f}\right] = \mathrm{Div}\left(\left[\frac{1}{2!}\boldsymbol{f} \mathbin{⨳} \boldsymbol{f}\right] : [\boldsymbol{V} \otimes \boldsymbol{K}] - \left[\frac{J}{3!}[\boldsymbol{f} \mathbin{⨳} \boldsymbol{f}] : \boldsymbol{f}\right] \otimes \boldsymbol{V}\right)$$

$$\mathrm{D}_t\left[\frac{J}{2!}\boldsymbol{f} \mathbin{⨳} \boldsymbol{f}\right] = \mathrm{Div}\left(\left[\frac{1}{1!}\boldsymbol{f}\right] \mathbin{⨳} [\boldsymbol{V} \otimes \boldsymbol{K}] - \left[\frac{J}{2!}\boldsymbol{f} \mathbin{⨳} \boldsymbol{f}\right] \otimes \boldsymbol{V}\right)$$

$$\mathrm{D}_t\left[\frac{J}{1!}\boldsymbol{f}\right] = \mathrm{Div}\left(\left[\frac{1}{0!}\boldsymbol{i}\right] \cdot [\boldsymbol{V} \otimes \boldsymbol{K}] - \left[\frac{J}{1!}\boldsymbol{f}\right] \otimes \boldsymbol{V}\right)$$

ii. Regular Points on the Boundary $\partial\mathcal{B}_{\mathrm{m}}$

$$\left[\left[\frac{J}{3!}[\boldsymbol{f} \mathbin{⨳} \boldsymbol{f}] : \boldsymbol{f}\right][\boldsymbol{W} - \boldsymbol{V}] + \left[\frac{1}{2!}\boldsymbol{f} \mathbin{⨳} \boldsymbol{f}\right] : [\boldsymbol{V} \otimes \boldsymbol{K}]\right] \cdot \boldsymbol{N} = \widehat{\jmath}_{\mathrm{m}}^{\mathrm{ext}}$$

$$\left[\left[\frac{J}{2!}\boldsymbol{f} \mathbin{⨳} \boldsymbol{f}\right][\boldsymbol{W} - \boldsymbol{V}] + \left[\frac{1}{1!}\boldsymbol{f}\right] \mathbin{⨳} [\boldsymbol{V} \otimes \boldsymbol{K}]\right] \cdot \boldsymbol{N} = \widehat{\boldsymbol{k}}_{\mathrm{m}}^{\mathrm{ext}}$$

$$\left[\left[\frac{J}{1!}\boldsymbol{f}\right][\boldsymbol{W} - \boldsymbol{V}] + \left[\frac{1}{0!}\boldsymbol{i}\right] \cdot [\boldsymbol{V} \otimes \boldsymbol{K}]\right] \cdot \boldsymbol{N} = \widehat{\boldsymbol{f}}_{\mathrm{m}}^{\mathrm{ext}}$$

ii. Regular Points at Singular Surfaces $\mathcal{S}_{\mathrm{m}}$

$$\left[\!\left[\left[\frac{J}{3!}[\boldsymbol{f} \mathbin{⨳} \boldsymbol{f}] : \boldsymbol{f}\right][\boldsymbol{W} - \boldsymbol{V}] + \left[\frac{1}{2!}\boldsymbol{f} \mathbin{⨳} \boldsymbol{f}\right] : [\boldsymbol{V} \otimes \boldsymbol{K}]\right]\!\right] \cdot \boldsymbol{M} = 0$$

$$\left[\!\left[\left[\frac{J}{2!}\boldsymbol{f} \mathbin{⨳} \boldsymbol{f}\right][\boldsymbol{W} - \boldsymbol{V}] + \left[\frac{1}{1!}\boldsymbol{f}\right] \mathbin{⨳} [\boldsymbol{V} \otimes \boldsymbol{K}]\right]\!\right] \cdot \boldsymbol{M} = \boldsymbol{0}$$

$$\left[\!\left[\left[\frac{J}{1!}\boldsymbol{f}\right][\boldsymbol{W} - \boldsymbol{V}] + \left[\frac{1}{0!}\boldsymbol{i}\right] \cdot [\boldsymbol{V} \otimes \boldsymbol{K}]\right]\!\right] \cdot \boldsymbol{M} = \boldsymbol{0}$$

6.8 Summary: 'Balances' of Material Maps

The above 'balances' of the material deformation measures, i.e. the material tangent, cotangent, and measure maps, display an illuminating structure and natural sequence, both in spatial and material control volumes, when recalling that $\boldsymbol{i} = \frac{1}{0!}\boldsymbol{f}^0$, $\boldsymbol{f} = \frac{1}{1!}\boldsymbol{f}$, $\boldsymbol{k} = \frac{1}{2!}\boldsymbol{f} \times\!\!\times \boldsymbol{f}$, and $j = \frac{1}{3!}[\boldsymbol{f} \times\!\!\times \boldsymbol{f}] : \boldsymbol{f}$ are, respectively, constant, linear, quadratic, and cubic in $\boldsymbol{f}$. To bring out these correlations more clearly and for convenience of comparison, Tables 6.3 and 6.4 re-express and juxtapose the local 'balances' for the material deformation measures.

References

1. Bonet J, Gil AJ, Lee CH, Aguirre M, Ortigosa R (2015) A first order hyperbolic framework for large strain computational solid dynamics. Part i: total lagrangian isothermal elasticity. Comput Methods Appl Mech Eng 283:689–732
2. Bonet J, Lee CH, Gil AJ, Ghavamian A (2021) A first order hyperbolic framework for large strain computational solid dynamics. Part iii: thermo-elasticity. Comput Methods Appl Mech Eng 373:113505
3. Gil AJ, Lee CH, Bonet J, Ortigosa R (2016) A first order hyperbolic framework for large strain computational solid dynamics. Part ii: total lagrangian compressible, nearly incompressible and truly incompressible elasticity. Comput Methods Appl Mech Eng 300:146–181

Chapter 7
Mechanical Balances

8,167 m 28°41’54"N 83°29’15"E

Abstract This chapter details the generic balances for the case of mechanical balances of mass, spatial momentum, and its vector and scalar moment, respectively, with emphasis on their global and local formats and the distinction between material and spatial control volumes.

Applying the procedure established previously for generic balances of volume extensive quantities to the *mass* and the *spatial momentum* as well as to the *vector* and *scalar moments of spatial momentum* renders the set of *mechanical balances*. Thereby, the balances of mass and spatial momentum, the latter commonly denoted as the balance of linear momentum, are independent balances for typically unquestioned fundamental quantities.

The balances of vector and scalar moments of spatial momentum, however, follow generically from taking corresponding moments of the accepted balance of spatial momentum. They only become truly *independent* balances under additional restrictive assumptions (spatial isotropy, positive homogeneity of the energy storage). This is a paradigm of a hen-egg problem: either the balances are postulated as independent and the above restrictions follow as requirements, or the above restrictions are postulated and the balances follow as independent.

Typically, the balance of vector moment of spatial momentum, commonly denoted as the balance of angular momentum, is accepted as an independent balance and the resulting restriction is satisfied by a priori postulating the symmetry of suited spatial stress measures.

Contrarily, the balance of scalar moment of spatial momentum, loosely related to the virial theorem that is instrumental to statistical mechanics, is of more formal nature and requires rather drastic restrictions, which are only satisfied in rare scenarios, for its independence.

P. Steinmann, *Spatial and Material Forces in Nonlinear Continuum Mechanics*,
Solid Mechanics and Its Applications 272,
https://doi.org/10.1007/978-3-030-89070-4_7

7.1 Mass

Mass is a fundamental property of matter that possesses a Janus-faced character as either *gravitational* or *inertial mass*. As such it contributes both to gravitation-induced body forces as well as to acceleration-induced inertial forces. The *balance of mass* is thus of utmost importance and relevance for all balances discussed subsequently. In particular for a closed system the balance of mass degenerates to the *conservation of mass*, a special case that shall be assumed exclusively in all what follows.

7.1.1 Global Format (Material Control Volume)

The *balance of mass* is stated generically in global format as

Global Balance of Mass (Material Control Volume):

$$\dot{\mathsf{m}}(\mathfrak{B}_{\mathrm{m}}) = \mathsf{m}^{\mathrm{ext}}(\mathfrak{B}_{\mathrm{m}}) \quad \forall \mathfrak{B}_{\mathrm{m}} \subseteq \mathcal{B}_{\mathrm{m}}. \tag{7.1}$$

Thereby m is expressed as the resultant over the material control volume $\mathfrak{B}_{\mathrm{m}}$ of the density of *mass* ρ_{m} per unit volume in the material configuration, i.e. as the *resultant mass* contained in the material control volume

$$\mathsf{m}(\mathfrak{B}_{\mathrm{m}}) := \int_{\mathfrak{B}_{\mathrm{m}}} \rho_{\mathrm{m}} \, \mathrm{d}V. \tag{7.2}$$

Moreover, $\mathsf{m}^{\mathrm{ext}}$ represents the resultant external mass supply and consists in general of volume and boundary integrals (any singular contributions are here ignored for the sake of presentation)

$$\mathsf{m}^{\mathrm{ext}}(\mathfrak{B}_{\mathrm{m}}) := \int_{\mathfrak{B}_{\mathrm{m}}} \overline{m}_{\mathrm{m}} \, \mathrm{d}V + \int_{\partial\mathfrak{B}_{\mathrm{m}}} \widehat{m}_{\mathrm{m}} \, \mathrm{d}A \tag{7.3}$$

in terms of the source density of mass $\overline{m}_{\mathrm{m}}$ per unit volume in the material configuration and the *extrinsic* flux density of mass $\widehat{m}_{\mathrm{m}}$ per unit area in the material configuration.

The extrinsic mass flux density along with the convective term stemming from the transport theorem are postulated to satisfy a Cauchy theorem for the *intrinsic* mass flux density (Kuhl and Steinmann [1–5])

$$m_{\mathrm{m}} := \boldsymbol{R} \cdot \boldsymbol{N} := \widehat{m}_{\mathrm{m}} - [\rho_{\mathrm{m}} \boldsymbol{W}] \cdot \boldsymbol{N} \quad \text{on} \quad \partial\mathfrak{B}_{\mathrm{m}}. \tag{7.4}$$

Thereby, $\boldsymbol{R} \in T\mathcal{B}_{\mathrm{m}}$ is introduced as the *Piola-type mass flux vector*, a material description vector mapping from the material cotangent space $T^*\mathcal{B}_{\mathrm{m}}$ to $\mathbb{R}$.

For non-living materials the balance of mass is conventionally formulated as the statement of *intrinsic mass conservation* (or simply mass conservation), a frequently occurring special case that shall be considered in the sequel. Thus for intrinsic mass conservation the mass source density $\overline{m}_{\mathrm{m}}$ and the intrinsic mass flux density m_{m} vanish identically

$$\overline{m}_{\mathrm{m}} \equiv 0 \quad \text{and} \quad m_{\mathrm{m}} \equiv 0. \tag{7.5}$$

Consequently, the extrinsic mass flux density $\widehat{m}_{\mathrm{m}} \doteq \rho_{\mathrm{m}} W_{\perp}$ equals the convective mass flux density $\rho_{\mathrm{m}} W_{\perp}$ and results in a corresponding total change of the resultant mass[1]

$$\dot{\mathsf{m}} = \int\limits_{\partial\mathfrak{B}_{\mathrm{m}}} \rho_{\mathrm{m}} W_{\perp} \, \mathrm{d}A. \tag{7.6}$$

As a conclusion, the global statement of intrinsic mass conservation (or simply mass conservation) expands as

$$\int\limits_{\mathfrak{B}_{\mathrm{m}}} \mathrm{D}_t \rho_{\mathrm{m}} \, \mathrm{d}V = \int\limits_{\mathfrak{S}_{\mathrm{m}}} [\![\rho_{\mathrm{m}} \boldsymbol{W}]\!] \cdot \boldsymbol{M} \, \mathrm{d}A. \tag{7.7}$$

7.1.2 Local Format (Material Control Volume)

The results of localizing the global statement of intrinsic mass conservation are summarized as

Local Mass Conservation (Material Control Volume):

i. Regular Points in the Domain

$$\mathrm{D}_t \rho_{\mathrm{m}} = 0 \quad \text{in} \quad \mathcal{B}_{\mathrm{m}}. \tag{7.8}$$

ii. Regular Points on the Boundary

$$[\rho_{\mathrm{m}} \boldsymbol{W}] \cdot \boldsymbol{N} = \widehat{m}_{\mathrm{m}}^{\mathrm{ext}} \quad \text{on} \quad \partial\mathcal{B}_{\mathrm{m}}. \tag{7.9}$$

iii. Regular Points at Singular Surfaces

$$[\![\rho_{\mathrm{m}} \boldsymbol{W}]\!] \cdot \boldsymbol{M} = 0 \quad \text{at} \quad \mathcal{S}_{\mathrm{m}}. \tag{7.10}$$

---

[1] Alternatively, the complementary case of *extrinsic mass conservation* could be defined such that the extrinsic mass flux density vanishes identically $\widehat{m}_{\mathrm{m}} \equiv 0$. Then, the intrinsic mass flux density $m_{\mathrm{m}} = \boldsymbol{R} \cdot \boldsymbol{N} \doteq -\rho_{\mathrm{m}} W_{\perp}$ cancels the convective mass flux density $\rho_{\mathrm{m}} W_{\perp}$, so that the boundary and jump conditions are trivially satisfied. If, in addition, the mass source density is set to $\overline{m}_{\mathrm{m}} \doteq \mathrm{Div}(\rho_{\mathrm{m}} \boldsymbol{W})$, the common local mass conservation $\mathrm{D}_t \rho_0 = 0$ results in regular points of the domain. Moreover, the corresponding total change of the resultant mass reads again as $\dot{\mathsf{m}} = \int_{\partial\mathfrak{B}_{\mathrm{m}}} \rho_{\mathrm{m}} W_{\perp} \, \mathrm{d}A$. Extrinsic mass conservation shall not further be considered in the sequel.

The local statement of mass conservation allows the following conclusions:

i. In a smooth domain $\mathcal{B}_m$, the statement of intrinsic mass conservation in Eq. 7.8 integrates into

$$\rho_m = \rho_m(\boldsymbol{X}) \quad \text{in} \quad \mathcal{B}_m. \tag{7.11}$$

Thus, for the special case of intrinsic mass conservation, the density of mass ρ_m per unit volume in the material configuration is exclusively parameterized in the material coordinates $\boldsymbol{X}$ and may display jump discontinuities $[\![\rho_m]\!] \neq 0$ at a *material* singular surface.

ii. The statement of intrinsic mass conservation on (the regular part of) the external boundary $\partial\mathcal{B}_m$ in Eq. 7.9 results, when restricted to singular surfaces, in the *cancelation* of the extrinsic mass flux density

$$\widehat{m}_m^+ + \widehat{m}_m^- = 0 \quad \text{at} \quad \mathcal{S}_m \tag{7.12}$$

and translates into the jump condition across singular surfaces

$$[\![\rho_m]\!] W_\perp = 0 \quad \text{at} \quad \mathcal{S}_m. \tag{7.13}$$

iii. At a (migrating) singular surface $\mathcal{S}_m$ the statement of intrinsic mass conservation in Eq. 7.10 requires for arbitrary $W_\perp \neq 0$ that

$$[\![\rho_m]\!] = 0 \quad \text{at} \quad \mathcal{S}_m. \tag{7.14}$$

Thus, for the special case of intrinsic mass conservation, the density of mass ρ_m per unit volume in the material configuration does not display jump discontinuities at singular surfaces $\mathfrak{S}_m$ migrating with $W_\perp \neq 0$. For material singular surfaces $W_\perp = 0$ holds trivially, thus here jump discontinuities $[\![\rho_m]\!] \neq 0$ are allowed.

7.1.3 Global Format (Spatial Control Volume)

Alternatively, the statement of intrinsic mass conservation is stated in global format as

Global Balance of Mass (Spatial Control Volume):

$$\dot{\mathsf{m}}(\mathfrak{B}_s) = \mathsf{m}^{ext}(\mathfrak{B}_s) \quad \forall \mathfrak{B}_s \subseteq \mathcal{B}_s. \tag{7.15}$$

Here, m is expressed as the resultant over the spatial control volume $\mathfrak{B}_s$ of the density of mass ρ_s per unit volume in the spatial configuration

$$\mathsf{m}(\mathfrak{B}_s) := \int_{\mathfrak{B}_s} \rho_s \, dv. \tag{7.16}$$

Moreover, for the case of intrinsic mass conservation m^{ext} represents the resultant external (convective) mass flux and merely consists in a boundary integral (any singular contributions are here ignored for the sake of presentation)

$$\mathsf{m}^{ext}(\mathfrak{B}_s) := \int_{\partial\mathfrak{B}_s} \widehat{m}_s \, da \tag{7.17}$$

in terms of the extrinsic flux density of mass $\widehat{m}_s = \widehat{j\,m}_m$ per unit area in the spatial configuration. Thereby the extrinsic mass flux density is related to the convective term stemming from the transport theorem via

$$\widehat{m}_s - [\rho_s \boldsymbol{w}] \cdot \boldsymbol{n} = -[\rho_s \boldsymbol{v}] \cdot \boldsymbol{n} \quad \text{on} \quad \partial\mathfrak{B}_s. \tag{7.18}$$

As a conclusion, the global statement of intrinsic mass conservation (or simply mass conservation) expands as

$$\int_{\mathfrak{B}_s} \mathrm{d}_t \rho_m \, \mathrm{d}V = -\int_{\partial\mathfrak{B}_s} [\rho_s \boldsymbol{v}] \cdot \boldsymbol{m} \, \mathrm{d}a + \int_{\mathfrak{S}_s} [\![\rho_s \boldsymbol{w}]\!] \cdot \boldsymbol{m} \, \mathrm{d}a. \tag{7.19}$$

7.1.4 Local Format (Spatial Control Volume)

The statements of local intrinsic mass conservation, expressed in terms of the density of mass ρ_s per unit volume in the spatial configuration and the extrinsic flux density of mass $\widehat{m}_s^{ext}$ per unit area in the spatial configuration, are summarized as

Local Mass Conservation (Spatial Control Volume):

i. Regular Points in the Domain

$$\mathrm{d}_t \rho_s + \mathrm{div}(\rho_s \boldsymbol{v}) = 0 \quad \text{in} \quad \mathcal{B}_s. \tag{7.20}$$

ii. Regular Points on the Boundary

$$\big[\rho_s[\boldsymbol{w} - \boldsymbol{v}]\big] \cdot \boldsymbol{n} = \widehat{m}_s^{ext} \quad \text{on} \quad \partial\mathcal{B}_s. \tag{7.21}$$

iii. Regular Points at Singular Surfaces

$$\big[\!\big[\rho_s[\boldsymbol{w} - \boldsymbol{v}]\big]\!\big] \cdot \boldsymbol{m} = 0 \quad \text{at} \quad \mathcal{S}_s. \tag{7.22}$$

Table 7.1 Tetragon of fully equivalent versions for the statement of local mass conservation at regular points in the domain, on the boundary and at singular surfaces

Regular Points in the Domain

$\mathrm{D}_t\rho_\mathrm{m} = \mathrm{Div}\boldsymbol{R}_\mathrm{D}$	⟷ PT	$j\,\mathrm{D}_t\rho_\mathrm{m} = \mathrm{div}\boldsymbol{\mu}_\mathrm{D}$
↕ TT		↕ TT
$J\,\mathrm{d}_t\rho_\mathrm{s} = \mathrm{Div}\boldsymbol{R}_\mathrm{d}$	⟷ PT	$\mathrm{d}_t\rho_\mathrm{s} = \mathrm{div}\boldsymbol{\mu}_\mathrm{d}$

Regular Points on the Boundary

$\left[\rho_\mathrm{m}\boldsymbol{W} + \boldsymbol{R}_\mathrm{D}\right]\cdot\boldsymbol{N} = \widehat{m}_\mathrm{m}^\mathrm{ext}$	⟷ PT	$\left[\rho_\mathrm{s}[\boldsymbol{w} - \boldsymbol{v}] + \boldsymbol{\mu}_\mathrm{D}\right]\cdot\boldsymbol{n} = \widehat{m}_\mathrm{s}^\mathrm{ext}$
↕ TT		↕ TT
$\left[\rho_\mathrm{m}[\boldsymbol{W} - \boldsymbol{V}] + \boldsymbol{R}_\mathrm{d}\right]\cdot\boldsymbol{N} = \widehat{m}_\mathrm{m}^\mathrm{ext}$	⟷ PT	$\left[\rho_\mathrm{s}\boldsymbol{w} + \boldsymbol{\mu}_\mathrm{d}\right]\cdot\boldsymbol{n} = \widehat{m}_\mathrm{s}^\mathrm{ext}$

Regular Points at Singular Surfaces

$[\![\rho_\mathrm{m}\boldsymbol{W} + \boldsymbol{R}_\mathrm{D}]\!]\cdot\boldsymbol{M} = 0$	⟷ PT	$[\![\rho_\mathrm{s}[\boldsymbol{w} - \boldsymbol{v}] + \boldsymbol{\mu}_\mathrm{D}]\!]\cdot\boldsymbol{m} = 0$
↕ TT		↕ TT
$[\![\rho_\mathrm{m}[\boldsymbol{W} - \boldsymbol{V}] + \boldsymbol{R}_\mathrm{d}]\!]\cdot\boldsymbol{M} = 0$	⟷ PT	$[\![\rho_\mathrm{s}\boldsymbol{w} + \boldsymbol{\mu}_\mathrm{d}]\!]\cdot\boldsymbol{m} = 0$

with *Dynamic Versions* of Intrinsic Mass Flux

$\boldsymbol{R}_\mathrm{D} := \boldsymbol{0}$	⟷ PT	$\boldsymbol{\mu}_\mathrm{D} := \boldsymbol{0}$
↕ TT		↕ TT
$\boldsymbol{R}_\mathrm{d} := \rho_\mathrm{m}\boldsymbol{V}$	⟷ PT	$\boldsymbol{\mu}_\mathrm{d} := -\rho_\mathrm{s}\boldsymbol{v}$

7.1.5 *Balance Tetragon*

The statement of local intrinsic mass conservation *at regular points in the domain*, *at regular points on the boundary* and *at regular points at singular surfaces* may be expressed in four different but fully equivalent versions, see the arrangement in the corresponding tetragons in Table 7.1. Here, *dynamic versions* of the intrinsic mass fluxes have been introduced as $\boldsymbol{\mu}_\mathrm{D} = \boldsymbol{R}_\mathrm{D} \cdot \operatorname{cof} \boldsymbol{f}$ and $\boldsymbol{\mu}_\mathrm{d} = \boldsymbol{R}_\mathrm{d} \cdot \operatorname{cof} \boldsymbol{f}$, respectively, whereby the subscripts D and d indicate that the corresponding intrinsic mass flux is consistent with either the material or the spatial time derivative D_t or d_t, respectively, of the mass density.

7.2 Spatial Momentum

Spatial momentum, commonly denoted as linear momentum, is a fundamental quantity characterizing the dynamics of a mechanical system under the action of spatial forces. In this regard, it serves as a measure of *non-equilibrium* of the spatial force system.

7.2.1 *Global Format (Material Control Volume)*

The *balance of spatial momentum* (aka the balance of spatial linear momentum) is stated in global format as

Global Balance of Spatial Momentum (Material Control Volume):

$$\dot{\mathsf{p}}(\mathfrak{B}_\mathrm{m}) = \mathsf{f}^\mathrm{ext}(\mathfrak{B}_\mathrm{m}) \quad \forall \mathfrak{B}_\mathrm{m} \subseteq \mathcal{B}_\mathrm{m}. \tag{7.23}$$

Thereby p is expressed as the resultant over the material control volume $\mathfrak{B}_\mathrm{m}$ of the density of *spatial momentum* $\boldsymbol{p}_\mathrm{m}$ per unit volume in the material configuration, i.e. as the *resultant spatial momentum* contained in the material control volume

$$\mathsf{p}(\mathfrak{B}_\mathrm{m}) := \int_{\mathfrak{B}_\mathrm{m}} \boldsymbol{p}_\mathrm{m} \, \mathrm{d}V. \tag{7.24}$$

Observe that the density of spatial momentum $\boldsymbol{p}_\mathrm{m}$ per unit volume in the material configuration is defined as conjugate to the spatial velocity $\boldsymbol{v}$

$$\boldsymbol{p}_{\rm m} := \frac{\partial \kappa_{\rm m}}{\partial \boldsymbol{v}} = \rho_{\rm m}\boldsymbol{v} \quad \text{with} \quad \kappa_{\rm m} := \frac{1}{2}\rho_{\rm m}\boldsymbol{v}\cdot\boldsymbol{v} \tag{7.25}$$

whereby $\kappa_{\rm m}$ is defined as the *kinetic energy density* per unit volume in the material configuration.

Moreover, $\mathbf{f}^{\rm ext}$ represents the resultant external spatial force (spatial momentum supply) and consists of volume and boundary integrals (any singular contributions are here ignored for the sake of presentation)

$$\mathbf{f}^{\rm ext}(\mathfrak{B}_{\rm m}) := \int_{\mathfrak{B}_{\rm m}} \bar{\boldsymbol{s}}_{\rm m}\,{\rm d}V + \int_{\partial\mathfrak{B}_{\rm m}} \widehat{\boldsymbol{s}}_{\rm m}\,{\rm d}A \tag{7.26}$$

in terms of the source density of spatial momentum $\bar{\boldsymbol{s}}_{\rm m}$ per unit volume in the material configuration (body force density, the paradigm being self-weight due to gravity) and the *extrinsic* flux density of spatial momentum $\widehat{\boldsymbol{s}}_{\rm m}$ per unit area in the material configuration (traction).

The extrinsic spatial momentum flux density along with the convective term stemming from the transport theorem are postulated to satisfy a Cauchy theorem for the *intrinsic* spatial momentum flux density

$$\boldsymbol{s}_{\rm m} := \boldsymbol{P}\cdot\boldsymbol{N} := \widehat{\boldsymbol{s}}_{\rm m} - [\boldsymbol{p}_{\rm m}\otimes\boldsymbol{W}]\cdot\boldsymbol{N} \quad \text{on} \quad \partial\mathfrak{B}_{\rm m}. \tag{7.27}$$

Thereby, $\boldsymbol{P} \in T^*\mathcal{B}_{\rm s} \times T\mathcal{B}_{\rm m}$ is introduced as the *Piola stress* (the transpose of the *nominal stress*), a spatial-material description (two-point, mixed-variant) tensor mapping from the material cotangent space $T^*\mathcal{B}_{\rm m}$ to its spatial counterpart $T^*\mathcal{B}_{\rm s}$.

As a consequence of the above definitions, the global format of the balance of spatial momentum expands as

$$\int_{\mathfrak{B}_{\rm m}} {\rm D}_t\boldsymbol{p}_{\rm m}\,{\rm d}V = \int_{\mathfrak{B}_{\rm m}} \bar{\boldsymbol{s}}_{\rm m}\,{\rm d}V + \int_{\partial\mathfrak{B}_{\rm m}} \underbrace{\boldsymbol{P}\cdot\boldsymbol{N}}_{\boldsymbol{s}_{\rm m}}\,{\rm d}A + \int_{\mathfrak{S}_{\rm m}} [\![\boldsymbol{p}_{\rm m}\otimes\boldsymbol{W}]\!]\cdot\boldsymbol{M}\,{\rm d}A. \tag{7.28}$$

7.2.2 *Local Format (Material Control Volume)*

The results of localizing the global balance of spatial momentum are summarized as

Local Balance of Spatial Momentum (Material Control Volume):

i. Regular Points in the Domain

$${\rm D}_t\boldsymbol{p}_{\rm m} - {\rm Div}\boldsymbol{P} = \bar{\boldsymbol{s}}_{\rm m} \quad \text{in} \quad \mathcal{B}_{\rm m}. \tag{7.29}$$

ii. Regular Points on the Boundary

$$[\boldsymbol{p}_{\mathrm{m}} \otimes \boldsymbol{W} + \boldsymbol{P}] \cdot \boldsymbol{N} = \widehat{\boldsymbol{s}}_{\mathrm{m}}^{\mathrm{ext}} \quad \text{on} \quad \partial\mathcal{B}_{\mathrm{m}}. \tag{7.30}$$

iii. Regular Points at Singular Surfaces

$$[\![\boldsymbol{p}_{\mathrm{m}} \otimes \boldsymbol{W} + \boldsymbol{P}]\!] \cdot \boldsymbol{M} = \boldsymbol{0} \quad \text{at} \quad \mathcal{S}_{\mathrm{m}}. \tag{7.31}$$

The local balance of spatial momentum allows the following conclusions:

i. In a smooth domain $\mathcal{B}_{\mathrm{m}}$, the statement of mass conservation $\mathrm{D}_t\rho_{\mathrm{m}} = 0$ in Eq. 7.8 simplifies the balance of spatial momentum in Eq. 7.29 to the more common '$F = ma$' type statement

$$\overline{\boldsymbol{s}}_{\mathrm{m}} + \mathrm{Div}\boldsymbol{P} = \rho_{\mathrm{m}}\boldsymbol{a} \quad \text{in} \quad \mathcal{B}_{\mathrm{m}}. \tag{7.32}$$

Here, for the special case of mass conservation, the *density of inertia* $\mathrm{D}_t\boldsymbol{p}_{\mathrm{m}}$ per unit volume in the material configuration is simply the spatial acceleration $\boldsymbol{a} := \mathrm{D}_t\boldsymbol{v}$ weighted by the mass density ρ_{m}.

ii. The balance of spatial momentum on (the regular part of) the external boundary $\partial\mathcal{B}_{\mathrm{m}}$ in Eq. 7.30 results, when restricted to singular surfaces, in the *cancelation* of the extrinsic spatial momentum flux density

$$\widehat{\boldsymbol{s}}_{\mathrm{m}}^{+} + \widehat{\boldsymbol{s}}_{\mathrm{m}}^{-} = \boldsymbol{0} \quad \text{at} \quad \mathcal{S}_{\mathrm{m}} \tag{7.33}$$

and translates into the jump condition across singular surfaces

$$[\![\boldsymbol{p}_{\mathrm{m}}]\!]W_{\perp} + [\![\boldsymbol{P}]\!] \cdot \boldsymbol{M} = \boldsymbol{0} \quad \text{at} \quad \mathcal{S}_{\mathrm{m}}. \tag{7.34}$$

iii. At singular surfaces $\mathcal{S}_{\mathrm{m}}$, the statement of mass conservation $[\![\rho_{\mathrm{m}}]\!] = 0$ resulting from Eq. 7.10 simplifies the balance of spatial momentum in Eq. 7.31 to

$$\rho_{\mathrm{m}}[\![\boldsymbol{v}]\!]W_{\perp} + [\![\boldsymbol{P}]\!] \cdot \boldsymbol{M} = \boldsymbol{0} \quad \text{at} \quad \mathcal{S}_{\mathrm{m}}. \tag{7.35}$$

Note that the Piola traction displays jump discontinuities $[\![\boldsymbol{P}]\!] \cdot \boldsymbol{M} \neq \boldsymbol{0}$ at singular surfaces $\mathfrak{S}_{\mathrm{m}}$ migrating with $W_{\perp} \neq 0$. For material singular surfaces with $W_{\perp} = 0$ no jump discontinuities of the Piola traction are allowed, i.e. $[\![\boldsymbol{P}]\!] \cdot \boldsymbol{M} = \boldsymbol{0}$.

7.2.3 Global Format (Spatial Control Volume)

Alternatively, the balance of spatial momentum is stated in global format as

Global Balance of Spatial Momentum (Spatial Control Volume):

$$\dot{\mathsf{p}}(\mathfrak{B}_s) = \mathsf{f}^{\text{ext}}(\mathfrak{B}_s) \quad \forall \mathfrak{B}_s \subseteq \mathcal{B}_s. \tag{7.36}$$

Here, p is expressed as the resultant over the spatial control volume $\mathfrak{B}_s$ of the density of spatial momentum $\boldsymbol{p}_s$ per unit volume in the spatial configuration

$$\mathsf{p}(\mathfrak{B}_s) := \int_{\mathfrak{B}_s} \boldsymbol{p}_s \,\mathrm{d}v. \tag{7.37}$$

Moreover, f^{ext} represents the resultant external spatial force and consists of volume and boundary integrals (any singular contributions are here ignored for the sake of presentation)

$$\mathsf{f}^{\text{ext}}(\mathfrak{B}_s) := \int_{\mathfrak{B}_s} \overline{\boldsymbol{s}}_s \,\mathrm{d}v + \int_{\partial\mathfrak{B}_s} \widehat{\boldsymbol{s}}_s \,\mathrm{d}a \tag{7.38}$$

in terms of the source density of spatial momentum $\overline{\boldsymbol{s}}_s = j\overline{\boldsymbol{s}}_m$ per unit volume in the spatial configuration and the extrinsic flux density of spatial momentum $\widehat{\boldsymbol{s}}_s = \widehat{j\boldsymbol{s}}_m$ per unit area in the spatial configuration.

The intrinsic spatial momentum flux density is defined as the Piola transformation $\boldsymbol{\sigma} := \boldsymbol{P} \cdot \operatorname{cof} \boldsymbol{f}$, thus the corresponding Cauchy theorem for $\boldsymbol{\sigma} \cdot \boldsymbol{n}$ reads in terms of the extrinsic spatial momentum flux density along with the convective term stemming from the transport theorem

$$\boldsymbol{s}_s := \boldsymbol{\sigma} \cdot \boldsymbol{n} := \widehat{\boldsymbol{s}}_s - \left[\boldsymbol{p}_s \otimes [\boldsymbol{w} - \boldsymbol{v}]\right] \cdot \boldsymbol{n} \quad \text{on} \quad \partial\mathfrak{B}_s. \tag{7.39}$$

Thereby, $\boldsymbol{\sigma} \in T^*\mathcal{B}_s \times T\mathcal{B}_s$ is introduced as the *Cauchy stress*, a spatial description (mixed-variant) tensor mapping from and to the spatial cotangent space $T^*\mathcal{B}_s$.

As a consequence of the above definitions, the global format of the balance of spatial momentum expands as

$$\int_{\mathfrak{B}_s} \mathrm{d}_t \boldsymbol{p}_s \,\mathrm{d}v = \int_{\mathfrak{B}_s} \overline{\boldsymbol{s}}_s \,\mathrm{d}v + \int_{\partial\mathfrak{B}_s} [\boldsymbol{\sigma} - \boldsymbol{p}_s \otimes \boldsymbol{v}] \cdot \boldsymbol{n} \,\mathrm{d}a + \int_{\mathfrak{S}_s} [\![\boldsymbol{p}_s \otimes \boldsymbol{w}]\!] \cdot \boldsymbol{m} \,\mathrm{d}a. \tag{7.40}$$

7.2.4 Local Format (Spatial Control Volume)

The local balance for the density of spatial momentum $\boldsymbol{p}_\mathrm{s}$ per unit volume in the spatial configuration, expressed in terms of the Cauchy stress $\boldsymbol{\sigma} = \boldsymbol{P} \cdot \operatorname{cof} \boldsymbol{f}$, the source density of spatial momentum $\overline{\boldsymbol{s}}_\mathrm{s}$ per unit volume in the spatial configuration and the extrinsic flux density of spatial momentum $\widehat{\boldsymbol{s}}_\mathrm{s}^\mathrm{ext}$ per unit area in the spatial configuration, is summarized as

Local Balance of Spatial Momentum (Spatial Control Volume):

i. Regular Points in the Domain

$$\mathrm{d}_t \boldsymbol{p}_\mathrm{s} - \operatorname{div}(\boldsymbol{\sigma} - \boldsymbol{p}_\mathrm{s} \otimes \boldsymbol{v}) = \overline{\boldsymbol{s}}_\mathrm{s} \quad \text{in} \quad \mathcal{B}_\mathrm{s}. \tag{7.41}$$

ii. Regular Points on the Boundary

$$\big[\boldsymbol{p}_\mathrm{s} \otimes \boldsymbol{w} + [\boldsymbol{\sigma} - \boldsymbol{p}_\mathrm{s} \otimes \boldsymbol{v}]\big] \cdot \boldsymbol{n} = \widehat{\boldsymbol{s}}_\mathrm{s}^\mathrm{ext} \quad \text{on} \quad \partial\mathcal{B}_\mathrm{s}. \tag{7.42}$$

iii. Regular Points at Singular Surfaces

$$[\![\boldsymbol{p}_\mathrm{s} \otimes \boldsymbol{w} + [\boldsymbol{\sigma} - \boldsymbol{p}_\mathrm{s} \otimes \boldsymbol{v}]]\!] \cdot \boldsymbol{m} = \boldsymbol{0} \quad \text{at} \quad \mathcal{S}_\mathrm{s}. \tag{7.43}$$

7.2.5 Spatial Stress Measures

In addition to the Piola transformation between the Piola and the Cauchy stress, it proves convenient to introduce the *Kirchhoff stress* $\boldsymbol{\tau}$ as

$$\boldsymbol{\tau} := J\boldsymbol{\sigma} = \boldsymbol{P} \cdot \boldsymbol{F}^t \quad \text{with} \quad \int_{\mathfrak{B}_\mathrm{m}} \boldsymbol{\tau} \cdots \mathrm{d}V = \int_{\mathfrak{B}_\mathrm{s}} \boldsymbol{\sigma} \cdots \mathrm{d}v.$$

The Kirchhoff stress might be considered either a contra-, a co-, or a co/contra- (mixed) variant spatial description tensor field (here no special notation shall be adopted to distinguish between these versions). Then, the contra-, co-, and co/contra- (mixed) variant pull-backs of the Kirchhoff stress follow as

$$\boldsymbol{S} := \boldsymbol{f} \cdot \boldsymbol{\tau} \cdot \boldsymbol{f}^t \quad \text{and} \quad \boldsymbol{T} := \boldsymbol{F}^t \cdot \boldsymbol{\tau} \cdot \boldsymbol{F} \quad \text{and} \quad \boldsymbol{M} := \boldsymbol{F}^t \cdot \boldsymbol{\tau} \cdot \boldsymbol{f}^t.$$

Table 7.2 Tetragon of fully equivalent versions for the local balance of spatial momentum at regular points in the domain, on the boundary, and at singular surfaces

Regular Points in the Domain

$\mathrm{D}_t \boldsymbol{p}_\mathrm{m} = \mathrm{Div} \boldsymbol{P}_\mathrm{D} + \overline{\boldsymbol{s}}_\mathrm{m}$	↔ PT	$j\, \mathrm{D}_t \boldsymbol{p}_\mathrm{m} = \mathrm{div} \boldsymbol{\sigma}_\mathrm{D} + \overline{\boldsymbol{s}}_\mathrm{s}$
↕ TT		TT ↕
$J\, \mathrm{d}_t \boldsymbol{p}_\mathrm{s} = \mathrm{Div} \boldsymbol{P}_\mathrm{d} + \overline{\boldsymbol{s}}_\mathrm{m}$	PT ↔	$\mathrm{d}_t \boldsymbol{p}_\mathrm{s} = \mathrm{div} \boldsymbol{\sigma}_\mathrm{d} + \overline{\boldsymbol{s}}_\mathrm{s}$

Regular Points on the Boundary

$\left[\boldsymbol{p}_\mathrm{m} \otimes \boldsymbol{W} + \boldsymbol{P}_\mathrm{D}\right] \cdot \boldsymbol{N} = \widehat{\boldsymbol{s}}_\mathrm{m}^\mathrm{ext}$	↔ PT	$\left[\boldsymbol{p}_\mathrm{s} \otimes [\boldsymbol{w} - \boldsymbol{v}] + \boldsymbol{\sigma}_\mathrm{D}\right] \cdot \boldsymbol{n} = \widehat{\boldsymbol{s}}_\mathrm{s}^\mathrm{ext}$
↕ TT		TT ↕
$\left[\boldsymbol{p}_\mathrm{m} \otimes [\boldsymbol{W} - \boldsymbol{V}] + \boldsymbol{P}_\mathrm{d}\right] \cdot \boldsymbol{N} = \widehat{\boldsymbol{s}}_\mathrm{m}^\mathrm{ext}$	PT ↔	$\left[\boldsymbol{p}_\mathrm{s} \otimes \boldsymbol{w} + \boldsymbol{\sigma}_\mathrm{d}\right] \cdot \boldsymbol{n} = \widehat{\boldsymbol{s}}_\mathrm{s}^\mathrm{ext}$

Regular Points at Singular Surfaces

$\llbracket \boldsymbol{p}_\mathrm{m} \otimes \boldsymbol{W} + \boldsymbol{P}_\mathrm{D} \rrbracket \cdot \boldsymbol{M} = \boldsymbol{0}$	↔ PT	$\llbracket \boldsymbol{p}_\mathrm{s} \otimes [\boldsymbol{w} - \boldsymbol{v}] + \boldsymbol{\sigma}_\mathrm{D} \rrbracket \cdot \boldsymbol{m} = \boldsymbol{0}$
↕ TT		TT ↕
$\llbracket \boldsymbol{p}_\mathrm{m} \otimes [\boldsymbol{W} - \boldsymbol{V}] + \boldsymbol{P}_\mathrm{d} \rrbracket \cdot \boldsymbol{M} = \boldsymbol{0}$	PT ↔	$\llbracket \boldsymbol{p}_\mathrm{s} \otimes \boldsymbol{w} + \boldsymbol{\sigma}_\mathrm{d} \rrbracket \cdot \boldsymbol{m} = \boldsymbol{0}$

with *Dynamic Versions* of Piola and Cauchy Stress

$\boldsymbol{P}_\mathrm{D} := \boldsymbol{P}$	↔ PT	$\boldsymbol{\sigma}_\mathrm{D} := \boldsymbol{\sigma}$
↕ TT		TT ↕
$\boldsymbol{P}_\mathrm{d} := \boldsymbol{P} + \boldsymbol{p}_\mathrm{m} \otimes \boldsymbol{V}$	PT ↔	$\boldsymbol{\sigma}_\mathrm{d} := \boldsymbol{\sigma} - \boldsymbol{p}_\mathrm{s} \otimes \boldsymbol{v}$

Here, $\boldsymbol{S}$, $\boldsymbol{T}$, and $\boldsymbol{M}$ denote a contravariantly transforming material description stress measure, i.e. the *Piola-Kirchhoff stress*, a covariantly transforming material description stress measure (sometimes denoted as the convected stress, [6]), and a co/contra-(mixed) variantly transforming material description stress measure, i.e. the *Mandel stress*, respectively. The relations between the various introduced stress measures are assembled in the following table:

	$\boldsymbol{\tau}$	$\boldsymbol{P}$	$\boldsymbol{S}$	$\boldsymbol{T}$	$\boldsymbol{M}$
$\boldsymbol{\tau}$	$\bullet$	$\boldsymbol{P}\cdot\boldsymbol{F}^t$	$\boldsymbol{F}\cdot\boldsymbol{S}\cdot\boldsymbol{F}^t$	$\boldsymbol{f}^t\cdot\boldsymbol{T}\cdot\boldsymbol{f}$	$\boldsymbol{f}^t\cdot\boldsymbol{M}\cdot\boldsymbol{F}^t$
$\boldsymbol{P}$	$\boldsymbol{\tau}\cdot\boldsymbol{f}^t$	$\bullet$	$\boldsymbol{F}\cdot\boldsymbol{S}$	$\boldsymbol{f}^t\cdot\boldsymbol{T}\cdot\boldsymbol{B}$	$\boldsymbol{f}^t\cdot\boldsymbol{M}$
$\boldsymbol{S}$	$\boldsymbol{f}\cdot\boldsymbol{\tau}\cdot\boldsymbol{f}^t$	$\boldsymbol{f}\cdot\boldsymbol{P}$	$\bullet$	$\boldsymbol{B}\cdot\boldsymbol{T}\cdot\boldsymbol{B}$	$\boldsymbol{B}\cdot\boldsymbol{M}$
$\boldsymbol{T}$	$\boldsymbol{F}^t\cdot\boldsymbol{\tau}\cdot\boldsymbol{F}$	$\boldsymbol{F}^t\cdot\boldsymbol{P}\cdot\boldsymbol{C}$	$\boldsymbol{C}\cdot\boldsymbol{S}\cdot\boldsymbol{C}$	$\bullet$	$\boldsymbol{M}\cdot\boldsymbol{C}$
$\boldsymbol{M}$	$\boldsymbol{F}^t\cdot\boldsymbol{\tau}\cdot\boldsymbol{f}^t$	$\boldsymbol{F}^t\cdot\boldsymbol{P}$	$\boldsymbol{C}\cdot\boldsymbol{S}$	$\boldsymbol{T}\cdot\boldsymbol{B}$	$\bullet$

Observe that $\boldsymbol{S}$ and $\boldsymbol{T}$ are *symmetric* if $\boldsymbol{\tau} = \boldsymbol{\tau}^t$ holds, however even in this case $\boldsymbol{M}$ is only symmetric if $\boldsymbol{C}$ and $\boldsymbol{S}$ or, likewise, $\boldsymbol{B}$ and $\boldsymbol{T}$ commute. The former is a consequence of the obligatory *spatial isotropy*, whereas the latter is an consequence of the optional *material isotropy*, see the discussions of these concepts in the below.

7.2.6 Balance Tetragon

The local balance of spatial momentum *at regular points in the domain, at regular points on the boundary* and *at regular points at singular surfaces* may be expressed in four different but fully equivalent versions, see the arrangement in the corresponding tetragons in Table 7.2. Here, *dynamic versions* of the Piola and the Cauchy stress have been introduced as $\boldsymbol{\sigma}_{\mathrm{D}} = \boldsymbol{P}_{\mathrm{D}} \cdot \operatorname{cof} \boldsymbol{f}$ and $\boldsymbol{\sigma}_{\mathrm{d}} = \boldsymbol{P}_{\mathrm{d}} \cdot \operatorname{cof} \boldsymbol{f}$, respectively, whereby the subscripts D and d indicate that the corresponding intrinsic spatial momentum flux is consistent with either the material or the spatial time derivative D_t or d_t, respectively, of the spatial momentum density.

7.3 Vector Moment of Spatial Momentum

Vector moments of spatial forces are denoted as *spatial couples* and serve as a measure of the *non-centrality* of the spatial force system. They contribute to the *balance of vector moment of spatial momentum* (aka the balance of angular spatial momentum).

7.3.1 Global Format (Material Control Volume)

The *balance of vector moment of spatial momentum* is stated in global format as

Global Balance of Vector Moment of Spatial Momentum (Material Control Volume):

$$\dot{\mathsf{l}}(\mathfrak{B}_{\mathrm{m}}) = \mathsf{c}^{\mathrm{ext}}(\mathfrak{B}_{\mathrm{m}}) \quad \forall \mathfrak{B}_{\mathrm{m}} \subseteq \mathcal{B}_{\mathrm{m}}. \tag{7.44}$$

Thereby, l is expressed as the resultant over the material control volume $\mathfrak{B}_{\mathrm{m}}$ of the density of *vector moment of spatial momentum* $\boldsymbol{l}_{\mathrm{m}}$ per unit volume in the material configuration, i.e. as the *resultant vector moment of spatial momentum* contained in the material control volume

$$\mathsf{l}(\mathfrak{B}_{\mathrm{m}}) := \int_{\mathfrak{B}_{\mathrm{m}}} \boldsymbol{l}_{\mathrm{m}} \, \mathrm{d}V. \tag{7.45}$$

For *non-polar* materials, i.e. for materials characterized by only translational degrees of freedom—a frequently occurring case that shall here be considered exclusively—the density of vector moment of spatial momentum $\boldsymbol{l}_{\mathrm{m}}$ per unit volume in the material configuration is defined as the *vector* (cross) product of the spatial distance vector

$$\boldsymbol{d}_0 := \boldsymbol{y} - \boldsymbol{y}_0 \tag{7.46}$$

to an arbitrary spatial reference point $\boldsymbol{y}_0$ (say $\boldsymbol{y}_0 \equiv \boldsymbol{0}$, i.e. the coordinate origin, for the sake of presentation) and the density of spatial momentum $\boldsymbol{p}_{\mathrm{m}}$ per unit volume in the material configuration

$$\boldsymbol{l}_{\mathrm{m}} := \boldsymbol{d}_0 \times \boldsymbol{p}_{\mathrm{m}}. \tag{7.47}$$

Moreover, $\mathsf{c}^{\mathrm{ext}}$ represents the resultant external spatial couple (supply of vector moment of spatial momentum) and consists of volume and boundary integrals (any singular contributions are here ignored for the sake of presentation)

$$\mathsf{c}^{\mathrm{ext}}(\mathfrak{B}_{\mathrm{m}}) := \int_{\mathfrak{B}_{\mathrm{m}}} \overline{\boldsymbol{t}}_{\mathrm{m}} \, \mathrm{d}V + \int_{\partial\mathfrak{B}_{\mathrm{m}}} \widehat{\boldsymbol{t}}_{\mathrm{m}} \, \mathrm{d}A \tag{7.48}$$

in terms of the source density of vector moment of spatial momentum $\overline{\boldsymbol{t}}_{\mathrm{m}}$ per unit volume in the material configuration (body couple density) and the *extrinsic* flux density of vector moment of spatial momentum $\widehat{\boldsymbol{t}}_{\mathrm{m}}$ per unit area in the material configuration (couple traction).

These are defined as the vector product of the spatial distance vector $\boldsymbol{d}_0$ and the corresponding source density of spatial momentum $\overline{\boldsymbol{s}}_{\mathrm{m}}$ and the extrinsic flux density of spatial momentum $\widehat{s}_{\text{m}}$, respectively, together with, in general, an autonomous contribution to the source density of vector moment of spatial momentum that will be identified later

$$\bar{t}_{\text{m}} := d_0 \times \bar{s}_{\text{m}} + \bar{t}_{\text{m}}^{\text{aut}} \quad \text{and} \quad \widehat{t}_{\text{m}} := d_0 \times \widehat{s}_{\text{m}}. \tag{7.49}$$

The extrinsic flux density of vector moment of spatial momentum along with the convective term stemming from the transport theorem are postulated to satisfy a Cauchy theorem for the *intrinsic* flux density of vector moment of spatial momentum

$$t_{\text{m}} := L \cdot N = \widehat{t}_{\text{m}} - [l_{\text{m}} \otimes W] \cdot N \quad \text{on} \quad \partial\mathfrak{B}_{\text{m}}. \tag{7.50}$$

Thereby, $L \in T^*\mathcal{B}_{\text{s}} \times T\mathcal{B}_{\text{m}}$ is introduced as the *Piola-type couple stress*, a spatial-material description (two-point, mixed-variant) tensor mapping from the material cotangent space $T^*\mathcal{B}_{\text{m}}$ to its spatial counterpart $T^*\mathcal{B}_{\text{s}}$.

Incorporating the specific formats for the flux and density terms, the intrinsic flux density of vector moment of spatial momentum and the corresponding Cauchy theorem particularize to

$$t_{\text{m}} := L \cdot N = d_0 \times \left[\widehat{s}_{\text{m}} - [p_{\text{m}} \otimes W] \cdot N\right] = d_0 \times s_{\text{m}} \quad \text{on} \quad \partial\mathfrak{B}_{\text{m}}, \tag{7.51}$$

from which, together with Eq. 7.27, the Piola-type couple stress is eventually identified as

$$L := d_0 \times P. \tag{7.52}$$

As a consequence of the above definitions, the global format of the balance of vector moment of spatial momentum expands as

$$\int_{\mathfrak{B}_{\text{m}}} \mathrm{D}_t l_{\text{m}} \, \mathrm{d}V = \int_{\mathfrak{B}_{\text{m}}} \bar{t}_{\text{m}} \, \mathrm{d}V + \int_{\partial\mathfrak{B}_{\text{m}}} \underbrace{L \cdot N}_{t_{\text{m}}} \, \mathrm{d}A + \int_{\partial\mathfrak{S}_{\text{m}}} [\![l_{\text{m}} \otimes W]\!] \cdot M \, \mathrm{d}A, \tag{7.53}$$

which, by incorporating the specific formats for the source and flux terms, specializes to

$$\begin{aligned} \int_{\mathfrak{B}_{\text{m}}} \mathrm{D}_t l_{\text{m}} \, \mathrm{d}V = \int_{\mathfrak{B}_{\text{m}}} [d_0 \times \bar{s}_{\text{m}} + \bar{t}_{\text{m}}^{\text{aut}}] \, \mathrm{d}V + \int_{\partial\mathfrak{B}_{\text{m}}} d_0 \times \underbrace{P \cdot N}_{s_{\text{m}}} \, \mathrm{d}A \\ + \int_{\partial\mathfrak{S}_{\text{m}}} [\![l_{\text{m}} \otimes W]\!] \cdot M \, \mathrm{d}A. \end{aligned} \tag{7.54}$$

The latter representation clearly reveals that only in case that the autonomous contribution to the source density of vector moment of spatial momentum vanishes, i.e. for $\bar{\boldsymbol{t}}_{\mathrm{m}}^{\mathrm{aut}} \equiv \mathbf{0}$, the resultant external spatial couple is exclusively due to the external spatial forces, thus making the balance of vector moment of spatial momentum an independent balance.

7.3.2 Local Format (Material Control Volume)

The results of localizing the global balance of vector moment of spatial momentum are summarized as

Local Balance of Vector Moment of Spatial Momentum (Material Control Volume):

i. Regular Points in the Domain

$$\mathrm{D}_t \boldsymbol{l}_{\mathrm{m}} - \mathrm{Div}\boldsymbol{L} = \bar{\boldsymbol{t}}_{\mathrm{m}} \quad \text{in} \quad \mathcal{B}_{\mathrm{m}}. \tag{7.55}$$

ii. Regular Points on the Boundary

$$[\boldsymbol{l}_{\mathrm{m}} \otimes \boldsymbol{W} + \boldsymbol{L}] \cdot \boldsymbol{N} = \widehat{\boldsymbol{t}}_{\mathrm{m}}^{\mathrm{ext}} \quad \text{on} \quad \partial\mathcal{B}_{\mathrm{m}}. \tag{7.56}$$

iii. Regular Points at Singular Surfaces

$$[\![\boldsymbol{l}_{\mathrm{m}} \otimes \boldsymbol{W} + \boldsymbol{L}]\!] \cdot \boldsymbol{M} = \mathbf{0} \quad \text{at} \quad \mathcal{S}_{\mathrm{m}}. \tag{7.57}$$

The local balance of vector moment of spatial momentum allows the following general conclusion:

i. The balance of vector moment of spatial momentum on (the regular part of) the external boundary $\partial\mathcal{B}_{\mathrm{m}}$ in Eq. 7.56 results, when restricted to singular surfaces, in the *cancelation* of the extrinsic flux density of vector moment of spatial momentum

$$\widehat{\boldsymbol{t}}_{\mathrm{m}}^{+} + \widehat{\boldsymbol{t}}_{\mathrm{m}}^{-} = \mathbf{0} \quad \text{at} \quad \mathcal{S}_{\mathrm{m}} \tag{7.58}$$

and translates into the jump condition across singular surfaces

$$[\![\boldsymbol{l}_{\mathrm{m}}]\!] W_{\perp} + [\![\boldsymbol{L}]\!] \cdot \boldsymbol{M} = \mathbf{0} \quad \text{at} \quad \mathcal{S}_{\mathrm{m}}. \tag{7.59}$$

Incorporating Specific Formats for Density, Source, and Flux Terms

Incorporating the specific formats for $\boldsymbol{l}_\mathrm{m}, \overline{\boldsymbol{t}}_\mathrm{m}, \widehat{\boldsymbol{t}}_\mathrm{m}$, and $\boldsymbol{L}$ in Eqs. 7.47, 7.49, and 7.52, respectively, while noting the identities

$$\mathrm{Div}(\boldsymbol{d}_0 \times \boldsymbol{P}) = \boldsymbol{d}_0 \times \mathrm{Div}\boldsymbol{P} + 2\mathrm{axl}(\boldsymbol{P} \cdot \boldsymbol{F}^t) \tag{7.60}$$

and

$$\mathrm{D}_t(\boldsymbol{d}_0 \times \boldsymbol{p}_\mathrm{m}) = \boldsymbol{d}_0 \times \mathrm{D}_t\boldsymbol{p}_\mathrm{m} + 2\mathrm{axl}(\boldsymbol{p}_\mathrm{m} \otimes \boldsymbol{v}), \tag{7.61}$$

whereby, e.g. $\mathrm{axl}(\boldsymbol{P} \cdot \boldsymbol{F}^t)$ denotes the *axial vector*[2] associated with the *skew symmetric* contribution $[\boldsymbol{P} \cdot \boldsymbol{F}^t]^{\mathrm{skw}}$ to the product $\boldsymbol{P} \cdot \boldsymbol{F}^t$, the local balance of vector moment of spatial momentum reads explicitly

Local Balance of Vector Moment of Spatial Momentum (Material Control Volume):

i. Regular Points in the Domain

$$\boldsymbol{d}_0 \times [\,\mathrm{D}_t\boldsymbol{p}_\mathrm{m} - \mathrm{Div}\boldsymbol{P}] - 2\mathrm{axl}(\boldsymbol{P} \cdot \boldsymbol{F}^t) = \boldsymbol{d}_0 \times \overline{\boldsymbol{s}}_\mathrm{m} + \overline{\boldsymbol{t}}_\mathrm{m}^{\mathrm{aut}} \quad \text{in} \quad \mathcal{B}_\mathrm{m}. \tag{7.62}$$

ii. Regular Points on the Boundary

$$\boldsymbol{d}_0 \times [\boldsymbol{p}_\mathrm{m} \otimes \boldsymbol{W} + \boldsymbol{P}] \cdot \boldsymbol{N} = \boldsymbol{d}_0 \times \widehat{\boldsymbol{s}}_\mathrm{m}^{\mathrm{ext}} \quad \text{on} \quad \partial\mathcal{B}_\mathrm{m}. \tag{7.63}$$

iii. Regular Points at Singular Surfaces

$$\boldsymbol{d}_0 \times [\![\boldsymbol{p}_\mathrm{m} \otimes \boldsymbol{W} + \boldsymbol{P}]\!] \cdot \boldsymbol{M} = \boldsymbol{0} \quad \text{at} \quad \mathcal{S}_\mathrm{m}. \tag{7.64}$$

Observe that in the above the orthogonality condition $2\mathrm{axl}(\boldsymbol{p}_\mathrm{m} \otimes \boldsymbol{v}) = \boldsymbol{v} \times \boldsymbol{p}_\mathrm{m} = \boldsymbol{0}$ and the coherence of the spatial deformation map $[\![\boldsymbol{d}_0]\!] = \boldsymbol{0}$ have been used.

Considering next the local balance of spatial momentum in Eqs. 7.29–7.31 establishes that the balance of vector moment of spatial momentum is trivially satisfied if

---

[2] For non-zero vectors $\boldsymbol{a} \in \mathbb{E}^3$ the *axial vector* $\boldsymbol{t} := \mathrm{axl}\,\boldsymbol{T} \in \mathbb{E}^3$ of an arbitrary second-order tensor $\boldsymbol{T} \in \mathbb{E}^3 \times \mathbb{E}^3$ is defined via the isomorphism

$$\boldsymbol{t} \times \boldsymbol{a} := \boldsymbol{T}^{\mathrm{skw}} \cdot \boldsymbol{a} \quad \forall \boldsymbol{a} \quad \in \mathbb{E}^3.$$

In Cartesian coordinates with coordinate representation for $\boldsymbol{T} = T_a{}^b\, \boldsymbol{e}^a \otimes \boldsymbol{e}_b$, the axial vector expands as $\boldsymbol{t} = t_c\, \boldsymbol{e}^c := -\frac{1}{2} T_a{}^b\, e^a{}_{bc}\, \boldsymbol{e}^c$. Vice versa $\boldsymbol{T}^{\mathrm{skw}} = -e_a{}^{bc}\, t_c\, \boldsymbol{e}^a \otimes \boldsymbol{e}_b$ holds. Thereby, the (mixed-variant) permutation symbols are denoted by $e^a{}_{bc} := [\boldsymbol{e}^a \times \boldsymbol{e}_b] \cdot \boldsymbol{e}_c$ and $e_a{}^{bc} := [\boldsymbol{e}_a \times \boldsymbol{e}^b] \cdot \boldsymbol{e}^c$, respectively.

the autonomous contribution to the corresponding source density of vector moment of spatial momentum is finally identified as

$$\bar{\boldsymbol{t}}_{\mathrm{m}}^{\mathrm{aut}} := -2\mathrm{axl}(\boldsymbol{P}\cdot\boldsymbol{F}^{t}). \tag{7.65}$$

Making the VMSM Balance Truly Independent (Spatial Isotropy)

The balance of vector moment of spatial momentum (in short VMSM) is an *independent* balance only in case that $\bar{\boldsymbol{t}}_{\mathrm{m}}^{\mathrm{aut}} \equiv \mathbf{0}$. Recall that in this situation the resultant external spatial couple is exclusively due to the external spatial forces.

The corresponding *symmetry constraint* $[\boldsymbol{P}\cdot\boldsymbol{F}^{t}]^{\mathrm{skw}} \doteq \mathbf{0}$, that is anyway a necessary consequence of the requirement of *spatial isotropy* (see remark below), can be regarded a constraint for the constitutive modeling of $\boldsymbol{P}$. Recall that $\boldsymbol{\tau} := J\,\boldsymbol{\sigma} = \boldsymbol{P}\cdot\boldsymbol{F}^{t}$ is denoted as the *Kirchhoff stress*, a spatial description (mixed-variant) tensor mapping from and to the spatial cotangent space $T^{*}\mathcal{B}_{\mathrm{s}}$. In the sequel, the Kirchhoff stress is assumed symmetric, thus automatically satisfying the balance of vector moment of spatial momentum.

Remark (Spatial Isotropy):

Constitutively, the symmetry of $\boldsymbol{\tau} = \boldsymbol{P}\cdot\boldsymbol{F}^{t}$ may be regarded a consequence of *spatial isotropy*. As will be shown later, for thermodynamically consistent constitutive modeling, $\boldsymbol{P}$ derives from the free energy density ψ_{m} per unit volume in the material configuration as $\boldsymbol{P} = \partial_{\boldsymbol{F}}\psi_{\mathrm{m}}$. Spatial isotropy, more commonly denoted as *objectivity* (or *frame indifference* in its passive version), requires that $\psi_{\mathrm{m}} = \psi_{\mathrm{m}}^{*}(\tau)$, with τ a time-like parameter, remains unaffected under *spatially superposed rigid body motions* that take $\boldsymbol{F}$ to $\boldsymbol{F}^{*}(\tau) = \boldsymbol{Q}(\tau)\cdot\boldsymbol{F}$, whereby $\boldsymbol{Q}$ is from the *special orthogonal group* $\mathrm{SO}(3) = \{\boldsymbol{Q}\,|\,\boldsymbol{Q}^{-1} = \boldsymbol{Q}^{t}, \det\boldsymbol{Q} = +1\}$. Then $\mathrm{D}_{\tau}\boldsymbol{F}^{*}(\tau) = \mathrm{D}_{\tau}\boldsymbol{Q}(\tau)\cdot\boldsymbol{Q}^{t}(\tau)\cdot\boldsymbol{F}^{*}(\tau)$ holds with $\mathrm{D}_{\tau}\boldsymbol{Q}(\tau)\cdot\boldsymbol{Q}^{t}(\tau) = -\boldsymbol{Q}(\tau)\cdot\mathrm{D}_{\tau}\boldsymbol{Q}^{t}(\tau)$ being skew-symmetric. Thus, spatial isotropy establishes eventually that $\mathrm{D}_{\tau}\psi_{\mathrm{m}}^{*} = \boldsymbol{\tau}^{*}(\tau) : [\mathrm{D}_{\tau}\boldsymbol{Q}(\tau)\cdot\boldsymbol{Q}^{t}(\tau)] \doteq 0$ and therefore especially that $\boldsymbol{\tau} = \boldsymbol{\tau}^{*}(\tau = 0)$ is necessarily symmetric. □

7.3.3 *Global Format (Spatial Control Volume)*

Alternatively, the balance of vector moment of spatial momentum is stated in global format as

Global Balance of Vector Moment of Spatial Momentum (Spatial Control Volume):

$$\dot{\mathsf{l}}(\mathfrak{B}_\mathrm{s}) = \mathsf{c}^\mathrm{ext}(\mathfrak{B}_\mathrm{s}) \quad \forall \mathfrak{B}_\mathrm{s} \subseteq \mathcal{B}_\mathrm{s}. \tag{7.66}$$

Here, l is expressed as the resultant over the spatial control volume $\mathfrak{B}_\mathrm{s}$ of the density of vector moment of spatial momentum $\boldsymbol{l}_\mathrm{s}$ per unit volume in the spatial configuration

$$\mathsf{l}(\mathfrak{B}_\mathrm{s}) := \int_{\mathfrak{B}_\mathrm{s}} \boldsymbol{l}_\mathrm{s} \,\mathrm{d}v. \tag{7.67}$$

The density of vector moment of spatial momentum $\boldsymbol{l}_\mathrm{s}$ per unit volume in the spatial configuration expands as

$$\boldsymbol{l}_\mathrm{s} := \boldsymbol{d}_0 \times \boldsymbol{p}_\mathrm{s}. \tag{7.68}$$

Moreover, c^ext represents the resultant external spatial couple and consists of volume and boundary integrals (any singular contributions are here ignored for the sake of presentation)

$$\mathsf{c}^\mathrm{ext}(\mathfrak{B}_\mathrm{s}) := \int_{\mathfrak{B}_\mathrm{s}} \bar{\boldsymbol{t}}_\mathrm{s} \,\mathrm{d}v + \int_{\partial\mathfrak{B}_\mathrm{s}} \widehat{\boldsymbol{t}}_\mathrm{s} \,\mathrm{d}a \tag{7.69}$$

in terms of the source density of vector moment of spatial momentum $\bar{\boldsymbol{t}}_\mathrm{s} = j\bar{\boldsymbol{t}}_\mathrm{m}$ per unit volume in the spatial configuration and the extrinsic flux density of vector moment of spatial momentum $\widehat{\boldsymbol{t}}_\mathrm{s} = \widehat{j\boldsymbol{t}}_\mathrm{m}$ per unit area in the spatial configuration. These expand as

$$\bar{\boldsymbol{t}}_\mathrm{s} := \boldsymbol{d}_0 \times \bar{\boldsymbol{s}}_\mathrm{s} + \bar{\boldsymbol{t}}_\mathrm{s}^\mathrm{aut} \quad \text{and} \quad \widehat{\boldsymbol{t}}_\mathrm{s} := \boldsymbol{d}_0 \times \widehat{\boldsymbol{s}}_\mathrm{s}. \tag{7.70}$$

The intrinsic flux density of vector moment of spatial momentum is defined as the Piola transformation $\boldsymbol{\theta} := \boldsymbol{L} \cdot \mathrm{cof}\, \boldsymbol{f}$, thus the corresponding Cauchy theorem for $\boldsymbol{\theta} \cdot \boldsymbol{n}$ reads in terms of the extrinsic flux density of vector moment of spatial momentum along with the convective term stemming from the transport theorem

$$\boldsymbol{t}_\mathrm{s} := \boldsymbol{\theta} \cdot \boldsymbol{n} := \widehat{\boldsymbol{t}}_\mathrm{s} - \left[\boldsymbol{l}_\mathrm{s} \otimes [\boldsymbol{w} - \boldsymbol{v}]\right] \cdot \boldsymbol{n} \quad \text{on} \quad \partial\mathfrak{B}_\mathrm{s}. \tag{7.71}$$

Thereby, $\boldsymbol{\theta} \in T^*\mathcal{B}_\mathrm{s} \times T\mathcal{B}_\mathrm{s}$ is introduced as the *Cauchy-type couple stress*, a spatial description (mixed-variant) tensor mapping from and to the spatial cotangent space $T^*\mathcal{B}_\mathrm{s}$.

Incorporating the specific format for the flux and density terms, the intrinsic flux density of vector moment of spatial momentum and the corresponding Cauchy theorem particularize to

$$\boldsymbol{t}_{\mathrm{s}} := \boldsymbol{\theta} \cdot \boldsymbol{n} = \boldsymbol{d}_0 \times \left[\widehat{\boldsymbol{s}}_{\mathrm{s}} - \left[\boldsymbol{p}_{\mathrm{s}} \otimes [\boldsymbol{w} - \boldsymbol{v}] \right] \right] \cdot \boldsymbol{n} = \boldsymbol{d}_0 \times \boldsymbol{s}_{\mathrm{s}} \quad \text{on} \quad \partial\mathfrak{B}_{\mathrm{s}}, \tag{7.72}$$

from which, together with Eq. 7.39, the Cauchy-type couple stress is eventually identified as

$$\boldsymbol{\theta} := \boldsymbol{d}_0 \times \boldsymbol{\sigma}. \tag{7.73}$$

As a consequence of the above definitions, the global format of the balance of vector moment of spatial momentum expands as

$$\int_{\mathfrak{B}_{\mathrm{s}}} \mathrm{d}_t \boldsymbol{l}_{\mathrm{s}} \,\mathrm{d}v = \int_{\mathfrak{B}_{\mathrm{s}}} \overline{\boldsymbol{t}}_{\mathrm{s}} \,\mathrm{d}v + \int_{\partial\mathfrak{B}_{\mathrm{s}}} [\boldsymbol{\theta} - \boldsymbol{l}_{\mathrm{s}} \otimes \boldsymbol{v}] \cdot \boldsymbol{n} \,\mathrm{d}a + \int_{\mathfrak{S}_{\mathrm{s}}} [\![\boldsymbol{l}_{\mathrm{s}} \otimes \boldsymbol{w}]\!] \cdot \boldsymbol{m} \,\mathrm{d}a, \tag{7.74}$$

which, by incorporating the specific formats for the source and flux terms, specializes to

$$\begin{aligned} \int_{\mathfrak{B}_{\mathrm{s}}} \mathrm{d}_t \boldsymbol{l}_{\mathrm{s}} \,\mathrm{d}v = \int_{\mathfrak{B}_{\mathrm{s}}} [\boldsymbol{d}_0 \times \overline{\boldsymbol{s}}_{\mathrm{s}} + \overline{\boldsymbol{t}}_{\mathrm{s}}^{\mathrm{aut}}] \,\mathrm{d}v + \int_{\partial\mathfrak{B}_{\mathrm{s}}} [\boldsymbol{d}_0 \times \boldsymbol{\sigma}_{\mathrm{d}}] \cdot \boldsymbol{n} \,\mathrm{d}a \\ + \int_{\partial\mathfrak{S}_{\mathrm{s}}} [\![\boldsymbol{l}_{\mathrm{s}} \otimes \boldsymbol{w}]\!] \cdot \boldsymbol{m} \,\mathrm{d}a. \end{aligned} \tag{7.75}$$

It becomes again obvious that the resultant external spatial couple is exclusively due to the external spatial forces only for $\overline{\boldsymbol{t}}_{\mathrm{s}}^{\mathrm{aut}} \equiv \boldsymbol{0}$.

7.3.4 *Local Format (Spatial Control Volume)*

The local balance for the density of vector moment of spatial momentum $\boldsymbol{l}_{\mathrm{s}}$ per unit volume in the spatial configuration, expressed in terms of the Cauchy-type couple stress $\boldsymbol{\theta} = \boldsymbol{L} \cdot \operatorname{cof} \boldsymbol{f}$, the source density of vector moment of spatial momentum $\overline{\boldsymbol{t}}_{\mathrm{s}}$ per unit volume in the spatial configuration and the extrinsic flux density of vector moment of spatial momentum $\widehat{\boldsymbol{t}}_{\mathrm{s}}^{\mathrm{ext}}$ per unit area in the spatial configuration, is summarized as

Local Balance of Vector Moment of Spatial Momentum (Spatial Control Volume):

i. Regular Points in the Domain

$$\mathrm{d}_t \boldsymbol{l}_{\mathrm{s}} - \operatorname{div}(\boldsymbol{\theta} - \boldsymbol{l}_{\mathrm{s}} \otimes \boldsymbol{v}) = \overline{\boldsymbol{t}}_{\mathrm{s}} \quad \text{in} \quad \mathcal{B}_{\mathrm{s}}. \tag{7.76}$$

ii. Regular Points on the Boundary

$$\left[\boldsymbol{l}_s \otimes \boldsymbol{w} + [\boldsymbol{\theta} - \boldsymbol{l}_s \otimes \boldsymbol{v}]\right] \cdot \boldsymbol{n} = \widehat{\boldsymbol{t}}_s^{\text{ext}} \quad \text{on} \quad \partial\mathcal{B}_s. \tag{7.77}$$

iii. Regular Points at Singular Surfaces

$$[\![\boldsymbol{l}_s \otimes \boldsymbol{w} + [\boldsymbol{\theta} - \boldsymbol{l}_s \otimes \boldsymbol{v}]]\!] \cdot \boldsymbol{m} = \boldsymbol{0} \quad \text{at} \quad \mathcal{S}_s. \tag{7.78}$$

Incorporating Specific Formats for Density, Source, and Flux Terms

Incorporating the specific formats for $\boldsymbol{l}_s$, $\overline{\boldsymbol{t}}_s$, $\widehat{\boldsymbol{t}}_s$ and $\boldsymbol{\theta}$ in Eqs. 7.68, 7.70, and 7.73, respectively, while noting the identity

$$\text{div}(\boldsymbol{d}_0 \times \boldsymbol{\sigma}_d) = \boldsymbol{d}_0 \times \text{div}\boldsymbol{\sigma}_d + 2\text{axl}\boldsymbol{\sigma}_d, \tag{7.79}$$

wherein $\nabla_x \boldsymbol{d}_0 = \boldsymbol{i}$ is used, the local balance of vector moment of spatial momentum reads explicitly

Local Balance of Vector Moment of Spatial Momentum (Spatial Control Volume):

i. Regular Points in the Domain

$$\boldsymbol{d}_0 \times [\,\mathrm{d}_t \boldsymbol{p}_s - \text{div}\boldsymbol{\sigma}_d] - 2\text{axl}\boldsymbol{\sigma}_d = \boldsymbol{d}_0 \times \overline{\boldsymbol{s}}_s + \overline{\boldsymbol{t}}_s^{\text{aut}} \quad \text{in} \quad \mathcal{B}_s. \tag{7.80}$$

ii. Regular Points on the Boundary

$$\boldsymbol{d}_0 \times [\boldsymbol{p}_s \otimes \boldsymbol{w} + \boldsymbol{\sigma}_d] \cdot \boldsymbol{n} = \boldsymbol{d}_0 \times \widehat{\boldsymbol{s}}_s^{\text{ext}} \quad \text{on} \quad \partial\mathcal{B}_s. \tag{7.81}$$

iii. Regular Points at Singular Surfaces

$$\boldsymbol{d}_0 \times [\![\boldsymbol{p}_s \otimes \boldsymbol{w} + \boldsymbol{\sigma}_d]\!] \cdot \boldsymbol{m} = \boldsymbol{0} \quad \text{at} \quad \mathcal{S}_s. \tag{7.82}$$

Observe that $\mathrm{d}_t \boldsymbol{d}_0 = \boldsymbol{0}$, thus $\mathrm{d}_t \boldsymbol{l}_s = \boldsymbol{d}_0 \times \mathrm{d}_t \boldsymbol{p}_s$, and $[\![\boldsymbol{d}_0]\!] = \boldsymbol{0}$, i.e. the coherence of the spatial deformation map, have been incorporated.

Considering next the local balance of spatial momentum in Eqs. 7.41–7.43 establishes that the balance of vector moment of spatial momentum is trivially satisfied if the autonomous contribution to the corresponding source density is finally identified as

Table 7.3 Tetragon of fully equivalent versions for the local balance of vector moment of spatial momentum at regular points in the domain, on the boundary, and at singular surfaces

Regular Points in the Domain

$\mathrm{D}_t \boldsymbol{l}_\mathrm{m} = \mathrm{Div} \boldsymbol{L}_\mathrm{D} + \overline{\boldsymbol{t}}_\mathrm{m}$	↔ PT	$j\, \mathrm{D}_t \boldsymbol{l}_\mathrm{m} = \mathrm{div} \boldsymbol{\theta}_\mathrm{D} + \overline{\boldsymbol{t}}_\mathrm{s}$
↕ TT		↕ TT
$J\, \mathrm{d}_t \boldsymbol{l}_\mathrm{s} = \mathrm{Div} \boldsymbol{L}_\mathrm{d} + \overline{\boldsymbol{t}}_\mathrm{m}$	↔ PT	$\mathrm{d}_t \boldsymbol{l}_\mathrm{s} = \mathrm{div} \boldsymbol{\theta}_\mathrm{d} + \overline{\boldsymbol{t}}_\mathrm{s}$

Regular Points on the Boundary

$\left[\boldsymbol{l}_\mathrm{m} \otimes \boldsymbol{W} + \boldsymbol{L}_\mathrm{D}\right] \cdot \boldsymbol{N} = \widehat{\boldsymbol{t}}_\mathrm{m}^\mathrm{ext}$	↔ PT	$\left[\boldsymbol{l}_\mathrm{s} \otimes [\boldsymbol{w} - \boldsymbol{v}] + \boldsymbol{\theta}_\mathrm{D}\right] \cdot \boldsymbol{n} = \widehat{\boldsymbol{t}}_\mathrm{s}^\mathrm{ext}$
↕ TT		↕ TT
$\left[\boldsymbol{l}_\mathrm{m} \otimes [\boldsymbol{W} - \boldsymbol{V}] + \boldsymbol{L}_\mathrm{d}\right] \cdot \boldsymbol{N} = \widehat{\boldsymbol{t}}_\mathrm{m}^\mathrm{ext}$	↔ PT	$\left[\boldsymbol{l}_\mathrm{s} \otimes \boldsymbol{w} + \boldsymbol{\theta}_\mathrm{d}\right] \cdot \boldsymbol{n} = \widehat{\boldsymbol{t}}_\mathrm{s}^\mathrm{ext}$

Regular Points at Singular Surfaces

$\llbracket \boldsymbol{l}_\mathrm{m} \otimes \boldsymbol{W} + \boldsymbol{L}_\mathrm{D} \rrbracket \cdot \boldsymbol{M} = \boldsymbol{0}$	↔ PT	$\llbracket \boldsymbol{l}_\mathrm{s} \otimes [\boldsymbol{w} - \boldsymbol{v}] + \boldsymbol{\theta}_\mathrm{D} \rrbracket \cdot \boldsymbol{m} = \boldsymbol{0}$
↕ TT		↕ TT
$\llbracket \boldsymbol{l}_\mathrm{m} \otimes [\boldsymbol{W} - \boldsymbol{V}] + \boldsymbol{L}_\mathrm{d} \rrbracket \cdot \boldsymbol{M} = \boldsymbol{0}$	↔ PT	$\llbracket \boldsymbol{l}_\mathrm{s} \otimes \boldsymbol{w} + \boldsymbol{\theta}_\mathrm{d} \rrbracket \cdot \boldsymbol{m} = \boldsymbol{0}$

with *Dynamic Versions* of Piola-type and Cauchy-type Couple Stress

$\boldsymbol{L}_\mathrm{D} := \boldsymbol{L}$	↔ PT	$\boldsymbol{\theta}_\mathrm{D} := \boldsymbol{\theta}$
↕ TT		↕ TT
$\boldsymbol{L}_\mathrm{d} := \boldsymbol{L} + \boldsymbol{l}_\mathrm{m} \otimes \boldsymbol{V}$	↔ PT	$\boldsymbol{\theta}_\mathrm{d} := \boldsymbol{\theta} - \boldsymbol{l}_\mathrm{s} \otimes \boldsymbol{v}$

$$\bar{\boldsymbol{t}}_{\mathrm{s}}^{\mathrm{aut}} := -2\operatorname{axl}\boldsymbol{\sigma}_{\mathrm{d}}. \tag{7.83}$$

Making the VMSM Balance Truly Independent (Spatial Isotropy)

Note that the *symmetry constraint* for (the dynamic version) of the Cauchy stress $\boldsymbol{\sigma}_{\mathrm{d}}^{\mathrm{skw}} := [\boldsymbol{\sigma} - \boldsymbol{p}_{\mathrm{s}} \otimes \boldsymbol{v}]^{\mathrm{skw}} \equiv \boldsymbol{\sigma}^{\mathrm{skw}} \doteq \mathbf{0}$ (due to $[\rho_{\mathrm{s}}\boldsymbol{v} \otimes \boldsymbol{v}]^{\mathrm{skw}} \equiv \mathbf{0}$) is fully equivalent to the symmetry constraint for the Kirchhoff stress $\boldsymbol{\tau}^{\mathrm{skw}} = [\boldsymbol{P} \cdot \boldsymbol{F}^{t}]^{\mathrm{skw}} \doteq \mathbf{0}$ as discussed earlier.

7.3.5 Balance Tetragon

The local balance of vector moment of spatial momentum *at regular points in the domain*, *at regular points on the boundary*, and *at regular points at singular surfaces* may be expressed in four different but fully equivalent versions, see the arrangement in the corresponding tetragons in Table 7.3. Here, *dynamic versions* of the Piola-type and the Cauchy-type couple stress have been introduced as $\boldsymbol{\theta}_{\mathrm{D}} = \boldsymbol{L}_{\mathrm{D}} \cdot \operatorname{cof} \boldsymbol{f}$ and $\boldsymbol{\theta}_{\mathrm{d}} = \boldsymbol{L}_{\mathrm{d}} \cdot \operatorname{cof} \boldsymbol{f}$, respectively, whereby the subscripts D and d indicate that the corresponding intrinsic flux of vector moment of spatial momentum is consistent with either the material or the spatial time derivative D_t or d_t, respectively, of the density of vector moment of spatial momentum.

7.4 Scalar Moment of Spatial Momentum*

Scalar moments of spatial forces are denoted as *spatial virials* and serve as a measure of the *centrality* of the spatial force system. They contribute to the *balance of scalar moment of spatial momentum*.

7.4.1 Global Format (Material Control Volume)

The *balance of scalar moment of spatial momentum* is stated in global format as

Global Balance of Scalar Moment of Spatial Momentum (Material Control Volume):

$$\dot{\mathsf{o}}(\mathfrak{B}_{\mathrm{m}}) = \mathsf{v}^{\mathrm{ext}}(\mathfrak{B}_{\mathrm{m}}) \quad \forall \mathfrak{B}_{\mathrm{m}} \subseteq \mathcal{B}_{\mathrm{m}}. \tag{7.84}$$

Thereby, o is expressed as the resultant over the material control volume $\mathfrak{B}_{\mathrm{m}}$ of the density of *scalar moment of spatial momentum* o_{m} per unit volume in the material configuration, i.e. as the *resultant scalar moment of spatial momentum* contained in the material control volume

$$\mathsf{o}(\mathfrak{B}_{\mathrm{m}}) := \int_{\mathfrak{B}_{\mathrm{m}}} o_{\mathrm{m}} \,\mathrm{d}V. \tag{7.85}$$

The density of scalar moment of spatial momentum o_{m} per unit volume in the material configuration is defined as the *scalar* product of the spatial distance vector $\boldsymbol{d}_0$ and the density of spatial momentum $\boldsymbol{p}_{\mathrm{m}}$ per unit volume in the material configuration

$$o_{\mathrm{m}} := \boldsymbol{d}_0 \cdot \boldsymbol{p}_{\mathrm{m}}. \tag{7.86}$$

Moreover, $\mathsf{v}^{\mathrm{ext}}$ represents the resultant external spatial virial (supply of scalar moment of spatial momentum) and consists of volume and boundary integrals (any singular contributions are here ignored for the sake of presentation)

$$\mathsf{v}^{\mathrm{ext}}(\mathfrak{B}_{\mathrm{m}}) := \int_{\mathfrak{B}_{\mathrm{m}}} \overline{p}_{\mathrm{m}} \,\mathrm{d}V + \int_{\partial\mathfrak{B}_{\mathrm{m}}} \widehat{p}_{\mathrm{m}} \,\mathrm{d}A \tag{7.87}$$

in terms of the source density of scalar moment of spatial momentum $\overline{p}_{\mathrm{m}}$ per unit volume in the material configuration and the *extrinsic* flux density of scalar moment of spatial momentum $\widehat{p}_{\mathrm{m}}$ per unit area in the material configuration.

These are defined as the scalar product of the spatial distance vector $\boldsymbol{d}_0$ and the corresponding source density of spatial momentum $\overline{\boldsymbol{s}}_{\mathrm{m}}$ and the extrinsic flux density of spatial momentum $\widehat{\boldsymbol{s}}_{\mathrm{m}}$, respectively, together with, in general, an autonomous contribution to the source density of scalar moment of spatial momentum that will be identified later

$$\overline{p}_{\mathrm{m}} := \boldsymbol{d}_0 \cdot \overline{\boldsymbol{s}}_{\mathrm{m}} + \overline{p}_{\mathrm{m}}^{\mathrm{aut}} \quad \text{and} \quad \widehat{p}_{\mathrm{m}} := \boldsymbol{d}_0 \cdot \widehat{\boldsymbol{s}}_{\mathrm{m}}. \tag{7.88}$$

The extrinsic flux density of scalar moment of spatial momentum along with the convective term stemming from the transport theorem are postulated to satisfy a Cauchy theorem for the *intrinsic* flux density of scalar moment of spatial momentum

$$p_{\mathrm{m}} := \boldsymbol{O} \cdot \boldsymbol{N} = \widehat{p}_{\mathrm{m}} - [o_{\mathrm{m}} \boldsymbol{W}] \cdot \boldsymbol{N} \quad \text{on} \quad \partial\mathfrak{B}_{\mathrm{m}}. \tag{7.89}$$

Thereby, $\boldsymbol{O} \in T\mathcal{B}_{\mathrm{m}}$ is introduced as the *Piola-type virial stress*, a material description vector mapping from the material cotangent space $T^*\mathcal{B}_{\mathrm{m}}$ to $\mathbb{R}$.

Incorporating the specific formats for the flux and density terms, the intrinsic flux density of scalar moment of spatial momentum and the corresponding Cauchy theorem particularize to

$$p_{\rm m} := \boldsymbol{O} \cdot \boldsymbol{N} = \boldsymbol{d}_0 \cdot \left[\widehat{\boldsymbol{s}}_{\rm m} - [\boldsymbol{p}_{\rm m} \otimes \boldsymbol{W}] \cdot \boldsymbol{N}\right] = \boldsymbol{d}_0 \cdot \boldsymbol{s}_{\rm m} \quad \text{on} \quad \partial\mathfrak{B}_{\rm m}, \tag{7.90}$$

from which, together with Eq. 7.27, the Piola-type virial stress is eventually identified as

$$\boldsymbol{O} := \boldsymbol{d}_0 \cdot \boldsymbol{P}. \tag{7.91}$$

As a consequence of the above definitions, the global format of the balance of scalar moment of spatial momentum expands as

$$\int_{\mathfrak{B}_{\rm m}} \mathrm{D}_t o_{\rm m} \,\mathrm{d}V = \int_{\mathfrak{B}_{\rm m}} \overline{p}_{\rm m} \,\mathrm{d}V + \int_{\partial\mathfrak{B}_{\rm m}} \underbrace{\boldsymbol{O} \cdot \boldsymbol{N}}_{p_{\rm m}} \,\mathrm{d}A + \int_{\partial\mathfrak{S}_{\rm m}} [\![o_{\rm m} \boldsymbol{W}]\!] \cdot \boldsymbol{M} \,\mathrm{d}A, \tag{7.92}$$

which, by incorporating the specific formats for the source and flux terms, specializes to

$$\int_{\mathfrak{B}_{\rm m}} \mathrm{D}_t o_{\rm m} \,\mathrm{d}V = \int_{\mathfrak{B}_{\rm m}} [\boldsymbol{d}_0 \cdot \overline{\boldsymbol{s}}_{\rm m} + \overline{p}_{\rm m}^{\rm aut}] \,\mathrm{d}V + \int_{\partial\mathfrak{B}_{\rm m}} \boldsymbol{d}_0 \cdot \underbrace{\boldsymbol{P} \cdot \boldsymbol{N}}_{\boldsymbol{s}_{\rm m}} \,\mathrm{d}A \tag{7.93}$$
$$+ \int_{\partial\mathfrak{S}_{\rm m}} [\![o_{\rm m} \boldsymbol{W}]\!] \cdot \boldsymbol{M} \,\mathrm{d}A.$$

The latter representation clearly reveals that only in case that the autonomous contribution to the source density of scalar moment of spatial momentum vanishes, i.e. for $\overline{p}_{\rm m}^{\rm aut} \equiv 0$, the resultant external spatial virial is exclusively due to the external spatial forces, thus making the balance of of scalar moment of spatial momentum an independent balance.

7.4.2 *Local Format (Material Control Volume)*

The results of localizing the global balance of scalar moment of spatial momentum are summarized as

Local Balance of Scalar Moment of Spatial Momentum (Material Control Volume):

i. Regular Points in the Domain

$$\mathrm{D}_t o_{\rm m} - \mathrm{Div}\boldsymbol{O} = \overline{p}_{\rm m} \quad \text{in} \quad \mathcal{B}_{\rm m}. \tag{7.94}$$

ii. Regular Points on the Boundary

$$[o_{\rm m} \boldsymbol{W} + \boldsymbol{O}] \cdot \boldsymbol{N} = \widehat{p}_{\rm m}^{\rm ext} \quad \text{on} \quad \partial\mathcal{B}_{\rm m}. \tag{7.95}$$

iii. Regular Points at Singular Surfaces

$$[\![o_{\mathrm{m}}\,\boldsymbol{W} + \boldsymbol{O}]\!] \cdot \boldsymbol{M} = 0 \quad \text{at} \quad \mathcal{S}_{\mathrm{m}}. \tag{7.96}$$

The local balance of scalar moment of spatial momentum allows the following general conclusion:

i. The balance of scalar moment of spatial momentum on (the regular part of) the external boundary $\partial\mathcal{B}_{\mathrm{m}}$ in Eq. 7.95 results, when restricted to singular surfaces, in the *cancelation* of the extrinsic flux density of scalar moment of spatial momentum

$$\widehat{p}_{\mathrm{m}}^{+} + \widehat{p}_{\mathrm{m}}^{-} = 0 \quad \text{at} \quad \mathcal{S}_{\mathrm{m}} \tag{7.97}$$

and translates into the jump condition across singular surfaces

$$[\![o_{\mathrm{m}}]\!] W_{\perp} + [\![\boldsymbol{O}]\!] \cdot \boldsymbol{M} = 0 \quad \text{at} \quad \mathcal{S}_{\mathrm{m}}. \tag{7.98}$$

Incorporating Specific Formats for Density, Source, and Flux Terms

Incorporating the specific formats for o_{m}, $\overline{p}_{\mathrm{m}}$, $\widehat{p}_{\mathrm{m}}$, and $\boldsymbol{O}$ in Eqs. 7.86, 7.88, and 7.91, respectively, while noting the identities

$$\mathrm{Div}(\boldsymbol{d}_0 \cdot \boldsymbol{P}) = \boldsymbol{d}_0 \cdot \mathrm{Div}\,\boldsymbol{P} + 3\mathrm{prs}(\boldsymbol{P} \cdot \boldsymbol{F}^t) \tag{7.99}$$

and

$$\mathrm{D}_t(\boldsymbol{d}_0 \cdot \boldsymbol{p}_{\mathrm{m}}) = \boldsymbol{d}_0 \cdot \mathrm{D}_t\boldsymbol{p}_{\mathrm{m}} + 3\mathrm{prs}(\boldsymbol{p}_{\mathrm{m}} \otimes \boldsymbol{v}), \tag{7.100}$$

whereby, e.g. $\mathrm{prs}(\boldsymbol{P} \cdot \boldsymbol{F}^t)$ denotes the *spatial pressure (scalar)*[3] associated with the *volumetric* contribution $[\boldsymbol{P} \cdot \boldsymbol{F}^t]^{\mathrm{vol}}$ to the product $\boldsymbol{P} \cdot \boldsymbol{F}^t$, the local balance of scalar moment of spatial momentum reads explicitly

---

[3] For non-zero vectors $\boldsymbol{a} \in \mathbb{E}^3$ the *pressure (scalar)* $t := \mathrm{prs}\,\boldsymbol{T} \in \mathbb{R}$ of an arbitrary second-order tensor $\boldsymbol{T} \in \mathbb{E}^3 \times \mathbb{E}^3$ is defined via the isomorphism

$$t\,\boldsymbol{a} := \boldsymbol{T}^{\mathrm{vol}} \cdot \boldsymbol{a} \quad \forall \boldsymbol{a} \quad \in \mathbb{E}^3.$$

In Cartesian coordinates with coordinate representation for $\boldsymbol{T} = T_a{}^b\,\boldsymbol{e}^a \otimes \boldsymbol{e}_b$, the pressure expands as $t := \frac{1}{3} T_a{}^b\,\delta^a{}_b$. Vice versa $\boldsymbol{T}^{\mathrm{vol}} = \delta_a{}^b\,t\,\boldsymbol{e}^a \otimes \boldsymbol{e}_b$ holds. Thereby, the (mixed-variant) Kronecker symbols are denoted by $\delta^a{}_b := \boldsymbol{e}^a \cdot \boldsymbol{e}_b$ and $\delta_a{}^b := \boldsymbol{e}_a \cdot \boldsymbol{e}^b$, respectively.

Local Balance of Scalar Moment of Spatial Momentum (Material Control Volume):

i. Regular Points in the Domain

$$\boldsymbol{d}_0 \cdot [\,\mathrm{D}_t \boldsymbol{p}_\mathrm{m} - \mathrm{Div}\boldsymbol{P}] - 3\mathrm{prs}\boldsymbol{\tau}_\mathrm{d} = \boldsymbol{d}_0 \cdot \bar{\boldsymbol{s}}_\mathrm{m} + \overline{p}_\mathrm{m}^\mathrm{aut} \quad \text{in} \quad \mathcal{B}_\mathrm{m}. \tag{7.101}$$

ii. Regular Points on the Boundary

$$\boldsymbol{d}_0 \cdot [\boldsymbol{p}_\mathrm{m} \otimes \boldsymbol{W} + \boldsymbol{P}] \cdot \boldsymbol{N} = \boldsymbol{d}_0 \cdot \widehat{\boldsymbol{s}}_\mathrm{m}^\mathrm{ext} \quad \text{on} \quad \partial\mathcal{B}_\mathrm{m}. \tag{7.102}$$

iii. Regular Points at Singular Surfaces

$$\boldsymbol{d}_0 \cdot [\![\boldsymbol{p}_\mathrm{m} \otimes \boldsymbol{W} + \boldsymbol{P}]\!] \cdot \boldsymbol{M} = 0 \quad \text{at} \quad \mathcal{S}_\mathrm{m}. \tag{7.103}$$

Observe that in the above the definition of the *dynamic* Kirchhoff stress $\boldsymbol{\tau}_\mathrm{d} := \boldsymbol{P} \cdot \boldsymbol{F}^t - \boldsymbol{p}_\mathrm{m} \otimes \boldsymbol{v}$ and the coherence of the spatial deformation map $[\![\boldsymbol{d}_0]\!] = \boldsymbol{0}$ have been used.

Considering next the local balance of spatial momentum in Eqs. 7.29–7.31 establishes that the balance of scalar moment of spatial momentum is trivially satisfied if the autonomous contribution to the corresponding source density of scalar moment of spatial momentum is finally identified as

$$\overline{p}_\mathrm{m}^\mathrm{aut} := -3\mathrm{prs}\boldsymbol{\tau}_\mathrm{d}. \tag{7.104}$$

Making the SMSM Balance Truly Independent (Spatial PHODT)

The balance of scalar moment of spatial momentum (in short SMSM) is an *independent* balance only in case that $\overline{p}_\mathrm{m}^\mathrm{aut} \equiv 0$. Recall that in this situation the resultant external spatial virial is exclusively due to the external spatial forces.

The corresponding *deviator constraint* $\boldsymbol{\tau}_\mathrm{d}^\mathrm{vol} \doteq \boldsymbol{0}$ for the dynamic Kirchhoff stress is a consequence of the special case of *spatial phodt* (see remark below) and can in general be regarded a constraint for the solution of an initial boundary value problem (similar to the classical *incompressibility constraint* $\mathrm{div}\boldsymbol{v} \doteq 0$ for the spatial velocity field). As a consequence, the balance of scalar moment of spatial momentum is in general not an independent balance equation!

Remark (Spatial PHODT):
For the special quasi-static, hyperelastic case, $\boldsymbol{\sigma} = \psi_\mathrm{s}\boldsymbol{i} - \boldsymbol{f}^t \cdot \partial_{\boldsymbol{f}}\psi_\mathrm{s}$ (this format for $\boldsymbol{\sigma}$ is derived in Sect. 12.7) is constitutively *trace-less* with $\mathrm{prs}\,\boldsymbol{\sigma} = \psi_\mathrm{s} - \frac{1}{3}\partial_{\boldsymbol{f}}\psi_\mathrm{s} : \boldsymbol{f} = 0$, if $\psi_\mathrm{s} = \psi_3(\boldsymbol{f})$ is *positively homogeneous of degree three* (spatial phodt) in $\boldsymbol{f}$, i.e. if $\partial_{\boldsymbol{f}}\psi_\mathrm{s} : \boldsymbol{f} = 3\psi_\mathrm{s}$. However, this constitutive scenario is rare to appear. □

7.4.3 Global Format (Spatial Control Volume)

Alternatively, the balance of scalar moment of spatial momentum is stated in global format as

Global Balance of Scalar Moment of Spatial Momentum (Spatial Control Volume):

$$\dot{\mathsf{o}}(\mathfrak{B}_s) = \mathsf{v}^{\text{ext}}(\mathfrak{B}_s) \quad \forall \mathfrak{B}_s \subseteq \mathcal{B}_s. \tag{7.105}$$

Here, o is expressed as the resultant over the spatial control volume $\mathfrak{B}_s$ of the density of scalar moment of spatial momentum o_s per unit volume in the spatial configuration

$$\mathsf{o}(\mathfrak{B}_s) := \int_{\mathfrak{B}_s} o_s \, \mathrm{d}v. \tag{7.106}$$

The density of scalar moment of spatial momentum o_s per unit volume in the spatial configuration expands as

$$o_s := \boldsymbol{d}_0 \cdot \boldsymbol{p}_s. \tag{7.107}$$

Moreover, v^{ext} represents the resultant external spatial virial and consists of volume and boundary integrals (any singular contributions are here ignored for the sake of presentation)

$$\mathsf{v}^{\text{ext}}(\mathfrak{B}_s) := \int_{\mathfrak{B}_s} \overline{p}_s \, \mathrm{d}v + \int_{\partial\mathfrak{B}_s} \widehat{p}_s \, \mathrm{d}a \tag{7.108}$$

in terms of the source density of scalar moment of spatial momentum $\overline{p}_s = j\,\overline{p}_m$ per unit volume in the spatial configuration and the extrinsic flux density of scalar moment of spatial momentum $\widehat{p}_s = \widehat{j}\,\widehat{p}_m$ per unit area in the spatial configuration. These expand as

$$\overline{p}_s := \boldsymbol{d}_0 \cdot \overline{\boldsymbol{s}}_s + \overline{p}_s^{\text{aut}} \quad \text{and} \quad \widehat{p}_s := \boldsymbol{d}_0 \cdot \widehat{\boldsymbol{s}}_s. \tag{7.109}$$

The intrinsic flux density of scalar moment of spatial momentum is defined as the Piola transformation $\boldsymbol{\phi} := \boldsymbol{O} \cdot \operatorname{cof} \boldsymbol{f}$, thus the corresponding Cauchy theorem for $\boldsymbol{\phi} \cdot \boldsymbol{n}$ reads in terms of the extrinsic flux density of scalar moment of spatial momentum along with the convective term stemming from the transport theorem

$$p_s := \boldsymbol{\phi} \cdot \boldsymbol{n} := \widehat{p}_s - \big[o_s\,[\boldsymbol{w} - \boldsymbol{v}]\big] \cdot \boldsymbol{n} \quad \text{on} \quad \partial\mathfrak{B}_s. \tag{7.110}$$

Thereby, $\boldsymbol{\phi} \in T^*\mathcal{B}_s \times T\mathcal{B}_s$ is introduced as the *Cauchy-type virial stress*, a spatial description (mixed-variant) tensor mapping from and to the spatial cotangent space $T^*\mathcal{B}_s$.

Incorporating the specific format for the flux and density terms, the intrinsic flux density of scalar moment of spatial momentum and the corresponding Cauchy theorem particularize to

$$p_s := \boldsymbol{\phi} \cdot \boldsymbol{n} = \boldsymbol{d}_0 \cdot \Big[\widehat{\boldsymbol{s}}_s - [\boldsymbol{p}_s \otimes [\boldsymbol{w} - \boldsymbol{v}]] \Big] \cdot \boldsymbol{n} = \boldsymbol{d}_0 \cdot \boldsymbol{s}_s \quad \text{on} \quad \partial \mathfrak{B}_s, \tag{7.111}$$

from which, together with Eq. 7.39, the Cauchy-type virial stress is eventually identified as

$$\boldsymbol{\phi} := \boldsymbol{d}_0 \cdot \boldsymbol{\sigma}. \tag{7.112}$$

As a consequence of the above definitions, the global format of the balance of spatial moment of spatial momentum expands as

$$\int_{\mathfrak{B}_s} \mathrm{d}_t o_s \, \mathrm{d}v = \int_{\mathfrak{B}_s} \overline{p}_s \, \mathrm{d}v + \int_{\partial \mathfrak{B}_s} [\boldsymbol{\phi} - o_s \, \boldsymbol{v}] \cdot \boldsymbol{n} \, \mathrm{d}a + \int_{\mathfrak{S}_s} [\![o_s \, \boldsymbol{w}]\!] \cdot \boldsymbol{m} \, \mathrm{d}a, \tag{7.113}$$

which, by incorporating the specific formats for the source and flux terms, specializes to

$$\begin{aligned} \int_{\mathfrak{B}_s} \mathrm{d}_t o_s \, \mathrm{d}v = & \int_{\mathfrak{B}_s} [\boldsymbol{d}_0 \cdot \overline{\boldsymbol{s}}_s + \overline{p}_s^{\mathrm{aut}}] \, \mathrm{d}v + \int_{\partial \mathfrak{B}_s} [\boldsymbol{d}_0 \cdot \boldsymbol{\sigma}_\mathrm{d}] \cdot \boldsymbol{n} \, \mathrm{d}a \\ & + \int_{\partial \mathfrak{S}_s} [\![o_s \, \boldsymbol{w}]\!] \cdot \boldsymbol{m} \, \mathrm{d}a. \end{aligned} \tag{7.114}$$

It becomes again obvious that the resultant external spatial virial is exclusively due to the external spatial forces only for $\overline{p}_s^{\mathrm{aut}} \equiv 0$.

7.4.4 *Local Format (Spatial Control Volume)*

The local balance for the density of scalar moment of spatial momentum o_s per unit volume in the spatial configuration, expressed in terms of the Cauchy-type virial stress $\boldsymbol{\phi} = \boldsymbol{O} \cdot \operatorname{cof} \boldsymbol{f}$, the source density of scalar moment of spatial momentum $\overline{p}_s$ per unit volume in the spatial configuration and the extrinsic flux density of scalar moment of spatial momentum $\widehat{p}_s^{\mathrm{ext}}$ per unit area in the spatial configuration, is summarized as

Local Balance of Spatial Moment of Spatial Momentum (Spatial Control Volume):

i. Regular Points in the Domain

$$\mathrm{d}_t o_\mathrm{s} - \mathrm{div}(\boldsymbol{\phi} - o_\mathrm{s}\,\boldsymbol{v}) = \overline{p}_\mathrm{s} \quad \text{in} \quad \mathcal{B}_\mathrm{s} \tag{7.115}$$

ii. Regular Points on the Boundary

$$\big[o_\mathrm{s}\,\boldsymbol{w} + [\boldsymbol{\phi} - o_\mathrm{s}\,\boldsymbol{v}]\big] \cdot \boldsymbol{n} = \widehat{p}_\mathrm{s}^\mathrm{ext} \quad \text{on} \quad \partial\mathcal{B}_\mathrm{s} \tag{7.116}$$

iii. Regular Points at Singular Surfaces

$$[\![o_\mathrm{s}\,\boldsymbol{w} + [\boldsymbol{\phi} - o_\mathrm{s}\,\boldsymbol{v}]]\!] \cdot \boldsymbol{m} = 0 \quad \text{at} \quad \mathcal{S}_\mathrm{s} \tag{7.117}$$

Incorporating Specific Formats for Density, Source, and Flux Terms

Incorporating the specific formats for o_s, $\overline{p}_\mathrm{s}$, $\widehat{p}_\mathrm{s}$, and $\boldsymbol{\phi}$ in Eqs. 7.107, 7.109, and 7.112, respectively, while noting the identity

$$\mathrm{div}(\boldsymbol{d}_0 \cdot \boldsymbol{\sigma}_\mathrm{d}) = \boldsymbol{d}_0 \cdot \mathrm{div}\boldsymbol{\sigma}_\mathrm{d} + 3\mathrm{prs}\boldsymbol{\sigma}_\mathrm{d}, \tag{7.118}$$

wherein $\nabla_x \boldsymbol{d}_0 = \boldsymbol{i}$ is used, the local balance of scalar moment of spatial momentum reads explicitly

Local Balance of Scalar Moment of Spatial Momentum (Spatial Control Volume):

i. Regular Points in the Domain

$$\boldsymbol{d}_0 \cdot [\,\mathrm{d}_t \boldsymbol{p}_\mathrm{s} - \mathrm{div}\boldsymbol{\sigma}_\mathrm{d}] - 3\mathrm{prs}\boldsymbol{\sigma}_\mathrm{d} = \boldsymbol{d}_0 \cdot \overline{\boldsymbol{s}}_\mathrm{s} + \overline{p}_\mathrm{s}^\mathrm{aut} \quad \text{in} \quad \mathcal{B}_\mathrm{s}. \tag{7.119}$$

ii. Regular Points on the Boundary

$$\boldsymbol{d}_0 \cdot [\boldsymbol{p}_\mathrm{s} \otimes \boldsymbol{w} + \boldsymbol{\sigma}_\mathrm{d}] \cdot \boldsymbol{n} = \boldsymbol{d}_0 \cdot \widehat{\boldsymbol{s}}_\mathrm{s}^\mathrm{ext} \quad \text{on} \quad \partial\mathcal{B}_\mathrm{s}. \tag{7.120}$$

iii. Regular Points at Singular Surfaces

$$\boldsymbol{d}_0 \cdot [\![\boldsymbol{p}_\mathrm{s} \otimes \boldsymbol{w} + \boldsymbol{\sigma}_\mathrm{d}]\!] \cdot \boldsymbol{m} = 0 \quad \text{at} \quad \mathcal{S}_\mathrm{s}. \tag{7.121}$$

Table 7.4 Tetragon of fully equivalent versions for the local balance of scalar moment of spatial momentum at regular points in the domain, on the boundary, and at singular surfaces

Regular Points in the Domain

	PT	
$\mathrm{D}_t o_\mathrm{m} = \mathrm{Div}\boldsymbol{O}_\mathrm{D} + \overline{p}_\mathrm{m}$	⟷ PT	$j\,\mathrm{D}_t o_\mathrm{m} = \mathrm{div}\boldsymbol{\phi}_\mathrm{D} + \overline{p}_\mathrm{s}$
↕ TT		↕ TT
$J\,\mathrm{d}_t o_\mathrm{s} = \mathrm{Div}\boldsymbol{O}_\mathrm{d} + \overline{p}_\mathrm{m}$	⟷ PT	$\mathrm{d}_t o_\mathrm{s} = \mathrm{div}\boldsymbol{\phi}_\mathrm{d} + \overline{p}_\mathrm{s}$

Regular Points on the Boundary

$\left[o_\mathrm{m}\,\boldsymbol{W} + \boldsymbol{O}_\mathrm{D}\right]\cdot\boldsymbol{N} = \widehat{p}_\mathrm{m}^\mathrm{ext}$	⟷ PT	$\left[o_\mathrm{s}\,[\boldsymbol{w}-\boldsymbol{v}] + \boldsymbol{\phi}_\mathrm{D}\right]\cdot\boldsymbol{n} = \widehat{p}_\mathrm{s}^\mathrm{ext}$
↕ TT		↕ TT
$\left[o_\mathrm{m}\,[\boldsymbol{W}-\boldsymbol{V}] + \boldsymbol{O}_\mathrm{d}\right]\cdot\boldsymbol{N} = \widehat{p}_\mathrm{m}^\mathrm{ext}$	⟷ PT	$\left[o_\mathrm{s}\,\boldsymbol{w} + \boldsymbol{\phi}_\mathrm{d}\right]\cdot\boldsymbol{n} = \widehat{p}_\mathrm{s}^\mathrm{ext}$

Regular Points at Singular Surfaces

$[\![o_\mathrm{m}\,\boldsymbol{W} + \boldsymbol{O}_\mathrm{D}]\!]\cdot\boldsymbol{M} = 0$	⟷ PT	$[\![o_\mathrm{s}\,[\boldsymbol{w}-\boldsymbol{v}] + \boldsymbol{\phi}_\mathrm{D}]\!]\cdot\boldsymbol{m} = 0$
↕ TT		↕ TT
$[\![o_\mathrm{m}\,[\boldsymbol{W}-\boldsymbol{V}] + \boldsymbol{O}_\mathrm{d}]\!]\cdot\boldsymbol{M} = 0$	⟷ PT	$[\![o_\mathrm{s}\,\boldsymbol{w} + \boldsymbol{\phi}_\mathrm{d}]\!]\cdot\boldsymbol{m} = 0$

with *Dynamic Versions* of Piola-type and Cauchy-type Virial Stress

$\boldsymbol{O}_\mathrm{D} := \boldsymbol{L}$	⟷ PT	$\boldsymbol{\phi}_\mathrm{D} := \boldsymbol{\phi}$
↕ TT		↕ TT
$\boldsymbol{O}_\mathrm{d} := \boldsymbol{L} + o_\mathrm{m}\,\boldsymbol{V}$	⟷ PT	$\boldsymbol{\phi}_\mathrm{d} := \boldsymbol{\phi} - o_\mathrm{s}\,\boldsymbol{v}$

Observe that $\mathrm{d}_t \boldsymbol{d}_0 = \mathbf{0}$, thus $\mathrm{d}_t o_\mathrm{s} = \boldsymbol{d}_0 \cdot \mathrm{d}_t \boldsymbol{p}_\mathrm{s}$, and $[\![\boldsymbol{d}_0]\!] = \mathbf{0}$, i.e. the coherence of the spatial deformation map, have been incorporated.

Considering next the local balance of spatial momentum in Eqs. 7.41–7.43 establishes that the balance of scalar moment of spatial momentum is trivially satisfied if the autonomous contribution to the corresponding source density is finally identified as

$$\overline{p}_\mathrm{s}^\mathrm{aut} := -3\mathrm{prs}\boldsymbol{\sigma}_\mathrm{d}. \tag{7.122}$$

Making the VMSM Balance Truly Independent

Note that the *deviator constraint* for the dynamic version of the Cauchy stress $\boldsymbol{\sigma}_\mathrm{d}^\mathrm{vol} := [\boldsymbol{\sigma} - \boldsymbol{p}_\mathrm{s} \otimes \boldsymbol{v}]^\mathrm{vol} \doteq \mathbf{0}$ is fully equivalent to the deviator constraint for the dynamic version of the Kirchhoff stress $\boldsymbol{\tau}_\mathrm{d}^\mathrm{vol} := [\boldsymbol{P} \cdot \boldsymbol{F}^t - \boldsymbol{p}_\mathrm{m} \otimes \boldsymbol{v}]^\mathrm{vol} \doteq \mathbf{0}$ as discussed earlier.

7.4.5 Balance Tetragon

The local balance of scalar moment of spatial momentum *at regular points in the domain*, *at regular points on the boundary*, and *at regular points at singular surfaces* may be expressed in four different but fully equivalent versions, see the arrangement in the corresponding tetragons in Table 7.4. Here, *dynamic versions* of the Piola-type and the Cauchy-type virial stress have been introduced as $\boldsymbol{\phi}_\mathrm{D} = \boldsymbol{O}_\mathrm{D} \cdot \mathrm{cof}\, \boldsymbol{f}$ and $\boldsymbol{\phi}_\mathrm{d} = \boldsymbol{O}_\mathrm{d} \cdot \mathrm{cof}\, \boldsymbol{f}$, respectively, whereby the subscripts D and d indicate that the corresponding intrinsic flux of scalar moment of spatial momentum is consistent with either the material or the spatial time derivative D_t or d_t, respectively, of the density of scalar moment of spatial momentum.

References

1. Kuhl E , Steinmann P (2003) Mass–and volume–specific views on thermodynamics for open systems. Proc R Soc London Ser A: Math, Phys Eng Sci 459:2547–2568
2. Kuhl E, Steinmann P (2003) On spatial and material settings of thermo-hyperelastodynamics for open systems. Acta Mech 160:179–217
3. Kuhl E, Steinmann P (2003) Theory and numerics of geometrically non-linear open system mechanics. Int J Numer Meth Eng 58:1593–1615
4. Kuhl E, Steinmann P (2004) Computational modeling of healing: an application of the material force method. Biomech Model Mechanobiol 2:187–203
5. Kuhl E, Steinmann P (2004) Material forces in open system mechanics. Comput Methods Appl Mech Eng 193:2357–2381
6. Marsden JE, Hughes TJR (1994) Mathematical foundations of elasticity. Dover, Illinois

Chapter 8
Consequences of Mechanical Balances

8,188 m 28°05'39"N 86°39'39"E

Abstract This chapter explores the consequences of the mechanical balances by elaborating on local and global formats of the balance of kinetic energy and the balance of material momentum and its vector and scalar moments, thereby differentiating between spatial and material control volumes.

The *balance of spatial momentum* is the fundamental vectorial statement that dictates the dynamics of a continuous mechanical system, i.e. of a *continuum body*. Interestingly, further insightful information on the dynamics of a continuum body is extractable from *manipulating* the balance of spatial momentum.
Its local version may be multiplied on the one hand by the *time gradient* of the spatial deformation map, aka the spatial velocity, and on the other hand by the *space gradient* of the spatial deformation map, aka the spatial deformation gradient. The former results eventually in the *balance of kinetic energy*, whereas the latter ultimately renders the *balance of material momentum*. However, these are not balances established in their own right but merely result as *consequences* of the balance of spatial momentum. In essence, the balances of kinetic energy and material momentum allow to observe the dynamics of a continuum body through a pair of polarized glasses, thereby elucidating features that are otherwise hidden. These 'derived' balances are of utmost support and importance when formulating and exploiting thermo-dynamical balances and their consequences in Chaps. 11 and 12.

Furthermore, balances of vector and scalar moments of material momentum follow from taking corresponding moments of the balance of material momentum. Collectively, these are related to the L-, and M-integrals of defect mechanics and the (vectorial) J-integral of fracture mechanics, respectively.

P. Steinmann, *Spatial and Material Forces in Nonlinear Continuum Mechanics*,
Solid Mechanics and Its Applications 272,
https://doi.org/10.1007/978-3-030-89070-4_8

8.1 Kinetic Energy

The *balance of kinetic energy* is not established in its own right but results as a *consequence* of the balance of spatial momentum.

8.1.1 Local Format (Material Control Volume)

The generic version of its local statement is summarized as

Generic Local Balance of Kinetic Energy (Material Control Volume):

i. Regular Points in the Domain

$$[2\,\mathrm{D}_t\kappa_\mathrm{m} - \boldsymbol{p}_\mathrm{m}\cdot\mathrm{D}_t\boldsymbol{v}] - \mathrm{Div}(\boldsymbol{v}\cdot\boldsymbol{P}) + \nabla_X\boldsymbol{v}:\boldsymbol{P} = \boldsymbol{v}\cdot\overline{\boldsymbol{s}}_\mathrm{m} \quad \text{in} \quad \mathcal{B}_\mathrm{m}. \tag{8.1}$$

ii. Regular Points on the Boundary

$$[2\kappa_\mathrm{m}\boldsymbol{W} + \boldsymbol{v}\cdot\boldsymbol{P}]\cdot\boldsymbol{N} = \boldsymbol{v}\cdot\widehat{\boldsymbol{s}}^\mathrm{ext}_\mathrm{m} \quad \text{on} \quad \partial\mathcal{B}_\mathrm{m}. \tag{8.2}$$

iii. Regular Points at Singular Surfaces

$$\Big[[\![2\kappa_\mathrm{m}]\!] - \{\boldsymbol{p}_\mathrm{m}\}\cdot[\![\boldsymbol{v}]\!]\big]\boldsymbol{W} + [\![\boldsymbol{v}\cdot\boldsymbol{P}]\!] - [\![\boldsymbol{v}]\!]\cdot\{\boldsymbol{P}\}\Big]\cdot\boldsymbol{M} = 0 \quad \text{at} \quad \mathcal{S}_\mathrm{m}. \tag{8.3}$$

For its derivation, starting from the multiplication of the local balance of spatial momentum with the *time gradient* of the spatial deformation map, i.e. with the spatial velocity, three different scenarios may be considered:

i. In a smooth domain $\mathcal{B}_\mathrm{m}$ multiplication of the balance of spatial momentum in Eq. 7.29 by the spatial velocity $\boldsymbol{v}$ renders

$$\boldsymbol{v}\cdot[\,\mathrm{D}_t\boldsymbol{p}_\mathrm{m} - \mathrm{Div}\boldsymbol{P}] = \boldsymbol{v}\cdot\overline{\boldsymbol{s}}_\mathrm{m} \quad \text{in} \quad \mathcal{B}_\mathrm{m}. \tag{8.4}$$

Then, based on the *product rule of differentiation*, the identities

$$\boldsymbol{v}\cdot\mathrm{D}_t\boldsymbol{p}_\mathrm{m} = 2\,\mathrm{D}_t\kappa_\mathrm{m} - \boldsymbol{p}_\mathrm{m}\cdot\mathrm{D}_t\boldsymbol{v},$$

since $\boldsymbol{v}\cdot\boldsymbol{p}_\mathrm{m} = 2\kappa_\mathrm{m}$, and

$$\boldsymbol{v}\cdot\mathrm{Div}\boldsymbol{P} = \mathrm{Div}(\boldsymbol{v}\cdot\boldsymbol{P}) - \nabla_X\boldsymbol{v}:\boldsymbol{P}$$

render the result in Eq. 8.1.

Note that, for the special case of mass conservation with $\mathrm{D}_t\rho_\mathrm{m} = 0$, the first two terms $2\,\mathrm{D}_t\kappa_\mathrm{m} - \boldsymbol{p}_\mathrm{m} \cdot \mathrm{D}_t\boldsymbol{v}$ in Eq. 8.1 simplify to $\mathrm{D}_t\kappa_\mathrm{m}$ since then $\boldsymbol{p}_\mathrm{m} \cdot \mathrm{D}_t\boldsymbol{v} = \mathrm{D}_t\kappa_\mathrm{m}$, whereby $\mathrm{D}_t[\boldsymbol{v}\cdot\boldsymbol{v}] = 2\boldsymbol{v}\cdot\mathrm{D}_t\boldsymbol{v}$ is used.

ii. On (the regular part of) the external boundary $\partial\mathcal{B}_\mathrm{m}$ multiplication of the balance of spatial momentum in Eq. 7.30 by the spatial velocity $\boldsymbol{v}$ renders

$$\boldsymbol{v}\cdot[\boldsymbol{p}_\mathrm{m}\otimes\boldsymbol{W}+\boldsymbol{P}]\cdot\boldsymbol{N} = \boldsymbol{v}\cdot\widehat{\boldsymbol{s}}^\mathrm{ext}_\mathrm{m} \quad \text{on} \quad \partial\mathcal{B}_\mathrm{m}. \tag{8.5}$$

Then $\boldsymbol{v}\cdot\boldsymbol{p}_\mathrm{m} = 2\kappa_\mathrm{m}$ immediately leads to the result in Eq. 8.2.

iii. At singular surfaces $\mathcal{S}_\mathrm{m}$ multiplication of the balance of spatial momentum in Eq. 7.31 by the average spatial velocity $\{\boldsymbol{v}\}$ renders

$$\{\boldsymbol{v}\}\cdot[\![\boldsymbol{p}_\mathrm{m}\otimes\boldsymbol{W}+\boldsymbol{P}]\!]\cdot\boldsymbol{M} = 0 \quad \text{on} \quad \mathcal{S}_\mathrm{m}. \tag{8.6}$$

Then, based on the *product rule of jumps*, the identities

$$\{\boldsymbol{v}\}\cdot[\![\boldsymbol{p}_\mathrm{m}]\!] = [\![2\kappa_\mathrm{m}]\!] - \{\boldsymbol{p}_\mathrm{m}\}\cdot[\![\boldsymbol{v}]\!],$$

since $[\![\boldsymbol{v}\cdot\boldsymbol{p}_\mathrm{m}]\!] = [\![2\kappa_\mathrm{m}]\!]$, and

$$\{\boldsymbol{v}\}\cdot[\![\boldsymbol{P}]\!] = [\![\boldsymbol{v}\cdot\boldsymbol{P}]\!] - [\![\boldsymbol{v}]\!]\cdot\{\boldsymbol{P}\}$$

render the result in Eq. 8.3.

Note that, for the special case of mass conservation with $[\![\rho_\mathrm{m}]\!] = 0$, the first two terms $[\![2\kappa_\mathrm{m}]\!] - \{\boldsymbol{p}_\mathrm{m}\}\cdot[\![\boldsymbol{v}]\!]$ in Eq. 8.3 simplify to $[\![\kappa_\mathrm{m}]\!]$ since then $\{\boldsymbol{p}_\mathrm{m}\}\cdot[\![\boldsymbol{v}]\!] = [\![\kappa_\mathrm{m}]\!]$, whereby $[\![\boldsymbol{v}\cdot\boldsymbol{v}]\!] = 2\{\boldsymbol{v}\}\cdot[\![\boldsymbol{v}]\!]$ is used.

Finally, the last term in Eq. 8.3 may be rewritten with the help of the coherence conditions VII and III in Eqs. 4.21 and 4.11 as

$$-[\![\boldsymbol{v}]\!]\cdot\{\boldsymbol{P}\}\cdot\boldsymbol{M} \overset{\text{(VII)}}{=} W_\perp\cdot[\![\boldsymbol{F}^t]\!]\cdot\{\boldsymbol{P}\}\cdot\boldsymbol{M} \overset{\text{(III)}}{=} W_\perp\,[\![\boldsymbol{F}]\!]:\{\boldsymbol{P}\}. \tag{8.7}$$

Thus, for the special case of mass conservation, Eq. 8.3 reads alternatively

$$\big[[\![\kappa_\mathrm{m}]\!] + [\![\boldsymbol{F}]\!]:\{\boldsymbol{P}\}\big]W_\perp + [\![\boldsymbol{v}\cdot\boldsymbol{P}]\!]\cdot\boldsymbol{M} = 0. \tag{8.8}$$

Internal Mechanical Power Generation at Singular Surfaces

Next, the jump condition across singular surfaces $\mathcal{S}_\mathrm{m}$ as generically given in Eq. 8.3 may be re-expressed in an interesting alternative format:

On the one hand, due to the cancelation of the extrinsic spatial momentum flux density $\widehat{\boldsymbol{s}}^+_\mathrm{m} + \widehat{\boldsymbol{s}}^-_\mathrm{m} = \boldsymbol{0}$ on $\mathcal{S}_\mathrm{m}$, the *product rule of averages* renders for the density of the (negative) internal mechanical power generation at singular surfaces

$$2\{\boldsymbol{v}\cdot\widehat{\boldsymbol{s}}_\mathrm{m}\} = \frac{1}{2}[\![\boldsymbol{v}]\!]\cdot[\![\widehat{\boldsymbol{s}}_\mathrm{m}]\!] =: -[\![\boldsymbol{v}]\!]\cdot\widetilde{\boldsymbol{s}}_\mathrm{m}, \tag{8.9}$$

whereby the *power generating traction* at singular surfaces is defined as

$$\widetilde{s}_{\mathrm{m}} := -\frac{1}{2}[\![\widehat{s}_{\mathrm{m}}]\!] = \{\boldsymbol{p}_{\mathrm{m}} \otimes \boldsymbol{W} + \boldsymbol{P}\} \cdot \boldsymbol{M}. \tag{8.10}$$

On the other hand, direct addition of the extrinsic kinetic energy flux densities in Eq. 8.2 at $\mathcal{S}_{\mathrm{m}}$ results in

$$\boldsymbol{v}^+ \cdot \widehat{\boldsymbol{s}}^+_{\mathrm{m}} + \boldsymbol{v}^- \cdot \widehat{\boldsymbol{s}}^-_{\mathrm{m}} = -[\![2\kappa_{\mathrm{m}} \boldsymbol{W} + \boldsymbol{v} \cdot \boldsymbol{P}]\!] \cdot \boldsymbol{M}. \tag{8.11}$$

Thus, the jump condition at singular surfaces as given in Eq. 8.3 may finally be re-expressed by

$$[\![2\kappa_{\mathrm{m}} \boldsymbol{W} + \boldsymbol{v} \cdot \boldsymbol{P}]\!] \cdot \boldsymbol{M} = [\![\boldsymbol{v}]\!] \cdot \widetilde{\boldsymbol{s}}_{\mathrm{m}} \quad \text{on} \quad \mathcal{S}_{\mathrm{m}}. \tag{8.12}$$

Observe that due to the coherence conditions VII and III in Eqs. 4.21 and 4.11 the term on the right-hand side of Eq. 8.12 expands into

$$[\![\boldsymbol{v}]\!] \cdot \widetilde{\boldsymbol{s}}_{\mathrm{m}} = \big[\{\boldsymbol{p}_{\mathrm{m}}\} \cdot [\![\boldsymbol{v}]\!] - [\![\boldsymbol{F}]\!] : \{\boldsymbol{P}\}\big] W_{\perp}. \tag{8.13}$$

Note finally that, for the special case of mass conservation, the term on the right-hand side of Eq. 8.12 simplifies further into

$$[\![\boldsymbol{v}]\!] \cdot \widetilde{\boldsymbol{s}}_{\mathrm{m}} = \big[[\![\kappa_{\mathrm{m}}]\!] - [\![\boldsymbol{F}]\!] : \{\boldsymbol{P}\}\big] W_{\perp} \tag{8.14}$$

thus rendering Eq. 8.12 to be consistent with Eq. 8.8.

Incorporating Mass Conservation

Consequently, for the special case of mass conservation, the local format of the balance of kinetic energy may be re-stated in an alternative, more intuitive special version that clearly identifies $\boldsymbol{E}^{\kappa} := \boldsymbol{v} \cdot \boldsymbol{P}$ as the flux and the right-hand sides as the sources of kinetic energy and that eventually alleviates its global interpretation:

Specific Local Balance of Kinetic Energy (Material Control Volume):

i. Regular Points in the Domain

$$\mathrm{D}_t \kappa_{\mathrm{m}} - \mathrm{Div}(\boldsymbol{v} \cdot \boldsymbol{P}) = \boldsymbol{v} \cdot \overline{\boldsymbol{s}}_{\mathrm{m}} - \nabla_X \boldsymbol{v} : \boldsymbol{P} \quad \text{in} \quad \mathcal{B}_{\mathrm{m}}. \tag{8.15}$$

ii. Regular Points on the Boundary

$$\begin{aligned} [\kappa_{\mathrm{m}} \boldsymbol{W} + \boldsymbol{v} \cdot \boldsymbol{P}] \cdot \boldsymbol{N} &= \boldsymbol{v} \cdot \widehat{\boldsymbol{s}}^{\mathrm{ext}}_{\mathrm{m}} - [\kappa_{\mathrm{m}} \boldsymbol{W}] \cdot \boldsymbol{N} \quad \text{on} \quad \partial\mathcal{B}_{\mathrm{m}} \\ \rightsquigarrow [\boldsymbol{v} \cdot \boldsymbol{P}] \cdot \boldsymbol{N} &= \boldsymbol{v} \cdot \boldsymbol{s}^{\mathrm{ext}}_{\mathrm{m}}. \end{aligned} \tag{8.16}$$

iii. Regular Points at Singular Surfaces

$$\begin{aligned}[\![\kappa_m \boldsymbol{W} + \boldsymbol{v} \cdot \boldsymbol{P}]\!] \cdot \boldsymbol{M} &= [\![\boldsymbol{v}]\!] \cdot \widetilde{\boldsymbol{s}}_m - [\![\kappa_m \boldsymbol{W}]\!] \cdot \boldsymbol{M} \quad \text{at} \quad \mathcal{S}_m \\ \rightsquigarrow [\![\boldsymbol{v} \cdot \boldsymbol{P}]\!] \cdot \boldsymbol{M} &= -\big[[\![\kappa_m]\!] + [\![\boldsymbol{F}]\!] : \{\boldsymbol{P}\}\big] W_\perp .\end{aligned} \tag{8.17}$$

8.1.2 *Global Format (Material Control Volume)*

Indeed, for the case of mass conservation and with these preliminaries at hand, the balance of kinetic energy may also be stated in global format as

Global Balance of Kinetic Energy (Material Control Volume):

$$\dot{\mathsf{K}}(\mathfrak{B}_m) = \mathsf{P}^{ext}(\mathfrak{B}_m) - \mathsf{P}^{int}(\mathfrak{B}_m) \quad \forall \mathfrak{B}_m \subseteq \mathcal{B}_m . \tag{8.18}$$

Thereby K is introduced as the resultant over the material control volume $\mathfrak{B}_m$ of the density of *kinetic energy* κ_m per unit volume in the material configuration, i.e. as the *resultant kinetic energy* contained in the material control volume

$$\mathsf{K}(\mathfrak{B}_m) := \int_{\mathfrak{B}_m} \kappa_m \, \mathrm{d}V . \tag{8.19}$$

Moreover, P^{ext} denotes the *resultant external mechanical power supply* (consisting of volume and boundary integrals)

$$\mathsf{P}^{ext}(\mathfrak{B}_m) := \int_{\mathfrak{B}_m} \boldsymbol{v} \cdot \bar{\boldsymbol{s}}_m \, \mathrm{d}V + \int_{\partial\mathfrak{B}_m} \boldsymbol{v} \cdot \widehat{\boldsymbol{s}}_m \, \mathrm{d}A \tag{8.20}$$

and P^{int} denotes the *resultant internal mechanical power generation* (consisting of volume, boundary *and* surface integrals)

$$\begin{aligned}\mathsf{P}^{int}(\mathfrak{B}_m) := &\int_{\mathfrak{B}_m} \nabla_X \boldsymbol{v} : \boldsymbol{P} \, \mathrm{d}V + \int_{\mathfrak{S}_m} [\![\boldsymbol{v}]\!] \cdot \widetilde{\boldsymbol{s}}_m \, \mathrm{d}A \\ &- \int_{\mathfrak{S}_m} [\![\kappa_m \boldsymbol{W}]\!] \cdot \boldsymbol{M} \, \mathrm{d}A + \int_{\partial\mathfrak{B}_m} [\kappa_m \boldsymbol{W}] \cdot \boldsymbol{N} \, \mathrm{d}A\end{aligned} \tag{8.21}$$

$$= \int_{\mathfrak{B}_m} \nabla_X \boldsymbol{v} : \boldsymbol{P} \, \mathrm{d}V - \int_{\mathfrak{S}_m} [\![\boldsymbol{F}]\!] : \{\boldsymbol{P}\} W_\perp \, \mathrm{d}A + \int_{\partial\mathfrak{B}_m} [\kappa_m \boldsymbol{W}] \cdot \boldsymbol{N} \, \mathrm{d}A.$$

Note the additional contributions to the resultant internal mechanical power generation in Eq. 8.21.1 due to the power generation at singular surfaces and the convection of the kinetic energy at singular surfaces and the boundary, respectively. In Eq. 8.21.2, the explicit representation for the power generation $[\![\boldsymbol{v}]\!] \cdot \widetilde{\boldsymbol{s}}_m = \big[[\![\kappa_m]\!] - [\![\boldsymbol{F}]\!] : \{\boldsymbol{P}\}\big] W_\perp$ at singular surfaces for the case of mass conservation has been incorporated. (In all of the above definitions any singular contributions are ignored for the sake of presentation.)

Consequently, the global statement of the balance of kinetic energy expands as

$$\begin{aligned} \int_{\mathfrak{B}_m} \mathrm{D}_t \kappa_m \, \mathrm{d}V &= \int_{\mathfrak{B}_m} \boldsymbol{v} \cdot \overline{\boldsymbol{s}}_m \, \mathrm{d}V + \int_{\partial\mathfrak{B}_m} \boldsymbol{v} \cdot \boldsymbol{s}_m \, \mathrm{d}A - \int_{\mathfrak{B}_m} \nabla_X \boldsymbol{v} : \boldsymbol{P} \, \mathrm{d}V \\ &\quad - \int_{\mathfrak{S}_m} [\![\boldsymbol{v}]\!] \cdot \widetilde{\boldsymbol{s}}_m \, \mathrm{d}A + \int_{\mathfrak{S}_m} [\![2\kappa_m \boldsymbol{W}]\!] \cdot \boldsymbol{M} \, \mathrm{d}A \\ &= \int_{\mathfrak{B}_m} \boldsymbol{v} \cdot \overline{\boldsymbol{s}}_m \, \mathrm{d}V + \int_{\partial\mathfrak{B}_m} \boldsymbol{v} \cdot \boldsymbol{s}_m \, \mathrm{d}A - \int_{\mathfrak{B}_m} \nabla_X \boldsymbol{v} : \boldsymbol{P} \, \mathrm{d}V \\ &\quad + \int_{\mathfrak{S}_m} \big[[\![\kappa_m]\!] + [\![\boldsymbol{F}]\!] : \{\boldsymbol{P}\}\big] W_\perp \, \mathrm{d}A. \end{aligned} \tag{8.22}$$

Here, the two convective terms stemming from the transport theorem and from the resultant internal mechanical power have been subsumed in Eq. 8.22.1 in the boundary integrand

$$\boldsymbol{v} \cdot \boldsymbol{s}_m = \boldsymbol{v} \cdot \boldsymbol{P} \cdot \boldsymbol{N} = \boldsymbol{v} \cdot \widehat{\boldsymbol{s}}_m - [2\kappa_m \boldsymbol{W}] \cdot \boldsymbol{N} \tag{8.23}$$

and the singular surface integrand $[\![2\kappa_m \boldsymbol{W}]\!] \cdot \boldsymbol{M}$. In Eq. 8.22.2 the explicit representation for the power generation $[\![\boldsymbol{v}]\!] \cdot \widetilde{\boldsymbol{s}}_m = \big[[\![\kappa_m]\!] - [\![\boldsymbol{F}]\!] : \{\boldsymbol{P}\}\big] W_\perp$ at singular surfaces for the case of mass conservation has additionally been incorporated. Note that upon localizing the global balance of kinetic energy in Eq. 8.22 indeed Eqs. 8.15–8.17 are obtained.

In summary the global balance of kinetic energy relates the total time derivative of the resultant kinetic energy to the difference between the resultant external and internal mechanical power under the precondition that the balance of spatial momentum is satisfied.

8.1.3 Local Format (Spatial Control Volume)

The generic version of the local statement of the balance of kinetic energy is summarized as

Generic Local Balance of Kinetic Energy (Spatial Control Volume):

i. Regular Points in the Domain

$$[2\,\mathrm{d}_t\kappa_s - \boldsymbol{p}_s \cdot \mathrm{d}_t\boldsymbol{v}] - \mathrm{div}(\boldsymbol{v}\cdot\boldsymbol{\sigma}_d) + \nabla_x\boldsymbol{v} : \boldsymbol{\sigma}_d = \boldsymbol{v}\cdot\overline{\boldsymbol{s}}_s \quad \text{in} \quad \mathcal{B}_s. \tag{8.24}$$

ii. Regular Points on the Boundary

$$[2\kappa_s\boldsymbol{w} + \boldsymbol{v}\cdot\boldsymbol{\sigma}_d]\cdot\boldsymbol{n} = \boldsymbol{v}\cdot\widehat{\boldsymbol{s}}_s^{\mathrm{ext}} \quad \text{on} \quad \partial\mathcal{B}_s. \tag{8.25}$$

iii. Regular Points at Singular Surfaces

$$\Big[[\![2\kappa_s]\!] - \{\boldsymbol{p}_s\}\cdot[\![\boldsymbol{v}]\!]\Big]\boldsymbol{w} + [\![\boldsymbol{v}\cdot\boldsymbol{\sigma}_d]\!] - [\![\boldsymbol{v}]\!]\cdot\{\boldsymbol{\sigma}_d\}\Big]\cdot\boldsymbol{m} = 0 \quad \text{at} \quad \mathcal{S}_s. \tag{8.26}$$

Again, for its derivation, starting from the multiplication of the local balance of spatial momentum with the *time gradient* of the spatial deformation map, i.e. with the spatial velocity, three different scenarios may be considered:

i. In a smooth domain $\mathcal{B}_s$ multiplication of the balance of spatial momentum in Eq. 7.41 by the spatial velocity $\boldsymbol{v}$ renders

$$\boldsymbol{v}\cdot[\,\mathrm{d}_t\boldsymbol{p}_s - \mathrm{div}\boldsymbol{\sigma}_d] = \boldsymbol{v}\cdot\overline{\boldsymbol{s}}_s \quad \text{in} \quad \mathcal{B}_s. \tag{8.27}$$

Then, based on the *product rule of differentiation*, the identities

$$\boldsymbol{v}\cdot\mathrm{d}_t\boldsymbol{p}_s = 2\,\mathrm{d}_t\kappa_s - \boldsymbol{p}_s\cdot\mathrm{d}_t\boldsymbol{v},$$

since $\boldsymbol{v}\cdot\boldsymbol{p}_s = 2\kappa_s$, and

$$\boldsymbol{v}\cdot\mathrm{div}\boldsymbol{\sigma}_d = \mathrm{div}(\boldsymbol{v}\cdot\boldsymbol{\sigma}_d) - \nabla_x\boldsymbol{v} : \boldsymbol{\sigma}_d$$

render the result in Eq. 8.24.

Note that, for the special case of mass conservation with $\mathrm{d}_t\rho_s = -\mathrm{div}\boldsymbol{p}_s$, the first two terms $2\,\mathrm{d}_t\kappa_s - \boldsymbol{p}_s\cdot\mathrm{d}_t\boldsymbol{v}$ in Eq. 8.24 simplify to $\mathrm{d}_t\kappa_s - \mathrm{div}(\kappa_s\boldsymbol{v}) + \boldsymbol{p}_s\cdot\nabla_x\boldsymbol{v}\cdot\boldsymbol{v}$ since then $\boldsymbol{p}_s\cdot\mathrm{d}_t\boldsymbol{v} = \mathrm{d}_t\kappa_s + \frac{1}{2}\boldsymbol{v}\cdot\boldsymbol{v}\,\mathrm{div}\boldsymbol{p}_s$, whereby $\mathrm{d}_t[\boldsymbol{v}\cdot\boldsymbol{v}] = 2\boldsymbol{v}\cdot\mathrm{d}_t\boldsymbol{v}$ is used. Moreover, $\frac{1}{2}\boldsymbol{v}\cdot\boldsymbol{v}\,\mathrm{div}\boldsymbol{p}_s = \mathrm{div}(\kappa_s\boldsymbol{v}) - \boldsymbol{p}_s\cdot\nabla_x\boldsymbol{v}\cdot\boldsymbol{v}$ holds. Together with the convective terms $\mathrm{div}(2\kappa_s\boldsymbol{v}) - \boldsymbol{p}_s\cdot\nabla_x\boldsymbol{v}\cdot\boldsymbol{v}$ contained in the last two terms

$-\text{div}(\boldsymbol{v} \cdot \boldsymbol{\sigma}_\text{d}) + \nabla_x \boldsymbol{v} : \boldsymbol{\sigma}_\text{d}$ of Eq. 8.24 eventually only $\text{d}_t \kappa_\text{s} + \text{div}(\kappa_\text{s} \boldsymbol{v}) - \text{div}(\boldsymbol{v} \cdot \boldsymbol{\sigma}) + \nabla_x \boldsymbol{v} : \boldsymbol{\sigma}$ remains on the left-hand side of Eq. 8.24.

ii. On (the regular part of) the external boundary $\partial\mathcal{B}_\text{s}$ multiplication of the balance of spatial momentum in Eq. 7.42 by the spatial velocity $\boldsymbol{v}$ renders

$$\boldsymbol{v} \cdot \big[\boldsymbol{p}_\text{s} \otimes [\boldsymbol{w} - \boldsymbol{v}] + \boldsymbol{\sigma}\big] \cdot \boldsymbol{n} = \boldsymbol{v} \cdot \widehat{\boldsymbol{s}}_\text{s}^\text{ext} \quad \text{on} \quad \partial\mathcal{B}_\text{s}. \tag{8.28}$$

Then $\boldsymbol{v} \cdot \boldsymbol{p}_\text{s} = 2\kappa_\text{s}$ immediately leads to the result in Eq. 8.25.

iii. At singular surfaces $\mathcal{S}_\text{s}$ multiplication of the balance of spatial momentum in Eq. 7.43 by the average spatial velocity $\{\boldsymbol{v}\}$ renders

$$\{\boldsymbol{v}\} \cdot [\![\boldsymbol{p}_\text{s} \otimes [\boldsymbol{w} - \boldsymbol{v}] + \boldsymbol{\sigma}]\!] \cdot \boldsymbol{m} = 0 \quad \text{on} \quad \mathcal{S}_\text{s}. \tag{8.29}$$

Then, based on the *product rule of jumps*, the identities

$$\{\boldsymbol{v}\} \cdot [\![\boldsymbol{p}_\text{s} \otimes [\boldsymbol{w} - \boldsymbol{v}]]\!] = [\![2\kappa_\text{s}[\boldsymbol{w} - \boldsymbol{v}]]\!] - [\![\boldsymbol{v}]\!] \cdot \{\boldsymbol{p}_\text{s} \otimes [\boldsymbol{w} - \boldsymbol{v}]\},$$

since $[\![\boldsymbol{v} \cdot \boldsymbol{p}_\text{s}]\!] = [\![2\kappa_\text{s}]\!]$, and

$$\{\boldsymbol{v}\} \cdot [\![\boldsymbol{\sigma}]\!] = [\![\boldsymbol{v} \cdot \boldsymbol{\sigma}]\!] - [\![\boldsymbol{v}]\!] \cdot \{\boldsymbol{\sigma}\}$$

render the result in Eq. 8.26.

Note that, for the special case of mass conservation, the first two terms $\big[[\![2\kappa_\text{s}[\boldsymbol{w} - \boldsymbol{v}]]\!] - [\![\boldsymbol{v}]\!] \cdot \{\boldsymbol{p}_\text{s} \otimes [\boldsymbol{w} - \boldsymbol{v}]\}\big] \cdot \boldsymbol{m}$ in Eq. 8.26 simplify to $[\![\kappa_\text{s}[\boldsymbol{w} - \boldsymbol{v}]]\!] \cdot \boldsymbol{m}$ since then $[\![\boldsymbol{v}]\!] \cdot \{\boldsymbol{p}_\text{s} \otimes [\boldsymbol{w} - \boldsymbol{v}]\} \cdot \boldsymbol{m} = [\![\kappa_\text{s}[\boldsymbol{w} - \boldsymbol{v}]]\!]$, whereby the product rule of averages, mass conservation $[\![\rho_\text{s}[\boldsymbol{w} - \boldsymbol{v}]]\!] \cdot \boldsymbol{m} = 0$, and $[\![\boldsymbol{v} \cdot \boldsymbol{v}]\!] = 2\{\boldsymbol{v}\} \cdot [\![\boldsymbol{v}]\!]$ are used.

Finally, the last term in Eq. 8.26 may be rewritten with the help of the coherence conditions VII and III in Eqs. 4.21 and 4.11 as

$$-[\![\boldsymbol{v}]\!] \cdot \{\boldsymbol{\sigma}\} \cdot \boldsymbol{m} \overset{\text{(VII)}}{=} \boldsymbol{W}_\perp \cdot [\![\boldsymbol{F}^t]\!] \cdot \{\boldsymbol{\sigma}\} \cdot \boldsymbol{m} \overset{\text{(III)}}{=} W_\perp \, \boldsymbol{j} \cdot \{\boldsymbol{\sigma}\} \cdot \boldsymbol{m}. \tag{8.30}$$

Thus, for the special case of mass conservation, Eq. 8.26 reads alternatively

$$\big[[\![\kappa_\text{s}[\boldsymbol{w} - \boldsymbol{v}]]\!] + W_\perp \, \boldsymbol{j} \cdot \{\boldsymbol{\sigma}\} + [\![\boldsymbol{v} \cdot \boldsymbol{\sigma}]\!]\big] \cdot \boldsymbol{m} = 0. \tag{8.31}$$

Note that $W_\perp \, [\boldsymbol{j} \otimes \boldsymbol{m}]$ has the flavor as the 'jump' of the spatial gradient of the spatial velocity $\nabla_x \boldsymbol{v} = \text{D}_t \boldsymbol{F} \cdot \boldsymbol{f}$, whereas $W_\perp \, [\boldsymbol{j} \otimes \boldsymbol{M}] = W_\perp \, [\![\boldsymbol{F}]\!]$ has the flavor as the 'jump' of the material gradient of the spatial velocity $\nabla_X \boldsymbol{v} = \text{D}_t \boldsymbol{F}$.

Internal Mechanical Power Generation at Singular Surfaces

Again, the jump condition across singular surfaces $\mathcal{S}_\text{s}$ as generically given in Eq. 8.26 may be re-expressed in an interesting alternative format:

On the one hand, due to the cancelation of the extrinsic spatial momentum flux density $\widehat{s}_s^+ + \widehat{s}_s^- = \mathbf{0}$ on $\mathcal{S}_m$, the *product rule of averages* renders for the density of the (negative) internal mechanical power generation at singular surfaces

$$2\{\boldsymbol{v} \cdot \widehat{\boldsymbol{s}}_s\} = \frac{1}{2}[\![\boldsymbol{v}]\!] \cdot [\![\widehat{\boldsymbol{s}}_s]\!] =: -[\![\boldsymbol{v}]\!] \cdot \widetilde{\boldsymbol{s}}_s, \tag{8.32}$$

whereby the *power generating traction* at singular surfaces is defined as

$$\widetilde{\boldsymbol{s}}_s := -\frac{1}{2}[\![\widehat{\boldsymbol{s}}_s]\!] = \{\boldsymbol{p}_s \otimes [\boldsymbol{w} - \boldsymbol{v}] + \boldsymbol{\sigma}\} \cdot \boldsymbol{m}. \tag{8.33}$$

On the other hand, direct addition of the extrinsic kinetic energy flux densities in Eq. 8.25 at $\mathcal{S}_s$ results in

$$\boldsymbol{v}^+ \cdot \widehat{\boldsymbol{s}}_s^+ + \boldsymbol{v}^- \cdot \widehat{\boldsymbol{s}}_s^- = -[\![2\kappa_s[\boldsymbol{w} - \boldsymbol{v}] + \boldsymbol{v} \cdot \boldsymbol{\sigma}]\!] \cdot \boldsymbol{m}. \tag{8.34}$$

Thus, the jump condition at singular surfaces as given in Eq. 8.26 may finally be re-expressed by

$$[\![2\kappa_s[\boldsymbol{w} - \boldsymbol{v}] + \boldsymbol{v} \cdot \boldsymbol{\sigma}]\!] \cdot \boldsymbol{m} = [\![\boldsymbol{v}]\!] \cdot \widetilde{\boldsymbol{s}}_s \quad \text{on} \quad \mathcal{S}_s. \tag{8.35}$$

Observe that due to the coherence conditions VII and III in Eqs. 4.21 and 4.11 the term on the right-hand side of Eq. 8.35 expands into

$$[\![\boldsymbol{v}]\!] \cdot \widetilde{\boldsymbol{s}}_s = [\![\boldsymbol{v}]\!] \cdot \{\boldsymbol{p}_s \otimes [\boldsymbol{w} - \boldsymbol{v}]\} \cdot \boldsymbol{m} - W_\perp\, \boldsymbol{j} \cdot \{\boldsymbol{\sigma}\} \cdot \boldsymbol{m}. \tag{8.36}$$

Note finally that, for the special case of mass conservation, the term on the right-hand side of Eq. 8.35 simplifies further into

$$[\![\boldsymbol{v}]\!] \cdot \widetilde{\boldsymbol{s}}_s = \Big[[\![\kappa_s[\boldsymbol{w} - \boldsymbol{v}]]\!] - W_\perp\, \boldsymbol{j} \cdot \{\boldsymbol{\sigma}\}\Big] \cdot \boldsymbol{m} \tag{8.37}$$

thus rendering Eq. 8.35 to be consistent with Eq. 8.31.

Incorporating Mass Conservation

Consequently, for the special case of mass conservation, the local format of the balance of kinetic energy may be re-stated in an alternative, more intuitive special version that clearly identifies $\boldsymbol{e}^\kappa := \boldsymbol{v} \cdot \boldsymbol{\sigma}$ as the flux and the right-hand sides as the sources of kinetic energy and that eventually alleviates its global interpretation:

Specific Local Balance of Kinetic Energy (Spatial Control Volume):

i. Regular Points in the Domain

$$\mathrm{d}_t \kappa_{\mathrm{s}} + \operatorname{div}(\kappa_{\mathrm{s}} \boldsymbol{v}) - \operatorname{div}(\boldsymbol{v} \cdot \boldsymbol{\sigma}) = \boldsymbol{v} \cdot \overline{\boldsymbol{s}}_{\mathrm{s}} - \nabla_x \boldsymbol{v} : \boldsymbol{\sigma} \quad \text{in} \quad \mathcal{B}_{\mathrm{s}}. \tag{8.38}$$

ii. Regular Points on the Boundary

$$\begin{aligned} \left[\kappa_{\mathrm{s}}[\boldsymbol{w} - \boldsymbol{v}] + \boldsymbol{v} \cdot \boldsymbol{\sigma}\right] \cdot \boldsymbol{n} &= \boldsymbol{v} \cdot \widehat{\boldsymbol{s}}_{\mathrm{s}}^{\mathrm{ext}} - \left[\kappa_{\mathrm{s}}[\boldsymbol{w} - \boldsymbol{v}]\right] \cdot \boldsymbol{n} \quad \text{on} \quad \partial\mathcal{B}_{\mathrm{s}} \\ \rightsquigarrow [\boldsymbol{v} \cdot \boldsymbol{\sigma}] \cdot \boldsymbol{n} &= \boldsymbol{v} \cdot \boldsymbol{s}_{\mathrm{s}}^{\mathrm{ext}}. \end{aligned} \tag{8.39}$$

iii. Regular Points at Singular Surfaces

$$\begin{aligned} \left[\!\left[\kappa_{\mathrm{s}}[\boldsymbol{w} - \boldsymbol{v}] + \boldsymbol{v} \cdot \boldsymbol{\sigma}\right]\!\right] \cdot \boldsymbol{m} &= [\![\boldsymbol{v}]\!] \cdot \widetilde{\boldsymbol{s}}_{\mathrm{s}} - \left[\!\left[\kappa_{\mathrm{s}}[\boldsymbol{w} - \boldsymbol{v}]\right]\!\right] \cdot \boldsymbol{m} \quad \text{at} \quad \mathcal{S}_{\mathrm{s}} \\ \rightsquigarrow [\![\boldsymbol{v} \cdot \boldsymbol{\sigma}]\!] \cdot \boldsymbol{m} &= -\left[\left[\!\left[\kappa_{\mathrm{s}}[\boldsymbol{w} - \boldsymbol{v}]\right]\!\right] + W_\perp \boldsymbol{j} \cdot \{\boldsymbol{\sigma}\}\right] \cdot \boldsymbol{m}. \end{aligned} \tag{8.40}$$

8.1.4 Global Format (Spatial Control Volume)

Indeed, for the case of mass conservation and with these preliminaries at hand, the balance of kinetic energy may also be stated in global format as

Global Balance of Kinetic Energy (Spatial Control Volume):

$$\dot{\mathsf{K}}(\mathfrak{B}_{\mathrm{s}}) = \mathsf{P}^{\mathrm{ext}}(\mathfrak{B}_{\mathrm{s}}) - \mathsf{P}^{\mathrm{int}}(\mathfrak{B}_{\mathrm{s}}) \quad \forall \mathfrak{B}_{\mathrm{s}} \subseteq \mathcal{B}_{\mathrm{s}}. \tag{8.41}$$

Thereby K is the resultant over the spatial control volume $\mathfrak{B}_{\mathrm{s}}$ of the density of *kinetic energy* κ_{s} per unit volume in the spatial configuration, i.e. as the *resultant kinetic energy* contained in the spatial control volume

$$\mathsf{K}(\mathfrak{B}_{\mathrm{s}}) := \int_{\mathfrak{B}_{\mathrm{s}}} \kappa_{\mathrm{s}} \, \mathrm{d}v. \tag{8.42}$$

Moreover, $\mathsf{P}^{\mathrm{ext}}$ denotes the *resultant external mechanical power supply* (consisting of volume and boundary integrals)

$$\mathsf{P}^{\mathrm{ext}}(\mathfrak{B}_{\mathrm{s}}) := \int_{\mathfrak{B}_{\mathrm{s}}} \boldsymbol{v} \cdot \overline{\boldsymbol{s}}_{\mathrm{s}} \, \mathrm{d}v + \int_{\partial\mathfrak{B}_{\mathrm{s}}} \boldsymbol{v} \cdot \widehat{\boldsymbol{s}}_{\mathrm{s}} \, \mathrm{d}a \tag{8.43}$$

and P^{int} denotes the *resultant internal mechanical power generation* (consisting of volume, boundary *and* surface integrals)

$$\begin{aligned}\mathsf{P}^{\text{int}}(\mathfrak{B}_s) := &\int_{\mathfrak{B}_s} \nabla_x \boldsymbol{v} : \boldsymbol{\sigma}\,\mathrm{d}v + \int_{\mathfrak{S}_s} [\![\boldsymbol{v}]\!] \cdot \widetilde{\boldsymbol{s}}_s\,\mathrm{d}a \\ &- \int_{\mathfrak{S}_s} [\![\kappa_s[\boldsymbol{w}-\boldsymbol{v}]]\!] \cdot \boldsymbol{m}\,\mathrm{d}a + \int_{\partial\mathfrak{B}_s} \big[\kappa_s[\boldsymbol{w}-\boldsymbol{v}]\big] \cdot \boldsymbol{n}\,\mathrm{d}a \\ = &\int_{\mathfrak{B}_s} \nabla_x \boldsymbol{v} : \boldsymbol{\sigma}\,\mathrm{d}v - \int_{\mathfrak{S}_s} W_\perp\,\boldsymbol{j} \cdot \{\boldsymbol{\sigma}\} \cdot \boldsymbol{m}\,\mathrm{d}a + \int_{\partial\mathfrak{B}_s} \big[\kappa_s[\boldsymbol{w}-\boldsymbol{v}]\big] \cdot \boldsymbol{n}\,\mathrm{d}a.\end{aligned} \tag{8.44}$$

Note the additional contributions to the resultant internal mechanical power generation in Eq. 8.44.1 due to the power generation at singular surfaces and the convection of the kinetic energy at singular surfaces and the boundary, respectively. In Eq. 8.44.2, the explicit representation for the power generation $[\![\boldsymbol{v}]\!] \cdot \widetilde{\boldsymbol{s}}_s = \big[[\![\kappa_m[\boldsymbol{w}-\boldsymbol{v}]]\!] - W_\perp\,\boldsymbol{j} \cdot \{\boldsymbol{\sigma}\}\big] \cdot \boldsymbol{m}$ at singular surfaces for the case of mass conservation has been incorporated. (In all of the above definitions any singular contributions are ignored for the sake of presentation.)

Consequently, the global statement of the balance of kinetic energy expands as

$$\begin{aligned}\int_{\mathfrak{B}_s} j\,\mathrm{D}_t\kappa_s\,\mathrm{d}v \equiv &\int_{\mathfrak{B}_s} \mathrm{d}_t\kappa_s + \operatorname{div}(\kappa_s\boldsymbol{v})\,\mathrm{d}v \\ = &\int_{\mathfrak{B}_s} \boldsymbol{v} \cdot \overline{\boldsymbol{s}}_s\,\mathrm{d}v + \int_{\partial\mathfrak{B}_s} \boldsymbol{v} \cdot \boldsymbol{s}_s\,\mathrm{d}a - \int_{\mathfrak{B}_s} \nabla_x \boldsymbol{v} : \boldsymbol{\sigma}\,\mathrm{d}v \\ &- \int_{\mathfrak{S}_s} [\![\boldsymbol{v}]\!] \cdot \widetilde{\boldsymbol{s}}_s\,\mathrm{d}a + \int_{\mathfrak{S}_s} [\![2\kappa_s[\boldsymbol{w}-\boldsymbol{v}]]\!] \cdot \boldsymbol{m}\,\mathrm{d}a \\ = &\int_{\mathfrak{B}_s} \boldsymbol{v} \cdot \overline{\boldsymbol{s}}_s\,\mathrm{d}v + \int_{\partial\mathfrak{B}_s} \boldsymbol{v} \cdot \boldsymbol{s}_s\,\mathrm{d}a - \int_{\mathfrak{B}_s} \nabla_x \boldsymbol{v} : \boldsymbol{\sigma}\,\mathrm{d}v \\ &+ \int_{\mathfrak{S}_s} \big[[\![\kappa_s[\boldsymbol{w}-\boldsymbol{v}]]\!] + W_\perp\,\boldsymbol{j} \cdot \{\boldsymbol{\sigma}\}\big] \cdot \boldsymbol{m}\,\mathrm{d}a.\end{aligned} \tag{8.45}$$

Here, the two convective terms stemming from the transport theorem and from the resultant internal mechanical power have been subsumed in Eq. 8.45.1 in the boundary integrand

$$\boldsymbol{v} \cdot \boldsymbol{s}_s = \boldsymbol{v} \cdot \boldsymbol{\sigma} \cdot \boldsymbol{n} = \boldsymbol{v} \cdot \widehat{\boldsymbol{s}}_s - \big[2\kappa_s[\boldsymbol{w}-\boldsymbol{v}]\big] \cdot \boldsymbol{n} \tag{8.46}$$

and the singular surface integrand $[\![2\kappa_s[\boldsymbol{w} - \boldsymbol{v}]]\!] \cdot \boldsymbol{m}$. In Eq. 8.45.2 the explicit representation for the power generation $[\![\boldsymbol{v}]\!] \cdot \widetilde{\boldsymbol{s}}_s = \left[[\![\kappa_s[\boldsymbol{w} - \boldsymbol{v}]]\!] - W_\perp\, \boldsymbol{j} \cdot \{\boldsymbol{\sigma}\}\right] \cdot \boldsymbol{m}$ at singular surfaces for the case of mass conservation has additionally been incorporated. Note that upon localizing the global balance of kinetic energy in Eq. 8.45 indeed Eqs. 8.38–8.40 are obtained.

8.1.5 Conjugated Spatial Stress and Strain Measures

The internal stress power density contributes to the balance of kinetic energy

$$\mathbb{P}^{\text{int}}_{\text{m}} := \boldsymbol{P} : \nabla_X \boldsymbol{v} = J\, \boldsymbol{\sigma} : \boldsymbol{l} =: J\, \mathbb{P}^{\text{int}}_{\text{s}} \tag{8.47}$$

i.e. as either the pairing of the Piola stress and the material gradient of the spatial velocity or the pairing of the Cauchy stress and the spatial gradient of the spatial velocity, respectively. Rendering the same internal stress power density, alternative pairings of spatial stress and (rates of) strain measures are denoted as (power) conjugated, e.g.

$$\mathbb{P}^{\text{int}}_{\text{m}} = \boldsymbol{\tau} : \boldsymbol{l} = \boldsymbol{M} : \boldsymbol{\Lambda} = \frac{1}{2}\boldsymbol{S} : \mathrm{D}_t \boldsymbol{C} = -\frac{1}{2}\boldsymbol{T} : \mathrm{D}_t \boldsymbol{B}. \tag{8.48}$$

The combination of conjugated pairings of stress and strain measures is of central importance for the determination of constitutive relations based on exploiting the dissipation power inequality in Sect. 12.3.

8.1.6 Balance Tetragon

The local balance of kinetic energy *at regular points in the domain*, *at regular points on the boundary* and *at regular points at singular surfaces* may be expressed in four different but fully equivalent versions, see the arrangement in the corresponding tetragons in Table 8.2. Here, *dynamic versions* of the kinetic energy fluxes have been introduced as $\boldsymbol{\varepsilon}^{\kappa}_{\text{D}} = \boldsymbol{E}^{\kappa}_{\text{D}} \cdot \text{cof}\, \boldsymbol{f}$ and $\boldsymbol{\varepsilon}^{\kappa}_{\text{d}} = \boldsymbol{E}^{\kappa}_{\text{d}} \cdot \text{cof}\, \boldsymbol{f}$, respectively. The subscripts D and d indicate that the corresponding kinetic energy flux is consistent with either the material or the spatial time derivative D_t or d_t, respectively, of the kinetic energy density.

Moreover, abbreviations for *effective* kinetic energy source terms in the domain, on the boundary, and at singular surfaces, respectively, have been introduced in Table 8.1.

Table 8.1 Effective kinetic energy source terms in regular points in the domain, on the boundary and at singular surfaces

Regular Points in the Domain

	PT	
$\overline{\varkappa}_{\mathrm{m}} := \boldsymbol{v} \cdot \overline{\boldsymbol{s}}_{\mathrm{m}} - \mathbb{P}^{\mathrm{int}}_{\mathrm{m}}$	$\longleftrightarrow$	$\overline{\varkappa}_{\mathrm{s}} := \boldsymbol{v} \cdot \overline{\boldsymbol{s}}_{\mathrm{s}} - \mathbb{P}^{\mathrm{int}}_{\mathrm{s}}$

Regular Points on the Boundary

	PT	
$\widehat{\varkappa}_{\mathrm{m}} := \boldsymbol{v} \cdot \widehat{\boldsymbol{s}}^{\mathrm{ext}}_{\mathrm{m}} - [\kappa_{\mathrm{m}} \boldsymbol{W}] \cdot \boldsymbol{N}$	$\longleftrightarrow$	$\widehat{\varkappa}_{\mathrm{s}} := \boldsymbol{v} \cdot \widehat{\boldsymbol{s}}^{\mathrm{ext}}_{\mathrm{s}} - \big[\kappa_{\mathrm{s}}[\boldsymbol{w} - \boldsymbol{v}]\big] \cdot \boldsymbol{n}$

Regular Points at Singular Surfaces

	PT	
$\widetilde{\varkappa}_{\mathrm{m}} := [\![\boldsymbol{v}]\!] \cdot \widetilde{\boldsymbol{s}}_{\mathrm{m}} - [\![\kappa_{\mathrm{m}} \boldsymbol{W}]\!] \cdot \boldsymbol{M}$	$\longleftrightarrow$	$\widetilde{\varkappa}_{\mathrm{s}} := [\![\boldsymbol{v}]\!] \cdot \widetilde{\boldsymbol{s}}_{\mathrm{s}} - \big[\!\big[\kappa_{\mathrm{s}}[\boldsymbol{w} - \boldsymbol{v}]\big]\!\big] \cdot \boldsymbol{n}$

8.2 Material Momentum

Configurational changes are due to *material forces*. They contribute to the *balance of material momentum* (aka the balance of material linear momentum), which, however, is not established in its own right but results as a consequence of the balance of spatial momentum.

8.2.1 Local Format (Material Control Volume)

Its local format is summarized as

Local Balance of Material Momentum (Material Control Volume):

i. Regular Points in the Domain

$$\mathrm{D}_t \boldsymbol{P}_{\mathrm{m}} - \mathrm{Div} \boldsymbol{\Sigma} =: \overline{\boldsymbol{S}}_{\mathrm{m}} \quad \text{in} \quad \mathcal{B}_{\mathrm{m}}. \tag{8.49}$$

ii. Regular Points on the Boundary

$$[\boldsymbol{P}_{\mathrm{m}} \otimes \boldsymbol{W} + \boldsymbol{\Sigma}] \cdot \boldsymbol{N} =: \widehat{\boldsymbol{S}}^{\mathrm{ext}}_{\mathrm{m}} \quad \text{on} \quad \partial\mathcal{B}_{\mathrm{m}} \setminus \partial^{3-\alpha}\mathcal{B}_{\mathrm{m}}. \tag{8.50}$$

Table 8.2 Tetragon of fully equivalent versions for the local balance of kinetic energy at regular points in the domain, on the boundary and at singular surfaces

Regular Points in the Domain

$\mathrm{D}_t\kappa_\mathrm{m} = \mathrm{Div}\boldsymbol{E}^\kappa_\mathrm{D} + \overline{\varkappa}_\mathrm{m}$	⟷ PT	$j\,\mathrm{D}_t\kappa_\mathrm{m} = \mathrm{div}\boldsymbol{\varepsilon}^\kappa_\mathrm{D} + \overline{\varkappa}_\mathrm{s}$
↕ TT		↕ TT
$J\,\mathrm{d}_t\kappa_\mathrm{s} = \mathrm{Div}\boldsymbol{E}^\kappa_\mathrm{d} + \overline{\varkappa}_\mathrm{m}$	⟷ PT	$\mathrm{d}_t\kappa_\mathrm{s} = \mathrm{div}\boldsymbol{\varepsilon}^\kappa_\mathrm{d} + \overline{\varkappa}_\mathrm{s}$

Regular Points on the Boundary

$\left[\kappa_\mathrm{m}\boldsymbol{W} + \boldsymbol{E}^\kappa_\mathrm{D}\right]\cdot\boldsymbol{N} = \widehat{\varkappa}^\mathrm{ext}_\mathrm{m}$	⟷ PT	$\left[\kappa_\mathrm{s}[\boldsymbol{w} - \boldsymbol{v}] + \boldsymbol{\varepsilon}^\kappa_\mathrm{D}\right]\cdot\boldsymbol{n} = \widehat{\varkappa}^\mathrm{ext}_\mathrm{s}$
↕ TT		↕ TT
$\left[\kappa_\mathrm{m}[\boldsymbol{W} - \boldsymbol{V}] + \boldsymbol{E}^\kappa_\mathrm{d}\right]\cdot\boldsymbol{N} = \widehat{\varkappa}^\mathrm{ext}_\mathrm{m}$	⟷ PT	$\left[\kappa_\mathrm{s}\boldsymbol{w} + \boldsymbol{\varepsilon}^\kappa_\mathrm{d}\right]\cdot\boldsymbol{n} = \widehat{\varkappa}^\mathrm{ext}_\mathrm{s}$

Regular Points at Singular Surfaces

$[\![\kappa_\mathrm{m}\boldsymbol{W} + \boldsymbol{E}^\kappa_\mathrm{D}]\!]\cdot\boldsymbol{M} = \widetilde{\varkappa}_\mathrm{m}$	⟷ PT	$[\![\kappa_\mathrm{s}[\boldsymbol{w} - \boldsymbol{v}] + \boldsymbol{\varepsilon}^\kappa_\mathrm{D}]\!]\cdot\boldsymbol{m} = \widetilde{\varkappa}_\mathrm{s}$
↕ TT		↕ TT
$[\![\kappa_\mathrm{m}[\boldsymbol{W} - \boldsymbol{V}] + \boldsymbol{E}^\kappa_\mathrm{d}]\!]\cdot\boldsymbol{M} = \widetilde{\varkappa}_\mathrm{m}$	⟷ PT	$[\![\kappa_\mathrm{s}\boldsymbol{w} + \boldsymbol{\varepsilon}^\kappa_\mathrm{d}]\!]\cdot\boldsymbol{m} = \widetilde{\varkappa}_\mathrm{s}$

with *Dynamic Versions* of Kinetic Energy Flux

$\boldsymbol{E}^\kappa_\mathrm{D} := \boldsymbol{E}^\kappa$	⟷ PT	$\boldsymbol{\varepsilon}^\kappa_\mathrm{D} := \boldsymbol{\varepsilon}^\kappa$
↕ TT		↕ TT
$\boldsymbol{E}^\kappa_\mathrm{d} := \boldsymbol{E}^\kappa + \kappa_\mathrm{m}\boldsymbol{V}$	⟷ PT	$\boldsymbol{\varepsilon}^\kappa_\mathrm{d} := \boldsymbol{\varepsilon}^\kappa - \kappa_\mathrm{s}\boldsymbol{v}$

iii. Regular Points at Singular Surfaces

$$[\![\boldsymbol{P}_{\rm m} \otimes \boldsymbol{W} + \boldsymbol{\Sigma}]\!] \cdot \boldsymbol{M} =: \widetilde{\mathsf{S}}_{\rm m} \quad \text{at} \quad \mathcal{S}_{\rm m}. \tag{8.51}$$

iv. Singular Points on the Boundary

$$\lim_{r\to 0} \int_{\partial\mathscr{B}_{\rm m}(r)} [\boldsymbol{P}_{\rm m} \otimes \boldsymbol{W} + \boldsymbol{\Sigma}] \cdot \boldsymbol{N}\, \mathrm{d}A =: \widehat{\mathsf{S}} \quad \text{on} \quad \partial^{3-\alpha}\mathcal{B}_{\rm m}. \tag{8.52}$$

For its derivation, starting from the multiplication of the local balance of spatial momentum with the *space gradient* of the spatial deformation map, i.e. with the spatial deformation gradient, three different scenarios may be considered:

i. In a smooth domain $\mathcal{B}_{\rm m}$ multiplication of the balance of spatial momentum in Eq. 7.29 by the transposed spatial deformation gradient $\boldsymbol{F}^t$ (Maugin [2, 3]) renders

$$\boldsymbol{F}^t \cdot [\,\mathrm{D}_t \boldsymbol{p}_{\rm m} - \mathrm{Div}\boldsymbol{P}] = \boldsymbol{F}^t \cdot \bar{\boldsymbol{s}}_{\rm m} \quad \text{in} \quad \mathcal{B}_{\rm m}. \tag{8.53}$$

Then, based on the *product rule of differentiation*, the identities

$$\boldsymbol{F}^t \cdot \mathrm{D}_t \boldsymbol{p}_{\rm m} = \mathrm{D}_t[\boldsymbol{F}^t \cdot \boldsymbol{p}_{\rm m}] - \mathrm{D}_t \boldsymbol{F} \cdot \boldsymbol{p}_{\rm m}$$

and

$$\boldsymbol{F}^t \cdot \mathrm{Div}\boldsymbol{P} = \mathrm{Div}(\boldsymbol{F}^t \cdot \boldsymbol{P}) - \nabla_X \boldsymbol{F}^t : \boldsymbol{P}$$

result in

$$\mathrm{D}_t[\boldsymbol{F}^t \cdot \boldsymbol{p}_{\rm m}] - \mathrm{D}_t \boldsymbol{F}^t \cdot \boldsymbol{p}_{\rm m} - \mathrm{Div}(\boldsymbol{F}^t \cdot \boldsymbol{P}) + \nabla_X \boldsymbol{F}^t : \boldsymbol{P} = \boldsymbol{F}^t \cdot \bar{\boldsymbol{s}}_{\rm m}. \tag{8.54}$$

Noting next, due to the *commutation rule of partial differentiation*, the identities

$$\mathrm{D}_t \boldsymbol{F}^t \cdot \boldsymbol{p}_{\rm m} = \boldsymbol{p}_{\rm m} \cdot \nabla_X \boldsymbol{v} \quad \text{and} \quad \nabla_X \boldsymbol{F}^t : \boldsymbol{P} = \boldsymbol{P} : \nabla_X \boldsymbol{F}$$

lead to

$$-\,\mathrm{D}_t \boldsymbol{P}_{\rm m} - \boldsymbol{p}_{\rm m} \cdot \nabla_X \boldsymbol{v} - \mathrm{Div}(\boldsymbol{F}^t \cdot \boldsymbol{P}) + \boldsymbol{P} : \nabla_X \boldsymbol{F} = \boldsymbol{F}^t \cdot \bar{\boldsymbol{s}}_{\rm m}. \tag{8.55}$$

Observe that here the density of *material momentum* $\boldsymbol{P}_{\rm m}$ per unit volume in the material configuration is introduced as conjugate to the material velocity $\boldsymbol{V}$

$$\boldsymbol{P}_{\mathrm{m}} := \frac{\partial \kappa_{\mathrm{m}}}{\partial \boldsymbol{V}} = \rho_{\mathrm{m}} \boldsymbol{C} \cdot \boldsymbol{V} \quad \text{with} \quad \kappa_{\mathrm{m}} := \frac{1}{2} \rho_{\mathrm{m}} \boldsymbol{V} \cdot \boldsymbol{C} \cdot \boldsymbol{V} \tag{8.56}$$

whereby κ_{m} is here understood as the *re-parameterized* kinetic energy density per unit volume in the material configuration. Based on the relation between the material and spatial velocities in Eq. 2.57 it then turns out that the material momentum density $\boldsymbol{P}_{\mathrm{m}}$ is the negative *pull-back* of its spatial counterpart

$$\boldsymbol{P}_{\mathrm{m}} = -\boldsymbol{F}^{t} \cdot \boldsymbol{p}_{\mathrm{m}}. \tag{8.57}$$

Next the *(interior)*[1] *free energy density* ψ_{m} per unit volume in the material configuration is introduced as

$$\psi_{\mathrm{m}} := \psi_{\mathrm{m}}(\boldsymbol{F}, \theta, \alpha; \boldsymbol{X}). \tag{8.58}$$

Thereby ψ_{m} *depends* in general on the spatial deformation gradient $\boldsymbol{F}$, the *absolute temperature* θ, and the generic (scalar-valued) *internal variable* α (the extension to sets of vector- and tensor-valued internal variables is straightforward and is thus here omitted for notational convenience). The *parametrization* of ψ_{m} in the material position vector $\boldsymbol{X}$ is included to capture any *inhomogeneities* in the material parameters. The more detailed introduction and discussion of the (interior) free energy density and its parametrization is postponed until the exploitation of the (interior) dissipation inequality as resulting from the consideration of the *balance of entropy*. Based on the (interior) free energy density the constitutive relations for the Piola stress $\boldsymbol{P}$, the *entropy density* σ_{m} and the *driving force* A_{m} (conjugate to the internal variable), both of the latter per unit volume in the material configuration, can then straightforwardly be shown to follow as

$$\boldsymbol{P} = \frac{\partial \psi_{\mathrm{m}}}{\partial \boldsymbol{F}} \quad \text{and} \quad \sigma_{\mathrm{m}} = -\frac{\partial \psi_{\mathrm{m}}}{\partial \theta} \quad \text{and} \quad \mathrm{A}_{\mathrm{m}} := -\frac{\partial \psi_{\mathrm{m}}}{\partial \alpha}. \tag{8.59}$$

Consequently, the material gradient of the (interior) free energy density ψ_{m} computes as

$$\nabla_X \psi_{\mathrm{m}} \equiv \mathrm{Div}(\psi_{\mathrm{m}} \boldsymbol{I}) = \boldsymbol{P} : \nabla_X \boldsymbol{F} - \sigma_{\mathrm{m}} \nabla_X \theta - \mathrm{A}_{\mathrm{m}} \nabla_X \alpha + \partial_X \psi_{\mathrm{m}}, \tag{8.60}$$

likewise, the material gradient of the kinetic energy density κ_{m} (in its original parametrization) is expressed as

$$\nabla_X \kappa_{\mathrm{m}} \equiv \mathrm{Div}(\kappa_{\mathrm{m}} \boldsymbol{I}) = \boldsymbol{p}_{\mathrm{m}} \cdot \nabla_X \boldsymbol{v} + \partial_X \kappa_{\mathrm{m}}. \tag{8.61}$$

[1] The terminology 'interior' versus 'exterior' will be elucidated in Chaps. 10 and 11 on the variational setting and on the thermo-dynamical balances, respectively. In a nutshell, 'exterior' refers to the inclusion of the external potential energy as part of the total energy.

In the above, *heterogeneities* in the fields $\boldsymbol{F}(\boldsymbol{X})$, $\theta(\boldsymbol{X})$, $\alpha(\boldsymbol{X})$, and $\boldsymbol{v}(\boldsymbol{X})$, i.e. their variations with respect to the material position vector $\boldsymbol{X}$, are captured by the material gradients $\nabla_X \boldsymbol{F}$, $\nabla_X \theta$, $\nabla_X \alpha$, and $\nabla_X \boldsymbol{v}$, respectively. Moreover, $\partial_X \psi_\text{m}$ and $\partial_X \kappa_\text{m}$ denote the *explicit gradients* of ψ_m and κ_m, respectively, with respect to their parametrization in the material position vector $\boldsymbol{X}$. As an example $\partial_X \kappa_\text{m}$ results from the possible dependence $\rho_\text{m} = \rho_\text{m}(\boldsymbol{X})$ that is consistent with the local balance of mass (mass conservation). In the sequel ψ_m and κ_m are combined into

$$\lambda_\text{m} := \kappa_\text{m} - \psi_\text{m}, \tag{8.62}$$

i.e. into a an energy density with the flavor of an *(interior) Lagrange-type free energy density*. Involving these relations, Eq. 8.55 finally transforms into the local format of the balance of material momentum in Eq. 8.49.

Therein $\boldsymbol{\Sigma} \in T^*\mathcal{B}_\text{m} \times T\mathcal{B}_\text{m}$ represents the celebrated *(interior) Eshelby* stress (that might also be denoted as the *material Cauchy stress* for terminological symmetry)

$$\boldsymbol{\Sigma} := -\lambda_\text{m} \boldsymbol{I} - \boldsymbol{F}^t \cdot \boldsymbol{P}. \tag{8.63}$$

Moreover, the (interior) source density of material momentum $\overline{\boldsymbol{S}}_\text{m}$ per unit volume in the material configuration has been introduced as an abbreviation

$$\overline{\boldsymbol{S}}_\text{m} := \sigma_\text{m} \nabla_X \theta + A_\text{m} \nabla_X \alpha + \partial_X \lambda_\text{m} - \boldsymbol{F}^t \cdot \overline{\boldsymbol{s}}_\text{m}. \tag{8.64}$$

The (interior) source density of material momentum $\overline{\boldsymbol{S}}_\text{m}$ captures essentially the field dependence (heterogeneity) of the absolute temperature θ and the generic internal variable α as well as the parametrization dependence (inhomogeneity) of the (interior) Lagrange-type free energy density λ_m on the material position vector $\boldsymbol{X}$.

Note that, for the special case of mass conservation, the first term in Eq. 8.49 simplifies to $\rho_\text{m} \, \mathrm{D}_t[\boldsymbol{C} \cdot \boldsymbol{V}]$, whereby $\boldsymbol{C} \cdot \boldsymbol{V}$ is a material covector.

ii. On the regular part $\partial\mathcal{B}_\text{m} \setminus \partial^{3-\alpha}\mathcal{B}_\text{m}$ of the external boundary $\partial\mathcal{B}_\text{m}$ multiplication of the balance of spatial momentum in Eq. 7.30 by the transposed spatial deformation gradient $\boldsymbol{F}^t$ renders

$$\boldsymbol{F}^t \cdot [\boldsymbol{p}_\text{m} \otimes \boldsymbol{W} + \boldsymbol{P}] \cdot \boldsymbol{N} = \boldsymbol{F}^t \cdot \widehat{\boldsymbol{s}}_\text{m}^\text{ext} \quad \text{on} \quad \partial\mathcal{B}_\text{m}. \tag{8.65}$$

Then, $\boldsymbol{F}^t \cdot \boldsymbol{p}_\text{m} = -\boldsymbol{P}_\text{m}$ and the definition of $\boldsymbol{\Sigma}$ in Eq. 8.63 immediately render the result in Eq. 8.50.

Therein $\widehat{\boldsymbol{S}}_\text{m}^\text{ext}$ represents the *extrinsic (interior) flux density of material momentum* per unit area of the external boundary in the material configuration

$$\widehat{\boldsymbol{S}}_\text{m}^\text{ext} := -\lambda_\text{m} \boldsymbol{N} - \boldsymbol{F}^t \cdot \widehat{\boldsymbol{s}}_\text{m}^\text{ext}. \tag{8.66}$$

The extrinsic (interior) material momentum flux density $\widehat{\boldsymbol{S}}_{\mathrm{m}}^{\mathrm{ext}}$ captures essentially the flow of the (interior) Lagrange-type free energy density λ_{m} across regular parts of the external boundary.

iii. At singular surfaces $\mathcal{S}_{\mathrm{m}}$ multiplication of the balance of spatial momentum in Eq. 7.31 by the average transposed spatial deformation gradient $\{\boldsymbol{F}^t\}$ renders

$$\{\boldsymbol{F}^t\} \cdot [\![\boldsymbol{p}_{\mathrm{m}} \otimes \boldsymbol{W} + \boldsymbol{P}]\!] \cdot \boldsymbol{M} = \boldsymbol{0} \quad \text{on} \quad \mathcal{S}_{\mathrm{m}}. \tag{8.67}$$

Then, based on the *product rule of jumps*, the identities

$$\{\boldsymbol{F}^t\} \cdot [\![\boldsymbol{p}_{\mathrm{m}}]\!] = -[\![\boldsymbol{P}_{\mathrm{m}}]\!] - [\![\boldsymbol{F}^t]\!] \cdot \{\boldsymbol{p}_{\mathrm{m}}\},$$

since $[\![\boldsymbol{F}^t \cdot \boldsymbol{p}_{\mathrm{m}}]\!] = -[\![\boldsymbol{P}_{\mathrm{m}}]\!]$, and

$$\{\boldsymbol{F}^t\} \cdot [\![\boldsymbol{P}]\!] = [\![\boldsymbol{F}^t \cdot \boldsymbol{P}]\!] - [\![\boldsymbol{F}^t]\!] \cdot \{\boldsymbol{P}\}$$

result in

$$[\![-\boldsymbol{F}^t \cdot \boldsymbol{P} + \boldsymbol{P}_{\mathrm{m}} \otimes \boldsymbol{W}]\!] \cdot \boldsymbol{M} = -[\![\boldsymbol{F}^t]\!] \cdot \{\boldsymbol{P} + \boldsymbol{p}_{\mathrm{m}} \otimes \boldsymbol{W}\} \cdot \boldsymbol{M}. \tag{8.68}$$

Incorporating finally the definitions of the (interior) Eshelby stress in Eq. 8.63 and the power generating traction at singular surfaces in Eq. 8.10 renders the result in Eq. 8.51.

Therein $\widetilde{\boldsymbol{S}}_{\mathrm{m}}$ represents the *(interior) flux density of material momentum* per unit area of singular surfaces in the material configuration

$$\widetilde{\boldsymbol{S}}_{\mathrm{m}} := -[\![\lambda_{\mathrm{m}}]\!]\boldsymbol{M} - [\![\boldsymbol{F}^t]\!] \cdot \widetilde{\boldsymbol{s}}_{\mathrm{m}}. \tag{8.69}$$

The (interior) material momentum flux density $\widetilde{\boldsymbol{S}}_{\mathrm{m}}$ captures essentially the flow of (interior) Lagrange-type free energy density λ_{m} to singular surfaces.

Based on the jump relation $[\![\boldsymbol{F}]\!]W_{\perp} = -[\![\boldsymbol{v}]\!] \otimes \boldsymbol{M}$ in Eq. 6.8 and the coherence condition III in Eq. 4.11, the second term on the right of Eq. 8.69 expands into

$$[\![\boldsymbol{F}^t]\!] \cdot \widetilde{\boldsymbol{s}}_{\mathrm{m}} = \big[[\![\boldsymbol{F}]\!] : \{\boldsymbol{P}\} - \{\boldsymbol{p}_{\mathrm{m}}\} \cdot [\![\boldsymbol{v}]\!]\big]\boldsymbol{M}. \tag{8.70}$$

Thus, the (interior) material momentum flux density in Eq. 8.69 is *purely normal* (Abeyaratne and Knowles [1]). Note finally that, for the special case of mass conservation, Eq. 8.69 simplifies further (Hill [4], Maugin [5])

$$\widetilde{\boldsymbol{S}}_{\mathrm{m}} := \big[[\![\psi_{\mathrm{m}}]\!] - [\![\boldsymbol{F}]\!] : \{\boldsymbol{P}\}\big]\boldsymbol{M}. \tag{8.71}$$

vi. On the singular part $\partial^{3-\alpha}\mathcal{B}_{\mathrm{m}}$ of the external boundary $\partial\mathcal{B}_{\mathrm{m}}$ multiplication of the limit of the balance of spatial momentum in Eq. 7.30 by the limit of transposed spatial deformation gradient $\boldsymbol{F}^t$ renders

$$\lim_{r\to 0} \boldsymbol{F}^t \cdot \lim_{r\to 0} \int\limits_{\partial\mathscr{B}_{\mathrm{m}}(r)} [\boldsymbol{p}_{\mathrm{m}} \otimes \boldsymbol{W} + \boldsymbol{P}] \cdot \boldsymbol{N} \,\mathrm{d}A = \tag{8.72}$$

$$\lim_{r\to 0} \boldsymbol{F}^t \cdot \lim_{r\to 0} \int\limits_{\partial\mathscr{B}_{\mathrm{m}}(r)} \widehat{\boldsymbol{s}}_{\mathrm{m}}^{\mathrm{ext}} \,\mathrm{d}A \qquad \text{on } \partial\mathcal{B}_{\mathrm{m}}^{3-\alpha}.$$

Pulling the constant $\lim_{r\to 0} \boldsymbol{F}^t$ into the integrals, noting that the original integrands can be substituted by $\lim_{r\to 0}[\boldsymbol{p}_{\mathrm{m}} \otimes \boldsymbol{W} + \boldsymbol{P}] \cdot \boldsymbol{N}$ and $\lim_{r\to 0} \widehat{\boldsymbol{s}}_{\mathrm{m}}^{\mathrm{ext}}$, respectively, and using a theorem equating the product of the limits of functions with the limit of the product of these functions renders eventually

$$\lim_{r\to 0} \int\limits_{\partial\mathscr{B}_{\mathrm{m}}(r)} \boldsymbol{F}^t \cdot [\boldsymbol{p}_{\mathrm{m}} \otimes \boldsymbol{W} + \boldsymbol{P}] \cdot \boldsymbol{N} \,\mathrm{d}A = \lim_{r\to 0} \int\limits_{\partial\mathscr{B}_{\mathrm{m}}(r)} \boldsymbol{F}^t \cdot \widehat{\boldsymbol{s}}_{\mathrm{m}}^{\mathrm{ext}} \,\mathrm{d}A, \tag{8.73}$$

which, by $\boldsymbol{F}^t \cdot \boldsymbol{p}_{\mathrm{m}} = -\boldsymbol{P}_{\mathrm{m}}$ and the definition of $\boldsymbol{\Sigma}$ in Eq. 8.63, may be rewritten as the result in Eq. 8.52.

Therein $\widehat{\mathbf{S}}$ represents the singular contribution to the extrinsic (interior) flux of material momentum at the external boundary in the material configuration

$$\widehat{\mathbf{S}} := -\lim_{r\to 0} \int\limits_{\partial\mathscr{B}_{\mathrm{m}}(r)} [\lambda_{\mathrm{m}} \boldsymbol{N} + \boldsymbol{F}^t \cdot \widehat{\boldsymbol{s}}_{\mathrm{m}}^{\mathrm{ext}}] \,\mathrm{d}A. \tag{8.74}$$

The singular contribution $\widehat{\mathbf{S}}$ captures essentially the singular flow of the (interior) Lagrange-type free energy density λ_{m} across singular parts of the external boundary.

8.2.2 *Global Format (Material Control Volume)*

With these preliminaries at hand, the balance of material momentum may formally also be stated in global format as

Global Balance of Material Momentum (Material Control Volume):

$$\dot{\mathbf{P}}(\mathfrak{B}_{\mathrm{m}}) = \mathbf{F}^{\mathrm{ext}}(\mathfrak{B}_{\mathrm{m}}) \quad \forall \mathfrak{B}_{\mathrm{m}} \subseteq \mathcal{B}_{\mathrm{m}}. \tag{8.75}$$

Thereby $\mathbf{P}$ is introduced as the resultant over the material control volume $\mathfrak{B}_{\mathrm{m}}$ of the density of *material momentum* $\boldsymbol{P}_{\mathrm{m}}$ per unit volume in the material configuration, i.e. as the *resultant material momentum* contained in the material control volume

$$\mathsf{P}(\mathfrak{B}_\mathrm{m}) := \int_{\mathfrak{B}_\mathrm{m}} \boldsymbol{P}_\mathrm{m}\,\mathrm{d}V. \tag{8.76}$$

Moreover, F^ext denotes the *resultant external material force* (consisting of volume, boundary *and* surface integrals *and* a singular contribution)

$$\mathsf{F}^\mathrm{ext}(\mathfrak{B}_\mathrm{m}) := \int_{\mathfrak{B}_\mathrm{m}} \overline{\boldsymbol{S}}_\mathrm{m}\,\mathrm{d}V + \int_{\partial\mathfrak{B}_\mathrm{m}\backslash\partial^{3-\alpha}\mathcal{B}_\mathrm{m}} \widehat{\boldsymbol{S}}_\mathrm{m}\,\mathrm{d}A + \widehat{\mathsf{S}} - \int_{\mathfrak{S}_\mathrm{m}} \widetilde{\boldsymbol{S}}_\mathrm{m}\,\mathrm{d}A. \tag{8.77}$$

Note the contributions to the resultant external material force due to the supply of material momentum to singular parts of the external boundary and to singular surfaces.

Consequently, the global statement of the balance of material momentum expands as

$$\int_{\mathfrak{B}_\mathrm{m}} \mathrm{D}_t \boldsymbol{P}_\mathrm{m}\,\mathrm{d}V = \tag{8.78}$$

$$\int_{\mathfrak{B}_\mathrm{m}} \overline{\boldsymbol{S}}_\mathrm{m}\,\mathrm{d}V + \int_{\partial\mathfrak{B}_\mathrm{m}\backslash\partial^{3-\alpha}\mathcal{B}_\mathrm{m}} \underbrace{\boldsymbol{\Sigma}\cdot\boldsymbol{N}}_{\boldsymbol{S}_\mathrm{m}}\,\mathrm{d}A + \mathsf{S} - \int_{\mathfrak{S}_\mathrm{m}} \widetilde{\boldsymbol{S}}_\mathrm{m}\,\mathrm{d}A + \int_{\mathfrak{S}_\mathrm{m}} [\![\boldsymbol{P}_\mathrm{m}\otimes\boldsymbol{W}]\!]\cdot\boldsymbol{M}\,\mathrm{d}A.$$

Here, the convective term stemming from the transport theorem has been subsumed in the boundary integrand, i.e. the *intrinsic* (interior) material momentum flux density

$$\boldsymbol{S}_\mathrm{m} := \boldsymbol{\Sigma}\cdot\boldsymbol{N} = \widehat{\boldsymbol{S}}_\mathrm{m} - [\boldsymbol{P}_\mathrm{m}\otimes\boldsymbol{W}]\cdot\boldsymbol{N}, \tag{8.79}$$

the singular *intrinsic* (interior) material momentum flux

$$\mathsf{S} := \lim_{r\to 0} \int_{\partial\mathscr{B}_\mathrm{m}(r)} \boldsymbol{\Sigma}\cdot\boldsymbol{N}\,\mathrm{d}A = \widehat{\mathsf{S}} - \lim_{r\to 0} \int_{\partial\mathscr{B}_\mathrm{m}(r)} [\boldsymbol{P}_\mathrm{m}\otimes\boldsymbol{W}]\cdot\boldsymbol{N}\,\mathrm{d}A \tag{8.80}$$

and the singular surface integrand $[\![\boldsymbol{P}_\mathrm{m}\otimes\boldsymbol{W}]\!]\cdot\boldsymbol{M}$. Note that upon localizing the global balance of material momentum in Eq. 8.78 indeed Eqs. 8.49–8.52 are obtained.

In summary the global balance of material momentum relates the total time derivative of the resultant material momentum to the resultant external material force under the precondition that the balance of spatial momentum is satisfied.

Identifying the Vectorial J-Integral

The balance of material momentum in Eq. 8.52 may be considered as the definition of a single material force acting on the singular part of the external boundary. In the classical terminology of fracture mechanics, this single material force is denoted as the *vectorial J-integral*

$$\mathcal{J} := -\lim_{r\to 0} \int_{\partial \mathcal{B}_m(r)} \boldsymbol{\Sigma} \cdot \boldsymbol{N} \, dA = -\mathbf{S}. \tag{8.81}$$

Here, $\mathcal{J}$ coincides with the classical definition of the vectorial J-integral. Observe the sign change that is a consequence of using the inward normal $-\boldsymbol{N}$ in the traditional definition of $\mathcal{J}$. The J-integral can be considered as the driving force for *translational* configurational changes of the material placement of, e.g. the tip of a mathematical crack.

Based on the global statement of the balance of material momentum in Eq. 8.78, conditions for the J-integral to be *path-independent* are straightforwardly established. These are as follows:

- $D_t \boldsymbol{P}_m = \boldsymbol{0}$, i.e. quasi-statics,
- $\overline{\boldsymbol{S}}_m = \boldsymbol{0}$, i.e. no heterogeneities or inhomogeneities,
- $\widetilde{\boldsymbol{S}}_m = \boldsymbol{0}$, i.e. no singular surfaces.

These conditions typically apply to homogeneous hyper-elasto-static specimens with a stationary pre-crack.

8.2.3 *Local Format (Spatial Control Volume)*

The local balance for the density of material momentum $\boldsymbol{P}_s := j\boldsymbol{P}_m$ per unit volume in the spatial configuration, expressed in terms of the material-spatial description (interior) Eshelby stress $\boldsymbol{p} := \boldsymbol{\Sigma} \cdot \operatorname{cof} \boldsymbol{f}$, the (interior) source density of material momentum $\overline{\boldsymbol{S}}_s := j\,\overline{\boldsymbol{S}}_m$ per unit volume in the spatial configuration and the extrinsic (interior) flux densities of material momentum $\widehat{\boldsymbol{S}}_s^{ext} := \widehat{j}\,\widehat{\boldsymbol{S}}_m^{ext}$ and $\widetilde{\boldsymbol{S}}_s := \widetilde{j}\,\widetilde{\boldsymbol{S}}_m$, respectively, per unit area in the spatial configuration, is summarized as

Local Balance of Material Momentum (Spatial Control Volume):

i. Regular Points in the Domain

$$d_t \boldsymbol{P}_s - \operatorname{div}(\boldsymbol{p} - \boldsymbol{P}_s \otimes \boldsymbol{v}) = \overline{\boldsymbol{S}}_s \quad \text{in} \quad \mathcal{B}_s. \tag{8.82}$$

ii. Regular Points on the Boundary

$$\big[\boldsymbol{P}_s \otimes \boldsymbol{w} + [\boldsymbol{p} - \boldsymbol{P}_s \otimes \boldsymbol{v}]\big] \cdot \boldsymbol{n} = \widehat{\boldsymbol{S}}_s^{ext} \quad \text{on} \quad \partial\mathcal{B}_s. \tag{8.83}$$

iii. Regular Points at Singular Surfaces

$$[\![\boldsymbol{P}_s \otimes \boldsymbol{w} + [\boldsymbol{p} - \boldsymbol{P}_s \otimes \boldsymbol{v}]]\!] \cdot \boldsymbol{m} = \widetilde{\boldsymbol{S}}_s \quad \text{at} \quad \mathcal{S}_s. \tag{8.84}$$

In the above definitions any singular contributions are ignored for the sake of presentation.

8.2.4 Global Format (Spatial Control Volume)

With these preliminaries at hand, the balance of material momentum may formally also be stated in global format as

Global Balance of Material Momentum (Spatial Control Volume):

$$\dot{\mathsf{P}}(\mathfrak{B}_s) = \mathsf{F}^{\text{ext}}(\mathfrak{B}_s) \quad \forall \mathfrak{B}_s \subseteq \mathcal{B}_s. \tag{8.85}$$

Thereby P is introduced as the resultant over the spatial control volume $\mathfrak{B}_s$ of the density of *material momentum* $\boldsymbol{P}_s$ per unit volume in the spatial configuration, i.e. as the *resultant material momentum* contained in the spatial control volume

$$\mathsf{P}(\mathfrak{B}_s) := \int_{\mathfrak{B}_s} \boldsymbol{P}_s \, \mathrm{d}v. \tag{8.86}$$

Moreover, F^{ext} denotes the *resultant external material force* (consisting of volume, boundary *and* surface integrals

$$\mathsf{F}^{\text{ext}}(\mathfrak{B}_s) := \int_{\mathfrak{B}_s} \overline{\boldsymbol{S}}_s \, \mathrm{d}v + \int_{\partial\mathfrak{B}_s} \widehat{\boldsymbol{S}}_s \, \mathrm{d}a - \int_{\mathfrak{S}_s} \widetilde{\boldsymbol{S}}_s \, \mathrm{d}a. \tag{8.87}$$

Note the contribution to the resultant external material force due to the supply of material momentum to singular surfaces. (In the above definition any singular contributions are ignored for the sake of presentation.)

Consequently, the global statement of the balance of material momentum expands as

$$\int_{\mathfrak{B}_s} \mathrm{d}_t \boldsymbol{P}_s \, \mathrm{d}v = \tag{8.88}$$

$$\int_{\mathfrak{B}_s} \overline{\boldsymbol{S}}_s \, \mathrm{d}v + \int_{\partial\mathfrak{B}_s} [\boldsymbol{p} - \boldsymbol{P}_s \otimes \boldsymbol{v}] \cdot \boldsymbol{n} \, \mathrm{d}a - \int_{\mathfrak{S}_s} \widetilde{\boldsymbol{S}}_s \, \mathrm{d}a + \int_{\mathfrak{S}_s} [\![\boldsymbol{P}_s \otimes \boldsymbol{w}]\!] \cdot \boldsymbol{m} \, \mathrm{d}a.$$

Here, the convective term stemming from the transport theorem has been subsumed in the boundary integrand, i.e. the *intrinsic* (interior) material momentum flux density

$$S_s := p \cdot n = \widehat{S}_s - \left[P_s \otimes [w - v]\right] \cdot n \qquad (8.89)$$

and the singular surface integrand $[\![P_s \otimes w]\!] \cdot m$. Note that upon localizing the global balance of material momentum in Eq. 8.88 indeed Eqs. 8.82–8.84 are obtained.

8.2.5 *Material Stress Measures*

In addition to the Piola transformation between the Eshelby-type stresses p and Σ it proves convenient to introduce the Eshelby-type stress $\mathcal{T}$ as

$$\mathcal{T} := j\Sigma = p \cdot f^t \quad \text{with} \quad \int_{\mathfrak{B}_s} \mathcal{T} \cdots \mathrm{d}v = \int_{\mathfrak{B}_m} \Sigma \cdots \mathrm{d}V.$$

The Eshelby-type stress $\mathcal{T}$ might be considered either a contra-, a co-, or a co/contra- (mixed) variant material description tensor field (again no special notation shall be adopted to distinguish between these versions). Then, the contra-, co-, and co/contra- (mixed) variant push-forwards of $\mathcal{T}$ follow as

$$s := F \cdot \mathcal{T} \cdot F^t \quad \text{and} \quad t := f^t \cdot \mathcal{T} \cdot f \quad \text{and} \quad m := f^t \cdot \mathcal{T} \cdot F^t.$$

Here s, t, and m denote contravariantly, covariantly and co/contra- (mixed) variantly transforming spatial description Eshelby-type stress measures, respectively. The relations between the various Eshelby-type stress measures are assembled in the following table:

	$\mathcal{T}$	p	s	t	m
$\mathcal{T}$	•	$p \cdot f^t$	$f \cdot s \cdot f^t$	$F^t \cdot t \cdot F$	$F^t \cdot m \cdot f^t$
p	$\mathcal{T} \cdot F^t$	•	$f \cdot s$	$F^t \cdot t \cdot b$	$F^t \cdot m$
s	$F \cdot \mathcal{T} \cdot F^t$	$F \cdot p$	•	$b \cdot t \cdot b$	$b \cdot m$
t	$f^t \cdot \mathcal{T} \cdot f$	$f^t \cdot p \cdot c$	$c \cdot s \cdot c$	•	$m \cdot c$
m	$f^t \cdot \mathcal{T} \cdot F^t$	$f^t \cdot p$	$c \cdot s$	$t \cdot b$	•

Note that $m = \psi_s i - \sigma$ is *symmetric*, thus $c \cdot s = s^t \cdot c$ or, likewise, $t \cdot b = b \cdot t^t$, however s and t are only symmetric (with c and s or, likewise, b and t commuting) if $\mathcal{T} = \mathcal{T}^t$ holds. The former is a consequence of the obligatory *spatial isotropy*, whereas the latter is a consequence of the optional *material isotropy*.

8.2.6 *Balance Tetragon*

The local balance of material momentum *at regular points in the domain, at regular points on the boundary* and *at regular points at singular surfaces* may then

Table 8.3 Tetragon of fully equivalent versions for the local balance of material momentum at regular points in the domain, on the boundary, and at singular surfaces

Regular Points in the Domain

$\mathrm{D}_t \boldsymbol{P}_\mathrm{m} = \mathrm{Div}\boldsymbol{\Sigma}_\mathrm{D} + \overline{\boldsymbol{S}}_\mathrm{m}$	⟷ PT	$j\,\mathrm{D}_t \boldsymbol{P}_\mathrm{m} = \mathrm{div}\boldsymbol{p}_\mathrm{D} + \overline{\boldsymbol{S}}_\mathrm{s}$
↕ TT		TT ↕
$J\,\mathrm{d}_t \boldsymbol{P}_\mathrm{s} = \mathrm{Div}\boldsymbol{\Sigma}_\mathrm{d} + \overline{\boldsymbol{S}}_\mathrm{m}$	PT ⟷	$\mathrm{d}_t \boldsymbol{P}_\mathrm{s} = \mathrm{div}\boldsymbol{p}_\mathrm{d} + \overline{\boldsymbol{S}}_\mathrm{s}$

Regular Points on the Boundary

$\big[\boldsymbol{P}_\mathrm{m} \otimes \boldsymbol{W} + \boldsymbol{\Sigma}_\mathrm{D}\big] \cdot \boldsymbol{N} = \widehat{\boldsymbol{S}}{}^\mathrm{ext}_\mathrm{m}$	⟷ PT	$\big[\boldsymbol{P}_\mathrm{s} \otimes [\boldsymbol{w} - \boldsymbol{v}] + \boldsymbol{p}_\mathrm{D}\big] \cdot \boldsymbol{n} = \widehat{\boldsymbol{S}}{}^\mathrm{ext}_\mathrm{s}$
↕ TT		TT ↕
$\big[\boldsymbol{P}_\mathrm{m} \otimes [\boldsymbol{W} - \boldsymbol{V}] + \boldsymbol{\Sigma}_\mathrm{d}\big] \cdot \boldsymbol{N} = \widehat{\boldsymbol{S}}{}^\mathrm{ext}_\mathrm{m}$	PT ⟷	$\big[\boldsymbol{P}_\mathrm{s} \otimes \boldsymbol{w} + \boldsymbol{p}_\mathrm{d}\big] \cdot \boldsymbol{n} = \widehat{\boldsymbol{S}}{}^\mathrm{ext}_\mathrm{s}$

Regular Points at Singular Surfaces

$[\![\boldsymbol{P}_\mathrm{m} \otimes \boldsymbol{W} + \boldsymbol{\Sigma}_\mathrm{D}]\!] \cdot \boldsymbol{M} = \widetilde{\boldsymbol{S}}_\mathrm{m}$	⟷ PT	$[\![\boldsymbol{P}_\mathrm{s} \otimes [\boldsymbol{w} - \boldsymbol{v}] + \boldsymbol{p}_\mathrm{D}]\!] \cdot \boldsymbol{m} = \widetilde{\boldsymbol{S}}_\mathrm{s}$
↕ TT		TT ↕
$[\![\boldsymbol{P}_\mathrm{m} \otimes [\boldsymbol{W} - \boldsymbol{V}] + \boldsymbol{\Sigma}_\mathrm{d}]\!] \cdot \boldsymbol{M} = \widetilde{\boldsymbol{S}}_\mathrm{m}$	PT ⟷	$[\![\boldsymbol{P}_\mathrm{s} \otimes \boldsymbol{w} + \boldsymbol{p}_\mathrm{d}]\!] \cdot \boldsymbol{m} = \widetilde{\boldsymbol{S}}_\mathrm{s}$

with *Dynamic Versions* of (Interior) Eshelby Stress

$\boldsymbol{\Sigma}_\mathrm{D} := \boldsymbol{\Sigma}$	⟷ PT	$\boldsymbol{p}_\mathrm{D} := \boldsymbol{p}$
↕ TT		TT ↕
$\boldsymbol{\Sigma}_\mathrm{d} := \boldsymbol{\Sigma} + \boldsymbol{P}_\mathrm{m} \otimes \boldsymbol{V}$	PT ⟷	$\boldsymbol{p}_\mathrm{d} := \boldsymbol{p} - \boldsymbol{P}_\mathrm{s} \otimes \boldsymbol{v}$

be expressed in four different but fully equivalent versions, see the arrangement in the corresponding tetragons in Table 8.3. Here, *dynamic versions* of the (interior) Eshelby stresses have been introduced as $\boldsymbol{\Sigma}_{\mathrm{D}} = \boldsymbol{p}_{\mathrm{D}} \cdot \mathrm{cof}\,\boldsymbol{F}$ and $\boldsymbol{\Sigma}_{\mathrm{d}} = \boldsymbol{p}_{\mathrm{d}} \cdot \mathrm{cof}\,\boldsymbol{F}$, respectively, whereby the subscripts D and d indicate that the corresponding (interior) Eshelby stress is consistent with either the material or the spatial time derivative D_t or d_t, respectively, of the material momentum density.

8.3 Vector Moment of Material Momentum*

Vector moments of material forces are denoted as *material couples* and serve as a measure of the *non-centrality* of the material force system. They contribute to the *balance of vector moment of material momentum* (aka the balance of angular material momentum), which, however, is not established in its own right but results as a consequence of the balance of material momentum.

8.3.1 Local Format (Material Control Volume)

Its local format is summarized as

Local Balance of Vector Moment of Material Momentum (Material Control Volume):

i. Regular Points in the Domain

$$\mathrm{D}_t \boldsymbol{L}_{\mathrm{m}} - \mathrm{Div}\boldsymbol{\Theta} =: \overline{\boldsymbol{T}}_{\mathrm{m}} \quad \text{in} \quad \mathcal{B}_{\mathrm{m}}. \tag{8.90}$$

ii. Regular Points on the Boundary

$$[\boldsymbol{L}_{\mathrm{m}} \otimes \boldsymbol{W} + \boldsymbol{\Theta}] \cdot \boldsymbol{N} =: \widehat{\boldsymbol{T}}_{\mathrm{m}}^{\mathrm{ext}} \quad \text{on} \quad \partial\mathcal{B}_{\mathrm{m}} \setminus \partial^{3-\alpha}\mathcal{B}_{\mathrm{m}}. \tag{8.91}$$

iii. Regular Points at Singular Surfaces

$$[\![\boldsymbol{L}_{\mathrm{m}} \otimes \boldsymbol{W} + \boldsymbol{\Theta}]\!] \cdot \boldsymbol{M} =: \widetilde{\boldsymbol{T}}_{\mathrm{m}} \quad \text{at} \quad \mathcal{S}_{\mathrm{m}}. \tag{8.92}$$

iv. Singular Points on the Boundary

$$\lim_{r \to 0} \int_{\partial\mathscr{B}_{\mathrm{m}}(r)} [\boldsymbol{L}_{\mathrm{m}} \otimes \boldsymbol{W} + \boldsymbol{\Theta}] \cdot \boldsymbol{N}\, \mathrm{d}A =: \widehat{\mathbf{T}} \quad \text{on} \quad \partial^{3-\alpha}\mathcal{B}_{\mathrm{m}}. \tag{8.93}$$

For its derivation, starting from the vector multiplication of the local balance of material momentum with the material distance vector

$$\boldsymbol{D}_0 := \boldsymbol{Y} - \boldsymbol{Y}_0 \tag{8.94}$$

to an arbitrary material reference point $\boldsymbol{Y}_0$ (whereby $\boldsymbol{Y}_0 = \boldsymbol{0}$ coincides with the origin for the sake of presentation), three different scenarios may be considered:

i. In a smooth domain $\mathcal{B}_\mathrm{m}$ vector multiplication of the balance of material momentum in Eq. 8.49 by the material distance vector $\boldsymbol{D}_0$ renders

$$\boldsymbol{D}_0 \times [\,\mathrm{D}_t \boldsymbol{P}_\mathrm{m} - \mathrm{Div}\,\boldsymbol{\Sigma}] = \boldsymbol{D}_0 \times \overline{\boldsymbol{S}}_\mathrm{m} \quad \text{in} \quad \mathcal{B}_\mathrm{m}. \tag{8.95}$$

Then, based on the product rule of differentiation, the identities

$$\boldsymbol{D}_0 \times \mathrm{D}_t \boldsymbol{P}_\mathrm{m} = \mathrm{D}_t[\boldsymbol{D}_0 \times \boldsymbol{P}_\mathrm{m}],$$

since $\mathrm{D}_t \boldsymbol{D}_0 = \boldsymbol{0}$, and

$$\boldsymbol{D}_0 \times \mathrm{Div}\,\boldsymbol{\Sigma} = \mathrm{Div}(\boldsymbol{D}_0 \times \boldsymbol{\Sigma}) - 2\mathrm{Axl}\,\boldsymbol{\Sigma},$$

wherein $\nabla_X \boldsymbol{D}_0 = \boldsymbol{I}$, hold. The *material axial vector*[2] is defined as

$$2\mathrm{Axl}\,\boldsymbol{\Sigma} := \boldsymbol{\Sigma} \rtimes \boldsymbol{I}.$$

As a result, the balance of vector moment of material momentum reads

$$\mathrm{D}_t[\boldsymbol{D}_0 \times \boldsymbol{P}_\mathrm{m}] - \mathrm{Div}(\boldsymbol{D}_0 \times \boldsymbol{\Sigma}) = \boldsymbol{D}_0 \times \overline{\boldsymbol{S}}_\mathrm{m} - 2\mathrm{Axl}\,\boldsymbol{\Sigma}. \tag{8.96}$$

Thus, the density of vector moment of material momentum $\boldsymbol{L}_\mathrm{m}$ per unit volume in the material configuration and the *(interior) Eshelby-type couple stress* $\boldsymbol{\Theta} \in T^*\mathcal{B}_\mathrm{m} \times T\mathcal{B}_\mathrm{m}$ are identified as

$$\boldsymbol{L}_\mathrm{m} := \boldsymbol{D}_0 \times \boldsymbol{P}_\mathrm{m} \quad \text{and} \quad \boldsymbol{\Theta} := \boldsymbol{D}_0 \times \boldsymbol{\Sigma}. \tag{8.97}$$

Finally, the (interior) source density of vector moment of material momentum $\overline{\boldsymbol{T}}_\mathrm{m}$ per unit volume in the material configuration follows as

$$\overline{\boldsymbol{T}}_\mathrm{m} := \boldsymbol{D}_0 \times \overline{\boldsymbol{S}}_\mathrm{m} - 2\mathrm{Axl}\,\boldsymbol{\Sigma}. \tag{8.98}$$

Observe the autonomous contribution $\overline{\boldsymbol{T}}^\mathrm{aut}_\mathrm{m}$ to $\overline{\boldsymbol{T}}_\mathrm{m}$

---

[2] In Cartesian coordinates with coordinate representation for $\boldsymbol{\Sigma} = \Sigma^a_b\, \boldsymbol{e}^b \otimes \boldsymbol{e}_a$, its (right) vector cross product with $\boldsymbol{I}$ expands as $\boldsymbol{\Sigma} \rtimes \boldsymbol{I} := \Sigma_m{}^a\, \delta^m{}_b\, e_a{}^b{}_c\, \boldsymbol{e}^c = \Sigma_b{}^a\, e_a{}^b{}_c\, \boldsymbol{e}^c$. Thereby, the (mixed-variant) permutation symbol is denoted by $e_a{}^b{}_c := [\boldsymbol{e}_a \times \boldsymbol{e}^b] \cdot \boldsymbol{e}_c$.

$$\overline{\boldsymbol{T}}_{\text{m}}^{\text{aut}} := -2\text{Axl}\boldsymbol{\Sigma}. \tag{8.99}$$

ii. On the regular part $\partial\mathcal{B}_{\text{m}} \setminus \partial^{3-\alpha}\mathcal{B}_{\text{m}}$ of the external boundary $\partial\mathcal{B}_{\text{m}}$ vector multiplication of the balance of material momentum in Eq. 8.50 by the material distance vector $\boldsymbol{D}_0$ renders

$$\boldsymbol{D}_0 \times [\boldsymbol{P}_{\text{m}} \otimes \boldsymbol{W} + \boldsymbol{\Sigma}] \cdot \boldsymbol{N} = \boldsymbol{D}_0 \times \widehat{\boldsymbol{S}}_{\text{m}}^{\text{ext}} \quad \text{on} \quad \partial\mathcal{B}_{\text{m}}. \tag{8.100}$$

Thus, 8.91 is trivial, whereby the extrinsic (interior) flux density of vector moment of material momentum $\widehat{\boldsymbol{T}}_{\text{m}}^{\text{ext}}$ per unit area of the external boundary in the material configuration follows as

$$\widehat{\boldsymbol{T}}_{\text{m}}^{\text{ext}} := \boldsymbol{D}_0 \times \widehat{\boldsymbol{S}}_{\text{m}}^{\text{ext}}. \tag{8.101}$$

iii. At singular surfaces $\mathcal{S}_{\text{m}}$ vector multiplication of the balance of material momentum in Eq. 8.51 by the average material distance vector $\{\boldsymbol{D}_0\} \equiv \boldsymbol{D}_0$ (where the identity follows from the coherence of singular surfaces) renders

$$\{\boldsymbol{D}_0\} \times [\![\boldsymbol{P}_{\text{m}} \otimes \boldsymbol{W} + \boldsymbol{\Sigma}]\!] \cdot \boldsymbol{M} = \{\boldsymbol{D}_0\} \times \widetilde{\boldsymbol{S}}_{\text{m}} \quad \text{on} \quad \mathcal{S}_{\text{m}}. \tag{8.102}$$

Thus, with $[\![\boldsymbol{D}_0]\!] \equiv \boldsymbol{0}$, 8.92 is trivial, whereby the (interior) flux density of vector moment of material momentum $\widetilde{\boldsymbol{T}}_{\text{m}}$ per unit area of singular surfaces in the material configuration follows as

$$\widetilde{\boldsymbol{T}}_{\text{m}} := \{\boldsymbol{D}_0\} \times \widetilde{\boldsymbol{S}}_{\text{m}}. \tag{8.103}$$

vi. On the singular part $\partial^{3-\alpha}\mathcal{B}_{\text{m}}$ of the external boundary $\partial\mathcal{B}_{\text{m}}$ vector multiplication of the limit of the balance of material momentum in Eq. 8.50 by the limit of the material distance $\boldsymbol{D}_0$ renders

$$\lim_{r\to 0} \boldsymbol{D}_0 \times \lim_{r\to 0} \int_{\partial\mathscr{B}_{\text{m}}(r)} [\boldsymbol{P}_{\text{m}} \otimes \boldsymbol{W} + \boldsymbol{\Sigma}] \cdot \boldsymbol{N}\, \mathrm{d}A = \tag{8.104}$$

$$\lim_{r\to 0} \boldsymbol{D}_0 \times \lim_{r\to 0} \int_{\partial\mathscr{B}_{\text{m}}(r)} \widehat{\boldsymbol{S}}_{\text{m}}^{\text{ext}}\, \mathrm{d}A \qquad \text{on} \quad \partial\mathcal{B}_{\text{m}}^{3-\alpha}.$$

Pulling the constant $\lim_{r\to 0} \boldsymbol{D}_0$ into the integrals, noting that the original integrands can be substituted by $\lim_{r\to 0}[\boldsymbol{P}_{\text{m}} \otimes \boldsymbol{W} + \boldsymbol{\Sigma}] \cdot \boldsymbol{N}$ and $\lim_{r\to 0} \widehat{\boldsymbol{S}}_{\text{m}}^{\text{ext}}$, respectively, and using a theorem equating the product of the limits of functions with the limit of the product of these functions renders eventually

$$\lim_{r\to 0} \int_{\partial\mathscr{B}_{\text{m}}(r)} \boldsymbol{D}_0 \times [\boldsymbol{P}_{\text{m}} \otimes \boldsymbol{W} + \boldsymbol{\Sigma}] \cdot \boldsymbol{N}\, \mathrm{d}A = \lim_{r\to 0} \int_{\partial\mathscr{B}_{\text{m}}(r)} \boldsymbol{D}_0 \times \widehat{\boldsymbol{S}}_{\text{m}}^{\text{ext}}\, \mathrm{d}A, \tag{8.105}$$

which, with $\boldsymbol{D}_0 \times \boldsymbol{P}_\mathrm{m} =: \boldsymbol{L}_\mathrm{m}$ and $\boldsymbol{D}_0 \times \boldsymbol{\Sigma} =: \boldsymbol{\Theta}$, may be rewritten as the result in Eq. 8.93.

Therein $\widehat{\mathsf{T}}$ represents the singular contribution to the extrinsic (interior) flux of vector moment of material momentum at the external boundary in the material configuration

$$\widehat{\mathsf{T}} := -\lim_{r\to 0} \int_{\partial\mathscr{B}_\mathrm{m}(r)} \boldsymbol{D}_0 \times [\lambda_\mathrm{m}\boldsymbol{N} + \boldsymbol{F}^t \cdot \widehat{\mathfrak{s}}_\mathrm{m}^\mathrm{ext}]\,\mathrm{d}A. \tag{8.106}$$

Making the VMMM Balance Truly Independent (Material Isotropy)

The balance of vector moment of material momentum (in short VMMM) is an *independent* balance only in case that $\overline{\boldsymbol{T}}_\mathrm{m}^\mathrm{aut} \equiv \boldsymbol{0}$. Only in this situation the resultant external material couple is exclusively due to the external material forces.

The corresponding *symmetry constraint* $\boldsymbol{\Sigma}^\mathrm{skw} \equiv -[\boldsymbol{F}^t \cdot \boldsymbol{P}]^\mathrm{skw} \doteq \boldsymbol{0}$ for the (interior) Eshelby stress is a consequence of the special case of *material isotropy* (see remark below) and can be regarded a constraint for the constitutive modeling of $\boldsymbol{P}$. Recall that $\boldsymbol{M} := \boldsymbol{F}^t \cdot \boldsymbol{P}$ is denoted as the *Mandel stress*, a material description (mixed-variant) tensor mapping from and to the material cotangent space $T^*\mathcal{B}_\mathrm{m}$. The Mandel stress is only symmetric in the case of material isotropy, in which case the balance of vector moment of material momentum is automatically satisfied.

Remark (Material Isotropy):

Constitutively, the symmetry of $\boldsymbol{M} = \boldsymbol{F}^t \cdot \boldsymbol{P}$ may be regarded a consequence of *material isotropy*. For thermodynamically consistent constitutive modeling, $\boldsymbol{P}$ derives from the free energy density ψ_m per unit volume in the material configuration as $\boldsymbol{P} = \partial_{\boldsymbol{F}}\psi_\mathrm{m}$. The special case of material isotropy requires that $\psi_\mathrm{m} = \psi_\mathrm{m}^*(\tau)$, with τ a time-like parameter, remains unaffected under *materially superposed rigid body motions* that take $\boldsymbol{F}$ to $\boldsymbol{F}^*(\tau) = \boldsymbol{F} \cdot \boldsymbol{Q}(\tau)$, whereby $\boldsymbol{Q}$ is from the *special orthogonal group* $\mathrm{SO}(3) = \{\boldsymbol{Q} | \boldsymbol{Q}^{-1} = \boldsymbol{Q}^t, \det \boldsymbol{Q} = +1\}$. Then $\mathrm{D}_\tau \boldsymbol{F}^*(\tau) = \boldsymbol{F}^*(\tau) \cdot \boldsymbol{Q}^t(\tau) \cdot \mathrm{D}_\tau \boldsymbol{Q}(\tau)$ holds with $\boldsymbol{Q}^t(\tau) \cdot \mathrm{D}_\tau \boldsymbol{Q}(\tau) = -\mathrm{D}_\tau \boldsymbol{Q}^t(\tau) \cdot \boldsymbol{Q}(\tau)$ being skew-symmetric. Thus, material isotropy establishes eventually that $\mathrm{D}_\tau \psi_\mathrm{m}^* = \boldsymbol{M}^*(\tau) : [\boldsymbol{Q}^t(\tau) \cdot \mathrm{D}_\tau \boldsymbol{Q}(\tau)] \doteq 0$ and therefore especially that $\boldsymbol{M} = \boldsymbol{M}^*(\tau = 0)$ is necessarily symmetric. □

8.3.2 Global Format (Material Control Volume)

With these preliminaries at hand, the balance of vector moment of material momentum may formally also be stated in global form as

Global Balance of Vector Moment of Material Momentum (Material Control Volume):

$$\dot{\mathsf{L}}(\mathfrak{B}_\mathrm{m}) = \mathsf{C}^\mathrm{ext}(\mathfrak{B}_\mathrm{m}) \quad \forall \mathfrak{B}_\mathrm{m} \subseteq \mathcal{B}_\mathrm{m}. \tag{8.107}$$

Thereby L is introduced as the resultant over the material control volume $\mathfrak{B}_\mathrm{m}$ of the density of *vector moment of material momentum* $\boldsymbol{L}_\mathrm{m}$ per unit volume in the material configuration, i.e. as the *resultant vector moment of material momentum* contained in the material control volume

$$\mathsf{L}(\mathfrak{B}_\mathrm{m}) := \int_{\mathfrak{B}_\mathrm{m}} \boldsymbol{L}_\mathrm{m}\,\mathrm{d}V. \tag{8.108}$$

Moreover, C^ext denotes the *resultant external material couple* (consisting of volume, boundary *and* surface integrals *and* a singular contribution)

$$\mathsf{C}^\mathrm{ext}(\mathfrak{B}_\mathrm{m}) := \int_{\mathfrak{B}_\mathrm{m}} \overline{\boldsymbol{T}}_\mathrm{m}\,\mathrm{d}V + \int_{\partial\mathfrak{B}_\mathrm{m}\setminus\partial^{3-\alpha}\mathcal{B}_\mathrm{m}} \widehat{\boldsymbol{T}}_\mathrm{m}\,\mathrm{d}A + \widehat{\mathsf{T}} - \int_{\mathfrak{S}_\mathrm{m}} \widetilde{\boldsymbol{T}}_\mathrm{m}\,\mathrm{d}A. \tag{8.109}$$

Note the contributions to the resultant external material couple due to the supply of vector moment of material momentum to singular parts of the external boundary and to singular surfaces.

Consequently, the global statement of the balance of vector moment of material momentum expands as

$$\int_{\mathfrak{B}_\mathrm{m}} \mathrm{D}_t\boldsymbol{L}_\mathrm{m}\,\mathrm{d}V = \tag{8.110}$$

$$\int_{\mathfrak{B}_\mathrm{m}} \overline{\boldsymbol{T}}_\mathrm{m}\,\mathrm{d}V + \int_{\partial\mathfrak{B}_\mathrm{m}\setminus\partial^{3-\alpha}\mathcal{B}_\mathrm{m}} \underbrace{\boldsymbol{\Theta}\cdot\boldsymbol{N}}_{\boldsymbol{T}_\mathrm{m}}\,\mathrm{d}A + \mathsf{T} - \int_{\mathfrak{S}_\mathrm{m}} \widetilde{\boldsymbol{T}}_\mathrm{m}\,\mathrm{d}A + \int_{\mathfrak{S}_\mathrm{m}} [\![\boldsymbol{L}_\mathrm{m}\otimes\boldsymbol{W}]\!]\cdot\boldsymbol{M}\,\mathrm{d}A.$$

Here, the convective term stemming from the transport theorem has been subsumed in the boundary integrand, i.e. the *intrinsic* (interior) flux density of vector moment of material momentum

$$\boldsymbol{T}_\mathrm{m} := \boldsymbol{\Theta}\cdot\boldsymbol{N} = \widehat{\boldsymbol{T}}_\mathrm{m} - [\boldsymbol{L}_\mathrm{m}\otimes\boldsymbol{W}]\cdot\boldsymbol{N}, \tag{8.111}$$

the singular *intrinsic* (interior) flux of vector moment of material momentum

$$\mathsf{T} := \lim_{r\to 0}\int_{\partial\mathscr{B}_\mathrm{m}(r)} \boldsymbol{\Theta}\cdot\boldsymbol{N}\,\mathrm{d}A = \widehat{\mathsf{T}} - \lim_{r\to 0}\int_{\partial\mathscr{B}_\mathrm{m}(r)} [\boldsymbol{L}_\mathrm{m}\otimes\boldsymbol{W}]\cdot\boldsymbol{N}\,\mathrm{d}A \tag{8.112}$$

and the singular surface integrand $[\![\boldsymbol{L}_\mathrm{m}\otimes\boldsymbol{W}]\!]\cdot\boldsymbol{M}$. Note that upon localizing the global balance of vector moment of material momentum in Eq. 8.110 indeed Eqs. 8.90–8.93 are obtained.

By incorporating the specific formats for the source and flux terms into Eq. 8.110, the global statement of the balance of vector moment of material momentum specializes to

$$\int_{\mathfrak{B}_m} \mathrm{D}_t \boldsymbol{L}_m \, \mathrm{d}V = \int_{\mathfrak{B}_m} [\boldsymbol{D}_0 \times \overline{\boldsymbol{S}}_m + \overline{\boldsymbol{T}}_m^{\mathrm{aut}}] \, \mathrm{d}V \tag{8.113}$$
$$+ \int_{\partial\mathfrak{B}_m \setminus \partial^{3-\alpha}\mathcal{B}_m} \boldsymbol{D}_0 \times \underbrace{\boldsymbol{\Sigma} \cdot \boldsymbol{N}}_{\boldsymbol{S}_m} \, \mathrm{d}A + \lim_{r \to 0} \int_{\partial\mathscr{B}_m(r)} \boldsymbol{D}_0 \times \boldsymbol{\Sigma} \cdot \boldsymbol{N} \, \mathrm{d}A$$
$$- \int_{\mathfrak{S}_m} \{\boldsymbol{D}_0\} \times \widetilde{\boldsymbol{S}}_m \, \mathrm{d}A + \int_{\mathfrak{S}_m} [\![\boldsymbol{L}_m \otimes \boldsymbol{W}]\!] \cdot \boldsymbol{M} \, \mathrm{d}A .$$

This format clearly reveals that only in case that the autonomous contribution to the source density of vector moment of material momentum vanishes, i.e. for $\overline{\boldsymbol{T}}_m^{\mathrm{aut}} \equiv \boldsymbol{0}$, the resultant external material couple is exclusively due to the external material forces, thus making the balance of vector moment of material momentum an independent balance.

In summary the global balance of vector moment of material momentum relates the total time derivative of the resultant vector moment of material momentum to the resultant external material couple under the precondition that the balance of material momentum (and thus the balance of spatial momentum) is satisfied.

Identifying the Vectorial L-Integral

The balance of vector moment of material momentum in Eq. 8.93 may be considered as the definition of a single material couple acting on the singular part of the external boundary. In the classical terminology of defect mechanics, this single material couple is denoted as the *vectorial L-integral*

$$\boldsymbol{\mathcal{L}} := - \lim_{r \to 0} \int_{\partial\mathscr{B}_m(r)} \boldsymbol{D}_0 \times \boldsymbol{\Sigma} \cdot \boldsymbol{N} \, \mathrm{d}A = -\mathsf{T}. \tag{8.114}$$

Here, $\boldsymbol{\mathcal{L}}$ coincides with the classical definition of the vectorial L-integral. Observe the sign change that is a consequence of using the inward normal $-\boldsymbol{N}$ in the traditional definition of $\boldsymbol{\mathcal{L}}$. The L-integral can be considered as the driving couple for *rotational* configurational changes of the material orientation of, e.g. a mathematical inclusion.

Based on the global statement of the balance of vector moment of material momentum in Eq. 8.110, conditions for the L-integral to be *path-independent* are straightforwardly established. These are

- $\mathrm{D}_t \boldsymbol{L}_m = \boldsymbol{0}$, i.e. quasi-statics,
- $\overline{\boldsymbol{T}}_m = \boldsymbol{0}$, i.e. no heterogeneities or inhomogeneities *and* material isotropy,
- $\widetilde{\boldsymbol{T}}_m = \boldsymbol{0}$, i.e. no singular surfaces.

These conditions typically apply to homogeneous, materially isotropic hyper-elasto-static specimens with a stationary inclusion.

8.3.3 Local Format (Spatial Control Volume)

The local balance for the density of vector moment of material momentum $\boldsymbol{L}_{\mathrm{s}} := j\boldsymbol{L}_{\mathrm{m}}$ per unit volume in the spatial configuration, expressed in terms of the material-spatial description (interior) Eshelby-type couple stress $\boldsymbol{l} := \boldsymbol{\Theta} \cdot \operatorname{cof} \boldsymbol{f}$, the (interior) source density of vector moment of material momentum $\overline{\boldsymbol{T}}_{\mathrm{s}} := j\,\overline{\boldsymbol{T}}_{\mathrm{m}}$ per unit volume in the spatial configuration and the extrinsic (interior) flux densities of vector moment of material momentum $\widehat{\boldsymbol{T}}^{\mathrm{ext}}_{\mathrm{s}} := \widehat{j}\,\widehat{\boldsymbol{T}}^{\mathrm{ext}}_{\mathrm{m}}$ and $\widetilde{\boldsymbol{T}}_{\mathrm{s}} := \widetilde{j}\,\widetilde{\boldsymbol{T}}_{\mathrm{m}}$, respectively, per unit area in the spatial configuration, is summarized as

Local Balance of Vector Moment of Material Momentum (Spatial Control Volume):

i. Regular Points in the Domain

$$\mathrm{d}_t \boldsymbol{L}_{\mathrm{s}} - \operatorname{div}(\boldsymbol{l} - \boldsymbol{L}_{\mathrm{s}} \otimes \boldsymbol{v}) = \overline{\boldsymbol{T}}_{\mathrm{s}} \quad \text{in} \quad \mathcal{B}_{\mathrm{s}}. \tag{8.115}$$

ii. Regular Points on the Boundary

$$\big[\boldsymbol{L}_{\mathrm{s}} \otimes \boldsymbol{w} + [\boldsymbol{l} - \boldsymbol{L}_{\mathrm{s}} \otimes \boldsymbol{v}]\big] \cdot \boldsymbol{n} = \widehat{\boldsymbol{T}}^{\mathrm{ext}}_{\mathrm{s}} \quad \text{on} \quad \partial\mathcal{B}_{\mathrm{s}}. \tag{8.116}$$

iii. Regular Points at Singular Surfaces

$$[\![\boldsymbol{L}_{\mathrm{s}} \otimes \boldsymbol{w} + [\boldsymbol{l} - \boldsymbol{L}_{\mathrm{s}} \otimes \boldsymbol{v}]]\!] \cdot \boldsymbol{m} = \widetilde{\boldsymbol{T}}_{\mathrm{s}} \quad \text{at} \quad \mathcal{S}_{\mathrm{s}}. \tag{8.117}$$

In the above definitions any singular contributions are ignored for the sake of presentation.

8.3.4 Global Format (Spatial Control Volume)

With these preliminaries at hand, the balance of vector moment of material momentum may formally also be stated in global format as

Global Balance of Vector Moment of Material Momentum (Spatial Control Volume):

$$\dot{\mathsf{L}}(\mathfrak{B}_{\mathrm{s}}) = \mathsf{C}^{\mathrm{ext}}(\mathfrak{B}_{\mathrm{s}}) \quad \forall \mathfrak{B}_{\mathrm{s}} \subseteq \mathcal{B}_{\mathrm{s}}. \tag{8.118}$$

Thereby L is introduced as the resultant over the spatial control volume $\mathfrak{B}_{\mathrm{s}}$ of the density of *vector moment of material momentum* $\boldsymbol{L}_{\mathrm{s}}$ per unit volume in the spatial configuration, i.e. as the *resultant vector moment of material momentum* contained in the spatial control volume

$$\mathsf{L}(\mathfrak{B}_s) := \int_{\mathfrak{B}_s} \boldsymbol{L}_s \,\mathrm{d}v. \tag{8.119}$$

Moreover, C^{ext} denotes the *resultant external material couple* (consisting of volume, boundary *and* surface integrals

$$\mathsf{C}^{\text{ext}}(\mathfrak{B}_s) := \int_{\mathfrak{B}_s} \overline{\boldsymbol{T}}_s \,\mathrm{d}v + \int_{\partial\mathfrak{B}_s} \widehat{\boldsymbol{T}}_s \,\mathrm{d}a - \int_{\mathfrak{S}_s} \widetilde{\boldsymbol{T}}_s \,\mathrm{d}a. \tag{8.120}$$

Note the contribution to the resultant external material couple due to the supply of vector moment of material momentum to singular surfaces. (In the above definition any singular contributions are ignored for the sake of presentation.)

Consequently, the global statement of the balance of vector moment of material momentum expands as

$$\int_{\mathfrak{B}_s} \mathrm{d}_t \boldsymbol{L}_s \,\mathrm{d}v = \tag{8.121}$$

$$\int_{\mathfrak{B}_s} \overline{\boldsymbol{T}}_s \,\mathrm{d}v + \int_{\partial\mathfrak{B}_s} [\boldsymbol{l} - \boldsymbol{L}_s \otimes \boldsymbol{v}] \cdot \boldsymbol{n} \,\mathrm{d}a - \int_{\mathfrak{S}_s} \widetilde{\boldsymbol{T}}_s \,\mathrm{d}a + \int_{\mathfrak{S}_s} [\![\boldsymbol{L}_s \otimes \boldsymbol{w}]\!] \cdot \boldsymbol{m} \,\mathrm{d}a.$$

Here, the convective term stemming from the transport theorem has been subsumed in the boundary integrand, i.e. the *intrinsic* (interior) flux density of vector moment of material momentum

$$\boldsymbol{T}_s := \boldsymbol{l} \cdot \boldsymbol{n} = \widehat{\boldsymbol{T}}_s - \big[\boldsymbol{L}_s \otimes [\boldsymbol{w} - \boldsymbol{v}]\big] \cdot \boldsymbol{n} \tag{8.122}$$

and the singular surface integrand $[\![\boldsymbol{L}_s \otimes \boldsymbol{w}]\!] \cdot \boldsymbol{m}$. Note that upon localizing the global balance of vector moment of material momentum in Eq. 8.121 indeed Eqs. 8.115–8.117 are obtained.

8.3.5 Balance Tetragon

The local balance of vector moment of material momentum *at regular points in the domain, at regular points on the boundary* and *at regular points at singular surfaces* may then be expressed in four different but fully equivalent versions, see the arrangement in the corresponding tetragons in Table 8.4. Here, *dynamic versions* of the (interior) Eshelby-type couple stresses have been introduced as $\boldsymbol{\Theta}_{\mathrm{D}} = \boldsymbol{l}_{\mathrm{D}} \cdot$

Table 8.4 Tetragon of fully equivalent versions for the local balance of vector moment of material momentum at regular points in the domain, on the boundary, and at singular surfaces

Regular Points in the Domain

$\mathrm{D}_t \boldsymbol{L}_\mathrm{m} = \mathrm{Div}\boldsymbol{\Theta}_\mathrm{D} + \overline{\boldsymbol{T}}_\mathrm{m}$	←PT→	$j\,\mathrm{D}_t \boldsymbol{L}_\mathrm{m} = \mathrm{div}\boldsymbol{l}_\mathrm{D} + \overline{\boldsymbol{T}}_\mathrm{s}$
↕ TT		↕ TT
$J\,\mathrm{d}_t \boldsymbol{L}_\mathrm{s} = \mathrm{Div}\boldsymbol{\Theta}_\mathrm{d} + \overline{\boldsymbol{T}}_\mathrm{m}$	←PT→	$\mathrm{d}_t \boldsymbol{L}_\mathrm{s} = \mathrm{div}\boldsymbol{l}_\mathrm{d} + \overline{\boldsymbol{T}}_\mathrm{s}$

Regular Points on the Boundary

$\left[\boldsymbol{L}_\mathrm{m} \otimes \boldsymbol{W} + \boldsymbol{\Theta}_\mathrm{D}\right] \cdot \boldsymbol{N} = \widehat{\boldsymbol{T}}^\mathrm{ext}_\mathrm{m}$	←PT→	$\left[\boldsymbol{L}_\mathrm{s} \otimes [\boldsymbol{w} - \boldsymbol{v}] + \boldsymbol{l}_\mathrm{D}\right] \cdot \boldsymbol{n} = \widehat{\boldsymbol{T}}^\mathrm{ext}_\mathrm{s}$
↕ TT		↕ TT
$\left[\boldsymbol{L}_\mathrm{m} \otimes [\boldsymbol{W} - \boldsymbol{V}] + \boldsymbol{\Theta}_\mathrm{d}\right] \cdot \boldsymbol{N} = \widehat{\boldsymbol{T}}^\mathrm{ext}_\mathrm{m}$	←PT→	$\left[\boldsymbol{L}_\mathrm{s} \otimes \boldsymbol{w} + \boldsymbol{l}_\mathrm{d}\right] \cdot \boldsymbol{n} = \widehat{\boldsymbol{T}}^\mathrm{ext}_\mathrm{s}$

Regular Points at Singular Surfaces

$[\![\boldsymbol{L}_\mathrm{m} \otimes \boldsymbol{W} + \boldsymbol{\Theta}_\mathrm{D}]\!] \cdot \boldsymbol{M} = \widetilde{\boldsymbol{T}}_\mathrm{m}$	←PT→	$[\![\boldsymbol{L}_\mathrm{s} \otimes [\boldsymbol{w} - \boldsymbol{v}] + \boldsymbol{l}_\mathrm{D}]\!] \cdot \boldsymbol{m} = \widetilde{\boldsymbol{T}}_\mathrm{s}$
↕ TT		↕ TT
$[\![\boldsymbol{L}_\mathrm{m} \otimes [\boldsymbol{W} - \boldsymbol{V}] + \boldsymbol{\Theta}_\mathrm{d}]\!] \cdot \boldsymbol{M} = \widetilde{\boldsymbol{T}}_\mathrm{m}$	←PT→	$[\![\boldsymbol{L}_\mathrm{s} \otimes \boldsymbol{w} + \boldsymbol{l}_\mathrm{d}]\!] \cdot \boldsymbol{m} = \widetilde{\boldsymbol{T}}_\mathrm{s}$

with *Dynamic Versions* of (Interior) Eshelby-Type Couple Stress

$\boldsymbol{\Theta}_\mathrm{D} := \boldsymbol{\Theta}$	←PT→	$\boldsymbol{l}_\mathrm{D} := \boldsymbol{l}$
↕ TT		↕ TT
$\boldsymbol{\Theta}_\mathrm{d} := \boldsymbol{\Theta} + \boldsymbol{L}_\mathrm{m} \otimes \boldsymbol{V}$	←PT→	$\boldsymbol{l}_\mathrm{d} := \boldsymbol{l} - \boldsymbol{L}_\mathrm{s} \otimes \boldsymbol{v}$

cof $\boldsymbol{F}$ and $\boldsymbol{\Theta}_\mathrm{d} = \boldsymbol{l}_\mathrm{d} \cdot \operatorname{cof} \boldsymbol{F}$, respectively, whereby the subscripts D and d indicate that the corresponding (interior) Eshelby-type couple stress is consistent with either the material or the spatial time derivative D_t or d_t, respectively, of the density of vector moment of material momentum.

8.4 Scalar Moment of Material Momentum*

Scalar moments of material forces are denoted as *material virials* and serve as a measure of the *centrality* of the material force system. They contribute to the *balance of scalar moment of material momentum*, which, however, is not established in its own right but results as a consequence of the balance of material momentum.

8.4.1 Local Format (Material Control Volume)

Its local format is summarized as

Local Balance of Scalar Moment of Material Momentum (Material Control Volume):

i. Regular Points in the Domain

$$\mathrm{D}_t O_\mathrm{m} - \operatorname{Div}\boldsymbol{\Phi} =: \overline{P}_\mathrm{m} \quad \text{in} \quad \mathcal{B}_\mathrm{m}. \tag{8.123}$$

ii. Regular Points on the Boundary

$$[O_\mathrm{m}\, \boldsymbol{W} + \boldsymbol{\Phi}] \cdot \boldsymbol{N} =: \widehat{P}_\mathrm{m}^\mathrm{ext} \quad \text{on} \quad \partial\mathcal{B}_\mathrm{m} \setminus \partial^{3-\alpha}\mathcal{B}_\mathrm{m}. \tag{8.124}$$

iii. Regular Points at Singular Surfaces

$$[\![O_\mathrm{m}\, \boldsymbol{W} + \boldsymbol{\Phi}]\!] \cdot \boldsymbol{M} =: \widetilde{P}_\mathrm{m} \quad \text{at} \quad \mathcal{S}_\mathrm{m}. \tag{8.125}$$

iv. Singular Points on the Boundary

$$\lim_{r\to 0} \int_{\partial\mathscr{B}_\mathrm{m}(r)} [O_\mathrm{m}\, \boldsymbol{W} + \boldsymbol{\Phi}] \cdot \boldsymbol{N}\, \mathrm{d}A =: \widehat{\mathsf{P}} \quad \text{on} \quad \partial^{3-\alpha}\mathcal{B}_\mathrm{m}. \tag{8.126}$$

For its derivation, starting from the scalar multiplication of the local balance of material momentum with the material distance vector $\boldsymbol{D}_0$, three different scenarios may be considered:

i. In a smooth domain $\mathcal{B}_m$ scalar multiplication of the balance of material momentum in Eq. 8.49 by the material distance vector $\boldsymbol{D}_0$ renders

$$\boldsymbol{D}_0 \cdot [\mathrm{D}_t \boldsymbol{P}_m - \mathrm{Div}\boldsymbol{\Sigma}] = \boldsymbol{D}_0 \cdot \overline{\boldsymbol{S}}_m \quad \text{in} \quad \mathcal{B}_m. \tag{8.127}$$

Then, based on the product rule of differentiation, the identities

$$\boldsymbol{D}_0 \cdot \mathrm{D}_t \boldsymbol{P}_m = \mathrm{D}_t[\boldsymbol{D}_0 \cdot \boldsymbol{P}_m],$$

since $\mathrm{D}_t \boldsymbol{D}_0 = \boldsymbol{0}$, and

$$\boldsymbol{D}_0 \cdot \mathrm{Div}\boldsymbol{\Sigma} = \mathrm{Div}(\boldsymbol{D}_0 \cdot \boldsymbol{\Sigma}) - 3\mathrm{Prs}\,\boldsymbol{\Sigma},$$

wherein $\nabla_X \boldsymbol{D}_0 = \boldsymbol{I}$, hold. The *material pressure (scalar)* is defined as

$$3\mathrm{Prs}\,\boldsymbol{\Sigma} := \boldsymbol{\Sigma} : \boldsymbol{I}.$$

As a result, the scalar moment of material momentum reads

$$\mathrm{D}_t[\boldsymbol{D}_0 \cdot \boldsymbol{P}_m] - \mathrm{Div}(\boldsymbol{D}_0 \cdot \boldsymbol{\Sigma}) = \boldsymbol{D}_0 \cdot \overline{\boldsymbol{S}}_m - 3\mathrm{Prs}\,\boldsymbol{\Sigma}. \tag{8.128}$$

Thus, the density of scalar moment of material momentum O_m per unit volume in the material configuration and the *(interior) Eshelby-type virial stress* $\boldsymbol{\Phi} \in T\mathcal{B}_m$ are identified as

$$O_m := \boldsymbol{D}_0 \cdot \boldsymbol{P}_m \quad \text{and} \quad \boldsymbol{\Phi} := \boldsymbol{D}_0 \cdot \boldsymbol{\Sigma}. \tag{8.129}$$

Finally, the (interior) source density of scalar moment of material momentum $\overline{P}_m$ per unit volume in the material configuration follows as

$$\overline{P}_m := \boldsymbol{D}_0 \cdot \overline{\boldsymbol{S}}_m - 3\mathrm{Prs}\,\boldsymbol{\Sigma}. \tag{8.130}$$

Observe the autonomous contribution $\overline{P}^{\mathrm{aut}}_m$ to $\overline{P}_m$

$$\overline{P}^{\mathrm{aut}}_m := -3\mathrm{Prs}\,\boldsymbol{\Sigma}. \tag{8.131}$$

ii. On the regular part $\partial\mathcal{B}_m \setminus \partial^{3-\alpha}\mathcal{B}_m$ of the external boundary $\partial\mathcal{B}_m$ scalar multiplication of the balance of material momentum in Eq. 8.50 by the material distance vector $\boldsymbol{D}_0$ renders

$$\boldsymbol{D}_0 \cdot [\boldsymbol{P}_m \otimes \boldsymbol{W} + \boldsymbol{\Sigma}] \cdot \boldsymbol{N} = \boldsymbol{D}_0 \cdot \widehat{\boldsymbol{S}}^{\mathrm{ext}}_m \quad \text{on} \quad \partial\mathcal{B}_m. \tag{8.132}$$

Thus, 8.124 is trivial, whereby the extrinsic (interior) flux density of scalar moment of material momentum $\widehat{P}^{\mathrm{ext}}_m$ per unit area of the external boundary in the material configuration follows as

$$\widehat{P}^{\text{ext}}_{\text{m}} := \boldsymbol{D}_0 \cdot \widehat{\boldsymbol{S}}^{\text{ext}}_{\text{m}}. \tag{8.133}$$

iii. At singular surfaces $\mathcal{S}_{\text{m}}$ scalar multiplication of the balance of material momentum in Eq. 8.51 by the average material distance vector $\{\boldsymbol{D}_0\} \equiv \boldsymbol{D}_0$ renders

$$\{\boldsymbol{D}_0\} \cdot [\![\boldsymbol{P}_{\text{m}} \otimes \boldsymbol{W} + \boldsymbol{\Sigma}]\!] \cdot \boldsymbol{M} = \{\boldsymbol{D}_0\} \cdot \widetilde{\boldsymbol{S}}_{\text{m}} \quad \text{on} \quad \mathcal{S}_{\text{m}}. \tag{8.134}$$

Thus 8.125 is trivial, whereby the (interior) flux density of scalar moment of material momentum $\widetilde{P}_{\text{m}}$ per unit area of singular surfaces in the material configuration follows as

$$\widetilde{P}_{\text{m}} := \{\boldsymbol{D}_0\} \cdot \widetilde{\boldsymbol{S}}_{\text{m}}. \tag{8.135}$$

vi. On the singular part $\partial^{3-\alpha}\mathcal{B}_{\text{m}}$ of the external boundary $\partial\mathcal{B}_{\text{m}}$ scalar multiplication of the limit of the balance of material momentum in Eq. 8.50 by the limit of the material distance $\boldsymbol{D}_0$ renders

$$\lim_{r\to 0} \boldsymbol{D}_0 \cdot \lim_{r\to 0} \int_{\partial\mathscr{B}_{\text{m}}(r)} [\boldsymbol{P}_{\text{m}} \otimes \boldsymbol{W} + \boldsymbol{\Sigma}] \cdot \boldsymbol{N}\, \text{d}A = \tag{8.136}$$
$$\lim_{r\to 0} \boldsymbol{D}_0 \cdot \lim_{r\to 0} \int_{\partial\mathscr{B}_{\text{m}}(r)} \widehat{\boldsymbol{S}}^{\text{ext}}_{\text{m}}\, \text{d}A \qquad \text{on} \quad \partial\mathcal{B}^{3-\alpha}_{\text{m}}.$$

Pulling the constant $\lim_{r\to 0} \boldsymbol{D}_0$ into the integrals, noting that the original integrands can be substituted by $\lim_{r\to 0}[\boldsymbol{P}_{\text{m}} \otimes \boldsymbol{W} + \boldsymbol{\Sigma}] \cdot \boldsymbol{N}$ and $\lim_{r\to 0} \widehat{\boldsymbol{S}}^{\text{ext}}_{\text{m}}$, respectively, and using a theorem equating the product of the limits of functions with the limit of the product of these functions renders eventually

$$\lim_{r\to 0} \int_{\partial\mathscr{B}_{\text{m}}(r)} \boldsymbol{D}_0 \cdot [\boldsymbol{P}_{\text{m}} \otimes \boldsymbol{W} + \boldsymbol{\Sigma}] \cdot \boldsymbol{N}\, \text{d}A = \lim_{r\to 0} \int_{\partial\mathscr{B}_{\text{m}}(r)} \boldsymbol{D}_0 \cdot \widehat{\boldsymbol{S}}^{\text{ext}}_{\text{m}}\, \text{d}A, \tag{8.137}$$

which, with $\boldsymbol{D}_0 \cdot \boldsymbol{P}_{\text{m}} =: O_{\text{m}}$ and $\boldsymbol{D}_0 \cdot \boldsymbol{\Sigma} =: \boldsymbol{\Phi}$, may be rewritten as the result in Eq. 8.126.
Therein $\widehat{\mathsf{P}}$ represents the singular contribution to the extrinsic (interior) flux of scalar moment of material momentum at the external boundary in the material configuration

$$\widehat{\mathsf{P}} := -\lim_{r\to 0} \int_{\partial\mathscr{B}_{\text{m}}(r)} \boldsymbol{D}_0 \cdot [\lambda_{\text{m}} \boldsymbol{N} + \boldsymbol{F}^t \cdot \widehat{\boldsymbol{s}}^{\text{ext}}_{\text{m}}]\, \text{d}A. \tag{8.138}$$

Making the SMMM Balance Truly Independent (Material PHODT)

The balance of scalar moment of material momentum (in short SMMM) is an *independent* balance only in case that $\overline{P}^{\text{aut}}_{\text{m}} \equiv 0$. Only in this situation the resultant external material virial is exclusively due to the external material forces.

The corresponding *deviator constraint* $\boldsymbol{\Sigma}^{\text{vol}} \doteq \mathbf{0}$ for the (interior) Eshelby stress is a consequence of the very special case of *material phodt* (see remark below) and can be regarded a constraint for the solution of an initial boundary value problem. As a consequence, the balance of scalar moment of material momentum is in general not an independent balance equation!

Remark (Material PHODT):

For the special quasi-static, hyperelastic case, $\boldsymbol{\Sigma} = \psi_{\text{m}} \boldsymbol{I} - \boldsymbol{F}^t \cdot \partial_F \psi_{\text{m}}$ is constitutively *traceless* with $\text{Prs}\, \boldsymbol{\Sigma} = \psi_{\text{m}} - \frac{1}{3} \partial_F \psi_{\text{m}} : \boldsymbol{F} = 0$, if $\psi_{\text{m}} = \psi_{\text{m}}(\boldsymbol{F})$ is *positively homogeneous of degree three* (material phodt) in $\boldsymbol{F}$, i.e. if $\partial_F \psi_{\text{m}} : \boldsymbol{F} = 3\psi_{\text{m}}$. □

8.4.2 Global Format (Material Control Volume)

With these preliminaries at hand, the balance of scalar moment of material momentum may formally also be stated in global form as

Global Balance of Scalar Moment of Material Momentum (Material Control Volume):

$$\dot{\mathsf{O}}(\mathfrak{B}_{\text{m}}) = \mathsf{V}^{\text{ext}}(\mathfrak{B}_{\text{m}}) \quad \forall \mathfrak{B}_{\text{m}} \subseteq \mathcal{B}_{\text{m}}. \tag{8.139}$$

Thereby O is introduced as the resultant over the material control volume $\mathfrak{B}_{\text{m}}$ of the density of *scalar moment of material momentum* O_{m} per unit volume in the material configuration, i.e. as the *resultant scalar moment of material momentum* contained in the material control volume

$$\mathsf{O}(\mathfrak{B}_{\text{m}}) := \int_{\mathfrak{B}_{\text{m}}} O_{\text{m}} \, \mathrm{d}V. \tag{8.140}$$

Moreover, V^{ext} denotes the *resultant external material virial* (consisting of volume, boundary *and* surface integrals *and* a singular contribution)

$$\mathsf{V}^{\text{ext}}(\mathfrak{B}_{\text{m}}) := \int_{\mathfrak{B}_{\text{m}}} \overline{P}_{\text{m}} \, \mathrm{d}V + \int_{\partial \mathfrak{B}_{\text{m}} \backslash \partial^{3-\alpha} \mathcal{B}_{\text{m}}} \widehat{P}_{\text{m}} \, \mathrm{d}A + \widehat{\mathsf{P}} - \int_{\mathfrak{S}_{\text{m}}} \widetilde{P}_{\text{m}} \, \mathrm{d}A. \tag{8.141}$$

Note the contributions to the resultant external material virial due to the supply of scalar moment of material momentum to singular parts of the external boundary and to singular surfaces.

Consequently, the global statement of the balance of scalar moment of material momentum expands as

$$\int_{\mathfrak{B}_m} \mathrm{D}_t O_m \, \mathrm{d}V = \tag{8.142}$$

$$\int_{\mathfrak{B}_m} \overline{P}_m \, \mathrm{d}V + \int_{\partial\mathfrak{B}_m \backslash \partial^{3-\alpha}\mathcal{B}_m} \underbrace{\boldsymbol{\Phi} \cdot \boldsymbol{N}}_{P_m} \, \mathrm{d}A + \mathsf{P} - \int_{\mathfrak{S}_m} \widetilde{P}_m \, \mathrm{d}A + \int_{\mathfrak{S}_m} [\![O_m \boldsymbol{W}]\!] \cdot \boldsymbol{M} \, \mathrm{d}A.$$

Here, the convective term stemming from the transport theorem has been subsumed in the boundary integrand, i.e. the *intrinsic* (interior) flux density of scalar moment of material momentum

$$P_m := \boldsymbol{\Phi} \cdot \boldsymbol{N} = \widehat{P}_m - [O_m \boldsymbol{W}] \cdot \boldsymbol{N}, \tag{8.143}$$

the singular *intrinsic* (interior) flux of scalar moment of material momentum

$$\mathsf{P} := \lim_{r \to 0} \int_{\partial\mathscr{B}_m(r)} \boldsymbol{\Phi} \cdot \boldsymbol{N} \, \mathrm{d}A = \widehat{\mathsf{P}} - \lim_{r \to 0} \int_{\partial\mathscr{B}_m(r)} [O_m \boldsymbol{W}] \cdot \boldsymbol{N} \, \mathrm{d}A \tag{8.144}$$

and the singular surface integrand $[\![O_m \boldsymbol{W}]\!] \cdot \boldsymbol{M}$. Note that upon localizing the global balance of scalar moment of material momentum in Eq. 8.142 indeed Eqs. 8.123–8.126 are obtained.

By incorporating the specific formats for the source and flux terms into Eq. 8.142, the global statement of the balance of scalar moment of material momentum specializes to

$$\begin{aligned} \int_{\mathfrak{B}_m} \mathrm{D}_t O_m \, \mathrm{d}V = & \int_{\mathfrak{B}_m} [\boldsymbol{D}_0 \cdot \overline{\boldsymbol{S}}_m + \overline{P}_m^{\mathrm{aut}}] \, \mathrm{d}V \\ & + \int_{\partial\mathfrak{B}_m \backslash \partial^{3-\alpha}\mathcal{B}_m} \boldsymbol{D}_0 \cdot \underbrace{\boldsymbol{\Sigma} \cdot \boldsymbol{N}}_{\boldsymbol{S}_m} \, \mathrm{d}A + \lim_{r \to 0} \int_{\partial\mathscr{B}_m(r)} \boldsymbol{D}_0 \cdot \boldsymbol{\Sigma} \cdot \boldsymbol{N} \, \mathrm{d}A \\ & - \int_{\mathfrak{S}_m} \{\boldsymbol{D}_0\} \cdot \widetilde{\boldsymbol{S}}_m \, \mathrm{d}A + \int_{\mathfrak{S}_m} [\![O_m \boldsymbol{W}]\!] \cdot \boldsymbol{M} \, \mathrm{d}A. \end{aligned} \tag{8.145}$$

This format clearly reveals that only in case that the autonomous contribution to the source density of scalar moment of material momentum vanishes, i.e. for $\overline{P}_m^{\mathrm{aut}} \equiv 0$, the resultant external material virial is exclusively due to the external material forces, thus making the balance of scalar moment of material momentum an independent balance.

In summary the global balance of scalar moment of material momentum relates the total time derivative of the resultant scalar moment of material momentum to the resultant external material virial under the precondition that the balance of material momentum (and thus the balance of spatial momentum) is satisfied.

Identifying the Scalar M-Integral

The balance of scalar moment of material momentum in Eq. 8.126 may be considered as the definition of a single material virial acting on the singular part of the external boundary. In the classical terminology of defect mechanics, this single material virial is denoted as the *scalar M-integral*

$$\mathcal{M} := -\lim_{r\to 0} \int_{\partial\mathcal{B}_{\mathrm{m}}(r)} \boldsymbol{D}_0 \cdot \boldsymbol{\Sigma} \cdot \boldsymbol{N} \,\mathrm{d}A = -\mathsf{P}. \tag{8.146}$$

Here, $\mathcal{M}$ relates to the classical definition of the scalar M-integral. Observe the sign change that is a consequence of using the inward normal $-\boldsymbol{N}$ in the traditional definition of $\mathcal{M}$. The M-integral can be considered as the driving virial for *dilatational* configurational changes of the material size of, e.g. a mathematical vacancy.

Based on the global statement of the balance of scalar moment of material momentum in Eq. 8.142, conditions for the M-integral to be *path-independent* are straightforwardly established. These are

- $\mathrm{D}_t O_{\mathrm{m}} = 0$, i.e. quasi-statics,
- $\overline{P}_{\mathrm{m}} = 0$, i.e. no heterogeneities or inhomogeneities *and* material phodt,
- $\widetilde{P}_{\mathrm{m}} = 0$, i.e. no singular surfaces.

These conditions typically apply to homogeneous, materially isotropic hyper-elasto-static specimens with a stationary vacancy.

8.4.3 Local Format (Spatial Control Volume)

The local balance for the density of scalar moment of material momentum $O_{\mathrm{s}} := j\, O_{\mathrm{m}}$ per unit volume in the spatial configuration, expressed in terms of the spatial description (interior) Eshelby-type virial stress $\boldsymbol{o} := \boldsymbol{\Phi} \cdot \mathrm{cof}\, \boldsymbol{f}$, the (interior) source density of scalar moment of material momentum $\overline{P}_{\mathrm{s}} := j\, \overline{P}_{\mathrm{m}}$ per unit volume in the spatial configuration and the extrinsic (interior) flux densities of scalar moment of material momentum $\widehat{P}^{\mathrm{ext}}_{\mathrm{s}} := \widehat{j}\, \widehat{P}^{\mathrm{ext}}_{\mathrm{m}}$ and $\widetilde{P}_{\mathrm{s}} := \widetilde{j}\, \widetilde{P}_{\mathrm{m}}$, respectively, per unit area in the spatial configuration, is summarized as

Local Balance of Scalar Moment of Material Momentum (Spatial Control Volume):

i. Regular Points in the Domain

$$\mathrm{d}_t O_{\mathrm{s}} - \mathrm{div}(\boldsymbol{o} - O_{\mathrm{s}}\, \boldsymbol{v}) = \overline{P}_{\mathrm{s}} \quad \text{in} \quad \mathcal{B}_{\mathrm{s}}. \tag{8.147}$$

ii. Regular Points on the Boundary

$$\big[O_{\mathrm{s}}\, \boldsymbol{w} + [\boldsymbol{o} - O_{\mathrm{s}}\, \boldsymbol{v}]\big] \cdot \boldsymbol{n} = \widehat{P}^{\mathrm{ext}}_{\mathrm{s}} \quad \text{on} \quad \partial\mathcal{B}_{\mathrm{s}}. \tag{8.148}$$

iii. Regular Points at Singular Surfaces

$$[\![O_s\, \boldsymbol{w} + [\boldsymbol{o} - O_s\, \boldsymbol{v}]]\!] \cdot \boldsymbol{m} = \widetilde{P}_s \quad \text{at} \quad \mathcal{S}_s. \tag{8.149}$$

In the above definitions any singular contributions are ignored for the sake of presentation.

8.4.4 Global Format (Spatial Control Volume)

With these preliminaries at hand, the balance of scalar moment of material momentum may formally also be stated in global format as

Global Balance of Scalar Moment of Material Momentum (Spatial Control Volume):

$$\dot{\mathsf{O}}(\mathfrak{B}_s) = \mathsf{V}^{\text{ext}}(\mathfrak{B}_s) \quad \forall \mathfrak{B}_s \subseteq \mathcal{B}_s. \tag{8.150}$$

Thereby O is introduced as the resultant over the spatial control volume $\mathfrak{B}_s$ of the density of *scalar moment of material momentum* O_s per unit volume in the spatial configuration, i.e. as the *resultant scalar moment of material momentum* contained in the spatial control volume

$$\mathsf{O}(\mathfrak{B}_s) := \int_{\mathfrak{B}_s} O_s\, \mathrm{d}v. \tag{8.151}$$

Moreover, V^{ext} denotes the *resultant external material virial* (consisting of volume, boundary *and* surface integrals

$$\mathsf{V}^{\text{ext}}(\mathfrak{B}_s) := \int_{\mathfrak{B}_s} \overline{P}_s\, \mathrm{d}v + \int_{\partial\mathfrak{B}_s} \widehat{P}_s\, \mathrm{d}a - \int_{\mathfrak{S}_s} \widetilde{P}_s\, \mathrm{d}a. \tag{8.152}$$

Note the contribution to the resultant external material virial due to the supply of scalar moment of material momentum to singular surfaces. (In the above definition any singular contributions are ignored for the sake of presentation.)

Consequently, the global statement of the balance of scalar moment of material momentum expands as

Table 8.5 Tetragon of fully equivalent versions for the local balance of scalar moment of material momentum at regular points in the domain, on the boundary, and at singular surfaces

Regular Points in the Domain

$\mathrm{D}_t O_\mathrm{m} = \mathrm{Div}\boldsymbol{\Phi}_\mathrm{D} + \overline{P}_\mathrm{m}$	⟷ PT	$j\,\mathrm{D}_t O_\mathrm{m} = \mathrm{div}\boldsymbol{o}_\mathrm{D} + \overline{P}_\mathrm{s}$
↕ TT		↕ TT
$J\,\mathrm{d}_t O_\mathrm{s} = \mathrm{Div}\boldsymbol{\Phi}_\mathrm{d} + \overline{P}_\mathrm{m}$	PT ⟷	$\mathrm{d}_t O_\mathrm{s} = \mathrm{div}\boldsymbol{o}_\mathrm{d} + \overline{P}_\mathrm{s}$

Regular Points on the Boundary

$\left[O_\mathrm{m}\,\boldsymbol{W} + \boldsymbol{\Phi}_\mathrm{D}\right]\cdot\boldsymbol{N} = \widehat{P}^\mathrm{ext}_\mathrm{m}$	⟷ PT	$\left[O_\mathrm{s}\,[\boldsymbol{w} - \boldsymbol{v}] + \boldsymbol{o}_\mathrm{D}\right]\cdot\boldsymbol{n} = \widehat{P}^\mathrm{ext}_\mathrm{s}$
↕ TT		↕ TT
$\left[O_\mathrm{m}\,[\boldsymbol{W} - \boldsymbol{V}] + \boldsymbol{\Phi}_\mathrm{d}\right]\cdot\boldsymbol{N} = \widehat{P}^\mathrm{ext}_\mathrm{m}$	PT ⟷	$\left[O_\mathrm{s}\,\boldsymbol{w} + \boldsymbol{o}_\mathrm{d}\right]\cdot\boldsymbol{n} = \widehat{P}^\mathrm{ext}_\mathrm{s}$

Regular Points at Singular Surfaces

$[\![O_\mathrm{m}\,\boldsymbol{W} + \boldsymbol{\Phi}_\mathrm{D}]\!]\cdot\boldsymbol{M} = \widetilde{P}_\mathrm{m}$	⟷ PT	$[\![O_\mathrm{s}\,[\boldsymbol{w} - \boldsymbol{v}] + \boldsymbol{o}_\mathrm{D}]\!]\cdot\boldsymbol{m} = \widetilde{P}_\mathrm{s}$
↕ TT		↕ TT
$[\![O_\mathrm{m}\,[\boldsymbol{W} - \boldsymbol{V}] + \boldsymbol{\Phi}_\mathrm{d}]\!]\cdot\boldsymbol{M} = \widetilde{P}_\mathrm{m}$	PT ⟷	$[\![O_\mathrm{s}\,\boldsymbol{w} + \boldsymbol{o}_\mathrm{d}]\!]\cdot\boldsymbol{m} = \widetilde{P}_\mathrm{s}$

with *Dynamic Versions* of (Interior) Eshelby-Type Virial Stress

$\boldsymbol{\Phi}_\mathrm{D} := \boldsymbol{\Phi}$	⟷ PT	$\boldsymbol{o}_\mathrm{D} := \boldsymbol{o}$
↕ TT		↕ TT
$\boldsymbol{\Phi}_\mathrm{d} := \boldsymbol{\Phi} + O_\mathrm{m}\,\boldsymbol{V}$	PT ⟷	$\boldsymbol{o}_\mathrm{d} := \boldsymbol{o} - O_\mathrm{s}\,\boldsymbol{v}$

$$\int_{\mathfrak{B}_s} \mathrm{d}_t O_s \,\mathrm{d}v = \tag{8.153}$$

$$\int_{\mathfrak{B}_s} \overline{P}_s \,\mathrm{d}v + \int_{\partial\mathfrak{B}_s} [\boldsymbol{o} - O_s \,\boldsymbol{v}] \cdot \boldsymbol{n} \,\mathrm{d}a - \int_{\mathfrak{S}_s} \widetilde{P}_s \,\mathrm{d}a + \int_{\mathfrak{S}_s} [\![O_s \,\boldsymbol{w}]\!] \cdot \boldsymbol{m} \,\mathrm{d}a.$$

Here, the convective term stemming from the transport theorem has been subsumed in the boundary integrand, i.e. the *intrinsic* (interior) flux density of scalar moment of material momentum

$$P_s := \boldsymbol{o} \cdot \boldsymbol{n} = \widehat{P}_s - \big[O_s \,[\boldsymbol{w} - \boldsymbol{v}]\big] \cdot \boldsymbol{n} \tag{8.154}$$

and the singular surface integrand $[\![O_s \,\boldsymbol{w}]\!] \cdot \boldsymbol{m}$. Note that upon localizing the global balance of scalar moment of material momentum in Eq. 8.153 indeed Eqs. 8.147–8.149 are obtained.

8.4.5 Balance Tetragon

The local balance of scalar moment of material momentum *at regular points in the domain, at regular points on the boundary* and *at regular points at singular surfaces* may then be expressed in four different but fully equivalent versions, see the arrangement in the corresponding tetragons in Table 8.5. Here, *dynamic versions* of the (interior) Eshelby-type virial stresses have been introduced as $\boldsymbol{\Phi}_\mathrm{D} = \boldsymbol{o}_\mathrm{D} \cdot \operatorname{cof} \boldsymbol{F}$ and $\boldsymbol{\Phi}_\mathrm{d} = \boldsymbol{o}_\mathrm{d} \cdot \operatorname{cof} \boldsymbol{F}$, respectively, whereby the subscripts D and d indicate that the corresponding (interior) Eshelby-type virial stress is consistent with either the material or the spatial time derivative D_t or d_t, respectively, of the density of scalar moment of material momentum.

References

1. Abeyaratne R, Knowles JK (1990) On the driving traction acting on a surface of strain discontinuity in a continuum. J Mech Phys Solids 38:345–360
2. Maugin GA (1993) Material inhomogeneities in elasticity. Chapman and Hall, London
3. Maugin GA (1995) Material forces: concepts and applications. ASME Appl Mech Rev 48:213–245
4. Hill R (1986) Energy-momentum tensors in elastostatics: some reflections on the general theory. J Mech Phys Solids 34:305–317
5. Maugin GA (2000) On the universality of the thermomechanics of forces driving singular sets. Arch Appl Mech 70:31–45

Chapter 9
Virtual Work

8,463 m 27°53'23"N 87°05'20"E

Abstract This chapter capitalizes on the referential setting when introducing the notions of spatial and material virtual displacements and discussing the accompanying spatial and material virtual work principles.

The local balances of momentum require sufficient smoothness of the stress measures in the domain. More specifically, the so-called *strong form* of the balances of momentum incorporates application of the *divergence (differential) operator* to the corresponding flux quantities, i.e. the stresses.
Testing the strong format of the momentum balances by suitably defined *test functions* and integrating over the (closed) domain results in the corresponding *weak form*. There the divergence operator as applied to the stresses in the strong form shifts to the *gradient operator* applied to the test function.

In the context of mechanics, the rather abstract concepts of test functions and weak forms take the more intuitive interpretations as *virtual displacements* and *virtual work principles*. The former, together with conditions for their admissibility, are best defined in terms of the *referential perspective* that is based on the idea of a stationary, i.e. time-independent (total) reference configuration. The latter serves a two-fold purpose: on the one hand as a prerequisite for discretization schemes such as the finite element method; on the other hand to rationally transit between representations of the local balances of momentum based on either material, spatial, or referential control volumes.

Virtual work principles hold generically for non-conservative and conservative cases. In the latter, the variational setting in terms of Hamilton and Dirichlet principles identifies the various terms involved energetically. This chapter shall thus pave the way for the variational setting by first establishing virtual displacements and corresponding virtual work principles.

P. Steinmann, *Spatial and Material Forces in Nonlinear Continuum Mechanics*,
Solid Mechanics and Its Applications 272,
https://doi.org/10.1007/978-3-030-89070-4_9

9.1 Referential Perspective

To set the stage for the subsequent discussions regarding the variational setting of continuum mechanics, the kinematics of the spatial and material motion are briefly reviewed from a *referential perspective*, i.e. from the viewpoint of the *stationary reference configuration* $\mathcal{B}_r$ with reference position vectors $\boldsymbol{\Xi}$, see Fig. 9.1. Then, the time-dependent spatial and material position vectors $\boldsymbol{x}(t)$ and $\boldsymbol{X}(t)$ follow from the *spatial* and *material placement maps* $\boldsymbol{\upsilon}(\boldsymbol{\Xi}, t)$ and $\boldsymbol{\Upsilon}(\boldsymbol{\Xi}, t)$, respectively, that collectively denote the *spatial-material configuration space*

$$\mathbb{Y} := \{\boldsymbol{x} = \boldsymbol{\upsilon}(\boldsymbol{\Xi}, t), \boldsymbol{X} = \boldsymbol{\Upsilon}(\boldsymbol{\Xi}, t)\}. \tag{9.1}$$

Concerning the geometry of the reference configuration $\mathcal{B}_r$, its boundary surface, assumed smooth almost everywhere for the sake of presentation, is denoted by $\partial\mathcal{B}_r$ and is equipped with the outwards pointing unit normal $\boldsymbol{\mathcal{N}}$. The smooth sub-parts of the boundary surface are enclosed by the set of boundary curves $\partial^2\mathcal{B}_r$ with outwards pointing unit normals $\widehat{\boldsymbol{\mathcal{N}}}$ tangent to $\partial\mathcal{B}_r$ and, at boundary points, outwards pointing unit normals $\widetilde{\boldsymbol{\mathcal{N}}}$ tangent to $\partial^2\mathcal{B}_r$.

The external loading at the boundary surface $\partial\mathcal{B}_{r,m,s} = \partial\mathcal{B}^D_{r,m,s} \cup \partial\mathcal{B}^N_{r,m,s}$ (with $\partial\mathcal{B}^D_{r,m,s} \cap \partial\mathcal{B}^N_{r,m,s} = \emptyset$) is applied as either Dirichlet data at the sub-parts of the boundary surface $\partial\mathcal{B}^D_{r,m,s}$ or Neumann data at the sub-parts of the boundary surface $\partial\mathcal{B}^N_{r,m,s}$, respectively.

Scalar-valued *referential* volume elements together with scalar- and vector-valued *referential* area, line and point elements are given by $\mathrm{d}\mathcal{V}$, $\mathrm{d}\mathcal{A}$, $\mathrm{d}\mathcal{L}$ and ($\mathrm{d}\mathcal{P} \equiv$) 1, and

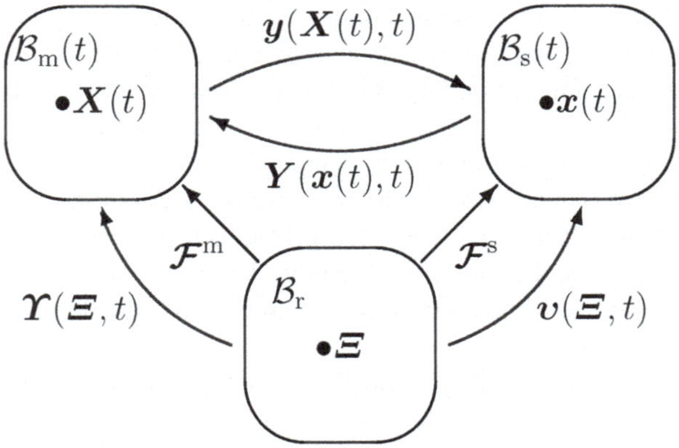

Fig. 9.1 The infinite-dimensional *spatial-material configuration space* defined as $\mathbb{Y} := \{\boldsymbol{\upsilon}, \boldsymbol{\Upsilon}\}$ consists collectively of the time-dependent placements $\boldsymbol{\upsilon}(\boldsymbol{\Xi}, t)$ and $\boldsymbol{\Upsilon}(\boldsymbol{\Xi}, t)$ that map from the stationary reference configuration $\mathcal{B}_r$ with reference position vectors $\boldsymbol{\Xi}$ into the time-dependent spatial and material configuration $\mathcal{B}_s(t)$ and $\mathcal{B}_m(t)$ with spatial and material position vectors $\boldsymbol{x}(t)$ and $\boldsymbol{X}(t)$, respectively,

$\mathrm{d}\boldsymbol{\mathcal{A}} := \mathcal{N}\,\mathrm{d}\mathcal{A}$, $\mathrm{d}\widehat{\boldsymbol{\mathcal{L}}} := \widehat{\mathcal{N}}\,\mathrm{d}\mathcal{L}$ and $(\,\mathrm{d}\widetilde{\boldsymbol{\mathcal{P}}} \equiv)\,\widetilde{\mathcal{N}}$, respectively. Finally, the unit normals at singular surfaces $\mathcal{S}_{\mathrm{r}}$ embedded into $\mathcal{B}_{\mathrm{r}}$ are denoted by $\boldsymbol{\mathcal{M}}$.

9.1.1 Kinematics in Bulk Volumes

In bulk volumes, the material and spatial line elements $\mathrm{d}\boldsymbol{X} \in T\mathcal{B}_{\mathrm{m}}$ and $\mathrm{d}\boldsymbol{x} \in T\mathcal{B}_{\mathrm{s}}$, respectively, are related to the *referential* line element $\mathrm{d}\boldsymbol{\Xi} \in T\mathcal{B}_{\mathrm{r}}$ by (referential-material and referential-spatial) *tangent maps*

$$\mathrm{d}\boldsymbol{X} = \boldsymbol{\mathcal{F}}^{\mathrm{m}} \cdot \mathrm{d}\boldsymbol{\Xi} \quad \text{and} \quad \mathrm{d}\boldsymbol{x} = \boldsymbol{\mathcal{F}}^{\mathrm{s}} \cdot \mathrm{d}\boldsymbol{\Xi}. \tag{9.2}$$

Consequently, the (referential-material and referential-spatial) tangent maps follow as the *referential* gradients of the material and spatial placement maps, respectively,

$$\boldsymbol{\mathcal{F}}^{\mathrm{m}} := \nabla_{\Xi}\boldsymbol{\Upsilon} \in T\mathcal{B}_{\mathrm{m}} \times T^*\mathcal{B}_{\mathrm{r}} \quad \text{and} \quad \boldsymbol{\mathcal{F}}^{\mathrm{s}} := \nabla_{\Xi}\boldsymbol{\upsilon} \in T\mathcal{B}_{\mathrm{s}} \times T^*\mathcal{B}_{\mathrm{r}}. \tag{9.3}$$

Hence, $\boldsymbol{\mathcal{F}}^{\mathrm{m}} : T\mathcal{B}_{\mathrm{r}} \to T\mathcal{B}_{\mathrm{m}}$ and $\boldsymbol{\mathcal{F}}^{\mathrm{s}} : T\mathcal{B}_{\mathrm{r}} \to T\mathcal{B}_{\mathrm{s}}$ are also denoted as the material and spatial *placement gradients*.

Referential vector-valued area elements $\mathrm{d}\boldsymbol{\mathcal{A}}$ are related to the material and spatial vector-valued area elements $\mathrm{d}\boldsymbol{A}$ and $\mathrm{d}\boldsymbol{a}$, respectively, via the cofactors of the (referential-material and referential-spatial) tangent maps

$$\mathrm{d}\boldsymbol{A} = \operatorname{cof}\boldsymbol{\mathcal{F}}^{\mathrm{m}} \cdot \mathrm{d}\boldsymbol{\mathcal{A}} \quad \text{and} \quad \mathrm{d}\boldsymbol{a} = \operatorname{cof}\boldsymbol{\mathcal{F}}^{\mathrm{s}} \cdot \mathrm{d}\boldsymbol{\mathcal{A}}. \tag{9.4}$$

The cofactors of the (referential-material and referential-spatial) tangent maps

$$\boldsymbol{\mathcal{K}}^{\mathrm{m}} := \operatorname{cof}\boldsymbol{\mathcal{F}}^{\mathrm{m}} \in T^*\mathcal{B}_{\mathrm{m}} \times T\mathcal{B}_{\mathrm{r}} \quad \text{and} \quad \boldsymbol{\mathcal{K}}^{\mathrm{s}} := \operatorname{cof}\boldsymbol{\mathcal{F}}^{\mathrm{s}} \in T^*\mathcal{B}_{\mathrm{s}} \times T\mathcal{B}_{\mathrm{r}} \tag{9.5}$$

are also denoted as the (referential-material and referential-spatial) *cotangent maps*.

Referential (scalar-valued) volume elements $\mathrm{d}\mathcal{V}$ are related to the material and spatial (scalar-valued) volume elements $\mathrm{d}V$ and $\mathrm{d}v$, respectively, via the determinants of the (referential-material and referential-spatial) tangent maps

$$\mathrm{d}V = \det\boldsymbol{\mathcal{F}}^{\mathrm{m}}\,\mathrm{d}\mathcal{V} \quad \text{and} \quad \mathrm{d}v = \det\boldsymbol{\mathcal{F}}^{\mathrm{s}}\,\mathrm{d}\mathcal{V}. \tag{9.6}$$

The determinants of the (referential-material and referential-spatial) tangent maps

$$\mathcal{J}^{\mathrm{m}} := \det\boldsymbol{\mathcal{F}}^{\mathrm{m}} \in \mathbb{R}^+ \quad \text{and} \quad \mathcal{J}^{\mathrm{s}} := \det\boldsymbol{\mathcal{F}}^{\mathrm{s}} \in \mathbb{R}^+ \tag{9.7}$$

are also denoted as the (referential-material and referential-spatial) *Jacobians* or rather the (referential-material and referential-spatial) *measure maps*.

9.1.2 Kinematics on Boundary Surfaces*

On boundary surfaces the material and spatial boundary surface line elements $\mathrm{d}\widehat{\boldsymbol{X}} \in T\partial\mathcal{B}_{\mathrm{m}}$ and $\mathrm{d}\widehat{\boldsymbol{x}} \in T\partial\mathcal{B}_{\mathrm{s}}$, respectively, are related to the *referential* boundary surface line element $\mathrm{d}\widehat{\boldsymbol{\Xi}} \in T\partial\mathcal{B}_{\mathrm{r}}$ by (referential-material and referential-spatial) *boundary surface tangent maps*

$$\mathrm{d}\widehat{\boldsymbol{X}} = \widehat{\boldsymbol{\mathcal{F}}}^{\mathrm{m}} \cdot \mathrm{d}\widehat{\boldsymbol{\Xi}} \quad \text{and} \quad \mathrm{d}\widehat{\boldsymbol{x}} = \widehat{\boldsymbol{\mathcal{F}}}^{\mathrm{s}} \cdot \mathrm{d}\widehat{\boldsymbol{\Xi}}. \tag{9.8}$$

Consequently, the (referential-material and referential-spatial) boundary surface tangent maps follow as the *referential* boundary surface gradients of the material and spatial boundary surface placement maps, respectively,

$$\widehat{\boldsymbol{\mathcal{F}}}^{\mathrm{m}} := \widehat{\nabla}_{\widehat{\Xi}} \widehat{\boldsymbol{\Upsilon}} \in T\partial\mathcal{B}_{\mathrm{m}} \times T^*\partial\mathcal{B}_{\mathrm{r}} \quad \text{and} \quad \widehat{\boldsymbol{\mathcal{F}}}^{\mathrm{s}} := \widehat{\nabla}_{\widehat{\Xi}} \widehat{\boldsymbol{\upsilon}} \in T\partial\mathcal{B}_{\mathrm{s}} \times T^*\partial\mathcal{B}_{\mathrm{r}}. \tag{9.9}$$

Hence, $\widehat{\boldsymbol{\mathcal{F}}}^{\mathrm{m}} : T\partial\mathcal{B}_{\mathrm{r}} \to T\partial\mathcal{B}_{\mathrm{m}}$ and $\widehat{\boldsymbol{\mathcal{F}}}^{\mathrm{s}} : T\partial\mathcal{B}_{\mathrm{r}} \to T\partial\mathcal{B}_{\mathrm{s}}$ are also denoted as the material and spatial *boundary surface placement gradients.*

Referential vector-valued line elements $\mathrm{d}\widehat{\boldsymbol{\mathcal{L}}}$ are related to the material and spatial vector-valued line elements $\mathrm{d}\widehat{\boldsymbol{L}}$ and $\mathrm{d}\widehat{\boldsymbol{l}}$, respectively, via the boundary surface cofactors of the (referential-material and referential-spatial) boundary surface tangent maps

$$\mathrm{d}\widehat{\boldsymbol{L}} = \widehat{\mathrm{cof}}\, \widehat{\boldsymbol{\mathcal{F}}}^{\mathrm{m}} \cdot \mathrm{d}\widehat{\boldsymbol{\mathcal{L}}} \quad \text{and} \quad \mathrm{d}\widehat{\boldsymbol{l}} = \widehat{\mathrm{cof}}\, \widehat{\boldsymbol{\mathcal{F}}}^{\mathrm{s}} \cdot \mathrm{d}\widehat{\boldsymbol{\mathcal{L}}}. \tag{9.10}$$

The boundary surface cofactors of the (referential-material and referential-spatial) boundary surface tangent maps

$$\widehat{\boldsymbol{\mathcal{K}}}^{\mathrm{m}} := \widehat{\mathrm{cof}}\, \widehat{\boldsymbol{\mathcal{F}}}^{\mathrm{m}} \in T^*\partial\mathcal{B}_{\mathrm{m}} \times T\partial\mathcal{B}_{\mathrm{r}} \quad \text{and} \quad \widehat{\boldsymbol{\mathcal{K}}}^{\mathrm{s}} := \widehat{\mathrm{cof}}\, \widehat{\boldsymbol{\mathcal{F}}}^{\mathrm{s}} \in T^*\partial\mathcal{B}_{\mathrm{s}} \times T\partial\mathcal{B}_{\mathrm{r}} \tag{9.11}$$

are also denoted as the (referential-material and referential-spatial) *boundary surface cotangent maps.*

Referential (scalar-valued) area elements $\mathrm{d}\mathcal{A}$ are related to the material and spatial (scalar-valued) area elements $\mathrm{d}A$ and $\mathrm{d}a$, respectively, via the boundary surface determinants of the (referential-material and referential-spatial) boundary surface tangent maps

$$\mathrm{d}A = \widehat{\det}\, \widehat{\boldsymbol{\mathcal{F}}}^{\mathrm{m}}\, \mathrm{d}\mathcal{A} \quad \text{and} \quad \mathrm{d}a = \widehat{\det}\, \widehat{\boldsymbol{\mathcal{F}}}^{\mathrm{s}}\, \mathrm{d}\mathcal{A}. \tag{9.12}$$

The boundary surface determinants of the (referential-material and referential-spatial) boundary surface tangent maps

$$\widehat{\mathcal{J}}^{\mathrm{m}} := \widehat{\det}\, \widehat{\boldsymbol{\mathcal{F}}}^{\mathrm{m}} \in \mathbb{R}^+ \quad \text{and} \quad \widehat{\mathcal{J}}^{\mathrm{s}} := \widehat{\det}\, \widehat{\boldsymbol{\mathcal{F}}}^{\mathrm{s}} \in \mathbb{R}^+ \tag{9.13}$$

are also denoted as the (referential-material and referential-spatial) *boundary surface Jacobians* or rather the (referential-material and referential-spatial) *boundary surface measure maps.*

9.1.3 Kinematics on Boundary Curves*

On boundary curves the material and spatial boundary curve line elements $\mathrm{d}\widetilde{X} \in T\partial^2\mathcal{B}_\mathrm{m}$ and $\mathrm{d}\widetilde{x} \in T\partial^2\mathcal{B}_\mathrm{s}$, respectively, are related to the *referential* boundary curve line element $\mathrm{d}\widetilde{\Xi} \in T\partial^2\mathcal{B}_\mathrm{r}$ by (referential-material and referential-spatial) *boundary curve tangent maps*

$$\mathrm{d}\widetilde{X} = \widetilde{\mathcal{F}}^\mathrm{m} \cdot \mathrm{d}\widetilde{\Xi} \quad \text{and} \quad \mathrm{d}\widetilde{x} = \widetilde{\mathcal{F}}^\mathrm{s} \cdot \mathrm{d}\widetilde{\Xi}. \tag{9.14}$$

Consequently, the (referential-material and referential-spatial) boundary curve tangent maps follow as the *referential* boundary curve gradients of the material and spatial boundary curve placement maps, respectively,

$$\widetilde{\mathcal{F}}^\mathrm{m} := \widetilde{\nabla}_{\widetilde{\Xi}}\widetilde{\Upsilon} \in T\partial^2\mathcal{B}_\mathrm{m} \times T^*\partial^2\mathcal{B}_\mathrm{r} \text{ and } \widetilde{\mathcal{F}}^\mathrm{s} := \widetilde{\nabla}_{\widetilde{\Xi}}\widetilde{\upsilon} \in T\partial^2\mathcal{B}_\mathrm{s} \times T^*\partial^2\mathcal{B}_\mathrm{r}. \tag{9.15}$$

Hence, $\widetilde{\mathcal{F}}^\mathrm{m} : T\partial^2\mathcal{B}_\mathrm{r} \to T\partial^2\mathcal{B}_\mathrm{m}$ and $\widetilde{\mathcal{F}}^\mathrm{s} : T\partial^2\mathcal{B}_\mathrm{r} \to T\partial^2\mathcal{B}_\mathrm{s}$ are also denoted as the material and spatial *boundary curve placement gradients*.

Referential vector-valued point elements $\widetilde{\mathcal{N}}$ are related to the material and spatial vector-valued point elements $\widetilde{N}$ and $\widetilde{n}$, respectively, via the boundary curve cofactors of the (referential-material and referential-spatial) boundary curve tangent maps

$$\widetilde{N} = \widetilde{\operatorname{cof}}\,\widetilde{\mathcal{F}}^\mathrm{m} \cdot \widetilde{\mathcal{N}} \quad \text{and} \quad \widetilde{n} = \widetilde{\operatorname{cof}}\,\widetilde{\mathcal{F}}^\mathrm{s} \cdot \widetilde{\mathcal{N}}. \tag{9.16}$$

The boundary curve cofactors of the (referential-material and referential-spatial) boundary curve tangent maps

$$\widetilde{\mathcal{K}}^\mathrm{m} := \widetilde{\operatorname{cof}}\,\widetilde{\mathcal{F}}^\mathrm{m} \in T^*\partial^2\mathcal{B}_\mathrm{m} \times T\partial^2\mathcal{B}_\mathrm{r} \text{ and } \widetilde{\mathcal{K}}^\mathrm{s} := \widetilde{\operatorname{cof}}\,\widetilde{\mathcal{F}}^\mathrm{s} \in T^*\partial^2\mathcal{B}_\mathrm{s} \times T\partial^2\mathcal{B}_\mathrm{r} \tag{9.17}$$

are also denoted as the (referential-material and referential-spatial) *boundary curve cotangent maps*.

Referential (scalar-valued) line elements $\mathrm{d}\mathcal{L}$ are related to the material and spatial (scalar-valued) line elements $\mathrm{d}L$ and $\mathrm{d}l$, respectively, via the boundary curve determinants of the (referential-material and referential-spatial) boundary curve tangent maps

$$\mathrm{d}L = \widetilde{\det}\,\widetilde{\mathcal{F}}^\mathrm{m}\,\mathrm{d}\mathcal{L} \quad \text{and} \quad \mathrm{d}l = \widetilde{\det}\,\widetilde{\mathcal{F}}^\mathrm{s}\,\mathrm{d}\mathcal{L}. \tag{9.18}$$

The boundary curve determinants of the (referential-material and referential-spatial) boundary curve tangent maps

$$\widetilde{\mathcal{J}}^\mathrm{m} := \widetilde{\det}\,\widetilde{\mathcal{F}}^\mathrm{m} \in \mathbb{R}^+ \quad \text{and} \quad \widetilde{\mathcal{J}}^\mathrm{s} := \widetilde{\det}\,\widetilde{\mathcal{F}}^\mathrm{s} \in \mathbb{R}^+ \tag{9.19}$$

are also denoted as the (referential-material and referential-spatial) *boundary curve Jacobians* or rather the (referential-material and referential-spatial) *boundary curve measure maps*.

9.1.4 Preliminaries

(Volume) Gauss Theorem (Referential Control Volume):
For smooth referential control volumes $\mathfrak{V}_r$ integrals extending over $\mathfrak{V}_r$ are related to integrals extending over $\partial\mathfrak{V}_r$ by *Gauss' theorem*, whereby $\mathcal{Z} = \mathcal{Z}(\Xi)$ is any (scalar-,) vector-, or tensor-valued field defined over the reference configuration $\mathcal{V}_r$

$$\int_{\mathfrak{V}_r} \mathrm{DIV}\mathcal{Z}\, \mathrm{d}\mathcal{V} = \int_{\partial\mathfrak{V}_r} \mathcal{Z} \cdot \mathcal{N}\, \mathrm{d}\mathcal{A}. \tag{9.20}$$

Here, $\mathrm{DIV}\mathcal{Z}$ denotes the properly defined *referential divergence* of $\mathcal{Z}$, whereby the *referential gradient operator* contracts with $\mathcal{Z}$ from the right, i.e. $\mathrm{DIV}\mathcal{Z} := \mathcal{Z} \cdot \nabla_\Xi$. Extending the Gauss theorem to cases with a singular surface $\mathfrak{S}_r$ contained in the referential control volume $\mathfrak{V}_r$ is straightforward and renders

$$\int_{\mathfrak{V}_r} \mathrm{DIV}\mathcal{Z}\, \mathrm{d}\mathcal{V} = \int_{\partial\mathfrak{V}_r} \mathcal{Z} \cdot \mathcal{N}\, \mathrm{d}\mathcal{A} - \int_{\mathfrak{S}_r} [\![\mathcal{Z}]\!] \cdot \mathcal{M}\, \mathrm{d}\mathcal{A}. \tag{9.21}$$

Here, $[\![\mathcal{Z}]\!] := \mathcal{Z}^+ - \mathcal{Z}^-$ denotes the finite jump discontinuity of $\mathcal{Z}$ across the singular surface. Thereby it is assumed that $\mathcal{Z}$ is smooth away from the singular surface and smooth up to the singular surface from either side where it takes values $\mathcal{Z}^\pm = \lim_{\epsilon\to 0} \mathcal{Z}(\Xi|_{\mathcal{S}_r} \pm \epsilon\mathcal{M})$ with $\epsilon \in \mathbb{R}^+$.

Global (Volume) Transport Theorem (Material/Spatial Control Volume):
The (volume) resultant Z of any time-dependent scalar-, vector-, or tensor-valued field $z_r = z_r(\Xi, t)$, a density per unit volume in the referential configuration, defined over the (smooth) referential control volume $\mathfrak{V}_r$ is given as

$$\mathsf{Z} := \int_{\mathfrak{V}_r} z_r\, \mathrm{d}\mathcal{V}. \tag{9.22}$$

Then, the (total) time derivative of the resultant Z is expressed as

$$\begin{aligned}\dot{\mathsf{Z}} = \int_{\mathfrak{V}_r} \dot{z}_r\, \mathrm{d}\mathcal{V} &= \int_{\mathfrak{V}_m} \mathrm{D}_t z_m\, \mathrm{d}V + \int_{\partial\mathfrak{V}_m} [z_m \otimes \boldsymbol{W}] \cdot \boldsymbol{N}\, \mathrm{d}A \\ &= \int_{\mathfrak{V}_s} \mathrm{d}_t z_s\, \mathrm{d}v + \int_{\partial\mathfrak{V}_s} [z_s \otimes \boldsymbol{w}] \cdot \boldsymbol{n}\, \mathrm{d}a\,.\end{aligned} \tag{9.23}$$

The proof, also extended to cases with a singular surface $\mathfrak{S}_r$ contained in the referential control volume $\mathfrak{V}_r$, follows in analogy to the steps leading, e.g. to Eq. 5.8.

Referential↔ Material/Spatial (Volume) Piola Transformations:
For any vector- or tensor-valued field $\mathcal{Z} \in \cdots \times T\mathcal{V}_r$, extending over the reference configuration $\mathcal{V}_r$ and mapping from the referential cotangent space $T^*\mathcal{V}_r$, the corresponding *Piola transformations* $\mathbf{Z} \in \cdots \times T\mathcal{V}_m$ and $z \in \cdots \times T\mathcal{V}_s$, extending over either the material or the spatial configurations $\mathcal{V}_m$ or $\mathcal{V}_s$, respectively, and mapping from either the material or the spatial cotangent spaces $T^*\mathcal{V}_m$ or $T^*\mathcal{V}_s$, respectively, are defined as

$$\mathbf{Z} := \mathcal{Z} \cdot \mathcal{K}^{m-1} \quad \text{and} \quad z := \mathcal{Z} \cdot \mathcal{K}^{s-1}. \tag{9.24}$$

Then, the relations between the properly defined *referential divergence* $\mathrm{DIV}\mathcal{Z}$, whereby the *referential gradient operator* contracts with $\mathcal{Z}$ from the right, i.e. $\mathrm{DIV}\mathcal{Z} := \mathcal{Z} \cdot \nabla_{\Xi}$, and the *material* or the *spatial divergence* of $\mathbf{Z}$ or z, respectively, follow as

$$\mathrm{DIV}\mathcal{Z} = \mathcal{J}^m \,\mathrm{Div}\mathbf{Z} = \mathcal{J}^s \,\mathrm{div} z. \tag{9.25}$$

Moreover, the relations between $\mathcal{Z} \cdot \mathcal{N}$ and $\mathbf{Z} \cdot \boldsymbol{N}$ or $z \cdot \boldsymbol{n}$, whereby $\mathcal{N}$ denotes the outwards pointing unit normal to the boundary surface $\partial\mathcal{V}_r$ of the referential configuration $\mathcal{V}_r$ follow as

$$\mathcal{Z} \cdot \mathcal{N} = \widehat{\mathcal{J}}^m \,\mathbf{Z} \cdot \boldsymbol{N} = \widehat{\mathcal{J}}^s \,z \cdot \boldsymbol{n}. \tag{9.26}$$

Likewise, the relations between $[\![\mathcal{Z}]\!] \cdot \mathcal{M}$ and $[\![\mathbf{Z}]\!] \cdot \boldsymbol{M}$ or $[\![z]\!] \cdot \boldsymbol{m}$, whereby $\mathcal{M}$ denotes the unit normal to the singular surface $\mathcal{S}_r$ within the referential configuration $\mathcal{V}_r$ follow as

$$[\![\mathcal{Z}]\!] \cdot \mathcal{M} = \widehat{\mathcal{J}}^m \,[\![\mathbf{Z}]\!] \cdot \boldsymbol{M} = \widehat{\mathcal{J}}^s \,[\![z]\!] \cdot \boldsymbol{m}. \tag{9.27}$$

Here, $\widehat{\mathcal{J}}^m := \mathrm{d}A/\mathrm{d}\mathcal{A} = \mathcal{J}^m \,|\mathcal{N} \cdot \mathcal{F}_m^{-1}|$ and $\widehat{\mathcal{J}}^s := \mathrm{d}a/\mathrm{d}\mathcal{A} = \mathcal{J}^s \,|\mathcal{N} \cdot \mathcal{F}_s^{-1}|$ denote the *(referential-material and referential-spatial) surface Jacobians.*

Referential↔Material/Spatial Local (Volume) Transport Theorems:
For any time-dependent scalar-, vector-, or tensor-valued field $z_r = z_r(\boldsymbol{\Xi}, t)$, a density per unit volume in the referential configuration, and its pendants $z_m = \mathcal{J}^{m-1} z_r \circ \boldsymbol{\Upsilon}^{-1}(\boldsymbol{X}, t) = z_m(\boldsymbol{X}, t)$ and $z_s = \mathcal{J}^{s-1} z_r \circ \boldsymbol{\upsilon}^{-1}(\boldsymbol{x}, t) = z_s(\boldsymbol{x}, t)$, densities per unit volume in the material and the spatial configuration, respectively, the local versions of the transport theorems read

$$\begin{aligned} \mathcal{J}^{m-1}\,\dot{z}_r &= \mathrm{D}_t z_m + \mathrm{Div}(z_m \otimes \boldsymbol{W}), \\ \mathcal{J}^{s\,-1}\,\dot{z}_r &= \mathrm{d}_t z_s \;+ \mathrm{div}(z_s \;\otimes\; \boldsymbol{w}\;) \end{aligned} \tag{9.28}$$

point-wise in smooth parts of any control volume. Moreover, for regular points on the boundary of the control volume, the (volume) Piola transformations (PT) (together with $\boldsymbol{\mathcal{F}}^s \cdot \boldsymbol{\mathcal{W}}^s := -\boldsymbol{w}$ and $\boldsymbol{\mathcal{F}}^m \cdot \boldsymbol{\mathcal{W}}^m := -\boldsymbol{W}$) render

$$-\,[z_r \otimes \boldsymbol{\mathcal{W}}^s] \cdot \boldsymbol{\mathcal{N}} \overset{\mathrm{PT}}{=} \widehat{\mathcal{J}}^s [z_s \otimes \boldsymbol{w}] \cdot \boldsymbol{n} \tag{9.29}$$

and

$$-[z_r \otimes \boldsymbol{\mathcal{W}}^m] \cdot \boldsymbol{\mathcal{N}} \overset{\mathrm{PT}}{=} \widehat{\mathcal{J}}^m [z_m \otimes \boldsymbol{W}] \cdot \boldsymbol{N}.$$

Likewise, for regular points at singular surfaces within the control volume the (volume) Piola transformations render

$$-\,[\![z_r \otimes \boldsymbol{\mathcal{W}}^s]\!] \cdot \boldsymbol{\mathcal{M}} \overset{\mathrm{PT}}{=} \widehat{\mathcal{J}}^s [\![z_s \otimes \boldsymbol{w}]\!] \cdot \boldsymbol{m} \tag{9.30}$$

and

$$-[\![z_r \otimes \boldsymbol{\mathcal{W}}^m]\!] \cdot \boldsymbol{\mathcal{M}} \overset{\mathrm{PT}}{=} \widehat{\mathcal{J}}^m [\![z_m \otimes \boldsymbol{W}]\!] \cdot \boldsymbol{N}.$$

Note that $-\boldsymbol{\mathcal{W}}^s \cdot \boldsymbol{\mathcal{N}} = \widehat{\mathcal{J}}^s \mathcal{J}^{s-1} \boldsymbol{w} \cdot \boldsymbol{n}$ and $-\boldsymbol{\mathcal{W}}^m \cdot \boldsymbol{\mathcal{N}} = \widehat{\mathcal{J}}^m \mathcal{J}^{m-1} \boldsymbol{W} \cdot \boldsymbol{N}$.

9.2 Virtual Displacements

In the context of mechanics, the concept of test functions takes the more intuitive interpretation as *virtual displacements*.

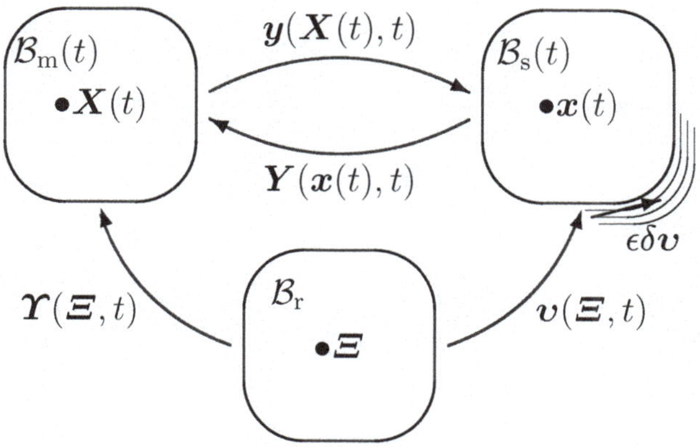

Fig. 9.2 Spatial virtual displacements as independent variations of the spatial placement map

9.2.1 Spatial Virtual Displacements

Spatial virtual displacements $\delta\boldsymbol{\upsilon}$, see Fig. 9.2, denote independent variations of the spatial placement map $\boldsymbol{\upsilon}(\Xi, t) = \boldsymbol{y} \circ \boldsymbol{\Upsilon}(\Xi, t)$ that is parameterized in referential coordinates and time. Thereby, the otherwise *arbitrary* $\delta\boldsymbol{\upsilon}$ need to be admissible, i.e. $\delta\boldsymbol{\upsilon} \in \mathbb{A}_{\delta\upsilon}$, whereby $\mathbb{A}_{\delta\upsilon}$ denotes the[1]:

Admissible Space for Spatial Virtual Displacements:

i. Regular Points in the Domain

$$\{\delta\boldsymbol{\upsilon}| \quad \delta\boldsymbol{\upsilon} \in \mathbb{H}_1 \quad \text{in} \quad \mathcal{B}_r\}. \tag{9.31}$$

ii. Regular Points on the Boundary

$$\{\delta\boldsymbol{\upsilon}| \quad \delta\boldsymbol{\upsilon} = \boldsymbol{0} \quad \text{on} \quad \partial\mathcal{B}_r^d\}. \tag{9.32}$$

iii. Regular Points at Singular Surfaces

$$\{\delta\boldsymbol{\upsilon}| \quad [\![\delta\boldsymbol{\upsilon}]\!] = \boldsymbol{0} \quad \text{on} \quad \mathcal{S}_r\}. \tag{9.33}$$

iv. End Points of Time Interval

---

[1] The Hilbert space $\mathbb{H}_1(\mathcal{D})$ consists of functions that, together with their (weak) partial derivatives, are square-integrable on the domain $\mathcal{D}$, e.g. a one-dimensional real-valued function $f(x) \in \mathbb{H}_1(\mathcal{D})$ with $x \in \mathcal{D}$ if

$$\int_{\mathcal{D}} \left[|f(x)|^2 + |f'(x)|^2\right] \mathrm{d}x < \infty.$$

.

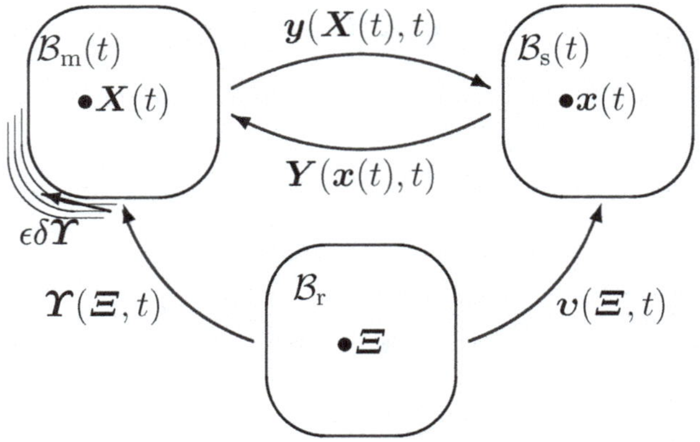

Fig. 9.3 Material virtual displacements as independent variations of the material placement map

$$\{\delta\boldsymbol{\upsilon}|\quad \delta\boldsymbol{\upsilon} = \mathbf{0} \quad \text{at} \quad \partial\mathcal{T}\}. \tag{9.34}$$

Here, $\partial\mathcal{B}_r^d$ denotes the spatial Dirichlet part of the boundary $\partial\mathcal{B}_r$ with prescribed $\boldsymbol{\upsilon} \doteq \boldsymbol{\upsilon}^p$ and $\mathcal{T} = [t_1, t_2]$ is the time interval of interest with (the set of) end points $\partial\mathcal{T} = \{t_1, t_2\}$. In the sequel, spatial virtual displacements on the boundary $\partial\mathcal{B}_r$ and at singular surfaces $\mathcal{S}_r$ shall be denoted as

$$\delta\widehat{\boldsymbol{\upsilon}} := \delta\boldsymbol{\upsilon}|_{\partial\mathcal{B}_r} \quad \text{and} \quad \delta\widetilde{\boldsymbol{\upsilon}} := \delta\boldsymbol{\upsilon}|_{\mathcal{S}_r}. \tag{9.35}$$

9.2.2 *Material Virtual Displacements*

Likewise, *material virtual displacements* $\delta\boldsymbol{\Upsilon}$, see Fig. 9.3, denote independent variations of the material placement map $\boldsymbol{\Upsilon}(\boldsymbol{\Xi}, t) = \boldsymbol{Y} \circ \boldsymbol{\upsilon}(\boldsymbol{\Xi}, t)$ that is parameterized in referential coordinates and time. Thereby, the otherwise *arbitrary* $\delta\boldsymbol{\Upsilon}$ need to be admissible, i.e. $\delta\boldsymbol{\Upsilon} \in \mathbb{A}_{\delta\Upsilon}$, whereby $\mathbb{A}_{\delta\Upsilon}$ denotes the

Admissible Space for Material Virtual Displacements:

i. Regular Points in the Domain

$$\{\delta\boldsymbol{\Upsilon}|\quad \delta\boldsymbol{\Upsilon} \in \mathbb{H}_1 \quad \text{in} \quad \mathcal{B}_r\}. \tag{9.36}$$

ii. Regular Points on the Boundary

$$\{\delta\boldsymbol{\Upsilon}|\quad \delta\boldsymbol{\Upsilon} = \mathbf{0} \quad \text{on} \quad \partial\mathcal{B}_r^D\}. \tag{9.37}$$

iii. Regular Points at Singular Surfaces

$$\{\delta \boldsymbol{\Upsilon} | \quad [\![\delta \boldsymbol{\Upsilon}]\!] = \mathbf{0} \quad \text{on} \quad \mathcal{S}_\mathrm{r}\}. \tag{9.38}$$

iv. End Points of Time Interval

$$\{\delta \boldsymbol{\Upsilon} | \quad \delta \boldsymbol{\Upsilon} = \mathbf{0} \quad \text{at} \quad \partial \mathcal{T}\}. \tag{9.39}$$

Here, $\partial \mathcal{B}_\mathrm{r}^\mathrm{D}$ denotes the material Dirichlet part of the boundary $\partial \mathcal{B}_\mathrm{r}$ with prescribed $\boldsymbol{\Upsilon} \doteq \boldsymbol{\Upsilon}^\mathrm{p}$. In the sequel, material virtual displacements on the boundary $\partial \mathcal{B}_\mathrm{r}$ and at singular surfaces $\mathcal{S}_\mathrm{r}$ shall be denoted as

$$\delta \widehat{\boldsymbol{\Upsilon}} := \delta \boldsymbol{\Upsilon}|_{\partial \mathcal{B}_\mathrm{r}} \quad \text{and} \quad \delta \widetilde{\boldsymbol{\Upsilon}} := \delta \boldsymbol{\Upsilon}|_{\mathcal{S}_\mathrm{r}}. \tag{9.40}$$

9.2.3 *Total Variations of Kinematic Quantities*

Total, Material and Spatial Variations:
The *total* variation of a generic field $z\big(\boldsymbol{\upsilon}(\boldsymbol{\Xi}), \boldsymbol{\Upsilon}(\boldsymbol{\Xi})\big)$ with space parametrization in the *spatial-material configuration space* $\mathbb{Y} := \{\boldsymbol{\upsilon}(\boldsymbol{\Xi}), \boldsymbol{\Upsilon}(\boldsymbol{\Xi})\}$ is defined as the directional derivative (at fixed time t)

$$\delta z := \frac{\mathrm{d}}{\mathrm{d}\epsilon} z(\boldsymbol{\upsilon}(\boldsymbol{\Xi}) + \epsilon \delta \boldsymbol{\upsilon}(\boldsymbol{\Xi}), \boldsymbol{\Upsilon}(\boldsymbol{\Xi}) + \epsilon \delta \boldsymbol{\Upsilon}(\boldsymbol{\Xi}))|_{\epsilon=0}. \tag{9.41}$$

Here, $\delta \boldsymbol{\upsilon} = \delta \boldsymbol{\upsilon}(\boldsymbol{\Xi})$ and $\delta \boldsymbol{\Upsilon} = \delta \boldsymbol{\Upsilon}(\boldsymbol{\Xi})$ are the spatial and the material virtual displacements as defined in the above. Moreover, D_δ and d_δ, at fixed $\boldsymbol{X}$ and $\boldsymbol{x}$, respectively, (and at fixed time t) denote the *material* and *spatial variation* of $z(\boldsymbol{\upsilon}(\boldsymbol{\Xi}), \boldsymbol{\Upsilon}(\boldsymbol{\Xi}))$ and are defined by

$$\mathrm{D}_\delta z := \delta z|_X = \frac{\mathrm{d}}{\mathrm{d}\epsilon} z(\boldsymbol{\upsilon}(\boldsymbol{X}) + \epsilon \delta \boldsymbol{\upsilon}(\boldsymbol{X}); \boldsymbol{X})|_{\epsilon=0} \tag{9.42}$$

and

$$\mathrm{d}_\delta z := \delta z|_x = \frac{\mathrm{d}}{\mathrm{d}\epsilon} z(\boldsymbol{\Upsilon}(\boldsymbol{x}) + \epsilon \delta \boldsymbol{\Upsilon}(\boldsymbol{x}); \boldsymbol{x})|_{\epsilon=0} \tag{9.43}$$

Note the slight abuse in notation for $\boldsymbol{\upsilon}(\boldsymbol{X}) \equiv \boldsymbol{y}(\boldsymbol{X})$ and $\boldsymbol{\Upsilon}(\boldsymbol{x}) \equiv \boldsymbol{Y}(\boldsymbol{x})$.

The following results will be particularly useful when stating variational principles in the sequel.

Variations in Bulk Volumes

The total variation of the *space gradients* of the deformation maps, i.e. of the deformation gradients, results in

$$\delta \boldsymbol{F} = \mathrm{D}_\delta \boldsymbol{F} - \boldsymbol{F} \cdot \mathrm{d}_\delta \boldsymbol{f} \cdot \boldsymbol{F} \quad \text{and} \quad \delta \boldsymbol{f} = \mathrm{d}_\delta \boldsymbol{f} - \boldsymbol{f} \cdot \mathrm{D}_\delta \boldsymbol{F} \cdot \boldsymbol{f}. \tag{9.44}$$

Thereby the material variation of the spatial deformation gradient $\mathrm{D}_\delta \boldsymbol{F}$ and the spatial variation of the material deformation gradient $\mathrm{d}_\delta \boldsymbol{f}$ are, respectively, defined as

$$\mathrm{D}_\delta \boldsymbol{F} := \nabla_X \delta \boldsymbol{\upsilon} \quad \text{and} \quad \mathrm{d}_\delta \boldsymbol{f} := \nabla_x \delta \boldsymbol{\Upsilon}. \tag{9.45}$$

Proof The proof follows from applying the product rule to the total variation of the deformation gradients $\boldsymbol{F} = \boldsymbol{\mathcal{F}}^{\mathrm{s}} \cdot \boldsymbol{\mathcal{F}}^{\mathrm{m}-1}$ and $\boldsymbol{f} = \boldsymbol{\mathcal{F}}^{\mathrm{m}} \cdot \boldsymbol{\mathcal{F}}^{\mathrm{s}-1}$, respectively, i.e.

$$\delta \boldsymbol{F} = \delta \boldsymbol{\mathcal{F}}^{\mathrm{s}} \cdot \boldsymbol{\mathcal{F}}^{\mathrm{m}-1} + \boldsymbol{\mathcal{F}}^{\mathrm{s}} \cdot \delta \boldsymbol{\mathcal{F}}^{\mathrm{m}-1} \quad \text{and} \quad \delta \boldsymbol{f} = \delta \boldsymbol{\mathcal{F}}^{\mathrm{m}} \cdot \boldsymbol{\mathcal{F}}^{\mathrm{s}-1} + \boldsymbol{\mathcal{F}}^{\mathrm{m}} \cdot \delta \boldsymbol{\mathcal{F}}^{\mathrm{s}-1}. \tag{9.46}$$

Here, the total variations of the placement gradients $\boldsymbol{\mathcal{F}}^{\mathrm{s}}$ and $\boldsymbol{\mathcal{F}}^{\mathrm{s}}$ result in

$$\begin{aligned} \delta \boldsymbol{\mathcal{F}}^{\mathrm{s}} \cdot \boldsymbol{\mathcal{F}}^{\mathrm{m}-1} &:= \frac{\mathrm{d}}{\mathrm{d}\epsilon} [\nabla_\Xi (\boldsymbol{\upsilon} + \epsilon \delta \boldsymbol{\upsilon})]_{\epsilon=0} \cdot \boldsymbol{\mathcal{F}}^{\mathrm{m}-1} = \nabla_X \delta \boldsymbol{\upsilon}, \\ \delta \boldsymbol{\mathcal{F}}^{\mathrm{m}} \cdot \boldsymbol{\mathcal{F}}^{\mathrm{s}\,-1} &:= \frac{\mathrm{d}}{\mathrm{d}\epsilon} [\nabla_\Xi (\boldsymbol{\Upsilon} + \epsilon \delta \boldsymbol{\Upsilon})]_{\epsilon=0} \cdot \boldsymbol{\mathcal{F}}^{\mathrm{s}\,-1} = \nabla_x \delta \boldsymbol{\Upsilon}. \end{aligned} \tag{9.47}$$

Moreover, the total variations of the inverse placement gradients $\boldsymbol{\mathcal{F}}^{\mathrm{m}-1}$ and $\boldsymbol{\mathcal{F}}^{\mathrm{s}-1}$ expand as

$$\begin{aligned} \boldsymbol{\mathcal{F}}^{\mathrm{s}} \cdot \delta \boldsymbol{\mathcal{F}}^{\mathrm{m}-1} &:= \boldsymbol{\mathcal{F}}^{\mathrm{s}} \cdot \frac{\mathrm{d}}{\mathrm{d}\epsilon} \left[\nabla_\Xi (\boldsymbol{\Upsilon} + \epsilon \delta \boldsymbol{\Upsilon})\right]^{-1} |_{\epsilon=0} = -\boldsymbol{F} \cdot \nabla_x \delta \boldsymbol{\Upsilon} \cdot \boldsymbol{F}, \\ \boldsymbol{\mathcal{F}}^{\mathrm{m}} \cdot \delta \boldsymbol{\mathcal{F}}^{\mathrm{s}\,-1} &:= \boldsymbol{\mathcal{F}}^{\mathrm{m}} \cdot \frac{\mathrm{d}}{\mathrm{d}\epsilon} \left[\nabla_\Xi (\boldsymbol{\upsilon} + \epsilon \delta \boldsymbol{\upsilon})\right]^{-1} |_{\epsilon=0} = -\boldsymbol{f} \cdot \nabla_X \delta \boldsymbol{\upsilon} \cdot \boldsymbol{f}. \end{aligned} \tag{9.48}$$

Combining these intermediate results renders the above expansions for $\delta \boldsymbol{F}$ and $\delta \boldsymbol{f}$, respectively. □

Remark 1 The total variations of the (referential-material and referential-spatial) measure maps compute as

$$\delta \mathcal{J}^{\mathrm{m}} = \mathcal{J}^{\mathrm{m}} \,\mathrm{Div} \delta \boldsymbol{\Upsilon} \quad \text{and} \quad \delta \mathcal{J}^{\mathrm{s}} = \mathcal{J}^{\mathrm{s}} \,\mathrm{div} \delta \boldsymbol{\upsilon}. \tag{9.49}$$

The proof follows from the definition of the measure maps as determinants of the deformation gradients. □

The total variation of the *time gradients* of the deformation maps, i.e. of the velocities, results in

$$\delta \boldsymbol{v} = \delta \boldsymbol{w} - \boldsymbol{F} \cdot \delta \boldsymbol{W} - \delta \boldsymbol{F} \cdot \boldsymbol{W} \quad \text{and} \quad \delta \boldsymbol{V} = \delta \boldsymbol{W} - \boldsymbol{f} \cdot \delta \boldsymbol{w} - \delta \boldsymbol{f} \cdot \boldsymbol{w}. \tag{9.50}$$

Here, the total variation of the total spatial and material velocity follow as

$$\delta \boldsymbol{w} = \frac{\mathrm{d}}{\mathrm{d}\epsilon}\left[\overline{\boldsymbol{v} + \epsilon \delta \boldsymbol{v}}^{\boldsymbol{\cdot}}\right]_{\epsilon=0} = \dot{\overline{\delta \boldsymbol{v}}} \quad \text{and} \quad \delta \boldsymbol{W} = \frac{\mathrm{d}}{\mathrm{d}\epsilon}\left[\overline{\boldsymbol{\Upsilon} + \epsilon \delta \boldsymbol{\Upsilon}}^{\boldsymbol{\cdot}}\right]_{\epsilon=0} = \dot{\overline{\delta \boldsymbol{\Upsilon}}}. \tag{9.51}$$

Variations on Boundary Surfaces*

The total variation of the *space gradients* of the boundary surface deformation maps, i.e. of the boundary surface deformation gradients, results in

$$\delta \widehat{\boldsymbol{F}} = \mathrm{D}_\delta \widehat{\boldsymbol{F}} - \widehat{\boldsymbol{F}} \cdot \mathrm{d}_\delta \widehat{\boldsymbol{f}} \cdot \widehat{\boldsymbol{F}} \quad \text{and} \quad \delta \widehat{\boldsymbol{f}} = \mathrm{d}_\delta \widehat{\boldsymbol{f}} - \widehat{\boldsymbol{f}} \cdot \mathrm{D}_\delta \widehat{\boldsymbol{F}} \cdot \widehat{\boldsymbol{f}}. \tag{9.52}$$

Thereby the material variation of the spatial boundary surface deformation gradient $\mathrm{D}_\delta \widehat{\boldsymbol{F}}$ and the spatial variation of the material boundary surface deformation gradient $\mathrm{d}_\delta \widehat{\boldsymbol{f}}$ are, respectively, defined as

$$\mathrm{D}_\delta \widehat{\boldsymbol{F}} := \widehat{\nabla}_{\widehat{X}} \delta \widehat{\boldsymbol{v}} \quad \text{and} \quad \mathrm{d}_\delta \widehat{\boldsymbol{f}} := \widehat{\nabla}_{\widehat{x}} \delta \widehat{\boldsymbol{\Upsilon}}. \tag{9.53}$$

The proof follows as in the case of the deformation gradients in bulk volumes.

Remark 2 The total variations of the (referential-material and referential-spatial) boundary surface measure maps result in

$$\delta \widehat{\mathcal{J}}^{\mathrm{m}} = \widehat{\mathcal{J}}^{\mathrm{m}}\, \widehat{\mathrm{Div} \delta \boldsymbol{\Upsilon}} \quad \text{and} \quad \delta \widehat{\mathcal{J}}^{\mathrm{s}} = \widehat{\mathcal{J}}^{\mathrm{s}}\, \widehat{\mathrm{div} \delta \boldsymbol{v}}. \tag{9.54}$$

The proof follows from the definition of the boundary surface measure maps as boundary surface determinants of the boundary surface deformation gradients. □

The total variation of the *time gradients* of the boundary surface deformation maps, i.e. of the boundary surface velocities, results in

$$\delta \widehat{\boldsymbol{v}} = \delta \widehat{\boldsymbol{w}} - \widehat{\boldsymbol{F}} \cdot \delta \widehat{\boldsymbol{W}} - \delta \widehat{\boldsymbol{F}} \cdot \widehat{\boldsymbol{W}} \quad \text{and} \quad \delta \widehat{\boldsymbol{V}} = \delta \widehat{\boldsymbol{W}} - \widehat{\boldsymbol{f}} \cdot \delta \widehat{\boldsymbol{w}} - \delta \widehat{\boldsymbol{f}} \cdot \widehat{\boldsymbol{w}}. \tag{9.55}$$

Variations on Boundary Curves*

The total variation of the *space gradients* of the boundary curve deformation maps, i.e. of the boundary curve deformation gradients, results in

$$\delta \widetilde{\boldsymbol{F}} = \mathrm{D}_\delta \widetilde{\boldsymbol{F}} - \widetilde{\boldsymbol{F}} \cdot \mathrm{d}_\delta \widetilde{\boldsymbol{f}} \cdot \widetilde{\boldsymbol{F}} \quad \text{and} \quad \delta \widetilde{\boldsymbol{f}} = \mathrm{d}_\delta \widetilde{\boldsymbol{f}} - \widetilde{\boldsymbol{f}} \cdot \mathrm{D}_\delta \widetilde{\boldsymbol{F}} \cdot \widetilde{\boldsymbol{f}}. \tag{9.56}$$

Thereby the material variation of the spatial boundary curve deformation gradient $\mathrm{D}_\delta \widetilde{\boldsymbol{F}}$ and the spatial variation of the material boundary curve deformation gradient $\mathrm{d}_\delta \widetilde{\boldsymbol{f}}$ are, respectively, defined as

$$\mathrm{D}_\delta \widetilde{\boldsymbol{F}} := \widetilde{\nabla}_{\widetilde{X}} \delta \widetilde{\boldsymbol{v}} \quad \text{and} \quad \mathrm{d}_\delta \widetilde{\boldsymbol{f}} := \widetilde{\nabla}_{\widetilde{x}} \delta \widetilde{\boldsymbol{\Upsilon}}. \tag{9.57}$$

The proof follows as in the case of the deformation gradients in bulk volumes.

Remark 3 The total variations of the (referential-material and referential-spatial) boundary curve measure maps result in

$$\delta\widetilde{\mathcal{J}}^{\mathrm{m}} = \widetilde{\mathcal{J}}^{\mathrm{m}}\,\widetilde{\mathrm{Div}}\delta\widetilde{\boldsymbol{\Upsilon}} \quad \text{and} \quad \delta\widetilde{\mathcal{J}}^{\mathrm{s}} = \widetilde{\mathcal{J}}^{\mathrm{s}}\,\widetilde{\mathrm{div}}\delta\widetilde{\boldsymbol{v}}. \tag{9.58}$$

The proof follows from the definition of the boundary curve measure maps as boundary curve determinants of the boundary curve deformation gradients. □

The total variation of the *time gradients* of the boundary curve deformation maps, i.e. of the boundary curve velocities, results in

$$\delta\widetilde{\boldsymbol{v}} = \delta\widetilde{\boldsymbol{w}} - \widetilde{\boldsymbol{F}}\cdot\delta\widetilde{\boldsymbol{W}} - \delta\widetilde{\boldsymbol{F}}\cdot\widetilde{\boldsymbol{W}} \quad \text{and} \quad \delta\widetilde{\boldsymbol{V}} = \delta\widetilde{\boldsymbol{W}} - \widetilde{\boldsymbol{f}}\cdot\delta\widetilde{\boldsymbol{w}} - \delta\widetilde{\boldsymbol{f}}\cdot\widetilde{\boldsymbol{w}}. \tag{9.59}$$

9.3 Spatial Virtual Work Principle

The *weak form* of the balance of spatial momentum (i.e. its reformulation that requires less regularity for the stresses) is stated as the

Spatial Virtual Work Principle:

$$\mathsf{P}_{\delta}^{\mathrm{ine}}(\delta\boldsymbol{v}) = \mathsf{P}_{\delta}^{\mathrm{ext}}(\delta\boldsymbol{v}) - \mathsf{P}_{\delta}^{\mathrm{int}}(\delta\boldsymbol{v}) \quad \forall\delta\boldsymbol{v}. \tag{9.60}$$

Here,

- $\mathsf{P}_{\delta}^{\mathrm{ine}}$ denotes the spatial virtual work of the inertial spatial forces, i.e. the *inertial spatial virtual work*,
- $\mathsf{P}_{\delta}^{\mathrm{ext}}$ denotes the spatial virtual work of the external spatial forces, i.e. the *external spatial virtual work*, and
- $\mathsf{P}_{\delta}^{\mathrm{int}}$ denotes the spatial virtual work of the spatial stresses (internal forces), i.e. the *internal spatial virtual work*.

9.3.1 Integrands in Material Configuration

Expressed as integrals extending over the material configuration $\mathcal{B}_{\mathrm{m}}$ and its boundary $\partial\mathcal{B}_{\mathrm{m}}$, these expand as the

Spatial Virtual Work (Material Configuration):

Inertial Spatial Virtual Work

$$\mathsf{P}^{\text{ine}}_{\delta}(\delta\boldsymbol{v}) := \int_{\mathcal{B}_{\text{m}}} \delta\boldsymbol{v} \cdot \left[\,\mathrm{D}_t\boldsymbol{p}_{\text{m}} + \mathrm{Div}(\boldsymbol{p}_{\text{m}} \otimes \boldsymbol{W})\right] \mathrm{d}V. \tag{9.61}$$

External Spatial Virtual Work

$$\mathsf{P}^{\text{ext}}_{\delta}(\delta\boldsymbol{v}) := \int_{\mathcal{B}_{\text{m}}} \delta\boldsymbol{v} \cdot \overline{\boldsymbol{s}}_{\text{m}}\, \mathrm{d}V + \int_{\partial\mathcal{B}_{\text{m}}} \delta\widehat{\boldsymbol{v}} \cdot \widehat{\boldsymbol{s}}^{\text{ext}}_{\text{m}}\, \mathrm{d}A. \tag{9.62}$$

Internal Spatial Virtual Work

$$\mathsf{P}^{\text{int}}_{\delta}(\delta\boldsymbol{v}) := \int_{\mathcal{B}_{\text{m}}} \nabla_X\delta\boldsymbol{v} : [\boldsymbol{P} + \boldsymbol{p}_{\text{m}} \otimes \boldsymbol{W}]\, \mathrm{d}V. \tag{9.63}$$

To derive the spatial virtual work principle in the material configuration, the local balance of spatial momentum is multiplied by the spatial virtual displacement, i.e. by the *spatial test function*, and the result integrated over the appropriate domain of the material configuration. Thereby, three different scenarios are considered:

i. The balance of spatial momentum as stated in Eq. 7.29 for regular points of the domain $\mathcal{B}_{\text{m}}$ is multiplied by the spatial virtual displacement $\delta\boldsymbol{v}$ and integrated over the domain $\mathcal{B}_{\text{m}}$ to render

$$\int_{\mathcal{B}_{\text{m}}} \delta\boldsymbol{v} \cdot [\,\mathrm{D}_t\boldsymbol{p}_{\text{m}} - \mathrm{Div}\boldsymbol{P}]\, \mathrm{d}V = \int_{\mathcal{B}_{\text{m}}} \delta\boldsymbol{v} \cdot \overline{\boldsymbol{s}}_{\text{m}}\, \mathrm{d}V. \tag{9.64}$$

Then, the identity $\delta\boldsymbol{v} \cdot \mathrm{Div}\boldsymbol{P} = \mathrm{Div}(\delta\boldsymbol{v} \cdot \boldsymbol{P}) - \nabla_X\delta\boldsymbol{v} : \boldsymbol{P}$ based on the *product rule of differentiation* leads to

$$\int_{\mathcal{B}_{\text{m}}} [\delta\boldsymbol{v} \cdot \mathrm{D}_t\boldsymbol{p}_{\text{m}} + \nabla_X\delta\boldsymbol{v} : \boldsymbol{P}]\, \mathrm{d}V = \int_{\mathcal{B}_{\text{m}}} [\delta\boldsymbol{v} \cdot \overline{\boldsymbol{s}}_{\text{m}} + \mathrm{Div}(\delta\boldsymbol{v} \cdot \boldsymbol{P})]\, \mathrm{d}V. \tag{9.65}$$

Applying finally the extended version of the Gauss theorem to the second term on the right-hand-side (any singular contributions are here ignored for the sake of presentation) results in

$$\int_{\mathcal{B}_m} [\delta \boldsymbol{v} \cdot \mathrm{D}_t \boldsymbol{p}_m + \nabla_X \delta \boldsymbol{v} : \boldsymbol{P}]\, \mathrm{d}V = \int_{\mathcal{B}_m} \delta \boldsymbol{v} \cdot \bar{\boldsymbol{s}}_m \, \mathrm{d}V + \int_{\partial \mathcal{B}_m} \delta \widehat{\boldsymbol{v}} \cdot \boldsymbol{P} \cdot \mathrm{d}\boldsymbol{A} - \int_{\mathcal{S}_m} [\![\delta \widetilde{\boldsymbol{v}} \cdot \boldsymbol{P}]\!] \cdot \mathrm{d}\boldsymbol{A}. \tag{9.66}$$

It now remains to re-express the last two terms on the right-hand side by the spatial extrinsic momentum flux density along with suited convective terms stemming from the transport theorem.

ii. The balance of spatial momentum as stated in Eq. 7.30 for regular points on the boundary $\partial \mathcal{B}_m$ is multiplied by the spatial virtual displacement $\delta \boldsymbol{v}$ and integrated over the boundary $\partial \mathcal{B}_m$ to render

$$\int_{\partial \mathcal{B}_m} \delta \widehat{\boldsymbol{v}} \cdot [\boldsymbol{p}_m \otimes \boldsymbol{W} + \boldsymbol{P}] \cdot \mathrm{d}\boldsymbol{A} = \int_{\partial \mathcal{B}_m} \delta \widehat{\boldsymbol{v}} \cdot \widehat{\boldsymbol{s}}_m^{\mathrm{ext}} \, \mathrm{d}A. \tag{9.67}$$

Applying the extended version of the Gauss theorem to the convective term on the left-hand side leads to

$$\int_{\partial \mathcal{B}_m} \delta \widehat{\boldsymbol{v}} \cdot \boldsymbol{P} \cdot \mathrm{d}\boldsymbol{A} + \int_{\mathcal{S}_m} \delta \widetilde{\boldsymbol{v}} \cdot [\![\boldsymbol{p}_m \otimes \boldsymbol{W}]\!] \, \mathrm{d}V = \int_{\partial \mathcal{B}_m} \delta \widehat{\boldsymbol{v}} \cdot \widehat{\boldsymbol{s}}_m^{\mathrm{ext}} \, \mathrm{d}A - \int_{\mathcal{B}_m} \mathrm{Div}\big(\delta \boldsymbol{v} \cdot [\boldsymbol{p}_m \otimes \boldsymbol{W}]\big) \, \mathrm{d}V. \tag{9.68}$$

Finally, the identity $\mathrm{Div}\big(\delta \boldsymbol{v} \cdot [\boldsymbol{p}_m \otimes \boldsymbol{W}]\big) = \delta \boldsymbol{v} \cdot \mathrm{Div}(\boldsymbol{p}_m \otimes \boldsymbol{W}) + \nabla_X \delta \boldsymbol{v} : [\boldsymbol{p}_m \otimes \boldsymbol{W}]$ based on the *product rule of differentiation* results in

$$\int_{\partial \mathcal{B}_m} \delta \widehat{\boldsymbol{v}} \cdot \boldsymbol{P} \cdot \mathrm{d}\boldsymbol{A} + \int_{\mathcal{S}_m} \delta \widetilde{\boldsymbol{v}} \cdot [\![\boldsymbol{p}_m \otimes \boldsymbol{W}]\!] \, \mathrm{d}V = \int_{\partial \mathcal{B}_m} \delta \widehat{\boldsymbol{v}} \cdot \widehat{\boldsymbol{s}}_m^{\mathrm{ext}} \, \mathrm{d}A - \int_{\mathcal{B}_m} \delta \boldsymbol{v} \cdot \mathrm{Div}(\boldsymbol{p}_m \otimes \boldsymbol{W}) \, \mathrm{d}V - \int_{\mathcal{B}_m} \nabla_X \delta \boldsymbol{v} : [\boldsymbol{p}_m \otimes \boldsymbol{W}] \, \mathrm{d}V. \tag{9.69}$$

It now remains to re-express the second term on the left-hand side by the virtual work of the Piola traction jump across singular surfaces.

iii. The balance of spatial momentum as stated in Eq. 7.31 for regular points at singular surfaces $\mathcal{S}_m$ is multiplied by the average spatial virtual displacement $\{\delta \boldsymbol{v}\} \equiv \delta \widetilde{\boldsymbol{v}}$ (where the identity follows from the coherence of singular surfaces) and integrated over the singular surface $\mathcal{S}_m$ to render

$$\int_{\mathcal{S}_m} \delta\widetilde{\boldsymbol{v}} \cdot [\![\boldsymbol{p}_m \otimes \boldsymbol{W}]\!] \cdot \mathrm{d}\boldsymbol{A} = -\int_{\mathcal{S}_m} \delta\widetilde{\boldsymbol{v}} \cdot [\![\boldsymbol{P}]\!] \cdot \mathrm{d}\boldsymbol{A}. \tag{9.70}$$

Inserting finally the result from Eqs. 9.70 and 9.69 into Eq. 9.66 renders the result.

9.3.2 Integrands in Spatial Configuration

The *spatial virtual work principle* may, likewise, be formulated in terms of integrals extending over the spatial configuration $\mathcal{B}_s$ and its boundary $\partial\mathcal{B}_s$ in terms of the

Spatial Virtual Work (Spatial Configuration):

Inertial Spatial Virtual Work

$$\mathsf{P}^{\text{ine}}_{\delta}(\delta\boldsymbol{v}) := \int_{\mathcal{B}_s} \delta\boldsymbol{v} \cdot \left[\mathrm{d}_t \boldsymbol{p}_s + \mathrm{div}(\boldsymbol{p}_s \otimes \boldsymbol{w})\right] \mathrm{d}v. \tag{9.71}$$

External Spatial Virtual Work

$$\mathsf{P}^{\text{ext}}_{\delta}(\delta\boldsymbol{v}) := \int_{\mathcal{B}_s} \delta\boldsymbol{v} \cdot \overline{\boldsymbol{s}}_s \,\mathrm{d}v + \int_{\partial\mathcal{B}_s} \delta\widehat{\boldsymbol{v}} \cdot \widehat{\boldsymbol{s}}^{\text{ext}}_s \,\mathrm{d}a. \tag{9.72}$$

Internal Spatial Virtual Work

$$\mathsf{P}^{\text{int}}_{\delta}(\delta\boldsymbol{v}) := \int_{\mathcal{B}_s} \nabla_x \delta\boldsymbol{v} : [\boldsymbol{\sigma}_d + \boldsymbol{p}_s \otimes \boldsymbol{w}] \,\mathrm{d}v. \tag{9.73}$$

The equivalence of each of the integrals with its counterpart in Eqs. 9.61–9.63 is justified as follows:

i. Based on the material variant of the local transport theorem

$$j\, \mathrm{D}_t \boldsymbol{p}_m = \mathrm{d}_t \boldsymbol{p}_s + \mathrm{div}(\boldsymbol{p}_s \otimes \boldsymbol{v}) \tag{9.74}$$

and the Piola identity jointly with the relation $\boldsymbol{W} \cdot \boldsymbol{F}^t = \boldsymbol{w} - \boldsymbol{v}$

$$j\, \mathrm{Div}(\boldsymbol{p}_m \otimes \boldsymbol{W}) = \mathrm{div}(\boldsymbol{p}_s \otimes [\boldsymbol{w} - \boldsymbol{v}]) \tag{9.75}$$

the corresponding term in the integrand of the *inertial spatial virtual work* $\mathsf{P}^{\text{ine}}_{\delta}(\delta\boldsymbol{v})$ is rewritten as

$$\int_{\mathcal{B}_m} \delta \boldsymbol{v} \cdot \left[\mathrm{D}_t \boldsymbol{p}_m + \mathrm{Div}(\boldsymbol{p}_m \otimes \boldsymbol{W}) \right] \mathrm{d}V \equiv \tag{9.76}$$

$$\int_{\mathcal{B}_s} \delta \boldsymbol{v} \cdot \left[\; \mathrm{d}_t \boldsymbol{p}_s \; + \; \mathrm{div}(\boldsymbol{p}_s \; \otimes \; \boldsymbol{w} \;) \right] \mathrm{d}v.$$

ii. The spatial momentum source density $\overline{\boldsymbol{s}}_s = j \, \overline{\boldsymbol{s}}_m$ per unit volume in the spatial configuration together with the spatial extrinsic momentum flux density $\widehat{\boldsymbol{s}}_s^{\mathrm{ext}} = \widehat{j} \, \widehat{\boldsymbol{s}}_m^{\mathrm{ext}}$ per unit area in the spatial configuration render the *external spatial virtual work* $\mathsf{P}_\delta^{\mathrm{ext}}(\delta \boldsymbol{v})$ as

$$\int_{\mathcal{B}_m} \delta \boldsymbol{v} \cdot \overline{\boldsymbol{s}}_m \, \mathrm{d}V + \int_{\partial \mathcal{B}_m} \delta \widehat{\boldsymbol{v}} \cdot \widehat{\boldsymbol{s}}_m^{\mathrm{ext}} \, \mathrm{d}A \equiv \tag{9.77}$$

$$\int_{\mathcal{B}_s} \delta \boldsymbol{v} \cdot \overline{\boldsymbol{s}}_s \; \mathrm{d}v \; + \int_{\partial \mathcal{B}_s} \delta \widehat{\boldsymbol{v}} \cdot \widehat{\boldsymbol{s}}_s^{\mathrm{ext}} \, \mathrm{d}a \; .$$

iii. Based on the appropriate (right-sided material→spatial) Piola transformation of the *dynamic* Piola stress

$$\boldsymbol{P} + \boldsymbol{p}_m \otimes \boldsymbol{W} \;\; \Longrightarrow \;\; \boldsymbol{\sigma} + \boldsymbol{p}_s \otimes [\boldsymbol{w} - \boldsymbol{v}] \tag{9.78}$$

and the incorporation of the convective term $\boldsymbol{p}_s \otimes \boldsymbol{v}$ into the definition of the *dynamic* (spatial→spatial) Cauchy stress $\boldsymbol{\sigma}_d := \boldsymbol{\sigma} - \boldsymbol{p}_s \otimes \boldsymbol{v}$, the *internal spatial virtual work* $\mathsf{P}_\delta^{\mathrm{int}}(\delta \boldsymbol{v})$ is expressed in terms of the spatial gradient $\nabla_x \delta \boldsymbol{v}$ of the spatial virtual displacement as

$$\int_{\mathcal{B}_m} \nabla_X \delta \boldsymbol{v} : [\boldsymbol{P} + \boldsymbol{p}_m \otimes \boldsymbol{W}] \, \mathrm{d}V \equiv \int_{\mathcal{B}_s} \nabla_x \delta \boldsymbol{v} : [\boldsymbol{\sigma}_d + \boldsymbol{p}_s \otimes \boldsymbol{w}] \, \mathrm{d}v. \tag{9.79}$$

Observe that these spatial virtual work expressions are fully consistent with the local format of the spatial momentum in the spatial configuration as listed in Eqs. 7.41–7.43.

9.3.3 *Integrands in Reference Configuration*

It is finally interesting to note that the *spatial virtual work principle* may also be reformulated in terms of integrals extending over the reference configuration $\mathcal{B}_r$ and its boundary $\partial \mathcal{B}_r$ in terms of the

Spatial Virtual Work (Reference Configuration):

Inertial Spatial Virtual Work

$$\mathsf{P}^{\rm ine}_{\delta}(\delta\boldsymbol{v}) := \int\limits_{\mathcal{B}_{\rm r}} \delta\boldsymbol{v} \cdot \dot{\boldsymbol{p}}_{\rm r}\, {\rm d}\mathcal{V}. \tag{9.80}$$

External Spatial Virtual Work

$$\mathsf{P}^{\rm ext}_{\delta}(\delta\boldsymbol{v}) := \int\limits_{\mathcal{B}_{\rm r}} \delta\boldsymbol{v} \cdot \overline{\boldsymbol{s}}_{\rm r}\, {\rm d}\mathcal{V} + \int\limits_{\partial\mathcal{B}_{\rm r}} \delta\widehat{\boldsymbol{v}} \cdot \widehat{\boldsymbol{s}}^{\rm ext}_{\rm r}\, {\rm d}\mathcal{A}. \tag{9.81}$$

Internal Spatial Virtual Work

$$\mathsf{P}^{\rm int}_{\delta}(\delta\boldsymbol{v}) := \int\limits_{\mathcal{B}_{\rm r}} \nabla_{\Xi}\delta\boldsymbol{v} : \boldsymbol{\mathcal{P}}^{\rm s}\, {\rm d}\mathcal{V}. \tag{9.82}$$

The rationale for the equivalence of each of the integrals with its counterpart in Eqs. 9.61–9.63 is as follows:

i. Based on the referential variant of the local transport theorem

$$\mathcal{J}^{{\rm m}-1}\dot{\boldsymbol{p}}_{\rm r} = {\rm D}_t\boldsymbol{p}_{\rm m} + {\rm Div}(\boldsymbol{p}_{\rm m} \otimes \boldsymbol{W}) \tag{9.83}$$

the corresponding term in the integrand of the *inertial spatial virtual work* $\mathsf{P}^{\rm ine}_{\delta}(\delta\boldsymbol{v})$ is recognized as the total time derivative (denoted by the superposed dot) of the spatial momentum density $\boldsymbol{p}_{\rm r} = \mathcal{J}^{\rm m}\,\boldsymbol{p}_{\rm m}$ per unit volume ${\rm d}\mathcal{V} = \mathcal{J}^{{\rm m}-1}\,{\rm d}V$ in the stationary reference configuration $\mathcal{B}_{\rm r}$, thus

$$\int\limits_{\mathcal{B}_{\rm m}} \delta\boldsymbol{v} \cdot \left[{\rm D}_t\boldsymbol{p}_{\rm m} + {\rm Div}(\boldsymbol{p}_{\rm m} \otimes \boldsymbol{W})\right]{\rm d}V \equiv \int\limits_{\mathcal{B}_{\rm r}} \delta\boldsymbol{v} \cdot \dot{\boldsymbol{p}}_{\rm r}\, {\rm d}\mathcal{V}. \tag{9.84}$$

ii. The spatial momentum source density $\overline{\boldsymbol{s}}_{\rm r} = \mathcal{J}^{\rm m}\,\overline{\boldsymbol{s}}_{\rm m}$ per unit volume in the reference configuration together with the spatial extrinsic momentum flux density $\widehat{\boldsymbol{s}}^{\rm ext}_{\rm r} = \widehat{\mathcal{J}}^{\rm m}\,\widehat{\boldsymbol{s}}^{\rm ext}_{\rm m}$ (with the area Jacobian $\widehat{\mathcal{J}}^{\rm m} := {\rm d}A/\,{\rm d}\mathcal{A}$) per unit area in the reference configuration render the *external spatial virtual work* $\mathsf{P}^{\rm ext}_{\delta}(\delta\boldsymbol{v})$ as

$$\int\limits_{\mathcal{B}_{\rm m}} \delta\boldsymbol{v} \cdot \overline{\boldsymbol{s}}_{\rm m}\, {\rm d}V + \int\limits_{\partial\mathcal{B}_{\rm m}} \delta\widehat{\boldsymbol{v}} \cdot \widehat{\boldsymbol{s}}^{\rm ext}_{\rm m}\, {\rm d}A \equiv \tag{9.85}$$

$$\int\limits_{\mathcal{B}_{\rm r}} \delta\boldsymbol{v} \cdot \overline{\boldsymbol{s}}_{\rm r}\, {\rm d}\mathcal{V} + \int\limits_{\partial\mathcal{B}_{\rm r}} \delta\widehat{\boldsymbol{v}} \cdot \widehat{\boldsymbol{s}}^{\rm ext}_{\rm r}\, {\rm d}\mathcal{A}.$$

iii. Based on the introduction of the *dynamic* Piola stress

$$\boldsymbol{P}_{\bullet} := \boldsymbol{P} + \boldsymbol{p}_{\mathrm{m}} \otimes \boldsymbol{W} \tag{9.86}$$

and its appropriate (right-sided material-referential) Piola transformation into the referential→spatial Piola stress

$$\mathcal{P}^{\mathrm{s}} := \mathcal{J}^{\mathrm{m}}\, \boldsymbol{P}_{\bullet} \cdot \mathcal{F}^{\mathrm{m}-t} = \boldsymbol{P}_{\bullet} \cdot \mathcal{K}^{\mathrm{m}}, \tag{9.87}$$

the *internal spatial virtual work* $\mathsf{P}^{\mathrm{int}}_{\delta}(\delta \boldsymbol{v})$ is expressed in terms of $\mathcal{P}^{\mathrm{s}}$ and the referential gradient $\nabla_{\Xi}\delta\boldsymbol{v} = \nabla_X \delta\boldsymbol{v} \cdot \mathcal{F}^{\mathrm{m}}$ of the spatial virtual displacement

$$\int_{\mathcal{B}_{\mathrm{m}}} \nabla_X \delta\boldsymbol{v} : [\boldsymbol{P} + \boldsymbol{p}_{\mathrm{m}} \otimes \boldsymbol{W}]\, \mathrm{d}V \equiv \int_{\mathcal{B}_{\mathrm{r}}} \nabla_{\Xi} \delta\boldsymbol{v} : \mathcal{P}^{\mathrm{s}}\, \mathrm{d}\mathcal{V}. \tag{9.88}$$

The dynamic Piola stress includes the corresponding convective term so as to refer to the total time derivative $\dot{\boldsymbol{p}}_{\mathrm{r}}$ of the spatial momentum density $\boldsymbol{p}_{\mathrm{r}}$ per unit volume in the reference configuration. The *spatial* momentum flux $\mathcal{P}^{\mathrm{s}}$ (referential→spatial Piola stress) is a two-point tensor that maps between the cotangent spaces to the reference and the *spatial* configuration $T^*\mathcal{B}_{\mathrm{r}}$ and $T^*\mathcal{B}_{\mathrm{s}}$, respectively, i.e.

$$\mathcal{P}^{\mathrm{s}} : T^*\mathcal{B}_{\mathrm{r}} \to T^*\mathcal{B}_{\mathrm{s}} \quad \text{with} \quad \mathcal{P}^{\mathrm{s}} \cdot \mathcal{N}\, \mathrm{d}\mathcal{A} \equiv \boldsymbol{P}_{\bullet} \cdot \boldsymbol{N}\, \mathrm{d}A. \tag{9.89}$$

Since the *spatial virtual work principle* holds for all admissible $\delta\boldsymbol{v}$ it is equivalent to the local balance of spatial momentum expressed in terms of referential quantities summarized as

Local Balance of Spatial Momentum (Referential Control Volume):

i. Regular Points in the Domain

$$\dot{\boldsymbol{p}}_{\mathrm{r}} - \mathrm{DIV}\mathcal{P}^{\mathrm{s}} = \overline{\boldsymbol{s}}_{\mathrm{r}} \quad \text{in} \quad \mathcal{B}_{\mathrm{r}}. \tag{9.90}$$

ii. Regular Points on the Boundary

$$\mathcal{P}^{\mathrm{s}} \cdot \mathcal{N} = \widehat{\boldsymbol{s}}^{\mathrm{ext}}_{\mathrm{r}} \quad \text{on} \quad \partial\mathcal{B}_{\mathrm{r}}. \tag{9.91}$$

iii. Regular Points at Singular Surfaces

$$[\![\mathcal{P}^{\mathrm{s}}]\!] \cdot \mathcal{M} = \boldsymbol{0} \quad \text{at} \quad \mathcal{S}_{\mathrm{r}}. \tag{9.92}$$

9.3.4 Balance Tetragons

The local balance of spatial momentum *at regular points in the domain, at regular points on the boundary* and *at regular points at singular surfaces* may thus be expressed in yet another set of different but fully equivalent versions, see the arrangements in the corresponding tetragons in Tables 9.1 and 9.2. Here, *dynamic versions* of the material→spatial Piola and the spatial→spatial Cauchy stress have been introduced as

$$\boldsymbol{P}_\bullet := \mathcal{P}^{\mathrm{s}} \cdot \mathcal{K}^{\mathrm{m}-1} \quad \text{and} \quad \boldsymbol{\sigma}_\bullet := \mathcal{P}^{\mathrm{s}} \cdot \mathcal{K}^{\mathrm{s}-1}, \tag{9.93}$$

respectively, whereby the subscript • indicates that the corresponding spatial intrinsic momentum flux is consistent with the total time derivative, denoted by a superposed dot, of the spatial momentum density. Moreover,

$$\mathcal{W}^{\mathrm{m}} := -\mathcal{F}^{\mathrm{m}-1} \cdot \boldsymbol{W} \quad \text{and} \quad \mathcal{W}^{\mathrm{s}} := -\mathcal{F}^{\mathrm{s}-1} \cdot \boldsymbol{w} \tag{9.94}$$

are defined as abbreviations for the (negative) referential pull-backs of the total material and spatial velocities, respectively.

9.4 Material Virtual Work Principle

The *weak form* of the balance of material momentum is stated as the

Material Virtual Work Principle:

$$\mathsf{p}_\delta^{\mathrm{ine}}(\delta\boldsymbol{\Upsilon}) = \mathsf{p}_\delta^{\mathrm{ext}}(\delta\boldsymbol{\Upsilon}) - \mathsf{p}_\delta^{\mathrm{int}}(\delta\boldsymbol{\Upsilon}) - \mathsf{p}_\delta^{\mathrm{ifc}}(\delta\boldsymbol{\Upsilon}) \quad \forall \delta\boldsymbol{\Upsilon}. \tag{9.95}$$

Here,

- $\mathsf{p}_\delta^{\mathrm{ine}}$ denotes the material virtual work of the inertial material forces, i.e. the *inertial material virtual work*,
- $\mathsf{p}_\delta^{\mathrm{ext}}$ denotes the material virtual work of the external material forces, i.e. the *external material virtual work*,
- $\mathsf{p}_\delta^{\mathrm{int}}$ denotes the material virtual work of the material stresses (internal forces), i.e. the *internal material virtual work*,

  and, in addition only for the material setting,
- $\mathsf{p}_\delta^{\mathrm{ifc}}$ denotes the material virtual work of the material forces at singular surfaces, i.e. the *interfacial material virtual work*.

Table 9.1 Tetragon of fully equivalent versions for the local balance of spatial momentum at regular points in the domain, on the boundary, and at singular surfaces

Regular Points in the Domain

$\dot{\boldsymbol{p}}_{\mathrm{r}} = \mathrm{DIV}\boldsymbol{\mathcal{P}}^{\mathrm{s}}_{\bullet} + \overline{\boldsymbol{s}}_{\mathrm{r}}$	⟷ PT	$\mathcal{J}^{\mathrm{m}-1}\dot{\boldsymbol{p}}_{\mathrm{r}} = \mathrm{Div}\boldsymbol{P}_{\bullet} + \overline{\boldsymbol{s}}_{\mathrm{m}}$
↕ TT		↕ TT
$\mathcal{J}^{\mathrm{m}}\,\mathrm{D}_t\boldsymbol{p}_{\mathrm{m}} = \mathrm{DIV}\boldsymbol{\mathcal{P}}^{\mathrm{s}}_{\mathrm{D}} + \overline{\boldsymbol{s}}_{\mathrm{r}}$	PT ⟷	$\mathrm{D}_t\boldsymbol{p}_{\mathrm{m}} = \mathrm{Div}\boldsymbol{P}_{\mathrm{D}} + \overline{\boldsymbol{s}}_{\mathrm{m}}$

Regular Points on the Boundary

$\boldsymbol{\mathcal{P}}^{\mathrm{s}}_{\bullet} \cdot \boldsymbol{\mathcal{N}} = \widehat{\boldsymbol{s}}^{\mathrm{ext}}_{\mathrm{r}}$	⟷ PT	$\boldsymbol{P}_{\bullet} \cdot \boldsymbol{N} = \widehat{\boldsymbol{s}}^{\mathrm{ext}}_{\mathrm{m}}$
↕ TT		↕ TT
$\left[\boldsymbol{\mathcal{P}}^{\mathrm{s}}_{\mathrm{D}} - \boldsymbol{p}_{\mathrm{r}} \otimes \boldsymbol{\mathcal{W}}^{\mathrm{m}}\right] \cdot \boldsymbol{\mathcal{N}} = \widehat{\boldsymbol{s}}^{\mathrm{ext}}_{\mathrm{r}}$	PT ⟷	$\left[\boldsymbol{P}_{\mathrm{D}} + \boldsymbol{p}_{\mathrm{m}} \otimes \boldsymbol{W}\right] \cdot \boldsymbol{N} = \widehat{\boldsymbol{s}}^{\mathrm{ext}}_{\mathrm{m}}$

Regular Points at Singular Surfaces

$[\![\boldsymbol{\mathcal{P}}^{\mathrm{s}}_{\bullet}]\!] \cdot \boldsymbol{\mathcal{M}} = \boldsymbol{0}$	⟷ PT	$[\![\boldsymbol{P}_{\bullet}]\!] \cdot \boldsymbol{M} = \boldsymbol{0}$
↕ TT		↕ TT
$[\![\boldsymbol{\mathcal{P}}^{\mathrm{s}}_{\mathrm{D}} - \boldsymbol{p}_{\mathrm{r}} \otimes \boldsymbol{\mathcal{W}}^{\mathrm{m}}]\!] \cdot \boldsymbol{\mathcal{M}} = \boldsymbol{0}$	PT ⟷	$[\![\boldsymbol{P}_{\mathrm{D}} + \boldsymbol{p}_{\mathrm{m}} \otimes \boldsymbol{W}]\!] \cdot \boldsymbol{M} = \boldsymbol{0}$

with *Dynamic Versions* of
Referential→Spatial Piola and Material→Spatial Piola Stress

$\boldsymbol{\mathcal{P}}^{\mathrm{s}}_{\bullet} := \boldsymbol{\mathcal{P}}^{\mathrm{s}}$	⟷ PT	$\boldsymbol{P}_{\bullet} := \boldsymbol{\mathcal{P}}^{\mathrm{s}}_{\bullet} \cdot \boldsymbol{\mathcal{K}}^{\mathrm{m}-1}$
↕ TT		↕ TT
$\boldsymbol{\mathcal{P}}^{\mathrm{s}}_{\mathrm{D}} := \boldsymbol{\mathcal{P}}^{\mathrm{s}}_{\bullet} + \boldsymbol{p}_{\mathrm{r}} \otimes \boldsymbol{\mathcal{W}}^{\mathrm{m}}$	PT ⟷	$\boldsymbol{P}_{\mathrm{D}} := \boldsymbol{P}_{\bullet} - \boldsymbol{p}_{\mathrm{m}} \otimes \boldsymbol{W}$

Table 9.2 Tetragon of fully equivalent versions for the local balance of spatial momentum at regular points in the domain, on the boundary, and at singular surfaces

Regular Points in the Domain

$\dot{\boldsymbol{p}}_{\mathrm{r}} = \mathrm{DIV}\boldsymbol{\mathcal{P}}^{\mathrm{s}}_{\bullet} + \overline{\boldsymbol{s}}_{\mathrm{r}}$	←PT→	$\mathcal{J}^{\mathrm{s}-1}\dot{\boldsymbol{p}}_{\mathrm{r}} = \mathrm{div}\boldsymbol{\sigma}_{\bullet} + \overline{\boldsymbol{s}}_{\mathrm{s}}$
↕ TT		↕ TT
$\mathcal{J}^{\mathrm{s}}\,\mathrm{d}_t\boldsymbol{p}_{\mathrm{s}} = \mathrm{DIV}\boldsymbol{\mathcal{P}}^{\mathrm{s}}_{\mathrm{d}} + \overline{\boldsymbol{s}}_{\mathrm{r}}$	←PT→	$\mathrm{d}_t\boldsymbol{p}_{\mathrm{s}} = \mathrm{div}\boldsymbol{\sigma}_{\mathrm{d}} + \overline{\boldsymbol{s}}_{\mathrm{s}}$

Regular Points on the Boundary

$\boldsymbol{\mathcal{P}}^{\mathrm{s}}_{\bullet} \cdot \boldsymbol{\mathcal{N}} = \widehat{\boldsymbol{s}}^{\mathrm{ext}}_{\mathrm{r}}$	←PT→	$\boldsymbol{\sigma}_{\bullet} \cdot \boldsymbol{n} = \widehat{\boldsymbol{s}}^{\mathrm{ext}}_{\mathrm{s}}$
↕ TT		↕ TT
$\left[\boldsymbol{\mathcal{P}}^{\mathrm{s}}_{\mathrm{d}} - \boldsymbol{p}_{\mathrm{r}} \otimes \boldsymbol{\mathcal{W}}^{\mathrm{s}}\right] \cdot \boldsymbol{\mathcal{N}} = \widehat{\boldsymbol{s}}^{\mathrm{ext}}_{\mathrm{r}}$	←PT→	$\left[\boldsymbol{\sigma}_{\mathrm{d}} + \boldsymbol{p}_{\mathrm{s}} \otimes \boldsymbol{w}\right] \cdot \boldsymbol{n} = \widehat{\boldsymbol{s}}^{\mathrm{ext}}_{\mathrm{s}}$

Regular Points at Singular Surfaces

$[\![\boldsymbol{\mathcal{P}}^{\mathrm{s}}_{\bullet}]\!] \cdot \boldsymbol{\mathcal{M}} = \boldsymbol{0}$	←PT→	$[\![\boldsymbol{\sigma}_{\bullet}]\!] \cdot \boldsymbol{m} = \boldsymbol{0}$
↕ TT		↕ TT
$[\![\boldsymbol{\mathcal{P}}^{\mathrm{s}}_{\mathrm{d}} - \boldsymbol{p}_{\mathrm{r}} \otimes \boldsymbol{\mathcal{W}}^{\mathrm{s}}]\!] \cdot \boldsymbol{\mathcal{M}} = \boldsymbol{0}$	←PT→	$[\![\boldsymbol{\sigma}_{\mathrm{d}} + \boldsymbol{p}_{\mathrm{s}} \otimes \boldsymbol{w}]\!] \cdot \boldsymbol{m} = \boldsymbol{0}$

with *Dynamic Versions* of
Referential→Spatial Piola and Spatial→Spatial Cauchy Stress

$\boldsymbol{\mathcal{P}}^{\mathrm{s}}_{\bullet} := \boldsymbol{\mathcal{P}}^{\mathrm{s}}$	←PT→	$\boldsymbol{\sigma}_{\bullet} := \boldsymbol{\mathcal{P}}^{\mathrm{s}}_{\bullet} \cdot \boldsymbol{\mathcal{K}}^{\mathrm{s}-1}$
↕ TT		↕ TT
$\boldsymbol{\mathcal{P}}^{\mathrm{s}}_{\mathrm{d}} := \boldsymbol{\mathcal{P}}^{\mathrm{s}}_{\bullet} + \boldsymbol{p}_{\mathrm{r}} \otimes \boldsymbol{\mathcal{W}}^{\mathrm{s}}$	←PT→	$\boldsymbol{\sigma}_{\mathrm{d}} := \boldsymbol{\sigma}_{\bullet} - \boldsymbol{p}_{\mathrm{s}} \otimes \boldsymbol{w}$

9.4.1 Integrands in Material Configuration

Expressed as integrals extending over the material configuration $\mathcal{B}_m$, its boundary $\partial\mathcal{B}_m$ and its interfaces $\mathcal{S}_m$, these expand as the

Material Virtual Work (Material Configuration):

Inertial Material Virtual Work

$$\mathsf{p}_\delta^{\text{ine}}(\delta\boldsymbol{\Upsilon}) := \int_{\mathcal{B}_m} \delta\boldsymbol{\Upsilon} \cdot \left[\mathrm{D}_t \boldsymbol{P}_m + \mathrm{Div}(\boldsymbol{P}_m \otimes \boldsymbol{W}) \right] \mathrm{d}V. \tag{9.96}$$

External Material Virtual Work

$$\mathsf{p}_\delta^{\text{ext}}(\delta\boldsymbol{\Upsilon}) := \int_{\mathcal{B}_m} \delta\boldsymbol{\Upsilon} \cdot \overline{\boldsymbol{S}}_m \, \mathrm{d}V + \int_{\partial\mathcal{B}_m} \delta\widehat{\boldsymbol{\Upsilon}} \cdot \widehat{\boldsymbol{S}}_m^{\text{ext}} \, \mathrm{d}A. \tag{9.97}$$

Internal Material Virtual Work

$$\mathsf{p}_\delta^{\text{int}}(\delta\boldsymbol{\Upsilon}) := \int_{\mathcal{B}_m} \nabla_X \delta\boldsymbol{\Upsilon} : [\boldsymbol{\Sigma} + \boldsymbol{P}_m \otimes \boldsymbol{W}] \, \mathrm{d}V. \tag{9.98}$$

Interfacial Material Virtual Work

$$\mathsf{p}_\delta^{\text{ifc}}(\delta\boldsymbol{\Upsilon}) := \int_{\mathcal{S}_m} \delta\widetilde{\boldsymbol{\Upsilon}} \cdot \widetilde{\boldsymbol{S}}_m \, \mathrm{d}A. \tag{9.99}$$

To derive the material virtual work principle in the material configuration, the local balance of material momentum is multiplied by the material virtual displacement, i.e. by the *material test function*, and the result integrated over the appropriate domain of the material configuration. Thereby, three different scenarios are considered:

i. The balance of material momentum as stated in Eq. 8.49 for regular points of the domain $\mathcal{B}_m$ is multiplied by the material virtual displacement $\delta\boldsymbol{\Upsilon}$ and integrated over the domain $\mathcal{B}_m$ to render

$$\int_{\mathcal{B}_m} \delta\boldsymbol{\Upsilon} \cdot [\mathrm{D}_t \boldsymbol{P}_m - \mathrm{Div}\boldsymbol{\Sigma}] \, \mathrm{d}V = \int_{\mathcal{B}_m} \delta\boldsymbol{\Upsilon} \cdot \overline{\boldsymbol{S}}_m \, \mathrm{d}V. \tag{9.100}$$

Then, the identity $\delta\boldsymbol{\Upsilon} \cdot \mathrm{Div}\boldsymbol{\Sigma} = \mathrm{Div}(\delta\boldsymbol{\Upsilon} \cdot \boldsymbol{\Sigma}) - \nabla_X \delta\boldsymbol{\Upsilon} : \boldsymbol{\Sigma}$ based on the *product rule of differentiation* leads to

$$\int_{\mathcal{B}_m} [\delta\boldsymbol{\Upsilon} \cdot \mathrm{D}_t \boldsymbol{P}_m + \nabla_X \delta\boldsymbol{\Upsilon} : \boldsymbol{\Sigma}]\,\mathrm{d}V = \int_{\mathcal{B}_m} [\delta\boldsymbol{\Upsilon} \cdot \overline{\boldsymbol{S}}_m + \mathrm{Div}(\delta\boldsymbol{\Upsilon} \cdot \boldsymbol{\Sigma})]\,\mathrm{d}V. \tag{9.101}$$

Applying finally the extended version of the Gauss theorem to the second term on the right-hand side (any singular contributions are here ignored for the sake of presentation) results in

$$\begin{aligned}\int_{\mathcal{B}_m} [\delta\boldsymbol{\Upsilon} \cdot \mathrm{D}_t \boldsymbol{P}_m + \nabla_X \delta\boldsymbol{\Upsilon} : \boldsymbol{\Sigma}]\,\mathrm{d}V = &\int_{\mathcal{B}_m} \delta\boldsymbol{\Upsilon} \cdot \overline{\boldsymbol{S}}_m\,\mathrm{d}V \\ &+ \int_{\partial\mathcal{B}_m} \delta\widehat{\boldsymbol{\Upsilon}} \cdot \boldsymbol{\Sigma} \cdot \mathrm{d}\boldsymbol{A} - \int_{\mathcal{S}_m} [\![\delta\widetilde{\boldsymbol{\Upsilon}} \cdot \boldsymbol{\Sigma}]\!] \cdot \mathrm{d}\boldsymbol{A}.\end{aligned} \tag{9.102}$$

It now remains to re-express the last two terms on the right-hand side by the (interior) material extrinsic momentum flux density along with suited convective terms stemming from the transport theorem.

ii. The balance of material momentum as stated in Eq. 8.50 for regular points on the boundary $\partial\mathcal{B}_m$ is multiplied by the material virtual displacement $\delta\boldsymbol{\Upsilon}$ and integrated over the boundary $\partial\mathcal{B}_m$ to render

$$\int_{\partial\mathcal{B}_m} \delta\widehat{\boldsymbol{\Upsilon}} \cdot [\boldsymbol{P}_m \otimes \boldsymbol{W} + \boldsymbol{\Sigma}] \cdot \mathrm{d}\boldsymbol{A} = \int_{\partial\mathcal{B}_m} \delta\widehat{\boldsymbol{\Upsilon}} \cdot \widehat{\boldsymbol{S}}_m^{\mathrm{ext}}\,\mathrm{d}A. \tag{9.103}$$

Applying the extended version of the Gauss theorem to the convective term on the left-hand side leads to

$$\begin{aligned}\int_{\partial\mathcal{B}_m} \delta\widehat{\boldsymbol{\Upsilon}} \cdot \boldsymbol{\Sigma} \cdot \mathrm{d}\boldsymbol{A} + \int_{\mathcal{S}_m} \delta\widetilde{\boldsymbol{\Upsilon}} \cdot [\![\boldsymbol{P}_m \otimes \boldsymbol{W}]\!]\,\mathrm{d}V = &\int_{\partial\mathcal{B}_m} \delta\widehat{\boldsymbol{\Upsilon}} \cdot \widehat{\boldsymbol{S}}_m^{\mathrm{ext}}\,\mathrm{d}A \\ &- \int_{\mathcal{B}_m} \mathrm{Div}\big(\delta\boldsymbol{\Upsilon} \cdot [\boldsymbol{P}_m \otimes \boldsymbol{W}]\big)\,\mathrm{d}V.\end{aligned} \tag{9.104}$$

Finally, the identity $\mathrm{Div}\big(\delta\boldsymbol{\Upsilon} \cdot [\boldsymbol{P}_m \otimes \boldsymbol{W}]\big) = \delta\boldsymbol{\Upsilon} \cdot \mathrm{Div}(\boldsymbol{P}_m \otimes \boldsymbol{W}) + \nabla_X \delta\boldsymbol{\Upsilon} : [\boldsymbol{P}_m \otimes \boldsymbol{W}]$ based on the *product rule of differentiation* results in

$$\begin{aligned}\int_{\partial\mathcal{B}_m} \delta\widehat{\boldsymbol{\Upsilon}} \cdot \boldsymbol{\Sigma} \cdot \mathrm{d}\boldsymbol{A} + \int_{\mathcal{S}_m} \delta\widetilde{\boldsymbol{\Upsilon}} \cdot [\![\boldsymbol{P}_m \otimes \boldsymbol{W}]\!]\,\mathrm{d}V = &\int_{\partial\mathcal{B}_m} \delta\widehat{\boldsymbol{\Upsilon}} \cdot \widehat{\boldsymbol{S}}_m^{\mathrm{ext}}\,\mathrm{d}A \\ &- \int_{\mathcal{B}_m} \delta\boldsymbol{\Upsilon} \cdot \mathrm{Div}(\boldsymbol{P}_m \otimes \boldsymbol{W})\,\mathrm{d}V \\ &- \int_{\mathcal{B}_m} \nabla_X \delta\boldsymbol{\Upsilon} : [\boldsymbol{P}_m \otimes \boldsymbol{W}]\,\mathrm{d}V.\end{aligned} \tag{9.105}$$

It now remains to re-express the second term on the left-hand-side by the virtual work of the (interior) Eshelby traction jump across singular surfaces.

iii. The balance of material momentum as stated in Eq. 8.51 for regular points at singular surfaces $\mathcal{S}_\text{m}$ is multiplied by the average material virtual displacement $\{\delta\boldsymbol{\Upsilon}\} \equiv \delta\widetilde{\boldsymbol{\Upsilon}}$ (were the identity follows from the coherence of singular surfaces) and integrated over the singular surface $\mathcal{S}_\text{m}$ to render

$$\int_{\mathcal{S}_\text{m}} \delta\widetilde{\boldsymbol{\Upsilon}} \cdot [\![\boldsymbol{P}_\text{m} \otimes \boldsymbol{W}]\!] \cdot \text{d}\boldsymbol{A} = -\int_{\mathcal{S}_\text{m}} \delta\widetilde{\boldsymbol{\Upsilon}} \cdot [\![\boldsymbol{\Sigma}]\!] \cdot \text{d}\boldsymbol{A} + \int_{\mathcal{S}_\text{m}} \delta\widetilde{\boldsymbol{\Upsilon}} \cdot \widetilde{\boldsymbol{S}}_\text{m} \cdot \text{d}\boldsymbol{A}. \tag{9.106}$$

Inserting finally the result from Eqs. 9.106 and 9.105 into Eq. 9.102 renders the result.

9.4.2 *Integrands in Spatial Configuration*

The *material virtual work principle* may, likewise, be formulated in terms of integrals extending over the spatial configuration $\mathcal{B}_\text{s}$, its boundary $\partial\mathcal{B}_\text{s}$ and its interfaces $\mathcal{S}_\text{s}$ in terms of the

Material Virtual Work (Spatial Configuration):

Inertial Material Virtual Work

$$\mathsf{p}_\delta^\text{ine}(\delta\boldsymbol{\Upsilon}) := \int_{\mathcal{B}_\text{s}} \delta\boldsymbol{\Upsilon} \cdot \left[\text{d}_t \boldsymbol{P}_\text{s} + \text{div}(\boldsymbol{P}_\text{s} \otimes \boldsymbol{w})\right] \text{d}v. \tag{9.107}$$

External Material Virtual Work

$$\mathsf{p}_\delta^\text{ext}(\delta\boldsymbol{\Upsilon}) := \int_{\mathcal{B}_\text{s}} \delta\boldsymbol{\Upsilon} \cdot \overline{\boldsymbol{S}}_\text{s}\, \text{d}v + \int_{\partial\mathcal{B}_\text{s}} \delta\widehat{\boldsymbol{\Upsilon}} \cdot \widehat{\boldsymbol{S}}_\text{s}^\text{ext}\, \text{d}a. \tag{9.108}$$

Internal Material Virtual Work

$$\mathsf{p}_\delta^\text{int}(\delta\boldsymbol{\Upsilon}) := \int_{\mathcal{B}_\text{s}} \nabla_x \delta\boldsymbol{\Upsilon} : [\boldsymbol{p}_\text{d} + \boldsymbol{P}_\text{s} \otimes \boldsymbol{w}]\, \text{d}v. \tag{9.109}$$

Interfacial Material Virtual Work

$$\mathsf{p}_\delta^\text{ifc}(\delta\boldsymbol{\Upsilon}) := \int_{\mathcal{S}_\text{s}} \delta\widetilde{\boldsymbol{\Upsilon}} \cdot \widetilde{\boldsymbol{S}}_\text{s}\, \text{d}a. \tag{9.110}$$

The equivalence of each of the integrals with its counterpart in Eqs. 9.96–9.99 is justified as follows:

i. Based on the material variant of the local transport theorem

$$j\,\mathrm{D}_t \boldsymbol{P}_\mathrm{m} = \mathrm{d}_t \boldsymbol{P}_\mathrm{s} + \mathrm{div}(\boldsymbol{P}_\mathrm{s} \otimes \boldsymbol{v}) \tag{9.111}$$

and the Piola identity jointly with the relation $\boldsymbol{W} \cdot \boldsymbol{F}^t = \boldsymbol{w} - \boldsymbol{v}$

$$j\,\mathrm{Div}(\boldsymbol{P}_\mathrm{m} \otimes \boldsymbol{W}) = \mathrm{div}(\boldsymbol{P}_\mathrm{s} \otimes [\boldsymbol{w} - \boldsymbol{v}]) \tag{9.112}$$

the corresponding term in the integrand of the *inertial material virtual work* $\mathsf{p}_\delta^\mathrm{ine}(\delta\boldsymbol{\Upsilon})$ is rewritten as

$$\int_{\mathcal{B}_\mathrm{m}} \delta\boldsymbol{\Upsilon} \cdot \big[\mathrm{D}_t \boldsymbol{P}_\mathrm{m} + \mathrm{Div}(\boldsymbol{P}_\mathrm{m} \otimes \boldsymbol{W})\big]\,\mathrm{d}V \equiv \tag{9.113}$$

$$\int_{\mathcal{B}_\mathrm{s}} \delta\boldsymbol{\Upsilon} \cdot \big[\ \mathrm{d}_t \boldsymbol{P}_\mathrm{s}\ +\ \mathrm{div}(\boldsymbol{P}_\mathrm{s}\ \otimes\ \boldsymbol{w}\,)\big]\,\mathrm{d}v.$$

ii. The (interior) material momentum source density $\overline{\boldsymbol{S}}_\mathrm{s} = j\,\overline{\boldsymbol{S}}_\mathrm{m}$ per unit volume in the spatial configuration together with the (interior) material extrinsic momentum flux density $\widehat{\boldsymbol{S}}_\mathrm{s}^\mathrm{ext} = \widehat{j\,\boldsymbol{S}}_\mathrm{m}^\mathrm{ext}$ per unit area in the spatial configuration render the *external material virtual work* $\mathsf{p}_\delta^\mathrm{ext}(\delta\boldsymbol{\Upsilon})$ as

$$\int_{\mathcal{B}_\mathrm{m}} \delta\boldsymbol{\Upsilon} \cdot \overline{\boldsymbol{S}}_\mathrm{m}\,\mathrm{d}V + \int_{\partial\mathcal{B}_\mathrm{m}} \delta\widehat{\boldsymbol{\Upsilon}} \cdot \widehat{\boldsymbol{S}}_\mathrm{m}^\mathrm{ext}\,\mathrm{d}A \equiv \tag{9.114}$$

$$\int_{\mathcal{B}_\mathrm{s}} \delta\boldsymbol{\Upsilon} \cdot \overline{\boldsymbol{S}}_\mathrm{s}\,\mathrm{d}v\ +\int_{\partial\mathcal{B}_\mathrm{s}} \delta\widehat{\boldsymbol{\Upsilon}} \cdot \widehat{\boldsymbol{S}}_\mathrm{s}^\mathrm{ext}\,\mathrm{d}a\,.$$

iii. Based on the appropriate (right-sided material→spatial) Piola transformation of the *dynamic* (material-material) Eshelby stress

$$\boldsymbol{\Sigma} + \boldsymbol{P}_\mathrm{m} \otimes \boldsymbol{W} \implies \boldsymbol{p} + \boldsymbol{P}_\mathrm{s} \otimes [\boldsymbol{w} - \boldsymbol{v}] \tag{9.115}$$

and the incorporation of the convective term $\boldsymbol{P}_\mathrm{s} \otimes \boldsymbol{v}$ into the definition of the *dynamic* (spatial→material) Eshelby stress $\boldsymbol{p}_\mathrm{d} := \boldsymbol{p} - \boldsymbol{P}_\mathrm{s} \otimes \boldsymbol{v}$, the *internal material virtual work* $\mathsf{p}_\delta^\mathrm{int}(\delta\boldsymbol{\Upsilon})$ is expressed in terms of the spatial gradient $\nabla_x \delta\boldsymbol{\Upsilon}$ of the material virtual displacement as

$$\int_{\mathcal{B}_\mathrm{m}} \nabla_X \delta\boldsymbol{\Upsilon} : [\boldsymbol{\Sigma} + \boldsymbol{P}_\mathrm{m} \otimes \boldsymbol{W}]\,\mathrm{d}V \equiv \int_{\mathcal{B}_\mathrm{s}} \nabla_x \delta\boldsymbol{\Upsilon} : [\boldsymbol{p}_\mathrm{d} + \boldsymbol{P}_\mathrm{s} \otimes \boldsymbol{w}]\,\mathrm{d}v. \tag{9.116}$$

iv. The (interior) material extrinsic momentum flux density $\widetilde{\boldsymbol{S}}_{\rm s} = \widetilde{j}\,\widetilde{\boldsymbol{S}}_{\rm m}$ per unit area of singular surfaces in the spatial configuration render the *interfacial material virtual work* $\mathsf{p}_\delta^{\rm sur}(\delta\boldsymbol{\Upsilon})$ as

$$\int_{\mathcal{S}_{\rm m}} \delta\widehat{\boldsymbol{\Upsilon}} \cdot \widetilde{\boldsymbol{S}}_{\rm m}\, \mathrm{d}A \equiv \int_{\mathcal{S}_{\rm s}} \delta\widehat{\boldsymbol{\Upsilon}} \cdot \widetilde{\boldsymbol{S}}_{\rm s}\, \mathrm{d}a. \tag{9.117}$$

Observe that these material virtual work expressions are fully consistent with the local format of the material momentum in the spatial configuration as listed in Eqs. 8.82–8.84.

9.4.3 Integrands in Reference Configuration

It is finally interesting to note that the *material virtual work principle* may also be reformulated in terms of integrals extending over the reference configuration $\mathcal{B}_{\rm r}$, its boundary $\partial\mathcal{B}_{\rm r}$ and its interfaces $\mathcal{S}_{\rm r}$ in terms of the

Material Virtual Work (Reference Configuration):

Inertial Material Virtual Work

$$\mathsf{p}_\delta^{\rm ine}(\delta\boldsymbol{\Upsilon}) := \int_{\mathcal{B}_{\rm r}} \delta\boldsymbol{\Upsilon} \cdot \dot{\boldsymbol{P}}_{\rm r}\, \mathrm{d}\mathcal{V}. \tag{9.118}$$

External Material Virtual Work

$$\mathsf{p}_\delta^{\rm ext}(\delta\boldsymbol{\Upsilon}) := \int_{\mathcal{B}_{\rm r}} \delta\boldsymbol{\Upsilon} \cdot \overline{\boldsymbol{S}}_{\rm r}\, \mathrm{d}\mathcal{V} + \int_{\partial\mathcal{B}_{\rm r}} \delta\widehat{\boldsymbol{\Upsilon}} \cdot \widehat{\boldsymbol{S}}_{\rm r}^{\rm ext}\, \mathrm{d}\mathcal{A}. \tag{9.119}$$

Internal Material Virtual Work

$$\mathsf{p}_\delta^{\rm int}(\delta\boldsymbol{\Upsilon}) := \int_{\mathcal{B}_{\rm r}} \nabla_{\Xi}\delta\boldsymbol{\Upsilon} : \boldsymbol{\mathcal{P}}^{\rm m}\, \mathrm{d}\mathcal{V}. \tag{9.120}$$

Interfacial Material Virtual Work

$$\mathsf{p}_\delta^{\rm ifc}(\delta\boldsymbol{\Upsilon}) := \int_{\mathcal{S}_{\rm r}} \delta\widehat{\boldsymbol{\Upsilon}} \cdot \widetilde{\boldsymbol{S}}_{\rm r}\, \mathrm{d}\mathcal{A}. \tag{9.121}$$

The rationale for the equivalence of each of the integrals with its counterpart in Eqs. 9.96–9.99 is as follows:

i. Based on the referential variant of the local transport theorem

$$\mathcal{J}^{m-1}\dot{\boldsymbol{P}}_r = D_t\boldsymbol{P}_m + \mathrm{Div}(\boldsymbol{P}_m \otimes \boldsymbol{W}) \tag{9.122}$$

the corresponding term in the integrand of the *inertial material virtual work* $\mathsf{p}_\delta^{ine}(\delta\boldsymbol{\Upsilon})$ is recognized as the total time derivative of the material momentum density $\boldsymbol{P}_r = \mathcal{J}^m\,\boldsymbol{P}_m$ per unit volume $d\mathcal{V} = \mathcal{J}^{m-1}\,dV$ in the stationary reference configuration $\mathcal{B}_r$, thus

$$\int_{\mathcal{B}_m} \delta\boldsymbol{\Upsilon}\cdot\left[D_t\boldsymbol{P}_m + \mathrm{Div}(\boldsymbol{P}_m \otimes \boldsymbol{W})\right]dv \equiv \int_{\mathcal{B}_r} \delta\boldsymbol{\Upsilon}\cdot\dot{\boldsymbol{P}}_r\,d\mathcal{V}. \tag{9.123}$$

ii. The (interior) material momentum source density $\overline{\mathbf{S}}_r = \mathcal{J}^m\,\overline{\mathbf{S}}_m$ per unit volume in the reference configuration together with the (interior) material extrinsic momentum flux density $\widehat{\mathbf{S}}_r^{ext} = \widehat{\mathcal{J}}^m\,\widehat{\mathbf{S}}_m^{ext}$ per unit area in the reference configuration render the *external material virtual work* $\mathsf{p}_\delta^{ext}(\delta\boldsymbol{\Upsilon})$ as

$$\int_{\mathcal{B}_m} \delta\boldsymbol{\Upsilon}\cdot\overline{\mathbf{S}}_m\,dV + \int_{\partial\mathcal{B}_m} \delta\widehat{\boldsymbol{\Upsilon}}\cdot\widehat{\mathbf{S}}_m^{ext}\,dA \equiv \tag{9.124}$$

$$\int_{\mathcal{B}_r} \delta\boldsymbol{\Upsilon}\cdot\overline{\mathbf{S}}_r\,d\mathcal{V} + \int_{\partial\mathcal{B}_r} \delta\widehat{\boldsymbol{\Upsilon}}\cdot\widehat{\mathbf{S}}_r^{ext}\,d\mathcal{A}.$$

iii. Based on the introduction of the *dynamic* Eshelby stress

$$\boldsymbol{\Sigma}_\bullet := \boldsymbol{\Sigma} + \boldsymbol{P}_m \otimes \boldsymbol{W} \tag{9.125}$$

and its appropriate (right-sided material-referential) Piola transformation into the (referential→material) Piola stress

$$\mathcal{P}^m := \mathcal{J}^m\,\boldsymbol{\Sigma}_\bullet\cdot\mathcal{F}^{m-t} = \boldsymbol{\Sigma}_\bullet\cdot\mathcal{K}^m, \tag{9.126}$$

the *internal material virtual work* $\mathsf{p}_\delta^{int}(\delta\boldsymbol{\Upsilon})$ is expressed in terms of $\mathcal{P}_m$ and the referential gradient $\nabla_\Xi\delta\boldsymbol{\Upsilon} = \nabla_X\delta\boldsymbol{\Upsilon}\cdot\mathcal{F}^m$ of the material virtual displacement

$$\int_{\mathcal{B}_m} \nabla_X\delta\boldsymbol{\Upsilon} : [\boldsymbol{\Sigma} + \boldsymbol{P}_m \otimes \boldsymbol{W}]\,dv \equiv \int_{\mathcal{B}_r} \nabla_\Xi\delta\boldsymbol{\Upsilon} : \mathcal{P}^m\,d\mathcal{V}. \tag{9.127}$$

The dynamic Eshelby stress includes the corresponding convective term so as to refer to the total time derivative $\dot{\boldsymbol{P}}_r$ of the material momentum density $\boldsymbol{P}_r$ per unit volume in the reference configuration. The *material* momentum flux

$\mathcal{P}^{\mathrm{m}}$ (referential→material Piola stress) is a two-point tensor that maps between the cotangent spaces to the reference and the *material* configuration $T^*\mathcal{B}_{\mathrm{r}}$ and $T^*\mathcal{B}_{\mathrm{m}}$, respectively, i.e.

$$\mathcal{P}^{\mathrm{m}} : T^*\mathcal{B}_{\mathrm{r}} \rightarrow T^*\mathcal{B}_{\mathrm{m}} \quad \text{with} \quad \mathcal{P}^{\mathrm{m}} \cdot \mathcal{N}\, \mathrm{d}\mathcal{A} \equiv \boldsymbol{\Sigma}_{\bullet} \cdot \boldsymbol{N}\, \mathrm{d}A. \tag{9.128}$$

vi. The (interior) material extrinsic momentum flux density $\widetilde{\mathbf{S}}_{\mathrm{r}} = \widetilde{\mathcal{J}}^{\mathrm{m}}\, \widetilde{\mathbf{S}}_{\mathrm{m}}$ (with the area Jacobian $\widetilde{\mathcal{J}}^{\mathrm{m}} := \mathrm{d}A / \mathrm{d}\mathcal{A}$) per unit area of singular surfaces in the reference configuration renders the *interfacial material virtual work* $\mathsf{p}^{\mathrm{sur}}_{\delta}(\delta\boldsymbol{\Upsilon})$ as

$$\int_{\mathcal{S}_{\mathrm{m}}} \delta\widehat{\boldsymbol{\Upsilon}} \cdot \widetilde{\mathbf{S}}_{\mathrm{m}}\, \mathrm{d}A \equiv \int_{\mathcal{S}_{\mathrm{r}}} \delta\widehat{\boldsymbol{\Upsilon}} \cdot \widetilde{\mathbf{S}}_{\mathrm{r}}\, \mathrm{d}\mathcal{A}. \tag{9.129}$$

Since the *material virtual work principle* holds for all admissible $\delta\boldsymbol{\Upsilon}$ it is equivalent to the local balance of material momentum expressed in terms of referential quantities summarized as

Local Balance of Material Momentum (Referential Control Volume):

i. Regular Points in the Domain

$$\dot{\boldsymbol{P}}_{\mathrm{r}} - \mathrm{DIV}\mathcal{P}^{\mathrm{m}} = \overline{\mathbf{S}}_{\mathrm{r}} \quad \text{in} \quad \mathcal{B}_{\mathrm{r}} \tag{9.130}$$

ii. Regular Points on the Boundary

$$\mathcal{P}^{\mathrm{m}} \cdot \mathcal{N} = \widehat{\mathbf{S}}^{\mathrm{ext}}_{\mathrm{r}} \quad \text{on} \quad \partial\mathcal{B}_{\mathrm{r}} \tag{9.131}$$

iii. Regular Points at Singular Surfaces

$$[\![\mathcal{P}^{\mathrm{m}}]\!] \cdot \mathcal{M} = \widetilde{\mathbf{S}}_{\mathrm{r}} \quad \text{at} \quad \mathcal{S}_{\mathrm{r}} \tag{9.132}$$

9.4.4 *Balance Tetragons*

The local balance of material momentum *at regular points in the domain*, *at regular points on the boundary* and *at regular points at singular surfaces* may thus be expressed in yet another set of different but fully equivalent versions, see the arrangements in the corresponding tetragons in Tables 9.3 and 9.4. Here *dynamic versions* of the spatial→material and the material→material (interior) Eshelby stress have been introduced as

$$\boldsymbol{p}_{\bullet} := \mathcal{P}^{\mathrm{m}} \cdot \mathcal{K}^{\mathrm{s}-1} \quad \text{and} \quad \boldsymbol{\Sigma}_{\bullet} := \mathcal{P}^{\mathrm{m}} \cdot \mathcal{K}^{\mathrm{m}-1} \tag{9.133}$$

Table 9.3 Tetragon of fully equivalent versions for the local balance of material momentum at regular points in the domain, on the boundary, and at singular surfaces

Regular Points in the Domain

$\dot{\boldsymbol{P}}_{\mathrm{r}} = \mathrm{DIV}\mathcal{P}^{\mathrm{m}}_{\bullet} + \overline{\boldsymbol{S}}_{\mathrm{r}}$	⟷ PT	$\mathcal{J}^{\mathrm{m}-1}\dot{\boldsymbol{P}}_{\mathrm{r}} = \mathrm{Div}\boldsymbol{\Sigma}_{\bullet} + \overline{\boldsymbol{S}}_{\mathrm{m}}$
↕ TT		TT ↕
$\mathcal{J}^{\mathrm{m}}\,\mathrm{D}_t\boldsymbol{P}_{\mathrm{m}} = \mathrm{DIV}\mathcal{P}^{\mathrm{m}}_{\mathrm{D}} + \overline{\boldsymbol{S}}_{\mathrm{r}}$	PT ⟷	$\mathrm{D}_t\boldsymbol{P}_{\mathrm{m}} = \mathrm{Div}\boldsymbol{\Sigma}_{\mathrm{D}} + \overline{\boldsymbol{S}}_{\mathrm{m}}$

Regular Points on the Boundary

$\mathcal{P}^{\mathrm{m}}_{\bullet}\cdot\mathcal{N} = \widehat{\boldsymbol{S}}^{\mathrm{ext}}_{\mathrm{r}}$	⟷ PT	$\boldsymbol{\Sigma}_{\bullet}\cdot\boldsymbol{N} = \widehat{\boldsymbol{S}}^{\mathrm{ext}}_{\mathrm{m}}$
↕ TT		TT ↕
$[\mathcal{P}^{\mathrm{m}}_{\mathrm{D}} - \boldsymbol{P}_{\mathrm{r}}\otimes\mathcal{W}^{\mathrm{m}}]\cdot\mathcal{N} = \widehat{\boldsymbol{S}}^{\mathrm{ext}}_{\mathrm{r}}$	PT ⟷	$[\boldsymbol{\Sigma}_{\mathrm{D}} + \boldsymbol{P}_{\mathrm{m}}\otimes\boldsymbol{W}]\cdot\boldsymbol{N} = \widehat{\boldsymbol{S}}^{\mathrm{ext}}_{\mathrm{m}}$

Regular Points at Singular Surfaces

$[\![\mathcal{P}^{\mathrm{m}}_{\bullet}]\!]\cdot\mathcal{M} = \widetilde{\boldsymbol{S}}_{\mathrm{r}}$	⟷ PT	$[\![\boldsymbol{\Sigma}_{\bullet}]\!]\cdot\boldsymbol{M} = \widetilde{\boldsymbol{S}}_{\mathrm{m}}$
↕ TT		TT ↕
$[\![\mathcal{P}^{\mathrm{m}}_{\mathrm{D}} - \boldsymbol{P}_{\mathrm{r}}\otimes\mathcal{W}^{\mathrm{m}}]\!]\cdot\mathcal{M} = \widetilde{\boldsymbol{S}}_{\mathrm{r}}$	PT ⟷	$[\![\boldsymbol{\Sigma}_{\mathrm{D}} + \boldsymbol{P}_{\mathrm{m}}\otimes\boldsymbol{W}]\!]\cdot\boldsymbol{M} = \widetilde{\boldsymbol{S}}_{\mathrm{m}}$

with *Dynamic Versions* of
Referential→Material Piola and Material→Material Eshelby Stresses

$\mathcal{P}^{\mathrm{m}}_{\bullet} := \mathcal{P}^{\mathrm{m}}$	⟷ PT	$\boldsymbol{\Sigma}_{\bullet} := \mathcal{P}^{\mathrm{m}}_{\bullet}\cdot\mathcal{K}^{\mathrm{m}-1}$
↕ TT		TT ↕
$\mathcal{P}^{\mathrm{m}}_{\mathrm{D}} := \mathcal{P}^{\mathrm{m}}_{\bullet} + \boldsymbol{P}_{\mathrm{r}}\otimes\mathcal{W}^{\mathrm{m}}$	PT ⟷	$\boldsymbol{\Sigma}_{\mathrm{D}} := \boldsymbol{\Sigma}_{\bullet} - \boldsymbol{P}_{\mathrm{m}}\otimes\boldsymbol{W}$

Table 9.4 Tetragon of fully equivalent versions for the local balance of material momentum at regular points in the domain, on the boundary, and at singular surfaces

Regular Points in the Domain

$\dot{\boldsymbol{P}}_{\rm r} = {\rm DIV}\boldsymbol{\mathcal{P}}^{\rm m}_{\bullet} + \overline{\boldsymbol{S}}_{\rm r}$	↔ PT	$\mathcal{J}^{{\rm s}-1}\dot{\boldsymbol{P}}_{\rm r} = {\rm div}\boldsymbol{p}_{\bullet} + \overline{\boldsymbol{S}}_{\rm s}$
↕ TT		TT ↕
$\mathcal{J}^{\rm s}\,{\rm d}_t\boldsymbol{P}_{\rm s} = {\rm DIV}\boldsymbol{\mathcal{P}}^{\rm m}_{\rm d} + \overline{\boldsymbol{S}}_{\rm r}$	PT ↔	${\rm d}_t\boldsymbol{P}_{\rm s} = {\rm div}\boldsymbol{p}_{\rm d} + \overline{\boldsymbol{S}}_{\rm s}$

Regular Points on the Boundary

$\boldsymbol{\mathcal{P}}^{\rm m}_{\bullet}\cdot\boldsymbol{\mathcal{N}} = \widehat{\boldsymbol{S}}^{\rm ext}_{\rm r}$	↔ PT	$\boldsymbol{p}_{\bullet}\cdot\boldsymbol{n} = \widehat{\boldsymbol{S}}^{\rm ext}_{\rm s}$
↕ TT		TT ↕
$\left[\boldsymbol{\mathcal{P}}^{\rm m}_{\rm d} - \boldsymbol{P}_{\rm r}\otimes\boldsymbol{\mathcal{W}}^{\rm s}\right]\cdot\boldsymbol{\mathcal{N}} = \widehat{\boldsymbol{S}}^{\rm ext}_{\rm r}$	PT ↔	$\left[\boldsymbol{p}_{\rm d} + \boldsymbol{P}_{\rm s}\otimes\boldsymbol{w}\right]\cdot\boldsymbol{n} = \widehat{\boldsymbol{S}}^{\rm ext}_{\rm s}$

Regular Points at Singular Surfaces

$[\![\boldsymbol{\mathcal{P}}^{\rm m}_{\bullet}]\!]\cdot\boldsymbol{\mathcal{M}} = \widetilde{\boldsymbol{S}}_{\rm r}$	↔ PT	$[\![\boldsymbol{p}_{\bullet}]\!]\cdot\boldsymbol{m} = \widetilde{\boldsymbol{S}}_{\rm s}$
↕ TT		TT ↕
$[\![\boldsymbol{\mathcal{P}}^{\rm m}_{\rm d} - \boldsymbol{P}_{\rm r}\otimes\boldsymbol{\mathcal{W}}^{\rm s}]\!]\cdot\boldsymbol{\mathcal{M}} = \widetilde{\boldsymbol{S}}_{\rm r}$	PT ↔	$[\![\boldsymbol{p}_{\rm d} + \boldsymbol{P}_{\rm s}\otimes\boldsymbol{w}]\!]\cdot\boldsymbol{m} = \widetilde{\boldsymbol{S}}_{\rm s}$

with *Dynamic Versions* of
Referential→Material Piola and Spatial→Material Eshelby Stress

$\boldsymbol{\mathcal{P}}^{\rm m}_{\bullet} := \boldsymbol{\mathcal{P}}^{\rm m}$	↔ PT	$\boldsymbol{p}_{\bullet} := \boldsymbol{\mathcal{P}}^{\rm m}_{\bullet}\cdot\boldsymbol{\mathcal{K}}^{{\rm s}-1}$
↕ TT		TT ↕
$\boldsymbol{\mathcal{P}}^{\rm m}_{\rm d} := \boldsymbol{\mathcal{P}}^{\rm m}_{\bullet} + \boldsymbol{P}_{\rm r}\otimes\boldsymbol{\mathcal{W}}^{\rm s}$	PT ↔	$\boldsymbol{p}_{\rm d} := \boldsymbol{p}_{\bullet} - \boldsymbol{P}_{\rm s}\otimes\boldsymbol{w}$

respectively, whereby the subscript • indicates that the corresponding material intrinsic momentum flux is consistent with the total time derivative of the material momentum density. Recall that

$$\mathcal{W}^{\mathrm{s}} := -\mathcal{F}^{\mathrm{s}-1} \cdot \boldsymbol{w} \quad \text{and} \quad \mathcal{W}^{\mathrm{m}} := -\mathcal{F}^{\mathrm{m}-1} \cdot \boldsymbol{W}$$

are defined as abbreviations for the (negative) referential pull-backs of the total spatial and material velocities, respectively.

Chapter 10
Variational Setting

8,516 m 27°57'42"N 86°56'00"E

Abstract This chapter expands on the variational setting in terms of extended Hamilton and Dirichlet principles for conservative elasto-dynamic and elasto-static cases, respectively, and carefully analyzes the resulting spatial and material Euler-Lagrange equations.

Before proceeding to general non-conservative cases that require the consideration of the thermo-dynamical balances of energy and entropy, it proves helpful to first consider purely conservative cases. These are characterized by internal and external potential energies (collectively denoted the *total* potential energy) in addition to the kinetic energy and commonly allow to determine the balance of spatial linear momentum as the Euler–Lagrange equation of a corresponding variational principle. Typical variational principles are the *Hamilton principle* in the case of elasto-dynamics and the *Dirichlet principle* in the case of elasto-statics.

In addition to delivering the spatial Euler–Lagrange equations from a stationarity condition under spatial variations in terms of spatial virtual displacements, the here proposed *extended* versions of the Hamilton and Dirichlet principles determine material body forces together with material boundary and surface tractions that contribute to additional material Euler–Lagrange equations. More importantly, these are power conjugated to material virtual displacements and thus capture the virtual energy release upon virtual configurational changes.

It is thus the aim of this chapter to determine the extended set of spatial *and* material Euler–Lagrange equations and in particular to identify the material or rather configurational forces that drive virtual changes in the material configuration. Thereby it proves most convenient if the referential perspective is adopted to compute total variations of the various energy contributions.

P. Steinmann, *Spatial and Material Forces in Nonlinear Continuum Mechanics*,
Solid Mechanics and Its Applications 272,
https://doi.org/10.1007/978-3-030-89070-4_10

10.1 Extended Hamilton Principle

Common Hamilton Principle

The common *Hamilton principle* or rather the *principle of least action* states that the *dynamics* of a system is described by the evolution of the common set of generalized coordinates

$$\boldsymbol{v}(\boldsymbol{\Xi})$$

in the common infinite dimensional configuration space

$$\mathbb{Y} := \{\boldsymbol{v}(\boldsymbol{\Xi})\}$$

between two given points in time, say t_1 and t_2, renders the common *action integral*

$$\mathcal{A}\{\boldsymbol{v}(\boldsymbol{\Xi})\},$$

i.e. a functional over the common configuration space, a *stationary value*, thus

$$\delta\mathcal{A} = 0$$

upon variations of the spatial placement map $\boldsymbol{v}(\boldsymbol{\Xi})$.

Rationale for the Extended Hamilton Principle

As an *extension*, the here defined *extended* Hamilton principle states that the dynamics of a system described by the evolution of the extended set of generalized coordinates

$$\boldsymbol{v}(\boldsymbol{\Xi}) \quad and \quad \boldsymbol{\Upsilon}(\boldsymbol{\Xi})$$

in the extended infinite dimensional configuration space

$$\mathbb{Y} := \{\boldsymbol{v}(\boldsymbol{\Xi}), \boldsymbol{\Upsilon}(\boldsymbol{\Xi})\}$$

between t_1 and t_2 renders the extended action integral

$$\mathcal{A}\{\boldsymbol{v}(\boldsymbol{\Xi}), \boldsymbol{\Upsilon}(\boldsymbol{\Xi})\},$$

i.e. a functional over the extended configuration space, a *non-negative value*

$$\delta\mathcal{A} \geq 0$$

upon configurational changes associated with variations of the material placement map $\boldsymbol{\Upsilon}(\boldsymbol{\Xi})$.

10.1.1 Variational Statement

Extended Hamilton Principle (Reference Configuration):
The *total* variation of the *extended action integral*

$$\mathcal{A}\{\boldsymbol{\upsilon}(\boldsymbol{\Xi}), \boldsymbol{\Upsilon}(\boldsymbol{\Xi})\} = \int_{t_1}^{t_2} \left[\int_{\mathcal{B}_r} \bar{l}_r \, \mathrm{d}\mathcal{V} + \int_{\partial\mathcal{B}_r} \widehat{l}_r \, \mathrm{d}\mathcal{A} \right] \mathrm{d}t \tag{10.1}$$

with exterior (see definition below) Lagrange-type potential energy density per unit volume in the reference configuration $\mathcal{B}_r$

$$\bar{l}_r = \bar{l}_r(\boldsymbol{\upsilon}, \nabla_{\Xi}\boldsymbol{\upsilon}, \dot{\boldsymbol{\upsilon}}, \boldsymbol{\Upsilon}, \nabla_{\Xi}\boldsymbol{\Upsilon}, \dot{\boldsymbol{\Upsilon}}; \boldsymbol{\Xi}) \tag{10.2}$$

and exterior Lagrange-type potential energy density per unit area on the boundary $\partial\mathcal{B}_r$ in the reference configuration

$$\widehat{l}_r = \widehat{l}_r(\widehat{\boldsymbol{\upsilon}}, \widehat{\nabla}_{\widehat{\Xi}}\widehat{\boldsymbol{\upsilon}}, \widehat{\boldsymbol{\Upsilon}}, \widehat{\nabla}_{\widehat{\Xi}}\widehat{\boldsymbol{\Upsilon}}; \widehat{\boldsymbol{\Xi}}) \tag{10.3}$$

defines the negative time integral of the *virtual energy release*

$$\delta\mathcal{A} =: -\int_{t_1}^{t_2} \mathcal{R}_\delta \, \mathrm{d}t \overset{!}{\geq} 0. \tag{10.4}$$

Thereby the virtual energy release is stated as

$$\mathcal{R}_\delta := \int_{\mathcal{B}_r} \delta\boldsymbol{\Upsilon} \cdot \overline{\boldsymbol{R}}_r \, \mathrm{d}\mathcal{V} + \int_{\partial\mathcal{B}_r} \delta\widehat{\boldsymbol{\Upsilon}} \cdot \widehat{\boldsymbol{R}}_r \, \mathrm{d}\mathcal{A} + \int_{\mathcal{S}_r} \delta\widetilde{\boldsymbol{\Upsilon}} \cdot \widetilde{\boldsymbol{R}}_r \, \mathrm{d}\mathcal{A} \overset{!}{\leq} 0, \tag{10.5}$$

with $\overline{\boldsymbol{R}}_r$ a *material* body force, and $\widehat{\boldsymbol{R}}_r$ and $\widetilde{\boldsymbol{R}}_r$ *material* boundary and surface tractions. These perform virtual power along *material virtual displacements* in the domain, on the boundary and at singular surfaces, respectively.

Only *non-negative* values of $\delta\mathcal{A} \geq 0$ are admissible for material virtual displacements to occur spontaneously with a virtual *release* of energy, i.e. with positive virtual dissipation of energy from the system.

10.1.2 Euler–Lagrange Equations: Summary

Upon total variation of the extended action integral $\mathcal{A}$, the corresponding local Euler–Lagrange equations are as follows:

Euler–Lagrange Equations of Extended Hamilton Principle (Referential Control Volume):

i. Regular Points in the Domain

$$-\frac{\partial \bar{l}_\mathrm{r}}{\partial \boldsymbol{v}} + \mathrm{DIV}\frac{\partial \bar{l}_\mathrm{r}}{\partial \nabla_\Xi \boldsymbol{v}} + \overline{\frac{\partial \bar{l}_\mathrm{r}}{\partial \dot{\boldsymbol{v}}}}^{\cdot} = \mathbf{0} \quad \text{in} \quad \mathcal{B}_\mathrm{r} \tag{10.6}$$

$$-\frac{\partial \bar{l}_\mathrm{r}}{\partial \boldsymbol{\Upsilon}} + \mathrm{DIV}\frac{\partial \bar{l}_\mathrm{r}}{\partial \nabla_\Xi \boldsymbol{\Upsilon}} + \overline{\frac{\partial \bar{l}_\mathrm{r}}{\partial \dot{\boldsymbol{\Upsilon}}}}^{\cdot} =: \overline{\boldsymbol{R}}_\mathrm{r}$$

ii. Regular Points on the Boundary

$$-\frac{\partial \widehat{l}_\mathrm{r}}{\partial \widehat{\boldsymbol{v}}} + \widehat{\mathrm{DIV}}\frac{\partial \widehat{l}_\mathrm{r}}{\partial \widehat{\nabla}_{\widehat{\Xi}} \widehat{\boldsymbol{v}}} - \left[\frac{\partial \bar{l}_\mathrm{r}}{\partial \nabla_\Xi \boldsymbol{v}}\right] \cdot \boldsymbol{\mathcal{N}} = \mathbf{0} \quad \text{on} \quad \partial\mathcal{B}_\mathrm{r} \tag{10.7}$$

$$-\frac{\partial \widehat{l}_\mathrm{r}}{\partial \widehat{\boldsymbol{\Upsilon}}} + \widehat{\mathrm{DIV}}\frac{\partial \widehat{l}_\mathrm{r}}{\partial \widehat{\nabla}_{\widehat{\Xi}} \widehat{\boldsymbol{\Upsilon}}} - \left[\frac{\partial \bar{l}_\mathrm{r}}{\partial \nabla_\Xi \boldsymbol{\Upsilon}}\right] \cdot \boldsymbol{\mathcal{N}} =: \widehat{\boldsymbol{R}}_\mathrm{r}$$

iii. Regular Points at Singular Surfaces

$$-\left[\!\left[\frac{\partial \bar{l}_\mathrm{r}}{\partial \nabla_\Xi \boldsymbol{v}}\right]\!\right] \cdot \boldsymbol{\mathcal{M}} = \mathbf{0} \quad \text{at} \quad \mathcal{S}_\mathrm{r} \tag{10.8}$$

$$-\left[\!\left[\frac{\partial \bar{l}_\mathrm{r}}{\partial \nabla_\Xi \boldsymbol{\Upsilon}}\right]\!\right] \cdot \boldsymbol{\mathcal{M}} =: \widetilde{\boldsymbol{R}}_\mathrm{r}$$

10.1.3 Euler–Lagrange Equations: Derivation

For the given parametrisation, the total variations of the various contributions to $\mathcal{A}$ are considered individually in the sequel:

Exterior Lagrange-Type Potential Energy in the Domain

The *exterior* Lagrange-type potential energy density $\overline{l}_r$ per unit volume in the reference configuration $\mathcal{B}_r$ is defined as the difference of the kinetic energy density κ_r and the *total* potential energy density $\overline{u}_r$, which in turn consists of the *internal* and the *external potential energy densities* w_r and $\overline{v}_r$, respectively

$$\overline{l}_r := \kappa_r - \overline{u}_r := \kappa_r - [w_r + \overline{v}_r]. \tag{10.9}$$

Recall the previous similar definitions of the *interior* Lagrange-type free and internal energy densities $\lambda_m := \kappa_m - \psi_m$ and $\ell_m := \kappa_m - \iota_m$, respectively, per unit volume in the material configuration $\mathcal{B}_m$, whereby the interior free and internal energies ψ_m and ι_m (see Chap. 11) collapse to the stored, i.e. *internal potential energy density* w_m for isothermal, conservative cases as considered here.

1. The kinetic energy density κ_r per unit volume in the reference configuration $\mathcal{B}_r$ expands as

$$\kappa_r := \frac{1}{2}\rho_r|\boldsymbol{v}|^2 \quad \text{with} \quad \rho_r = \mathcal{J}^m\,\rho_m(\boldsymbol{X}). \tag{10.10}$$

With $\boldsymbol{v} = \boldsymbol{w} - \boldsymbol{F}\cdot\boldsymbol{W}$ its total variation computes as

$$\begin{aligned}\delta\kappa_r = {} & \mathcal{J}^m\big[\,\boldsymbol{p}_m\cdot\dot{\overline{\delta\boldsymbol{v}}} + [\qquad - \boldsymbol{p}_m\otimes\boldsymbol{W}] : \nabla_X\,\delta\boldsymbol{v}\big] \\ & + \mathcal{J}^m\big[\boldsymbol{P}_m\cdot\dot{\overline{\delta\boldsymbol{\Upsilon}}} + [\kappa_m\boldsymbol{I} - \boldsymbol{P}_m\otimes\boldsymbol{W}] : \nabla_X\delta\boldsymbol{\Upsilon}\big] \\ & + \mathcal{J}^m\big[\partial_X\kappa_m\cdot\delta\boldsymbol{\Upsilon}\big].\end{aligned} \tag{10.11}$$

Here, the spatial and material linear momentum densities $\boldsymbol{p}_m := \partial_v\kappa_m = \partial_v\overline{l}_m$ and $\boldsymbol{P}_m := \partial_V\kappa_m = \partial_V\overline{l}_m$ have been incorporated as abbreviations

$$\boldsymbol{p}_m = \rho_m\boldsymbol{v} \quad \text{and} \quad \boldsymbol{P}_m = -\boldsymbol{F}^t\cdot\boldsymbol{p}_m = \rho_m\boldsymbol{C}\cdot\boldsymbol{V}. \tag{10.12}$$

2. The internal potential (stored) energy density w_r per unit volume in the reference configuration $\mathcal{B}_r$ is parameterized as

$$w_r := \mathcal{J}^m\,w_m \quad \text{with} \quad w_m = w_m(\boldsymbol{F};\boldsymbol{X}). \tag{10.13}$$

Its total variation computes as

$$\delta w_r = \tag{10.14}$$

$$\mathcal{J}^m\Big[\boldsymbol{P}' : \nabla_X\delta\boldsymbol{v} + \partial_X w_m\cdot\delta\boldsymbol{\Upsilon} + [w_m\boldsymbol{I} - \boldsymbol{F}^t\cdot\boldsymbol{P}'] : \nabla_X\delta\boldsymbol{\Upsilon}\Big].$$

Here the variational definition of the of the common material→spatial (exterior) Piola stress $\boldsymbol{P}'$ has been introduced as abbreviation

$$\boldsymbol{P}' := \partial_{F} w_{\mathrm{m}} = -\partial_{F} \bar{l}_{\mathrm{m}}. \tag{10.15}$$

3. The external potential energy density $\overline{v}_{\mathrm{r}}$ per unit volume in the reference configuration $\mathcal{B}_{\mathrm{r}}$ is parameterized as

$$\overline{v}_{\mathrm{r}} := \mathcal{J}^{\mathrm{m}}\, \overline{v}_{\mathrm{m}} \quad \text{with} \quad \overline{v}_{\mathrm{m}} = \overline{v}_{\mathrm{m}}(\boldsymbol{v}; \boldsymbol{X}). \tag{10.16}$$

Its total variation computes as

$$\delta \overline{v}_{\mathrm{r}} = \mathcal{J}^{\mathrm{m}} \Big[-\overline{\boldsymbol{s}}'_{\mathrm{m}} \cdot \delta \boldsymbol{v} + \partial_{X} \overline{v}_{\mathrm{m}} \cdot \delta \boldsymbol{\Upsilon} + [\overline{v}_{\mathrm{m}} \boldsymbol{I}] : \nabla_{X} \delta \boldsymbol{\Upsilon} \Big]. \tag{10.17}$$

Here the variational definition for the (exterior) spatial linear momentum source density has been introduced as an abbreviation

$$\overline{\boldsymbol{s}}'_{\mathrm{m}} := -\partial_{x} \overline{v}_{\mathrm{m}} = \partial_{x} \bar{l}_{\mathrm{m}}. \tag{10.18}$$

Taken together, the total variation of the exterior Lagrange-type potential energy density $\bar{l}_{\mathrm{r}}$ per unit volume in the reference configuration $\mathcal{B}_{\mathrm{r}}$ results in

$$\begin{aligned} \delta \bar{l}_{\mathrm{r}} &= \mathcal{J}^{\mathrm{m}} \left[\overline{\boldsymbol{s}}'_{\mathrm{m}} \cdot \delta \boldsymbol{v} + \boldsymbol{p}_{\mathrm{m}} \cdot \dot{\overline{\delta \boldsymbol{v}}} - [\boldsymbol{P}' + \boldsymbol{p}_{\mathrm{m}} \otimes \boldsymbol{W}] : \nabla_{X} \delta \boldsymbol{v} \right] \\ &+ \mathcal{J}^{\mathrm{m}} \left[\overline{\boldsymbol{S}}'_{\mathrm{m}} \cdot \delta \boldsymbol{\Upsilon} + \boldsymbol{P}_{\mathrm{m}} \cdot \dot{\overline{\delta \boldsymbol{\Upsilon}}} - [\boldsymbol{\Sigma}' + \boldsymbol{P}_{\mathrm{m}} \otimes \boldsymbol{W}] : \nabla_{X} \delta \boldsymbol{\Upsilon} \right] \\ &\equiv \overline{\boldsymbol{s}}'_{\mathrm{r}} \cdot \delta \boldsymbol{v} + \boldsymbol{p}_{\mathrm{r}} \cdot \dot{\overline{\delta \boldsymbol{v}}} - \mathcal{P}^{\mathrm{s}\prime} : \nabla_{\Xi} \delta \boldsymbol{v} \\ &+ \overline{\boldsymbol{S}}'_{\mathrm{r}} \cdot \delta \boldsymbol{\Upsilon} + \boldsymbol{P}_{\mathrm{r}} \cdot \dot{\overline{\delta \boldsymbol{\Upsilon}}} - \mathcal{P}^{\mathrm{m}\prime} : \nabla_{\Xi} \delta \boldsymbol{\Upsilon}. \end{aligned} \tag{10.19}$$

Here the variational definitions for the exterior Eshelby stress $\boldsymbol{\Sigma}'$ and the exterior material linear momentum source density $\overline{\boldsymbol{S}}'_{\mathrm{m}}$ have been introduced as

$$\boldsymbol{\Sigma}' := -\bar{l}_{\mathrm{m}} \boldsymbol{I} - \boldsymbol{F}^{t} \cdot \boldsymbol{P}' \quad \text{and} \quad \overline{\boldsymbol{S}}'_{\mathrm{m}} := \partial_{X} \bar{l}_{\mathrm{m}}. \tag{10.20}$$

Moreover, $\mathcal{P}^{\mathrm{s}\prime}$ and $\mathcal{P}^{\mathrm{m}\prime}$ are the right-sided material→referential Piola transformations of the *dynamic* exterior Piola and Eshelby stresses and variationally compute as

$$\mathcal{P}^{\mathrm{s}\prime} := -\partial_{\mathcal{F}^{\mathrm{s}}} \bar{l}_{\mathrm{r}} \quad \text{and} \quad \mathcal{P}^{\mathrm{m}\prime} := -\partial_{\mathcal{F}^{\mathrm{m}}} \bar{l}_{\mathrm{r}}. \tag{10.21}$$

Note that all variational quantities in the above are expressed in terms of the exterior Lagrange-type potential energy density $\bar{l}_{\mathrm{m}}$. □

Exterior Lagrange-Type Potential Energy on the Boundary

The *exterior* Lagrange-type boundary potential energy density $\widehat{l}_{\rm r}$ per unit area on the boundary $\partial\mathcal{B}_{\rm r}$ in the reference configuration is in general defined as the difference of the boundary kinetic energy density $\widehat{\kappa}_{\rm r}$ and the total boundary potential energy density $\widehat{u}_{\rm r}$, which in turn consists of the *internal* and the *external boundary potential energy densities* $\widehat{w}_{\rm r}$ and $\widehat{v}_{\rm r}$, respectively

$$\widehat{l}_{\rm r} := \widehat{\kappa}_{\rm r} - \widehat{u}_{\rm r} = \widehat{\kappa}_{\rm r} - [\widehat{w}_{\rm r} + \widehat{v}_{\rm r}]. \tag{10.22}$$

The boundary kinetic energy density $\widehat{\kappa}_{\rm r}$, capturing superficial contributions from boundary velocities, and the internal boundary potential energy $\widehat{w}_{\rm r}$, capturing superficial contributions from boundary deformation gradients, are here explicitly introduced for the sake of completeness. These terms, however, are typically absent in standard applications.

1. The boundary kinetic energy density $\widehat{\kappa}_{\rm r}$ per unit area on the boundary $\partial\mathcal{B}_{\rm r}$ shall here be neglected entirely, thus
$$\widehat{\kappa}_{\rm r} \equiv 0. \tag{10.23}$$

2. The internal boundary potential (stored) energy density $\widehat{w}_{\rm r}$ per unit area on the boundary $\partial\mathcal{B}_{\rm r}$ is parameterized as
$$\widehat{w}_{\rm r} := \widehat{\mathcal{J}}^{\rm m}\,\widehat{w}_{\rm m} \quad \text{with} \quad \widehat{w}_{\rm m} = \widehat{w}_{\rm m}(\widehat{\boldsymbol{F}}; \widehat{\boldsymbol{X}}). \tag{10.24}$$
Any dependence of $\widehat{w}_{\rm r}$ on the outwards pointing material boundary normal $\boldsymbol{N}$, needed to capture anisotropic response of boundaries, is here neglected for the sake of simplifying the presentation. Then, the total variation of $\widehat{w}_{\rm r}$ computes as
$$\delta\widehat{w}_{\rm r} = \tag{10.25}$$
$$\widehat{\mathcal{J}}^{\rm m}\left[\widehat{\boldsymbol{P}}' : \widehat{\nabla}_{\widehat{X}}\delta\widehat{\boldsymbol{v}} + \partial_{\widehat{X}}\widehat{w}_{\rm m} \cdot \delta\widehat{\boldsymbol{\Upsilon}} + [\widehat{w}_{\rm m}\widehat{\boldsymbol{I}} - \widehat{\boldsymbol{F}}^t \cdot \widehat{\boldsymbol{P}}'] : \widehat{\nabla}_{\widehat{X}}\delta\widehat{\boldsymbol{\Upsilon}}\right].$$
Here the variational definition of the (exterior) boundary Piola stress $\widehat{\boldsymbol{P}}'$ has been introduced as abbreviation
$$\widehat{\boldsymbol{P}}' := \partial_{\widehat{F}}\widehat{w}_{\rm m} = -\partial_{\widehat{F}}\widehat{l}_{\rm m}. \tag{10.26}$$

3. The external boundary potential energy density $\widehat{v}_{\rm r}$ per unit area on the boundary $\partial\mathcal{B}_{\rm r}$ in the reference configuration is parameterized as
$$\widehat{v}_{\rm r} := \widehat{\mathcal{J}}^{\rm m}\,\widehat{v}_{\rm m} \quad \text{with} \quad \widehat{v}_{\rm m} = \widehat{v}_{\rm m}(\widehat{\boldsymbol{v}}; \widehat{\boldsymbol{X}}). \tag{10.27}$$

Any dependence of $\widehat{\upsilon}_r$ on the outwards pointing spatial boundary normal $\boldsymbol{n}$, needed to capture, e.g. conservative follower loads, is here neglected for the sake of simplifying the presentation. Its total variation computes as

$$\delta\widehat{\upsilon}_r = \widehat{\mathcal{J}}^m \Big[-\widehat{\boldsymbol{s}}_m^{\text{ext}\prime} \cdot \delta\widehat{\boldsymbol{\upsilon}} + \partial_{\widehat{\boldsymbol{X}}}\widehat{\upsilon}_m \cdot \delta\widehat{\boldsymbol{\Upsilon}} + [\widehat{\upsilon}_m \widehat{\boldsymbol{I}}] : \widehat{\nabla}_{\widehat{X}}\delta\widehat{\boldsymbol{\Upsilon}} \Big]. \tag{10.28}$$

Here the variational definition for the (exterior) spatial extrinsic linear momentum flux density has been introduced as an abbreviation

$$\widehat{\boldsymbol{s}}_m^{\text{ext}\prime} := -\partial_{\widehat{\boldsymbol{x}}}\widehat{\upsilon}_m = \partial_{\widehat{\boldsymbol{x}}}\widehat{l}_m. \tag{10.29}$$

Taken together, the total variation of the external Lagrange-type boundary potential energy density $\widehat{l}_r$ per unit area on the boundary $\partial\mathcal{B}_r$ results in

$$\begin{aligned} \delta\widehat{l}_r &= \widehat{\mathcal{J}}^m \, [\widehat{\boldsymbol{s}}_m^{\text{ext}\prime} \cdot \delta\widehat{\boldsymbol{\upsilon}} \; - \widehat{\boldsymbol{P}}' : \widehat{\nabla}_{\widehat{X}}\delta\widehat{\boldsymbol{\upsilon}} \;] \\ &+ \widehat{\mathcal{J}}^m \, [\widehat{\boldsymbol{S}}_m^{\text{ext}\prime} \cdot \delta\widehat{\boldsymbol{\Upsilon}} - \widehat{\boldsymbol{\Sigma}}' : \widehat{\nabla}_{\widehat{X}}\delta\widehat{\boldsymbol{\Upsilon}}] \\ &\equiv \widehat{\boldsymbol{s}}_r^{\text{ext}\prime} \cdot \delta\widehat{\boldsymbol{\upsilon}} \; - \widehat{\boldsymbol{\mathcal{P}}}^{s\prime} \; : \widehat{\nabla}_{\widehat{\Xi}}\delta\widehat{\boldsymbol{\upsilon}} \\ &+ \widehat{\boldsymbol{S}}_r^{\text{ext}\prime} \cdot \delta\widehat{\boldsymbol{\Upsilon}} - \widehat{\boldsymbol{\mathcal{P}}}^{m\prime} : \widehat{\nabla}_{\widehat{\Xi}}\delta\widehat{\boldsymbol{\Upsilon}}. \end{aligned} \tag{10.30}$$

Here the variational definitions for the exterior boundary Eshelby stress $\widehat{\boldsymbol{\Sigma}}'$ and the exterior material extrinsic linear momentum flux density $\widehat{\boldsymbol{S}}_m^{\text{ext}\prime}$ have been introduced as

$$\widehat{\boldsymbol{\Sigma}}' := -\widehat{l}_m\widehat{\boldsymbol{I}} - \widehat{\boldsymbol{F}}^t \cdot \widehat{\boldsymbol{P}}' \quad \text{and} \quad \widehat{\boldsymbol{S}}_m^{\text{ext}\prime} := \partial_{\widehat{\boldsymbol{X}}}\widehat{l}_m. \tag{10.31}$$

Moreover, $\widehat{\boldsymbol{\mathcal{P}}}^{s\prime}$ and $\widehat{\boldsymbol{\mathcal{P}}}^{m\prime}$ are the right-sided material→referential boundary Piola transformations of the exterior boundary Piola and Eshelby stresses and variationally compute as

$$\widehat{\boldsymbol{\mathcal{P}}}^{s\prime} := -\partial_{\widehat{\mathcal{F}}^s}\widehat{l}_r \quad \text{and} \quad \widehat{\boldsymbol{\mathcal{P}}}^{m\prime} := -\partial_{\widehat{\mathcal{F}}^m}\widehat{l}_r. \tag{10.32}$$

Note that all the above variational quantities are expressed in terms of the exterior Lagrange-type boundary potential energy density $\widehat{l}_m$. □

Summary of Total Variation

Next, partial integration in time of the terms involving the spatial and material momentum densities

$$\boldsymbol{p}_r \cdot \dot{\overline{\delta\boldsymbol{\upsilon}}} = \dot{\overline{\boldsymbol{p}_r \cdot \delta\boldsymbol{\upsilon}}} - \dot{\boldsymbol{p}}_r \cdot \delta\boldsymbol{\upsilon} \quad \text{and} \quad \boldsymbol{P}_r \cdot \dot{\overline{\delta\boldsymbol{\Upsilon}}} = \dot{\overline{\boldsymbol{P}_r \cdot \delta\boldsymbol{\Upsilon}}} - \dot{\boldsymbol{P}}_r \cdot \delta\boldsymbol{\Upsilon} \tag{10.33}$$

and explicit incorporation of the constraints on the spatial and material virtual displacements at the end points of the time interval

$$\delta\boldsymbol{\upsilon}|_{t_1} = \delta\boldsymbol{\upsilon}|_{t_2} = \delta\boldsymbol{\Upsilon}|_{t_1} = \delta\boldsymbol{\Upsilon}|_{t_2} \equiv \boldsymbol{0} \tag{10.34}$$

renders eventually the total variation of the action integral $\mathcal{A}$ expanded over the reference configuration as given in Eq. 10.35.

$$\delta\mathcal{A} = \int_{t_1}^{t_2} \Big[\int_{\mathcal{B}_r} [[\overline{\boldsymbol{s}}'_r - \dot{\boldsymbol{p}}_r] \cdot \delta\boldsymbol{v} - \boldsymbol{\mathcal{P}}^{s\prime} : \nabla_{\Xi}\delta\boldsymbol{v}]\,\mathrm{d}\mathcal{V} \tag{10.35}$$

$$+ \int_{\mathcal{B}_r} [[\overline{\boldsymbol{S}}'_r - \dot{\boldsymbol{P}}_r] \cdot \delta\boldsymbol{\Upsilon} - \boldsymbol{\mathcal{P}}^{m\prime} : \nabla_{\Xi}\delta\boldsymbol{\Upsilon}]\,\mathrm{d}\mathcal{V}$$

$$+ \int_{\partial\mathcal{B}_r} \left[\widehat{\boldsymbol{s}}_r^{\mathrm{ext}\prime} \cdot \delta\widehat{\boldsymbol{v}} - \widehat{\boldsymbol{\mathcal{P}}}^{s\prime} : \widehat{\nabla}_{\widehat{\Xi}}\delta\widehat{\boldsymbol{v}}\right]\mathrm{d}\mathcal{A}$$

$$+ \int_{\partial\mathcal{B}_r} \left[\widehat{\boldsymbol{S}}_r^{\mathrm{ext}\prime} \cdot \delta\widehat{\boldsymbol{\Upsilon}} - \widehat{\boldsymbol{\mathcal{P}}}^{m\prime} : \widehat{\nabla}_{\widehat{\Xi}}\delta\widehat{\boldsymbol{\Upsilon}}\right]\mathrm{d}\mathcal{A}\Big]\,\mathrm{d}t$$

Likewise, the total variation of the action integral $\mathcal{A}$ expanded over the material configuration reads as in Eq. 10.36.

$$\delta\mathcal{A} = \int_{t_1}^{t_2} \Big[\int_{\mathcal{B}_m} [[\overline{\boldsymbol{s}}'_m - \underline{\mathcal{J}_m^{-1}\,\dot{\boldsymbol{p}}_r}] \cdot \delta\boldsymbol{v} - [\boldsymbol{P}' + \boldsymbol{p}_m \otimes \boldsymbol{W}] : \nabla_X\delta\boldsymbol{v}]\,\mathrm{d}V \tag{10.36}$$

$$+ \int_{\mathcal{B}_m} [[\overline{\boldsymbol{S}}'_m - \underline{\mathcal{J}_m^{-1}\dot{\boldsymbol{P}}_r}] \cdot \delta\boldsymbol{\Upsilon} - [\boldsymbol{\Sigma}' + \boldsymbol{P}_m \otimes \boldsymbol{W}] : \nabla_X\delta\boldsymbol{\Upsilon}]\,\mathrm{d}V$$

$$+ \int_{\partial\mathcal{B}_m} \left[\widehat{\boldsymbol{s}}_m^{\mathrm{ext}\prime} \cdot \delta\widehat{\boldsymbol{v}} - \underline{\underline{\widehat{\boldsymbol{P}}' : \widehat{\nabla}_{\widehat{X}}\delta\widehat{\boldsymbol{v}}}}\right]\mathrm{d}A$$

$$+ \int_{\partial\mathcal{B}_m} \left[\widehat{\boldsymbol{S}}_m^{\mathrm{ext}\prime} \cdot \delta\widehat{\boldsymbol{\Upsilon}} - \underline{\underline{\widehat{\boldsymbol{\Sigma}}' : \widehat{\nabla}_{\widehat{X}}\delta\widehat{\boldsymbol{\Upsilon}}}}\right]\mathrm{d}A\Big]\,\mathrm{d}t.$$

Toward Euler–Lagrange Equations

Observe the local transport theorem in the domain to substitute the underlined terms in Eq. 10.36 by

$$\mathrm{D}_t\boldsymbol{p}_m + \mathrm{Div}(\boldsymbol{p}_m \otimes \boldsymbol{W}) \quad \text{and} \quad \mathrm{D}_t\boldsymbol{P}_m + \mathrm{Div}(\boldsymbol{P}_m \otimes \boldsymbol{W})$$

as well as the product rule of differentiation on the boundary that here expands as

$$\begin{aligned}\widehat{\text{Div}}(\delta\widehat{\boldsymbol{v}}\cdot\widehat{\boldsymbol{P}}') &= [\delta\widehat{\boldsymbol{v}}\cdot\widehat{\boldsymbol{P}}']_{,\alpha}\cdot\widehat{\boldsymbol{A}}^{\alpha}\\ &= [\delta\widehat{\boldsymbol{v}}_{,\alpha}\cdot\widehat{\boldsymbol{P}}' + \delta\widehat{\boldsymbol{v}}\cdot\widehat{\boldsymbol{P}}'_{,\alpha}]\cdot\widehat{\boldsymbol{A}}^{\alpha}\\ &= \widehat{\boldsymbol{P}}' : [\delta\widehat{\boldsymbol{v}}_{,\alpha}\otimes\widehat{\boldsymbol{A}}^{\alpha}] + \delta\widehat{\boldsymbol{v}}\cdot\widehat{\boldsymbol{P}}'_{,\alpha}\cdot\widehat{\boldsymbol{A}}^{\alpha}\\ &= \widehat{\boldsymbol{P}}' : \widehat{\nabla}_{\widehat{X}}\delta\widehat{\boldsymbol{v}} + \delta\widehat{\boldsymbol{v}}\cdot\widehat{\text{Div}}\widehat{\boldsymbol{P}}'\end{aligned}$$

and

$$\begin{aligned}\widehat{\text{Div}}(\delta\widehat{\boldsymbol{\Upsilon}}\cdot\widehat{\boldsymbol{\Sigma}}') &= [\delta\widehat{\boldsymbol{\Upsilon}}\cdot\widehat{\boldsymbol{\Sigma}}']_{,\alpha}\cdot\widehat{\boldsymbol{A}}^{\alpha}\\ &= [\delta\widehat{\boldsymbol{v}}_{,\alpha}\cdot\widehat{\boldsymbol{\Sigma}}' + \delta\widehat{\boldsymbol{v}}\cdot\widehat{\boldsymbol{\Sigma}}'_{,\alpha}]\cdot\widehat{\boldsymbol{A}}^{\alpha}\\ &= \widehat{\boldsymbol{\Sigma}}' : [\delta\widehat{\boldsymbol{\Upsilon}}_{,\alpha}\otimes\widehat{\boldsymbol{A}}^{\alpha}] + \delta\widehat{\boldsymbol{\Upsilon}}\cdot\widehat{\boldsymbol{\Sigma}}'_{,\alpha}\cdot\widehat{\boldsymbol{A}}^{\alpha}\\ &= \widehat{\boldsymbol{\Sigma}}' : \widehat{\nabla}_{\widehat{X}}\delta\widehat{\boldsymbol{\Upsilon}} + \delta\widehat{\boldsymbol{\Upsilon}}\cdot\widehat{\text{Div}}\widehat{\boldsymbol{\Sigma}}',\end{aligned}$$

respectively, to substitute the double underlined terms in Eq. 10.36 by

$$\widehat{\text{Div}}(\delta\widehat{\boldsymbol{v}}\cdot\widehat{\boldsymbol{P}}') - \delta\widehat{\boldsymbol{v}}\cdot\widehat{\text{Div}}\widehat{\boldsymbol{P}}' \quad\text{and}\quad \widehat{\text{Div}}(\delta\widehat{\boldsymbol{\Upsilon}}\cdot\widehat{\boldsymbol{\Sigma}}') - \delta\widehat{\boldsymbol{\Upsilon}}\cdot\widehat{\text{Div}}\widehat{\boldsymbol{\Sigma}}',$$

respectively. Thereby the integrals of $\widehat{\text{Div}}(\delta\widehat{\boldsymbol{v}}\cdot\widehat{\boldsymbol{P}}')$ and $\widehat{\text{Div}}(\delta\widehat{\boldsymbol{\Upsilon}}\cdot\widehat{\boldsymbol{\Sigma}}')$ over the closed boundary $\partial\mathcal{B}_{\text{m}}$, here assumed smooth for the sake of simplicity, vanish since $\widehat{\boldsymbol{P}}'\cdot\boldsymbol{N} = \boldsymbol{0}$ and $\widehat{\boldsymbol{\Sigma}}'\cdot\boldsymbol{N} = \boldsymbol{0}$, respectively (compare to the boundary Gauss theorem). Thus, integration by parts and application of the extended Gauss theorem renders eventually the total variation of the action integral $\mathcal{A}$ expanded over the reference configuration as given in Eq. 10.37.

$$\begin{aligned}\delta\mathcal{A} = \int_{t_1}^{t_2}\Bigg[&\int_{\mathcal{B}_\text{r}}[\bar{\boldsymbol{s}}_\text{r}' + \text{DIV}\boldsymbol{\mathcal{P}}^{\text{s}\prime} - \dot{\boldsymbol{p}}_\text{r}]\cdot\delta\boldsymbol{v}\,\text{d}\mathcal{V}\\ &+\int_{\mathcal{B}_\text{r}}[\bar{\boldsymbol{S}}_\text{r}' + \text{DIV}\boldsymbol{\mathcal{P}}^{\text{m}\prime} - \dot{\boldsymbol{P}}_\text{r}]\cdot\delta\boldsymbol{\Upsilon}\,\text{d}\mathcal{V}\\ &+\int_{\partial\mathcal{B}_\text{r}}\Big[\widehat{\boldsymbol{s}}_\text{r}^{\text{ext}\prime} + \widehat{\text{DIV}}\widehat{\boldsymbol{\mathcal{P}}}^{\text{s}\prime} - \boldsymbol{\mathcal{P}}^{\text{s}\prime}\cdot\boldsymbol{\mathcal{N}}\Big]\cdot\delta\widehat{\boldsymbol{v}}\,\text{d}\mathcal{A}\\ &+\int_{\partial\mathcal{B}_\text{r}}\Big[\widehat{\boldsymbol{S}}_\text{r}^{\text{ext}\prime} + \widehat{\text{DIV}}\widehat{\boldsymbol{\mathcal{P}}}^{\text{m}\prime} - \boldsymbol{\mathcal{P}}^{\text{m}\prime}\cdot\boldsymbol{\mathcal{N}}\Big]\cdot\delta\widehat{\boldsymbol{\Upsilon}}\,\text{d}\mathcal{A}\\ &+\int_{\mathcal{S}_\text{r}}\Big[[\![\boldsymbol{\mathcal{P}}^{\text{s}\prime}]\!]\cdot\boldsymbol{\mathcal{M}}\Big]\cdot\delta\widetilde{\boldsymbol{v}}\,\text{d}\mathcal{A}\end{aligned} \tag{10.37}$$

$$+\int_{\mathcal{S}_r} [[\![\mathcal{P}^{m'}]\!] \cdot \mathcal{M}] \cdot \delta\widetilde{\boldsymbol{\Upsilon}}\, \mathrm{d}\mathcal{A}\Bigg]\, \mathrm{d}t$$

Likewise, the total variation of the action integral $\mathcal{A}$ expanded over the material configuration reads as in Eq. 10.38. The bracketed terms preceding the entirely arbitrary spatial and material virtual displacements in the integrands of the individual integrals (a) to (f) constitute the Euler–Lagrange equations. These are pair-wise not entirely independent and their fulfillment due to the satisfaction of the balances of spatial and material linear momentum in the domain, on the boundary, and at singular surfaces will be analyzed in the sequel.

$$\delta\mathcal{A} = \int_{t_1}^{t_2} \Bigg[(a) \int_{\mathcal{B}_m} [\, \overline{\boldsymbol{s}}'_m + \mathrm{Div}\boldsymbol{P}' - \mathrm{D}_t\, \boldsymbol{p}_m] \cdot \delta\boldsymbol{v}\, \mathrm{d}V \tag{10.38}$$

$$(b) + \int_{\mathcal{B}_m} [\overline{\boldsymbol{S}}'_m + \mathrm{Div}\boldsymbol{\Sigma}' - \mathrm{D}_t \boldsymbol{P}_m] \cdot \delta\boldsymbol{\Upsilon}\, \mathrm{d}V$$

$$(c) + \int_{\partial\mathcal{B}_m} \big[\, \widehat{\boldsymbol{s}}^{\mathrm{ext}\prime}_m + \widehat{\mathrm{Div}}\widehat{\boldsymbol{P}}' - [\boldsymbol{P}' + \boldsymbol{p}_m \otimes \boldsymbol{W}] \cdot \boldsymbol{N}\big] \cdot \delta\widehat{\boldsymbol{v}}\, \mathrm{d}A$$

$$(d) + \int_{\partial\mathcal{B}_m} \big[\widehat{\boldsymbol{S}}^{\mathrm{ext}\prime}_m + \widehat{\mathrm{Div}}\widehat{\boldsymbol{\Sigma}}' - [\boldsymbol{\Sigma}' + \boldsymbol{P}_m \otimes \boldsymbol{W}] \cdot \boldsymbol{N}\big] \cdot \delta\widehat{\boldsymbol{\Upsilon}}\, \mathrm{d}A$$

$$(e) + \int_{\mathcal{S}_m} \big[[\![\boldsymbol{P}' + \boldsymbol{p}_m \otimes \boldsymbol{W}]\!] \cdot \boldsymbol{M}\big] \cdot \delta\widetilde{\boldsymbol{v}}\, \mathrm{d}A$$

$$(f) + \int_{\mathcal{S}_m} \big[[\![\boldsymbol{\Sigma}' + \boldsymbol{P}_m \otimes \boldsymbol{W}]\!] \cdot \boldsymbol{M}\big] \cdot \delta\widetilde{\boldsymbol{\Upsilon}}\, \mathrm{d}A\Bigg]\, \mathrm{d}t.$$

10.1.4 Euler–Lagrange Equations: Analysis

The integrands of the integrals (a) to (f) in Eq. 10.38 are further analyzed in the following:

(*a*) The variational definition for the (exterior) Piola stress $\boldsymbol{P}'$ coincides with $\boldsymbol{P} \equiv \boldsymbol{P}'$ and, likewise, the (exterior) spatial linear momentum source density $\overline{\boldsymbol{s}}'_{\mathrm{m}}$ coincides with $\overline{\boldsymbol{s}}_{\mathrm{m}} \equiv \overline{\boldsymbol{s}}'_{\mathrm{m}}$.

In conclusion, the bracketed term in the integrand of (*a*) vanishes due to the local balance of spatial linear momentum at regular points in the domain in Eq. 7.29.

(*b*) The local balance of material linear momentum at regular points in the domain in Eq. 8.49 is expressed in terms of the interior Eshelby stress $\boldsymbol{\Sigma}$ and the interior material linear momentum source density $\overline{\boldsymbol{S}}_{\mathrm{m}}$ that have been introduced earlier in Eqs. 8.63 and 8.64 and result for the current isothermal, conservative case in

$$\boldsymbol{\Sigma} := -\lambda_{\mathrm{m}} \boldsymbol{I} - \boldsymbol{F}^t \cdot \boldsymbol{P} \quad \text{and} \quad \overline{\boldsymbol{S}}_{\mathrm{m}} := \partial_X \lambda_{\mathrm{m}} - \boldsymbol{F}^t \cdot \overline{\boldsymbol{s}}_{\mathrm{m}}. \tag{10.39}$$

Note that rather being expressed in terms of the exterior Lagrange-type potential energy density $\bar{l}_{\mathrm{m}}$ the interior $\boldsymbol{\Sigma}$ and $\overline{\boldsymbol{S}}_{\mathrm{m}}$ are expressed in terms of the interior Lagrange-type free energy density λ_{m} that, for the here considered isothermal, conservative case, relates to the exterior Lagrange-type potential energy density $\bar{l}_{\mathrm{m}}$ as

$$\lambda_{\mathrm{m}} := \kappa_{\mathrm{m}} - w_{\mathrm{m}} \equiv \bar{l}_{\mathrm{m}} + \overline{v}_{\mathrm{m}} \quad \text{with} \quad \bar{l}_{\mathrm{m}} := \kappa_{\mathrm{m}} - [w_{\mathrm{m}} + \overline{v}_{\mathrm{m}}]. \tag{10.40}$$

Thus, the exterior and interior Eshelby stress and the exterior and interior material linear momentum source density are connected by

$$\boldsymbol{\Sigma}' = \boldsymbol{\Sigma} + \overline{v}_{\mathrm{m}} \boldsymbol{I} \quad \text{and} \quad \overline{\boldsymbol{S}}'_{\mathrm{m}} = \overline{\boldsymbol{S}}_{\mathrm{m}} - \nabla_X \overline{v}_{\mathrm{m}} \tag{10.41}$$

with $\nabla_X \overline{v}_{\mathrm{m}} = \partial_X \overline{v}_{\mathrm{m}} + \boldsymbol{F}^t \cdot \partial_x \overline{v}_{\mathrm{m}}$. As a result it holds that

$$\operatorname{Div} \boldsymbol{\Sigma}' + \overline{\boldsymbol{S}}'_{\mathrm{m}} \equiv \operatorname{Div} \boldsymbol{\Sigma} + \overline{\boldsymbol{S}}_{\mathrm{m}}, \tag{10.42}$$

whereby $\operatorname{Div}(\overline{v}_{\mathrm{m}} \boldsymbol{I}) = \nabla_X \overline{v}_{\mathrm{m}} = \partial_X \overline{v}_{\mathrm{m}} - \boldsymbol{F}^t \cdot \overline{\boldsymbol{s}}_{\mathrm{m}}$ has been used.

In conclusion, the bracketed term in the integrand of (*b*) vanishes due to the local balance of material linear momentum at regular points in the domain in Eq. 8.49.

(*c*) The variational definition for the (exterior) boundary Piola stress $\widehat{\boldsymbol{P}}'$ coincides with $\widehat{\boldsymbol{P}} \equiv \widehat{\boldsymbol{P}}'$ and, likewise, the (exterior) spatial extrinsic linear momentum flux density $\widehat{\boldsymbol{s}}^{\mathrm{ext}\prime}_{\mathrm{m}}$ coincides with $\widehat{\boldsymbol{s}}^{\mathrm{ext}}_{\mathrm{m}} \equiv \widehat{\boldsymbol{s}}^{\mathrm{ext}\prime}_{\mathrm{m}}$. Thus, the integrand of (*c*) represents the local balance of spatial linear momentum at regular points on the boundary when Eq. 7.30 is properly extended by the additional term $\widehat{\operatorname{Div}} \widehat{\boldsymbol{P}}$ due to boundary stresses

$$\widehat{\operatorname{Div}} \widehat{\boldsymbol{P}} + \widehat{\boldsymbol{s}}^{\mathrm{ext}}_{\mathrm{m}} = [\boldsymbol{P} + \boldsymbol{p}_{\mathrm{m}} \otimes \boldsymbol{W}] \cdot \boldsymbol{N}. \tag{10.43}$$

In conclusion, the bracketed term in the integrand of (c) vanishes due to the local balance of spatial linear momentum at regular points on the boundary.

(d) Due to the definition of the exterior Eshelby stress and jointly with Eq. 10.43 the bracketed term in the integrand of (d) is re-expressed as

$$\widehat{\boldsymbol{S}}^{\text{ext}\prime}_{\text{m}} + \widehat{\text{Div}}\,\widehat{\boldsymbol{\Sigma}}' + \bar{l}_{\text{m}}\boldsymbol{N} + \boldsymbol{F}^t \cdot [\widehat{\boldsymbol{s}}^{\text{ext}\prime}_{\text{m}} + \widehat{\text{Div}}\,\widehat{\boldsymbol{P}}']. \tag{10.44}$$

Next, the surface divergence $\widehat{\text{Div}}(-\widehat{l}_{\text{m}}\widehat{\boldsymbol{I}})$ of the spherical part of $\widehat{\boldsymbol{\Sigma}}'$ expands as

$$-\widehat{\nabla}_{\widehat{X}}\widehat{l}_{\text{m}} - \widehat{l}_{\text{m}}\,\widehat{C}\,\boldsymbol{N} = -\nabla_{\widehat{X}}\widehat{l}_{\text{m}} + \left[\nabla_N\widehat{l}_{\text{m}} - [\widehat{l}_{\text{m}}\,\widehat{\boldsymbol{I}}] : \widehat{\boldsymbol{C}}\right]\boldsymbol{N}. \tag{10.45}$$

with the gradient $-\nabla_{\widehat{X}}\widehat{l}_{\text{m}}$ unfolding as

$$-\partial_{\widehat{X}}\widehat{l}_{\text{m}} - \boldsymbol{F}^t \cdot \partial_{\widehat{x}}\widehat{l}_{\text{m}} - \partial_{\widehat{F}}\widehat{l}_{\text{m}} : \nabla_{\widehat{X}}\widehat{\boldsymbol{F}} = -\widehat{\boldsymbol{S}}^{\text{ext}\prime}_{\text{m}} - \boldsymbol{F}^t \cdot \widehat{\boldsymbol{s}}^{\text{ext}\prime}_{\text{m}} + \widehat{\boldsymbol{P}}' : \nabla_{\widehat{X}}\widehat{\boldsymbol{F}}. \tag{10.46}$$

Finally, the surface divergence of the non-spherical part of $\widehat{\boldsymbol{\Sigma}}'$ reads[1]

$$\widehat{\text{Div}}(-\widehat{\boldsymbol{F}}^t \cdot \widehat{\boldsymbol{P}}') = -\widehat{\boldsymbol{F}}^t \cdot \widehat{\text{Div}}\,\widehat{\boldsymbol{P}}' - \widehat{\boldsymbol{P}}' : \widehat{\nabla}_{\widehat{X}}\widehat{\boldsymbol{F}} - [\widehat{\boldsymbol{F}}^t \cdot \widehat{\boldsymbol{P}}'] : \widehat{\boldsymbol{C}}\,\boldsymbol{N}. \tag{10.47}$$

Taken together, due to the local balance of material linear momentum at regular points on the boundary in Eq. 8.50, the bracketed term in the integrand of (d) is purely normal

$$[\bar{l}_{\text{m}} + \nabla_N\widehat{l}_{\text{m}} + \widehat{\boldsymbol{\Sigma}}' : \widehat{\boldsymbol{C}} + \nabla_N\boldsymbol{y} \cdot \widehat{\text{Div}}\,\widehat{\boldsymbol{P}}' + \widehat{\boldsymbol{P}}' : \nabla_N\widehat{\boldsymbol{F}}]\boldsymbol{N}. \tag{10.48}$$

(e) The variational definition for the (exterior) boundary Piola stress $\widehat{\boldsymbol{P}}'$ coincides with $\widehat{\boldsymbol{P}} \equiv \widehat{\boldsymbol{P}}'$.

---

[1] A step-by-step derivation is as follows:

$$\begin{aligned}
&\widehat{\text{Div}}(\widehat{\boldsymbol{F}}^t \cdot \widehat{\boldsymbol{P}}) = \left[[\widehat{\boldsymbol{G}}^\alpha \otimes \widehat{\boldsymbol{g}}_\alpha] \cdot \widehat{\boldsymbol{P}}\right]_{,\beta} \cdot \widehat{\boldsymbol{G}}^\beta \\
&= [\widehat{\boldsymbol{G}}^\alpha \otimes \widehat{\boldsymbol{g}}_\alpha] \cdot \widehat{\boldsymbol{P}}_{,\beta} \cdot \widehat{\boldsymbol{G}}^\beta + [\widehat{\boldsymbol{G}}^\alpha \otimes \widehat{\boldsymbol{g}}_{\alpha,\beta} + \widehat{\boldsymbol{G}}^\alpha{}_{,\beta} \otimes \widehat{\boldsymbol{g}}_\alpha] \cdot \widehat{\boldsymbol{P}} \cdot \widehat{\boldsymbol{G}}^\beta \\
&= \widehat{\boldsymbol{F}}^t \cdot \widehat{\text{Div}}\,\widehat{\boldsymbol{P}} + \widehat{\boldsymbol{P}} : \left[\widehat{\boldsymbol{g}}_{\alpha,\beta} \otimes \widehat{\boldsymbol{G}}^\beta \otimes \widehat{\boldsymbol{G}}^\alpha + \widehat{\boldsymbol{g}}_\alpha \otimes \widehat{\boldsymbol{G}}^\beta \otimes \widehat{\boldsymbol{G}}^\alpha{}_{,\beta}\right] \\
&= \widehat{\boldsymbol{F}}^t \cdot \widehat{\text{Div}}\,\widehat{\boldsymbol{P}} + \widehat{\boldsymbol{P}} : \left[\widehat{\boldsymbol{g}}_{\alpha,\beta} \otimes \widehat{\boldsymbol{G}}^\alpha \otimes \widehat{\boldsymbol{G}}^\beta + \widehat{\boldsymbol{g}}_\alpha \otimes \widehat{\boldsymbol{G}}^\beta \otimes [-\widehat{\Gamma}^\alpha{}_{\beta\gamma}\,\widehat{\boldsymbol{G}}^\gamma + \widehat{C}^\alpha{}_\beta\,\boldsymbol{N}]\right] \\
&= \widehat{\boldsymbol{F}}^t \cdot \widehat{\text{Div}}\,\widehat{\boldsymbol{P}} + \widehat{\boldsymbol{P}} : \left[[\widehat{\boldsymbol{g}}_{\alpha,\gamma} \otimes \widehat{\boldsymbol{G}}^\alpha - \widehat{\Gamma}^\alpha{}_{\beta\gamma}\,\widehat{\boldsymbol{g}}_\alpha \otimes \widehat{\boldsymbol{G}}^\beta] \otimes \widehat{\boldsymbol{G}}^\gamma + \widehat{C}^\alpha{}_\beta\,\widehat{\boldsymbol{g}}_\alpha \otimes \widehat{\boldsymbol{G}}^\beta \otimes \boldsymbol{N}\right] \\
&= \widehat{\boldsymbol{F}}^t \cdot \widehat{\text{Div}}\,\widehat{\boldsymbol{P}} + \widehat{\boldsymbol{P}} : \left[[\widehat{\boldsymbol{g}}_{\alpha,\gamma} \otimes \widehat{\boldsymbol{G}}^\alpha - \widehat{\Gamma}^\alpha{}_{\gamma\beta}\,\widehat{\boldsymbol{g}}_\alpha \otimes \widehat{\boldsymbol{G}}^\beta] \otimes \widehat{\boldsymbol{G}}^\gamma + \widehat{C}^\alpha{}_\beta\,\widehat{\boldsymbol{F}} \cdot \widehat{\boldsymbol{G}}_\alpha \otimes \widehat{\boldsymbol{G}}^\beta \otimes \boldsymbol{N}\right] \\
&= \widehat{\boldsymbol{F}}^t \cdot \widehat{\text{Div}}\,\widehat{\boldsymbol{P}} + \widehat{\boldsymbol{P}} : \left[[\widehat{\boldsymbol{g}}_{\alpha,\gamma} \otimes \widehat{\boldsymbol{G}}^\alpha - \widehat{\Gamma}^\alpha{}_{\gamma\beta}\,\widehat{\boldsymbol{g}}_\alpha \otimes \widehat{\boldsymbol{G}}^\beta + \widehat{C}^\alpha{}_\gamma\,\widehat{\boldsymbol{g}}_\alpha \otimes \boldsymbol{N}] \otimes \widehat{\boldsymbol{G}}^\gamma + \widehat{\boldsymbol{F}} \cdot \widehat{\boldsymbol{C}} \otimes \boldsymbol{N}\right] \\
&= \widehat{\boldsymbol{F}}^t \cdot \widehat{\text{Div}}\,\widehat{\boldsymbol{P}} + \widehat{\boldsymbol{P}} : \left[\widehat{\boldsymbol{g}}_{\alpha,\gamma} \otimes \widehat{\boldsymbol{G}}^\alpha + \widehat{\boldsymbol{g}}_\alpha \otimes [-\widehat{\Gamma}^\alpha{}_{\gamma\beta}\,\widehat{\boldsymbol{G}}^\beta + \widehat{C}^\alpha{}_\gamma\,\boldsymbol{N}]\right] \otimes \widehat{\boldsymbol{G}}^\gamma + [\widehat{\boldsymbol{F}}^t \cdot \widehat{\boldsymbol{P}}] : \widehat{\boldsymbol{C}}\,\boldsymbol{N} \\
&= \widehat{\boldsymbol{F}}^t \cdot \widehat{\text{Div}}\,\widehat{\boldsymbol{P}} + \widehat{\boldsymbol{P}} : [\widehat{\boldsymbol{g}}_{\alpha,\gamma} \otimes \widehat{\boldsymbol{G}}^\alpha + \widehat{\boldsymbol{g}}_\alpha \otimes \widehat{\boldsymbol{G}}^\alpha{}_{,\gamma}] \otimes \widehat{\boldsymbol{G}}^\gamma + [\widehat{\boldsymbol{F}}^t \cdot \widehat{\boldsymbol{P}}] : \widehat{\boldsymbol{C}}\,\boldsymbol{N} \\
&= \widehat{\boldsymbol{F}}^t \cdot \widehat{\text{Div}}\,\widehat{\boldsymbol{P}} + \widehat{\boldsymbol{P}} : \widehat{\nabla}_{\widehat{X}}\widehat{\boldsymbol{F}} + [\widehat{\boldsymbol{F}}^t \cdot \widehat{\boldsymbol{P}}] : \widehat{\boldsymbol{C}}\,\boldsymbol{N}.
\end{aligned}$$

In conclusion, the bracketed term in the integrand of (e) vanishes due to the local balance of spatial linear momentum at regular points at singular surfaces in Eq. 7.31.

(f) Due to the local balance of material linear momentum at regular points at singular surfaces in Eq. 8.51, the bracketed term in the integrand of (f), that is abbreviated as $\widetilde{\boldsymbol{S}}'_{\mathrm{m}} := \widetilde{\boldsymbol{S}}_{\mathrm{m}} + [\![\overline{v}_{\mathrm{m}}]\!]\,\boldsymbol{M}$, is purely normal (as is $\widetilde{\boldsymbol{S}}_{\mathrm{m}}$)

$$\widetilde{\boldsymbol{S}}'_{\mathrm{m}} = \big[[\![\overline{u}_{\mathrm{m}}]\!] - [\![\boldsymbol{F}]\!] : \{\boldsymbol{P}\}\big]\boldsymbol{M}. \tag{10.49}$$

In summary, the total variation of the action integral reduces eventually to

$$\begin{aligned}\delta\mathcal{A} = \int_{t_1}^{t_2} \Bigg[\int_{\partial\mathcal{B}_{\mathrm{m}}} [\overline{l}_{\mathrm{m}} + \nabla_N \widehat{l}_{\mathrm{m}} + \widehat{\boldsymbol{\Sigma}}' : \widehat{\boldsymbol{C}} + \nabla_N \boldsymbol{y} \cdot \widehat{\mathrm{Div}\boldsymbol{P}'} + \widehat{\boldsymbol{P}}' : \nabla_N \widehat{\boldsymbol{F}}]\, \delta\widehat{\Upsilon}_\perp \,\mathrm{d}A \\ + \int_{\mathcal{S}_{\mathrm{m}}} \big[[\![\overline{u}_{\mathrm{m}}]\!] - [\![\boldsymbol{F}]\!] : \{\boldsymbol{P}\}\big]\, \delta\widetilde{\Upsilon}_\perp \,\mathrm{d}A \Bigg] \mathrm{d}t \end{aligned} \tag{10.50}$$

and compares to the negative virtual energy release

$$\delta\mathcal{A} =: -\int_{t_1}^{t_2} \Bigg[\int_{\mathcal{B}_{\mathrm{m}}} \delta\boldsymbol{\Upsilon} \cdot \overline{\boldsymbol{R}}_{\mathrm{m}} \,\mathrm{d}V + \int_{\partial\mathcal{B}_{\mathrm{m}}} \delta\widehat{\boldsymbol{\Upsilon}} \cdot \widehat{\boldsymbol{R}}_{\mathrm{m}} \,\mathrm{d}A + \int_{\mathcal{S}_{\mathrm{m}}} \delta\widetilde{\boldsymbol{\Upsilon}} \cdot \widetilde{\boldsymbol{R}}_{\mathrm{m}} \,\mathrm{d}A \Bigg] \mathrm{d}t \overset{!}{\geq} 0. \tag{10.51}$$

By comparison of contributions, it is thus concluded that the material body force vanishes identically

$$\overline{\boldsymbol{R}}_{\mathrm{m}} \equiv \boldsymbol{0}, \tag{10.52}$$

whereas the material tractions driving material virtual displacements on the boundary and at singular surfaces read as

$$\begin{aligned}\widehat{\boldsymbol{R}}_{\mathrm{m}} &\equiv -[\overline{l}_{\mathrm{m}} + \nabla_N \widehat{l}_{\mathrm{m}} + \widehat{\boldsymbol{\Sigma}}' : \widehat{\boldsymbol{C}} + \nabla_N \boldsymbol{y} \cdot \widehat{\mathrm{Div}\boldsymbol{P}'} + \widehat{\boldsymbol{P}}' : \nabla_N \widehat{\boldsymbol{F}}]\, \boldsymbol{N}, \\ \widetilde{\boldsymbol{R}}_{\mathrm{m}} &\equiv -\big[[\![\overline{u}_{\mathrm{m}}]\!] - [\![\boldsymbol{F}]\!] : \{\boldsymbol{P}\}\big]\, \boldsymbol{M}.\end{aligned} \tag{10.53}$$

Observe that the material tractions $\widehat{\boldsymbol{R}}_{\mathrm{m}}$ and $\widetilde{\boldsymbol{R}}_{\mathrm{m}}$ are purely normal.

In the above, the material virtual displacement $\delta\widehat{\boldsymbol{\Upsilon}}$ and $\delta\widetilde{\boldsymbol{\Upsilon}}$ on the boundary $\partial\mathcal{B}_{\mathrm{m}}$ and at singular surfaces $\mathcal{S}_{\mathrm{m}}$ have been decomposed into orthogonal contributions, i.e. tangential and normal to the boundary and to singular surfaces, respectively, with the vector-valued $\delta\widehat{\boldsymbol{\Upsilon}}_{\parallel} := \widehat{\boldsymbol{I}} \cdot \delta\widehat{\boldsymbol{\Upsilon}}$ and $\delta\widetilde{\boldsymbol{\Upsilon}}_{\parallel} := \widetilde{\boldsymbol{I}} \cdot \delta\widetilde{\boldsymbol{\Upsilon}}$ and the scalar-valued $\delta\widehat{\Upsilon}_\perp := \delta\widehat{\boldsymbol{\Upsilon}} \cdot \boldsymbol{N}$ and $\delta\widetilde{\Upsilon}_\perp := \delta\widetilde{\boldsymbol{\Upsilon}} \cdot \boldsymbol{M}$, respectively.

10.1.5 Euler–Lagrange Equations: PB/PF Operations

Alternatively, the integrands of the integrals (a) to (f) in Eq. 10.38 may be analyzed by considering appropriate pull-back/push-forward operations:

$(a) \leftrightarrow (b)$ When expressed in terms of the exterior Lagrange-type potential energy density per unit volume in the material configuration $\mathcal{B}_\mathrm{m}$, i.e.

$$\bar{l}_\mathrm{m} = \bar{l}_\mathrm{m}(\boldsymbol{y}, \boldsymbol{v}, \boldsymbol{F}; \boldsymbol{X}),$$

with its partial derivatives defined as

$$\bar{\boldsymbol{s}}'_\mathrm{m} := \partial_x \bar{l}_\mathrm{m}, \quad \boldsymbol{p}_\mathrm{m} := \partial_v \bar{l}_\mathrm{m} \quad \text{and} \quad \boldsymbol{P}' := -\partial_F \bar{l}_\mathrm{m}, \tag{10.54}$$

the integrand (a), representing the local balance of spatial linear momentum at regular points in the domain, reads as

$$\partial_x \bar{l}_\mathrm{m} - \mathrm{Div}\partial_F \bar{l}_\mathrm{m} - \mathrm{D}_t \partial_v \bar{l}_\mathrm{m} = \boldsymbol{0}. \tag{10.55}$$

As a vectorial statement, this is a spatial covariant vector, i.e. a covector, consequently its material counterpart follows from a (negative) covariant pull-back with the spatial deformation gradient as

$$\boldsymbol{F}^t \cdot [-\partial_x \bar{l}_\mathrm{m} + \mathrm{Div}\partial_F \bar{l}_\mathrm{m} + \mathrm{D}_t \partial_v \bar{l}_\mathrm{m}] = \boldsymbol{0}. \tag{10.56}$$

Due to the product rule of differentiation and the compatibility of $\boldsymbol{F}$ the second term in the above rewrites as

$$\boldsymbol{F}^t \cdot \mathrm{Div}\partial_F \bar{l}_\mathrm{m} = \mathrm{Div}(\boldsymbol{F}^t \cdot \partial_F \bar{l}_\mathrm{m}) - \partial_F \bar{l}_\mathrm{m} : \nabla_X \boldsymbol{F}. \tag{10.57}$$

Next, with the explicit expansion of the (negative) gradient of $\bar{l}_\mathrm{m}$, i.e.

$$-\nabla_X \bar{l}_\mathrm{m} = \mathrm{Div}(-\bar{l}_\mathrm{m} \boldsymbol{I}),$$

the last term in the above reads as

$$-\partial_F \bar{l}_\mathrm{m} : \nabla_X \boldsymbol{F} = \mathrm{Div}(-\bar{l}_\mathrm{m} \boldsymbol{I}) + \boldsymbol{F}^t \cdot \partial_x \bar{l}_\mathrm{m} + \partial_v \bar{l}_\mathrm{m} \cdot \nabla_X \boldsymbol{v} + \partial_X \bar{l}_\mathrm{m}. \tag{10.58}$$

Together with the identity (that follows from $-\partial_V \bar{l}_\mathrm{m} = \boldsymbol{F}^t \cdot \partial_v \bar{l}_\mathrm{m}$)

$$\partial_v \bar{l}_\mathrm{m} \cdot \nabla_X \boldsymbol{v} = -\mathrm{D}_t \partial_V \bar{l}_\mathrm{m} - \boldsymbol{F}^t \cdot \mathrm{D}_t \partial_v \bar{l}_\mathrm{m},$$

the (negative) covariant pull-back of (a) results in

$$\partial_X \bar{l}_m + \mathrm{Div}(-\bar{l}_m \boldsymbol{I} + \boldsymbol{F}^t \cdot \partial_F \bar{l}_m) - \mathrm{D}_t \partial_V \bar{l}_m = \boldsymbol{0}. \tag{10.59}$$

Clearly, this is the local balance of material linear momentum at regular points in the domain expressed in terms of the exterior Lagrange-type potential energy density per unit volume in the material configuration $\mathcal{B}_m$ parameterized as

$$\bar{l}_m = \bar{l}_m(\boldsymbol{Y}, \boldsymbol{V}, \boldsymbol{F}; \boldsymbol{x}),$$

and its partial derivatives defined as

$$\overline{\boldsymbol{S}}'_m := \partial_X \bar{l}_m, \quad \boldsymbol{P}_m := \partial_V \bar{l}_m \quad \text{and} \quad \boldsymbol{\Sigma}' := -\bar{l}_m \boldsymbol{I} + \boldsymbol{F}^t \cdot \partial_F \bar{l}_m. \tag{10.60}$$

Consequently, the local balances of spatial and material linear momentum at regular points in the domain are simultaneously satisfied, i.e. the integrals (a) *and* (b) *vanish concurrently!*

Observe finally that the exterior Eshelby stress $\boldsymbol{\Sigma}' = \boldsymbol{p}' \cdot \mathrm{cof}\, \boldsymbol{F}$ follows as the Piola transformation of the stress measure

$$\boldsymbol{p}' := -\partial_f \bar{l}_s = -j\, [\bar{l}_m \boldsymbol{F}^t + \partial_f \bar{l}_m] = -j\, [\bar{l}_m \boldsymbol{I} - \boldsymbol{F}^t \cdot \partial_F \bar{l}_m] \cdot \boldsymbol{F}^t, \tag{10.61}$$

thus $\boldsymbol{p}' = \boldsymbol{\Sigma}' \cdot \mathrm{cof}\, \boldsymbol{f}$, whereby

$$\bar{l}_s := j\, \bar{l}_m(\boldsymbol{Y}, \boldsymbol{V}, \boldsymbol{f}; \boldsymbol{x})$$

is the exterior Lagrange-type potential energy density per unit volume in the spatial configuration $\mathcal{B}_s$. Thus

$$\partial_X \bar{l}_s - \mathrm{div} \partial_f \bar{l}_s - j\, \mathrm{D}_t \partial_V \bar{l}_m = \boldsymbol{0} \tag{10.62}$$

represents an alternative for the local balance of material linear momentum at regular points in the domain.

$(c) \leftrightarrow (d)$ When expressed in terms of the exterior Lagrange-type potential energy density per unit area on the boundary $\partial\mathcal{B}_m$ in the material configuration, i.e.

$$\widehat{l}_m = \widehat{l}_m(\widehat{\boldsymbol{y}}, \widehat{\boldsymbol{F}}; \widehat{\boldsymbol{X}}),$$

with its partial derivatives defined as

$$\widehat{\boldsymbol{s}}^{\mathrm{ext}\prime}_m := \partial_{\widehat{x}} \widehat{l}_m \quad \text{and} \quad \widehat{\boldsymbol{P}}' := -\partial_{\widehat{F}} \widehat{l}_m, \tag{10.63}$$

the *tangential* part of the integrand (c), representing the tangential part of the local balance of spatial linear momentum at regular points on the boundary, reads as

$$\widehat{\boldsymbol{i}} \cdot \big[\partial_{\widehat{x}} \widehat{l}_{\mathrm{m}} - \widehat{\mathrm{Div}} \partial_{\widehat{F}} \widehat{l}_{\mathrm{m}} + [\partial_F \bar{l}_{\mathrm{m}} - \partial_v \bar{l}_{\mathrm{m}} \otimes \boldsymbol{W}] \cdot \boldsymbol{N}\big] = \mathbf{0}. \tag{10.64}$$

As a vectorial statement, this is a spatial covariant vector, i.e. a covector, tangential to the boundary, consequently its material counterpart follows from a (negative) covariant pull-back with the spatial boundary deformation gradient as

$$\widehat{\boldsymbol{F}}^t \cdot \big[- \partial_{\widehat{x}} \widehat{l}_{\mathrm{m}} + \widehat{\mathrm{Div}} \partial_{\widehat{F}} \widehat{l}_{\mathrm{m}} - [\partial_F \bar{l}_{\mathrm{m}} - \partial_v \bar{l}_{\mathrm{m}} \otimes \boldsymbol{W}] \cdot \boldsymbol{N}\big] = \mathbf{0}. \tag{10.65}$$

Due to the product rule of differentiation and the compatibility of $\widehat{\boldsymbol{F}}$ the second term in the above rewrites as[2]

$$\widehat{\boldsymbol{F}}^t \cdot \widehat{\mathrm{Div}} \partial_{\widehat{F}} \widehat{l}_{\mathrm{m}} = \widehat{\boldsymbol{I}} \cdot \widehat{\mathrm{Div}}(\widehat{\boldsymbol{F}}^t \cdot \partial_{\widehat{F}} \widehat{l}_{\mathrm{m}}) - \partial_{\widehat{F}} \widehat{l}_{\mathrm{m}} : \widehat{\nabla}_{\widehat{X}} \widehat{\boldsymbol{F}}. \tag{10.66}$$

Next, with the explicit expansion of the (negative) boundary gradient of $\widehat{l}_{\mathrm{m}}$, i.e.

$$-\widehat{\nabla}_{\widehat{X}} \widehat{l}_{\mathrm{m}} = \widehat{\boldsymbol{I}} \cdot \widehat{\mathrm{Div}}(-\widehat{l}_{\mathrm{m}} \widehat{\boldsymbol{I}}),$$

the last term in the above reads as

$$- \partial_{\widehat{F}} \widehat{l}_{\mathrm{m}} : \widehat{\nabla}_{\widehat{X}} \widehat{\boldsymbol{F}} = \widehat{\boldsymbol{I}} \cdot \widehat{\mathrm{Div}}(-\widehat{l}_{\mathrm{m}} \widehat{\boldsymbol{I}}) + \widehat{\boldsymbol{F}}^t \cdot \partial_{\widehat{x}} \widehat{l}_{\mathrm{m}} + \widehat{\boldsymbol{I}} \cdot \partial_{\widehat{X}} \widehat{l}_{\mathrm{m}}. \tag{10.67}$$

Together with the identities

$$\widehat{\boldsymbol{F}}^t \cdot \partial_F \bar{l}_{\mathrm{m}} = \widehat{\boldsymbol{I}} \cdot \boldsymbol{F}^t \cdot \partial_F \bar{l}_{\mathrm{m}} \quad \text{and} \quad \widehat{\boldsymbol{F}}^t \cdot \partial_v \bar{l}_{\mathrm{m}} = -\widehat{\boldsymbol{I}} \cdot \partial_V \bar{l}_{\mathrm{m}}$$

the (negative) covariant tangential pull-back of (c) results in

$$\widehat{\boldsymbol{I}} \cdot \big[\partial_{\widehat{X}} \widehat{l}_{\mathrm{m}} + \widehat{\mathrm{Div}}(-\widehat{l}_{\mathrm{m}} \widehat{\boldsymbol{I}} + \widehat{\boldsymbol{F}}^t \cdot \partial_{\widehat{F}} \widehat{l}_{\mathrm{m}}) - [\boldsymbol{F}^t \cdot \partial_F \bar{l}_{\mathrm{m}} + \partial_V \bar{l}_{\mathrm{m}} \otimes \boldsymbol{W}] \cdot \boldsymbol{N}\big] = \mathbf{0}. \tag{10.68}$$

[2] From the established result

$$\widehat{\mathrm{Div}}(\widehat{\boldsymbol{F}}^t \cdot \widehat{\boldsymbol{P}}) = \widehat{\boldsymbol{F}}^t \cdot \widehat{\mathrm{Div}} \widehat{\boldsymbol{P}} + \widehat{\boldsymbol{P}} : \widehat{\nabla}_{\widehat{X}} \widehat{\boldsymbol{F}} + [\widehat{\boldsymbol{F}}^t \cdot \widehat{\boldsymbol{P}}] : \widehat{\boldsymbol{C}}\, \boldsymbol{N}$$

it follows for its tangential and normal parts, respectively, that

$$\widehat{\boldsymbol{I}} \cdot \widehat{\mathrm{Div}}(\widehat{\boldsymbol{F}}^t \cdot \widehat{\boldsymbol{P}}) = \widehat{\boldsymbol{F}}^t \cdot \widehat{\mathrm{Div}} \widehat{\boldsymbol{P}} + \widehat{\boldsymbol{P}} : \widehat{\nabla}_{\widehat{X}} \widehat{\boldsymbol{F}} \quad \text{and} \quad \boldsymbol{N} \cdot \widehat{\mathrm{Div}}(\widehat{\boldsymbol{F}}^t \cdot \widehat{\boldsymbol{P}}) = [\widehat{\boldsymbol{F}}^t \cdot \widehat{\boldsymbol{P}}] : \widehat{\boldsymbol{C}}.$$

.

Clearly, by adding the identity $\widehat{\boldsymbol{I}} \cdot [\bar{l}_{\mathrm{m}} \boldsymbol{I}] \cdot \boldsymbol{N} = \boldsymbol{0}$, this is the tangential part of the local balance of material linear momentum at regular points on the boundary expressed in terms of the exterior Lagrange-type potential energy density per unit area on the boundary $\partial \mathcal{B}_{\mathrm{m}}$ in the material configuration parameterized as

$$\widehat{l}_{\mathrm{m}} = \widehat{l}_{\mathrm{m}}(\widehat{\boldsymbol{Y}}, \widehat{\boldsymbol{F}}; \widehat{\boldsymbol{x}}),$$

and its partial derivatives defined as

$$\widehat{\boldsymbol{S}}^{\mathrm{ext}\,\prime}_{\mathrm{m}} := \partial_{\widehat{X}} \widehat{l}_{\mathrm{m}} \quad \text{and} \quad \widehat{\boldsymbol{\Sigma}}' := -\widehat{l}_{\mathrm{m}} \widehat{\boldsymbol{I}} + \widehat{\boldsymbol{F}}^t \cdot \partial_{\widehat{F}} \widehat{l}_{\mathrm{m}}. \tag{10.69}$$

Consequently, the <u>*tangential*</u> *parts of the local balances of spatial and material linear momentum at regular points on the boundary are simultaneously satisfied, i.e. the tangential parts of the integrals* (c) *and* (d) *vanish concurrently!*

Observe, moreover, that the exterior boundary Eshelby stress $\widehat{\boldsymbol{\Sigma}}' = \widehat{\boldsymbol{p}}' \cdot \widehat{\operatorname{cof} \boldsymbol{F}}$ follows as the boundary Piola transformation of the stress measure

$$\widehat{\boldsymbol{p}}' := -\partial_{\widehat{f}} \widehat{l}_{\mathrm{s}} = -\widehat{j}\,[\widehat{l}_{\mathrm{m}} \widehat{\boldsymbol{F}}^t + \partial_{\widehat{f}} \widehat{l}_{\mathrm{m}}] = -\widehat{j}\,[\widehat{l}_{\mathrm{m}} \widehat{\boldsymbol{I}} - \widehat{\boldsymbol{F}}^t \cdot \partial_{\widehat{F}} \widehat{l}_{\mathrm{m}}] \cdot \widehat{\boldsymbol{F}}^t, \tag{10.70}$$

thus $\widehat{\boldsymbol{p}}' = \widehat{\boldsymbol{\Sigma}}' \cdot \widehat{\operatorname{cof} \boldsymbol{f}}$, whereby

$$\widehat{l}_{\mathrm{s}} := \widehat{j}\,\widehat{l}_{\mathrm{m}}(\widehat{\boldsymbol{Y}}, \widehat{\boldsymbol{f}}; \widehat{\boldsymbol{x}})$$

is the exterior Lagrange-type potential energy density per unit area on the boundary $\partial \mathcal{B}_{\mathrm{s}}$ in the spatial configuration. Thus

$$\widehat{\boldsymbol{I}} \cdot \Big[\partial_{\widehat{X}} \widehat{l}_{\mathrm{s}} - \widehat{\operatorname{div}} \partial_{\widehat{f}} \widehat{l}_{\mathrm{s}} + \big[\partial_f \bar{l}_{\mathrm{s}} - \partial_V \bar{l}_{\mathrm{s}} \otimes [\boldsymbol{w} - \boldsymbol{v}] \big] \cdot \boldsymbol{n} \Big] = \boldsymbol{0} \tag{10.71}$$

represents an alternative for the tangential part of the local balance of material linear momentum at regular points on the boundary.

Eventually, the *normal* part[3] of the local balance of material linear momentum at regular points on the boundary reads as

[3] A step-by-step derivation is as follows:

$$\begin{aligned} \boldsymbol{N} \cdot \widehat{\operatorname{Div}} \widehat{\boldsymbol{\Sigma}}' = \boldsymbol{N} \cdot \widehat{\boldsymbol{\Sigma}}'_{,\alpha} \cdot \widehat{\boldsymbol{A}}^{\alpha} &= -\boldsymbol{N}_{,\alpha} \cdot \widehat{\boldsymbol{\Sigma}}' \cdot \widehat{\boldsymbol{A}}^{\alpha} + \big[\boldsymbol{N} \cdot \widehat{\boldsymbol{\Sigma}}' \big]_{,\alpha} \cdot \widehat{\boldsymbol{A}}^{\alpha} \\ &= -\big[\boldsymbol{N}_{,\alpha} \otimes \widehat{\boldsymbol{A}}^{\alpha} \big] : \widehat{\boldsymbol{\Sigma}}' + \widehat{\operatorname{Div}}(\boldsymbol{N} \cdot \widehat{\boldsymbol{\Sigma}}') \\ &= -\widehat{\nabla}_{\widehat{X}} \boldsymbol{N} : \widehat{\boldsymbol{\Sigma}}' = \widehat{\boldsymbol{C}} : \widehat{\boldsymbol{\Sigma}}'. \end{aligned}$$

$$\boldsymbol{N} \cdot [\boldsymbol{\Sigma}' + \boldsymbol{P}_\mathrm{m} \otimes \boldsymbol{W}] \cdot \boldsymbol{N} - [\boldsymbol{N} \cdot \widehat{\boldsymbol{S}}^{\mathrm{ext}\prime}_\mathrm{m} + \widehat{\boldsymbol{\Sigma}}' : \widehat{\boldsymbol{C}}] := \boldsymbol{N} \cdot \widehat{\boldsymbol{R}}_\mathrm{m}. \tag{10.72}$$

Here, the first part expands as

$$\boldsymbol{N} \cdot [\boldsymbol{\Sigma}' + \boldsymbol{P}_\mathrm{m} \otimes \boldsymbol{W}] \cdot \boldsymbol{N} = -\bar{l}_\mathrm{m} - \boldsymbol{N} \cdot \boldsymbol{F}^t \cdot [\widehat{\boldsymbol{s}}^{\mathrm{ext}\prime}_\mathrm{m} + \widehat{\mathrm{Div}\boldsymbol{P}'}]. \tag{10.73}$$

Thus, with the normal gradient of $\widehat{l}_\mathrm{m} = \widehat{l}_\mathrm{m}(\widehat{\boldsymbol{X}}, \widehat{\boldsymbol{y}}, \widehat{\boldsymbol{F}})$, i.e.

$$\nabla_N \widehat{l}_\mathrm{m} = \boldsymbol{N} \cdot \widehat{\boldsymbol{S}}^{\mathrm{ext}\prime}_\mathrm{m} + \boldsymbol{N} \cdot \boldsymbol{F}^t \cdot \widehat{\boldsymbol{s}}^{\mathrm{ext}\prime}_\mathrm{m} - \widehat{\boldsymbol{P}}' : \nabla_N \widehat{\boldsymbol{F}},$$

the final result reads as

$$-\bar{l}_\mathrm{m} - \nabla_N \widehat{l}_\mathrm{m} - \widehat{\boldsymbol{\Sigma}}' : \widehat{\boldsymbol{C}} - \nabla_N \boldsymbol{y} \cdot \widehat{\mathrm{Div}\boldsymbol{P}'} - \widehat{\boldsymbol{P}}' : \nabla_N \widehat{\boldsymbol{F}} := \boldsymbol{N} \cdot \widehat{\boldsymbol{R}}_\mathrm{m}. \tag{10.74}$$

$(e) \leftrightarrow (f)$ The *tangential* part of the integrand (e), representing the tangential part of the local balance of spatial linear momentum at regular points at singular surfaces, reads as

$$\widehat{\boldsymbol{i}} \cdot [\![\partial_F \bar{l}_\mathrm{m} - \partial_v \bar{l}_\mathrm{m} \otimes \boldsymbol{W}]\!] \cdot \boldsymbol{M} = \boldsymbol{0}. \tag{10.75}$$

As a vectorial statement, this is a spatial covariant vector, i.e. a covector, tangential to singular surfaces, consequently its material counterpart follows from a (negative) covariant pull-back with the tangential part of the spatial deformation gradient as

$$\widehat{\boldsymbol{F}}^t \cdot [\![-\partial_F \bar{l}_\mathrm{m} + \partial_v \bar{l}_\mathrm{m} \otimes \boldsymbol{W}]\!] \cdot \boldsymbol{M} = \boldsymbol{0}. \tag{10.76}$$

Note that $\widehat{\boldsymbol{F}} \equiv \{\widehat{\boldsymbol{F}}\}$ due to the tangential continuity of $\widehat{\boldsymbol{F}}$. Together with the identities

$$\widehat{\boldsymbol{F}}^t \cdot [\![\partial_F \bar{l}_\mathrm{m}]\!] = \widehat{\boldsymbol{I}} \cdot [\![\boldsymbol{F}^t \cdot \partial_F \bar{l}_\mathrm{m}]\!] \quad \text{and} \quad \widehat{\boldsymbol{F}}^t \cdot [\![\partial_v \bar{l}_\mathrm{m}]\!] = -\widehat{\boldsymbol{I}} \cdot [\![\partial_V \bar{l}_\mathrm{m}]\!]$$

the (negative) covariant tangential pull-back of (e) results in

$$\widehat{\boldsymbol{I}} \cdot [\![-\boldsymbol{F}^t \cdot \partial_F \bar{l}_\mathrm{m} - \partial_V \bar{l}_\mathrm{m} \otimes \boldsymbol{W}]\!] \cdot \boldsymbol{M} = \boldsymbol{0}. \tag{10.77}$$

Clearly, by adding the identity $\widehat{\boldsymbol{I}} \cdot [\![\bar{l}_\mathrm{m} \boldsymbol{I}]\!] \cdot \boldsymbol{M} = \boldsymbol{0}$, this is the tangential part of the local balance of material linear momentum at regular points at singular surfaces.

Consequently, the tangential parts of the local balances of spatial and material linear momentum at regular points at singular surfaces are simultaneously satisfied, i.e. the tangential parts of the integrals (e) *and* (f) *vanish concurrently!*

Observe that

$$\widehat{\boldsymbol{I}} \cdot \left[\!\left[\partial_f \bar{l}_s - \partial_V \bar{l}_s \otimes [\boldsymbol{w} - \boldsymbol{v}] \right]\!\right] \cdot \boldsymbol{m} = \boldsymbol{0} \tag{10.78}$$

represents an alternative for the tangential part of the local balance of material linear momentum at regular points at singular surfaces.

Eventually, the *normal* part of the local balance of material linear momentum at regular points at singular surfaces reads as

$$-\boldsymbol{M} \cdot [\![\boldsymbol{\Sigma}' + \boldsymbol{P}_{\mathrm{m}} \otimes \boldsymbol{W}]\!] \cdot \boldsymbol{M} := \boldsymbol{M} \cdot \widetilde{\boldsymbol{R}}_{\mathrm{m}} \tag{10.79}$$

and, with the balance of material linear momentum at singular surfaces

$$[\![\boldsymbol{\Sigma} + \boldsymbol{P}_{\mathrm{m}} \otimes \boldsymbol{W}]\!] \cdot \boldsymbol{M} =: \widetilde{\boldsymbol{S}}_{\mathrm{m}}$$

in Eq. 8.51 together with the definition

$$\widetilde{\boldsymbol{S}}_{\mathrm{m}} := \left[[\![\overline{w}_{\mathrm{m}}]\!] - [\![\boldsymbol{F}]\!] : \{\boldsymbol{P}\} \right] \boldsymbol{M}$$

according to Eq. 8.71 and the relation $\boldsymbol{\Sigma}' = \boldsymbol{\Sigma} + \overline{v}_{\mathrm{m}} \boldsymbol{I}$, expands as

$$-\left[[\![\overline{u}_{\mathrm{m}}]\!] - [\![\boldsymbol{F}]\!] : \{\boldsymbol{P}\} \right] := \boldsymbol{M} \cdot \widetilde{\boldsymbol{R}}_{\mathrm{m}}. \tag{10.80}$$

The pertinent pull-back/push-forward relations for fully equivalent variants of the local balances of spatial and material linear momentum are finally summarized for convenience in Table 10.1.

Table 10.1 Pull-back/push-forward relations for fully equivalent variants of the local balances of spatial and material linear momentum at regular points in the domain, on the boundary, and at singular surfaces

Regular Points in the Domain

$\mathrm{D}_t\boldsymbol{p}_{\mathrm{m}} = \mathrm{Div}\boldsymbol{P}' + \overline{\boldsymbol{s}}'_{\mathrm{m}}$	⟷ PT	$j\,\mathrm{D}_t\boldsymbol{p}_{\mathrm{m}} = \mathrm{div}\boldsymbol{\sigma}' + \overline{\boldsymbol{s}}'_{\mathrm{s}}$
PB ↕ PF		PB ↕ PF
$\mathrm{D}_t\boldsymbol{P}_{\mathrm{m}} = \mathrm{Div}\boldsymbol{\Sigma}' + \overline{\boldsymbol{S}}'_{\mathrm{m}}$	PT ⟷	$j\,\mathrm{D}_t\boldsymbol{P}_{\mathrm{s}} = \mathrm{div}\boldsymbol{p}' + \overline{\boldsymbol{S}}'_{\mathrm{s}}$

Regular Points on the Boundary

$\widehat{\boldsymbol{i}} \cdot \boldsymbol{P}'_{\bullet} \cdot \boldsymbol{N} = \widehat{\boldsymbol{i}} \cdot \left[\widehat{\mathrm{Div}\widehat{\boldsymbol{P}}'} + \widehat{\boldsymbol{s}}^{\mathrm{ext}\prime}_{\mathrm{m}}\right]$	⟷ PT	$\widehat{\boldsymbol{i}} \cdot \boldsymbol{\sigma}'_{\bullet} \cdot \boldsymbol{n} = \widehat{\boldsymbol{i}} \cdot \left[\widehat{\mathrm{div}\widehat{\boldsymbol{\sigma}}'} + \widehat{\boldsymbol{s}}^{\mathrm{ext}\prime}_{\mathrm{s}}\right]$
PB ↕ PF		PB ↕ PF
$\widehat{\boldsymbol{I}} \cdot \boldsymbol{\Sigma}'_{\bullet} \cdot \boldsymbol{N} = \widehat{\boldsymbol{I}} \cdot \left[\widehat{\mathrm{Div}\widehat{\boldsymbol{\Sigma}}'} + \widehat{\boldsymbol{S}}^{\mathrm{ext}\prime}_{\mathrm{m}}\right]$	PT ⟷	$\widehat{\boldsymbol{I}} \cdot \boldsymbol{p}'_{\bullet} \cdot \boldsymbol{n} = \widehat{\boldsymbol{I}} \cdot \left[\widehat{\mathrm{div}\widehat{\boldsymbol{p}}'} + \widehat{\boldsymbol{S}}^{\mathrm{ext}\prime}_{\mathrm{s}}\right]$

Regular Points at Singular Surfaces

$\widehat{\boldsymbol{i}} \cdot [\![\boldsymbol{P}'_{\bullet}]\!] \cdot \boldsymbol{M} = \boldsymbol{0}$	⟷ PT	$\widehat{\boldsymbol{i}} \cdot [\![\boldsymbol{\sigma}'_{\bullet}]\!] \cdot \boldsymbol{m} = \boldsymbol{0}$
PB ↕ PF		PB ↕ PF
$\widehat{\boldsymbol{I}} \cdot [\![\boldsymbol{\Sigma}'_{\bullet}]\!] \cdot \boldsymbol{M} = \boldsymbol{0}$	PT ⟷	$\widehat{\boldsymbol{I}} \cdot [\![\boldsymbol{p}'_{\bullet}]\!] \cdot \boldsymbol{m} = \boldsymbol{0}$

with *Dynamic Versions* of Spatial and Material Stresses

$\boldsymbol{P}'_{\bullet} := \boldsymbol{P}' + \boldsymbol{p}_{\mathrm{m}} \otimes \boldsymbol{W}$	⟷ PT	$\boldsymbol{\sigma}'_{\bullet} := \boldsymbol{\sigma}' + \boldsymbol{p}_{\mathrm{s}} \otimes [\boldsymbol{w} - \boldsymbol{v}]$
PB ↕ PF		PB ↕ PF
$\boldsymbol{\Sigma}'_{\bullet} := \boldsymbol{\Sigma}' + \boldsymbol{P}_{\mathrm{m}} \otimes \boldsymbol{W}$	PT ⟷	$\boldsymbol{p}'_{\bullet} := \boldsymbol{p}' + \boldsymbol{P}_{\mathrm{s}} \otimes [\boldsymbol{w} - \boldsymbol{v}]$

10.2 Extended Dirichlet Principle

Common Dirichlet Principle

The common *Dirichlet principle* or rather the *principle of minimum potential energy* states that the *equilibrium* of a system in terms of the common set of generalized coordinates

$$\boldsymbol{v}(\Xi)$$

in the common infinite dimensional configuration space

$$\mathbb{Y} := \{\boldsymbol{v}(\Xi)\},$$

renders the common *total potential energy functional*

$$\mathcal{U}\{\boldsymbol{v}(\Xi)\},$$

i.e. a functional over the common configuration space, a *stationary value*, thus

$$\delta\mathcal{U} = 0$$

upon variations of the spatial placement map $\boldsymbol{v}(\Xi)$.

Rationale for the Extended Dirichlet Principle

As an *extension*, the here defined *extended* Dirichlet principle states that the equilibrium of a system in terms of the extended set of generalized coordinates

$$\boldsymbol{v}(\Xi) \quad and \quad \boldsymbol{\Upsilon}(\Xi)$$

in the extended infinite dimensional configuration space

$$\mathbb{Y} := \{\boldsymbol{v}(\Xi), \boldsymbol{\Upsilon}(\Xi)\},$$

renders the extended total potential energy functional

$$\mathcal{U}\{\boldsymbol{v}(\Xi), \boldsymbol{\Upsilon}(\Xi)\},$$

i.e. a functional over the extended configuration space, a *non-positive value*

$$\delta\mathcal{U} \leq 0$$

upon configurational changes associated with variations of the material placement map $\boldsymbol{\Upsilon}(\Xi)$.

10.2.1 Variational Statement

Extended Dirichlet Principle (Reference Configuration):
The *total* variation of the *total potential energy functional*

$$\mathcal{U}\{\boldsymbol{v}(\boldsymbol{\Xi}), \boldsymbol{\Upsilon}(\boldsymbol{\Xi})\} = \int_{\mathcal{B}_r} \overline{u}_r \, d\mathcal{V} + \int_{\partial\mathcal{B}_r} \widehat{u}_r \, d\mathcal{A} \tag{10.81}$$

with total potential energy density per unit volume in the reference configuration $\mathcal{B}_r$

$$\overline{u}_r = \overline{u}_r(\boldsymbol{v}, \nabla_{\Xi}\boldsymbol{v}, \boldsymbol{\Upsilon}, \nabla_{\Xi}\boldsymbol{\Upsilon}; \boldsymbol{\Xi}) \tag{10.82}$$

and total potential energy density per unit area on the boundary $\partial\mathcal{B}_r$ in the reference configuration

$$\widehat{u}_r = \widehat{u}_r(\widehat{\boldsymbol{v}}, \widehat{\nabla}_{\widehat{\Xi}}\widehat{\boldsymbol{v}}, \widehat{\boldsymbol{\Upsilon}}, \widehat{\nabla}_{\widehat{\Xi}}\widehat{\boldsymbol{\Upsilon}}; \widehat{\boldsymbol{\Xi}}) \tag{10.83}$$

defines the *virtual energy release*

$$\delta\mathcal{U} =: \mathcal{R}_\delta \overset{!}{\leq} 0. \tag{10.84}$$

Thereby the virtual energy release is stated as

$$\mathcal{R}_\delta := \int_{\mathcal{B}_r} \delta\boldsymbol{\Upsilon} \cdot \overline{\boldsymbol{R}}_r \, d\mathcal{V} + \int_{\partial\mathcal{B}_r} \delta\widehat{\boldsymbol{\Upsilon}} \cdot \widehat{\boldsymbol{R}}_r \, d\mathcal{A} + \int_{\mathcal{S}_r} \delta\widetilde{\boldsymbol{\Upsilon}} \cdot \widetilde{\boldsymbol{R}}_r \, d\mathcal{A} \overset{!}{\leq} 0, \tag{10.85}$$

with $\overline{\boldsymbol{R}}_r$ a *material* body force, and $\widehat{\boldsymbol{R}}_r$ and $\widetilde{\boldsymbol{R}}_r$ *material* boundary and surface tractions. These perform virtual power along *material virtual displacements* in the domain, on the boundary and at singular surfaces, respectively.

Only *non-positive* values of $\delta\mathcal{U} \leq 0$ are admissible for material virtual displacements to occur spontaneously with a virtual *release* of energy, i.e. with positive virtual dissipation of energy from the system.

10.2.2 Euler–Lagrange Equations: Summary

Upon total variation of the extended total potential energy functional $\mathcal{U}$, the corresponding local Euler–Lagrange equations are as follows:

Euler–Lagrange Equations of Extended Dirichlet Principle (Referential Control Volume):

i. Regular Points in the Domain

$$\frac{\partial \overline{u}_r}{\partial \boldsymbol{v}} - \text{DIV}\frac{\partial \overline{u}_r}{\partial \nabla_{\Xi}\boldsymbol{v}} = \mathbf{0} \quad \text{in} \quad \mathcal{B}_r \tag{10.86}$$
$$\frac{\partial \overline{u}_r}{\partial \boldsymbol{\Upsilon}} - \text{DIV}\frac{\partial \overline{u}_r}{\partial \nabla_{\Xi}\boldsymbol{\Upsilon}} =: \overline{\boldsymbol{R}}_r$$

ii. Regular Points on the Boundary

$$\frac{\partial \widehat{u}_r}{\partial \widehat{\boldsymbol{v}}} - \widehat{\text{DIV}}\frac{\partial \widehat{u}_r}{\partial \widehat{\nabla}_{\Xi}\widehat{\boldsymbol{v}}} + \left[\frac{\partial \overline{u}_r}{\partial \nabla_{\Xi}\boldsymbol{v}}\right]\cdot \mathcal{N} = \mathbf{0} \quad \text{on} \quad \partial\mathcal{B}_r \tag{10.87}$$
$$\frac{\partial \widehat{u}_r}{\partial \widehat{\boldsymbol{\Upsilon}}} - \widehat{\text{DIV}}\frac{\partial \widehat{u}_r}{\partial \widehat{\nabla}_{\Xi}\widehat{\boldsymbol{\Upsilon}}} + \left[\frac{\partial \overline{u}_r}{\partial \nabla_{\Xi}\boldsymbol{\Upsilon}}\right]\cdot \mathcal{N} =: \widehat{\boldsymbol{R}}_r$$

iii. Regular Points at Singular Surfaces

$$\left[\!\left[\frac{\partial \overline{u}_r}{\partial \nabla_{\Xi}\boldsymbol{v}}\right]\!\right]\cdot \mathcal{M} = \mathbf{0} \quad \text{at} \quad \mathcal{S}_r \tag{10.88}$$
$$\left[\!\left[\frac{\partial \overline{u}_r}{\partial \nabla_{\Xi}\boldsymbol{\Upsilon}}\right]\!\right]\cdot \mathcal{M} =: \widetilde{\boldsymbol{R}}_r$$

10.2.3 Euler–Lagrange Equations: Derivation

For the given parametrisation, the total variations of the various contributions to $\mathcal{U}$ are considered individually in the sequel:

Total Potential Energy in the Domain

The *total* potential energy density $\overline{u}_r$ per unit volume in the reference configuration $\mathcal{B}_r$ consists of the *internal* and the *external potential energy densities* w_r and $\overline{v}_r$, respectively

$$\overline{u}_\mathrm{r} := w_\mathrm{r} + \overline{v}_\mathrm{r}. \tag{10.89}$$

1. The internal potential (stored) energy density w_r per unit volume in the reference configuration $\mathcal{B}_\mathrm{r}$ is parameterized as $w_\mathrm{r} := \mathcal{J}^\mathrm{m}\, w_\mathrm{m}(\boldsymbol{F}; \boldsymbol{X})$. Its total variation computes as in Eq. 10.14, whereby the variational definition of the common material→spatial (exterior) Piola stress $\boldsymbol{P}'$ now reads as

$$\boldsymbol{P}' := \partial_{\boldsymbol{F}} w_\mathrm{m} = \partial_{\boldsymbol{F}} \overline{u}_\mathrm{m}. \tag{10.90}$$

2. The external potential energy density $\overline{v}_\mathrm{r}$ per unit volume in the reference configuration $\mathcal{B}_\mathrm{r}$ is parameterized as $\overline{v}_\mathrm{r} := \mathcal{J}^\mathrm{m}\, \overline{v}_\mathrm{m}(\boldsymbol{v}; \boldsymbol{X})$. Its total variation computes as in Eq. 10.17, whereby the variational definition for the (exterior) spatial linear momentum source density now reads as

$$\overline{\boldsymbol{s}}'_\mathrm{m} := -\partial_{\boldsymbol{x}} \overline{v}_\mathrm{m} = -\partial_{\boldsymbol{x}} \overline{u}_\mathrm{m}. \tag{10.91}$$

Taken together, the total variation of the total potential energy density $\overline{u}_\mathrm{r}$ per unit volume in the reference configuration $\mathcal{B}_\mathrm{r}$ results in

$$\delta \overline{u}_\mathrm{r} = \mathcal{J}^\mathrm{m}\big[-\, \overline{\boldsymbol{s}}'_\mathrm{m} \cdot \delta \boldsymbol{v} + \boldsymbol{P}' : \nabla_{\boldsymbol{X}} \delta \boldsymbol{v} - \overline{\boldsymbol{S}}'_\mathrm{m} \cdot \delta \boldsymbol{\Upsilon} + \boldsymbol{\Sigma}' : \nabla_{\boldsymbol{X}} \delta \boldsymbol{\Upsilon} \big]. \tag{10.92}$$

Here the variational definitions for the exterior Eshelby stress $\boldsymbol{\Sigma}'$ and the exterior material linear momentum source density $\overline{\boldsymbol{S}}'_\mathrm{m}$ now read as

$$\boldsymbol{\Sigma}' := \overline{u}_\mathrm{m} \boldsymbol{I} - \boldsymbol{F}^t \cdot \boldsymbol{P}' \quad \text{and} \quad \overline{\boldsymbol{S}}'_\mathrm{m} := -\partial_{\boldsymbol{X}} \overline{u}_\mathrm{m}. \tag{10.93}$$

Note that all variational quantities in the above are expressed in terms of the total potential energy density $\overline{u}_\mathrm{m}$. □

Total Potential Energy on the Boundary

The *total* boundary potential energy density $\widehat{u}_\mathrm{r}$ per unit area on the boundary $\partial\mathcal{B}_\mathrm{r}$ in the reference configuration consists of the *internal* and the *external boundary potential energy densities* $\widehat{w}_\mathrm{r}$ and $\widehat{v}_\mathrm{r}$, respectively

$$\widehat{u}_\mathrm{r} = \widehat{w}_\mathrm{r} + \widehat{v}_\mathrm{r}. \tag{10.94}$$

1. The internal boundary potential (stored) energy density $\widehat{w}_\mathrm{r}$ per unit area on the boundary $\partial\mathcal{B}_\mathrm{r}$ is parameterized as $\widehat{w}_\mathrm{r} := \widehat{\mathcal{J}}^\mathrm{m}\, \widehat{w}_\mathrm{m}(\widehat{\boldsymbol{F}}; \widehat{\boldsymbol{X}})$. Its total variation computes as in Eq. 10.25, whereby the variational definition of the (exterior) boundary Piola stress $\widehat{\boldsymbol{P}}'$ now reads as

$$\widehat{\boldsymbol{P}}' := \partial_{\widehat{\boldsymbol{F}}} \widehat{w}_\mathrm{m} = \partial_{\widehat{\boldsymbol{F}}} \widehat{u}_\mathrm{m}. \tag{10.95}$$

2. The external boundary potential energy density $\widehat{v}_\mathrm{r}$ per unit area on the boundary $\partial\mathcal{B}_\mathrm{r}$ in the reference configuration is parameterized as $\widehat{v}_\mathrm{r} := \widehat{\mathcal{J}}^\mathrm{m}\, \widehat{v}_\mathrm{m}(\widehat{\boldsymbol{v}}; \widehat{\boldsymbol{X}})$. Its

total variation computes as in Eq. 10.28, whereby the variational definition for the (exterior) spatial extrinsic linear momentum flux density now reads as

$$\widehat{\boldsymbol{s}}_{\mathrm{m}}^{\mathrm{ext}\prime} := -\partial_{\widehat{\boldsymbol{x}}}\widehat{v}_{\mathrm{m}} = -\partial_{\widehat{\boldsymbol{x}}}\widehat{u}_{\mathrm{m}}. \tag{10.96}$$

Taken together, the total variation of the total boundary potential energy density $\widehat{u}_{\mathrm{r}}$ per unit area on the boundary $\partial\mathcal{B}_{\mathrm{r}}$ results in

$$\delta\widehat{u}_{\mathrm{r}} = \widehat{\mathcal{J}}^{\mathrm{m}}\,[-\widehat{\boldsymbol{s}}_{\mathrm{m}}^{\mathrm{ext}\prime}\cdot\delta\widehat{\boldsymbol{v}} + \widehat{\boldsymbol{P}}' : \widehat{\nabla}_{\widehat{X}}\delta\widehat{\boldsymbol{v}} - \widehat{\boldsymbol{S}}_{\mathrm{m}}^{\mathrm{ext}\prime}\cdot\delta\widehat{\boldsymbol{\Upsilon}} + \widehat{\boldsymbol{\Sigma}}' : \widehat{\nabla}_{\widehat{X}}\delta\widehat{\boldsymbol{\Upsilon}}]. \tag{10.97}$$

Here the variational definitions for the exterior boundary Eshelby stress $\widehat{\boldsymbol{\Sigma}}'$ and the exterior material extrinsic linear momentum flux density $\widehat{\boldsymbol{S}}_{\mathrm{m}}^{\mathrm{ext}\prime}$ now read as

$$\widehat{\boldsymbol{\Sigma}}' := \widehat{u}_{\mathrm{m}}\widehat{\boldsymbol{I}} - \widehat{\boldsymbol{F}}^{t}\cdot\widehat{\boldsymbol{P}}' \quad\text{and}\quad \widehat{\boldsymbol{S}}_{\mathrm{m}}^{\mathrm{ext}\prime} := -\partial_{\widehat{X}}\widehat{u}_{\mathrm{m}}. \tag{10.98}$$

Note that all variational quantities in the above are expressed in terms of the total boundary potential energy density $\widehat{u}_{\mathrm{m}}$. □

Summary of Total Variation

The total variation of the total potential energy functional $\mathcal{U}$ expanded over the material configuration reads as in Eq. 10.99.

$$\begin{aligned}\delta\mathcal{U} =& \int_{\mathcal{B}_{\mathrm{m}}}[\boldsymbol{P}' : \nabla_X\delta\boldsymbol{v} - \overline{\boldsymbol{s}}_{\mathrm{m}}'\cdot\delta\boldsymbol{v}]\mathrm{d}V + \int_{\partial\mathcal{B}_{\mathrm{m}}}\Big[\widehat{\boldsymbol{P}}' : \widehat{\nabla}_{\widehat{X}}\delta\widehat{\boldsymbol{v}} - \widehat{\boldsymbol{s}}_{\mathrm{m}}^{\mathrm{ext}\prime}\cdot\delta\widehat{\boldsymbol{v}}\Big]\mathrm{d}A \\ &+ \int_{\mathcal{B}_{\mathrm{m}}}[\boldsymbol{\Sigma}' : \nabla_X\delta\boldsymbol{\Upsilon} - \overline{\boldsymbol{S}}_{\mathrm{m}}'\cdot\delta\boldsymbol{\Upsilon}]\mathrm{d}V + \int_{\partial\mathcal{B}_{\mathrm{m}}}\Big[\widehat{\boldsymbol{\Sigma}}' : \widehat{\nabla}_{\widehat{X}}\delta\widehat{\boldsymbol{\Upsilon}} - \widehat{\boldsymbol{S}}_{\mathrm{m}}^{\mathrm{ext}\prime}\cdot\delta\widehat{\boldsymbol{\Upsilon}}\Big]\mathrm{d}A.\end{aligned} \tag{10.99}$$

Toward Euler–Lagrange Equations

Integration by parts and application of the extended Gauss theorem renders eventually the total variation of the total potential energy functional $\mathcal{U}$ expanded over the material configuration as in Eq. 10.100.

$$\delta\mathcal{U} = \underbrace{-\int_{\mathcal{B}_\mathrm{m}} [\overline{\boldsymbol{s}}'_\mathrm{m} + \mathrm{Div}\boldsymbol{P}'] \cdot \delta\boldsymbol{v}\,\mathrm{d}V}_{(a)} \underbrace{-\int_{\mathcal{B}_\mathrm{m}} [\overline{\boldsymbol{S}}'_\mathrm{m} + \mathrm{Div}\boldsymbol{\Sigma}'] \cdot \delta\boldsymbol{\Upsilon}\,\mathrm{d}V}_{(b)} \qquad (10.100)$$

$$\underbrace{-\int_{\partial\mathcal{B}_\mathrm{m}} \left[\widehat{\boldsymbol{s}}^{\mathrm{ext}\prime}_\mathrm{m} + \widehat{\mathrm{Div}}\widehat{\boldsymbol{P}}' - \boldsymbol{P}' \cdot \boldsymbol{N}\right] \cdot \delta\widehat{\boldsymbol{v}}\,\mathrm{d}A}_{(c)}$$

$$\underbrace{-\int_{\partial\mathcal{B}_\mathrm{m}} \left[\widehat{\boldsymbol{S}}^{\mathrm{ext}\prime}_\mathrm{m} + \widehat{\mathrm{Div}}\widehat{\boldsymbol{\Sigma}}' - \boldsymbol{\Sigma}' \cdot \boldsymbol{N}\right] \cdot \delta\widehat{\boldsymbol{\Upsilon}}\,\mathrm{d}A}_{(d)}$$

$$\underbrace{-\int_{\mathcal{S}_\mathrm{m}} \left[[\![\boldsymbol{P}']\!] \cdot \boldsymbol{M}\right] \cdot \delta\widetilde{\boldsymbol{v}}\,\mathrm{d}A}_{(e)} \underbrace{-\int_{\mathcal{S}_\mathrm{m}} \left[[\![\boldsymbol{\Sigma}']\!] \cdot \boldsymbol{M}\right] \cdot \delta\widetilde{\boldsymbol{\Upsilon}}\,\mathrm{d}A}_{(f)}.$$

The bracketed terms preceding the entirely arbitrary spatial and material virtual displacements in the integrands of the individual integrals (a) to (f) constitute the Euler–Lagrange equations.

10.2.4 Euler–Lagrange Equations: Analysis

The integrands of the integrals (a) to (f) in Eq. 10.100 are further analyzed in the following:

(a) Following the analysis as before, the bracketed term in the integrand of (a) vanishes due to the quasi-static version of the local balance of spatial linear momentum at regular points in the domain in Eq. 7.29.

(b) Following the analysis as before, the bracketed term in the integrand of (b) vanishes due to the quasi-static version of the local balance of material linear momentum at regular points in the domain in Eq. 8.49.

(c) Following the analysis as before, the bracketed term in the integrand of (c) vanishes due to the quasi-static version of the local balance of spatial linear momentum at regular points on the boundary.

(d) Following the analysis as before, due to the quasi-static version of the local balance of material linear momentum at regular points on the boundary in Eq. 8.50, the bracketed term in the integrand of (d) is purely normal

$$[-\overline{u}_\mathrm{m} - \nabla_N \widehat{u}_\mathrm{m} + \widehat{\boldsymbol{\Sigma}}' : \widehat{\boldsymbol{C}} + \nabla_N \boldsymbol{y} \cdot \widehat{\mathrm{Div}}\widehat{\boldsymbol{P}}' + \widehat{\boldsymbol{P}}' : \nabla_N \widehat{\boldsymbol{F}}]\boldsymbol{N}. \qquad (10.101)$$

(*e*) Following the analysis as before, the bracketed term in the integrand of (*e*) vanishes due to the quasi-static version of the local balance of spatial linear momentum at regular points at singular surfaces in Eq. 7.31.

(*f*) Following the analysis as before, due to the quasi-static version of the local balance of material linear momentum at regular points at singular surfaces in Eq. 8.51, the bracketed term in the integrand of (*f*) is purely normal

$$\big[[\![\overline{u}_{\mathrm{m}}]\!] - [\![\boldsymbol{F}]\!] : \{\boldsymbol{P}\}\big]\boldsymbol{M}. \tag{10.102}$$

In summary, the total variation of the total potential energy functional reduces eventually to

$$\begin{aligned}\delta\mathcal{U} = &\int_{\partial\mathcal{B}_{\mathrm{m}}} [\overline{u}_{\mathrm{m}} + \nabla_N \widehat{u}_{\mathrm{m}} - \widehat{\boldsymbol{\Sigma}}' : \widehat{\boldsymbol{C}} - \nabla_N \boldsymbol{y} \cdot \widehat{\mathrm{Div}\boldsymbol{P}'} - \widehat{\boldsymbol{P}}' : \nabla_N \widehat{\boldsymbol{F}}]\,\delta\widehat{\Upsilon}_\perp \,\mathrm{d}A \\ &- \int_{\mathcal{S}_{\mathrm{m}}} \big[[\![\overline{u}_{\mathrm{m}}]\!] - [\![\boldsymbol{F}]\!] : \{\boldsymbol{P}\}\big]\,\delta\widetilde{\Upsilon}_\perp \,\mathrm{d}A\end{aligned} \tag{10.103}$$

and compares to the virtual energy release

$$\delta\mathcal{U} =: \int_{\mathcal{B}_{\mathrm{m}}} \delta\boldsymbol{\Upsilon} \cdot \overline{\boldsymbol{R}}_{\mathrm{m}} \,\mathrm{d}V + \int_{\partial\mathcal{B}_{\mathrm{m}}} \delta\widehat{\boldsymbol{\Upsilon}} \cdot \widehat{\boldsymbol{R}}_{\mathrm{m}} \,\mathrm{d}A + \int_{\mathcal{S}_{\mathrm{m}}} \delta\widetilde{\boldsymbol{\Upsilon}} \cdot \widetilde{\boldsymbol{R}}_{\mathrm{m}} \,\mathrm{d}A \overset{!}{\leq} 0. \tag{10.104}$$

By comparison of contributions, it is thus concluded that the material body force vanishes identically

$$\overline{\boldsymbol{R}}_{\mathrm{m}} \equiv \boldsymbol{0}, \tag{10.105}$$

whereas the material tractions driving material virtual displacements on the boundary and at singular surfaces read as

$$\begin{aligned}\widehat{\boldsymbol{R}}_{\mathrm{m}} &\equiv \quad [\overline{u}_{\mathrm{m}} + \nabla_N \widehat{u}_{\mathrm{m}} - \widehat{\boldsymbol{\Sigma}}' : \widehat{\boldsymbol{C}} - \nabla_N \boldsymbol{y} \cdot \widehat{\mathrm{Div}\boldsymbol{P}'} - \widehat{\boldsymbol{P}}' : \nabla_N \widehat{\boldsymbol{F}}]\,\boldsymbol{N}, \\ \widetilde{\boldsymbol{R}}_{\mathrm{m}} &\equiv -\big[[\![\overline{u}_{\mathrm{m}}]\!] - [\![\boldsymbol{F}]\!] : \{\boldsymbol{P}\}\big]\,\boldsymbol{M}.\end{aligned} \tag{10.106}$$

Observe again that the material tractions $\widehat{\boldsymbol{R}}_{\mathrm{m}}$ and $\widetilde{\boldsymbol{R}}_{\mathrm{m}}$ are purely normal.

Example: Zero Boundary Potential

For *zero* boundary potential with $\widehat{u}_{\mathrm{m}} = 0$, the exterior boundary Eshelby stress $\widehat{\boldsymbol{\Sigma}}' = \boldsymbol{0}$ and the exterior boundary Piola stress $\widehat{\boldsymbol{P}}' \equiv \boldsymbol{0}$ vanish. In this situation, the material boundary traction and the corresponding virtual energy release, which here merely occurs from the total potential energy in the domain, read as

$$\widehat{\boldsymbol{R}}_{\mathrm{m}} \equiv \overline{u}_{\mathrm{m}}\,\boldsymbol{N} \quad \text{and} \quad \mathcal{R}_\delta = \int_{\partial\mathcal{B}_{\mathrm{m}}} [\overline{u}_{\mathrm{m}}]\,\delta\widehat{\Upsilon}_\perp \,\mathrm{d}A \overset{!}{\leq} 0.$$

Taken together, with *positive* total potential energy density in the domain, the bracketed term in the integrand is entirely *positive*. □

Example: Material Boundary Tension
For material boundary *tension* with $\widehat{u}_{\rm m} = \widehat{\sigma}_{\rm m} = $ constant the exterior boundary Eshelby stress is purely spherical with $\widehat{\boldsymbol{\Sigma}}' = \widehat{\sigma}_{\rm m}\boldsymbol{I}$, while the exterior boundary Piola stress vanishes $\widehat{\boldsymbol{P}}' \equiv \boldsymbol{0}$. In this situation, the material boundary traction and the corresponding virtual energy release, which here occurs from both the total potential energy in the domain and on the boundary, read as

$$\widehat{\boldsymbol{R}}_{\rm m} \equiv [\overline{u}_{\rm m} - \widehat{\sigma}_{\rm m}\,\widehat{C}]\,\boldsymbol{N} \quad \text{and} \quad \mathcal{R}_\delta = \int\limits_{\partial\mathcal{B}_{\rm m}} [\overline{u}_{\rm m} - \widehat{\sigma}_{\rm m}\,\widehat{C}]\,\delta\widehat{\Upsilon}_\perp\,{\rm d}A \overset{!}{\leq} 0.$$

Material boundary tension applies, e.g. to grain boundaries in crystalline materials.

Analyzing the various terms, it appears that virtual energy release due to the total potential energy in the bulk is competing with the effects of material boundary tension and curvature. Thus for a sufficiently large positive boundary curvature $\widehat{C} > 0$ so that $\overline{u}_{\rm m} - \widehat{\sigma}_{\rm m}\,\widehat{C} < 0$ an admissible material variation has to point toward the outward normal direction with $\delta\widehat{\Upsilon}_\perp \geq 0$. A positive boundary curvature corresponds to a locally concave body as, for example, in the case of a cracked body with a rounded crack tip. Then, due to the contribution $-\widehat{\sigma}_{\rm m}\,\widehat{C}$, material boundary tension may act as an obstacle to further crack extension.

Taken together, with *positive* total potential energy density in the domain together with *positive* material boundary tension and *negative* boundary curvature $\widehat{C}$, the bracketed term in the integrand is entirely *positive*. □

Example: Spatial Boundary Tension
For spatial boundary *tension* with $\widehat{u}_{\rm s} = \widehat{\sigma}_{\rm s} = $ constant, the exterior boundary Eshelby stress vanishes $\widehat{\boldsymbol{\Sigma}}' \equiv \boldsymbol{0}$, while the exterior boundary Piola stress reads as $\widehat{\boldsymbol{P}}' \equiv \widehat{\sigma}_{\rm s}\,\widehat{\operatorname{cof}\boldsymbol{F}}$, thus here $\nabla_N \widehat{J}\,\widehat{\sigma}_{\rm s} \equiv \widehat{\boldsymbol{P}}' : \nabla_N \widehat{\boldsymbol{F}}$ holds. In this situation, since furthermore $\widehat{\operatorname{Div}(\operatorname{cof}\boldsymbol{F})} = \widehat{J}\,\widehat{c}\,\boldsymbol{n}$ and $\nabla_N \boldsymbol{y}\cdot\boldsymbol{n} = \widehat{j}\,J$ hold, the material boundary traction and the corresponding virtual energy release, which here occurs from both the total potential energy in the domain and on the boundary, read

$$\widehat{\boldsymbol{R}}_{\rm m} \equiv [\overline{u}_{\rm m} - J\,\widehat{\sigma}_{\rm s}\,\widehat{c}]\,\boldsymbol{N} \quad \text{and} \quad \mathcal{R}_\delta = \int\limits_{\partial\mathcal{B}_{\rm m}} [\overline{u}_{\rm m} - J\,\widehat{\sigma}_{\rm s}\,\widehat{c}]\,\delta\widehat{\Upsilon}_\perp\,{\rm d}A \overset{!}{\leq} 0.$$

Spatial boundary tension applies, e.g. to uncontaminated interfaces between immiscible fluids where it complements the celebrated Laplace–Young relation $\boldsymbol{\sigma}\cdot\boldsymbol{n} = \widehat{\sigma}_{\rm s}\,\widehat{c}\,\boldsymbol{n}$.

Taken together, with *positive* total potential energy density in the domain together with *positive* spatial boundary tension and *negative* boundary curvature $\widehat{C}$, the bracketed term in the integrand is entirely *positive*. □

Example: Spatial Boundary Traction
For spatial boundary *traction* with $\widehat{u}_{\rm m} = \widehat{v}_{\rm m}(\widehat{\boldsymbol{v}}; \widehat{\boldsymbol{X}})$, the exterior boundary Eshelby stress is purely spherical with $\widehat{\boldsymbol{\Sigma}}' = \widehat{v}_{\rm m}\boldsymbol{I}$, while the exterior boundary Piola stress vanishes $\widehat{\boldsymbol{P}}' \equiv \boldsymbol{0}$. In this situation, the material boundary traction reads as

$$\widehat{\boldsymbol{R}}_{\rm m} \equiv [\overline{u}_{\rm m} + \nabla_N \widehat{v}_{\rm m} - \widehat{v}_{\rm m}\,\widehat{C}]\,\boldsymbol{N}$$

with correspondingly virtual energy release

$$\mathcal{R}_\delta = \int\limits_{\partial\mathcal{B}_{\rm m}} [\overline{u}_{\rm m} + \nabla_N \widehat{v}_{\rm m} - \widehat{v}_{\rm m}\,\widehat{C}]\,\delta\widehat{\Upsilon}_\perp\,{\rm d}A \overset{!}{\leq} 0. \tag{10.107}$$

Taken together, with *positive* total potential energy density in the domain and for *positive* normal gradient of the external boundary potential energy density together with *positive* external boundary potential energy density and *negative* boundary curvature $\widehat{C}$, the bracketed term in the integrand is entirely *positive*. □

Conclusion
For the above cases with *positive* bracketed term in the integrand of the virtual energy release, the normal part of the virtual material displacement $\delta\widehat{\Upsilon}_\perp$ is necessarily *negative*. Thus, in order to virtually release energy, the variation of the material placement map $\delta\widehat{\boldsymbol{\Upsilon}}$ and the outwards pointing normal $\boldsymbol{N}$ have to form an obtuse angle.

Clearly, in this situation, the volume of the material configuration, i.e. the measure vol($\mathcal{B}_{\rm m}$) decreases, thus $\overline{u}_{\rm m} \geq 0$ as well as $\nabla_N \widehat{v}_{\rm m} \geq 0$ are released. Likewise, in this situation, the area of the boundary to the material configuration, i.e. the measure area($\partial\mathcal{B}_{\rm m}$) decreases for *negative* boundary curvature $\widehat{C}$, i.e. a locally convex body, (which explains the negative sign in front of the curvature term), thus $\widehat{v}_{\rm m}$ is also released. □

10.2.5 Euler–Lagrange Equations: PB/PF Operations

Alternatively, the integrands of the integrals (a) to (f) in Eq. 10.100 may be analyzed by considering appropriate pull-back/push-forward operations:

$(a) \leftrightarrow (b)$ The quasi-static version of the local balance of spatial linear momentum at regular points in the domain is a spatial covariant vector. Consequently, its material counterpart follows from a (negative) covariant pull-back with the spatial deformation gradient as

$$\boldsymbol{F}^t \cdot [-\text{Div}\partial_F \overline{u}_\text{m} + \partial_x \overline{u}_\text{m}] = \boldsymbol{0} \tag{10.108}$$

and results in the quasi-static version of the local balance of material linear momentum at regular points in the domain

$$\text{Div}(\overline{u}_\text{m}\boldsymbol{I} - \boldsymbol{F}^t \cdot \partial_F \overline{u}_\text{m}) - \partial_X \overline{u}_\text{m} = \boldsymbol{0}. \tag{10.109}$$

Consequently, the quasi-static versions of the local balances of spatial and material linear momentum at regular points in the domain are simultaneously satisfied, i.e. the integrals (*a*) *and* (*b*) *vanish concurrently!*

Observe that an alternative for the quasi-static version of the local balance of material linear momentum at regular points in the domain is represented by

$$\text{div}\partial_f \overline{u}_\text{s} - \partial_X \overline{u}_\text{s} = \boldsymbol{0}. \tag{10.110}$$

(*c*) ↔ (*d*) The tangential part of the quasi-static version of the local balance of spatial linear momentum at regular points on the boundary is a spatial covariant vector, tangential to the boundary. Consequently, its material counterpart follows from a (negative) covariant pull-back with the spatial boundary deformation gradient as

$$\widehat{\boldsymbol{F}}^t \cdot \left[-\widehat{\text{Div}}\partial_{\widehat{F}} \widehat{u}_\text{m} + \partial_{\widehat{x}} \widehat{u}_\text{m} + \partial_F \overline{u}_\text{m} \cdot \boldsymbol{N} \right] = \boldsymbol{0} \tag{10.111}$$

and results in the tangential part of the quasi-static version of the local balance of material linear momentum at regular points on the boundary

$$\widehat{\boldsymbol{I}} \cdot \left[\widehat{\text{Div}}(\widehat{u}_\text{m}\widehat{\boldsymbol{I}} - \widehat{\boldsymbol{F}}^t \cdot \partial_{\widehat{F}} \widehat{u}_\text{m}) - \partial_{\widehat{X}} \widehat{u}_\text{m} - [\overline{u}_\text{m}\boldsymbol{I} - \boldsymbol{F}^t \cdot \partial_F \overline{u}_\text{m}] \cdot \boldsymbol{N} \right] = \boldsymbol{0}. \tag{10.112}$$

Consequently, the tangential parts of the quasi-static version of the local balances of spatial and material linear momentum at regular points on the boundary are simultaneously satisfied, i.e. the tangential parts of the integrals (*c*) *and* (*d*) *vanish concurrently!*

Observe that an alternative for the tangential part of the quasi-static version of the local balance of material linear momentum at regular points on the boundary is represented by

$$\widehat{\boldsymbol{I}} \cdot \left[\widehat{\text{div}}\partial_{\widehat{f}} \widehat{u}_\text{s} - \partial_{\widehat{X}} \widehat{u}_\text{s} - \partial_f \overline{u}_\text{s} \cdot \boldsymbol{n} \right] = \boldsymbol{0}. \tag{10.113}$$

Eventually, the *normal* part of the quasi-static version of the local balance of material linear momentum at regular points on the boundary reads as

$$\overline{u}_{\mathrm{m}} + \nabla_N \widehat{u}_{\mathrm{m}} - \widehat{\boldsymbol{\Sigma}}' : \widehat{\boldsymbol{C}} - \nabla_N \boldsymbol{y} \cdot \widehat{\mathrm{Div}\boldsymbol{P}'} - \widehat{\boldsymbol{P}}' : \nabla_N \widehat{\boldsymbol{F}} := \boldsymbol{N} \cdot \widehat{\boldsymbol{R}}_{\mathrm{m}}. \quad (10.114)$$

$(e) \leftrightarrow (f)$ The tangential part of the quasi-static version of the local balance of spatial linear momentum at regular points at singular surfaces is a spatial covariant vector, tangential to singular surfaces. Consequently, its material counterpart follows from a (negative) covariant pull-back with the tangential part of the spatial deformation gradient as

$$\widehat{\boldsymbol{F}}^t \cdot [\![-\partial_F \overline{u}_{\mathrm{m}}]\!] \cdot \boldsymbol{M} = \boldsymbol{0} \quad (10.115)$$

and results in the tangential part of the quasi-static version of the local balance of material linear momentum at regular points at singular surfaces

$$\widehat{\boldsymbol{I}} \cdot [\![\overline{u}_{\mathrm{m}} \boldsymbol{I} - \boldsymbol{F}^t \cdot \partial_F \overline{u}_{\mathrm{m}}]\!] \cdot \boldsymbol{M} = \boldsymbol{0}. \quad (10.116)$$

Consequently, the tangential parts of the quasi-static version of the local balances of spatial and material linear momentum at regular points at singular surfaces are simultaneously satisfied, i.e. the tangential parts of the integrals (e) and (f) vanish concurrently!

Observe that an alternative for the tangential part of the quasi-static version of the local balance of material linear momentum at regular points at singular surfaces is represented by

$$\widehat{\boldsymbol{I}} \cdot \left[\!\left[\partial_f \overline{u}_{\mathrm{s}} \right]\!\right] \cdot \boldsymbol{m} = \boldsymbol{0}. \quad (10.117)$$

Eventually, the *normal* part of the quasi-static version of the local balance of material linear momentum at regular points at singular surfaces reads as

$$-\left[[\![\overline{u}_{\mathrm{m}}]\!] - [\![\boldsymbol{F}]\!] : \{\boldsymbol{P}\} \right] := \boldsymbol{M} \cdot \widetilde{\boldsymbol{R}}_{\mathrm{m}}. \quad (10.118)$$

The pertinent pull-back/push-forward relations for fully equivalent variants of the quasi-static versions of the local balances of spatial and material linear momentum are finally summarized for convenience in Table 10.2.

Table 10.2 Pull-back/push-forward relations for fully equivalent variants of the local balances of spatial and material linear momentum at regular points in the domain, on the boundary, and at singular surfaces

Regular Points in the Domain

$0 = \mathrm{Div}\boldsymbol{P}' + \overline{\boldsymbol{s}}'_{\mathrm{m}}$	←PT→	$0 = \mathrm{div}\boldsymbol{\sigma}' + \overline{\boldsymbol{s}}'_{\mathrm{s}}$
PB ↕ PF		PB ↕ PF
$0 = \mathrm{Div}\boldsymbol{\Sigma}' + \overline{\boldsymbol{S}}'_{\mathrm{m}}$	←PT→	$0 = \mathrm{div}\boldsymbol{p}' + \overline{\boldsymbol{S}}'_{\mathrm{s}}$

Regular Points on the Boundary

$\widehat{\boldsymbol{i}} \cdot \boldsymbol{P}' \cdot \boldsymbol{N} = \widehat{\boldsymbol{i}} \cdot \left[\widehat{\mathrm{Div}\widehat{\boldsymbol{P}}'} + \widehat{\boldsymbol{s}}^{\mathrm{ext}\prime}_{\mathrm{m}}\right]$	←PT→	$\widehat{\boldsymbol{i}} \cdot \boldsymbol{\sigma}' \cdot \boldsymbol{n} = \widehat{\boldsymbol{i}} \cdot \left[\widehat{\mathrm{div}\widehat{\boldsymbol{\sigma}}'} + \widehat{\boldsymbol{s}}^{\mathrm{ext}\prime}_{\mathrm{s}}\right]$
PB ↕ PF		PB ↕ PF
$\widehat{\boldsymbol{I}} \cdot \boldsymbol{\Sigma}' \cdot \boldsymbol{N} = \widehat{\boldsymbol{I}} \cdot \left[\widehat{\mathrm{Div}\widehat{\boldsymbol{\Sigma}}'} + \widehat{\boldsymbol{S}}^{\mathrm{ext}\prime}_{\mathrm{m}}\right]$	←PT→	$\widehat{\boldsymbol{I}} \cdot \boldsymbol{p}' \cdot \boldsymbol{n} = \widehat{\boldsymbol{I}} \cdot \left[\widehat{\mathrm{div}\widehat{\boldsymbol{p}}'} + \widehat{\boldsymbol{S}}^{\mathrm{ext}\prime}_{\mathrm{s}}\right]$

Regular Points at Singular Surfaces

$\widehat{\boldsymbol{i}} \cdot [\![\boldsymbol{P}']\!] \cdot \boldsymbol{M} = 0$	←PT→	$\widehat{\boldsymbol{i}} \cdot [\![\boldsymbol{\sigma}']\!] \cdot \boldsymbol{m} = 0$
PB ↕ PF		PB ↕ PF
$\widehat{\boldsymbol{I}} \cdot [\![\boldsymbol{\Sigma}']\!] \cdot \boldsymbol{M} = 0$	←PT→	$\widehat{\boldsymbol{I}} \cdot [\![\boldsymbol{p}']\!] \cdot \boldsymbol{m} = 0$

with *Quasi-Static Versions* of Spatial and Material Stresses

$\boldsymbol{P}' := \partial_{\boldsymbol{F}}\overline{u}_{\mathrm{m}} \qquad \widehat{\boldsymbol{P}}' := \partial_{\widehat{\boldsymbol{F}}}\widehat{u}_{\mathrm{m}}$	←PT→	$\boldsymbol{\sigma}' := \overline{u}_{\mathrm{s}}\boldsymbol{i} - \boldsymbol{f}^t \cdot \partial_{\boldsymbol{f}}\overline{u}_{\mathrm{s}}$ $\widehat{\boldsymbol{\sigma}}' := \widehat{u}_{\mathrm{s}}\widehat{\boldsymbol{i}} - \widehat{\boldsymbol{f}}^t \cdot \partial_{\widehat{\boldsymbol{f}}}\widehat{u}_{\mathrm{s}}$
PB ↕ PF		PB ↕ PF
$\boldsymbol{\Sigma}' := \overline{u}_{\mathrm{m}}\boldsymbol{I} - \boldsymbol{F}^t \cdot \partial_{\boldsymbol{F}}\overline{u}_{\mathrm{m}}$ $\widehat{\boldsymbol{\Sigma}}' := \widehat{u}_{\mathrm{m}}\widehat{\boldsymbol{I}} - \widehat{\boldsymbol{F}}^t \cdot \partial_{\widehat{\boldsymbol{F}}}\widehat{u}_{\mathrm{m}}$	←PT→	$\boldsymbol{p}' := \partial_{\boldsymbol{f}}\overline{u}_{\mathrm{s}} \qquad \widehat{\boldsymbol{p}}' := \partial_{\widehat{\boldsymbol{f}}}\widehat{u}_{\mathrm{s}}$

Chapter 11
Thermodynamical Balances

8,586 m 27°42'09"N 88°08'48"E

Abstract This chapter specifies the generic balances for the case of thermodynamical balances of energy and entropy, whereby the case of exterior energy is demarcated from the common case of (interior) energy by the formal incorporation of the external potential energy into the notion of internal energy.

Thermodynamical balances of relevance for the subsequent discussions are the balance of (either *interior* or *exterior*) *total energy* and the balance of *entropy* (first and second law of thermodynamics).
The (*interior*) *internal* energy captures the (interior) total energy minus the macroscopic kinetic energy, i.e. the energy internal to a system due to microscopic velocity fluctuations (temperature), internal (interatomic or inter- and intramolecular) interaction potential energies, and other microscopic energetic contributions. If external potential energies for the external mechanical loads exist, these may be considered in addition to the common notion of (interior) internal energy to render the notion of *exterior* internal energy (admittedly an awkward terminology, however, a useful one in light of subsequent discussions and comparison with the variational setting).

The balance of total energy captures transformation of kinetic and internal energy under the supply of external power (we shall here restrict to external mechanical and external thermal power as paradigms of external power supply). By incorporating the balance of kinetic energy, which holds thanks to the balance of spatial momentum, the balance of internal energy results.

Sloppily speaking, entropy relates to the thermodynamics probability of macroscopically undistinguishable microscopic states (recall Boltzmann's gravestone engraving $S = k \ln w$). Then, the balance of entropy dictates the direction of macroscopic thermodynamics processes, necessitating the introduction of *entropy source*, *entropy flux*, and *positive entropy production*.

P. Steinmann, *Spatial and Material Forces in Nonlinear Continuum Mechanics*,
Solid Mechanics and Its Applications 272,
https://doi.org/10.1007/978-3-030-89070-4_11

11.1 Interior Total Energy

For the generic case of *non-conservative external mechanical loading*, external potential energies for the external mechanical loads neither exist in the domain nor on the external boundary. Consequently, it is only the *interior* total energy E, defined as the sum of the kinetic and *(interior) internal* energy, that may be balanced. Note that all later conclusions regarding the release of energy then indeed refer to the definition of interior total energy. Quantities deriving from the balance of interior total energy E are denoted as their *interior definition*. The alternative case of exterior total energy E' characterized by the existence of external potential energies is subsequently discussed in Sect. 11.2.

11.1.1 Global Format (Material Control Volume)

The *balance of interior total energy* is stated in global format for a material control volume as

Global Balance of Interior Total Energy (Material Control Volume):

$$\dot{\mathsf{E}}(\mathfrak{B}_{\rm m}) = \mathsf{E}^{\rm ext}(\mathfrak{B}_{\rm m}) \quad \forall \mathfrak{B}_{\rm m} \subseteq \mathcal{B}_{\rm m} \tag{11.1}$$

Thereby E is expressed as the resultant of the *interior total energy density* $\varepsilon_{\rm m}$ per unit volume in the material configuration over the material control volume $\mathfrak{B}_{\rm m}$, i.e. as the *resultant interior total energy* contained in the material control volume

$$\mathsf{E}(\mathfrak{B}_{\rm m}) := \int_{\mathfrak{B}_{\rm m}} \varepsilon_{\rm m} \, \mathrm{d}V. \tag{11.2}$$

The interior total energy E is defined as the summation of the kinetic and the *(interior) internal* energy

$$\mathsf{E}(\mathfrak{B}_{\rm m}) := \mathsf{K}(\mathfrak{B}_{\rm m}) + \mathsf{I}(\mathfrak{B}_{\rm m}) \quad \text{with} \quad \mathsf{I}(\mathfrak{B}_{\rm m}) := \int_{\mathfrak{B}_{\rm m}} \iota_{\rm m} \, \mathrm{d}V. \tag{11.3}$$

whereby $\iota_{\rm m}$ is defined as the *(interior) internal energy density* per unit volume in the material configuration. Thus, the interior total energy density $\varepsilon_{\rm m} := \kappa_{\rm m} + \iota_{\rm m}$ has the flavor of an *interior Hamilton-type internal energy density*. Note carefully that there are *no* exterior contributions to the interior total energy density $\varepsilon_{\rm m}$, thus the annoying addition *interior*. Instead, external power supplies are accounted for by $\mathsf{E}^{\rm ext}$ representing the resultant external (mechanical and non-mechanical) power supply.

E^{ext} consists of volume and boundary integrals (any singular contributions are here ignored for the sake of presentation)

$$\mathsf{E}^{\text{ext}}(\mathfrak{B}_{\text{m}}) := \int_{\mathfrak{B}_{\text{m}}} \overline{e}_{\text{m}}\, \text{d}V + \int_{\partial\mathfrak{B}_{\text{m}}} \widehat{e}_{\text{m}}\, \text{d}A \tag{11.4}$$

in terms of the *total energy source density* $\overline{e}_{\text{m}}$ per unit volume in the material configuration and the *extrinsic total energy flux density* $\widehat{e}_{\text{m}}$ per unit area in the material configuration. In the sequel, thermal power supply as the paradigm for non-mechanical power supply is considered in addition to mechanical power supply, thus

$$\overline{e}_{\text{m}} := \boldsymbol{v} \cdot \overline{\boldsymbol{s}}_{\text{m}} + \overline{h}_{\text{m}} \quad \text{and} \quad \widehat{e}_{\text{m}} := \boldsymbol{v} \cdot \widehat{\boldsymbol{s}}_{\text{m}} + \widehat{h}_{\text{m}}, \tag{11.5}$$

with $\overline{h}_{\text{m}}$ and $\widehat{h}_{\text{m}}$ denoting the *heat source density* per unit volume in the material configuration and the *extrinsic heat flux density* per unit area in the material configuration, respectively. Thus, the resultant external power supply consists of the resultant external mechanical and the *resultant external thermal power supply*

$$\mathsf{E}^{\text{ext}}(\mathfrak{B}_{\text{m}}) := \mathsf{P}^{\text{ext}}(\mathfrak{B}_{\text{m}}) + \mathsf{H}^{\text{ext}}(\mathfrak{B}_{\text{m}}) \tag{11.6}$$

whereby the resultant external thermal power supply H^{ext} consists of volume and boundary integrals

$$\mathsf{H}^{\text{ext}}(\mathfrak{B}_{\text{m}}) := \int_{\mathfrak{B}_{\text{m}}} \overline{h}_{\text{m}}\, \text{d}V + \int_{\partial\mathfrak{B}_{\text{m}}} \widehat{h}_{\text{m}}\, \text{d}A. \tag{11.7}$$

The extrinsic total energy flux density along with the convective term stemming from the transport theorem are postulated to satisfy a Cauchy theorem for the *intrinsic interior total energy flux density*

$$e_{\text{m}} := \boldsymbol{E} \cdot \boldsymbol{N} := \widehat{e}_{\text{m}} - [\varepsilon_{\text{m}} \boldsymbol{W}] \cdot \boldsymbol{N} \quad \text{on} \quad \partial\mathfrak{B}_{\text{m}}. \tag{11.8}$$

Thereby $\boldsymbol{E} \in T\mathcal{B}_{\text{m}}$ is introduced as the *Piola-type intrinsic interior total energy flux* consisting of a mechanical and a thermal contribution

$$\boldsymbol{E} := \boldsymbol{v} \cdot \boldsymbol{P} - \boldsymbol{H}. \tag{11.9}$$

The thermal contribution $\boldsymbol{H} \in T\mathcal{B}_{\text{m}}$ denotes the *Piola-type intrinsic heat flux* (the minus sign indicating the *system's egoistic point of view*, i.e. heat and thus energy is flowing *into* the control volume if $\boldsymbol{H}$ and the outwards pointing normal $\boldsymbol{N}$ form an obtuse angle such that $\boldsymbol{H} \cdot \boldsymbol{N} < 0$). Both $\boldsymbol{E}$ and $\boldsymbol{H}$ are material description vectors mapping from the material cotangent space $T^*\mathcal{B}_{\text{m}}$ to $\mathbb{R}$.

Inserting the previous definitions $\boldsymbol{v} \cdot \boldsymbol{P} \cdot \boldsymbol{N} =: \boldsymbol{v} \cdot \widehat{\boldsymbol{s}}_{\text{m}} - [2\kappa_{\text{m}} \boldsymbol{W}] \cdot \boldsymbol{N}$ and $\widehat{e}_{\text{m}} := \boldsymbol{v} \cdot \widehat{\boldsymbol{s}}_{\text{m}} + \widehat{h}_{\text{m}}$ into Eq. 11.8 identifies the relation between the intrinsic and the extrinsic heat flux density as

$$h_{\rm m} := -\boldsymbol{H} \cdot \boldsymbol{N} = \widehat{h}_{\rm m} + [\ell_{\rm m} \boldsymbol{W}] \cdot \boldsymbol{N} \quad \text{on} \quad \partial\mathfrak{B}_{\rm m}. \tag{11.10}$$

Here $\ell_{\rm m}$ denotes an energy density with the flavor of an *interior Lagrange-type internal energy density*

$$\ell_{\rm m} := \kappa_{\rm m} - \iota_{\rm m}. \tag{11.11}$$

Finally, as a consequence of the above definitions, the global format of the balance of interior total energy expands as

$$\int_{\mathfrak{B}_{\rm m}} {\rm D}_t \varepsilon_{\rm m}\, {\rm d}V = \int_{\mathfrak{B}_{\rm m}} \overline{e}_{\rm m}\, {\rm d}V + \int_{\partial\mathfrak{B}_{\rm m}} \underbrace{\boldsymbol{E} \cdot \boldsymbol{N}}_{e_{\rm m}}\, {\rm d}A + \int_{\mathfrak{S}_{\rm m}} [\![\varepsilon_{\rm m} \boldsymbol{W}]\!] \cdot \boldsymbol{M}\, {\rm d}A. \tag{11.12}$$

11.1.2 Local Format (Material Control Volume)

The results of localizing the global balance of interior total energy are summarized as:

Local Balance of Interior Total Energy (Material Control Volume):

i. Regular Points in the Domain

$$ {\rm D}_t \varepsilon_{\rm m} - {\rm Div} \boldsymbol{E} = \overline{e}_{\rm m} \quad \text{in} \quad \mathcal{B}_{\rm m} \tag{11.13}$$

ii. Regular Points on the Boundary

$$[\varepsilon_{\rm m} \boldsymbol{W} + \boldsymbol{E}] \cdot \boldsymbol{N} = \widehat{e}_{\rm m}^{\rm ext} \quad \text{on} \quad \partial\mathcal{B}_{\rm m} \tag{11.14}$$

iii. Regular Points at Singular Surfaces

$$[\![\varepsilon_{\rm m} \boldsymbol{W} + \boldsymbol{E}]\!] \cdot \boldsymbol{M} = 0 \quad \text{at} \quad \mathcal{S}_{\rm m} \tag{11.15}$$

The local balance of interior total energy allows the following conclusions:

i. In a smooth domain $\mathcal{B}_{\rm m}$, the formats of the Piola-type intrinsic interior total energy flux $\boldsymbol{E}$ in Eq. 11.9 and the total energy source density $\overline{e}_{\rm m}$ in Eq. 11.5.1 particularize the balance of interior total energy in Eq. 11.13 to the more common format

$$ {\rm D}_t \varepsilon_{\rm m} = \boldsymbol{v} \cdot \overline{\boldsymbol{s}}_{\rm m} + {\rm Div}(\boldsymbol{v} \cdot \boldsymbol{P}) + \overline{h}_{\rm m} - {\rm Div} \boldsymbol{H} \quad \text{in} \quad \mathcal{B}_{\rm m}. \tag{11.16}$$

ii. The balance of interior total energy on (the regular part of) the boundary $\partial\mathfrak{B}_{\rm m}$ of a smooth material control volume $\mathfrak{B}_{\rm m}$ in Eq. 11.14, results, when restricted to a singular surface, in *extrinsic total energy flux density cancelation*

$$\widehat{e}_{\mathrm{m}}^{+} + \widehat{e}_{\mathrm{m}}^{-} = 0 \quad \text{on} \quad \mathcal{S}_{\mathrm{m}} \tag{11.17}$$

and translates into the jump condition across the singular surface

$$[\![\varepsilon_{\mathrm{m}}]\!] W_{\perp} + [\![\boldsymbol{E}]\!] \cdot \boldsymbol{M} = 0 \quad \text{on} \quad \mathcal{S}_{\mathrm{m}}. \tag{11.18}$$

iii. At a singular surface $\mathcal{S}_{\mathrm{m}}$, the format of the Piola-type intrinsic interior total energy flux $\boldsymbol{E}$ in Eq. 11.9 simplifies the balance of interior total energy in Eq. 11.15 to the more common format

$$[\![\varepsilon_{\mathrm{m}}]\!] W_{\perp} = -[\![\boldsymbol{v} \cdot \boldsymbol{P}]\!] \cdot \boldsymbol{M} + [\![\boldsymbol{H}]\!] \cdot \boldsymbol{M} \quad \text{on} \quad \mathcal{S}_{\mathrm{m}}. \tag{11.19}$$

Note for material singular surfaces with $W_{\perp} = 0$ (and thus $[\![\boldsymbol{v}]\!] = \boldsymbol{0}$), the jump discontinuities in the mechanical contribution to the intrinsic interior total energy flux density $[\![\boldsymbol{v} \cdot \boldsymbol{P}]\!] \cdot \boldsymbol{M}$ equal the jump discontinuities in the thermal contribution to the intrinsic interior total energy flux density $[\![\boldsymbol{H}]\!] \cdot \boldsymbol{M}$, however, under these conditions, both terms vanish identically due to the balance of kinetic energy in 8.3, i.e. $[\![\boldsymbol{v} \cdot \boldsymbol{P}]\!] \cdot \boldsymbol{M} \equiv 0$, and thus consequently $[\![\boldsymbol{H}]\!] \cdot \boldsymbol{M} \equiv 0$.

Incorporating Generic Balance of Kinetic Energy

By subtracting the generic version of the local balance of kinetic energy from the local balance of interior total energy, the local balance of *interior Lagrange-type internal energy* (with density $\ell_{\mathrm{m}} = \kappa_{\mathrm{m}} - \iota_{\mathrm{m}}$) is obtained:

Local Balance of Interior Lagrange-Type Internal Energy (Material Control Volume):

i. Regular Points in the Domain

$$[-\mathrm{D}_t \ell_{\mathrm{m}} + \boldsymbol{p}_{\mathrm{m}} \cdot \mathrm{D}_t \boldsymbol{v}] - \nabla_X \boldsymbol{v} : \boldsymbol{P} + \mathrm{Div} \boldsymbol{H} = \overline{h}_{\mathrm{m}} \quad \text{in} \quad \mathcal{B}_{\mathrm{m}} \tag{11.20}$$

ii. Regular Points on the Boundary

$$[-\ell_{\mathrm{m}} \boldsymbol{W} - \boldsymbol{H}] \cdot \boldsymbol{N} = \widehat{h}_{\mathrm{m}}^{\mathrm{ext}} \quad \text{on} \quad \partial \mathcal{B}_{\mathrm{m}} \tag{11.21}$$

iii. Regular Points at Singular Surfaces

$$\Big[[-[\![\ell_{\mathrm{m}}]\!] + \{\boldsymbol{p}_{\mathrm{m}}\} \cdot [\![\boldsymbol{v}]\!]] \boldsymbol{W} + [\![\boldsymbol{v}]\!] \cdot \{\boldsymbol{P}\} - [\![\boldsymbol{H}]\!]\Big] \cdot \boldsymbol{M} = 0 \quad \text{at} \quad \mathcal{S}_{\mathrm{m}} \tag{11.22}$$

Incorporating Specific Balance of Kinetic Energy

Alternatively, by subtracting the specific version of the local balance of kinetic energy, that holds for the case of mass conservation, from the local balance of interior total energy, the more common local balance of *(interior) internal energy* (with density ι_{m}) is obtained:

Local Balance of (Interior) Internal Energy (Material Control Volume):

i. Regular Points in the Domain

$$\mathrm{D}_t \iota_{\text{m}} + \mathrm{Div}\boldsymbol{H} = \overline{h}_{\text{m}} + \nabla_X \boldsymbol{v} : \boldsymbol{P} \quad \text{in} \quad \mathcal{B}_{\text{m}} \tag{11.23}$$

ii. Regular Points on the Boundary

$$-\boldsymbol{H} \cdot \boldsymbol{N} = \widehat{h}^{\text{ext}}_{\text{m}} + \big[[\kappa_{\text{m}} - \iota_{\text{m}}]\boldsymbol{W}\big] \cdot \boldsymbol{N} \quad \text{on} \quad \partial\mathcal{B}_{\text{m}} \tag{11.24}$$

$$\rightsquigarrow -\boldsymbol{H} \cdot \boldsymbol{N} = h^{\text{ext}}_{\text{m}}$$

iii. Regular Points at Singular Surfaces

$$-[\![\boldsymbol{H}]\!] \cdot \boldsymbol{M} = \widetilde{h}_{\text{m}} + \big[\![[\kappa_{\text{m}} - \iota_{\text{m}}]\boldsymbol{W}]\!\big] \cdot \boldsymbol{M} \quad \text{at} \quad \mathcal{S}_{\text{m}} \tag{11.25}$$

$$\rightsquigarrow -[\![\boldsymbol{H}]\!] \cdot \boldsymbol{M} = -\big[[\![\iota_{\text{m}}]\!] - [\![\boldsymbol{F}]\!] : \{\boldsymbol{P}\}\big] W_{\perp}$$

Internal Thermal Power Generation at Singular Surfaces

Here the heat flux density generated at the singular surface per unit area in the material configuration has been defined as

$$\widetilde{h}_{\text{m}} := -[\![\boldsymbol{v}]\!] \cdot \widetilde{\boldsymbol{s}}_{\text{m}} \overset{\text{Eq. 8.14}}{\equiv} -\big[[\![\kappa_{\text{m}}]\!] - [\![\boldsymbol{F}]\!] : \{\boldsymbol{P}\}\big] W_{\perp} \tag{11.26}$$

based on (1) the extrinsic total energy flux density cancelation $\widetilde{e}_{\text{m}} := \widehat{e}^{+}_{\text{m}} + \widehat{e}^{-}_{\text{m}} = 0$ and (2) the postulation of a format for the total energy flux density generated at the singular surface

$$\widetilde{e}_{\text{m}} := [\![\boldsymbol{v}]\!] \cdot \widetilde{\boldsymbol{s}}_{\text{m}} + \widetilde{h}_{\text{m}} \tag{11.27}$$

that mimics the format for the extrinsic total energy flux density $\widehat{e}_{\text{m}} := \boldsymbol{v} \cdot \widehat{\boldsymbol{s}}^{\text{ext}}_{\text{m}} + \widehat{h}_{\text{m}}$ on the external boundary as introduced earlier.

Note finally that $[\![\boldsymbol{F}]\!] : \{\boldsymbol{P}\} W_{\perp} = -[\![\boldsymbol{v}]\!] \cdot \{\boldsymbol{P}\} \cdot \boldsymbol{M}$ holds due to the result in Eq. 8.7, thus the local balance of interior internal energy across a singular surface reads alternatively

$$[\![\iota_{\text{m}}]\!] W_{\perp} + \big[[\![\boldsymbol{v}]\!] \cdot \{\boldsymbol{P}\} - [\![\boldsymbol{H}]\!]\big] \cdot \boldsymbol{M} = 0. \tag{11.28}$$

Observe that for material singular surfaces with $W_\perp = 0$ (and thus $[\![\boldsymbol{v}]\!] = \boldsymbol{0}$) the jump discontinuities in the intrinsic heat flux density $[\![\boldsymbol{H}]\!] \cdot \boldsymbol{M} \equiv 0$ vanish identically.

11.1.3 Global Format (Spatial Control Volume)

Alternatively, the *balance of interior total energy* is stated in global format for a spatial control volume as

Global Balance of Interior Total Energy (Spatial Control Volume):

$$\dot{\mathsf{E}}(\mathfrak{B}_s) = \mathsf{E}^{\text{ext}}(\mathfrak{B}_s) \quad \forall \mathfrak{B}_s \subseteq \mathcal{B}_s \tag{11.29}$$

Here E is expressed as the resultant of the *interior total energy density* $\varepsilon_s = \kappa_s + \iota_s$ per unit volume in the spatial configuration over the spatial control volume $\mathfrak{B}_s$

$$\mathsf{E}(\mathfrak{B}_s) := \int_{\mathfrak{B}_s} \varepsilon_s \, \mathrm{d}v. \tag{11.30}$$

Moreover, E^{ext} represents the resultant external (mechanical and non-mechanical) power supply and consists of volume and boundary integrals (any singular contributions are here ignored for the sake of presentation)

$$\mathsf{E}^{\text{ext}}(\mathfrak{B}_s) := \int_{\mathfrak{B}_s} \overline{e}_s \, \mathrm{d}v + \int_{\partial\mathfrak{B}_s} \widehat{e}_s \, \mathrm{d}a \tag{11.31}$$

in terms of the *total energy source density* $\overline{e}_s = j\overline{e}_m := \boldsymbol{v} \cdot \overline{\boldsymbol{s}}_s + \overline{h}_s$ per unit volume in the spatial configuration and the *extrinsic total energy flux density* $\widehat{e}_s = \widehat{je}_m := \boldsymbol{v} \cdot \widehat{\boldsymbol{s}}_s + \widehat{h}_s$ per unit area in the spatial configuration. The intrinsic interior total energy flux is defined as the Piola transformation $\boldsymbol{\varepsilon} := \boldsymbol{E} \cdot \operatorname{cof} \boldsymbol{f}$, thus the corresponding Cauchy theorem for $\boldsymbol{\varepsilon} \cdot \boldsymbol{n}$ reads in terms of the extrinsic total energy flux density along with the convective term stemming from the transport theorem

$$e_s := \boldsymbol{\varepsilon} \cdot \boldsymbol{n} := \widehat{e}_s - \big[\varepsilon_s[\boldsymbol{w} - \boldsymbol{v}]\big] \cdot \boldsymbol{n} \quad \text{on} \quad \partial\mathfrak{B}_s. \tag{11.32}$$

Thereby $\boldsymbol{\varepsilon} \in T\mathcal{B}_s$ is introduced as the *Cauchy-type intrinsic interior total energy flux* consisting of a mechanical and a thermal contribution

$$\boldsymbol{\varepsilon} := \boldsymbol{v} \cdot \boldsymbol{\sigma} - \boldsymbol{\eta}. \tag{11.33}$$

The thermal contribution $\boldsymbol{\eta} := \boldsymbol{H} \cdot \operatorname{cof} \boldsymbol{f} \in T\mathcal{B}_s$ denotes the *Cauchy-type intrinsic heat flux*. Both $\boldsymbol{\varepsilon}$ and $\boldsymbol{\eta}$ are spatial description vectors mapping from the spatial cotangent space $T^*\mathcal{B}_s$ to $\mathbb{R}$.

As a consequence of the above definitions, the global format of the balance of interior total energy expands as

$$\int_{\mathfrak{B}_s} \mathrm{d}_t \varepsilon_s \,\mathrm{d}v = \int_{\mathfrak{B}_s} \overline{e}_s \,\mathrm{d}v + \int_{\partial\mathfrak{B}_s} [\boldsymbol{\varepsilon} - \varepsilon_s \boldsymbol{v}] \cdot \boldsymbol{n} \,\mathrm{d}a + \int_{\mathfrak{S}_s} [\![\varepsilon_s \boldsymbol{w}]\!] \cdot \boldsymbol{m} \,\mathrm{d}a. \tag{11.34}$$

11.1.4 Local Format (Spatial Control Volume)

The local balance for the interior total energy density ε_s per unit volume in the spatial configuration, expressed in terms of the Cauchy-type interior total energy flux $\boldsymbol{\varepsilon} = \boldsymbol{E} \cdot \operatorname{cof} \boldsymbol{f}$, the total energy source density $\overline{e}_s$ per unit volume in the spatial configuration and the extrinsic total energy flux density $\widehat{e}_s^{\text{ext}}$ per unit area in the spatial configuration, is summarized as:

Local Balance of Interior Total Energy (Spatial Control Volume):

i. Regular Points in the Domain

$$\mathrm{d}_t \varepsilon_s - \operatorname{div}(\boldsymbol{\varepsilon} - \varepsilon_s \boldsymbol{v}) = \overline{e}_s \quad \text{in} \quad \mathcal{B}_s \tag{11.35}$$

ii. Regular Points on the Boundary

$$\big[\varepsilon_s \boldsymbol{w} + [\boldsymbol{\varepsilon} - \varepsilon_s \boldsymbol{v}]\big] \cdot \boldsymbol{n} = \widehat{e}_s^{\text{ext}} \quad \text{on} \quad \partial\mathcal{B}_s \tag{11.36}$$

iii. Regular Points at Singular Surfaces

$$[\![\varepsilon_s \boldsymbol{w} + [\boldsymbol{\varepsilon} - \varepsilon_s \boldsymbol{v}]]\!] \cdot \boldsymbol{m} = 0 \quad \text{at} \quad \mathcal{S}_s \tag{11.37}$$

Incorporating Generic Balance of Kinetic Energy

By subtracting the generic version of the local balance of kinetic energy from the local balance of interior total energy, the local balance of *interior Lagrange-type internal energy* (with density $\ell_s = \kappa_s - \iota_s$) is obtained:

Local Balance of Interior Lagrange-Type Internal Energy (Spatial Control Volume):

i. Regular Points in the Domain

$$[-\mathrm{d}_t \ell_s + \boldsymbol{p}_s \cdot \mathrm{d}_t \boldsymbol{v}] - \nabla_x \boldsymbol{v} : \boldsymbol{\sigma}_d + \operatorname{div}(\boldsymbol{\eta} - \ell_s \boldsymbol{v}) = \overline{h}_s \quad \text{in} \quad \mathcal{B}_s \tag{11.38}$$

ii. Regular Points on the Boundary

$$\big[-\ell_s \boldsymbol{w} - [\boldsymbol{\eta} - \ell_s \boldsymbol{v}]\big] \cdot \boldsymbol{n} = \widehat{h}_s^{\text{ext}} \quad \text{on} \quad \partial\mathcal{B}_s \tag{11.39}$$

iii. Regular Points at Singular Surfaces

$$\Big[[-[\![\ell_s]\!] + \{\boldsymbol{p}_s\} \cdot [\![\boldsymbol{v}]\!]]\boldsymbol{w} + [\![\boldsymbol{v}]\!] \cdot \{\boldsymbol{\sigma}_d\} - [\![\boldsymbol{\eta} - \ell_s \boldsymbol{v}]\!]\Big] \cdot \boldsymbol{m} = 0 \quad \text{at} \quad \mathcal{S}_s \tag{11.40}$$

Incorporating Specific Balance of Kinetic Energy

By subtracting the specific version of the local balance of kinetic energy, that holds for the case of mass conservation, from the local balance of interior total energy, the more common local balance of *(interior) internal energy* (with density ι_m) is obtained:

Local Balance of (Interior) Internal Energy (Spatial Control Volume):

i. Regular Points in the Domain

$$\mathrm{d}_t \iota_s + \operatorname{div}(\boldsymbol{\eta} + \iota_s \boldsymbol{v}) = \overline{h}_s + \nabla_x \boldsymbol{v} : \boldsymbol{\sigma} \quad \text{in} \quad \mathcal{B}_s \tag{11.41}$$

ii. Regular Points on the Boundary

$$-\boldsymbol{\eta} \cdot \boldsymbol{n} = \widehat{h}_s^{\text{ext}} + \big[[\kappa_s - \iota_s][\boldsymbol{w} - \boldsymbol{v}]\big] \cdot \boldsymbol{n} \quad \text{on} \quad \partial\mathcal{B}_s \tag{11.42}$$

$$\rightsquigarrow -\boldsymbol{\eta} \cdot \boldsymbol{n} = h_s^{\text{ext}}$$

iii. Regular Points at Singular Surfaces

$$-[\![\boldsymbol{\eta}]\!] \cdot \boldsymbol{m} = \widetilde{h}_s + [\![[\kappa_s - \iota_s][\boldsymbol{w} - \boldsymbol{v}]]\!] \cdot \boldsymbol{m} \quad \text{at} \quad \mathcal{S}_s \tag{11.43}$$

$$\rightsquigarrow -[\![\boldsymbol{\eta}]\!] \cdot \boldsymbol{m} = -\Big[[\![\iota_s[\boldsymbol{w} - \boldsymbol{v}]]\!] - W_\perp \boldsymbol{j} \cdot \{\boldsymbol{\sigma}\}\Big] \cdot \boldsymbol{m}$$

Internal Thermal Power Generation at Singular Surfaces

Here the heat flux density generated at the singular surface per unit area in the spatial configuration has been defined as

$$\widetilde{h}_s := -[\![\boldsymbol{v}]\!] \cdot \widetilde{\boldsymbol{s}}_s \overset{\text{Eq. 8.37}}{\equiv} -\Big[[\![\kappa_s[\boldsymbol{w} - \boldsymbol{v}]]\!] - W_\perp \boldsymbol{j} \cdot \{\boldsymbol{\sigma}\}\Big] \cdot \boldsymbol{m}. \tag{11.44}$$

Note finally that $W_\perp \boldsymbol{j} \cdot \{\boldsymbol{\sigma}\} \cdot \boldsymbol{m} = -[\![\boldsymbol{v}]\!] \cdot \{\boldsymbol{\sigma}\} \cdot \boldsymbol{m}$ holds due to the result in Eq. 8.30, thus the local balance of interior internal energy across a singular surface reads alternatively

$$\Big[[\![\iota_s[\boldsymbol{w} - \boldsymbol{v}]]\!] + [\![\boldsymbol{v}]\!] \cdot \{\boldsymbol{\sigma}\} - [\![\boldsymbol{\eta}]\!]\Big] \cdot \boldsymbol{m} = 0. \tag{11.45}$$

11.1.5 Balance Tetragons

The local balance of interior total energy *at regular points in the domain, at regular points on the boundary*, and *at regular points at singular surfaces* may be expressed in four different but fully equivalent versions, see the arrangement in the corresponding tetragons in Table 11.2. Here *dynamic versions* of the interior total energy fluxes have been introduced as $\boldsymbol{\varepsilon}_D = \boldsymbol{E}_D \cdot \operatorname{cof} \boldsymbol{f}$ and $\boldsymbol{\varepsilon}_d = \boldsymbol{E}_d \cdot \operatorname{cof} \boldsymbol{f}$, respectively. The subscripts D and d indicate that the corresponding interior total energy flux is consistent with either the material or the spatial time derivative D_t or d_t, respectively, of the interior total energy density.

Likewise, the local balance of (interior) internal energy *at regular points in the domain, at regular points on the boundary*, and *at regular points at singular surfaces* may be expressed in four different but fully equivalent versions, see the arrangement in the corresponding tetragons in Table 11.3. Here *dynamic versions* of the heat fluxes have been introduced as $\boldsymbol{\eta}_D = \boldsymbol{H}_D \cdot \operatorname{cof} \boldsymbol{f}$ and $\boldsymbol{\eta}_d = \boldsymbol{H}_d \cdot \operatorname{cof} \boldsymbol{f}$, respectively. The subscripts D and d indicate that the corresponding heat flux is consistent with either the material or the spatial time derivative D_t or d_t, respectively, of the (interior) internal energy density.

Moreover, abbreviations for *effective* internal energy source terms in the domain, on the boundary, and at singular surfaces, respectively, have been introduced in Table 11.1.

11.2 Exterior Total Energy

For cases of *conservative external mechanical loading*, the interior total energy E may be upgraded to the exterior total energy E' by the contributions of the external potential energy densities $\overline{v}_m$ and $\widehat{v}_m$ in the domain and on the external boundary, respectively. However, when balancing the exterior total energy E' instead of the interior total energy E, one should carefully note that all resulting conclusions regarding the release of energy refer indeed to two different types of energy. To highlight this distinction, quantities deriving eventually from starting with the previous balance of interior total energy E are denoted as their *interior definition*, whereas quantities deriving eventually from starting with the balance of exterior total energy E' are denoted as their *exterior definition* and are highlighted by a prime.

11.2.1 Global Format (Material Control Volume)

Then the *balance of exterior total energy* is stated in global format for a material control volume as

Global Balance of Exterior Total Energy (Material Control Volume):

$$\dot{\mathsf{E}}'(\mathfrak{B}_\mathrm{m}) = \mathsf{E}'^\mathrm{ext}(\mathfrak{B}_\mathrm{m}) \quad \forall \mathfrak{B}_\mathrm{m} \subseteq \mathcal{B}_\mathrm{m} \tag{11.46}$$

The exterior total energy E' is defined as the summation of the kinetic and the *interior internal* and the *external potential* energies

$$\mathsf{E}'(\mathfrak{B}_\mathrm{m}) := \mathsf{E}(\mathfrak{B}_\mathrm{m}) + \mathsf{V}(\mathfrak{B}_\mathrm{m}) \quad \text{with} \quad \mathsf{V}(\mathfrak{B}_\mathrm{m}) := \int_{\mathfrak{B}_\mathrm{m}} \overline{v}_\mathrm{m} \, \mathrm{d}V + \int_{\partial\mathfrak{B}_\mathrm{m} \cap \partial\mathcal{B}_\mathrm{m}} \widehat{v}_\mathrm{m} \, \mathrm{d}A. \tag{11.47}$$

whereby $\overline{v}_\mathrm{m}$ and $\widehat{v}_\mathrm{m}$ are defined as the *external potential energy density* per unit volume in the domain and per unit area on the external boundary, respectively, of the material configuration. These are defined so as to render

$$\mathrm{D}_t \overline{v}_\mathrm{m} := -\boldsymbol{v} \cdot \overline{\boldsymbol{s}}_\mathrm{m} \quad \text{in} \quad \mathcal{B}_\mathrm{m} \quad \text{and} \quad \mathrm{D}_t \widehat{v}_\mathrm{m} := -\boldsymbol{v} \cdot \widehat{\boldsymbol{s}}^\mathrm{ext}_\mathrm{m} \quad \text{on} \quad \partial\mathcal{B}_\mathrm{m}. \tag{11.48}$$

Table 11.1 Effective internal energy source terms in regular points in the domain, on the boundary and at singular surfaces

Regular Points in the Domain

$$\overline{k}_\mathrm{m} := \overline{h}_\mathrm{m} + \nabla_X \boldsymbol{v} : \boldsymbol{P} \quad \overset{\mathrm{PT}}{\longleftrightarrow} \quad \overline{k}_\mathrm{s} := \overline{h}_\mathrm{s} + \nabla_x \boldsymbol{v} : \boldsymbol{\sigma}$$

Regular Points on the Boundary

$$\widehat{k}_\mathrm{m} := \widehat{h}^\mathrm{ext}_\mathrm{m} + [\kappa_\mathrm{m} \boldsymbol{W}] \cdot \boldsymbol{N} \quad \overset{\mathrm{PT}}{\longleftrightarrow} \quad \widehat{k}_\mathrm{s} := \widehat{h}^\mathrm{ext}_\mathrm{s} + \big[\kappa_\mathrm{s}[\boldsymbol{w} - \boldsymbol{v}]\big] \cdot \boldsymbol{n}$$

Regular Points at Singular Surfaces

$$\widetilde{k}_\mathrm{m} := \widetilde{h}_\mathrm{m} + [\![\kappa_\mathrm{m} \boldsymbol{W}]\!] \cdot \boldsymbol{M} \quad \overset{\mathrm{PT}}{\longleftrightarrow} \quad \widetilde{k}_\mathrm{s} := \widetilde{h}_\mathrm{s} + \big[\!\big[\kappa_\mathrm{s}[\boldsymbol{w} - \boldsymbol{v}]\big]\!\big] \cdot \boldsymbol{n}$$

Table 11.2 Tetragon of fully equivalent versions for the local balance of interior total energy at regular points in the domain, on the boundary and at singular surfaces

Regular Points in the Domain

$\mathrm{D}_t\varepsilon_\mathrm{m} = \mathrm{Div}\boldsymbol{E}_\mathrm{D} + \overline{e}_\mathrm{m}$	$\longleftrightarrow$ PT	$j\,\mathrm{D}_t\varepsilon_\mathrm{m} = \mathrm{div}\boldsymbol{\varepsilon}_\mathrm{D} + \overline{e}_\mathrm{s}$
$\updownarrow$ TT		TT $\updownarrow$
$J\,\mathrm{d}_t\varepsilon_\mathrm{s} = \mathrm{Div}\boldsymbol{E}_\mathrm{d} + \overline{e}_\mathrm{m}$	PT $\longleftrightarrow$	$\mathrm{d}_t\varepsilon_\mathrm{s} = \mathrm{div}\boldsymbol{\varepsilon}_\mathrm{d} + \overline{e}_\mathrm{s}$

Regular Points on the Boundary

$\big[\varepsilon_\mathrm{m}\boldsymbol{W} + \boldsymbol{E}_\mathrm{D}\big]\cdot\boldsymbol{N} = \widehat{e}_\mathrm{m}^\mathrm{ext}$	$\longleftrightarrow$ PT	$\big[\varepsilon_\mathrm{s}[\boldsymbol{w}-\boldsymbol{v}] + \boldsymbol{\varepsilon}_\mathrm{D}\big]\cdot\boldsymbol{n} = \widehat{e}_\mathrm{s}^\mathrm{ext}$
$\updownarrow$ TT		TT $\updownarrow$
$\big[\varepsilon_\mathrm{m}[\boldsymbol{W}-\boldsymbol{V}] + \boldsymbol{E}_\mathrm{d}\big]\cdot\boldsymbol{N} = \widehat{e}_\mathrm{m}^\mathrm{ext}$	PT $\longleftrightarrow$	$\big[\varepsilon_\mathrm{s}\boldsymbol{w} + \boldsymbol{\varepsilon}_\mathrm{d}\big]\cdot\boldsymbol{n} = \widehat{e}_\mathrm{s}^\mathrm{ext}$

Regular Points at Singular Surfaces

$[\![\varepsilon_\mathrm{m}\boldsymbol{W} + \boldsymbol{E}_\mathrm{D}]\!]\cdot\boldsymbol{M} = 0$	$\longleftrightarrow$ PT	$[\![\varepsilon_\mathrm{s}[\boldsymbol{w}-\boldsymbol{v}] + \boldsymbol{\varepsilon}_\mathrm{D}]\!]\cdot\boldsymbol{m} = 0$
$\updownarrow$ TT		TT $\updownarrow$
$[\![\varepsilon_\mathrm{m}[\boldsymbol{W}-\boldsymbol{V}] + \boldsymbol{E}_\mathrm{d}]\!]\cdot\boldsymbol{M} = 0$	PT $\longleftrightarrow$	$[\![\varepsilon_\mathrm{s}\boldsymbol{w} + \boldsymbol{\varepsilon}_\mathrm{d}]\!]\cdot\boldsymbol{m} = 0$

with *Dynamic Versions* of Intrinsic Interior Total Energy Flux

$\boldsymbol{E}_\mathrm{D} := \boldsymbol{E}$	$\longleftrightarrow$ PT	$\boldsymbol{\varepsilon}_\mathrm{D} := \boldsymbol{\varepsilon}$
$\updownarrow$ TT		TT $\updownarrow$
$\boldsymbol{E}_\mathrm{d} := \boldsymbol{E} + \varepsilon_\mathrm{m}\boldsymbol{V}$	PT $\longleftrightarrow$	$\boldsymbol{\varepsilon}_\mathrm{d} := \boldsymbol{\varepsilon} - \varepsilon_\mathrm{s}\boldsymbol{v}$

Table 11.3 Tetragon of fully equivalent versions for the local balance of (interior) internal energy at regular points in the domain, on the boundary and at singular surfaces

Regular Points in the Domain

$\mathrm{D}_t \iota_\mathrm{m} = -\mathrm{Div}\boldsymbol{H}_\mathrm{D} + \overline{k}_\mathrm{m}$	⟷ PT	$j\,\mathrm{D}_t \iota_\mathrm{m} = -\mathrm{div}\boldsymbol{\eta}_\mathrm{D} + \overline{k}_\mathrm{s}$
↕ TT		↕ TT
$J\,\mathrm{d}_t \iota_\mathrm{s} = -\mathrm{Div}\boldsymbol{H}_\mathrm{d} + \overline{k}_\mathrm{m}$	⟷ PT	$\mathrm{d}_t \iota_\mathrm{s} = -\mathrm{div}\boldsymbol{\eta}_\mathrm{d} + \overline{k}_\mathrm{s}$

Regular Points on the Boundary

$[\iota_\mathrm{m}\boldsymbol{W} - \boldsymbol{H}_\mathrm{D}] \cdot \boldsymbol{N} = \widehat{k}_\mathrm{m}$	⟷ PT	$[\iota_\mathrm{s}[\boldsymbol{w} - \boldsymbol{v}] - \boldsymbol{\eta}_\mathrm{D}] \cdot \boldsymbol{n} = \widehat{k}_\mathrm{s}$
↕ TT		↕ TT
$[\iota_\mathrm{m}[\boldsymbol{W} - \boldsymbol{V}] - \boldsymbol{H}_\mathrm{d}] \cdot \boldsymbol{N} = \widehat{k}_\mathrm{m}$	⟷ PT	$[\iota_\mathrm{s}\boldsymbol{w} - \boldsymbol{\eta}_\mathrm{d}] \cdot \boldsymbol{n} = \widehat{k}_\mathrm{s}$

Regular Points at Singular Surfaces

$[\![\iota_\mathrm{m}\boldsymbol{W} - \boldsymbol{H}_\mathrm{D}]\!] \cdot \boldsymbol{M} = \widetilde{k}_\mathrm{m}$	⟷ PT	$[\![\iota_\mathrm{s}[\boldsymbol{w} - \boldsymbol{v}] - \boldsymbol{\eta}_\mathrm{D}]\!] \cdot \boldsymbol{m} = \widetilde{k}_\mathrm{s}$
↕ TT		↕ TT
$[\![\iota_\mathrm{m}[\boldsymbol{W} - \boldsymbol{V}] - \boldsymbol{H}_\mathrm{d}]\!] \cdot \boldsymbol{M} = \widetilde{k}_\mathrm{m}$	⟷ PT	$[\![\iota_\mathrm{s}\boldsymbol{w} - \boldsymbol{\eta}_\mathrm{d}]\!] \cdot \boldsymbol{m} = \widetilde{k}_\mathrm{s}$

with *Dynamic Versions* of Intrinsic Heat Flux

$\boldsymbol{H}_\mathrm{D} := \boldsymbol{H}$	⟷ PT	$\boldsymbol{\eta}_\mathrm{D} := \boldsymbol{\eta}$
↕ TT		↕ TT
$\boldsymbol{H}_\mathrm{d} := \boldsymbol{H} - \iota_\mathrm{m}\boldsymbol{V}$	⟷ PT	$\boldsymbol{\eta}_\mathrm{d} := \boldsymbol{\eta} + \iota_\mathrm{s}\boldsymbol{v}$

Note that the last integral appears *only* at the external boundary where the *external traction* $\widehat{s}_m^{ext}$ is applied (thus the mechanical contact interactions at the material control volume boundary not intersecting with the external boundary are accounted for by E'^{ext}, see below). Furthermore, the following exterior total energy densities or rather *exterior Hamilton-type internal energy densities* are introduced as abbreviations

$$\overline{\varepsilon}_m := \varepsilon_m + \overline{\upsilon}_m \quad \text{in/at} \quad \mathcal{B}_m/\mathcal{S}_m \quad \text{and} \quad \varepsilon'_m := \overline{\varepsilon}_m + \upsilon'_m \quad \text{on} \quad \partial\mathcal{B}_m \tag{11.49}$$

whereby υ'_m is defined so as to account for the *curved* external boundary

$$\upsilon'_m := \nabla_N \widehat{\upsilon}_m - \widehat{\upsilon}_m \widehat{C} \quad \text{on} \quad \partial\mathcal{B}_m. \tag{11.50}$$

Then, with $W_\| = 0$, the total time derivative of E' expands as

$$\begin{aligned}\dot{\mathsf{E}}'(\mathfrak{B}_m) = &\int_{\mathfrak{B}_m} \mathrm{D}_t\overline{\varepsilon}_m \, \mathrm{d}V + \int_{\partial\mathfrak{B}_m \not\cap \partial\mathcal{B}_m} [\overline{\varepsilon}_m \boldsymbol{W}] \cdot \boldsymbol{N} \, \mathrm{d}A - \int_{\mathfrak{S}_m} [\![\overline{\varepsilon}_m \boldsymbol{W}]\!] \cdot \boldsymbol{M} \, \mathrm{d}A \\ &+ \int_{\partial\mathfrak{B}_m \cap \partial\mathcal{B}_m} \mathrm{D}_t \widehat{\upsilon}_m \, \mathrm{d}A + \int_{\partial\mathfrak{B}_m \cap \partial\mathcal{B}_m} [\varepsilon'_m \boldsymbol{W}] \cdot \boldsymbol{N} \, \mathrm{d}A.\end{aligned} \tag{11.51}$$

Moreover, E'^{ext} represents the resultant external (non-mechanical and mechanical) power supply, whereby the mechanical contact interactions are only accounted for at the material control volume boundary not intersecting with the external boundary

$$\mathsf{E}^{ext\prime}(\mathfrak{B}_m) := \mathsf{H}^{ext}(\mathfrak{B}_m) + \int_{\partial\mathfrak{B}_m \not\cap \partial\mathcal{B}_m} \boldsymbol{v} \cdot \widehat{\boldsymbol{s}}_m \, \mathrm{d}A. \tag{11.52}$$

The extrinsic total energy flux density $\widehat{e}_m := \widehat{h}_m + \boldsymbol{v} \cdot \widehat{\boldsymbol{s}}_m$ at the material control volume boundary not intersecting with the external boundary and $\widehat{e}_m^{ext} := \widehat{h}_m^{ext} - \mathrm{D}_t \widehat{\upsilon}_m$ on the external boundary, respectively, along with the respective convective terms stemming from the transport theorem are postulated to satisfy a Cauchy theorem for the *intrinsic exterior total energy flux density*

$$\begin{aligned} e_m &:= \boldsymbol{E} \cdot \boldsymbol{N} := \widehat{h}_m + \boldsymbol{v} \cdot \widehat{\boldsymbol{s}}_m - [\overline{\varepsilon}_m \boldsymbol{W}] \cdot \boldsymbol{N} \quad \text{on} \quad \partial\mathfrak{B}_m \not\cap \partial\mathcal{B}_m, \\ e_m^{ext} &:= \boldsymbol{E} \cdot \boldsymbol{N} := \widehat{h}_m^{ext} - \mathrm{D}_t \widehat{\upsilon}_m - [\varepsilon'_m \boldsymbol{W}] \cdot \boldsymbol{N} \quad \text{on} \quad \partial\mathfrak{B}_m \cap \partial\mathcal{B}_m. \end{aligned} \tag{11.53}$$

Thereby $\boldsymbol{E} \in T\mathcal{B}_m$ is introduced as *Piola-type intrinsic exterior total energy flux* consisting of a mechanical and a thermal contribution

$$\boldsymbol{E} := \boldsymbol{v} \cdot \boldsymbol{P} - \boldsymbol{H}. \tag{11.54}$$

The thermal contribution $\boldsymbol{H} \in T\mathcal{B}_m$ denotes the *Piola-type intrinsic heat flux*.

Inserting next the results $\boldsymbol{v} \cdot \widehat{\boldsymbol{s}}_m = \boldsymbol{v} \cdot \boldsymbol{P} \cdot \boldsymbol{N} + [2\kappa_m \boldsymbol{W}] \cdot \boldsymbol{N}$ at the material control volume boundary not intersecting with the external boundary and $-\mathrm{D}_t \widehat{\upsilon}_m =$

$\boldsymbol{v} \cdot \boldsymbol{P} \cdot \boldsymbol{N} + [2\kappa_{\mathrm{m}} \boldsymbol{W}] \cdot \boldsymbol{N}$ on the external boundary into Eq. 11.53 identifies the relation between the intrinsic and the extrinsic heat flux density defined as

$$\begin{aligned} h_{\mathrm{m}} &:= -\boldsymbol{H} \cdot \boldsymbol{N} = \widehat{h}_{\mathrm{m}} + [\overline{\ell}_{\mathrm{m}} \boldsymbol{W}] \cdot \boldsymbol{N} \quad \text{on} \quad \partial\mathfrak{B}_{\mathrm{m}} \not\cap \partial\mathcal{B}_{\mathrm{m}}, \\ h^{\mathrm{ext}}_{\mathrm{m}} &:= -\boldsymbol{H} \cdot \boldsymbol{N} = \widehat{h}^{\mathrm{ext}}_{\mathrm{m}} + [\ell'_{\mathrm{m}} \boldsymbol{W}] \cdot \boldsymbol{N} \quad \text{on} \quad \partial\mathfrak{B}_{\mathrm{m}} \cap \partial\mathcal{B}_{\mathrm{m}}. \end{aligned} \tag{11.55}$$

Here $\overline{\ell}_{\mathrm{m}}$ and ℓ'_{m} denote energy densities with the flavor of *exterior Lagrange-type internal energy densities*, i.e. including the *exterior internal energy densities* $\overline{\iota}_{\mathrm{m}}$ and ι'_{m} as

$$\begin{aligned} \overline{\ell}_{\mathrm{m}} &:= \kappa_{\mathrm{m}} - \overline{\iota}_{\mathrm{m}} := \kappa_{\mathrm{m}} - [\iota_{\mathrm{m}} + \overline{\upsilon}_{\mathrm{m}}], \\ \ell'_{\mathrm{m}} &:= \kappa_{\mathrm{m}} - \iota'_{\mathrm{m}} := \kappa_{\mathrm{m}} - [\overline{\iota}_{\mathrm{m}} + \upsilon'_{\mathrm{m}}]. \end{aligned} \tag{11.56}$$

Finally, as a consequence of the above definitions, the global format of the balance of exterior total energy expands as

$$\begin{aligned} \int_{\mathfrak{B}_{\mathrm{m}}} \mathrm{D}_t \overline{\varepsilon}_{\mathrm{m}} \, \mathrm{d}V = \int_{\mathfrak{B}_{\mathrm{m}}} \overline{h}_{\mathrm{m}} \, \mathrm{d}V + \int_{\partial\mathfrak{B}_{\mathrm{m}} \not\cap \partial\mathcal{B}_{\mathrm{m}}} \underbrace{\boldsymbol{E} \cdot \boldsymbol{N}}_{e_{\mathrm{m}}} \, \mathrm{d}A + \int_{\mathfrak{S}_{\mathrm{m}}} [\![\overline{\varepsilon}_{\mathrm{m}} \boldsymbol{W}]\!] \cdot \boldsymbol{M} \, \mathrm{d}A \\ + \int_{\partial\mathfrak{B}_{\mathrm{m}} \cap \partial\mathcal{B}_{\mathrm{m}}} \underbrace{\boldsymbol{E} \cdot \boldsymbol{N}}_{e^{\mathrm{ext}}_{\mathrm{m}}} \, \mathrm{d}A. \end{aligned} \tag{11.57}$$

11.2.2 Local Format (Material Control Volume)

The results of localizing the global balance of exterior total energy are summarized as:

Local Balance of Exterior Total Energy (Material Control Volume):

i. Regular Points in the Domain

$$\mathrm{D}_t \overline{\varepsilon}_{\mathrm{m}} - \mathrm{Div} \boldsymbol{E} = \overline{h}_{\mathrm{m}} \quad \text{in} \quad \mathcal{B}_{\mathrm{m}} \tag{11.58}$$

ii. Regular Points on the Boundary

$$\mathrm{D}_t \widehat{\upsilon}_{\mathrm{m}} + [\varepsilon'_{\mathrm{m}} \boldsymbol{W} + \boldsymbol{E}] \cdot \boldsymbol{N} = \widehat{h}^{\mathrm{ext}}_{\mathrm{m}} \quad \text{on} \quad \partial\mathcal{B}_{\mathrm{m}} \tag{11.59}$$

iii. Regular Points at Singular Surfaces

$$[\![\overline{\varepsilon}_{\mathrm{m}} \boldsymbol{W} + \boldsymbol{E}]\!] \cdot \boldsymbol{M} = 0 \quad \text{at} \quad \mathcal{S}_{\mathrm{m}} \tag{11.60}$$

The local balance of exterior total energy allows the following conclusions:

i. In a smooth domain $\mathcal{B}_m$ the format of the Piola-type exterior intrinsic total energy flux vector $\boldsymbol{E}$ in Eq. 11.54.1 particularizes the balance of exterior total energy in Eq. 11.58 to a more common format

$$\mathrm{D}_t\overline{\varepsilon}_m = \mathrm{Div}(\boldsymbol{v}\cdot\boldsymbol{P}) + \overline{h}_m - \mathrm{Div}\boldsymbol{H} \quad \text{in} \quad \mathcal{B}_m. \tag{11.61}$$

Note that identifying $-\mathrm{D}_t\overline{v}_m \equiv \boldsymbol{v}\cdot\overline{\boldsymbol{s}}_m$, shows the balance of exterior total energy to *coincide* identically with the balance of interior total energy in the domain $\mathcal{B}_m$.

ii. On the external boundary $\partial\mathcal{B}_m$, the format of the Piola-type intrinsic exterior total energy flux vector $\boldsymbol{E}$ in Eq. 11.54.2 particularizes the balance of exterior total energy in Eq. 11.59 to a more common format

$$\varepsilon'_m W_\perp + \boldsymbol{v}\cdot\boldsymbol{P}\cdot\boldsymbol{N} - \boldsymbol{H}\cdot\boldsymbol{N} = \widehat{h}^{\mathrm{ext}}_m - \mathrm{D}_t\widehat{v}_m \quad \text{on} \quad \partial\mathcal{B}_m. \tag{11.62}$$

Note that due to the ε'_m term as opposed to the ε_m term, the balance of exterior total energy *differs* from the balance of interior total energy on the external boundary $\mathcal{B}_m$.

iii. The balance of exterior total energy on (the regular part of) the boundary $\partial\mathfrak{B}_m$ of a smooth material control volume $\mathfrak{B}_m$ not intersecting with the external boundary, as resulting from particularizing Eq. 11.59 by setting v_m to $\overline{v}_m$, $\widehat{h}^{\mathrm{ext}}_m$ to $\widehat{h}_m$, and $\mathrm{D}_t\widehat{v}_m$ to $-\boldsymbol{v}\cdot\widehat{\boldsymbol{s}}_m$, renders, when restricted to a singular surface, the *extrinsic total energy flux density cancelation*

$$\widehat{h}^+_m + \boldsymbol{v}^+\cdot\widehat{\boldsymbol{s}}^+_m + \widehat{h}^-_m + \boldsymbol{v}^-\cdot\widehat{\boldsymbol{s}}^-_m = 0 \quad \text{on} \quad \mathcal{S}_m \tag{11.63}$$

and translates into the jump condition across the singular surface

$$[\![\overline{\varepsilon}_m]\!]W_\perp + [\![\boldsymbol{E}]\!]\cdot\boldsymbol{M} = 0 \quad \text{on} \quad \mathcal{S}_m. \tag{11.64}$$

iv. At a singular surface $\mathcal{S}_m$, the format of the Piola-type intrinsic exterior total energy flux vector $\boldsymbol{E}$ in Eq. 11.54.1 simplifies the balance of exterior total energy in Eq. 11.60 to a more common format

$$[\![\overline{\varepsilon}_m]\!]W_\perp = -[\![\boldsymbol{v}\cdot\boldsymbol{P}]\!]\cdot\boldsymbol{M} + [\![\boldsymbol{H}]\!]\cdot\boldsymbol{M} \quad \text{on} \quad \mathcal{S}_m. \tag{11.65}$$

Note that due to the $\overline{\varepsilon}_m$ term as opposed to the ε_m term, the balance of exterior total energy *differs* from the balance of interior total energy at singular surfaces $\mathcal{S}_m$.

Incorporating Generic Balance of Kinetic Energy

By subtracting the generic version of the local balance of kinetic energy from the local balance of exterior total energy, whereby $\mathrm{D}_t\overline{v}_m$ and $\mathrm{D}_t\widehat{v}_m$ are identified with

$-\boldsymbol{v}\cdot\overline{\boldsymbol{s}}_{\rm m}$ and $-\boldsymbol{v}\cdot\widehat{\boldsymbol{s}}^{\rm ext}_{\rm m}$, respectively, the local balance of *exterior Lagrange-type internal energy* (with densities $\ell_{\rm m}=\kappa_{\rm m}-\iota_{\rm m}$, $\overline{\ell}_{\rm m}=\ell_{\rm m}-\overline{\upsilon}_{\rm m}$ and $\ell'_{\rm m}=\overline{\ell}_{\rm m}-\upsilon'_{\rm m}$) is obtained:

Local Balance of Exterior Lagrange-Type Internal Energy (Material Control Volume):

i. Regular Points in the Domain

$$[-\mathrm{D}_t\ell_{\rm m}+\boldsymbol{p}_{\rm m}\cdot\mathrm{D}_t\boldsymbol{v}]-\nabla_X\boldsymbol{v}:\boldsymbol{P}+\mathrm{Div}\boldsymbol{H}=\overline{h}_{\rm m}\quad\text{in}\quad\mathcal{B}_{\rm m}\tag{11.66}$$

ii. Regular Points on the Boundary

$$[-\ell'_{\rm m}\boldsymbol{W}-\boldsymbol{H}]\cdot\boldsymbol{N}=\widehat{h}^{\rm ext}_{\rm m}\quad\text{on}\quad\partial\mathcal{B}_{\rm m}\tag{11.67}$$

iii. Regular Points at Singular Surfaces

$$\Big[\big[-[\![\overline{\ell}_{\rm m}]\!]+\{\boldsymbol{p}_{\rm m}\}\cdot[\![\boldsymbol{v}]\!]\big]\boldsymbol{W}+[\![\boldsymbol{v}]\!]\cdot\{\boldsymbol{P}\}-[\![\boldsymbol{H}]\!]\Big]\cdot\boldsymbol{M}=0\quad\text{at}\quad\mathcal{S}_{\rm m}\tag{11.68}$$

Incorporating Specific Balance of Kinetic Energy

Alternatively, by subtracting the specific version of the local balance of kinetic energy, that holds for the case of mass conservation, from the local balance of exterior total energy, a more common local balance of *exterior internal energy* (with densities $\iota_{\rm m}$, $\overline{\iota}_{\rm m}=\iota_{\rm m}+\overline{\upsilon}_{\rm m}$ and $\iota'_{\rm m}=\overline{\iota}_{\rm m}+\upsilon'_{\rm m}$) is obtained:

Local Balance of Exterior Internal Energy (Material Control Volume):

i. Regular Points in the Domain

$$\mathrm{D}_t\iota_{\rm m}+\mathrm{Div}\boldsymbol{H}=\overline{h}_{\rm m}+\nabla_X\boldsymbol{v}:\boldsymbol{P}\quad\text{in}\quad\mathcal{B}_{\rm m}\tag{11.69}$$

ii. Regular Points on the Boundary

$$-\boldsymbol{H}\cdot\boldsymbol{N}=\widehat{h}^{\rm ext}_{\rm m}+\big[[\kappa_{\rm m}-\iota'_{\rm m}]\boldsymbol{W}\big]\cdot\boldsymbol{N}\quad\text{on}\quad\partial\mathcal{B}_{\rm m}\tag{11.70}$$

iii. Regular Points at Singular Surfaces

$$-[\![\boldsymbol{H}]\!]\cdot\boldsymbol{M}=\widetilde{h}_{\rm m}+\big[\!\big[[\kappa_{\rm m}-\overline{\iota}_{\rm m}]\boldsymbol{W}\big]\!\big]\cdot\boldsymbol{M}\quad\text{at}\quad\mathcal{S}_{\rm m}\tag{11.71}$$

11.2.3 Global Format (Spatial Control Volume)

Alternatively, the *balance of exterior total energy* is stated in global format for a spatial control volume as

Global Balance of Exterior Total Energy (Spatial Control Volume):

$$\dot{\mathsf{E}}'(\mathfrak{B}_s) = \mathsf{E}'^{\text{ext}}(\mathfrak{B}_s) \quad \forall \mathfrak{B}_s \subseteq \mathcal{B}_s \tag{11.72}$$

The exterior total energy E' is defined as the summation of the kinetic and the *interior internal* and the *external* energies

$$\mathsf{E}'(\mathfrak{B}_s) := \mathsf{E}(\mathfrak{B}_s) + \mathsf{V}(\mathfrak{B}_s) \quad \text{with} \quad \mathsf{V}(\mathfrak{B}_s) := \int_{\mathfrak{B}_s} \overline{v}_s \, \mathrm{d}v + \int_{\partial\mathfrak{B}_s \cap \partial\mathcal{B}_s} \widehat{v}_s \, \mathrm{d}a. \tag{11.73}$$

with $\overline{v}_s$ and $\widehat{v}_s$ the *external energy density* per unit volume in the domain and per unit area on the external boundary, respectively, of the spatial configuration. These are defined so as to render

$$j \, \mathrm{D}_t \overline{v}_m := -\boldsymbol{v} \cdot \overline{\boldsymbol{s}}_s \quad \text{in} \quad \mathcal{B}_s \quad \text{and} \quad \widehat{j} \, \mathrm{D}_t \widehat{v}_m := -\boldsymbol{v} \cdot \widehat{\boldsymbol{s}}_s^{\text{ext}} \quad \text{on} \quad \partial\mathcal{B}_s. \tag{11.74}$$

Furthermore, the following *exterior Hamilton-type internal energy densities* are introduced as abbreviations

$$\overline{\varepsilon}_s := \varepsilon_s + \overline{v}_s \quad \text{in/at} \quad \mathcal{B}_s/\mathcal{S}_s \quad \text{and} \quad \varepsilon_s' := \overline{\varepsilon}_s + v_s' \quad \text{on} \quad \partial\mathcal{B}_s \tag{11.75}$$

whereby v_s' is defined so as to account for the *curved* external boundary

$$v_s' := \nabla_n \widehat{v}_s - \widehat{v}_s \widehat{c} \quad \text{on} \quad \partial\mathcal{B}_s. \tag{11.76}$$

Then, with $w_\parallel - v_\parallel = 0$, the total time derivative of E' expands as

$$\begin{aligned} \dot{\mathsf{E}}'(\mathfrak{B}_s) = & \int_{\mathfrak{B}_s} \mathrm{d}_t \overline{\varepsilon}_s \, \mathrm{d}v + \int_{\partial\mathfrak{B}_s \not\cap \partial\mathcal{B}_s} [\overline{\varepsilon}_s \boldsymbol{w}] \cdot \boldsymbol{n} \, \mathrm{d}a - \int_{\mathfrak{S}_s} [\![\overline{\varepsilon}_s \boldsymbol{w}]\!] \cdot \boldsymbol{m} \, \mathrm{d}a \\ & + \int_{\partial\mathfrak{B}_s \cap \partial\mathcal{B}_s} \widehat{j} \, \mathrm{D}_t \widehat{v}_s \, \mathrm{d}a + \int_{\partial\mathfrak{B}_s \cap \partial\mathcal{B}_s} [\varepsilon_s' \boldsymbol{w}] \cdot \boldsymbol{n} \, \mathrm{d}a - \int_{\partial\mathfrak{B}_s \cap \partial\mathcal{B}_s} [\, v_s' \boldsymbol{v} \,] \cdot \boldsymbol{n} \, \mathrm{d}a. \end{aligned} \tag{11.77}$$

Moreover, E'^{ext} represents the resultant external (non-mechanical and mechanical) power supply, whereby the mechanical contact interactions are only accounted for at the spatial control volume boundary not intersecting with the external boundary

$$\mathsf{E}^{\text{ext}\prime}(\mathfrak{B}_{\text{s}}) := \mathsf{H}^{\text{ext}}(\mathfrak{B}_{\text{s}}) + \int\limits_{\partial\mathfrak{B}_{\text{s}} \not\cap \partial\mathcal{B}_{\text{s}}} \boldsymbol{v} \cdot \widehat{\boldsymbol{s}}_{\text{s}} \, da. \tag{11.78}$$

The extrinsic total energy flux densities $\widehat{e}_{\text{s}} := \widehat{h}_{\text{s}} + \boldsymbol{v} \cdot \widehat{\boldsymbol{s}}_{\text{s}}$ at the spatial control volume boundary not intersecting with the external boundary and $\widehat{e}^{\text{ext}}_{\text{s}} := \widehat{h}^{\text{ext}}_{\text{s}} - \widehat{j}\, \mathrm{D}_t \widehat{v}_{\text{s}}$ on the external boundary, respectively, along with the respective convective terms stemming from the transport theorem are postulated to satisfy a Cauchy theorem for the *intrinsic exterior total energy flux density*

$$\begin{aligned} e_{\text{s}} \quad &:= \boldsymbol{\varepsilon} \cdot \boldsymbol{n} := \widehat{h}_{\text{s}} \;\; + \boldsymbol{v} \cdot \widehat{\boldsymbol{s}}_{\text{s}} - \left[\overline{\varepsilon}_{\text{s}}[\boldsymbol{w} - \boldsymbol{v}]\right] \cdot \boldsymbol{n} \quad \text{on} \quad \partial\mathfrak{B}_{\text{m}} \not\cap \partial\mathcal{B}_{\text{m}}, \\ e^{\text{ext}}_{\text{s}} &:= \boldsymbol{\varepsilon} \cdot \boldsymbol{n} := \widehat{h}^{\text{ext}}_{\text{s}} - \widehat{j}\, \mathrm{D}_t \widehat{v}_{\text{m}} - \left[\varepsilon'_{\text{s}}[\boldsymbol{w} - \boldsymbol{v}]\right] \cdot \boldsymbol{n} \quad \text{on} \quad \partial\mathfrak{B}_{\text{m}} \cap \partial\mathcal{B}_{\text{m}}. \end{aligned} \tag{11.79}$$

Thereby $\boldsymbol{\varepsilon} \in T\mathcal{B}_{\text{m}}$ is introduced as *Cauchy-type intrinsic exterior total energy flux* consisting of a mechanical and a thermal contribution

$$\boldsymbol{\varepsilon} := \boldsymbol{v} \cdot \boldsymbol{\sigma} - \boldsymbol{\eta}. \tag{11.80}$$

The thermal contribution $\boldsymbol{\eta} \in T\mathcal{B}_{\text{ms}}$ denotes the *Cauchy-type intrinsic heat flux*.

Finally, as a consequence of the above definitions, the global format of the balance of exterior total energy expands as

$$\begin{aligned} \int\limits_{\mathfrak{B}_{\text{s}}} \mathrm{d}_t \overline{\varepsilon}_{\text{s}} \, dv = \int\limits_{\mathfrak{B}_{\text{s}}} \overline{h}_{\text{s}} \, dv + \int\limits_{\partial\mathfrak{B}_{\text{s}} \not\cap \partial\mathcal{B}_{\text{s}}} [\boldsymbol{\varepsilon} - \overline{\varepsilon}_{\text{s}} \boldsymbol{v}] \cdot \boldsymbol{n} \, da + \int\limits_{\mathfrak{S}_{\text{s}}} [\![\overline{\varepsilon}_{\text{s}} \boldsymbol{w}]\!] \cdot \boldsymbol{m} \, da \\ + \int\limits_{\partial\mathfrak{B}_{\text{s}} \cap \partial\mathcal{B}_{\text{s}}} [\boldsymbol{\varepsilon} - \overline{\varepsilon}_{\text{s}} \boldsymbol{v}] \cdot \boldsymbol{n} \, da. \end{aligned} \tag{11.81}$$

11.2.4 Local Format (Spatial Control Volume)

The local balance for the exterior total energy density $\overline{\varepsilon}_{\text{s}}$ per unit volume in the spatial configuration, expressed in terms of the Cauchy-type exterior total energy flux $\boldsymbol{\varepsilon} = \boldsymbol{E} \cdot \operatorname{cof} \boldsymbol{f}$, the heat source density $\overline{h}_{\text{s}}$ per unit volume in the spatial configuration, and the extrinsic heat flux density $\widehat{h}^{\text{ext}}_{\text{s}}$ per unit area in the spatial configuration, is summarized as:

> **Local Balance of Exterior Total Energy (Spatial Control Volume):**
>
> **i. Regular Points in the Domain**
>
> $$\mathrm{d}_t \overline{\varepsilon}_{\text{s}} - \operatorname{div}(\boldsymbol{\varepsilon} - \overline{\varepsilon}_{\text{s}} \boldsymbol{v}) = \overline{h}_{\text{s}} \quad \text{in} \quad \mathcal{B}_{\text{s}} \tag{11.82}$$

ii. Regular Points on the Boundary

$$\widehat{j}\,\mathrm{D}_t\widehat{v}_{\mathrm{m}} + \big[\varepsilon'_{\mathrm{s}}\boldsymbol{w} + [\boldsymbol{\varepsilon} - \varepsilon'_{\mathrm{s}}\boldsymbol{v}]\big]\cdot\boldsymbol{n} = \widehat{h}^{\mathrm{ext}}_{\mathrm{s}} \quad \text{on} \quad \partial\mathcal{B}_{\mathrm{s}} \tag{11.83}$$

iii. Regular Points at Singular Surfaces

$$\big[\!\big[\overline{\varepsilon}_{\mathrm{s}}\boldsymbol{w} + [\boldsymbol{\varepsilon} - \overline{\varepsilon}_{\mathrm{s}}\boldsymbol{v}]\big]\!\big]\cdot\boldsymbol{m} = 0 \quad \text{at} \quad \mathcal{S}_{\mathrm{s}} \tag{11.84}$$

Incorporating Generic Balance of Kinetic Energy

By subtracting the generic version of the local balance of kinetic energy from the local balance of exterior total energy, whereby $j\,\mathrm{D}_t\overline{v}_{\mathrm{m}}$ and $\widehat{j}\,\mathrm{D}_t\widehat{v}_{\mathrm{m}}$ are identified with $-\boldsymbol{v}\cdot\widehat{\boldsymbol{s}}_{\mathrm{s}}$ and $-\boldsymbol{v}\cdot\widehat{\boldsymbol{s}}^{\mathrm{ext}}_{\mathrm{s}}$, respectively, the local balance of *exterior Lagrange-type internal energy* (with density $\ell_{\mathrm{s}} = \kappa_{\mathrm{s}} - \iota_{\mathrm{s}}$, $\overline{\ell}_{\mathrm{s}} = \ell_{\mathrm{s}} - \overline{v}_{\mathrm{s}}$ and $\ell'_{\mathrm{s}} = \overline{\ell}_{\mathrm{s}} - v'_{\mathrm{s}}$) is obtained:

Local Balance of Exterior Lagrange-Type Internal Energy (Spatial Control Volume):

i. Regular Points in the Domain

$$[-\mathrm{d}_t\ell_{\mathrm{s}} + \boldsymbol{p}_{\mathrm{s}}\cdot\mathrm{d}_t\boldsymbol{v}] - \nabla_x\boldsymbol{v} : \boldsymbol{\sigma}_{\mathrm{d}} + \mathrm{div}(\boldsymbol{\eta} - \ell_{\mathrm{s}}\boldsymbol{v}) = \overline{h}_{\mathrm{s}} \quad \text{in} \quad \mathcal{B}_{\mathrm{s}} \tag{11.85}$$

ii. Regular Points on the Boundary

$$\big[-\ell'_{\mathrm{s}}\boldsymbol{w} - [\boldsymbol{\eta} - \ell'_{\mathrm{s}}\boldsymbol{v}]\big]\cdot\boldsymbol{n} = \widehat{h}^{\mathrm{ext}}_{\mathrm{s}} \quad \text{on} \quad \partial\mathcal{B}_{\mathrm{s}} \tag{11.86}$$

iii. Regular Points at Singular Surfaces

$$\Big[\big[-[\![\overline{\ell}_{\mathrm{s}}]\!] + \{\boldsymbol{p}_{\mathrm{s}}\}\cdot[\![\boldsymbol{v}]\!]\big]\boldsymbol{w} + [\![\boldsymbol{v}]\!]\cdot\{\boldsymbol{\sigma}_{\mathrm{d}}\} - [\![\boldsymbol{\eta} - \overline{\ell}_{\mathrm{s}}\boldsymbol{v}]\!]\Big]\cdot\boldsymbol{m} = 0 \quad \text{at} \quad \mathcal{S}_{\mathrm{s}} \tag{11.87}$$

Incorporating Specific Balance of Kinetic Energy

Alternatively, by subtracting the specific version of the local balance of kinetic energy, that holds for the case of mass conservation, from the local balance of exterior total energy, the local balance of *exterior internal energy* (with densities ι_{s}, $\overline{\iota}_{\mathrm{s}} = \iota_{\mathrm{s}} + \overline{v}_{\mathrm{s}}$ and $\iota'_{\mathrm{s}} = \overline{\iota}_{\mathrm{s}} + v'_{\mathrm{s}}$) is obtained:

Local Balance of Exterior Internal Energy (Spatial Control Volume):

i. Regular Points in the Domain

$$\mathrm{d}_t \iota_s + \mathrm{div}(\boldsymbol{\eta} + \iota_s \boldsymbol{v}) = \overline{h}_s + \nabla_x \boldsymbol{v} : \boldsymbol{\sigma} \quad \text{in} \quad \mathcal{B}_s \tag{11.88}$$

ii. Regular Points on the Boundary

$$-\boldsymbol{\eta} \cdot \boldsymbol{n} = \widehat{h}_s^{\mathrm{ext}} + \big[[\kappa_s - \iota_s'][\boldsymbol{w} - \boldsymbol{v}]\big] \cdot \boldsymbol{n} \quad \text{on} \quad \partial\mathcal{B}_s \tag{11.89}$$

iii. Regular Points at Singular Surfaces

$$-[\![\boldsymbol{\eta}]\!] \cdot \boldsymbol{m} = \widetilde{h}_s + \big[\!\big[[\kappa_s - \overline{\iota}_s][\boldsymbol{w} - \boldsymbol{v}]\big]\!\big] \cdot \boldsymbol{m} \quad \text{at} \quad \mathcal{S}_s \tag{11.90}$$

11.2.5 Balance Tetragon

The local balance of exterior total energy *at regular points in the domain*, *at regular points on the boundary*, and *at regular points at singular surfaces* may be expressed in four different but fully equivalent versions, see the arrangement in the corresponding tetragons in Table 11.4. Here *dynamic versions* of the intrinsic exterior total energy fluxes have been introduced as $\boldsymbol{\varepsilon}_\mathrm{D} = \boldsymbol{E}_\mathrm{D} \cdot \mathrm{cof}\, \boldsymbol{f}$ and $\boldsymbol{\varepsilon}_\mathrm{d} = \boldsymbol{E}_\mathrm{d} \cdot \mathrm{cof}\, \boldsymbol{f}$, respectively. The subscripts D and d indicate that the corresponding exterior total energy flux is consistent with either the material or the spatial time derivative D_t or d_t, respectively, of the exterior total energy density.

Likewise, the local balance of exterior internal energy *at regular points in the domain*, *at regular points on the boundary*, and *at regular points at singular surfaces* may be expressed in four different but fully equivalent versions, see the arrangement in the corresponding tetragons in Table 11.5. Here *dynamic versions* of the heat fluxes have been introduced as $\boldsymbol{\eta}_\mathrm{D} = \boldsymbol{H}_\mathrm{D} \cdot \mathrm{cof}\, \boldsymbol{f}$ and $\boldsymbol{\eta}_\mathrm{d} = \boldsymbol{H}_\mathrm{d} \cdot \mathrm{cof}\, \boldsymbol{f}$, respectively. The subscripts D and d indicate that the corresponding heat flux is consistent with either the material or the spatial time derivative D_t or d_t, respectively, of the internal energy density.

The abbreviations for *effective* internal energy source terms in the domain, on the boundary, and at singular surfaces, respectively, have already been introduced in Table 11.1.

Table 11.4 Tetragon of fully equivalent versions for the local balance of exterior total energy at regular points in the domain, on the boundary and at singular surfaces

Regular Points in the Domain

	PT	
$\mathrm{D}_t\bar{\varepsilon}_\mathrm{m} = \mathrm{Div}\boldsymbol{E}_\mathrm{D} + \bar{h}_\mathrm{m}$	⟷ PT	$j\,\mathrm{D}_t\bar{\varepsilon}_\mathrm{m} = \mathrm{div}\boldsymbol{\varepsilon}_\mathrm{D} + \bar{h}_\mathrm{s}$
↕ TT		↕ TT
$J\,\mathrm{d}_t\bar{\varepsilon}_\mathrm{s} = \mathrm{Div}\boldsymbol{E}_\mathrm{d} + \bar{h}_\mathrm{m}$	⟷ PT	$\mathrm{d}_t\bar{\varepsilon}_\mathrm{s} = \mathrm{div}\boldsymbol{\varepsilon}_\mathrm{d} + \bar{h}_\mathrm{s}$

Regular Points on the Boundary

$\left[\varepsilon'_\mathrm{m}\boldsymbol{W} + \boldsymbol{E}_\mathrm{D}\right]\cdot\boldsymbol{N} = \widehat{e}^\mathrm{ext}_\mathrm{m}$	⟷ PT	$\left[\varepsilon'_\mathrm{s}[\boldsymbol{w}-\boldsymbol{v}] + \boldsymbol{\varepsilon}_\mathrm{D}\right]\cdot\boldsymbol{n} = \widehat{e}^\mathrm{ext}_\mathrm{s}$
↕ TT		↕ TT
$\left[\varepsilon'_\mathrm{m}[\boldsymbol{W}-\boldsymbol{V}] + \boldsymbol{E}_\mathrm{d}\right]\cdot\boldsymbol{N} = \widehat{e}^\mathrm{ext}_\mathrm{m}$	⟷ PT	$\left[\varepsilon'_\mathrm{s}\boldsymbol{w} + \boldsymbol{\varepsilon}_\mathrm{d}\right]\cdot\boldsymbol{n} = \widehat{e}^\mathrm{ext}_\mathrm{s}$

Regular Points at Singular Surfaces

$[\![\bar{\varepsilon}_\mathrm{m}\boldsymbol{W} + \boldsymbol{E}_\mathrm{D}]\!]\cdot\boldsymbol{M} = 0$	⟷ PT	$[\![\bar{\varepsilon}_\mathrm{s}[\boldsymbol{w}-\boldsymbol{v}] + \boldsymbol{\varepsilon}_\mathrm{D}]\!]\cdot\boldsymbol{m} = 0$
↕ TT		↕ TT
$[\![\bar{\varepsilon}_\mathrm{m}[\boldsymbol{W}-\boldsymbol{V}] + \boldsymbol{E}_\mathrm{d}]\!]\cdot\boldsymbol{M} = 0$	⟷ PT	$[\![\bar{\varepsilon}_\mathrm{s}\boldsymbol{w} + \boldsymbol{\varepsilon}_\mathrm{d}]\!]\cdot\boldsymbol{m} = 0$

with *Dynamic Versions* of Intrinsic Exterior Total Energy Fluxes

$\boldsymbol{E}_\mathrm{D} := \boldsymbol{E}$	⟷ PT	$\boldsymbol{\varepsilon}_\mathrm{D} := \boldsymbol{\varepsilon}$
↕ TT		↕ TT
$\boldsymbol{E}_\mathrm{d} := \boldsymbol{E} + \bar{\varepsilon}_\mathrm{m}\boldsymbol{V} \mid \boldsymbol{E}_\mathrm{d} := \boldsymbol{E} + \varepsilon'_\mathrm{m}\boldsymbol{V}$	⟷ PT	$\boldsymbol{\varepsilon}_\mathrm{d} := \boldsymbol{\varepsilon} - \bar{\varepsilon}_\mathrm{s}\boldsymbol{v} \mid \boldsymbol{\varepsilon}_\mathrm{d} := \boldsymbol{\varepsilon} - \varepsilon'_\mathrm{s}\boldsymbol{v}$

Table 11.5 Tetragon of fully equivalent versions for the local balance of exterior internal energy at regular points in the domain, on the boundary and at singular surfaces

Regular Points in the Domain

$\mathrm{D}_t\iota_\mathrm{m} = -\mathrm{Div}\boldsymbol{H}_\mathrm{D} + \overline{k}_\mathrm{m}$	⟷ PT	$j\,\mathrm{D}_t\iota_\mathrm{m} = -\mathrm{div}\boldsymbol{\eta}_\mathrm{D} + \overline{k}_\mathrm{s}$
↕ TT		TT ↕
$J\,\mathrm{d}_t\iota_\mathrm{s} = -\mathrm{Div}\boldsymbol{H}_\mathrm{d} + \overline{k}_\mathrm{m}$	PT ⟷	$\mathrm{d}_t\iota_\mathrm{s} = -\mathrm{div}\boldsymbol{\eta}_\mathrm{d} + \overline{k}_\mathrm{s}$

Regular Points on the Boundary

$[\iota'_\mathrm{m}\boldsymbol{W} - \boldsymbol{H}_\mathrm{D}]\cdot\boldsymbol{N} = \widehat{k}_\mathrm{m}$	⟷ PT	$[\iota'_\mathrm{s}[\boldsymbol{w} - \boldsymbol{v}] - \boldsymbol{\eta}_\mathrm{D}]\cdot\boldsymbol{n} = \widehat{k}_\mathrm{s}$
↕ TT		TT ↕
$[\iota'_\mathrm{m}[\boldsymbol{W} - \boldsymbol{V}] - \boldsymbol{H}_\mathrm{d}]\cdot\boldsymbol{N} = \widehat{k}_\mathrm{m}$	PT ⟷	$[\iota'_\mathrm{s}\boldsymbol{w} - \boldsymbol{\eta}_\mathrm{d}]\cdot\boldsymbol{n} = \widehat{k}_\mathrm{s}$

Regular Points at Singular Surfaces

$[\![\bar{\iota}_\mathrm{m}\boldsymbol{W} - \boldsymbol{H}_\mathrm{D}]\!]\cdot\boldsymbol{M} = \widetilde{k}_\mathrm{m}$	⟷ PT	$[\![\bar{\iota}_\mathrm{s}[\boldsymbol{w} - \boldsymbol{v}] - \boldsymbol{\eta}_\mathrm{D}]\!]\cdot\boldsymbol{m} = \widetilde{k}_\mathrm{s}$
↕ TT		TT ↕
$[\![\bar{\iota}_\mathrm{m}[\boldsymbol{W} - \boldsymbol{V}] - \boldsymbol{H}_\mathrm{d}]\!]\cdot\boldsymbol{M} = \widetilde{k}_\mathrm{m}$	PT ⟷	$[\![\bar{\iota}_\mathrm{s}\boldsymbol{w} - \boldsymbol{\eta}_\mathrm{d}]\!]\cdot\boldsymbol{m} = \widetilde{k}_\mathrm{s}$

with *Dynamic Versions* of Intrinsic Heat Fluxes

$\boldsymbol{H}_\mathrm{D} := \boldsymbol{H}$	⟷ PT	$\boldsymbol{\eta}_\mathrm{D} := \boldsymbol{\eta}$
↕ TT		TT ↕
$\boldsymbol{H}_\mathrm{d} := \boldsymbol{H} - \bar{\iota}_\mathrm{m}\boldsymbol{V} \mid \boldsymbol{H}_\mathrm{d} := \boldsymbol{H} - \iota'_\mathrm{m}\boldsymbol{V}$	PT ⟷	$\boldsymbol{\eta}_\mathrm{d} := \boldsymbol{\eta} + \bar{\iota}_\mathrm{s}\boldsymbol{v} \mid \boldsymbol{\eta}_\mathrm{d} := \boldsymbol{\eta} + \iota'_\mathrm{s}\boldsymbol{v}$

11.3 Entropy

The balance of entropy dictates the direction of macroscopic thermodynamic processes. Its continuum formulation necessitates introduction of the generic concepts *entropy source, entropy flux*, and *positive entropy production.*

11.3.1 Global Format (Material Control Volume)

The *balance of entropy* is stated in global format for a material control volume

Global Balance of Entropy (Material Control Volume):

$$\dot{\mathsf{S}}(\mathfrak{B}_{\mathrm{m}}) = \mathsf{S}^{\mathrm{ext}}(\mathfrak{B}_{\mathrm{m}}) + \mathsf{S}^{\mathrm{pro}}(\mathfrak{B}_{\mathrm{m}}) \quad \forall \mathfrak{B}_{\mathrm{m}} \subseteq \mathcal{B}_{\mathrm{m}} \tag{11.91}$$

Thereby S is expressed as the resultant of the *entropy density* σ_{m} per unit volume in the material configuration over the material control volume $\mathfrak{B}_{\mathrm{m}}$, i.e. as the *resultant entropy* contained in the material control volume

$$\mathsf{S}(\mathfrak{B}_{\mathrm{m}}) := \int_{\mathfrak{B}_{\mathrm{m}}} \sigma_{\mathrm{m}} \, \mathrm{d}V. \tag{11.92}$$

Moreover, $\mathsf{S}^{\mathrm{ext}}$ represents the resultant external entropy input and consists of volume and boundary integrals (any singular contributions are here ignored for the sake of presentation)

$$\mathsf{S}^{\mathrm{ext}}(\mathfrak{B}_{\mathrm{m}}) := \int_{\mathfrak{B}_{\mathrm{m}}} \overline{n}_{\mathrm{m}} \, \mathrm{d}V + \int_{\partial\mathfrak{B}_{\mathrm{m}}} \widehat{n}_{\mathrm{m}} \, \mathrm{d}A \tag{11.93}$$

in terms of the *entropy source density* $\overline{n}_{\mathrm{m}}$ per unit volume in the material configuration and the *extrinsic entropy flux density* $\widehat{n}_{\mathrm{m}}$ per unit area in the material configuration (both to be specified later).

Finally, as the unique feature of the balance of entropy, $\mathsf{S}^{\mathrm{pro}}$ represents the resultant entropy *production* that is (globally) constrained by

$$\mathsf{S}^{\mathrm{pro}} \geq 0. \tag{11.94}$$

Thereby, remarkably, the entropy production consists of volume, boundary *and* surface integrals

$$\mathsf{S}^{\mathrm{pro}}(\mathfrak{B}_{\mathrm{m}}) := \int_{\mathfrak{B}_{\mathrm{m}}} \overline{\pi}_{\mathrm{m}} \, \mathrm{d}V + \int_{\partial\mathfrak{B}_{\mathrm{m}} \cap \partial\mathcal{B}_{\mathrm{m}}} \widehat{\pi}_{\mathrm{m}} \, \mathrm{d}A + \int_{\mathfrak{S}_{\mathrm{m}}} \widetilde{\pi}_{\mathrm{m}} \, \mathrm{d}A \tag{11.95}$$

in terms of the *entropy production source density* $\overline{\pi}_{\mathrm{m}}$ per unit volume in the material configuration and the *extrinsic entropy production flux densities* $\widehat{\pi}_{\mathrm{m}}$ and $\widetilde{\pi}_{\mathrm{m}}$ per unit area in the material configuration, i.e. on $\partial\mathfrak{B}_{\mathrm{m}} \cap \partial\mathcal{B}_{\mathrm{m}}$ and $\mathfrak{S}_{\mathrm{m}}$, respectively.

Observe (i) that (due to reasons motivated below) the extrinsic entropy production flux density $\widehat{\pi}_{\mathrm{m}}$ only arises at the external boundary $\partial\mathcal{B}_{\mathrm{m}}$ and (ii) the appearance of the additional extrinsic entropy production flux density $\widetilde{\pi}_{\mathrm{m}}$ at the singular surface $\mathcal{S}_{\mathrm{m}}$ (which will be demonstrated in the sequel to result from the addition of the respective extrinsic entropy production fluxes $\widetilde{\pi}_{\mathrm{m}} := \widehat{\pi}^{+}_{\mathrm{m}} + \widehat{\pi}^{-}_{\mathrm{m}}$ at both sides of the singular surface $\mathcal{S}_{\mathrm{m}}$).

Then, as the key aspect of the balance of entropy, the entropy production has to satisfy the (point-wise) *positivity constraints*

$$\overline{\pi}_{\mathrm{m}} \geq 0 \quad \text{and} \quad \widehat{\pi}_{\mathrm{m}} \geq 0 \quad \text{and} \quad \widetilde{\pi}_{\mathrm{m}} \geq 0 \tag{11.96}$$

so as to provide irreversible processes with a direction. As a consequence of the positivity constraint $\widehat{\pi}_{\mathrm{m}} \geq 0$, the extrinsic entropy production flux density cannot satisfy a corresponding cancelation condition at any arbitrary internal (regular) cut-surface (i.e. at any empty intersection $\partial\mathfrak{B}_{\mathrm{m}} \cap \partial\mathcal{B}_{\mathrm{m}} = \emptyset$ of a smooth material control volume boundary with the external boundary), thus resulting in the above observation (i).

Note that it is only the explicit incorporation of the entropy production that allows to formulate a *balance* of entropy rather than the more common *imbalance* of entropy, that is given in global format as

Global Imbalance of Entropy (Material Control Volume):

$$\dot{\mathsf{S}}(\mathfrak{B}_{\mathrm{m}}) - \mathsf{S}^{\mathrm{ext}}(\mathfrak{B}_{\mathrm{m}}) = \mathsf{S}^{\mathrm{pro}}(\mathfrak{B}_{\mathrm{m}}) \geq 0 \quad \forall \mathfrak{B}_{\mathrm{m}} \subseteq \mathcal{B}_{\mathrm{m}} \tag{11.97}$$

The imbalance of entropy states that the total rate of change of the resultant entropy contained in a material control volume $\mathfrak{B}_{\mathrm{m}}$ *is not less than the resultant external entropy input supplied to* $\mathfrak{B}_{\mathrm{m}}$.

The extrinsic entropy and entropy production flux densities along with the convective term stemming from the transport theorem are postulated to satisfy a Cauchy theorem for the *intrinsic entropy flux density* for any intersection $\partial\mathfrak{B}_{\mathrm{m}} \cap \partial\mathcal{B}_{\mathrm{m}}$ of a smooth material control volume boundary with the external boundary

$$n^{\mathrm{ext}}_{\mathrm{m}} := -\boldsymbol{S} \cdot \boldsymbol{N} := \widehat{n}^{\mathrm{ext}}_{\mathrm{m}} + \widehat{\pi}_{\mathrm{m}} - [\sigma_{\mathrm{m}} \boldsymbol{W}] \cdot \boldsymbol{N} \quad \text{on} \quad \partial\mathfrak{B}_{\mathrm{m}} \cap \partial\mathcal{B}_{\mathrm{m}}. \tag{11.98}$$

Note, however, that for any empty intersection $\partial\mathfrak{B}_{\mathrm{m}} \cap \partial\mathcal{B}_{\mathrm{m}} = \emptyset$ of a smooth material control volume boundary with the external boundary, the extrinsic entropy production flux density vanishes, i.e.

$$\widehat{\pi}_{\mathrm{m}} = 0 \quad \text{on} \quad \partial\mathfrak{B}_{\mathrm{m}} \not\cap \partial\mathcal{B}_{\mathrm{m}}, \tag{11.99}$$

thus in this case, the intrinsic entropy flux density is as follows:

$$n_{\mathrm{m}} := -\boldsymbol{S}\cdot\boldsymbol{N} := \widehat{n}_{\mathrm{m}} - [\sigma_{\mathrm{m}}\boldsymbol{W}]\cdot\boldsymbol{N} \quad \text{on} \quad \partial\mathfrak{B}_{\mathrm{m}} \not\cap \partial\mathcal{B}_{\mathrm{m}}. \tag{11.100}$$

In the above, $\boldsymbol{S} \in T\mathcal{B}_{\mathrm{m}}$ is introduced as the *Piola-type intrinsic entropy flux* (the minus sign indicating the *system's egoistic point of view*, i.e. entropy is flowing *into* the material control volume if $\boldsymbol{S}$ and the outwards pointing normal $\boldsymbol{N}$ form an obtuse angle such that $\boldsymbol{S}\cdot\boldsymbol{N} < 0$).

Finally, as a consequence of the above definitions, the global format of the balance of entropy expands as

$$\begin{aligned}\int\limits_{\mathfrak{B}_{\mathrm{m}}} \mathrm{D}_t\sigma_{\mathrm{m}}\,\mathrm{d}V = \int\limits_{\mathfrak{B}_{\mathrm{m}}} \overline{n}_{\mathrm{m}}\,\mathrm{d}V - \int\limits_{\partial\mathfrak{B}_{\mathrm{m}}\not\cap\partial\mathcal{B}_{\mathrm{m}}} \underbrace{\boldsymbol{S}\cdot\boldsymbol{N}}_{-n_{\mathrm{m}}}\,\mathrm{d}A + \int\limits_{\mathfrak{S}_{\mathrm{m}}} [\![\sigma_{\mathrm{m}}\boldsymbol{W}]\!]\cdot\boldsymbol{M}\,\mathrm{d}A \\ + \int\limits_{\mathfrak{B}_{\mathrm{m}}} \overline{\pi}_{\mathrm{m}}\,\mathrm{d}V - \int\limits_{\partial\mathfrak{B}_{\mathrm{m}}\cap\partial\mathcal{B}_{\mathrm{m}}} \underbrace{\boldsymbol{S}\cdot\boldsymbol{N}}_{-n^{\mathrm{ext}}_{\mathrm{m}}}\,\mathrm{d}A + \int\limits_{\mathfrak{S}_{\mathrm{m}}} \widetilde{\pi}_{\mathrm{m}}\,\mathrm{d}A.\end{aligned} \tag{11.101}$$

11.3.2 Local Format (Material Control Volume)

The results of localizing the global balance of entropy are summarized as:

Local Balance of Entropy (Material Control Volume):

i. Regular Points in the Domain

$$\mathrm{D}_t\sigma_{\mathrm{m}} + \mathrm{Div}\boldsymbol{S} = \overline{n}_{\mathrm{m}} + \overline{\pi}_{\mathrm{m}} \quad \text{in} \quad \mathcal{B}_{\mathrm{m}} \tag{11.102}$$

ii. Regular Points on the Boundary

$$[\sigma_{\mathrm{m}}\boldsymbol{W} - \boldsymbol{S}]\cdot\boldsymbol{N} = \widehat{n}^{\mathrm{ext}}_{\mathrm{m}} + \widehat{\pi}_{\mathrm{m}} \quad \text{on} \quad \partial\mathcal{B}_{\mathrm{m}} \tag{11.103}$$

iii. Regular Points at Singular Surfaces

$$[\![\sigma_{\mathrm{m}}\boldsymbol{W} - \boldsymbol{S}]\!]\cdot\boldsymbol{M} = -\widetilde{\pi}_{\mathrm{m}} \quad \text{at} \quad \mathcal{S}_{\mathrm{m}} \tag{11.104}$$

Observe that the *extrinsic entropy flux density cancelation* $\widehat{n}^{+}_{\mathrm{m}} + \widehat{n}^{-}_{\mathrm{m}} = 0$ on $\mathcal{S}_{\mathrm{m}}$ together with the definition for the extrinsic entropy production flux density at the singular surface

$$\widetilde{\pi}_{\mathrm{m}} := \widehat{\pi}^{+}_{\mathrm{m}} + \widehat{\pi}^{-}_{\mathrm{m}} \quad \text{at} \quad \mathcal{S}_{\mathrm{m}} \tag{11.105}$$

is consistent with the jump condition $[\![\sigma_{\rm m}]\!] W_\perp - [\![\boldsymbol{S}]\!] \cdot \boldsymbol{M} = -\widetilde{\pi}_{\rm m}$ across $\mathcal{S}_{\rm m}$.

Note that since $\widehat{\pi}_{\rm m}^{\pm} \geq 0$ and thus $\widetilde{\pi}_{\rm m} \geq 0$ no cancelation for the extrinsic entropy *production* flux density may be formulated.

11.3.3 Global Format (Spatial Control Volume)

Alternatively, the *balance of entropy* is stated in global format for a spatial control volume as

Global Balance of Entropy (Spatial Control Volume):

$$\dot{\mathsf{S}}(\mathfrak{B}_{\rm s}) = \mathsf{S}^{\rm ext}(\mathfrak{B}_{\rm s}) + \mathsf{S}^{\rm pro}(\mathfrak{B}_{\rm s}) \quad \forall \mathfrak{B}_{\rm s} \subseteq \mathcal{B}_{\rm s} \tag{11.106}$$

Here S is expressed as the resultant of the *entropy density* $\sigma_{\rm s}$ per unit volume in the spatial configuration over the spatial control volume $\mathfrak{B}_{\rm s}$

$$\mathsf{S}(\mathfrak{B}_{\rm s}) := \int_{\mathfrak{B}_{\rm s}} \sigma_{\rm s} \, {\rm d}v. \tag{11.107}$$

Moreover, $\mathsf{S}^{\rm ext}$ represents the resultant external entropy input and consists of volume and boundary integrals (any singular contributions are here ignored for the sake of presentation)

$$\mathsf{S}^{\rm ext}(\mathfrak{B}_{\rm s}) := \int_{\mathfrak{B}_{\rm s}} \overline{n}_{\rm s} \, {\rm d}v + \int_{\partial \mathfrak{B}_{\rm s}} \widehat{n}_{\rm s} \, {\rm d}a \tag{11.108}$$

in terms of the *entropy source density* $\overline{n}_{\rm s}$ per unit volume in the spatial configuration and the *extrinsic entropy flux density* $\widehat{n}_{\rm s}$ per unit area in the spatial configuration.

Finally $\mathsf{S}^{\rm pro} \geq 0$ represents the resultant external entropy *production* consisting of volume, boundary *and* surface integrals

$$\mathsf{S}^{\rm pro}(\mathfrak{B}_{\rm s}) := \int_{\mathfrak{B}_{\rm s}} \overline{\pi}_{\rm s} \, {\rm d}v + \int_{\partial \mathfrak{B}_{\rm s} \cap \partial \mathcal{B}_{\rm s}} \widehat{\pi}_{\rm s} \, {\rm d}a + \int_{\mathfrak{S}_{\rm s}} \widetilde{\pi}_{\rm s} \, {\rm d}a \tag{11.109}$$

in terms of the *entropy production source density* $\overline{\pi}_{\rm s} \geq 0$ per unit volume in the spatial configuration and the *extrinsic entropy production flux densities* $\widehat{\pi}_{\rm s} \geq 0$ and $\widetilde{\pi}_{\rm s} \geq 0$ per unit area in the spatial configuration, i.e. on $\partial \mathfrak{B}_{\rm s} \cap \partial \mathcal{B}_{\rm s}$ and $\mathfrak{S}_{\rm s}$, respectively.

The intrinsic entropy flux is defined as the Piola transformation $\boldsymbol{v} := \boldsymbol{S} \cdot {\rm cof}\, \boldsymbol{f}$, thus the corresponding Cauchy theorem for the *intrinsic entropy flux density* reads in terms of the extrinsic entropy and entropy production flux densities along with the convective term stemming from the transport theorem

$$n_s^{\text{ext}} := -\boldsymbol{\nu} \cdot \boldsymbol{n} := \widehat{n}_s^{\text{ext}} + \widehat{\pi}_s - \big[\sigma_m[\boldsymbol{w} - \boldsymbol{v}]\big] \cdot \boldsymbol{n} \quad \text{on} \quad \partial\mathfrak{B}_s \cap \partial\mathcal{B}_s, \tag{11.110}$$
$$n_s \;\; := -\boldsymbol{\nu} \cdot \boldsymbol{n} := \widehat{n}_s \qquad\qquad - \big[\sigma_m[\boldsymbol{w} - \boldsymbol{v}]\big] \cdot \boldsymbol{n} \quad \text{on} \quad \partial\mathfrak{B}_s \not\cap \partial\mathcal{B}_s.$$

Thereby $\boldsymbol{\nu} \in T\mathcal{B}_s$ is introduced as the *Cauchy-type intrinsic entropy flux*.

Finally, as a consequence of the above definitions, the global format of the balance of entropy expands as

$$\int_{\mathfrak{B}_s} \mathrm{d}_t\sigma_s \,\mathrm{d}v = \int_{\mathfrak{B}_s} \overline{n}_s \,\mathrm{d}v - \int_{\partial\mathfrak{B}_s \not\cap \partial\mathcal{B}_s} [\boldsymbol{\nu} + \sigma_s \boldsymbol{v}] \cdot \boldsymbol{n} \,\mathrm{d}a + \int_{\mathfrak{S}_s} [\![\sigma_s \boldsymbol{w}]\!] \cdot \boldsymbol{m} \,\mathrm{d}a \tag{11.111}$$
$$+ \int_{\mathfrak{B}_s} \overline{\pi}_s \,\mathrm{d}v - \int_{\partial\mathfrak{B}_s \cap \partial\mathcal{B}_s} [\boldsymbol{\nu} + \sigma_s \boldsymbol{v}] \cdot \boldsymbol{n} \,\mathrm{d}a + \int_{\mathfrak{S}_s} \widetilde{\pi}_s \,\mathrm{d}a.$$

11.3.4 Local Format (Spatial Control Volume)

The local balance for the entropy density σ_s per unit volume in the spatial configuration, expressed in terms of the Cauchy-type total entropy flux $\boldsymbol{\nu} = \boldsymbol{S} \cdot \operatorname{cof} \boldsymbol{f}$, the entropy and entropy production source densities $\overline{n}_s$ and $\overline{\pi}_s \geq 0$ per unit volume in the spatial configuration and the extrinsic entropy and entropy production flux densities $\widehat{n}_s^{\text{ext}}$, $\widehat{\pi}_s \geq 0$ and $\widetilde{\pi}_s \geq 0$ per unit area in the spatial configuration, is summarized as:

Local Balance of Entropy (Spatial Control Volume):

i. Regular Points in the Domain

$$\mathrm{d}_t\sigma_s + \operatorname{div}(\boldsymbol{\nu} + \sigma_s \boldsymbol{v}) = \overline{n}_s + \overline{\pi}_s \quad \text{in} \quad \mathcal{B}_s \tag{11.112}$$

ii. Regular Points on the Boundary

$$\big[\sigma_s \boldsymbol{w} - [\boldsymbol{\nu} + \sigma_s \boldsymbol{v}]\big] \cdot \boldsymbol{n} = \widehat{n}_s^{\text{ext}} + \widehat{\pi}_s \quad \text{on} \quad \partial\mathcal{B}_s \tag{11.113}$$

iii. Regular Points at Singular Surfaces

$$\big[\!\big[\sigma_s \boldsymbol{w} - [\boldsymbol{\nu} + \sigma_s \boldsymbol{v}]\big]\!\big] \cdot \boldsymbol{m} = -\widetilde{\pi}_s \quad \text{at} \quad \mathcal{S}_s \tag{11.114}$$

Table 11.6 Effective entropy source terms in regular points in the domain, on the boundary and at singular surfaces

Regular Points in the Domain

$$\overline{s}_{\mathrm{m}} := \overline{n}_{\mathrm{m}} + \overline{\pi}_{\mathrm{m}} \quad \overset{\text{PT}}{\longleftrightarrow} \quad \overline{s}_{\mathrm{s}} := \overline{n}_{\mathrm{s}} + \overline{\pi}_{\mathrm{s}}$$

Regular Points on the Boundary

$$\widehat{s}_{\mathrm{m}} := \widehat{n}_{\mathrm{m}}^{\mathrm{ext}} + \widehat{\pi}_{\mathrm{m}} \quad \overset{\text{PT}}{\longleftrightarrow} \quad \widehat{s}_{\mathrm{s}} := \widehat{n}_{\mathrm{s}}^{\mathrm{ext}} + \widehat{\pi}_{\mathrm{s}}$$

Regular Points at Singular Surfaces

$$\widetilde{s}_{\mathrm{m}} := -\widetilde{\pi}_{\mathrm{m}} \quad \overset{\text{PT}}{\longleftrightarrow} \quad \widetilde{s}_{\mathrm{s}} := -\widetilde{\pi}_{\mathrm{s}}$$

11.3.5 Balance Tetragon

The local balance of entropy *at regular points in the domain*, *at regular points on the boundary*, and *at regular points at singular surfaces* may be expressed in four different but fully equivalent versions, see the arrangement in the corresponding tetragons in Table 11.7. Here *dynamic versions* of the entropy fluxes have been introduced as $\boldsymbol{\nu}_{\mathrm{D}} = \boldsymbol{S}_{\mathrm{D}} \cdot \operatorname{cof} \boldsymbol{f}$ and $\boldsymbol{\nu}_{\mathrm{d}} = \boldsymbol{S}_{\mathrm{d}} \cdot \operatorname{cof} \boldsymbol{f}$, respectively, whereby the subscripts D and d indicate that the corresponding heat flux is consistent with either the material or the spatial time derivative D_t or d_t, respectively, of the entropy density.

Moreover, abbreviations for *effective* entropy source terms in the domain, on the boundary, and at singular surfaces, respectively, have been introduced in Table 11.6.

Table 11.7 Tetragon of fully equivalent versions for the local balance of entropy at regular points in the domain, on the boundary and at singular surfaces

Regular Points in the Domain

$\mathrm{D}_t\sigma_\mathrm{m} = -\mathrm{Div}\boldsymbol{S}_\mathrm{D} + \overline{s}_\mathrm{m}$	⟷ PT	$j\,\mathrm{D}_t\sigma_\mathrm{m} = -\mathrm{div}\boldsymbol{\nu}_\mathrm{D} + \overline{s}_\mathrm{s}$
↕ TT		↕ TT
$J\,\mathrm{d}_t\sigma_\mathrm{s} = -\mathrm{Div}\boldsymbol{S}_\mathrm{d} + \overline{s}_\mathrm{m}$	PT ⟷	$\mathrm{d}_t\sigma_\mathrm{s} = -\mathrm{div}\boldsymbol{\nu}_\mathrm{d} + \overline{s}_\mathrm{s}$

Regular Points on the Boundary

$\left[\sigma_\mathrm{m}\boldsymbol{W} - \boldsymbol{S}_\mathrm{D}\right] \cdot \boldsymbol{N} = \widehat{s}_\mathrm{m}$	⟷ PT	$\left[\sigma_\mathrm{s}[\boldsymbol{w} - \boldsymbol{v}] - \boldsymbol{\nu}_\mathrm{D}\right] \cdot \boldsymbol{n} = \widehat{s}_\mathrm{s}$
↕ TT		↕ TT
$\left[\sigma_\mathrm{m}[\boldsymbol{W} - \boldsymbol{V}] - \boldsymbol{S}_\mathrm{d}\right] \cdot \boldsymbol{N} = \widehat{s}_\mathrm{m}$	PT ⟷	$\left[\sigma_\mathrm{s}\boldsymbol{w} - \boldsymbol{\nu}_\mathrm{d}\right] \cdot \boldsymbol{n} = \widehat{s}_\mathrm{s}$

Regular Points at Singular Surfaces

$[\![\sigma_\mathrm{m}\boldsymbol{W} - \boldsymbol{S}_\mathrm{D}]\!] \cdot \boldsymbol{M} = \widetilde{s}_\mathrm{m}$	⟷ PT	$[\![\sigma_\mathrm{s}[\boldsymbol{w} - \boldsymbol{v}] - \boldsymbol{\nu}_\mathrm{D}]\!] \cdot \boldsymbol{m} = \widetilde{s}_\mathrm{s}$
↕ TT		↕ TT
$[\![\sigma_\mathrm{m}[\boldsymbol{W} - \boldsymbol{V}] - \boldsymbol{S}_\mathrm{d}]\!] \cdot \boldsymbol{M} = \widetilde{s}_\mathrm{m}$	PT ⟷	$[\![\sigma_\mathrm{s}\boldsymbol{w} - \boldsymbol{\nu}_\mathrm{d}]\!] \cdot \boldsymbol{m} = \widetilde{s}_\mathrm{s}$

with *Dynamic Versions* of Intrinsic Entropy Flux

$\boldsymbol{S}_\mathrm{D} := \boldsymbol{S}$	⟷ PT	$\boldsymbol{\nu}_\mathrm{D} := \boldsymbol{\nu}$
↕ TT		↕ TT
$\boldsymbol{S}_\mathrm{d} := \boldsymbol{S} - \sigma_\mathrm{m}\boldsymbol{V}$	PT ⟷	$\boldsymbol{\nu}_\mathrm{d} := \boldsymbol{\nu} + \sigma_\mathrm{s}\boldsymbol{v}$

Chapter 12
Consequences of Thermodynamical Balances

8,610 m 35°52'57"N 76°30'48"E

Abstract This chapter exploits the consequences of the thermodynamical balances and the resulting formats of the dissipation power inequalities by identifying the forces driving material (configurational) changes on the boundary and at singular surfaces as the appropriate contributions to the balance of material momentum. It is the here advocated thermodynamical derivation of material forces that qualifies the current approach as being dissipation-consistent.

Typically, the Clausius–Duhem assumption specifies the entropy flux and source densities in terms of the corresponding heat flux and source densities, and the absolute temperature. More general expansions in terms of extra entropy flux and source densities are possible, however, for the sake of presentation, these shall not be considered here.
Incorporating a suited form of the (interior or exterior) internal energy balance renders the dissipation power inequality in the domain, on the boundary, and at singular surfaces. The dissipation power inequality may be expressed in three equivalent Versions I to III. Version I is especially beneficial when comparing with the dissipation power inequality expressed in terms of the material (Eshelby) stress in Version III. Exploiting Version II is instrumental when determining thermodynamically or rather dissipation consistent constitutive relations in the domain (including the spatial stress) that automatically respect positive dissipation power as a constraint. Finally, Version III identifies constitutive relations for the material stress in the domain together with thermodynamically conjugated forces, expressed in terms of the material stress, that drive material (configurational) changes on the boundary and at singular surfaces. The latter may eventually be exploited to determine corresponding evolution equations based on a convex analysis setting.

Interestingly, the spatial (Piola and Cauchy) and material (Eshelby) stresses can be cast in dual formats and a four-dimensional space-time formalism.

P. Steinmann, *Spatial and Material Forces in Nonlinear Continuum Mechanics*,
Solid Mechanics and Its Applications 272,
https://doi.org/10.1007/978-3-030-89070-4_12

12.1 Clausius–Duhem Assumption

In the sequel, a classical Ansatz according to the Clausius–Duhem assumption is made for the entropy source density as well as for the extrinsic and intrinsic entropy flux densities

$$\overline{\eta}_{\mathrm{m}} := \frac{\overline{h}_{\mathrm{m}}}{\theta} \quad \text{and} \quad \widehat{\eta}_{\mathrm{m}} := \frac{\widehat{h}_{\mathrm{m}}}{\theta} \quad \text{with} \quad \boldsymbol{S} := \frac{\boldsymbol{H}}{\theta} \tag{12.1}$$

whereby θ denotes the absolute temperature (the integrating denominator in classical pVT-thermodynamics), assumed spatially C_0-continuous across singular surfaces

$$\theta > 0 \quad \text{with} \quad [\![\theta]\!] = 0. \tag{12.2}$$

Singular surfaces with vanishing temperature jump $[\![\theta]\!] = 0$ are denoted as *highly conductive* interfaces (Javili et al. [1], Esmaeili et al. [2]). It shall be noted that generalizations to cases with $[\![\theta]\!] \neq 0$ are possible and fall under the umbrella of *Kapitza* interfaces, the case $[\![\theta]\!] \neq 0$ is investigated elsewhere (for example Javili et al. [3, 4], Kaessmair et al. [5], Esmaeili et al. [6, 7]).

Moreover, multiplying the entropy production source and extrinsic flux densities $\overline{\pi}_{\mathrm{m}}$, $\widehat{\pi}_{\mathrm{m}}$ and $\widetilde{\pi}_{\mathrm{m}}$, respectively, by the absolute temperature θ, the *dissipation power source density* $\overline{\delta}_{\mathrm{m}}$ per unit volume in the material configuration, i.e. in $\mathcal{B}_{\mathrm{m}}$, and the *extrinsic dissipation power flux densities* $\widehat{\delta}_{\mathrm{m}}$ and $\widetilde{\delta}_{\mathrm{m}}$ per unit area in the material configuration, i.e. on $\partial\mathcal{B}_{\mathrm{m}}$ and at $\mathcal{S}_{\mathrm{m}}$, respectively, are introduced

$$0 \leq \overline{\delta}_{\mathrm{m}} := \theta\,\overline{\pi}_{\mathrm{m}} \quad \text{and} \quad 0 \leq \widehat{\delta}_{\mathrm{m}} := \theta\,\widehat{\pi}_{\mathrm{m}} \quad \text{and} \quad 0 \leq \widetilde{\delta}_{\mathrm{m}} := \theta\,\widetilde{\pi}_{\mathrm{m}}. \tag{12.3}$$

Finally, the product of the entropy density σ_{m} and the absolute temperature θ is denoted, as a novel terminology, the *entropergy density* and is abbreviated as

$$\tau_{\mathrm{m}} := \theta\,\sigma_{\mathrm{m}}. \tag{12.4}$$

Here the terminology *entropergy density* (the *entrop*ic contribution to the en*ergy* density) is artificially constructed from the ancient Greek expressions $\acute{\epsilon}\nu$ (= in), $\tau\rho o\pi\acute{\eta}$ (= transformation), and $\acute{\epsilon}\rho\gamma o\nu$ (= work), which are also the etymological roots for the terminology *energy* and *entropy*.

With these preliminaries at hand, the local format of the global balance of entropy for a material control volume is rewritten as the *local balance of entropergy*:

Local Balance of Entropergy (Material Control Volume):

i. Regular Points in the Domain

$$\mathrm{D}_t\tau_{\mathrm{m}} + \mathrm{Div}\boldsymbol{H} - \sigma_{\mathrm{m}}\,\mathrm{D}_t\theta = \overline{h}_{\mathrm{m}} + \overline{\delta}_{\mathrm{m}}^{\mathrm{loc}} \quad \text{in} \quad \mathcal{B}_{\mathrm{m}} \tag{12.5}$$

ii. Regular Points on the Boundary

$$[\tau_{\mathrm{m}} \boldsymbol{W} - \boldsymbol{H}] \cdot \boldsymbol{N} = \widehat{h}_{\mathrm{m}}^{\mathrm{ext}} + \widehat{\delta}_{\mathrm{m}} \quad \text{on} \quad \partial\mathcal{B}_{\mathrm{m}} \tag{12.6}$$

iii. Regular Points at Singular Surfaces

$$[\![\tau_{\mathrm{m}} \boldsymbol{W} - \boldsymbol{H}]\!] \cdot \boldsymbol{M} = -\widetilde{\delta}_{\mathrm{m}} \quad \text{at} \quad \mathcal{S}_{\mathrm{m}} \tag{12.7}$$

Here, following standard convention, the dissipation power source density $\overline{\delta}_{\mathrm{m}}$ is decomposed into its *local* and *conductive* contributions

$$\overline{\delta}_{\mathrm{m}} = \overline{\delta}_{\mathrm{m}}^{\mathrm{loc}} + \overline{\delta}_{\mathrm{m}}^{\mathrm{con}} \quad \text{with} \quad \overline{\delta}_{\mathrm{m}}^{\mathrm{con}} := -\nabla_X \ln\theta \cdot \boldsymbol{H}, \tag{12.8}$$

whereby, as a sufficient condition for $\overline{\delta}_{\mathrm{m}} \geq 0$, the positivity constraint is typically postulated separately for both contributions

$$\overline{\delta}_{\mathrm{m}}^{\mathrm{loc}} \geq 0 \quad \text{and} \quad \overline{\delta}_{\mathrm{m}}^{\mathrm{con}} \geq 0. \tag{12.9}$$

Observe furthermore that the *extrinsic heat flux density cancelation* $\widehat{h}_{\mathrm{m}}^{+} + \widehat{h}_{\mathrm{m}}^{-} = 0$ on $\mathcal{S}_{\mathrm{m}}$ together with the definition for the extrinsic dissipation power flux density

$$\widetilde{\delta}_{\mathrm{m}} := \widehat{\delta}_{\mathrm{m}}^{+} + \widehat{\delta}_{\mathrm{m}}^{-} \quad \text{at} \quad \mathcal{S}_{\mathrm{m}} \tag{12.10}$$

is consistent with the jump condition $[\![\tau_{\mathrm{m}}]\!] W_{\perp} - [\![\boldsymbol{H}]\!] \cdot \boldsymbol{M} = -\widetilde{\delta}_{\mathrm{m}}$ across the singular surface.

Note that since $\widehat{\delta}_{\mathrm{m}}^{\pm} \geq 0$ and thus $\widetilde{\delta}_{\mathrm{m}} \geq 0$ no cancelation for the extrinsic dissipation power flux density may be formulated.

Finally, in passing the local balance of entropergy is also re-stated for a spatial control volume:

Local Balance of Entropergy (Spatial Control Volume):

i. Regular Points in the Domain

$$\mathrm{d}_t \tau_{\mathrm{s}} + \operatorname{div}(\boldsymbol{\eta} + \tau_{\mathrm{s}} \boldsymbol{v}) - \sigma_{\mathrm{s}}\, \mathrm{D}_t \theta = \overline{h}_{\mathrm{s}} + \overline{\delta}_{\mathrm{s}}^{\mathrm{loc}} \quad \text{in} \quad \mathcal{B}_{\mathrm{s}} \tag{12.11}$$

ii. Regular Points on the Boundary

$$\big[\tau_{\mathrm{s}} \boldsymbol{w} - [\boldsymbol{\eta} + \tau_{\mathrm{s}} \boldsymbol{v}]\big] \cdot \boldsymbol{n} = \widehat{h}_{\mathrm{s}}^{\mathrm{ext}} + \widehat{\delta}_{\mathrm{s}} \quad \text{on} \quad \partial\mathcal{B}_{\mathrm{s}} \tag{12.12}$$

iii. Regular Points at Singular Surfaces

$$\big[\!\big[\tau_{\mathrm{s}} \boldsymbol{w} - [\boldsymbol{\eta} + \tau_{\mathrm{s}} \boldsymbol{v}]\big]\!\big] \cdot \boldsymbol{m} = -\widetilde{\delta}_{\mathrm{s}} \quad \text{at} \quad \mathcal{S}_{\mathrm{s}} \tag{12.13}$$

12.2 Dissipation Power Inequality (DPI) I, II

Introducing the (interior or exterior) free energy and incorporating a suited form of the (interior or exterior) internal energy balance, renders equivalent (interior or exterior) Versions I and II of the dissipation power inequality in the domain, on the boundary, and at singular surfaces. Interestingly, only the dissipation power inequality in the domain as expressed in Version II coincides for interior and exterior cases. On the boundary and at singular surfaces, for interior and exterior cases, the dissipation power inequalities clearly differ.

12.2.1 Dissipation of Interior Total Energy

To proceed, the *interior free energy density* is defined as the *Legendre transformation* of the *interior internal energy density*

$$\psi_{\mathrm{m}} := \iota_{\mathrm{m}} - \theta\sigma_{\mathrm{m}} = \iota_{\mathrm{m}} - \tau_{\mathrm{m}}, \tag{12.14}$$

thus revealing the flavor of an energy contribution for the entropergy τ_{m} as introduced earlier in Eq. 12.4. As a side-result

$$\lambda_{\mathrm{m}} = \tau_{\mathrm{m}} + \ell_{\mathrm{m}} \tag{12.15}$$

holds for the relation between the interior Lagrange-type free and internal energy densities $\lambda_{\mathrm{m}} = \kappa_{\mathrm{m}} - \psi_{\mathrm{m}}$ and $\ell_{\mathrm{m}} = \kappa_{\mathrm{m}} - \iota_{\mathrm{m}}$ that were already introduced earlier.

Incorporating Balance of Interior Lagrange-Type Internal Energy

Then subtracting the local balance of interior Lagrange-type internal energy from the local balance of entropergy renders eventually the statements:

Local Interior Dissipation Power Inequality (Version I) (Material Control Volume):

i. Regular Points in the Domain

$$\mathbb{P}^{\mathrm{int}}_{\mathrm{m}} + [\,\mathrm{D}_t\lambda_{\mathrm{m}} - \boldsymbol{p}_{\mathrm{m}} \cdot \mathrm{D}_t\boldsymbol{v}] - \sigma_{\mathrm{m}}\,\mathrm{D}_t\theta = \overline{\delta}^{\mathrm{loc}}_{\mathrm{m}} \geq 0 \quad \text{in} \quad \mathcal{B}_{\mathrm{m}} \tag{12.16}$$

with

$$\mathbb{P}^{\mathrm{int}}_{\mathrm{m}} := \nabla_X\boldsymbol{v} : \boldsymbol{P}$$

ii. Regular Points on the Boundary

$$\lambda_{\mathrm{m}}\boldsymbol{W} \cdot \boldsymbol{N} = \widehat{\delta}_{\mathrm{m}} \geq 0 \quad \text{on} \quad \partial\mathcal{B}_{\mathrm{m}} \tag{12.17}$$

iii. Regular Points at Singular Surfaces

$$\Big[- [\![\boldsymbol{v}]\!] \cdot \{\boldsymbol{P}\} + \big[[\![\lambda_{\mathrm{m}}]\!] - \{\boldsymbol{p}_{\mathrm{m}}\} \cdot [\![\boldsymbol{v}]\!] \big] \boldsymbol{W} \Big] \cdot \boldsymbol{M} = -\widetilde{\delta}_{\mathrm{m}} \le 0 \quad \text{at} \quad \mathcal{S}_{\mathrm{m}} \qquad (12.18)$$

In passing the local interior dissipation power inequality is also re-stated for a spatial control volume:

Local Interior Dissipation Power Inequality (Version I) (Spatial Control Volume):

i. Regular Points in the Domain

$$\mathbb{P}^{\mathrm{int}}_{\mathrm{s}} + [j\, \mathrm{D}_t \lambda_{\mathrm{m}} - \boldsymbol{p}_{\mathrm{s}} \cdot \mathrm{D}_t \boldsymbol{v}] - \sigma_{\mathrm{s}}\, \mathrm{D}_t \theta = \overline{\delta}^{\mathrm{loc}}_{\mathrm{s}} \ge 0 \quad \text{in} \quad \mathcal{B}_{\mathrm{s}} \qquad (12.19)$$

with

$$\mathbb{P}^{\mathrm{int}}_{\mathrm{s}} := \nabla_x \boldsymbol{v} : \boldsymbol{\sigma}$$

ii. Regular Points on the Boundary

$$\lambda_{\mathrm{s}}[\boldsymbol{w} - \boldsymbol{v}] \cdot \boldsymbol{n} = \widehat{\delta}_{\mathrm{s}} \ge 0 \quad \text{on} \quad \partial\mathcal{B}_{\mathrm{s}} \qquad (12.20)$$

iii. Regular Points at Singular Surfaces

$$\Big[- [\![\boldsymbol{v}]\!] \cdot \{\boldsymbol{\sigma}_{\mathrm{d}}\} + [\![\lambda_{\mathrm{s}}[\boldsymbol{w} - \boldsymbol{v}]]\!] - \{\boldsymbol{p}_{\mathrm{s}}\} \cdot [\![\boldsymbol{v}]\!] \boldsymbol{w} \Big] \cdot \boldsymbol{m} = -\widetilde{\delta}_{\mathrm{s}} \le 0 \quad \text{at} \quad \mathcal{S}_{\mathrm{s}} \qquad (12.21)$$

The format in Eqs. 12.16 and 12.19 is especially beneficial when the formal similarity for the expression of $\overline{\delta}^{\mathrm{loc}}_{\mathrm{m}}$ in terms of the interior Eshelby stress will be highlighted later in Eqs. 12.54 and 12.72.

Incorporating Balance of Interior Internal Energy

However, for the Coleman-Noll-type exploitation of the interior dissipation power inequality $\overline{\delta}^{\mathrm{loc}}_{\mathrm{m}} \ge 0$ that aims in the thermodynamically consistent derivation of constitutive relations, the following fully equivalent format is preferable. Thereby subtracting the local balance of interior internal energy from the local balance of entropergy renders the statements:

Local Interior Dissipation Power Inequality (Version II) (Material Control Volume):

i. Regular Points in the Domain

$$\mathbb{P}^{\text{int}}_{\text{m}} - \mathrm{D}_t\psi_{\text{m}} - \sigma_{\text{m}}\,\mathrm{D}_t\theta = \overline{\delta}^{\text{loc}}_{\text{m}} \geq 0 \quad \text{in} \quad \mathcal{B}_{\text{m}} \tag{12.22}$$

ii. Regular Points on the Boundary

$$\lambda_{\text{m}}\boldsymbol{W}\cdot\boldsymbol{N} = \widehat{\delta}_{\text{m}} \geq 0 \quad \text{on} \quad \partial\mathcal{B}_{\text{m}} \tag{12.23}$$

iii. Regular Points at Singular Surfaces

$$\widetilde{h}_{\text{m}} + [\![\lambda_{\text{m}}\boldsymbol{W}]\!]\cdot\boldsymbol{M} = -\widetilde{\delta}_{\text{m}} \leq 0 \quad \text{at} \quad \mathcal{S}_{\text{m}} \tag{12.24}$$
$$\rightsquigarrow -\big[[\![\psi_{\text{m}}]\!] - [\![\boldsymbol{F}]\!] : \{\boldsymbol{P}\}\big]W_{\perp} = -\widetilde{\delta}_{\text{m}}$$

The local interior dissipation power inequality as stated in the above gives rise to the following observations:

i. In a smooth domain $\mathcal{B}_{\text{m}}$, the interior dissipation power density depends exclusively on material time derivatives $\mathrm{D}_t\{\bullet\}$, i.e. on time derivatives at fixed material coordinates. Thus, it is not at all affected by any changes of the material configuration as captured by the total material velocity $\boldsymbol{W}$.
ii. On (the regular part of) the external boundary $\partial\mathcal{B}_{\text{m}}$, the interior dissipation power density depends exclusively on the total material velocity $\boldsymbol{W}$, thus it is only affected by changes of the material configuration. It shall also be noted carefully that $\widehat{\delta}_{\text{m}}$ contains no contributions from external tractions, (i.e. neither from the spatial extrinsic nor spatial intrinsic linear momentum flux density), thus it exclusively captures the *interior* contribution to the dissipation power density at the external boundary $\partial\mathcal{B}_{\text{m}}$.
iii. At a singular surface $\mathcal{S}_{\text{m}}$, the extrinsic heat flux density generated at the singular surface in Eq. 12.24 is expressed in the format

$$\widetilde{h}_{\text{m}} = -\big[[\![\kappa_{\text{m}}]\!] - [\![\boldsymbol{F}]\!] : \{\boldsymbol{P}\}\big]W_{\perp}.$$

Note again that

$$[\![\boldsymbol{F}]\!] : \{\boldsymbol{P}\}W_{\perp} = -[\![\boldsymbol{v}]\!]\cdot\{\boldsymbol{P}\}\cdot\boldsymbol{M}$$

holds due to the result in Eq. 8.7, thus the local interior dissipation power inequality across a singular surface reads alternatively

$$[\![\psi_{\text{m}}]\!]W_{\perp} + [\![\boldsymbol{v}]\!]\cdot\{\boldsymbol{P}\}\cdot\boldsymbol{M} = \widetilde{\delta}_{\text{m}} \geq 0. \tag{12.25}$$

By subtracting the corresponding local balance of interior internal energy across a singular surface renders alternatively

$$- [\![\tau_m]\!] W_\perp + [\![\boldsymbol{H}]\!] \cdot \boldsymbol{M} = \widetilde{\delta}_m \geq 0. \tag{12.26}$$

Note that for material singular surfaces with $W_\perp = 0$ (and $[\![\boldsymbol{H}]\!] \cdot \boldsymbol{M} = 0$), the corresponding interior dissipation power flux density $\widetilde{\delta}_m \equiv 0$ clearly vanishes identically.

In passing the local interior dissipation power inequality is also re-stated for a spatial control volume:

Local Interior Dissipation Power Inequality (Version II) (Spatial Control Volume):

i. Regular Points in the Domain

$$\mathbb{P}_s^{int} - j\,\mathrm{D}_t\psi_m - \sigma_s\,\mathrm{D}_t\theta = \overline{\delta}_s^{loc} \geq 0 \quad \text{in} \quad \mathcal{B}_s \tag{12.27}$$

ii. Regular Points on the Boundary

$$\lambda_s[\boldsymbol{w} - \boldsymbol{v}] \cdot \boldsymbol{n} = \widehat{\delta}_s \geq 0 \quad \text{on} \quad \partial\mathcal{B}_s \tag{12.28}$$

iii. Regular Points at Singular Surfaces

$$\widetilde{h}_s + [\![\lambda_s[\boldsymbol{w} - \boldsymbol{v}]]\!] \cdot \boldsymbol{m} = -\widetilde{\delta}_s \leq 0 \quad \text{at} \quad \mathcal{S}_s \tag{12.29}$$

$$\rightsquigarrow -\Big[[\![\psi_s[\boldsymbol{w} - \boldsymbol{v}]]\!] - W_\perp \boldsymbol{j} \cdot \{\boldsymbol{\sigma}\}\Big] \cdot \boldsymbol{m} = -\widetilde{\delta}_s$$

12.2.2 Dissipation of Exterior Total Energy

Likewise, the *exterior free energy densities* are defined as the *Legendre transformations* of the *exterior internal energy densities*

$$\overline{\psi}_m := \overline{\iota}_m - \theta\sigma_m = \overline{\iota}_m - \tau_m \quad \text{and} \quad \psi'_m := \iota'_m - \theta\sigma_m = \iota'_m - \tau_m. \tag{12.30}$$

Recall that $\overline{\iota}_m = \iota_m + \overline{\upsilon}_m$ and $\iota'_m = \overline{\iota}_m + \upsilon'_m$ with $\upsilon'_m := \nabla_N\widehat{\upsilon}_m - \widehat{\upsilon}_m\widehat{C}$ take into account the external potential energy on the curved external boundary. As a side-result

$$\overline{\lambda}_m = \tau_m + \overline{\ell}_m \quad \text{and} \quad \lambda'_m = \tau_m + \ell'_m \tag{12.31}$$

hold for the relation between the exterior Lagrange-type free and internal energy densities $\overline{\lambda}_{\rm m} := \kappa_{\rm m} - \overline{\psi}_{\rm m}$ and $\overline{\ell}_{\rm m} := \kappa_{\rm m} - \overline{\iota}_{\rm m}$ as well as $\lambda'_{\rm m} := \kappa_{\rm m} - \psi'_{\rm m}$ and $\ell'_{\rm m} := \kappa_{\rm m} - \iota'_{\rm m}$.

Incorporating Balance of Exterior Lagrange-Type Internal Energy

Subtracting alternatively the local balance of exterior Lagrange-type internal energy from the local balance of entropergy renders the statements:

Local Exterior Dissipation Power Inequality (Version I) (Material Control Volume):

i. Regular Points in the Domain

$$\mathbb{P}^{\rm int}_{\rm m} + [\,{\rm D}_t\lambda_{\rm m} - \boldsymbol{p}_{\rm m}\cdot{\rm D}_t\boldsymbol{v}] - \sigma_{\rm m}\,{\rm D}_t\theta = \overline{\delta}^{\rm loc}_{\rm m} \geq 0 \quad \text{in} \quad \mathcal{B}_{\rm m} \tag{12.32}$$

ii. Regular Points on the Boundary

$$\lambda'_{\rm m}\boldsymbol{W}\cdot\boldsymbol{N} = \widehat{\delta}_{\rm m} \geq 0 \quad \text{on} \quad \partial\mathcal{B}_{\rm m} \tag{12.33}$$

iii. Regular Points at Singular Surfaces

$$\Big[-[\![\boldsymbol{v}]\!]\cdot\{\boldsymbol{P}\} + \big[[\![\overline{\lambda}_{\rm m}]\!] - \{\boldsymbol{p}_{\rm m}\}\cdot[\![\boldsymbol{v}]\!]\big]\boldsymbol{W}\Big]\cdot\boldsymbol{M} = -\widetilde{\delta}_{\rm m} \leq 0 \quad \text{at} \quad \mathcal{S}_{\rm m} \tag{12.34}$$

In passing the local exterior dissipation power inequality is also re-stated for a spatial control volume:

Local Exterior Dissipation Power Inequality (Version I) (Spatial Control Volume):

i. Regular Points in the Domain

$$\mathbb{P}^{\rm int}_{\rm s} + [\,j\,{\rm D}_t\lambda_{\rm m} - \boldsymbol{p}_{\rm s}\cdot{\rm D}_t\boldsymbol{v}] - \sigma_{\rm s}\,{\rm D}_t\theta = \overline{\delta}^{\rm loc}_{\rm s} \geq 0 \quad \text{in} \quad \mathcal{B}_{\rm s} \tag{12.35}$$

ii. Regular Points on the Boundary

$$\lambda'_{\rm s}[\boldsymbol{w} - \boldsymbol{v}]\cdot\boldsymbol{n} = \widehat{\delta}_{\rm s} \geq 0 \quad \text{on} \quad \partial\mathcal{B}_{\rm s} \tag{12.36}$$

iii. Regular Points at Singular Surfaces

$$\Big[-[\![\boldsymbol{v}]\!]\cdot\{\boldsymbol{\sigma}_{\rm d}\} + \big[\!\big[\overline{\lambda}_{\rm s}[\boldsymbol{w} - \boldsymbol{v}]\big]\!\big] - \{\boldsymbol{p}_{\rm s}\}\cdot[\![\boldsymbol{v}]\!]\boldsymbol{w}\Big]\cdot\boldsymbol{m} = -\widetilde{\delta}_{\rm s} \leq 0 \quad \text{at} \quad \mathcal{S}_{\rm s} \tag{12.37}$$

The format in Eqs. 12.32 and 12.35 is especially beneficial when the formal similarity for the expression of $\overline{\delta}_{\rm m}^{\rm loc}$ in terms of the exterior Eshelby stress will be highlighted later in Eqs. 12.83 and 12.90.

Incorporating Balance of Exterior Internal Energy

Subtracting alternatively the local balance of exterior internal energy from the local balance of entropergy renders the statements:

Local Exterior Dissipation Power Inequality (Version II) (Material Control Volume):

i. Regular Points in the Domain

$$\mathbb{P}_{\rm m}^{\rm int} - {\rm D}_t \psi_{\rm m} - \sigma_{\rm m}\, {\rm D}_t \theta = \overline{\delta}_{\rm m}^{\rm loc} \geq 0 \quad \text{in} \quad \mathcal{B}_{\rm m} \tag{12.38}$$

ii. Regular Points on the Boundary

$$\lambda'_{\rm m} \boldsymbol{W} \cdot \boldsymbol{N} = \widehat{\delta}_{\rm m} \geq 0 \quad \text{on} \quad \partial\mathcal{B}_{\rm m} \tag{12.39}$$

iii. Regular Points at Singular Surfaces

$$\widetilde{h}_{\rm m} + [\![\overline{\lambda}_{\rm m} \boldsymbol{W}]\!] \cdot \boldsymbol{M} = -\widetilde{\delta}_{\rm m} \leq 0 \quad \text{at} \quad \mathcal{S}_{\rm m} \tag{12.40}$$

$$\rightsquigarrow -\big[[\![\overline{\psi}_{\rm m}]\!] - [\![\boldsymbol{F}]\!] : \{\boldsymbol{P}\}\big] W_\perp = -\widetilde{\delta}_{\rm m}$$

The local exterior dissipation power inequality as stated in the above gives rise to the following observations:

i. In a smooth domain $\mathcal{B}_{\rm m}$, the interior and exterior dissipation power densities do entirely coincide. Thus, no differentiation between the interior and exterior dissipation power density is needed in the domain.
ii. On (the regular part of) the external boundary $\partial\mathcal{B}_{\rm m}$, the interior and exterior dissipation power densities differ by the term $\lambda_{\rm m} - \lambda'_{\rm m} = \overline{\upsilon}_{\rm m} + \upsilon'_{\rm m}$.
iii. At a singular surface $\mathcal{S}_{\rm m}$, the interior and exterior dissipation power densities differ by the term $-\psi_{\rm m} + \overline{\psi}_{\rm m} = \overline{\upsilon}_{\rm m}$.

In passing the local exterior dissipation power inequality is also re-stated for a spatial control volume:

Local Exterior Dissipation Power Inequality (Version II) (Spatial Control Volume):

i. Regular Points in the Domain

$$\mathbb{P}_{\rm s}^{\rm int} - j\, {\rm D}_t \psi_{\rm m} - \sigma_{\rm s}\, {\rm D}_t \theta = \overline{\delta}_{\rm s}^{\rm loc} \geq 0 \quad \text{in} \quad \mathcal{B}_{\rm s} \tag{12.41}$$

ii. Regular Points on the Boundary

$$\lambda'_s[\boldsymbol{w}-\boldsymbol{v}]\cdot\boldsymbol{n}=\widehat{\delta}_s\geq 0 \quad \text{on} \quad \partial\mathcal{B}_s \tag{12.42}$$

iii. Regular Points at Singular Surfaces

$$\widetilde{h}_s+[\![\overline{\lambda}_s[\boldsymbol{w}-\boldsymbol{v}]]\!]\cdot\boldsymbol{m}=-\widetilde{\delta}_s\leq 0 \quad \text{at} \quad \mathcal{S}_s \tag{12.43}$$

$$\rightsquigarrow -\Big[[\![\overline{\psi}_s[\boldsymbol{w}-\boldsymbol{v}]]\!]-W_\perp\boldsymbol{j}\cdot\{\boldsymbol{\sigma}\}\Big]\cdot\boldsymbol{m}=-\widetilde{\delta}_s$$

12.3 Exploitation of DPI in the Domain

In the domain $\mathcal{B}_m$, the Coleman-Noll-type exploitation of the inequality for the local dissipation power source density in Eq. 12.9.1 leading to Eq. 12.22 or Eq. 12.38 alike and the conductive dissipation power density in Eq. 12.9.2 renders the constitutive relations between the (primal) *state variables* and their resulting (dual) *state functions* (here state variables and state functions shall collectively denote the *state quantities* that fully characterize the *state of a system*). These constitutive relations are needed to close the system of kinematic and kinetic equations describing a continuum problem.

In a Boltzmann-type (local) continuum description, the state variables (collected in the set $\mathfrak{S}_\mathfrak{v}$) consist of:

- the spatial deformation map $\boldsymbol{y}$,
- the absolute temperature θ,
- a generic (scalar-valued) internal variable α (the extension to sets of vector- and tensor-valued internal variables is straightforward),
- and their material gradients $\boldsymbol{F}$ and $\boldsymbol{\Gamma}:=\nabla_X\theta$,
- (and potentially $\nabla_X\alpha$, $\nabla_X\nabla_X\alpha,\cdots$).

Thus

$$\mathfrak{S}_\mathfrak{v}:=\{\underbrace{\boldsymbol{y},\boldsymbol{F},\theta,\boldsymbol{\Gamma}}_{\mathfrak{S}_\mathfrak{v}^{\text{ext}}},\underbrace{\alpha,\nabla_X\alpha,\nabla_X\nabla_X\alpha,\cdots}_{\mathfrak{S}_\mathfrak{v}^{\text{int}}}\}. \tag{12.44}$$

Correspondingly, the state functions (collected in the set $\mathfrak{S}_\mathfrak{f}$) are:

- the interior free energy density ψ_m,
- the Piola stress $\boldsymbol{P}$,
- the entropy density σ_m,
- the intrinsic entropy flux vector $\boldsymbol{S}$,
- and the driving force A_m (conjugate to the internal variable).

Thus

$$\mathfrak{S}_\mathfrak{f} := \{\psi_\mathrm{m}, \boldsymbol{P}, \sigma_\mathrm{m}, \boldsymbol{S}, A_\mathrm{m}\}. \tag{12.45}$$

With these preliminaries at hand, the Coleman-Noll-type exploitation proceeds along the following three steps:

1. **Principle of Equipresence:** Initially the complete list of state variables has to be assumed as arguments of the state functions, i.e.

$$\mathfrak{S}_\mathfrak{f} \doteq \mathfrak{S}_\mathfrak{f}(\mathfrak{S}_\mathfrak{v}). \tag{12.46}$$

It is remarked that possible higher-order gradients of the spatial deformation map $\boldsymbol{y}$ and the absolute temperature θ drop out from the list of arguments for the state functions in the current Boltzmann-type (local) continuum description. Thus, they are not considered here from the onset for the sake of conciseness of exposition.

As a consequence of the principle of equipresence, especially the material time derivative of the interior free energy density reads as

$$\begin{aligned} \mathrm{D}_t\psi_\mathrm{m} &= \frac{\partial\psi_\mathrm{m}}{\partial\mathfrak{S}_\mathfrak{v}} \star \mathrm{D}_t\mathfrak{S}_\mathfrak{v} \\ &= \frac{\partial\psi_\mathrm{m}}{\partial\boldsymbol{y}} \cdot \mathrm{D}_t\boldsymbol{y} + \frac{\partial\psi_\mathrm{m}}{\partial\boldsymbol{F}} : \mathrm{D}_t\boldsymbol{F} + \frac{\partial\psi_\mathrm{m}}{\partial\theta}\mathrm{D}_t\theta + \frac{\partial\psi_\mathrm{m}}{\partial\boldsymbol{\Gamma}} \cdot \mathrm{D}_t\boldsymbol{\Gamma} \\ &+ \frac{\delta\psi_\mathrm{m}}{\delta\alpha}\mathrm{D}_t\alpha. \end{aligned} \tag{12.47}$$

Here, $\star$ is a suited scalar product and $\delta_\alpha\psi_\mathrm{m}$ denotes the *variational derivative* of the free energy density ψ_m with respect to the internal variable α as a short-hand notation to capture the possible presence of $\nabla_X\alpha$, $\nabla_X\nabla_X\alpha, \cdots$ in the list $\mathfrak{S}_\mathfrak{v}$ of state variables. In the case that $\mathfrak{S}_\mathfrak{v}$ depends only on α, the corresponding derivative trivially degenerates to the *partial derivative* $\partial_\alpha\psi_\mathrm{m}$.

2. **Positive Dissipation Power Density:** $\overline{\delta}_\mathrm{m} \geq 0$ shall hold for all admissible thermo-mechanical processes $\boldsymbol{x} = \boldsymbol{y}(\boldsymbol{X}, t)$ and $\theta = \theta(\boldsymbol{X}, t)$ (whereby admissible thermo-mechanical processes $\boldsymbol{x} = \boldsymbol{y}(\boldsymbol{X}, t)$ and $\theta = \theta(\boldsymbol{X}, t)$ may be realized by appropriately prescribing external source terms in the balances of linear momentum and energy) under the constraints imposed by the principle of equipresence. Then, after re-grouping, the dissipation power density is re-expressed as

$$\begin{aligned}\overline{\delta}_{\mathrm{m}} = &-\frac{\partial \psi_{\mathrm{m}}}{\partial \boldsymbol{y}} \cdot \mathrm{D}_t \boldsymbol{y} \\ &+\left[\boldsymbol{P} - \frac{\partial \psi_{\mathrm{m}}}{\partial \boldsymbol{F}}\right] : \mathrm{D}_t \boldsymbol{F} \\ &-\left[\sigma_{\mathrm{m}} + \frac{\partial \psi_{\mathrm{m}}}{\partial \theta}\right] \mathrm{D}_t \theta \\ &-\frac{\partial \psi_{\mathrm{m}}}{\partial \boldsymbol{\Gamma}} \cdot \mathrm{D}_t \boldsymbol{\Gamma} \\ &-\frac{\delta \psi_{\mathrm{m}}}{\delta \alpha} \mathrm{D}_t \alpha \\ &-\boldsymbol{S} \cdot \boldsymbol{\Gamma} \qquad \geq 0.\end{aligned} \tag{12.48}$$

Here it shall be remarked that each of the *external* state variables in $\mathfrak{S}_{\mathfrak{v}}^{\mathrm{ext}}$ and in particular each of their material time derivatives in $\mathrm{D}_t \mathfrak{S}_{\mathfrak{v}}^{\mathrm{ext}}$ are externally (and individually) *controllable*. This is in contrast to the *internal* state variables in $\mathfrak{S}_{\mathfrak{v}}^{\mathrm{int}}$ whose value and evolution are *not* externally controllable. It shall also be noted that the state functions $\boldsymbol{P}$, σ_{m} and A_{m} are in the sequel exclusively assumed *energetic*, i.e. they shall here not depend on the material time derivatives of their respective conjugate state variables for the sake of presentation. The incorporation and consideration of *dissipative* contributions to $\boldsymbol{P}$, σ_{m}, and A_{m} is however straightforward.

3. **Constitutive Relations, Reduced Dissipation Power Density:** Since $\mathfrak{S}_{\mathfrak{v}}^{\mathrm{ext}}$ and in particular $\mathrm{D}_t \mathfrak{S}_{\mathfrak{v}}^{\mathrm{ext}}$ are externally controllable, all contributions to $\overline{\delta}_{\mathrm{m}}$ that are linear in $\mathrm{D}_t \mathfrak{S}_{\mathfrak{v}}^{\mathrm{ext}}$ have to vanish individually, thus rendering the constitutive relations

$$\begin{aligned}\boldsymbol{0} &= \frac{\partial \psi_{\mathrm{m}}}{\partial \boldsymbol{y}} \implies \psi_{\mathrm{m}} \neq \psi_{\mathrm{m}}(\boldsymbol{y}, \cdots) \\ \boldsymbol{P} &= \frac{\partial \psi_{\mathrm{m}}}{\partial \boldsymbol{F}} \\ -\sigma_{\mathrm{m}} &= \frac{\partial \psi_{\mathrm{m}}}{\partial \theta} \\ \boldsymbol{0} &= \frac{\partial \psi_{\mathrm{m}}}{\partial \boldsymbol{\Gamma}} \implies \psi_{\mathrm{m}} \neq \psi_{\mathrm{m}}(\cdots, \boldsymbol{\Gamma}, \cdots).\end{aligned} \tag{12.49}$$

As a conclusion the interior free energy density ψ_{m} can not depend on $\boldsymbol{y}$ and $\boldsymbol{\Gamma}$ (indeed the dependence of ψ_{m} on $\boldsymbol{y}$ is also ruled out by an alternative argument of spatial objectivity[1]), thus

---

[1] *Spatial objectivity* is here particularized as invariance of the free energy density ψ_{m} under superposed rigid body motions (srbm). These follow as

$$\boldsymbol{y}(\boldsymbol{X}, t) \quad \rightarrow \quad \boldsymbol{Q}(t) \cdot \boldsymbol{y}(\boldsymbol{X}, t) + \boldsymbol{c}(t)$$

with $\boldsymbol{Q} \in \mathrm{SO}(3)$ and $\boldsymbol{c} \in \mathrm{T}(3)$ arbitrary elements from the special orthogonal and the translational (three-dimensional) groups, respectively. Thus, spatial objectivity requires

$$\psi_{\mathrm{m}} = \psi_{\mathrm{m}}(\mathfrak{S}_{\mathfrak{v}}^{\mathrm{ext,red}}, \mathfrak{S}_{\mathfrak{v}}^{\mathrm{int}}) = \psi_{\mathrm{m}}(\underbrace{\boldsymbol{F}, \theta}_{\mathfrak{S}_{\mathfrak{v}}^{\mathrm{ext,red}}}, \underbrace{\alpha, \nabla_X \alpha, \nabla_X \nabla_X \alpha, \cdots}_{\mathfrak{S}_{\mathfrak{v}}^{\mathrm{int}}}; \boldsymbol{X}). \tag{12.50}$$

Finally, based on the abbreviation for the driving force

$$A_{\mathrm{m}} := -\frac{\delta \psi_{\mathrm{m}}}{\delta \alpha} \tag{12.51}$$

the *reduced dissipation inequality* reads as

$$\bar{\delta}_{\mathrm{m}}^{\mathrm{red}} = \bar{\delta}_{\mathrm{m}}^{\mathrm{loc,red}} + \bar{\delta}_{\mathrm{m}}^{\mathrm{con}} \geq 0 \quad \text{with} \quad \bar{\delta}_{\mathrm{m}}^{\mathrm{loc,red}} := A_{\mathrm{m}} \, \mathrm{D}_t \alpha \geq 0 \tag{12.52}$$

It shall finally be noted that the intrinsic entropy flux $\boldsymbol{S}$ may in general still depend on the (almost) full set of state variables (except $\boldsymbol{y}$ for reasons of objectivity), i.e. $\boldsymbol{S} = \boldsymbol{S}(\boldsymbol{F}, \theta, \boldsymbol{\Gamma}, \alpha)$.

An example is the generalized Fourier law with

$$\boldsymbol{H} = \theta \boldsymbol{S} = -k_{\mathrm{m}}(\theta, \alpha) \boldsymbol{C}^{-1} \cdot \boldsymbol{\Gamma} \tag{12.53}$$

with a temperature and internal variable dependent heat conductivity $k_{\mathrm{m}}(\theta, \alpha) > 0$.

Summarizing, constitutive relations are denoted as *thermodynamically consistent*, i.e. they do not violate the second law of thermodynamics in the domain (by producing negative dissipation and thus allowing for a perpetuum mobile), whenever the Piola stress $\boldsymbol{P}$ and the entropy density σ_{m} derive from an interior free energy density ψ_{m} that is parameterized in the deformation gradient $\boldsymbol{F}$, the absolute temperature θ, and the internal variable in $\mathfrak{S}_{\mathfrak{v}}^{\mathrm{int}}$; and if, moreover, the reduced dissipation inequality $\bar{\delta}_{\mathrm{m}}^{\mathrm{red}} \geq 0$ is satisfied. A thermodynamically consistent evolution equation for the internal variable α is, for example, obtained if $\mathrm{D}_t \alpha$ is positively proportional to A_{m}, however, other more sophisticated formats for $\mathrm{D}_t \alpha$ are also possible (Steinmann and Runesson [8]).

$$\psi_{\mathrm{m}}(\boldsymbol{y}, \cdots) \doteq \psi_{\mathrm{m}}(\boldsymbol{Q} \cdot \boldsymbol{y} + \boldsymbol{c}, \cdots) \quad \forall \boldsymbol{Q} \in \mathrm{SO}(3) \quad \text{and} \quad \forall \boldsymbol{c} \in \mathrm{T}(3).$$

Obviously, this can only be satisfied for $\psi_{\mathrm{m}} \neq \psi_{\mathrm{m}}(\boldsymbol{y})$. It is also recalled that spatial objectivity results in the dependence of ψ_{m} on $\boldsymbol{F}$ through $\boldsymbol{C}$ (under srbm $\boldsymbol{F} \to \boldsymbol{Q} \cdot \boldsymbol{F}$, whereas $\boldsymbol{C}$ remains invariant, thus $\psi_{\mathrm{m}}(\boldsymbol{C})$ remains invariant too) and thus, in agreement with the balance of spatial angular momentum, eventually also in the symmetry of the spatial description Kirchhoff stress $\boldsymbol{\tau} = \boldsymbol{\tau}^t$.

12.4 Dissipation Power Inequality (DPI) III

The importance of the *interior* and *exterior* Eshelby stresses as driving changes in the material configuration (with non-zero total material velocity $\boldsymbol{W}$) will be elucidated in terms of the interior and exterior dissipation power inequality, respectively.

12.4.1 Dissipation of Interior Total Energy

To this end, first the interior dissipation power densities in the bulk, on the boundary and at singular surfaces, are re-stated in terms of the interior Eshelby stress:

Local Interior Dissipation Power Inequality (Version III) (Material Control Volume):

i. Regular Points in the Domain

$$\mathbb{P}_{\mathrm{m}}^{\mathrm{int}} + [J\,\mathrm{d}_t\lambda_{\mathrm{s}} - \boldsymbol{P}_{\mathrm{m}} \cdot \mathrm{d}_t\boldsymbol{V}] - \sigma_{\mathrm{m}}\,\mathrm{d}_t\theta = \overline{\delta}_{\mathrm{m}}^{\mathrm{loc}} \geq 0 \quad \text{in} \quad \mathcal{B}_{\mathrm{m}} \tag{12.54}$$

with

$$\mathbb{P}_{\mathrm{m}}^{\mathrm{int}} := \nabla_X \boldsymbol{V} : \boldsymbol{\Sigma}_{\mathrm{d}} - \boldsymbol{V} \cdot \overline{\boldsymbol{S}}_{\mathrm{m}}^{\mathrm{mec}}$$

ii. Regular Points on the Boundary

$$-\boldsymbol{W} \cdot \widehat{\boldsymbol{S}}_{\mathrm{m}}^{\mathrm{int}} = \widehat{\delta}_{\mathrm{m}} \geq 0 \quad \text{on} \quad \partial\mathcal{B}_{\mathrm{m}} \tag{12.55}$$

iii. Regular Points at Singular Surfaces

$$\boldsymbol{W} \cdot \widetilde{\boldsymbol{S}}_{\mathrm{m}} = \widetilde{\delta}_{\mathrm{m}} \geq 0 \quad \text{at} \quad \mathcal{S}_{\mathrm{m}} \tag{12.56}$$

For the derivation of these results, clearly highlighting the role of the interior Eshelby stress in the interior dissipation power densities, three different scenarios may be considered:

i. In a smooth domain $\mathcal{B}_{\mathrm{m}}$ the stress power $\mathbb{P}_{\mathrm{m}}^{\mathrm{int}} := \nabla_X \boldsymbol{v} : \boldsymbol{P}$ in Eq. 12.16 is re-expressed via the balance of kinetic energy in Eq. 8.15 as

$$\nabla_X \boldsymbol{v} : \boldsymbol{P} = \mathrm{Div}(\boldsymbol{v} \cdot \boldsymbol{P}) + \boldsymbol{v} \cdot \overline{\boldsymbol{s}}_{\mathrm{m}} - \mathrm{D}_t\kappa_{\mathrm{m}}. \tag{12.57}$$

Incorporating further the relation $\boldsymbol{v} = -\boldsymbol{F} \cdot \boldsymbol{V}$ between the spatial and material velocities leads to

$$\nabla_X \boldsymbol{v} : \boldsymbol{P} = \mathrm{Div}(-\boldsymbol{V} \cdot \boldsymbol{F}^t \cdot \boldsymbol{P}) - \boldsymbol{V} \cdot \boldsymbol{F}^t \cdot \overline{\boldsymbol{s}}_{\mathrm{m}} - \mathrm{D}_t\kappa_{\mathrm{m}}. \tag{12.58}$$

Then the interior Eshelby stress

$$\boldsymbol{\Sigma} := -\lambda_{\mathrm{m}}\boldsymbol{I} - \boldsymbol{F}^t \cdot \boldsymbol{P}$$

already introduced in Eq. 8.63 is incorporated to render

$$\nabla_X \boldsymbol{v} : \boldsymbol{P} = \mathrm{Div}(\boldsymbol{V} \cdot \boldsymbol{\Sigma}) - \boldsymbol{V} \cdot \boldsymbol{F}^t \cdot \overline{\boldsymbol{s}}_{\mathrm{m}} + \mathrm{Div}(\lambda_{\mathrm{m}} \boldsymbol{V}) - \mathrm{D}_t \kappa_{\mathrm{m}}. \tag{12.59}$$

Involving next the balance of material linear momentum in Eq. 8.49 transforms the above into

$$\nabla_X \boldsymbol{v} : \boldsymbol{P} = \nabla_X \boldsymbol{V} : \boldsymbol{\Sigma} - \boldsymbol{V} \cdot [\overline{\boldsymbol{S}}^{\mathrm{int}}_{\mathrm{m}} - \mathrm{D}_t \boldsymbol{P}_{\mathrm{m}}] + \mathrm{Div}(\lambda_{\mathrm{m}} \boldsymbol{V}) - \mathrm{D}_t \kappa_{\mathrm{m}}. \tag{12.60}$$

Here the *internal* part $\overline{\boldsymbol{S}}^{\mathrm{int}}_{\mathrm{m}}$ of the interior material linear momentum source density $\overline{\boldsymbol{S}}_{\mathrm{m}}$ in Eq. 8.64 is defined as

$$\overline{\boldsymbol{S}}^{\mathrm{int}}_{\mathrm{m}} := \overline{\boldsymbol{S}}_{\mathrm{m}} + \boldsymbol{F}^t \cdot \overline{\boldsymbol{s}}_{\mathrm{m}} = \sigma_{\mathrm{m}} \nabla_X \theta + A_{\mathrm{m}} \nabla_X \alpha + \partial_X \lambda_{\mathrm{m}} \tag{12.61}$$

and captures the dependence on the material position of the temperature field and the internal variable (heterogeneity) and the explicit dependence on the material position of the interior Lagrange-type free energy density (inhomogeneity). Next, the material time derivative may be exchanged for the spatial time derivative by applying the Euler formula to the material velocity $\boldsymbol{V}$ and the absolute temperature θ, i.e.

$$\mathrm{D}_t \boldsymbol{V} = \mathrm{d}_t \boldsymbol{V} - \nabla_X \boldsymbol{V} \cdot \boldsymbol{V} \quad \text{and} \quad \mathrm{D}_t \theta = \mathrm{d}_t \theta - \nabla_X \theta \cdot \boldsymbol{V}. \tag{12.62}$$

Then incorporating the identities

$$\boldsymbol{V} \cdot \mathrm{D}_t \boldsymbol{P}_{\mathrm{m}} = 2\,\mathrm{D}_t \kappa_{\mathrm{m}} - \boldsymbol{P}_{\mathrm{m}} \cdot \mathrm{D}_t \boldsymbol{V} \quad \text{and} \quad \mathrm{D}_t \kappa_{\mathrm{m}} = \boldsymbol{p}_{\mathrm{m}} \cdot \mathrm{D}_t \boldsymbol{v},$$

whereby the former holds due to $\boldsymbol{V} \cdot \boldsymbol{P}_{\mathrm{m}} = 2\kappa_{\mathrm{m}}$, together with the local transport theorem for the interior Lagrange-type free energy density

$$\mathrm{D}_t \lambda_{\mathrm{m}} = J\, \mathrm{d}_t \lambda_{\mathrm{s}} - \mathrm{Div}(\lambda_{\mathrm{m}} \boldsymbol{V}) \tag{12.63}$$

results after rearrangement in

$$\begin{aligned} \nabla_X \boldsymbol{v} : \boldsymbol{P} = {} & [J\, \mathrm{d}_t \lambda_{\mathrm{s}} - \mathrm{D}_t \lambda_{\mathrm{m}}] \\ & + [\boldsymbol{p}_{\mathrm{m}} \cdot \mathrm{D}_t \boldsymbol{v} - \boldsymbol{P}_{\mathrm{m}} \cdot \mathrm{d}_t \boldsymbol{V}] \\ & + \nabla_X \boldsymbol{V} : [\boldsymbol{\Sigma} + \boldsymbol{P}_{\mathrm{m}} \otimes \boldsymbol{V}] - \boldsymbol{V} \cdot \overline{\boldsymbol{S}}^{\mathrm{mec}}_{\mathrm{m}} \\ & + \sigma_{\mathrm{m}} [\mathrm{D}_t \theta - \mathrm{d}_t \theta]. \end{aligned} \tag{12.64}$$

Here the *internal mechanical* part $\overline{\boldsymbol{S}}_{\mathrm{m}}^{\mathrm{mec}}$ of the interior material linear momentum source density $\overline{\boldsymbol{S}}_{\mathrm{m}}$ in Eq. 8.64 is defined as

$$\overline{\boldsymbol{S}}_{\mathrm{m}}^{\mathrm{mec}} := \overline{\boldsymbol{S}}_{\mathrm{m}}^{\mathrm{int}} - \sigma_{\mathrm{m}} \nabla_X \theta = A_{\mathrm{m}} \nabla_X \alpha + \partial_X \lambda_{\mathrm{m}} \tag{12.65}$$

and captures the dependence on the material position of the internal variable (heterogeneity) and the explicit dependence on the material position of the interior Lagrange-type free energy density (inhomogeneity). Inserting the re-expressed stress power $\nabla_X \boldsymbol{v} : \boldsymbol{P}$ into Eq. 12.16 and abbreviating

$$\mathbb{p}_{\mathrm{m}}^{\mathrm{int}} := \nabla_X \boldsymbol{V} : \boldsymbol{\Sigma}_{\mathrm{d}} - \boldsymbol{V} \cdot \overline{\boldsymbol{S}}_{\mathrm{m}}^{\mathrm{mec}} \quad \text{with} \quad \boldsymbol{\Sigma}_{\mathrm{d}} := [\boldsymbol{\Sigma} + \boldsymbol{P}_{\mathrm{m}} \otimes \boldsymbol{V}] \tag{12.66}$$

renders finally the result in Eq. 12.54.

ii. On (the regular part of) the external boundary $\partial \mathcal{B}_{\mathrm{m}}$ the relation

$$\boldsymbol{F}^t \cdot \boldsymbol{P} \cdot \boldsymbol{N} = [\boldsymbol{P}_{\mathrm{m}} \otimes \boldsymbol{W}] \cdot \boldsymbol{N} + \boldsymbol{F}^t \cdot \widehat{\boldsymbol{s}}_{\mathrm{m}}^{\mathrm{ext}} \tag{12.67}$$

results from the definition of the spatial linear momentum flux densities in Eq. 7.27. Then incorporating the definition of the interior Eshelby stress

$$\boldsymbol{\Sigma} := -\lambda_{\mathrm{m}} \boldsymbol{I} - \boldsymbol{F}^t \cdot \boldsymbol{P}$$

as already introduced in Eq. 8.63 renders

$$-\lambda_{\mathrm{m}} \boldsymbol{N} = [\boldsymbol{\Sigma} + \boldsymbol{P}_{\mathrm{m}} \otimes \boldsymbol{W}] \cdot \boldsymbol{N} + \boldsymbol{F}^t \cdot \widehat{\boldsymbol{s}}_{\mathrm{m}}^{\mathrm{ext}}. \tag{12.68}$$

Involving finally the balance of material linear momentum in Eq. 8.50 transforms the above into

$$-\lambda_{\mathrm{m}} \boldsymbol{N} = \widehat{\boldsymbol{S}}_{\mathrm{m}}^{\mathrm{ext}} + \boldsymbol{F}^t \cdot \widehat{\boldsymbol{s}}_{\mathrm{m}}^{\mathrm{ext}}, \tag{12.69}$$

which, when introducing the abbreviation

$$\widehat{\boldsymbol{S}}_{\mathrm{m}}^{\mathrm{int}} := \widehat{\boldsymbol{S}}_{\mathrm{m}}^{\mathrm{ext}} + \boldsymbol{F}^t \cdot \widehat{\boldsymbol{s}}_{\mathrm{m}}^{\mathrm{ext}}, \tag{12.70}$$

for the *internal* part of the interior material extrinsic linear momentum flux density establishes the result in Eq. 12.55.

iii. At a singular surface $\mathcal{S}_{\mathrm{m}}$, the extrinsic heat flux density generated at the singular surface $\widetilde{h}_{\mathrm{m}} = -[\![\boldsymbol{v}]\!] \cdot \widetilde{\boldsymbol{s}}_{\mathrm{m}}$ in Eq. 12.24.1, due to the coherence condition $[\![\boldsymbol{v}]\!] = -\boldsymbol{W} \cdot [\![\boldsymbol{F}^t]\!]$, is related to the interior material extrinsic linear momentum flux density $\widetilde{\boldsymbol{S}}_{\mathrm{m}}$ in Eq. 8.69 via

$$\widetilde{h}_{\mathrm{m}} := \boldsymbol{W} \cdot [\![\boldsymbol{F}^t]\!] \cdot \widetilde{\boldsymbol{s}}_{\mathrm{m}} = -\boldsymbol{W} \cdot \big[[\![\lambda_{\mathrm{m}}]\!] \boldsymbol{M} + \widetilde{\boldsymbol{S}}_{\mathrm{m}}\big]. \tag{12.71}$$

Combining with the term $[\![\lambda_{\mathrm{m}} \boldsymbol{W}]\!] \cdot \boldsymbol{M}$ from Eq. 12.24.1 leads to the result in Eq. 12.56.

In passing the local interior dissipation power inequality is also re-stated for a spatial control volume:

Local Interior Dissipation Power Inequality (Version III) (Spatial Control Volume):

i. Regular Points in the Domain

$$\mathbb{p}_s^{\rm int} + [\,{\rm d}_t\lambda_s - \boldsymbol{P}_s \cdot {\rm d}_t\boldsymbol{V}] - \sigma_s\,{\rm d}_t\theta = \overline{\delta}_s^{\rm loc} \geq 0 \quad \text{in} \quad \mathcal{B}_s \tag{12.72}$$

with

$$\mathbb{p}_s^{\rm int} := \nabla_x\boldsymbol{V} : \boldsymbol{p}_{\rm d} - \boldsymbol{V} \cdot \overline{\boldsymbol{S}}_s^{\rm mec}$$

ii. Regular Points on the Boundary

$$-\,\boldsymbol{W} \cdot \widehat{\boldsymbol{S}}_s^{\rm int} = \widehat{\delta}_s \geq 0 \quad \text{on} \quad \partial\mathcal{B}_s \tag{12.73}$$

iii. Regular Points at Singular Surfaces

$$\boldsymbol{W} \cdot \widetilde{\boldsymbol{S}}_s = \widetilde{\delta}_s \geq 0 \quad \text{at} \quad \mathcal{S}_s \tag{12.74}$$

Observe that the local interior dissipation power inequalities in Eqs. 12.54 and 12.72 establish a *formal similarity* to those in Eqs. 12.19 and 12.16. Indeed, since $\boldsymbol{v} = -\boldsymbol{F} \cdot \boldsymbol{V}$, Eqs. 12.16, 12.22, and 12.54, as well as Eqs. 12.19, 12.27, and 12.72 are entirely equivalent, however, the format in Eq. 12.22 is most common since it serves as the starting point for deriving the usual constitutive relations via the Coleman-Noll-type exploitation of the interior dissipation power inequality.

Discussion of DPI Expressed in Interior Eshelby Stress

The local interior dissipation power inequalities in Eqs. 12.54–12.56 and 12.72 to 12.90, as stated in terms of the interior Eshelby stress, allow the following discussions:

i. In a smooth domain $\mathcal{B}_s$, despite its equivalence to Eqs. 12.16, 12.19, 12.22, 12.27, and 12.54, the inequality for the local interior dissipation power source density $\overline{\delta}_s^{\rm loc} \geq 0$ in Eq. 12.72 is best suited to derive constitutive relations for the interior Eshelby stress. To this end, it is recalled that the interior Lagrange-type free energy $\lambda_s = \kappa_s - \psi_s$ per unit volume in the spatial configuration $\mathcal{B}_s$ is parameterized as

$$\begin{aligned} \lambda_s &= \kappa_s(\boldsymbol{Y}, \boldsymbol{f}, \boldsymbol{V}) \\ &\quad - \psi_s(\boldsymbol{Y}, \boldsymbol{f}, \qquad \theta, \alpha, \nabla_X\alpha, \nabla_X\nabla_X\alpha, \cdots) \\ &= \lambda_s(\boldsymbol{Y}, \boldsymbol{f}, \boldsymbol{V}, \theta, \alpha, \nabla_X\alpha, \nabla_X\nabla_X\alpha, \cdots). \end{aligned} \tag{12.75}$$

Consequently, its spatial time derivative expands as

$$\begin{aligned} \mathrm{d}_t\lambda_s = &\frac{\partial\lambda_s}{\partial \boldsymbol{Y}}\cdot \mathrm{d}_t\boldsymbol{Y} + \frac{\mathrm{d}\lambda_s}{\mathrm{d}\boldsymbol{f}} : \mathrm{d}_t\boldsymbol{f} + \frac{\partial\lambda_s}{\partial \boldsymbol{V}}\cdot \mathrm{d}_t\boldsymbol{V} \\ &+ \frac{\partial\lambda_s}{\partial\theta}\mathrm{d}_t\theta + \frac{\delta\lambda_s}{\delta\alpha}[\mathrm{D}_t\alpha + \nabla_X\alpha\cdot\boldsymbol{V}], \end{aligned} \tag{12.76}$$

that, when inserted into Eq. 12.72, allows the following identifications:

$$\begin{aligned} \boldsymbol{p}_\mathrm{d} &= -\frac{\mathrm{d}\lambda_s}{\mathrm{d}\boldsymbol{f}}, \\ \overline{\boldsymbol{S}}_s^{\mathrm{mec}} &= \frac{\delta\lambda_s}{\delta\alpha}\nabla_X\alpha + \frac{\partial\lambda_s}{\partial\boldsymbol{Y}}, \\ \boldsymbol{P}_s &= \frac{\partial\lambda_s}{\partial\boldsymbol{V}}, \\ \sigma_s &= \frac{\partial\lambda_s}{\partial\theta}, \\ \bar{\delta}_s^{\mathrm{loc}} &= \frac{\delta\lambda_s}{\delta\alpha}\mathrm{D}_t\alpha. \end{aligned} \tag{12.77}$$

In the above, the partial derivative of λ_s with respect to $\boldsymbol{f}$ is denoted by $\mathrm{d}\lambda_s/\mathrm{d}\boldsymbol{f}$ to indicate that the kinetic energy density κ_s contained in $\lambda_s = \kappa_s - \psi_s$ is here expressed as $\kappa_s = \frac{1}{2}\rho_s\boldsymbol{V}\cdot\boldsymbol{C}\cdot\boldsymbol{V}$ (otherwise, for the common expression $\kappa_s = \frac{1}{2}\rho_s\boldsymbol{v}\cdot\boldsymbol{v}$ a capital D is used for partial derivative). Note, however, that no such distinction is necessary for the interior free energy density ψ_s. Then incorporating the abbreviation $A_s = jA_m$ for the driving force in Eq. 12.51, introducing the short-hand notation ∂_X for the explicit derivative with respect to the material position vector $\boldsymbol{X} = \boldsymbol{Y}(\boldsymbol{x}, t)$, noting that $\partial_V\kappa_s = \rho_s\boldsymbol{C}\cdot\boldsymbol{V}$ and $\mathrm{d}_f\kappa_s = \kappa_s\boldsymbol{F}^t + \boldsymbol{P}_s\otimes\boldsymbol{v}$, and observing that $\kappa_s \neq \kappa_s(\theta, \alpha, \nabla_X\alpha, \cdots)$ and $\psi_s \neq \psi_s(\boldsymbol{V})$, renders eventually the more explicit representation

$$\begin{aligned} \boldsymbol{p}_\mathrm{d} &= \frac{\partial\psi_s}{\partial\boldsymbol{f}} - \kappa_s\boldsymbol{F}^t - \boldsymbol{P}_s\otimes\boldsymbol{v}, \\ \overline{\boldsymbol{S}}_s^{\mathrm{mec}} &= A_s\nabla_X\alpha + \partial_X\lambda_s, \\ \boldsymbol{P}_s &= \rho_s\boldsymbol{C}\cdot\boldsymbol{V}, \\ \sigma_s &= -\frac{\partial\psi_s}{\partial\theta}, \\ \bar{\delta}_s^{\mathrm{loc}} &= A_s\,\mathrm{D}_t\alpha. \end{aligned} \tag{12.78}$$

It remains to relate the above format for $\boldsymbol{p}_\mathrm{d}$ to the interior Eshelby stress $\boldsymbol{\Sigma}$. To this end, it is recalled that $\psi_s = j\psi_m$, $\partial_f j = \mathrm{cof}\,\boldsymbol{f}$, and $\partial_f\psi_m = -\boldsymbol{F}^t\cdot\partial_F\psi_m\cdot\boldsymbol{F}^t$ with the constitutive relation $\boldsymbol{P} = \partial_F\psi_m$. Taken together

$$\frac{\partial\psi_s}{\partial\boldsymbol{f}} = [\psi_m\boldsymbol{I} - \boldsymbol{F}^t\cdot\boldsymbol{P}]\cdot\mathrm{cof}\,\boldsymbol{f} \tag{12.79}$$

holds, that, when inserted into the above expression for $\boldsymbol{p}_{\mathrm{d}}$, renders

$$\boldsymbol{p}_{\mathrm{d}} = [-\lambda_{\mathrm{m}} - \boldsymbol{F}^t \cdot \boldsymbol{P} + \boldsymbol{P}_{\mathrm{m}} \otimes \boldsymbol{V}] \cdot \mathrm{cof}\, \boldsymbol{f} = \boldsymbol{\Sigma}_{\mathrm{d}} \cdot \mathrm{cof}\, \boldsymbol{f} \qquad (12.80)$$

Recalling that $\boldsymbol{\Sigma}_{\mathrm{d}} = \boldsymbol{\Sigma} + \boldsymbol{P}_{\mathrm{m}} \otimes \boldsymbol{V}$ establishes eventually the sought-for relation.

ii. On (the regular part of) the external boundary $\partial\mathcal{B}_{\mathrm{m}}$ the inequality for the interior extrinsic dissipation power flux density $\widehat{\delta}_{\mathrm{m}} \geq 0$ in Eq. 12.55 reads alternatively as

$$\widehat{\delta}_{\mathrm{m}} = \lambda_{\mathrm{m}} W_{\perp} = -\boldsymbol{W} \cdot \widehat{\boldsymbol{S}}^{\mathrm{int}}_{\mathrm{m}} \geq 0 \quad \text{on} \quad \partial\mathcal{B}_{\mathrm{m}}. \qquad (12.81)$$

Thus, the total material velocity $\boldsymbol{W}$ has to form an *obtuse* angle, i.e. $\boldsymbol{W} \cdot \widehat{\boldsymbol{S}}^{\mathrm{int}}_{\mathrm{m}} \leq 0$, with the *internal part* $\widehat{\boldsymbol{S}}^{\mathrm{int}}_{\mathrm{m}} = \widehat{\boldsymbol{S}}^{\mathrm{ext}}_{\mathrm{m}} + \boldsymbol{F}^t \cdot \widehat{\boldsymbol{s}}^{\mathrm{ext}}_{\mathrm{m}}$ of the interior material extrinsic linear momentum flux density (or rather the interior Eshelby traction) $\widehat{\boldsymbol{S}}^{\mathrm{ext}}_{\mathrm{m}} = [\boldsymbol{\Sigma} + \boldsymbol{P}_{\mathrm{m}} \otimes \boldsymbol{W}] \cdot \boldsymbol{N}$, in order to produce a positive interior extrinsic dissipation power flux density $\widehat{\delta}_{\mathrm{m}} \geq 0$ on the external boundary.

Stated sloppily, material configurational changes with total velocity $\boldsymbol{W}$ on the external boundary are only possible *opposite* to the *internal part* of the interior Eshelby traction $\widehat{\boldsymbol{S}}^{\mathrm{int}}_{\mathrm{m}} = -\lambda_{\mathrm{m}}\boldsymbol{N}$, (Eq. 8.66).

In particular for the *quasi-static* case with $\lambda_{\mathrm{m}} = -\psi_{\mathrm{m}} < 0$, it has to hold that $\boldsymbol{W} \cdot \boldsymbol{N} < 0$ for $\widehat{\delta}_{\mathrm{m}} > 0$, thus interior free energy can only *spontaneously* (that is without any further external stimulus) be released if the external boundary moves inwards, i.e. opposite to the outwards pointing normal.

iii. At singular surfaces $\mathcal{S}_{\mathrm{m}}$, the inequality for the interior extrinsic dissipation power flux density $\widetilde{\delta}_{\mathrm{m}}$ in Eq. 12.56 reads alternatively as

$$\widetilde{\delta}_{\mathrm{m}} = \big[[\![\psi_{\mathrm{m}}]\!] - [\![\boldsymbol{F}]\!] : \{\boldsymbol{P}\}\big] W_{\perp} = \boldsymbol{W} \cdot \widetilde{\boldsymbol{S}}_{\mathrm{m}} \geq 0 \quad \text{at} \quad \mathcal{S}_{\mathrm{m}}. \qquad (12.82)$$

Thus, the total material velocity $\boldsymbol{W}$ has to form an *acute* angle, i.e. $\boldsymbol{W} \cdot \widetilde{\boldsymbol{S}}_{\mathrm{m}} \geq 0$, with the interior material extrinsic linear momentum flux density (or rather the jump in the interior Eshelby traction) $\widetilde{\boldsymbol{S}}_{\mathrm{m}} = [\![\boldsymbol{\Sigma} + \boldsymbol{P}_{\mathrm{m}} \otimes \boldsymbol{W}]\!] \cdot \boldsymbol{M}$, in order to produce a positive interior extrinsic dissipation power flux density $\widetilde{\delta}_{\mathrm{m}} \geq 0$ at singular surfaces.

Stated sloppily, material configurational changes with total velocity $\boldsymbol{W}$ of the singular surface are only possible *in* the direction of the jump in the interior Eshelby traction $\widetilde{\boldsymbol{S}}_{\mathrm{m}} = \big[[\![\psi_{\mathrm{m}}]\!] - [\![\boldsymbol{F}]\!] : \{\boldsymbol{P}\}\big]\boldsymbol{M}$, (Eq. 8.71).

Thus, the migration of the singular surface in the direction of the normal $\boldsymbol{M}$ is driven by $[\![\psi_{\mathrm{m}}]\!] > 0$, i.e. the sub-domain with the lower interior free energy $\psi^-_{\mathrm{m}} < \psi^+_{\mathrm{m}}$ is extending, and by $[\![\boldsymbol{F}]\!] : \{\boldsymbol{P}\} < 0$, i.e. sloppily speaking for a stress state of average tension $\{\boldsymbol{P}\} > 0$ the 'softer' sub-domain with $\boldsymbol{F}^- > \boldsymbol{F}^+$ is extending, likewise, for a stress state of average compression $\{\boldsymbol{P}\} < 0$ the 'stiffer' sub-domain with $\boldsymbol{F}^- < \boldsymbol{F}^+$ is extending.

12.4.2 Dissipation of Exterior Total Energy

Next, the exterior dissipation power densities in the bulk, on the boundary and at singular surfaces are re-stated in terms of the exterior Eshelby stress:

Local Exterior Dissipation Power Inequality (Version III) (Material Control Volume):

i. Regular Points in the Domain

$$\mathbb{p}_{\mathrm{m}}^{\mathrm{int}} + [J\,\mathrm{d}_t\lambda_{\mathrm{s}} - \boldsymbol{P}_{\mathrm{m}} \cdot \mathrm{d}_t\boldsymbol{V}] - \sigma_{\mathrm{m}}\,\mathrm{d}_t\theta = \overline{\delta}_{\mathrm{m}}^{\mathrm{loc}} \geq 0 \quad \text{in} \quad \mathcal{B}_{\mathrm{m}} \tag{12.83}$$

ii. Regular Points on the Boundary

$$-\boldsymbol{W} \cdot \widehat{\boldsymbol{S}}_{\mathrm{m}}^{\mathrm{int}\prime} = \widehat{\delta}_{\mathrm{m}} \geq 0 \quad \text{on} \quad \partial\mathcal{B}_{\mathrm{m}} \tag{12.84}$$

iii. Regular Points at Singular Surfaces

$$\boldsymbol{W} \cdot \widetilde{\boldsymbol{S}}_{\mathrm{m}}' = \widetilde{\delta}_{\mathrm{m}} \geq 0 \quad \text{at} \quad \mathcal{S}_{\mathrm{m}} \tag{12.85}$$

For the derivation of these results, clearly highlighting the role of the exterior Eshelby stress in the exterior dissipation power densities, three different scenarios may be considered:

i. In a smooth domain $\mathcal{B}_{\mathrm{m}}$, the interior and the exterior dissipation power inequalities do entirely coincide. Thus, the result in Eq. 12.83 is established exactly as that in Eq. 12.54.
ii. Combining the definitions for the exterior Eshelby stress $\boldsymbol{\Sigma}'$ (with $-\overline{\lambda}_{\mathrm{m}} = -\lambda_{\mathrm{m}} + \overline{\upsilon}_{\mathrm{m}}$) and the exterior boundary Eshelby stress $\widehat{\boldsymbol{\Sigma}}'$ as

$$\boldsymbol{\Sigma}' := \boldsymbol{\Sigma} + \overline{\upsilon}_{\mathrm{m}}\boldsymbol{I} = -\overline{\lambda}_{\mathrm{m}}\boldsymbol{I} - \boldsymbol{F}^t \cdot \boldsymbol{P} \quad \text{and} \quad \widehat{\boldsymbol{\Sigma}}' := \widehat{\upsilon}_{\mathrm{m}}\widehat{\boldsymbol{I}}$$

(similar to the definitions in Eqs. 10.20.1 and 10.31.1 of the variational setting in Chap. 10) and the identity for $\upsilon_{\mathrm{m}}' := \nabla_N\widehat{\upsilon}_{\mathrm{m}} - \widehat{\upsilon}_{\mathrm{m}}\widehat{C}$ (compare to Eqs. 10.45 and 10.46)

$$\upsilon_{\mathrm{m}}'\boldsymbol{N} = -\widehat{\boldsymbol{S}}_{\mathrm{m}}^{\mathrm{ext}\prime} - \widehat{\mathrm{Div}}\,\widehat{\boldsymbol{\Sigma}}' - \boldsymbol{F}^t \cdot \widehat{\boldsymbol{s}}_{\mathrm{m}}^{\mathrm{ext}}$$

(where $\widehat{\boldsymbol{S}}_{\mathrm{m}}^{\mathrm{ext}\prime}$ is defined in Eq. 10.31.2) with the definition of the spatial linear momentum flux densities on (the regular part of) the external boundary $\partial\mathcal{B}_{\mathrm{m}}$ in Eq. 7.27 and $-\lambda_{\mathrm{m}}' = -\overline{\lambda}_{\mathrm{m}} + \upsilon_{\mathrm{m}}' = -\lambda_{\mathrm{m}} + \overline{\upsilon}_{\mathrm{m}} + \upsilon_{\mathrm{m}}'$ renders

$$-\lambda_{\mathrm{m}}'\boldsymbol{N} = -\widehat{\boldsymbol{S}}_{\mathrm{m}}^{\mathrm{ext}\prime} - \widehat{\mathrm{Div}}\,\widehat{\boldsymbol{\Sigma}}' + [\boldsymbol{\Sigma}' + \boldsymbol{P}_{\mathrm{m}} \otimes \boldsymbol{W}] \cdot \boldsymbol{N}. \tag{12.86}$$

Involving the balance of material linear momentum in Eq. 8.50 and the definition for $\widehat{\boldsymbol{S}}^{\text{int}}_{\text{m}} := \widehat{\boldsymbol{S}}^{\text{ext}}_{\text{m}} + \boldsymbol{F}^{t} \cdot \widehat{\boldsymbol{s}}^{\text{ext}}_{\text{m}}$ in Eq. 12.70 transforms the above alternatively into

$$-\lambda'_{\text{m}}\boldsymbol{N} = -\widehat{\boldsymbol{S}}^{\text{ext}\prime}_{\text{m}} - \widehat{\text{Div}\,\boldsymbol{\Sigma}}' + \widehat{\boldsymbol{S}}^{\text{ext}}_{\text{m}} + \overline{v}_{\text{m}}\boldsymbol{N} = \widehat{\boldsymbol{S}}^{\text{int}}_{\text{m}} + [\overline{v}_{\text{m}} + v'_{\text{m}}]\boldsymbol{N}. \quad (12.87)$$

Taken together, when introducing the abbreviation

$$\begin{aligned} \widehat{\boldsymbol{S}}^{\text{int}\prime}_{\text{m}} &:= -\widehat{\boldsymbol{S}}^{\text{ext}\prime}_{\text{m}} - \widehat{\text{Div}\,\boldsymbol{\Sigma}}' + [\boldsymbol{\Sigma}' + \boldsymbol{P}_{\text{m}} \otimes \boldsymbol{W}] \cdot \boldsymbol{N} & (12.88)\\ &= -\widehat{\boldsymbol{S}}^{\text{ext}\prime}_{\text{m}} - \widehat{\text{Div}\,\boldsymbol{\Sigma}}' + \widehat{\boldsymbol{S}}^{\text{ext}}_{\text{m}} + \overline{v}_{\text{m}}\boldsymbol{N} = \widehat{\boldsymbol{S}}^{\text{int}}_{\text{m}} + [\overline{v}_{\text{m}} + v'_{\text{m}}]\boldsymbol{N}, \end{aligned}$$

for the *internal* part of the exterior material extrinsic linear momentum density establishes the result in Eq. 12.84. Observe that for the conservative case of the variational setting in Chap. 10, $\widehat{\boldsymbol{S}}^{\text{int}\prime}_{\text{m}}$ coincides with the material traction $\widehat{\boldsymbol{R}}_{\text{m}}$ driving material virtual displacements on the boundary (compare to Eq. 10.53.1).

iii. At a singular surface $\mathcal{S}_{\text{m}}$, similar to Eq. 10.49 of the variational setting in Chap. 10, the exterior material extrinsic linear momentum flux density is defined as

$$\widetilde{\boldsymbol{S}}'_{\text{m}} := \widetilde{\boldsymbol{S}}_{\text{m}} + [\![\overline{v}_{\text{m}}]\!]\boldsymbol{M}.$$

With this definition, the extrinsic heat flux density $\widetilde{h}_{\text{m}}$ generated at the singular surface in Eq. 12.40.1 expands as

$$\widetilde{h}_{\text{m}} := \boldsymbol{W} \cdot [\![\boldsymbol{F}^{t}]\!] \cdot \widetilde{\boldsymbol{s}}_{\text{m}} = -\boldsymbol{W} \cdot \left[[\![\overline{\lambda}_{\text{m}}]\!]\boldsymbol{M} + \widetilde{\boldsymbol{S}}'_{\text{m}}\right]. \quad (12.89)$$

Combining with the term $[\![\overline{\lambda}_{\text{m}}\boldsymbol{W}]\!] \cdot \boldsymbol{M}$ from Eq. 12.40.1 leads to the result in Eq. 12.85.

In passing the local exterior dissipation power inequality is also re-stated for a spatial control volume:

Local Exterior Dissipation Power Inequality (Version III) (Spatial Control Volume):

i. Regular Points in the Domain

$$\mathbb{p}^{\text{int}}_{\text{s}} + [\,\text{d}_t\lambda_{\text{s}} - \boldsymbol{P}_{\text{s}} \cdot \text{d}_t\boldsymbol{V}] - \sigma_{\text{s}}\,\text{d}_t\theta = \overline{\delta}^{\text{loc}}_{\text{s}} \geq 0 \quad \text{in} \quad \mathcal{B}_{\text{s}} \quad (12.90)$$

ii. Regular Points on the Boundary

$$-\boldsymbol{W} \cdot \widehat{\boldsymbol{S}}^{\text{int}\prime}_{\text{s}} = \widehat{\delta}_{\text{s}} \geq 0 \quad \text{on} \quad \partial\mathcal{B}_{\text{s}} \quad (12.91)$$

iii. Regular Points at Singular Surfaces

$$\boldsymbol{W} \cdot \widetilde{\boldsymbol{S}}'_{\text{s}} = \widetilde{\delta}_{\text{s}} \geq 0 \quad \text{at} \quad \mathcal{S}_{\text{s}} \quad (12.92)$$

Discussion of DPI Expressed in Exterior Eshelby Stress

The local interior dissipation power inequalities in Eqs. 12.83 to 12.85 and 12.90 to 12.92, as stated in terms of the exterior Eshelby stress, allow the following discussions:

i. In a smooth domain $\mathcal{B}_s$ the constitutive relations for the interior and the exterior Eshelby stress do entirely coincide. Thus, no differentiation between the notions of interior and the exterior Eshelby stress is needed in the domain.
ii. On (the regular part of) the external boundary $\partial\mathcal{B}_m$, the inequality for the exterior extrinsic dissipation power flux density $\widehat{\delta}_m \geq 0$ in Eq. 12.84 reads alternatively as

$$\widehat{\delta}_m = \lambda'_m W_\perp = -\boldsymbol{W} \cdot \widehat{\boldsymbol{S}}_m^{\text{int}\prime} \geq 0 \quad \text{on} \quad \partial\mathcal{B}_m. \tag{12.93}$$

iii. At singular surfaces $\mathcal{S}_m$, the inequality for the exterior extrinsic dissipation power flux density $\widetilde{\delta}_m$ in Eq. 12.85 reads alternatively as

$$\widetilde{\delta}_m = \big[[\![\bar{\psi}_m]\!] - [\![\boldsymbol{F}]\!] : \{\boldsymbol{P}\}\big] W_\perp = \boldsymbol{W} \cdot \widetilde{\boldsymbol{S}}'_m \geq 0 \quad \text{at} \quad \mathcal{S}_m. \tag{12.94}$$

12.5 Exploitation of DPI on the Boundary

The inequality for the dissipation power density on (the regular part of) the external boundary

$$\widehat{\delta}_m = \widehat{\varsigma}_m \, W_\perp \geq 0 \quad \text{with} \quad \widehat{\varsigma}_m = \begin{cases} \lambda_m & \text{interior DPI} \\ \lambda'_m & \text{exterior DPI} \end{cases} \tag{12.95}$$

may be used to determine the normal part $W_\perp = \boldsymbol{W} \cdot \boldsymbol{N}$ of its total material velocity $\boldsymbol{W}$.

To this end, a convex but possibly non-smooth boundary dissipation potential

$$\widehat{\Delta} = \widehat{\Delta}(W_\perp) \tag{12.96}$$

and its Legendre transformation

$$\widehat{\Delta}^* = \widehat{\Delta}^*(\widehat{\varsigma}_m) \tag{12.97}$$

are introduced so that

$$\widehat{\Delta} + \widehat{\Delta}^* = \widehat{\delta}_m \tag{12.98}$$

(indeed, for the rate-independent case $\widehat{\Delta}^*$ is non-smooth, see below). Positive dissipation power is then assured by the convexity and positivity of the dual boundary dissipation potential $\widehat{\Delta}^*(\widehat{\varsigma}_m)$ together with $\widehat{\Delta}^*(0) = 0$.

Then the normal part $W_\perp$ of the total material velocity $\boldsymbol{W}$ at the external boundary follows as the *sub-differential* $\partial(\bullet)$, i.e. the set of sub-derivatives, of the convex but possibly non-smooth dual boundary dissipation potential

$$W_\perp \in \partial_{\widehat{\varsigma}_{\rm m}} \widehat{\Delta}^*. \tag{12.99}$$

For the rate-independent case the convex and non-smooth dual boundary dissipation potential $\widehat{\Delta}^*$ is the *indicator function* I of the *admissible domain*

$$\widehat{\Delta}^* = I\big(\widehat{\omega}(\widehat{\varsigma}_{\rm m})\big) \quad \text{with} \quad I(\widehat{\omega}) := \begin{cases} 0 & \widehat{\omega} \leq 0 \\ & \text{for} \\ \infty & \text{else} \end{cases} \tag{12.100}$$

that is expressed in terms of a convex *configurational external-boundary-evolution condition*

$$\widehat{\omega}(\widehat{\varsigma}_{\rm m}) \leq 0. \tag{12.101}$$

Then the evolution of the external boundary is alternatively expressed as

$$W_\perp = \widehat{\gamma}\, \partial_{\widehat{\varsigma}_{\rm m}} \widehat{\omega}, \tag{12.102}$$

whereby $\widehat{\gamma} \geq 0$ denotes a positive multiplier that follows from the Karush–Kuhn–Tucker complementary conditions

$$\widehat{\omega} \leq 0, \qquad \widehat{\gamma} \geq 0, \qquad \widehat{\gamma}\widehat{\omega} = 0. \tag{12.103}$$

A possible example for a configurational external-boundary-evolution condition resembling a frictional slider rheological element is given by

$$\widehat{\omega} = \widehat{\omega}(\widehat{\varsigma}_{\rm m}) = |\widehat{\varsigma}_{\rm m}| - \widehat{\varsigma}_0 \leq 0. \tag{12.104}$$

Here $\widehat{\varsigma}_0$ is a given threshold value. The corresponding associated evolution law for the external boundary is then as follows:

$$W_\perp = \widehat{\gamma} \frac{\widehat{\varsigma}_{\rm m}}{|\widehat{\varsigma}_{\rm m}|} \quad \text{with} \quad \widehat{\gamma} = |W_\perp| \quad \text{and} \quad |\widehat{\varsigma}_{\rm m}| = \widehat{\varsigma}_0. \tag{12.105}$$

It is illuminating to note that the above evolution law for the external boundary is consistent with the sub-differential

$$\widehat{\varsigma}_{\rm m} \in \partial_{W_\perp} \widehat{\Delta} \tag{12.106}$$

of the boundary dissipation potential

$$\widehat{\Delta} = \widehat{\varsigma}_0\, |W_\perp| \tag{12.107}$$

that is obtained as the Legendre transformation of the dual boundary dissipation potential $\widehat{\Delta}^*$.

This elementary modeling framework can straightforwardly be extended to hardening and/or modified to the rate-dependent case (in which the boundary dissipation potentials are smooth).

12.6 Exploitation of DPI at Singular Surfaces

The inequality for the dissipation power density at singular surfaces

$$\widetilde{\delta}_{\mathrm{m}} = \widetilde{\varsigma}_{\mathrm{m}} \, W_\perp \geq 0 \quad \text{with} \quad \widetilde{\varsigma}_{\mathrm{m}} = \begin{cases} [\![\psi_{\mathrm{m}}]\!] - [\![\boldsymbol{F}]\!] : \{\boldsymbol{P}\} & \text{interior DPI} \\ \\ [\![\overline{\psi}_{\mathrm{m}}]\!] - [\![\boldsymbol{F}]\!] : \{\boldsymbol{P}\} & \text{exterior DPI} \end{cases} \tag{12.108}$$

may be used, in full analogy to the exploitation of the dissipation inequality on the external boundary, to determine the normal part $W_\perp = \boldsymbol{W} \cdot \boldsymbol{M}$ of its total material velocity $\boldsymbol{W}$.

A possible example for a configurational singular-surface-evolution condition resembling a frictional slider rheological element is given by

$$\widetilde{\omega} = \widehat{\omega}(\widetilde{\varsigma}_{\mathrm{m}}) = |\widetilde{\varsigma}_{\mathrm{m}}| - \widetilde{\varsigma}_0 \leq 0. \tag{12.109}$$

Here $\widetilde{\varsigma}_0$ is a given threshold value. The corresponding associated evolution law for the singular surface is then as follows:

$$W_\perp = \widetilde{\gamma} \frac{\widetilde{\varsigma}_{\mathrm{m}}}{|\widetilde{\varsigma}_{\mathrm{m}}|} \quad \text{with} \quad \widetilde{\gamma} = |W_\perp| \quad \text{and} \quad |\widetilde{\varsigma}_{\mathrm{m}}| = \widetilde{\varsigma}_0, \tag{12.110}$$

whereby $\widetilde{\gamma} \geq 0$ denotes a positive multiplier that follows from the Karush–Kuhn–Tucker complementary conditions

$$\widetilde{\omega} \leq 0, \qquad \widetilde{\gamma} \geq 0, \qquad \widetilde{\gamma}\,\widetilde{\omega} = 0. \tag{12.111}$$

This elementary modeling framework can straightforwardly be extended to hardening and/or modified to the rate-dependent case.

12.7 Duality of Spatial and Material Stress

It is interesting to observe that the constitutive relations for the interior stresses entering the balances of spatial and material linear momentum (aka the spatial and material stress) display striking formal similarities.

To highlight these, first the partial derivatives of the two different parameterizations for the kinetic energy density

$$\kappa_{\mathrm{m}} = \frac{1}{2}\rho_{\mathrm{m}}(\boldsymbol{X})|\boldsymbol{v}(\boldsymbol{X},t)|^2 \quad \text{versus} \quad \kappa_{\mathrm{m}} = \frac{1}{2}\rho_{\mathrm{m}}\big(\boldsymbol{Y}(\boldsymbol{x},t)\big)|\boldsymbol{V}(\boldsymbol{x},t)|^2_{\boldsymbol{C}} \tag{12.112}$$

need to be determined with respect to $\boldsymbol{F}$ and $\boldsymbol{f}$, respectively.

To this end, $\kappa_{\mathrm{s}} = j\kappa_{\mathrm{m}}$ and $\partial_{\boldsymbol{f}} j = \operatorname{cof} \boldsymbol{f}$ together with

$$\partial_{\boldsymbol{f}}\boldsymbol{C} = -\boldsymbol{C}\overline{\otimes}\boldsymbol{F}^t - \boldsymbol{F}^t\underline{\otimes}\boldsymbol{C} \quad \text{and} \quad \partial_{\boldsymbol{F}}\boldsymbol{C} = \boldsymbol{F}^t\overline{\otimes}\boldsymbol{I} + \boldsymbol{I}\underline{\otimes}\boldsymbol{F}^t$$

are used to obtain the results assembled in Table 12.1.

Here the partial derivatives of the kinetic energy density are denoted by either upper case D or lower case d to indicate that the kinetic energy is understood to be parameterized in either $\boldsymbol{X}$ or $\boldsymbol{x}$, respectively.

Then the constitutive relations for the interior material→spatial and spatial→material Piola-type stresses from Eqs. 12.49.2 and 12.77.1 render, while noting that

$$\lambda_{\mathrm{m}} = J\lambda_{\mathrm{s}}, \quad \partial_{\boldsymbol{F}} J = \operatorname{cof}\boldsymbol{F} \quad \text{and} \quad \mathrm{D}_{\boldsymbol{F}}\lambda_{\mathrm{s}} = -\boldsymbol{f}^t \cdot \mathrm{D}_{\boldsymbol{f}}\lambda_{\mathrm{s}} \cdot \boldsymbol{f}^t,$$

and vice versa that

$$\lambda_{\mathrm{s}} = j\lambda_{\mathrm{m}}, \quad \partial_{\boldsymbol{f}} j = \operatorname{cof}\boldsymbol{f} \quad \text{and} \quad \mathrm{d}_{\boldsymbol{f}}\lambda_{\mathrm{m}} = -\boldsymbol{F}^t \cdot \mathrm{d}_{\boldsymbol{F}}\lambda_{\mathrm{m}} \cdot \boldsymbol{F}^t,$$

the constitutive relations assembled in Table 12.2.

Thereby, the interior spatial→spatial Cauchy and material→material Eshelby stresses display a so-called *energy-momentum format*. The energy-momentum format consists of:

- a spherical part in terms of an *energy* density, i.e. here the interior Lagrange-type free energy density, times the unit tensor, plus
- a general part in terms of the primary solution field gradient, i.e. here a deformation gradient, times the partial derivative of the energy density with respect

Table 12.1 Partial derivatives of the kinetic energy density with respect to the deformation gradients

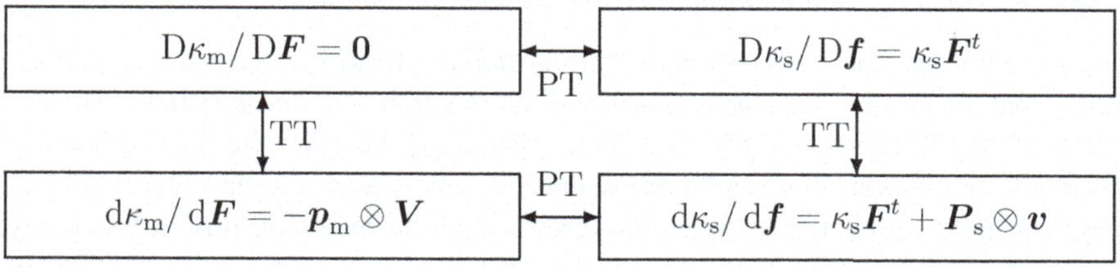

Table 12.2 Duality in the constitutive relations for the interior stresses entering the spatial and material linear momentum balances

Stresses in Spatial Linear Momentum Balance

$\boldsymbol{P}_{\mathrm{D}} = -\,\mathrm{D}\lambda_{\mathrm{m}}/\,\mathrm{D}\boldsymbol{F}$	⟷ PT	$\boldsymbol{\sigma}_{\mathrm{D}} = -\lambda_{\mathrm{s}}\boldsymbol{i} + \boldsymbol{f}^{t} \cdot \mathrm{D}\lambda_{\mathrm{s}}/\,\mathrm{D}\boldsymbol{f}$
↕ TT		↕ TT
$\boldsymbol{P}_{\mathrm{d}} = -\,\mathrm{d}\lambda_{\mathrm{m}}/\,\mathrm{d}\boldsymbol{F}$	⟷ PT	$\boldsymbol{\sigma}_{\mathrm{d}} = -\lambda_{\mathrm{s}}\boldsymbol{i} + \boldsymbol{f}^{t} \cdot \mathrm{d}\lambda_{\mathrm{s}}/\,\mathrm{d}\boldsymbol{f}$

Stresses in Material Linear Momentum Balance

$\boldsymbol{\Sigma}_{\mathrm{D}} = -\lambda_{\mathrm{m}}\boldsymbol{I} + \boldsymbol{F}^{t} \cdot \mathrm{D}\lambda_{\mathrm{m}}/\,\mathrm{D}\boldsymbol{F}$	⟷ PT	$\boldsymbol{p}_{\mathrm{D}} = -\,\mathrm{D}\lambda_{\mathrm{s}}/\,\mathrm{D}\boldsymbol{f}$
↕ TT		↕ TT
$\boldsymbol{\Sigma}_{\mathrm{d}} = -\lambda_{\mathrm{m}}\boldsymbol{I} + \boldsymbol{F}^{t} \cdot \mathrm{d}\lambda_{\mathrm{m}}/\,\mathrm{d}\boldsymbol{F}$	⟷ PT	$\boldsymbol{p}_{\mathrm{d}} = -\,\mathrm{d}\lambda_{\mathrm{s}}/\,\mathrm{d}\boldsymbol{f}$

Table 12.3 Energy-momentum format of the interior spatial→spatial Cauchy and material→material Eshelby stresses entering the spatial and material linear momentum balances

$\boldsymbol{\Sigma}_{\mathrm{D}} = -\lambda_{\mathrm{m}}\boldsymbol{I} + \boldsymbol{F}^{t} \cdot \boldsymbol{P}_{\mathrm{D}}$	$\boldsymbol{\sigma}_{\mathrm{D}} = -\lambda_{\mathrm{s}}\boldsymbol{i} + \boldsymbol{f}^{t} \cdot \boldsymbol{p}_{\mathrm{D}}$
$\boldsymbol{\Sigma}_{\mathrm{d}} = -\lambda_{\mathrm{m}}\boldsymbol{I} + \boldsymbol{F}^{t} \cdot \boldsymbol{P}_{\mathrm{d}}$	$\boldsymbol{\sigma}_{\mathrm{d}} = -\lambda_{\mathrm{s}}\boldsymbol{i} + \boldsymbol{f}^{t} \cdot \boldsymbol{p}_{\mathrm{d}}$

to the primary solution field gradient, a *momentum* flux, i.e. here the interior spatial→material and material→spatial Piola-type stresses.

The terminology energy-momentum format derives from the formal resemblance of the spatial→spatial Cauchy stresses and the material→material Eshelby stresses to the Maxwell stress of electro-magneto-dynamics, whereby the electric and the magnetic energy densities constitute the energy density term and the electric and the magnetic field, respectively, times the electric and the magnetic flux, respectively, constitute the momentum term. The duality of the interior spatial→spatial Cauchy

Table 12.4 Explicit constitutive relations for the interior stresses entering the spatial and material linear momentum balances

Stresses in Spatial Linear Momentum Balance

$\boldsymbol{P}_{\mathrm{D}} = \partial\psi_{\mathrm{m}}/\partial\boldsymbol{F}$	⟷ PT	$\boldsymbol{\sigma}_{\mathrm{D}} = \psi_{\mathrm{s}}\boldsymbol{i} - \boldsymbol{f}^t \cdot \partial\psi_{\mathrm{s}}/\partial\boldsymbol{f}$
↕ TT		↕ TT
$\boldsymbol{P}_{\mathrm{d}} = \partial\psi_{\mathrm{m}}/\partial\boldsymbol{F} + \boldsymbol{p}_{\mathrm{m}} \otimes \boldsymbol{V}$	⟷ PT	$\boldsymbol{\sigma}_{\mathrm{d}} = \psi_{\mathrm{s}}\boldsymbol{i} - \boldsymbol{f}^t \cdot \partial\psi_{\mathrm{s}}/\partial\boldsymbol{f} - \boldsymbol{p}_{\mathrm{s}} \otimes \boldsymbol{v}$

Stresses in Material Linear Momentum Balance

$\boldsymbol{\Sigma}_{\mathrm{D}} = -\lambda_{\mathrm{m}}\boldsymbol{I} - \boldsymbol{F}^t \cdot \partial\psi_{\mathrm{m}}/\partial\boldsymbol{F}$	⟷ PT	$\boldsymbol{p}_{\mathrm{D}} = \partial\psi_{\mathrm{s}}/\partial\boldsymbol{f} - \kappa_{\mathrm{s}}\boldsymbol{F}^t$
↕ TT		↕ TT
$\boldsymbol{\Sigma}_{\mathrm{d}} = -\lambda_{\mathrm{m}}\boldsymbol{I} - \boldsymbol{F}^t \cdot \partial\psi_{\mathrm{m}}/\partial\boldsymbol{F} + \boldsymbol{P}_{\mathrm{m}} \otimes \boldsymbol{V}$	⟷ PT	$\boldsymbol{p}_{\mathrm{d}} = \partial\psi_{\mathrm{s}}/\partial\boldsymbol{f} - \kappa_{\mathrm{s}}\boldsymbol{F}^t - \boldsymbol{P}_{\mathrm{s}} \otimes \boldsymbol{v}$

and material→material Eshelby stresses in energy-momentum format becomes especially apparent in Table 12.3, where the constitutive relations for the interior spatial→material and material→spatial Piola-type stresses from Table 12.2 have been explicitly incorporated.

More detailed representations of the interior stresses entering the balances of spatial and material linear momentum in terms of the kinetic energy density and the interior free energy density are the explicit constitutive relations finally assembled in Table 12.4. These, however, do not highlight too much anymore their duality.

Duality of Spatial and Material Stress in Case of Polyconvexity

In hyperelasticity, polyconvexity of the internal potential energy density $w_{\mathrm{m}}(\boldsymbol{F})$ (together with the requirement of coercivity, i.e. $w_{\mathrm{m}}(\boldsymbol{F}) \to +\infty$ for $\|\boldsymbol{F}\| \to \infty$ with $\|\boldsymbol{F}\|$ a suited norm) assures the existence of minimizers for the total potential energy. Thereby polyconvexity of a stored energy density $w_{\mathrm{m}}(\boldsymbol{F})$ is defined by the existence of a function $w^{\mathrm{p}}_{\mathrm{m}}(\boldsymbol{F}, \boldsymbol{K}, J)$, convex in each of its arguments $\boldsymbol{F}$, $\boldsymbol{K}$, and J, such that $w_{\mathrm{m}}(\boldsymbol{F})$ can be represented as $w_{\mathrm{m}}(\boldsymbol{F}) = w^{\mathrm{p}}_{\mathrm{m}}(\boldsymbol{F}, \boldsymbol{K}, J)$ with $\boldsymbol{K} := \operatorname{cof} \boldsymbol{F} := \frac{1}{2!}\boldsymbol{F} \times\!\!\times \boldsymbol{F}$ and $J := \det \boldsymbol{F} := \frac{1}{3!}[\boldsymbol{F} \times\!\!\times \boldsymbol{F}] : \boldsymbol{F}$. Polyconvexity implies quasiconvexity (that assures homogeneous deformations at minimum potential energy for a homogeneous body under applied affine Dirichlet boundary conditions), which in turn implies rank-one convexity (that assures real wave speeds in the sense of the Legendre-Hadarmard condition), whereby the converse implications are not true.

For non-conservative, i.e. inelastic material behavior, polyconvexity is required for the interior free energy density $\psi_{\mathrm{m}}(\boldsymbol{F}, \cdots)$ rather than for $w_{\mathrm{m}}(\boldsymbol{F})$, with the result that, in this case, $\psi_{\mathrm{m}}(\boldsymbol{F}, \cdots)$ can be represented as $\psi_{\mathrm{m}}(\boldsymbol{F}, \cdots) = \psi^{\mathrm{p}}_{\mathrm{m}}(\boldsymbol{F}, \boldsymbol{K}, J, \cdots)$

with the function $\psi^{\rm p}_{\rm m}(\boldsymbol{F},\boldsymbol{K},J,\cdots)$ convex in its first three arguments.

The duality of the stress measures entering the balances of spatial and material linear momentum shall next be detailed further for the case that the interior free energy density contained in the interior Lagrange-type free energy density is polyconvex in the deformation gradient, i.e.

$$\lambda_{\rm m}(\boldsymbol{F},\cdots)=\lambda^{\rm p}_{\rm m}(\boldsymbol{F},\boldsymbol{K},J,\cdots)\ \text{ and }\ \lambda_{\rm s}(\boldsymbol{f},\cdots)=\lambda^{\rm p}_{\rm s}(\boldsymbol{f},\boldsymbol{k},j,\cdots).\qquad(12.113)$$

Then, with $\mathfrak{d}$ representing either the partial derivative $\mathfrak{d}=\mathrm{D}$ or $\mathfrak{d}=\mathrm{d}$, the partial derivatives of the polyconvex interior Lagrange-type free energy density with respect to its first three arguments shall be denoted as

$$\boldsymbol{\Pi}^{\rm F}_{\mathfrak{d}}:=-\frac{\mathfrak{d}\lambda^{\rm p}_{\rm m}}{\mathfrak{d}\boldsymbol{F}}\ \text{ and }\ \boldsymbol{\Pi}^{\rm K}_{\mathfrak{d}}:=-\frac{\mathfrak{d}\lambda^{\rm p}_{\rm m}}{\mathfrak{d}\boldsymbol{K}}\ \text{ and }\ \Pi^{\rm J}_{\mathfrak{d}}:=-\frac{\mathfrak{d}\lambda^{\rm p}_{\rm m}}{\mathfrak{d}J}.\qquad(12.114)$$

and

$$\boldsymbol{\pi}^{\rm f}_{\mathfrak{d}}:=-\frac{\mathfrak{d}\lambda^{\rm p}_{\rm s}}{\mathfrak{d}\boldsymbol{f}}\ \text{ and }\ \boldsymbol{\pi}^{\rm k}_{\mathfrak{d}}:=-\frac{\mathfrak{d}\lambda^{\rm p}_{\rm s}}{\mathfrak{d}\boldsymbol{k}}\ \text{ and }\ \pi^{\rm j}_{\mathfrak{d}}:=-\frac{\mathfrak{d}\lambda^{\rm p}_{\rm s}}{\mathfrak{d}j}.\qquad(12.115)$$

The interior material→spatial and spatial→material Piola stresses are as follows:

$$\boldsymbol{P}_{\mathfrak{d}}=\boldsymbol{P}^{\rm F}_{\mathfrak{d}}+\boldsymbol{P}^{\rm K}_{\mathfrak{d}}+\boldsymbol{P}^{\rm J}_{\mathfrak{d}}\ \text{ and }\ \boldsymbol{p}_{\mathfrak{d}}=\boldsymbol{p}^{\rm f}_{\mathfrak{d}}+\boldsymbol{p}^{\rm k}_{\mathfrak{d}}+\boldsymbol{p}^{\rm j}_{\mathfrak{d}},\qquad(12.116)$$

whereby the following definitions for the various contributions to the interior material→spatial and spatial→material Piola stresses have been introduced

$$\boldsymbol{P}^{\rm F}_{\mathfrak{d}}:=\boldsymbol{\Pi}^{\rm F}_{\mathfrak{d}}\ \text{ and }\ \boldsymbol{P}^{\rm K}_{\mathfrak{d}}:=\boldsymbol{\Pi}^{\rm K}_{\mathfrak{d}}\times\!\!\times\boldsymbol{F}\ \text{ and }\ \boldsymbol{P}^{\rm J}_{\mathfrak{d}}:=\Pi^{\rm J}_{\mathfrak{d}}\boldsymbol{K}\qquad(12.117)$$

as well as

$$\boldsymbol{p}^{\rm f}_{\mathfrak{d}}:=\boldsymbol{\pi}^{\rm f}_{\mathfrak{d}}\ \text{ and }\ \boldsymbol{p}^{\rm k}_{\mathfrak{d}}:=\boldsymbol{\pi}^{\rm k}_{\mathfrak{d}}\times\!\!\times\boldsymbol{f}\ \text{ and }\ \boldsymbol{p}^{\rm j}_{\mathfrak{d}}:=\pi^{\rm j}_{\mathfrak{d}}\boldsymbol{k}.\qquad(12.118)$$

Then the interior spatial→spatial Chauchy stresses reads in polyconvex energy-momentum format as

$$\begin{aligned}\boldsymbol{\sigma}_{\mathfrak{d}}&=-\lambda^{\rm p}_{\rm s}\boldsymbol{i}-\boldsymbol{f}^t\cdot\boldsymbol{p}_{\mathfrak{d}}\qquad(12.119)\\&=-\lambda^{\rm p}_{\rm s}\boldsymbol{i}-\boldsymbol{f}^t\cdot[\boldsymbol{p}^{\rm f}_{\mathfrak{d}}+\boldsymbol{p}^{\rm k}_{\mathfrak{d}}+\boldsymbol{p}^{\rm j}_{\mathfrak{d}}]\\&=-\lambda^{\rm p}_{\rm s}\boldsymbol{i}-\boldsymbol{f}^t\cdot[\boldsymbol{\pi}^{\rm f}_{\mathfrak{d}}+\boldsymbol{\pi}^{\rm k}_{\mathfrak{d}}\times\!\!\times\boldsymbol{f}+\pi^{\rm j}_{\mathfrak{d}}\boldsymbol{k}]\\&=-\lambda^{\rm p}_{\rm s}\boldsymbol{i}-\boldsymbol{f}^t\cdot\boldsymbol{\pi}^{\rm f}_{\mathfrak{d}}-\mathbf{k}^T:\boldsymbol{\pi}^{\rm k}_{\mathfrak{d}}-\boldsymbol{j}^t\pi^{\rm j}_{\mathfrak{d}}\end{aligned}$$

with the fourth- and second-order spatial cofactor and determinant tensors

$$\mathbf{k}:=\boldsymbol{k}\otimes\boldsymbol{i}-\boldsymbol{k}\,\underline{\otimes}\,\boldsymbol{i}\implies\mathbf{k}^T=\boldsymbol{i}\otimes\boldsymbol{k}-\boldsymbol{i}\,\underline{\otimes}\,\boldsymbol{k}^t\ \text{ and }\ \boldsymbol{j}:=j\boldsymbol{i}.\qquad(12.120)$$

Observe the interesting format of the momentum terms in the last line of Eq. 12.119 with the 'gradients of the primary solution fields' $\boldsymbol{f}^t$, $\mathbf{k}^T$ and $\boldsymbol{j}^t$ pre-multiplying the 'fluxes' $\pi^{\mathrm{f}}_{\partial}$, $\pi^{\mathrm{k}}_{\partial}$, and $\pi^{\mathrm{j}}_{\partial}$, respectively.

Likewise, the interior material→material Eshelby stresses reads in polyconvex energy-momentum format as

$$\begin{aligned}\boldsymbol{\Sigma}_{\partial} &= -\lambda^{\mathrm{p}}_{\mathrm{m}}\boldsymbol{I} - \boldsymbol{F}^t \cdot \boldsymbol{P}_{\partial} \\ &= -\lambda^{\mathrm{p}}_{\mathrm{m}}\boldsymbol{I} - \boldsymbol{F}^t \cdot [\boldsymbol{P}^{\mathrm{F}}_{\partial} + \boldsymbol{P}^{\mathrm{K}}_{\partial} + \boldsymbol{P}^{\mathrm{J}}_{\partial}] \\ &= -\lambda^{\mathrm{p}}_{\mathrm{m}}\boldsymbol{I} - \boldsymbol{F}^t \cdot [\boldsymbol{\Pi}^{\mathrm{F}}_{\partial} + \boldsymbol{\Pi}^{\mathrm{K}}_{\partial} \times\!\!\times \boldsymbol{F} + \Pi^{\mathrm{J}}_{\partial}\boldsymbol{K}] \\ &= -\lambda^{\mathrm{p}}_{\mathrm{m}}\boldsymbol{I} - \boldsymbol{F}^t \cdot \boldsymbol{\Pi}^{\mathrm{F}}_{\partial} - \mathbf{K}^T : \boldsymbol{\Pi}^{\mathrm{K}}_{\partial} - \boldsymbol{J}^t \Pi^{\mathrm{J}}_{\partial}\end{aligned} \tag{12.121}$$

with the fourth- and second-order material cofactor and determinant tensors

$$\mathbf{K} := \boldsymbol{K} \otimes \boldsymbol{I} - \boldsymbol{K} \underline{\otimes} \boldsymbol{I} \implies \mathbf{K}^T = \boldsymbol{I} \otimes \boldsymbol{K} - \boldsymbol{I} \underline{\otimes} \boldsymbol{K}^t \quad \text{and} \quad \boldsymbol{J} := J\boldsymbol{I}. \tag{12.122}$$

Observe again the interesting format of the momentum terms in the last line of Eq. 12.121 with the 'gradients of the primary solution fields' $\boldsymbol{F}^t$, $\mathbf{K}^T$ and $\boldsymbol{J}^t$ pre-multiplying the 'fluxes' $\boldsymbol{\Pi}^{\mathrm{F}}_{\partial}$, $\boldsymbol{\Pi}^{\mathrm{K}}_{\partial}$, and $\Pi^{\mathrm{J}}_{\partial}$, respectively.

12.8 Four-Dimensional Formalism

It is an interesting to observe that the pertinent balance relations may be re-cast in a *four-dimensional formalism* in terms of appropriately defined *four-vectors* and *-tensors* in the case of elasto-dynamics (Maugin [9], Kienzler and Herrmann [10, 11]). The four-dimensional formalism results in a unified representation of the balances of spatial momentum and mass or the balances of material momentum and (interior total) *mechanical* energy, respectively. It thus highlights the intimate relation between mass and energy known from relativity. The four-vectors of spatial and material coordinates $\underline{\boldsymbol{x}}$ and $\underline{\boldsymbol{X}}$ consisting of the ordinary space-like coordinates $\boldsymbol{x}$ and $\boldsymbol{X}$ in the first three entries and a time-like coordinate $\Xi = \xi := ct$ in the fourth entry are related by suited *four-deformation maps*

$$\underline{\boldsymbol{x}} := \begin{bmatrix} \boldsymbol{x} \\ \xi \end{bmatrix} = \underline{\boldsymbol{y}}(\underline{\boldsymbol{X}}) \quad \text{and} \quad \underline{\boldsymbol{X}} := \begin{bmatrix} \boldsymbol{X} \\ \Xi \end{bmatrix} = \underline{\boldsymbol{Y}}(\underline{\boldsymbol{x}}). \tag{12.123}$$

Here c denotes a proper reference velocity that needs to be introduced for dimensional consistency. The corresponding *four-deformation gradients* follow from application of appropriate *four-gradients* to the four-deformation maps

$$\underline{\boldsymbol{F}} := \underline{\nabla_X \boldsymbol{y}} = \begin{bmatrix} \boldsymbol{F} & \boldsymbol{v}/c \\ \boldsymbol{0} & 1 \end{bmatrix} \quad \text{and} \quad \underline{\boldsymbol{f}} := \underline{\nabla_x \boldsymbol{Y}} = \begin{bmatrix} \boldsymbol{f} & \boldsymbol{V}/c \\ \boldsymbol{0} & 1 \end{bmatrix}. \tag{12.124}$$

Next, the *Piola four-stress* is introduced via the Lagrange-type potential energy density $l_{\rm m} := \kappa_{\rm m}(\boldsymbol{v}/c, 1) - w_{\rm m}(\boldsymbol{F})$ with $2\kappa_{\rm m} := \rho_{\rm m}c^2\big[|\boldsymbol{v}/c|^2 + 1^2\big]$ the *four-kinetic energy density* and $w_{\rm m} = w_{\rm m}(\boldsymbol{F})$ the elastically stored (potential) energy density as

$$\underline{\boldsymbol{P}} := -\frac{\partial l_{\rm m}}{\partial \underline{\boldsymbol{F}}} = \begin{bmatrix} \boldsymbol{P} & -c\,\boldsymbol{p}_{\rm m} \\ \boldsymbol{0} & -\rho_{\rm m}c^2 \end{bmatrix} \Longrightarrow \tag{12.125}$$

$$\underline{\boldsymbol{\sigma}} := \underline{\boldsymbol{P}} \cdot \mathrm{cof}\,\underline{\boldsymbol{f}} = \begin{bmatrix} \boldsymbol{\sigma} - \boldsymbol{p}_{\rm s} \otimes \boldsymbol{v} & -c\,\boldsymbol{p}_{\rm s} \\ -c\,\boldsymbol{p}_{\rm s} & -\rho_{\rm s}c^2 \end{bmatrix}.$$

Here, any possible external potential energy density is neglected for the sake of presentation, thus no source or (Neumann) boundary terms appear. Note the symmetry of the *Cauchy four-stress* $\underline{\boldsymbol{\sigma}}$ introduced as the *Piola four-transformation* of the Piola four-stress. Applying the corresponding material and spatial *four-divergence operators* to either the Piola or the Cauchy four-stress renders collectively the appropriate versions of the (source-free) balances of spatial momentum and mass

$$\underline{\mathrm{Div}\boldsymbol{P}} = \begin{bmatrix} \mathrm{Div}\boldsymbol{P} - \mathrm{D}_t\boldsymbol{p}_{\rm m} \\ -c\,\mathrm{D}_t\rho_{\rm m} \end{bmatrix} \equiv \tag{12.126}$$

$$\equiv J\,\underline{\mathrm{div}\boldsymbol{\sigma}} = J\begin{bmatrix} \mathrm{div}(\boldsymbol{\sigma} - \boldsymbol{p}_{\rm s} \otimes \boldsymbol{v}) - \mathrm{d}_t\boldsymbol{p}_{\rm s} \\ -c\,[\mathrm{div}\boldsymbol{p}_{\rm s} + \mathrm{d}_t\rho_{\rm s}] \end{bmatrix} \doteq \underline{\boldsymbol{0}}.$$

These balance equations may equivalently be expressed in terms of a pull-back by the transposed four-deformation gradient to the material manifold

$$-\,\underline{\boldsymbol{F}}^t \cdot \underline{\mathrm{Div}\boldsymbol{P}} = \underline{\mathrm{Div}}\left(-l_{\rm m}\underline{\boldsymbol{I}} + \underline{\boldsymbol{F}}^t \cdot \frac{\partial l_{\rm m}}{\partial \underline{\boldsymbol{F}}}\right) =: \underline{\mathrm{Div}\boldsymbol{\Sigma}} \doteq \underline{\boldsymbol{0}}. \tag{12.127}$$

Thereby, the *Eshelby four-stress* $\underline{\boldsymbol{\Sigma}}$ expands in terms of the Eshelby stress $\boldsymbol{\Sigma} := -l_{\rm m}\boldsymbol{I} - \boldsymbol{F}^t \cdot \boldsymbol{P}$, the material momentum density $\boldsymbol{P}_{\rm m} := -\boldsymbol{F}^t \cdot \boldsymbol{p}_{\rm m}$, the Piola-type mechanical energy flux $\boldsymbol{E} := \boldsymbol{v} \cdot \boldsymbol{P}$ and the Hamiltonian-type potential energy density $\varepsilon_{\rm m} := \kappa_{\rm m} + w_{\rm m}$ as

$$\underline{\boldsymbol{\Sigma}} = \begin{bmatrix} \boldsymbol{\Sigma} & -c\,\boldsymbol{P}_{\rm m} \\ -\boldsymbol{E}/c & \varepsilon_{\rm m} \end{bmatrix} \Longrightarrow \tag{12.128}$$

$$\underline{\boldsymbol{p}} := \underline{\boldsymbol{\Sigma}} \cdot \mathrm{cof}\,\underline{\boldsymbol{f}} = \begin{bmatrix} \boldsymbol{p} - \boldsymbol{P}_{\rm s} \otimes \boldsymbol{v} & -c\,\boldsymbol{P}_{\rm s} \\ -[\boldsymbol{\varepsilon} - \varepsilon_{\rm s}\boldsymbol{v}]/c & \varepsilon_{\rm s} \end{bmatrix}.$$

Here, $\underline{\boldsymbol{p}}$ is the two-point *Eshelby-type four-stress*, and $\boldsymbol{\varepsilon} := \boldsymbol{E} \cdot \mathrm{cof}\,\boldsymbol{f} = \boldsymbol{v} \cdot \boldsymbol{\sigma}$ denotes the Cauchy-type mechanical energy flux. Applying the corresponding material and spatial four-divergence operators renders

$$\underline{\mathrm{Div}\boldsymbol{\Sigma}} = \begin{bmatrix} \mathrm{Div}\boldsymbol{\Sigma} - \mathrm{D}_t\boldsymbol{P}_\mathrm{m} \\ -[\mathrm{Div}\boldsymbol{E} - \mathrm{D}_t\varepsilon_\mathrm{m}]/c \end{bmatrix} \equiv \tag{12.129}$$

$$\equiv J\underline{\mathrm{div}\boldsymbol{p}} = J\begin{bmatrix} \mathrm{div}(\boldsymbol{p} - \boldsymbol{P}_\mathrm{s}\otimes\boldsymbol{v}) - \mathrm{d}_t\boldsymbol{P}_\mathrm{s} \\ -[\mathrm{div}(\boldsymbol{\varepsilon} - \varepsilon_\mathrm{s}\boldsymbol{v}) - \mathrm{d}_t\varepsilon_\mathrm{s}]/c \end{bmatrix} \doteq \underline{\boldsymbol{0}}.$$

The former are alternative versions of the balance of material momentum, the latter of the balance of mechanical energy. Noteworthy, a pull-back of the four-balance equation by the transposed four-deformation gradient to the material manifold exchanges the balance of mass for the balance of mechanical energy. The Cauchy four-stress $\underline{\boldsymbol{\sigma}}$, the Piola four-stress $\underline{\boldsymbol{P}}$, the Eshelby four-stress $\underline{\boldsymbol{\Sigma}}$, and its Piola four-transformation $\underline{\boldsymbol{p}}$ are alternatives for the four-dimensional energy-momentum tensor of elastodynamics.

For large c (e.g. the speed of light) the scaled energy-momentum tensor, $-\underline{\boldsymbol{\sigma}}/c^2$ degenerates to only the entry ρ_s different from zero (Newtonian limit). The density is related to Newton's gravitational potential via $\Delta\gamma = 4\pi\rho_\mathrm{s}$, thus identifying the gravitational potential $\gamma = \gamma(\boldsymbol{x})$ as the key contribution to the space-time metric. The gravito-elastically coupled problem for deformable heavy masses of finite volume may be recovered from a variational setting in terms of the deformation map and Newton's gravitational potential (Steinmann [12]).

References

1. Javili A, McBride A, Steinmann P (2013) Numerical modelling of thermomechanical solids with highly conductive energetic interfaces. Int J Numer Methods Eng 93:551–574
2. Esmaeili A, Javili A, Steinmann P (2017) Highly-conductive energetic coherent interfaces subject to in-plane degradation. Math Mech Solids 22:1696–1716
3. Javili A, Kaessmair S, Steinmann P (2014) General imperfect interfaces. Comput Methods Appl Mech Eng 275:76–97
4. Javili A, McBride A, Steinmann P (2012) Numerical modelling of thermomechanical solids with mechanically energetic (generalised) kapitza interfaces. Comput Mat Sci 65:542–551
5. Kaessmair S, Javili A, Steinmann P (2014) Thermomechanics of solids with general imperfect coherent interfaces. Arch Appl Mech 84:1409–1426
6. Esmaeili A, Javili A, Steinmann P (2016) A thermo-mechanical cohesive zone model accounting for mechanically energetic kapitza interfaces. Int J Solids Struct 92:29–44
7. Esmaeili A, Steinmann P, Javili A (2017) Coupled thermally general imperfect and mechanically coherent energetic interfaces subject to in-plane degradation. J Mech Mat Struct 12:289–312
8. Steinmann P, Runesson K (2021) The catalogue of computational material models: basic geometrically linear models in 1D. Springer Nature, Berlin
9. Maugin GA (2000) On the universality of the thermomechanics of forces driving singular sets. Arch Appl Mech 70:31–45
10. Kienzler R, Herrmann G (2003) On the four-dimensional formalism in continuum mechanics. Acta Mech 161:103–125
11. Kienzler R, Herrmann G (2004) On conservation laws in elastodynamics. Int J Solids Struct 41:3595–3606
12. Steinmann P (2011) Geometrically nonlinear gravito-elasticity: hyperelastostatics coupled to newtonian gravitation. Int J Eng Sci 49:1452–1460

Chapter 13
Computational Setting

8,848 m 27°59'17"N 86°55'31"E

Abstract This chapter sketches the consequences for computational mechanics by outlining the material force method based on finite element discretization of the material virtual work principle and highlights its applicability to geometrically nonlinear fracture mechanics by some computational examples.

Nonlinear (as well as linear) continuum mechanics problems are typically too involved to allow for exact analytical solution, thus asking for computational approaches approximating the true solution with controllable accuracy. To this end, the virtual work formats of the balances of spatial and material momentum in Eqs. 9.60 and 9.95 lend themselves readily for a straightforward Galerkin discretization resulting in the celebrated *Finite Element Method*. Here, for the sake of demonstration, we shall restrict ourselves to the *quasi-static* and *conservative* case. In the *quasi-static* case, the external loading is so slow that inertia effects are neglectable, thus the balances of momentum degenerate to their equilibrium version. Then time is merely a parameter ordering the sequence of external loading. Moreover, the *conservative* case is characterized by hyperelastic constitutive response and external loads that derive from a potential and that are often independent of the deformation.

Discrete nodal *spatial* and *material forces* arise from the discretization. The former contribute to the residual statement representing equilibrium of the discretized spatial configuration. This is, in general, a nonlinear algebraic system that may be solved iteratively for the discrete nodal spatial deformation, e.g. by the Newton method. The latter result from post-processing and are energetically conjugate to changes of the discretized material configuration. Discrete nodal material forces allow analyzing in a uniform computational framework the propagation of cracks, the movement of interfaces and inclusions, the flow of discrete (and continuous) distributions of dislocations, and shape sensitivities, among other challenges from defect mechanics. This chapter will focus on computational examples from fracture mechanics.

P. Steinmann, *Spatial and Material Forces in Nonlinear Continuum Mechanics*,
Solid Mechanics and Its Applications 272,
https://doi.org/10.1007/978-3-030-89070-4_13

13.1 Continuous Spatial Virtual Work

For the *quasi-static* case, the principle of *spatial virtual work* is equivalent to the equilibrium of resultant spatial volume and surface forces exerted on a continuum body. It is stated in terms of spatial virtual displacements as

Quasi-Static Spatial Virtual Work Principle (Material Configuration):

$$\mathsf{P}_\delta^{\mathrm{ext}}(\delta \boldsymbol{v}) = \mathsf{P}_\delta^{\mathrm{int}}(\delta \boldsymbol{v}) \quad \forall \delta \boldsymbol{v} \tag{13.1}$$

with:

External Spatial Virtual Work

$$\mathsf{P}_\delta^{\mathrm{ext}}(\delta \boldsymbol{v}) := \int_{\mathcal{B}_{\mathrm{m}}} \delta \boldsymbol{v} \cdot \overline{\boldsymbol{s}}_{\mathrm{m}} \, \mathrm{d}V + \int_{\partial \mathcal{B}_{\mathrm{m}}} \delta \widehat{\boldsymbol{v}} \cdot \widehat{\boldsymbol{s}}_{\mathrm{m}}^{\mathrm{ext}} \, \mathrm{d}A \tag{13.2}$$

Internal Spatial Virtual Work

$$\mathsf{P}_\delta^{\mathrm{int}}(\delta \boldsymbol{v}) := \int_{\mathcal{B}_{\mathrm{m}}} \nabla_X \delta \boldsymbol{v} : \boldsymbol{P} \, \mathrm{d}V \tag{13.3}$$

13.2 Discretized Spatial Virtual Work

For general boundary value problems with given external volume and surface loading and given boundary support, the analytical solution for the unknown spatial deformation map is typically unavailable. Thus, within the *finite element method*, the continuous setting of the spatial virtual work principle is algebraizised through suited discretization of the geometry and the deformation map. The resulting nonlinear algebraic equation system, i.e. the residual that shall equate to zero, is then in principle solvable by a Newton-type iteration and renders an approximation to the true solution. It typically differs from the true solution mainly due to errors in the geometry description, the approximation quality, and the integration accuracy. Upon sufficient refinement of the discretization, these errors approach zero asymptotically.

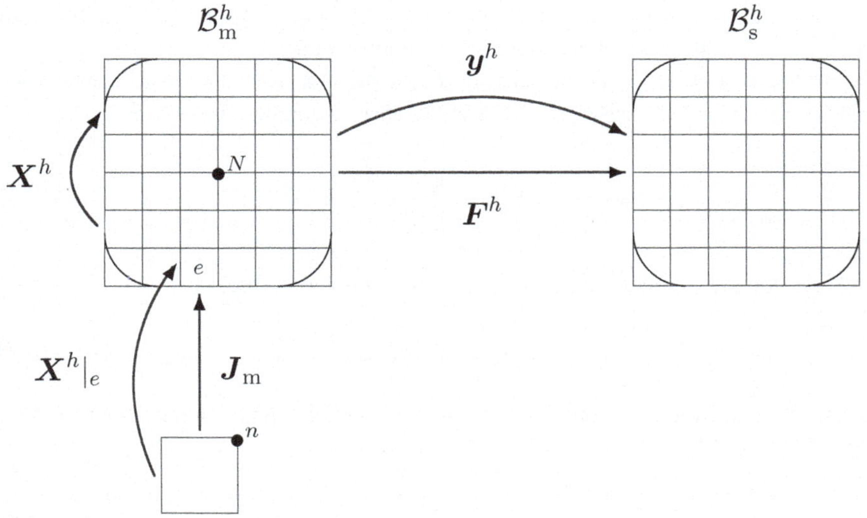

Fig. 13.1 Discretized spatial virtual work principle: discretization of material configuration into finite elements (element-wise and patch-wise views)

13.2.1 Finite Element Discretization

Element-Wise View

Commonly, an element-wise view is adopted toward finite element discretization. Here, the solution domain, i.e. the material configuration, is discretized into n_{el} (*number of elements*) non-overlapping finite elements, see Fig. 13.1

$$\mathcal{B}^h_{\rm m} = \bigcup_{e=1}^{n_{el}} \mathcal{B}^e_{\rm m} \quad \text{with} \quad \emptyset = \bigcap_{e=1}^{n_{el}} \mathcal{B}^e_{\rm m} \tag{13.4}$$

with $e \in [1, n_{el}]$ denoting the (global) element numbering. On each element $\mathcal{B}^e_{\rm m}$ the discretized spatial deformation map $\boldsymbol{y}^h|_e$ and the material coordinates $\boldsymbol{X}^h|_e$ are interpolated from the positions $\boldsymbol{y}_n$ and $\boldsymbol{X}_n$ of the n_{en} (*number of element nodes*) local node points by local shape functions

$$N^n = N^n(\boldsymbol{\xi}) \tag{13.5}$$

with $n \in [1, n_{en}]$ denoting the local node numbering and $\boldsymbol{\xi} \in [-1, 1]^{n_{dm}}$ the coordinates in the n_{dm} (*number of dimensions*) dimensional isoparametric domain, thus

$$\boldsymbol{y}^h|_e = \sum_{n=1}^{n_{en}} N^n(\boldsymbol{\xi})\, \boldsymbol{y}_n \quad \text{and} \quad \boldsymbol{X}^h|_e = \sum_{n=1}^{n_{en}} N^n(\boldsymbol{\xi})\, \boldsymbol{X}_n. \tag{13.6}$$

The local shape functions N^n satisfy the local $\delta^n_m = N^n(\boldsymbol{\xi}_m)$ property and obey the local partition of unity condition $\sum_{n=1}^{n_{en}} N^n(\boldsymbol{\xi}) = 1$, furthermore the global assembly of the element-wise expansions (see below) ensures a globally C^0-continuous interpolation of the discretized spatial deformation map.

The element-wise material Jacobian $\boldsymbol{J}_\mathrm{m}$ of the map $\boldsymbol{X}^h|_e : \square \to \mathcal{B}^e_\mathrm{m}$, where the symbol $\square = [-1, 1]^{n_{dm}}$ denotes the isoparametric domain, computes as

$$\boldsymbol{J}_\mathrm{m} = \nabla_\xi \boldsymbol{X}^h|_e = \sum_{n=1}^{n_{en}} \boldsymbol{X}_n \otimes \nabla_\xi N^n(\boldsymbol{\xi}) \tag{13.7}$$

and is needed to compute material gradients by the chain rule, i.e.

$$\nabla_X\{\bullet\} = \nabla_\xi\{\bullet\} \cdot \boldsymbol{J}_\mathrm{m}^{-1}. \tag{13.8}$$

With these prerequisites at hand, the discretized spatial deformation gradient on each element is as follows:

$$\boldsymbol{F}^h|_e = \nabla_X \boldsymbol{y}^h|_e = \sum_{n=1}^{n_{en}} \boldsymbol{y}_n \otimes \nabla_X N^n(\boldsymbol{\xi}). \tag{13.9}$$

The element-wise discretization of the *spatial virtual displacement* field $\delta\boldsymbol{v}^h|_e$ into nodal values $\delta\boldsymbol{v}_n$, which are interpolated as well by the local shape functions $N^n = N^n(\boldsymbol{\xi})$ in the spirit of a Bubnov-Galerkin expansion, renders the representation

$$\delta\boldsymbol{v}^h|_e = \sum_{n=1}^{n_{en}} N^n(\boldsymbol{\xi})\, \delta\boldsymbol{v}_n. \tag{13.10}$$

Finally, the material gradient of the spatial virtual displacement field is element-wise given by

$$\nabla_X \delta\boldsymbol{v}^h|_e = \sum_{n=1}^{n_{en}} \delta\boldsymbol{v}_n \otimes \nabla_X N^n(\boldsymbol{\xi}). \tag{13.11}$$

Remark In passing, it is noted that the corresponding spatial gradient is as follows:

$$\nabla_x \delta\boldsymbol{v}^h|_e = \sum_{n=1}^{n_{en}} \delta\boldsymbol{v}_n \otimes \nabla_x N^n(\boldsymbol{\xi}),$$

whereby $\nabla_x N^n$ computes from the element-wise relation

$$\nabla_X N^n = \nabla_x N^n \cdot \boldsymbol{F}^h|_e$$

or likewise as $\nabla_x N^n = \nabla_\xi N^n \cdot \boldsymbol{J}_\mathrm{s}^{-1}$, with the corresponding spatial Jacobian $\boldsymbol{J}_\mathrm{s}^{-1}$, see Fig. 13.2 and below. $\square$

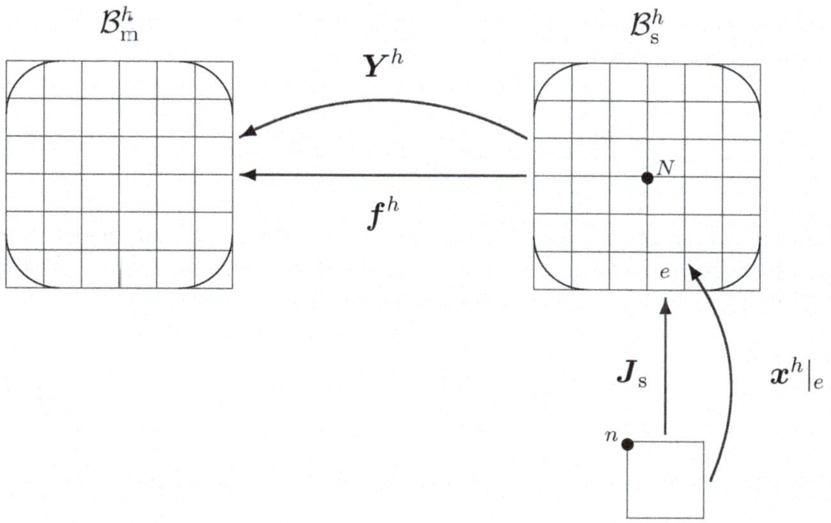

Fig. 13.2 Discretized material virtual work principle: discretization of spatial configuration into finite elements (element-wise and patch-wise views)

Patch-Wise View

It is sometimes convenient to alternatively adopt a patch-wise view. Here, the solution domain, i.e. the material configuration, is discretized into N_{np} (*number of node points*) global node points with overlapping finite patches defined by their attached elements, see Fig. 13.1

$$\mathcal{B}^h_m = \bigcup_{N=1}^{N_{np}} \mathcal{P}^N_m \quad \text{with} \quad \mathcal{B}^h_m = \bigcap_{N=1}^{N_{np}} \mathcal{P}^N_m \tag{13.12}$$

with $N \in [1, N_{np}]$ denoting the global node numbering. On the discretized solution domain $\mathcal{B}^h_m$, the discretized spatial deformation map $\boldsymbol{y}^h$ and the material coordinates $\boldsymbol{X}^h$ are interpolated from the positions $\boldsymbol{y}_N$ and $\boldsymbol{X}_N$ of the N_{np} global node points by global shape functions

$$N^N = N^N(\boldsymbol{X}), \tag{13.13}$$

thus

$$\boldsymbol{y}^h = \sum_{N=1}^{N_{np}} N^N(\boldsymbol{X})\,\boldsymbol{y}_N \quad \text{and} \quad \boldsymbol{X}^h = \sum_{N=1}^{N_{np}} N^N(\boldsymbol{X})\,\boldsymbol{X}_N. \tag{13.14}$$

The hat-like global shape functions N^N satisfy the global $\delta^N_M = N^N(\boldsymbol{X}_M)$ property, ensuring a globally C^0-continuous interpolation of the discretized spatial deformation map, and obey the global partition of unity condition $\sum_{N=1}^{N_{np}} N^N(\boldsymbol{X}) = 1$.

The Jacobian of the map $\boldsymbol{y}^h : \mathcal{B}_{\rm m} \to \mathcal{B}_{\rm s}^h$, i.e. the discretized spatial deformation gradient $\boldsymbol{F}^h$, is as follows:

$$\boldsymbol{F}^h = \nabla_X \boldsymbol{y}^h = \sum_{N=1}^{N_{np}} \boldsymbol{y}_N \otimes \nabla_X N^N(\boldsymbol{X}). \tag{13.15}$$

The discretization of the *spatial virtual displacement* field $\delta\boldsymbol{v}^h$ into nodal values $\delta\boldsymbol{v}_N$, which are interpolated as well by the global shape functions $N^N = N^N(\boldsymbol{X})$ in the spirit of a Bubnov-Galerkin expansion, renders the representation

$$\delta\boldsymbol{v}^h = \sum_{N=1}^{N_{np}} N^N(\boldsymbol{X})\, \delta\boldsymbol{v}_N. \tag{13.16}$$

Finally, the material gradient of the spatial virtual displacement field follows by

$$\nabla_X \delta\boldsymbol{v}^h = \sum_{N=1}^{N_{np}} \delta\boldsymbol{v}_N \otimes \nabla_X N^N(\boldsymbol{X}). \tag{13.17}$$

Matrix Arrangement

It proves convenient to assign the global vector-valued nodal values $\delta\boldsymbol{v}_N \in \mathbb{E}^{N_{dm}}$ to corresponding (column) matrices $\delta\mathbf{y}_N \in \mathbb{R}^{N_{dm}}$ (whereby the redundant equality $N_{dm} \equiv n_{dm}$ is only introduced for notational symetry), i.e.

$$\delta\boldsymbol{v}_N \in \mathbb{E}^{N_{dm}} \quad \mapsto \quad \delta\mathbf{y}_N \in \mathbb{R}^{N_{dm}} \tag{13.18}$$

and to assemble these into a global (column) matrix $\delta\mathbf{y} \in \mathbb{R}^{N_{np}N_{dm}}$ with

$$\delta\mathbf{y} := \left[[\delta\mathbf{y}_1]^T \cdots [\delta\mathbf{y}_N]^T \cdots [\delta\mathbf{y}_{N_{np}}]^T\right]^T. \tag{13.19}$$

Here, ^{T} denotes transposition of matrices (thus, e.g. $\delta\mathbf{y}^T$ denotes a row matrix). Likewise, we assign the local vector-valued nodal values $\delta\boldsymbol{v}_n \in \mathbb{E}^{n_{dm}}$ to corresponding (column) matrices $\delta\mathbf{y}_n \in \mathbb{R}^{n_{dm}}$, i.e.

$$\delta\boldsymbol{v}_n \in \mathbb{E}^{n_{dm}} \quad \mapsto \quad \delta\mathbf{y}_n \in \mathbb{R}^{n_{dm}} \tag{13.20}$$

and assemble these into an element (column) matrix $\delta\mathbf{y}_e \in \mathbb{R}^{n_{en}n_{dm}}$ with

$$\delta\mathbf{y}_e := \left[[\delta\mathbf{y}_1]^T \cdots [\delta\mathbf{y}_n]^T \cdots [\delta\mathbf{y}_{n_{en}}]^T\right]^T. \tag{13.21}$$

Remark In passing, it is noted that, e.g. the local shape functions N^n may accordingly be arranged into an element (rectangular) matrix $\mathbf{N}^e \in \mathbb{R}^{n_{dm} \times n_{en} n_{dm}}$ with

$$\mathbf{N}^e(\boldsymbol{\xi}) := \left[\mathbf{N}^1(\boldsymbol{\xi}) \cdots \mathbf{N}^n(\boldsymbol{\xi}) \cdots \mathbf{N}^{n_{en}}(\boldsymbol{\xi})\right]$$

and the (diagonal) nodal shape function matrices $\mathbf{N}^n \in \mathbb{R}^{n_{dm} \times n_{dm}}$ defined as

$$\mathbf{N}^n(\boldsymbol{\xi}) := \left\lceil N^n(\boldsymbol{\xi}) \right\rfloor.$$

Thus, the approximation of, e.g. the element-wise spatial virtual displacement field $\delta \boldsymbol{v}^h|_e$ takes the alternative, more abbreviated matrix representation

$$\delta \boldsymbol{v}^h|_e = \mathbf{N}^e(\boldsymbol{\xi})\, \delta \mathbf{y}_e.$$

Similar relations hold of course also for the global shape functions N^N. □

Furthermore, the $n_{en} n_{dm}$ local (element-wise) degrees of freedom assembled in $\mathbf{y}_e$ are assigned to the $N_{np} N_{dm}$ global degrees of freedom assembled in $\mathbf{y}$ by the help of Boolean (assembly) matrices $\mathbf{a}_e \in \mathbb{R}^{n_{en} n_{dm} \times N_{np} N_{dm}}$ that capture the local versus global numbering of degrees of freedom, thus

$$\delta \mathbf{y}_e := \mathbf{a}_e\, \delta \mathbf{y}. \tag{13.22}$$

The Boolean (assembly) matrices $\mathbf{a}_e$ consist of mostly zeros and only very few ones and are thus only used to conceptually represent the process of assembly.

13.2.2 Finite Element Algebraization

Element-Wise View

For the element-wise view, the discretized internal and external contributions to the spatial virtual work follow as a sum over all elements

$$\mathsf{P}_\delta^{\mathrm{int}}(\delta \boldsymbol{v}^h) = \sum_{e=1}^{n_{el}} \mathsf{P}_\delta^{\mathrm{int}}(\delta \boldsymbol{v}^h|_e) \quad \text{and} \quad \mathsf{P}_\delta^{\mathrm{ext}}(\delta \boldsymbol{v}^h) = \sum_{e=1}^{n_{el}} \mathsf{P}_\delta^{\mathrm{ext}}(\delta \boldsymbol{v}^h|_e). \tag{13.23}$$

The external contribution to the spatial virtual work decomposes furthermore into bulk volume and boundary surface contributions $\mathsf{P}_\delta^{\mathrm{ext}} = \mathsf{P}_\delta^{\mathrm{vol}} + \mathsf{P}_\delta^{\mathrm{sur}}$ with

$$\mathsf{P}_\delta^{\mathrm{vol}}(\delta \boldsymbol{v}^h) = \sum_{e=1}^{n_{el}} \mathsf{P}_\delta^{\mathrm{vol}}(\delta \boldsymbol{v}^h|_e) \quad \text{and} \quad \mathsf{P}_\delta^{\mathrm{sur}}(\delta \boldsymbol{v}^h) = \sum_{e=1}^{n_{el}} \mathsf{P}_\delta^{\mathrm{sur}}(\delta \boldsymbol{v}^h|_e). \tag{13.24}$$

The element-wise expansion for the internal contribution reads

$$\mathsf{P}_\delta^{\text{int}}(\delta\boldsymbol{v}^h|_e) := \sum_{n=1}^{n_{en}} \delta\boldsymbol{v}_n \cdot \boldsymbol{f}_{\text{int}}^n = \delta\mathbf{y}_e^T\, \mathbf{f}_{\text{int}}^e. \tag{13.25}$$

Thereby, we assigned the (local) vector-valued spatial internal nodal forces

$$\boldsymbol{f}_{\text{int}}^n := \int_{\mathcal{B}_{\text{m}}^e} \boldsymbol{P}^h \cdot \nabla_X N^n \,\mathrm{d}V \equiv \int_{\mathcal{B}_{\text{s}}^e} \boldsymbol{\sigma}^h \cdot \nabla_x N^n \,\mathrm{d}v \quad \mapsto \quad \mathbf{f}_{\text{int}}^n \tag{13.26}$$

to corresponding (column) matrices $\mathbf{f}_{\text{int}}^n \in \mathbb{R}^{n_{dm}}$ and assembled these into a local element (column) matrix of spatial internal nodal forces $\mathbf{f}_{\text{int}}^e \in \mathbb{R}^{n_{en}n_{dm}}$, i.e.

$$\mathbf{f}_{\text{int}}^e := \left[[\mathbf{f}_{\text{int}}^1]^T, \cdots, [\mathbf{f}_{\text{int}}^n]^T, \cdots, [\mathbf{f}_{\text{int}}^{n_{en}}]^T\right]^T. \tag{13.27}$$

Note that the spatial internal nodal forces $\boldsymbol{f}_{\text{int}}^n$ depend, usually in a nonlinear fashion, on the discretized spatial deformation map $\boldsymbol{y}^h|_e$ (or rather its material gradient $\boldsymbol{F}^h|_e$) through the discretized Piola stress $\boldsymbol{P}^h := \boldsymbol{P}(\boldsymbol{F}^h|_e)$. It sometimes proves convenient to express $\boldsymbol{f}_{\text{int}}^n$ alternatively through the discretized Cauchy stress $\boldsymbol{\sigma}^h := \boldsymbol{P}^h(\boldsymbol{F}^h|_e) \cdot [\operatorname{cof} \boldsymbol{F}^h|_e]^{-1}$.

Next, the element-wise expansion for the external bulk volume contribution reads

$$\mathsf{P}_\delta^{\text{vol}}(\delta\boldsymbol{v}^h|_e) := \sum_{n=1}^{n_{en}} \delta\boldsymbol{v}_n \cdot \boldsymbol{f}_{\text{vol}}^n = \delta\mathbf{y}_e^T\, \mathbf{f}_{\text{vol}}^e. \tag{13.28}$$

Thereby, we assigned the (local) vector-valued spatial external bulk volume nodal forces

$$\boldsymbol{f}_{\text{vol}}^n := \int_{\mathcal{B}_{\text{m}}^e} \bar{s}_{\text{m}}\, N^n \,\mathrm{d}V \equiv \int_{\mathcal{B}_{\text{s}}^e} \bar{s}_{\text{s}}\, N^n \,\mathrm{d}v \quad \mapsto \quad \mathbf{f}_{\text{vol}}^n \tag{13.29}$$

to corresponding (column) matrices $\mathbf{f}_{\text{vol}}^n \in \mathbb{R}^{n_{dm}}$ and assembled these into a local element (column) matrix of spatial external bulk volume nodal forces $\mathbf{f}_{\text{vol}}^e \in \mathbb{R}^{n_{en}n_{dm}}$, i.e.

$$\mathbf{f}_{\text{vol}}^e := \left[[\mathbf{f}_{\text{vol}}^1]^T, \cdots, [\mathbf{f}_{\text{vol}}^n]^T, \cdots, [\mathbf{f}_{\text{vol}}^{n_{en}}]^T\right]^T. \tag{13.30}$$

Note that the body (bulk volume) force density $\bar{s}_{\text{m}}$ is typically a given quantity and independent of the spatial deformation map.

Finally, the element-wise expansion for the external boundary surface contribution reads as

$$\mathsf{P}_{\delta}^{\text{sur}}(\delta\boldsymbol{v}^h|_e) := \sum_{n=1}^{n_{en}} \delta\boldsymbol{v}_n \cdot \boldsymbol{f}_{\text{sur}}^n = \delta\mathbf{y}_e^T\, \mathbf{f}_{\text{sur}}^e. \tag{13.31}$$

Thereby we assigned the (local) vector-valued spatial external boundary surface nodal forces

$$\boldsymbol{f}_{\text{sur}}^n := \int_{\partial\mathcal{B}_{\text{m}}^e \cap \partial\mathcal{B}_{\text{m}}} \widehat{\boldsymbol{s}}_{\text{m}}^{\text{ext}} N^n \,\text{d}A \equiv \int_{\partial\mathcal{B}_{\text{s}}^e \cap \partial\mathcal{B}_{\text{s}}} \widehat{\boldsymbol{s}}_{\text{s}}^{\text{ext}} N^n \,\text{d}a \quad \mapsto \quad \mathbf{f}_{\text{sur}}^n \tag{13.32}$$

to corresponding (column) matrices $\mathbf{f}_{\text{sur}}^n \in \mathbb{R}^{n_{dm}}$ and assembled these into a local element (column) matrix of spatial external boundary surface nodal forces $\mathbf{f}_{\text{sur}}^e \in \mathbb{R}^{n_{en}n_{dm}}$, i.e.

$$\mathbf{f}_{\text{sur}}^e := \left[[\mathbf{f}_{\text{sur}}^1]^T, \cdots, [\mathbf{f}_{\text{sur}}^n]^T, \cdots, [\mathbf{f}_{\text{sur}}^{n_{en}}]^T\right]^T. \tag{13.33}$$

Note that the external traction (boundary surface force density) $\widehat{\boldsymbol{s}}_{\text{m}}^{\text{ext}}$ is typically a given quantity and for conservative external loading independent of the spatial deformation map.

Taken together, the quasi-static spatial virtual work principle re-expresses in element-wise matrix representation as

$$\sum_{e=1}^{n_{el}} \delta\mathbf{y}_e^T\, \mathbf{f}_{\text{int}}^e = \sum_{e=1}^{n_{el}} \delta\mathbf{y}_e^T\, [\mathbf{f}_{\text{vol}}^e + \mathbf{f}_{\text{sur}}^e] \quad \forall \delta\mathbf{y}_e \quad \text{with continuity enforced.} \tag{13.34}$$

Here, the continuity of the global discretized spatial virtual displacement field $\delta\boldsymbol{v}^h$ needs separate enforcement. This is straightforwardly achieved by the help of the Boolean (assembly) matrices $\mathbf{a}_e$ relating element-wise to global degrees of freedom as $\delta\mathbf{y}_e = \mathbf{a}_e\, \delta\mathbf{y}$. Thus, the quasi-static spatial virtual work principle reads in element-wise matrix format as

$$\delta\mathbf{y}^T \sum_{e=1}^{n_{el}} \mathbf{a}_e^T\, \mathbf{f}_{\text{int}}^e = \delta\mathbf{y}^T \sum_{e=1}^{n_{el}} \mathbf{a}_e^T\, [\mathbf{f}_{\text{vol}}^e + \mathbf{f}_{\text{sur}}^e] \quad \forall \delta\mathbf{y}. \tag{13.35}$$

In finite element notation, also in order to highlight that in actual implementation, the assembly is effected rather in terms of connectivity lists for the sake of efficiency and reduced storage requirements, the element-wise assembly into global matrices is typically represented in terms of the *assembly operator* $\mathbf{A}_{e=1}^{n_{el}}$ resulting in global (column) matrices of spatial internal, bulk volume, and boundary surface nodal forces

$$\mathbf{f}_{\text{int}} := \mathop{\mathsf{A}}_{e=1}^{n_{el}} \mathbf{f}_{\text{int}}^{e} \equiv \sum_{e=1}^{n_{el}} \mathbf{a}_e^T \, \mathbf{f}_{\text{int}}^{e}, \tag{13.36}$$

$$\mathbf{f}_{\text{vol}} := \mathop{\mathsf{A}}_{e=1}^{n_{el}} \mathbf{f}_{\text{vol}}^{e} \equiv \sum_{e=1}^{n_{el}} \mathbf{a}_e^T \, \mathbf{f}_{\text{vol}}^{e},$$

$$\mathbf{f}_{\text{sur}} := \mathop{\mathsf{A}}_{e=1}^{n_{el}} \mathbf{f}_{\text{sur}}^{e} \equiv \sum_{e=1}^{n_{el}} \mathbf{a}_e^T \, \mathbf{f}_{\text{sur}}^{e}.$$

Patch-Wise View

For the patch-wise view, the internal contribution to the spatial virtual work reads

$$\mathsf{P}_{\delta}^{\text{int}}(\delta \boldsymbol{v}^h) := \sum_{N=1}^{N_{np}} \delta \boldsymbol{v}_N \cdot \boldsymbol{f}_{\text{int}}^N = \delta \mathbf{y}^T \, \mathbf{f}_{\text{int}}. \tag{13.37}$$

Thereby, we assigned the (global) vector-valued spatial internal nodal forces

$$\boldsymbol{f}_{\text{int}}^N := \int_{\mathcal{B}_\text{m}} \boldsymbol{P}^h \cdot \nabla_X N^N \, \text{d}V \equiv \int_{\mathcal{B}_\text{s}} \boldsymbol{\sigma}^h \cdot \nabla_x N^N \, \text{d}v \quad \mapsto \quad \mathbf{f}_{\text{int}}^N \tag{13.38}$$

to corresponding (column) matrices $\mathbf{f}_{\text{int}}^N \in \mathbb{R}^{N_{dm}}$ and assembled these into a global (column) matrix of spatial internal nodal forces $\mathbf{f}_{\text{int}} \in \mathbb{R}^{N_{np} N_{dm}}$, i.e.

$$\mathbf{f}_{\text{int}} := \big[[\mathbf{f}_{\text{int}}^1]^T, \cdots, [\mathbf{f}_{\text{int}}^N]^T, \cdots, [\mathbf{f}_{\text{int}}^{N_{np}}]^T \big]^T. \tag{13.39}$$

Next, the external bulk volume contribution to the spatial virtual work reads as

$$\mathsf{P}_{\delta}^{\text{vol}}(\delta \boldsymbol{v}^h) := \sum_{N=1}^{N_{np}} \delta \boldsymbol{v}_N \cdot \boldsymbol{f}_{\text{vol}}^N = \delta \mathbf{y}^T \, \mathbf{f}_{\text{vol}}. \tag{13.40}$$

Thereby we assigned the (global) vector-valued spatial external bulk volume nodal forces

$$\boldsymbol{f}_{\text{vol}}^N := \int_{\mathcal{B}_\text{m}} \bar{s}_\text{m} \, N^N \, \text{d}V \equiv \int_{\mathcal{B}_\text{s}} \bar{s}_\text{s} \, N^N \, \text{d}v \quad \mapsto \quad \mathbf{f}_{\text{vol}}^N \tag{13.41}$$

to corresponding (column) matrices $\mathbf{f}_{\text{vol}}^N \in \mathbb{R}^{N_{dm}}$ and assembled these into a global (column) matrix of spatial external bulk volume nodal forces $\mathbf{f}_{\text{vol}} \in \mathbb{R}^{N_{np} N_{dm}}$, i.e.

$$\mathbf{f}_{\text{vol}} := \big[[\mathbf{f}_{\text{vol}}^1]^T, \cdots, [\mathbf{f}_{\text{vol}}^N]^T, \cdots, [\mathbf{f}_{\text{vol}}^{N_{np}}]^T \big]^T. \tag{13.42}$$

Finally, the external boundary surface contribution to the spatial virtual work reads as

$$\mathsf{P}_\delta^{\text{sur}}(\delta\boldsymbol{v}^h) := \sum_{N=1}^{N_{np}} \delta\boldsymbol{v}_N \cdot \boldsymbol{f}_{\text{sur}}^N = \delta\mathbf{y}^T\,\mathbf{f}_{\text{sur}}. \tag{13.43}$$

Thereby we assigned the (global) vector-valued spatial external boundary surface nodal forces

$$\boldsymbol{f}_{\text{sur}}^N := \int_{\partial\mathcal{B}_\text{m}} \widehat{\boldsymbol{s}}_\text{m}^{\text{ext}} N^N \,\mathrm{d}A \equiv \int_{\partial\mathcal{B}_\text{s}} \widehat{\boldsymbol{s}}_\text{s}^{\text{ext}} N^N \,\mathrm{d}a \quad \mapsto \quad \mathbf{f}_{\text{sur}}^N \tag{13.44}$$

to corresponding (column) matrices $\mathbf{f}_{\text{sur}}^N \in \mathbb{R}^{N_{dm}}$ and assembled these into a global (column) matrix of spatial external boundary surface nodal forces $\mathbf{f}_{\text{sur}} \in \mathbb{R}^{N_{np}N_{dm}}$, i.e.

$$\mathbf{f}_{\text{sur}} := \left[[\mathbf{f}_{\text{sur}}^1]^T, \cdots, [\mathbf{f}_{\text{sur}}^N]^T, \cdots, [\mathbf{f}_{\text{sur}}^{N_{np}}]^T\right]^T. \tag{13.45}$$

Residual Statement

Consequently, in terms of the global (column) matrices of spatial internal and external nodal forces, the quasi-static spatial virtual work principle takes the concise format

$$\delta\mathbf{y}^T\,\mathbf{f}_{\text{int}} = \delta\mathbf{y}^T\,[\mathbf{f}_{\text{vol}} + \mathbf{f}_{\text{sur}}] \quad \forall\delta\mathbf{y}. \tag{13.46}$$

Observe that the discrete spatial nodal forces are energetically conjugated to variations of the spatial nodal positions, i.e. the spatial nodal virtual displacements $\delta\mathbf{y}$. Considering the arbitrariness of $\delta\mathbf{y}$, the global discrete spatial nodal residual reads eventually as

$$\mathbf{r} := \mathbf{f}_{\text{vol}} + \mathbf{f}_{\text{sur}} - \mathbf{f}_{\text{int}} \doteq \mathbf{0}. \tag{13.47}$$

The above residual statement for $\mathbf{r} = \mathbf{r}(\mathbf{y}) \in \mathbb{R}^{N_{np}nN_{dm}}$ is a nonlinear algebraic system for the discrete, i.e. nodal spatial deformation map $\mathbf{y} \in \mathbb{R}^{N_{np}N_{dm}}$ that may be solved iteratively by Newton's method.

13.2.3 Finite Element Linearization

Restricting to conservative external loading, only the global (column) matrix of spatial nodal internal forces depends on the discrete spatial deformation map, i.e. only $\mathbf{f}_{\text{int}} = \mathbf{f}_{\text{int}}(\mathbf{y})$.

Element-Wise View

Then, for the element-wise view, the linearization of the discretized internal contribution to the spatial virtual work follows as a sum over all elements

$$\mathrm{d}\mathsf{P}_\delta^{\text{int}}(\delta\boldsymbol{v}^h, \mathrm{d}\boldsymbol{y}^h) = \sum_{e=1}^{n_{el}} \mathrm{d}\mathsf{P}_\delta^{\text{int}}(\delta\boldsymbol{v}^h|_e, \mathrm{d}\boldsymbol{y}^h|_e) \tag{13.48}$$

with element-wise expansion

$$\mathsf{dP}^{\text{int}}_{\delta}(\delta\boldsymbol{v}^h|_e,\, \mathrm{d}\boldsymbol{y}^h|_e) := \sum_{n=1}^{n_{en}} \sum_{m=1}^{n_{en}} \delta\boldsymbol{v}_n \cdot \boldsymbol{k}^{nm} \cdot \mathrm{d}\boldsymbol{y}_m = \delta\mathbf{y}_e^T\, \mathbf{k}^e\, \mathrm{d}\mathbf{y}_e. \tag{13.49}$$

Thereby we assigned the (local) tensor-valued spatial nodal tangent stiffnesses

$$\boldsymbol{k}^{nm} := \int_{\mathcal{B}^e_{\mathrm{m}}} \frac{\partial \boldsymbol{P}^h \cdot \nabla_X N^n}{\partial \boldsymbol{F}^h|_e} \cdot \nabla_X N^m \,\mathrm{d}V \quad \mapsto \quad \mathbf{k}^{nm} \tag{13.50}$$

to corresponding (square) matrices $\mathbf{k}^{nm} \in \mathbb{R}^{n_{dm} \times n_{dm}}$ and assembled these into a local element (square) matrix of spatial tangent stiffnesses $\mathbf{k}^e \in \mathbb{R}^{\bar{n} n_{dm} \times \bar{m} n_{dm}}$ (with $\bar{n} = \bar{m} := n_{en}$ used as abbreviation)

$$\mathbf{k}^e := \begin{bmatrix} \mathbf{k}^{11} & \dots & \mathbf{k}^{1m} & \dots & \mathbf{k}^{1\bar{m}} \\ \vdots & \ddots & \vdots & \ddots & \vdots \\ \mathbf{k}^{n1} & \dots & \mathbf{k}^{nm} & \dots & \mathbf{k}^{n\bar{m}} \\ \vdots & \ddots & \vdots & \ddots & \vdots \\ \mathbf{k}^{\bar{n}1} & \dots & \mathbf{k}^{\bar{n}m} & \dots & \mathbf{k}^{\bar{n}\bar{m}} \end{bmatrix}. \tag{13.51}$$

Finally, $\mathbf{k} \in \mathbb{R}^{N_{np} N_{dm} \times N_{np} N_{dm}}$ denotes the global tangent stiffness (square) matrix

$$\mathbf{k} := \mathop{\mathsf{A}}_{e=1}^{n_{el}} \mathbf{k}^e \equiv \sum_{e=1}^{n_{el}} \mathbf{a}_e^T\, \mathbf{k}^e\, \mathbf{a}_e. \tag{13.52}$$

Patch-Wise View

Alternatively, for the patch-wise view, the linearization of the discretized internal contribution to the spatial virtual work is as follows:

$$\mathsf{dP}^{\text{int}}_{\delta}(\delta\boldsymbol{v}^h,\, \mathrm{d}\boldsymbol{y}^h) := \sum_{N=1}^{N_{np}} \sum_{M=1}^{N_{np}} \delta\boldsymbol{v}_N \cdot \boldsymbol{k}^{NM} \cdot \mathrm{d}\boldsymbol{y}_M = \delta\mathbf{y}^T\, \mathbf{k}\, \mathrm{d}\mathbf{y}. \tag{13.53}$$

Thereby we assigned the (global) tensor-valued spatial nodal tangent stiffnesses

$$\boldsymbol{k}^{NM} := \int_{\mathcal{B}_{\mathrm{m}}} \frac{\partial \boldsymbol{P}^h \cdot \nabla_X N^N}{\partial \boldsymbol{F}^h} \cdot \nabla_X N^M \,\mathrm{d}V \quad \mapsto \quad \mathbf{k}^{NM} \tag{13.54}$$

to corresponding (square) matrices $\mathbf{k}^{NM} \in \mathbb{R}^{N_{dm} \times N_{dm}}$ and assembled these into the global (square) matrix of spatial tangent stiffnesses $\mathbf{k} \in \mathbb{R}^{\bar{N} N_{dm} \times \bar{M} N_{dm}}$ (with $\bar{N} = \bar{M} := N_{np}$ used as abbreviation)

$$\mathbf{k} := \begin{bmatrix} \mathbf{k}^{11} & \dots & \mathbf{k}^{1M} & \dots & \mathbf{k}^{1\bar{M}} \\ \vdots & \ddots & \vdots & \ddots & \vdots \\ \mathbf{k}^{N1} & \dots & \mathbf{k}^{NM} & \dots & \mathbf{k}^{N\bar{M}} \\ \vdots & \ddots & \vdots & \ddots & \vdots \\ \mathbf{k}^{\bar{N}1} & \dots & \mathbf{k}^{\bar{N}M} & \dots & \mathbf{k}^{\bar{N}\bar{M}} \end{bmatrix}. \tag{13.55}$$

Newton Update

With these preliminaries, each Newton iteration, in a nutshell, is as follows (from $\mathbf{r} + \mathrm{d}\mathbf{r} \doteq \mathbf{0}$):

$$\mathbf{y} + \mathbf{k}^{-1}\mathbf{r} \mapsto \mathbf{y} \quad \text{with} \quad \mathbf{k} := -\frac{\partial \mathbf{r}}{\partial \mathbf{y}} \tag{13.56}$$

until the norm of the residual falls under a given tolerance

$$|\mathbf{r}(\mathbf{y})| \leq \varepsilon. \tag{13.57}$$

Newton's method enjoys the computationally beneficial property of quadratic convergence, i.e. in the vicinity of the solution the norm of the residual reduces as $0 > \log|\mathbf{r}| \mapsto 2\log|\mathbf{r}|$ in each iteration.

13.3 Continuous Material Virtual Work

For the *quasi-static* case, the principle of *material virtual work* is equivalent to the 'equilibrium' of the resultant material volume, surface and interfacial forces acting on a continuum body. It is stated in terms of material virtual displacements as

Quasi-Static Material Virtual Work Principle (Spatial Configuration):

$$\mathsf{p}^{\text{ext}}_{\delta}(\delta\boldsymbol{\Upsilon}) = \mathsf{p}^{\text{int}}_{\delta}(\delta\boldsymbol{\Upsilon}) + \mathsf{p}^{\text{ifc}}_{\delta}(\delta\boldsymbol{\Upsilon}) \quad \forall \delta\boldsymbol{\Upsilon} \tag{13.58}$$

with:

External Material Virtual Work

$$\mathsf{p}^{\text{ext}}_{\delta}(\delta\boldsymbol{\Upsilon}) := \int_{\mathcal{B}_s} \delta\boldsymbol{\Upsilon} \cdot \overline{\boldsymbol{S}}_s \, \mathrm{d}v + \int_{\partial\mathcal{B}_s} \delta\widehat{\boldsymbol{\Upsilon}} \cdot \widehat{\boldsymbol{S}}^{\text{ext}}_s \, \mathrm{d}a \tag{13.59}$$

Internal Material Virtual Work

$$\mathsf{p}_\delta^{\text{int}}(\delta\boldsymbol{\Upsilon}) := \int_{\mathcal{B}_s} \nabla_x \delta\boldsymbol{\Upsilon} : \boldsymbol{p}\, \mathrm{d}v \tag{13.60}$$

Interfacial Material Virtual Work

$$\mathsf{p}_\delta^{\text{ifc}}(\delta\boldsymbol{\Upsilon}) := \int_{\mathcal{S}_s} \delta\widetilde{\boldsymbol{\Upsilon}} \cdot \widetilde{\boldsymbol{S}}_s\, \mathrm{d}a \tag{13.61}$$

Observe the additional interfacial contribution $\mathsf{p}_\delta^{\text{ifc}}(\delta\boldsymbol{\Upsilon})$ with the corresponding term lacking in the spatial virtual work principle. It originates in the non-vanishing jump of the material traction at singular surfaces as expressed most conveniently in the material configuration

$$[\![\boldsymbol{\Sigma}]\!] \cdot \boldsymbol{M} =: \widetilde{\boldsymbol{S}}_{\mathrm{m}} = \left[[\![\psi_{\mathrm{m}}]\!] - [\![\boldsymbol{F}]\!] : \{\boldsymbol{P}\}\right] \boldsymbol{M} \neq \boldsymbol{0}. \tag{13.62}$$

Recall that $[\![\bullet]\!] := \{\bullet\}^+ - \{\bullet\}^-$ with $\boldsymbol{M}$ pointing from $\mathcal{B}_{\mathrm{m}}^-$ to $\mathcal{B}_{\mathrm{m}}^+$. Note also that a discontinuous $\boldsymbol{F}$ at $\mathcal{S}_{\mathrm{m}}$ results both in $[\![\psi_{\mathrm{m}}]\!] \neq 0$ as well as $[\![\boldsymbol{F}]\!] \neq \boldsymbol{0}$.

13.4 Discretized Material Virtual Work

Once the spatial virtual work statement has been evaluated in its discretized format, the discretized material virtual work amounts to a mere post-processing of the discrete spatial solution to determine material boundary surface and interfacial nodal forces. Jointly, these will be denoted as discrete material nodal forces in the sequel. It is recalled that the Eshelby traction $\boldsymbol{\Sigma} \cdot \boldsymbol{N}$ at the (here considered unloaded part of the) external boundary $\partial\mathcal{B}_{\mathrm{m}}$ and $[\![\boldsymbol{\Sigma}]\!] \cdot \boldsymbol{M}$ at singular surfaces $\mathcal{S}_{\mathrm{m}}$ are the driving forces for configurational changes. Their discrete determination is thus of utmost importance to assess the tendency of all kind of defects (cracks, interfaces, inclusions, dislocations, etc.) to propagate relative to the material.

In a nutshell, the determination of discrete material forces is essentially a matter of looking at the pre-determined solution of the spatial balance equations with a pair of glasses filtering out important, however, otherwise unappreciated aspects, in particular, the driving forces acting on generic defects.

13.4.1 Finite Element Discretization

In the sequel, we shall adapt the patch-wise view for the sake of brevity.

Patch-Wise View

Here, the post-processing domain, e.g. the spatial configuration, is considered discretized into the N_{np} global node points with overlapping finite patches defined by their attached elements, see Fig. 13.2

$$\mathcal{B}_s^h = \bigcup_{N=1}^{N_{np}} \mathcal{P}_s^N \quad \text{with} \quad \mathcal{B}_s^h = \bigcap_{N=1}^{N_{np}} \mathcal{P}_s^N. \tag{13.63}$$

On the discretized post-processing domain $\mathcal{B}_s^h$, the discretized material deformation map $\boldsymbol{Y}^h$ and the spatial coordinates $\boldsymbol{x}^h$ interpolate from the positions $\boldsymbol{Y}_N$ and $\boldsymbol{x}_N$ of the N_{np} global node points by the global shape functions

$$N^N = N^N(\boldsymbol{x}), \tag{13.64}$$

now parameterized in $\boldsymbol{x}$ (i.e. $N^N(\boldsymbol{x}) = N^N(\boldsymbol{X}) \circ \boldsymbol{Y}^h(\boldsymbol{x})$), thus

$$\boldsymbol{Y}^h = \sum_{N=1}^{N_{np}} N^N(\boldsymbol{x})\, \boldsymbol{Y}_N \quad \text{and} \quad \boldsymbol{x}^h = \sum_{N=1}^{N_{np}} N^N(\boldsymbol{x})\, \boldsymbol{x}_N. \tag{13.65}$$

Consequently, the Jacobian of the map $\boldsymbol{Y}^h : \mathcal{B}_s \to \mathcal{B}_m^h$, i.e. the discretized material deformation gradient $\boldsymbol{f}^h$, expands as

$$\boldsymbol{f}^h = \nabla_x \boldsymbol{Y}^h = \sum_{N=1}^{N_{np}} \boldsymbol{Y}_N \otimes \nabla_x N^N(\boldsymbol{x}). \tag{13.66}$$

The discretization of the *material virtual displacement* field $\delta\boldsymbol{\Upsilon}^h$ into nodal values $\delta\boldsymbol{\Upsilon}_N$, interpolated as well by the global shape functions $N^N = N^N(\boldsymbol{x})$, renders the representation

$$\delta\boldsymbol{\Upsilon}^h = \sum_{N=1}^{N_{np}} N^N(\boldsymbol{x})\, \delta\boldsymbol{\Upsilon}_N. \tag{13.67}$$

Thus, the spatial gradient of the material virtual displacement field reads as

$$\nabla_x \delta\boldsymbol{\Upsilon}^h = \sum_{N=1}^{N_{np}} \delta\boldsymbol{\Upsilon}_N \otimes \nabla_x N^N(\boldsymbol{x}). \tag{13.68}$$

Remark In passing, it is noted that the corresponding material gradient is as follows:

$$\nabla_X \delta\boldsymbol{\Upsilon}^h = \sum_{N=1}^{N_{np}} \delta\boldsymbol{\Upsilon}_N \otimes \nabla_X N^N(\boldsymbol{X}),$$

whereby $\nabla_X N^N$ computes from the relation

$$\nabla_x N^N = \nabla_X N^N \cdot \boldsymbol{f}^h$$

and a re-parametrization from $\nabla_X N^N(\boldsymbol{X}) \circ \boldsymbol{Y}^h(\boldsymbol{x}) = \nabla_X N^N(\boldsymbol{x})$. □

13.4.2 Finite Element Algebraization

Patch-Wise View

Here, the internal contribution to the material virtual work reads as

$$\mathsf{p}_\delta^{\text{int}}(\delta\boldsymbol{\Upsilon}^h) := \sum_{N=1}^{N_{np}} \delta\boldsymbol{\Upsilon}_N \cdot \boldsymbol{F}_{\text{int}}^N = \delta\mathbf{Y}^T\,\mathbf{F}_{\text{int}}. \tag{13.69}$$

Thereby we assigned the (global) vector-valued material internal nodal forces

$$\boldsymbol{F}_{\text{int}}^N := \int_{\mathcal{B}_s} \boldsymbol{p}^h \cdot \nabla_x N^N \,\mathrm{d}v \equiv \int_{\mathcal{B}_m} \boldsymbol{\Sigma}^h \cdot \nabla_X N^N \,\mathrm{d}V \quad \mapsto \quad \mathbf{F}_{\text{int}}^N \tag{13.70}$$

to corresponding (column) matrices $\mathbf{F}_{\text{int}}^N \in \mathbb{R}^{N_{dm}}$ and assembled these into a global (column) matrix of material internal nodal forces $\mathbf{F}_{\text{int}} \in \mathbb{R}^{N_{np}N_{dm}}$, i.e.

$$\mathbf{F}_{\text{int}} := \big[[\mathbf{F}_{\text{int}}^1]^T, \cdots, [\mathbf{F}_{\text{int}}^N]^T, \cdots, [\mathbf{F}_{\text{int}}^{N_{np}}]^T\big]^T. \tag{13.71}$$

Note that $\boldsymbol{p}^h = \boldsymbol{p}^h(\boldsymbol{f}^h)$ is easily computable in a post-processing step once the spatial equilibrium problem is solved. Observe furthermore the alternative, more convenient expression for $\mathbf{F}_{\text{int}}^N$ in terms of the discretized Eshelby stress $\boldsymbol{\Sigma}^h := \boldsymbol{p}^h(\boldsymbol{f}^h) \cdot [\operatorname{cof} \boldsymbol{f}^h]^{-1}$.

Next, the bulk volume contribution to the material virtual work reads

$$\mathsf{p}_\delta^{\text{vol}}(\delta\boldsymbol{\Upsilon}^h) := \sum_{N=1}^{N_{np}} \delta\boldsymbol{\Upsilon}_N \cdot \boldsymbol{F}_{\text{vol}}^N = \delta\mathbf{Y}^T\,\mathbf{F}_{\text{vol}}. \tag{13.72}$$

Thereby, we assigned the (global) vector-valued material bulk volume nodal forces

$$\boldsymbol{F}_{\text{vol}}^N := \int_{\mathcal{B}_s} \overline{\boldsymbol{S}}_s^h\, N^N \,\mathrm{d}v \equiv \int_{\partial\mathcal{B}_m} \overline{\boldsymbol{S}}_m^h N^N \,\mathrm{d}V \quad \mapsto \quad \mathbf{F}_{\text{vol}}^N \tag{13.73}$$

to corresponding (column) matrices $\mathbf{F}_{\text{vol}}^N \in \mathbb{R}^{N_{dm}}$ and assembled these into a global (column) matrix of material bulk volume nodal forces $\mathbf{F}_{\text{vol}} \in \mathbb{R}^{N_{np}N_{dm}}$, i.e.

$$\mathsf{F}_{\text{vol}} := \left[[\mathsf{F}^1_{\text{vol}}]^T, \cdots, [\mathsf{F}^N_{\text{vol}}]^T, \cdots, [\mathsf{F}^{N_{np}}_{\text{vol}}]^T\right]^T. \tag{13.74}$$

Note that $\overline{S}^h_{\text{s}} = \overline{S}_{\text{s}}(f^h)$ is easily computable in a post-processing step once the spatial equilibrium problem is solved. Observe furthermore the alternative, more convenient expression for $\mathsf{F}^N_{\text{vol}}$ in terms of $\overline{S}^h_{\text{m}} := \overline{S}^h_{\text{s}}(f^h)/\det f^h$.

For the sake of completeness, the boundary surface contribution to the material virtual work reads as

$$\mathsf{p}^{\text{sur}}_\delta(\delta\boldsymbol{\Upsilon}^h) := \sum_{N=1}^{N_{np}} \delta\boldsymbol{\Upsilon}_N \cdot \boldsymbol{F}^N_{\text{sur}} = \delta\mathsf{Y}^T\,\mathsf{F}_{\text{sur}}. \tag{13.75}$$

Thereby, conceptually, we assigned the (global) vector-valued material boundary surface nodal forces

$$\boldsymbol{F}^N_{\text{sur}} := \int_{\partial\mathcal{B}_{\text{s}}} \widehat{S}^{\text{ext}}_{\text{s}} N^N \,\mathrm{d}a \equiv \int_{\partial\mathcal{B}_{\text{m}}} \widehat{S}^{\text{ext}}_{\text{m}} N^N \,\mathrm{d}A \quad\mapsto\quad \mathsf{F}^N_{\text{sur}} \tag{13.76}$$

to corresponding (column) matrices $\mathsf{F}^N_{\text{sur}} \in \mathbb{R}^{N_{dm}}$ and assembled these into a global (column) matrix of material boundary surface nodal forces $\mathsf{F}_{\text{sur}} \in \mathbb{R}^{N_{np}N_{dm}}$, i.e.

$$\mathsf{F}_{\text{sur}} := \left[[\mathsf{F}^1_{\text{sur}}]^T, \cdots, [\mathsf{F}^N_{\text{sur}}]^T, \cdots, [\mathsf{F}^{N_{np}}_{\text{sur}}]^T\right]^T. \tag{13.77}$$

Note that the evaluation of $\boldsymbol{F}^N_{\text{sur}}$ is only conceptual, since the material nodal forces F_{sur} will eventually be determined from material 'equilibrium'.

Likewise for the sake of completeness, finally, the interfacial contribution to the material virtual work reads as

$$\mathsf{p}^{\text{ifc}}_\delta(\delta\boldsymbol{\Upsilon}^h) := \sum_{N=1}^{N_{np}} \delta\boldsymbol{\Upsilon}_N \cdot \boldsymbol{F}^N_{\text{ifc}} = \delta\mathsf{Y}^T\,\mathsf{F}_{\text{ifc}}. \tag{13.78}$$

Thereby, again conceptually, we assigned the (global) vector-valued material interfacial nodal forces

$$\boldsymbol{F}^N_{\text{ifc}} := \int_{\partial\mathcal{B}_{\text{s}}} \widetilde{S}_{\text{s}} N^N \,\mathrm{d}a \equiv \int_{\partial\mathcal{B}_{\text{m}}} \widetilde{S}_{\text{m}} N^N \,\mathrm{d}A \quad\mapsto\quad \mathsf{F}^N_{\text{ifc}} \tag{13.79}$$

to corresponding (column) matrices $\mathsf{F}^N_{\text{ifc}} \in \mathbb{R}^{N_{dm}}$ and assembled these into a global (column) matrix of material interfacial nodal forces $\mathsf{F}_{\text{sur}} \in \mathbb{R}^{N_{np}N_{dm}}$, i.e.

$$\mathsf{F}_{\text{ifc}} := \left[[\mathsf{F}^1_{\text{ifc}}]^T, \cdots, [\mathsf{F}^N_{\text{ifc}}]^T, \cdots, [\mathsf{F}^{N_{np}}_{\text{ifc}}]^T\right]^T. \tag{13.80}$$

Note that the evaluation of $\boldsymbol{F}^N_{\text{ifc}}$ is likewise only conceptual, since the material nodal forces F_{ifc} will eventually be determined from material 'equilibrium'.

Material Force Method

Consequently, in terms of the global (column) matrices of material internal, volume, surface and interfacial nodal forces, the quasi-static material virtual work principle takes the concise format

$$\delta\mathbf{Y}^T\,[\mathbf{F}_{\text{int}} - \mathbf{F}_{\text{vol}}] =: \delta\mathbf{Y}^T\,[\mathbf{F}_{\text{sur}} - \mathbf{F}_{\text{ifc}}] \quad \forall\delta\mathbf{Y}. \tag{13.81}$$

Observe that the discrete material nodal forces are energetically conjugated to variations of the material nodal positions, i.e. the material nodal virtual displacements $\delta\mathbf{Y}$. Considering the arbitrariness of $\delta\mathbf{Y}$, the global discrete, i.e. nodal material equilibrium reads eventually

$$\mathbf{F}_{\text{int}} - \mathbf{F}_{\text{vol}} = \mathbf{F}_{\text{mat}} \quad \text{with} \quad \mathbf{F}_{\text{mat}} := \mathbf{F}_{\text{sur}} - \mathbf{F}_{\text{ifc}}. \tag{13.82}$$

Note the negative sign associated with the material interfacial nodal forces $\mathbf{F}_{\text{ifc}}$. The above material equilibrium statement is a *definition* for the discrete material nodal forces $\mathbf{F}_{\text{mat}}$ at the boundary surface and at singular surfaces, respectively, that are driving *actual* discrete configurational changes $\dot{\mathbf{Y}}$. Recall that the discrete dissipation due to configurational changes follows collectively as

$$-\,\dot{\mathbf{Y}}^T\,\mathbf{F}_{\text{mat}} \geq 0, \tag{13.83}$$

thus actual discrete configurational changes $\dot{\mathbf{Y}}$ are only possible *opposite* to the discrete material nodal forces in order to satisfy second law's requirement of positive dissipation!

On the one hand, in the case of a singularity in the spatial equilibrium problem, e.g. for a crack, the length of the material force opposite to the direction of the possible crack extension corresponds to the value of the celebrated J-integral.

On the other hand, we do not expect any discrete material nodal forces in an otherwise homogeneous material. The presence of such *spurious* material forces indicates that a change of the material node point positions of the discretization renders an improved mesh regarding its approximation quality, e.g. a mesh rendering less stored energy in the case of a conservative problem.

Lastly, due to the interpretation of material forces as being energetically conjugate to configurational changes, discrete material nodal forces at the boundary may be considered as a measure of the shape sensitivity of a solution domain.

Thus, in anticipation of the results obtained from the computational examples below, we may thus state that besides the usual sensitivity of the results with respect to the mesh design, the *Material Force Method* (Ackermann et al. [1], Steinmann et al. [2, 3]) has the advantages that it:

1. requires no additional FE data structure,
2. is extremely versatile,
3. renders additional indicators for geometrical shape sensitivity, and
4. renders additional indicators for the mesh quality.

Table 13.1 Computational Steps for the Material Force Method

(1) Determine Eshelby stress and material body force at quadrature points
(2) Perform standard numerical quadrature

Indeed, very few additional computational steps pertaining to the material force method are listed in Table 13.1.

Thus, the material force method simply consists in the determination of the discrete material nodal forces corresponding to the material (Eshelby) stress and body force which are trivially *computable* after the spatial equilibrium problem has been solved. It will be demonstrated in the sequel that the discrete material nodal forces may effectively be used, e.g., for the assessment of hyperelastic configurations with defects such as cracks and interfaces.

Remark The computation of the discrete material nodal forces involves the same operations as the computation of the discrete spatial nodal forces which are already available for the solution of the spatial equilibrium problem and does, therefore, not contribute significantly to the overall computational costs. □

Remark Recall that the discretized version of the spatial equilibrium is solved for given data contained in $\mathsf{P}_\delta^{\text{sur}}$ and $\mathsf{P}_\delta^{\text{vol}}$. Then, the discretized version of the material 'equilibrium' is subsequently used in a mere post-processing computation to evaluate the discrete material nodal forces from $\mathsf{p}_\delta^{\text{int}}$ and $\mathsf{p}_\delta^{\text{vol}}$ a posteriori. □

Remark It is interesting to note that the finite dimensional discrete spatial and material nodal forces preserve the properties of the corresponding equilibrium equations of the underlying infinite-dimensional continuous system. As an example, the equilibrium of spatial and material forces is trivially satisfied element-wise. For verification, it suffices to consider the body force free situation. Then, the internal contributions to the discrete nodal forces of one element only satisfy

$$\sum_{n=1}^{n_{en}} \int_{\mathcal{B}_{\text{s}}^e} \boldsymbol{\sigma} \cdot \nabla_x N^n \, \mathrm{d}v = \int_{\mathcal{B}_{\text{s}}^e} \boldsymbol{\sigma} \cdot \left[\sum_{n=1}^{n_{en}} \nabla_x N^n \right] \mathrm{d}v$$

and likewise

$$\sum_{n=1}^{n_{en}} \int_{\mathcal{B}_{\text{m}}^e} \boldsymbol{\Sigma} \cdot \nabla_X N^n \, \mathrm{d}V = \int_{\mathcal{B}_{\text{m}}^e} \boldsymbol{\Sigma} \cdot \left[\sum_{n=1}^{n_{en}} \nabla_X N^n \right] \mathrm{d}V.$$

Obviously

$$\sum_{n=1}^{n_{en}} \nabla_x N^n = \mathbf{0} \quad \text{and} \quad \sum_{n=1}^{n_{en}} \nabla_X N^n = \mathbf{0}$$

holds due to the (local) partition of unity condition $\sum_{n=1}^{n_{en}} N^n = 1$. Likewise arguments hold in the case of distributed volume forces. □

13.5 Computational Examples

We will finally exemplify the material force method by computationally analyzing mode I crack problems. To this end, we consider two-dimensional unit Pacman-shaped domains in the $\boldsymbol{e}_1 - \boldsymbol{e}_2$ plane representing the vicinity to the tip of a crack that extents in $\boldsymbol{e}_1$-direction. We apply Dirichlet boundary conditions everywhere at the external boundary of the Pacman-shaped domains (except at the crack faces, which remain unconstrained) that correspond to the linear elastic near crack tip solution in plane strain under mode I loading in $\boldsymbol{e}_2$-direction

$$\begin{bmatrix} u_1(R,\Theta) \\ u_2(R,\Theta) \end{bmatrix} = \frac{K_I}{E}\sqrt{\frac{R}{8\pi}}[1+\nu]\begin{bmatrix} [2\gamma-1]\cos(\Theta/2)-\cos(3\Theta/2) \\ [2\gamma+1]\sin(\Theta/2)-\sin(3\Theta/2) \end{bmatrix}, \tag{13.84}$$

with abbreviation $\gamma := 3 - 4\nu$. Here, E and ν, respectively, denote the elasticity modulus and Poisson ratio of isotropic elasticity. These relate to the Lamé constants λ and μ as used below via

$$E = \mu\frac{3\lambda + 2\mu}{\lambda + \mu} \quad \text{and} \quad \nu = \frac{1}{2}\frac{\lambda}{\lambda + \mu}. \tag{13.85}$$

Note that multiplying λ and μ by the same factor, say κ, results in an elasticity modulus multiplied by κ, while the Poisson ratio remains unchanged. In the case of mode I loading, the stress intensity K_I relates to the value of the here prescribed J as

$$K_I^2 = J\frac{E}{1-\nu^2}. \tag{13.86}$$

Finally, R denotes the radius (here $R = 1$ at the boundary of a unit Pacman domain) and Θ is the anti-clock-wise angle from the horizontal ligament.

Regarding the hyperelastic constitutive model, we choose the isotropic compressible Neo-Hooke (interior) free energy density

$$\psi_{\rm m} = \frac{1}{2}\mu[I_1 - 3] + \frac{1}{2}\lambda \ln^2 J - \mu \ln J. \tag{13.87}$$

The compressible Neo-Hooke model agrees with linear Hooke elasticity in the limit of infinitesimal deformations. Of course, only in this limit, we can expect that the computed resultant material force $\mathsf{F}_{\rm mat}$ in the crack direction (see below) coincides with the prescribed J (modulo the sign) in the case of homogeneous material properties. However, importantly, it also captures the case of finite deformation for which typically $\mathsf{F}^{\text{mat}} > J$ holds.

Then, based on the free energy density ψ_{m}, the Piola and the Eshelby stress[1] compute explicitly as

$$\boldsymbol{P} = \mu[\boldsymbol{F} - \boldsymbol{f}^t] + \lambda \ln J \boldsymbol{f}^t \quad \text{and} \quad \boldsymbol{\Sigma} = \psi_{\text{m}} \boldsymbol{I} - \mu[\boldsymbol{C} - \boldsymbol{I}] - \lambda \ln J \boldsymbol{I}. \tag{13.88}$$

We uniformly discretize a unit square in the $\Xi_1, \Xi_2 \in [0, R]$ domain and map the resulting mesh into the material configuration of the Pacman-shaped domains by the coordinate transformation

$$X_1 = -[R - \Xi_1] \cos(2\pi\, \Xi_2/R) \quad \text{and} \quad X_2 = [R - \Xi_1] \sin(2\pi\, \Xi_2/R). \tag{13.89}$$

For the loading, we apply a prescribed J ranging over five orders of magnitude in the above Dirichlet boundary conditions, i.e.

$$J = 10^{-1}, \quad J = 10^{0}, \quad J = 10^{1}, \quad J = 10^{2}, \quad J = 10^{3}. \tag{13.90}$$

In the sequel, we shall investigate two different problem classes, a homogeneous and a heterogeneous Pacman-shaped domain under mode I loading.

13.5.1 Homogeneous Pacman-Shaped Domain

For a homogeneous Pacman-shaped domain, we set the Lamé constants for the sake of demonstration to $\lambda = 1800$ and $\mu = 3800$, and use uniform mesh refinements with **(a)** 10×10, **(b)** 20×20, **(c)** 30×30, **(d)** 40×40, **(e)** 50×50, and **(f)** 60×60 (bi-linear) $Q1$ elements. The resulting spatial configurations for $J = 10^{-1}$ to $J = 10^{3}$ are highlighted together with the nodal material forces (normalized to their respective summed-up value in a mesh) in Figs. 13.4, 13.5, 13.6, 13.7, and 13.8, respectively (for $J = 10^{-1}$ the spatial and material configurations are almost not distinguishable in the eye-ball-norm).

- As a first observation, note that larger deformations and thus geometrical non-linearities are obtained for increased loading by J, whereas loading by small J approaches the limit of infinitesimal deformations (with the classical crack tip

[1] Moreover, for example, the Piola-Kirchhoff and Mandel stress read concretely as

$$\boldsymbol{S} = \mu[\boldsymbol{I} - \boldsymbol{B}] + \lambda \ln J \boldsymbol{B} \quad \text{and} \quad \boldsymbol{M} = \mu[\boldsymbol{C} - \boldsymbol{I}] + \lambda \ln J \boldsymbol{I},$$

whereas the Cauchy stress and the Piola-type Eshelby stress expand as

$$J\boldsymbol{\sigma} = \mu[\boldsymbol{b} - \boldsymbol{i}] + \lambda \ln J \boldsymbol{i} \quad \text{and} \quad J\boldsymbol{p} = \psi_{\text{m}} \boldsymbol{F}^t - \mu \boldsymbol{F}^t \cdot [\boldsymbol{b} - \boldsymbol{i}] - \lambda \ln J \boldsymbol{F}^t.$$

Δ	a)	b)	c)	d)	e)	f)
$J = 10^{-1}$	9.9	3.5	2.0	1.4	1.1	0.9
$J = 10^{+0}$	10.25	3.83	2.41	1.83	1.53	1.34
$J = 10^{+1}$	11.26	4.96	3.58	3.01	2.71	2.53
$J = 10^{+2}$	14.12	8.15	6.82	6.26	5.96	5.78
$J = 10^{+3}$	18.78	13.64	12.43	11.91	11.62	11.45

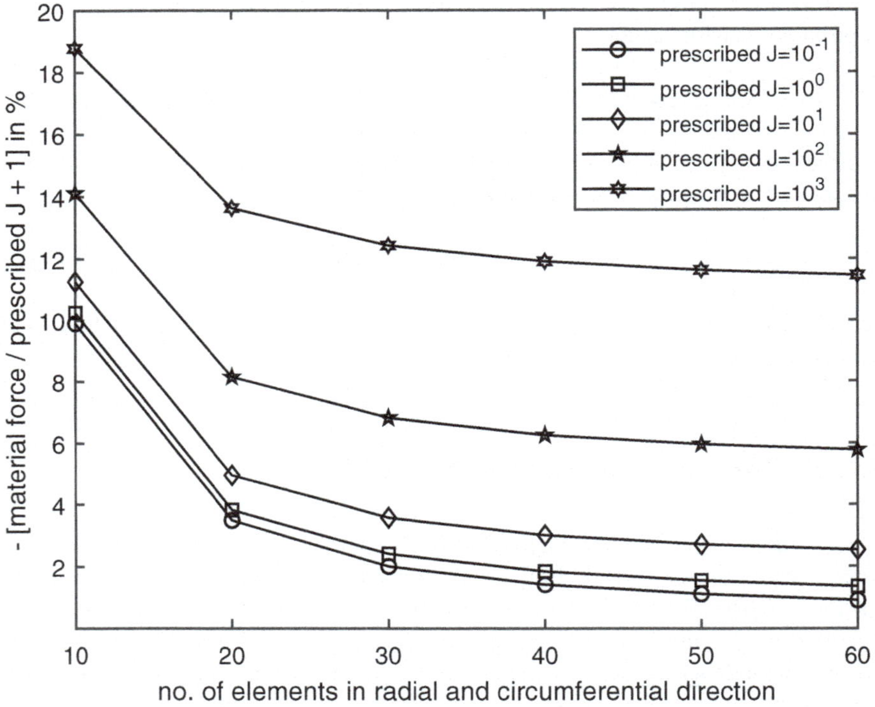

Fig. 13.3 Percentaged deviation $\Delta := [|\mathsf{F}_{\text{mat}}| - J]/J$ for prescribed Dirichlet BCs corresponding to the classical mode I solution with $J = 10^{-1}, 10^{0}, 10^{1}, 10^{2}, 10^{3}$ ($\lambda = 1800$, $\mu = 3800$). Uniform mesh refinement with **a** 10×10, **b** 20×20, **c** 30×30, **d** 40×40, **e** 50×50, and **f** 60×60 (bilinear) $Q1$ elements

solutions being valid exactly only in this limit). However, the close vicinity to the crack tip obviously always suffers extreme deformations, which is in agreement with the typical crack tip singularity in 'stress' and 'strain' and thus justifies a geometrically exact, i.e. nonlinear approach.

- As a second observation, note that 'spurious' nodal material forces show up in the discretization in the close vicinity to the crack tip, thus indicating insufficient mesh resolution of the ensuing singular 'stress' and 'strain' fields. Clearly, the crack tip

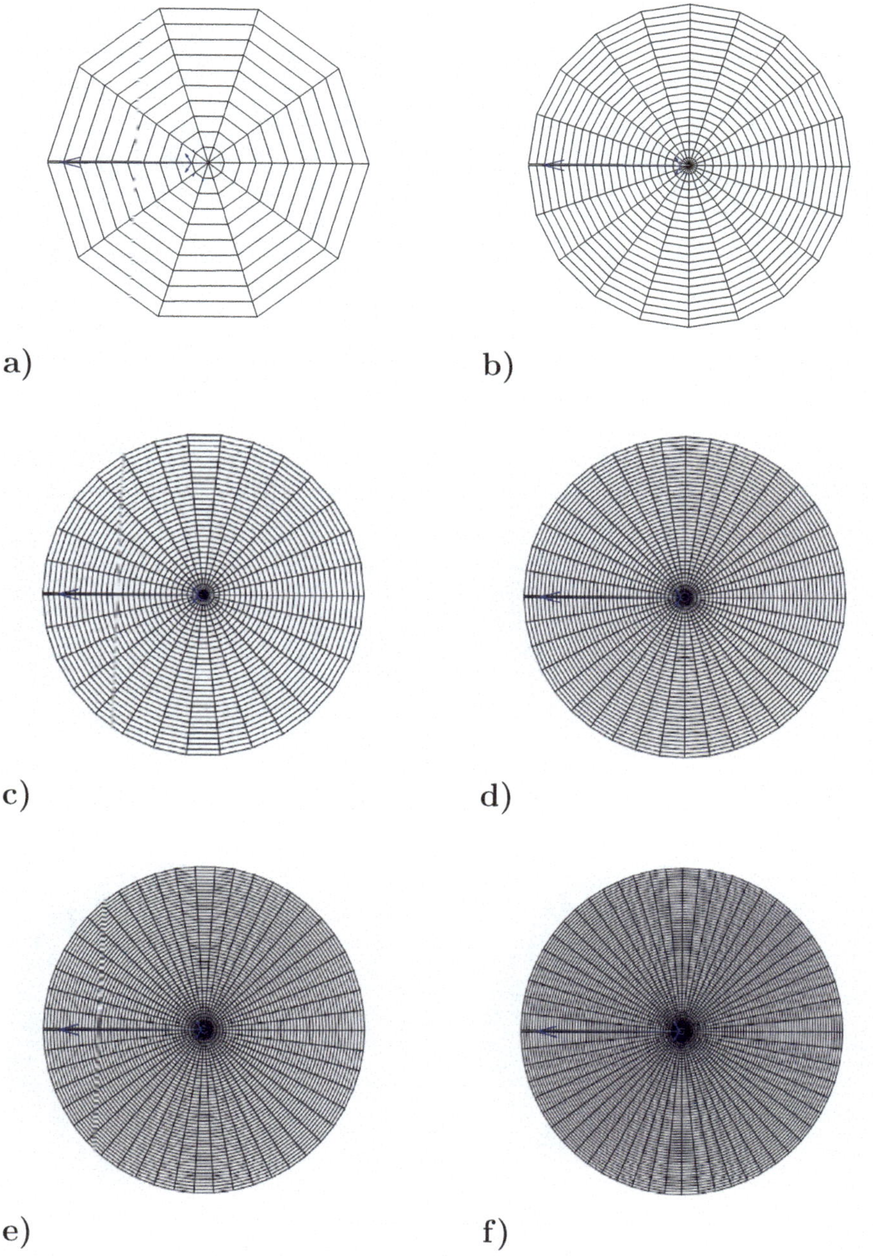

Fig. 13.4 Spatial Pacman configuration and material forces for prescribed Dirichlet BCs corresponding to the classical mode I solution with $J = 10^{-1}$ ($\lambda = 1800$, $\mu = 3800$). Uniform mesh refinement with **a** 10×10, **b** 20×20, **c** 30×30, **d** 40×40, **e** 50×50, and **f** 60×60 (bi-linear) $Q1$ elements

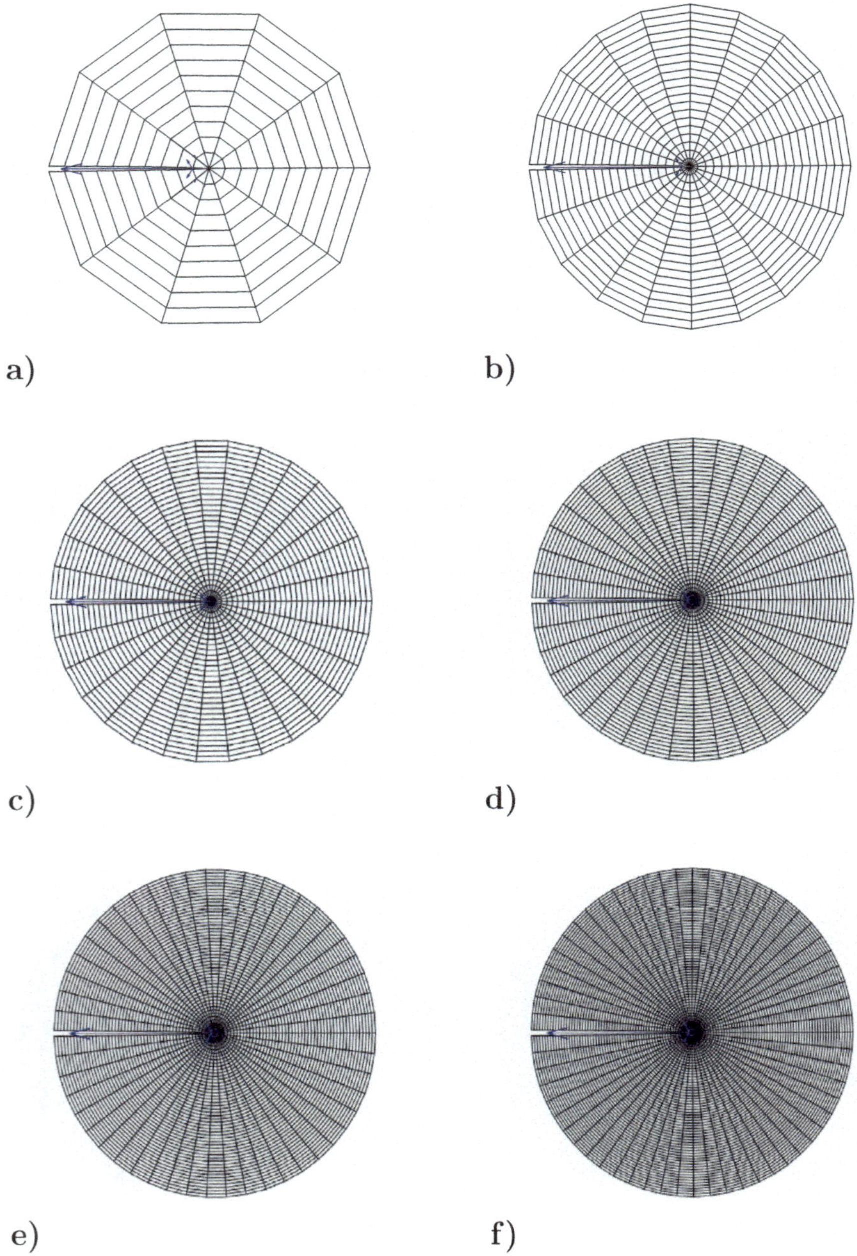

Fig. 13.5 Spatial Pacman configuration and material forces for prescribed Dirichlet BCs corresponding to the classical mode I solution with $J = 10^0$ ($\lambda = 1800$, $\mu = 3800$). Uniform mesh refinement with **a** 10×10, **b** 20×20, **c** 30×30, **d** 40×40, **e** 50×50, and **f** 60×60 (bi-linear) $Q1$ elements

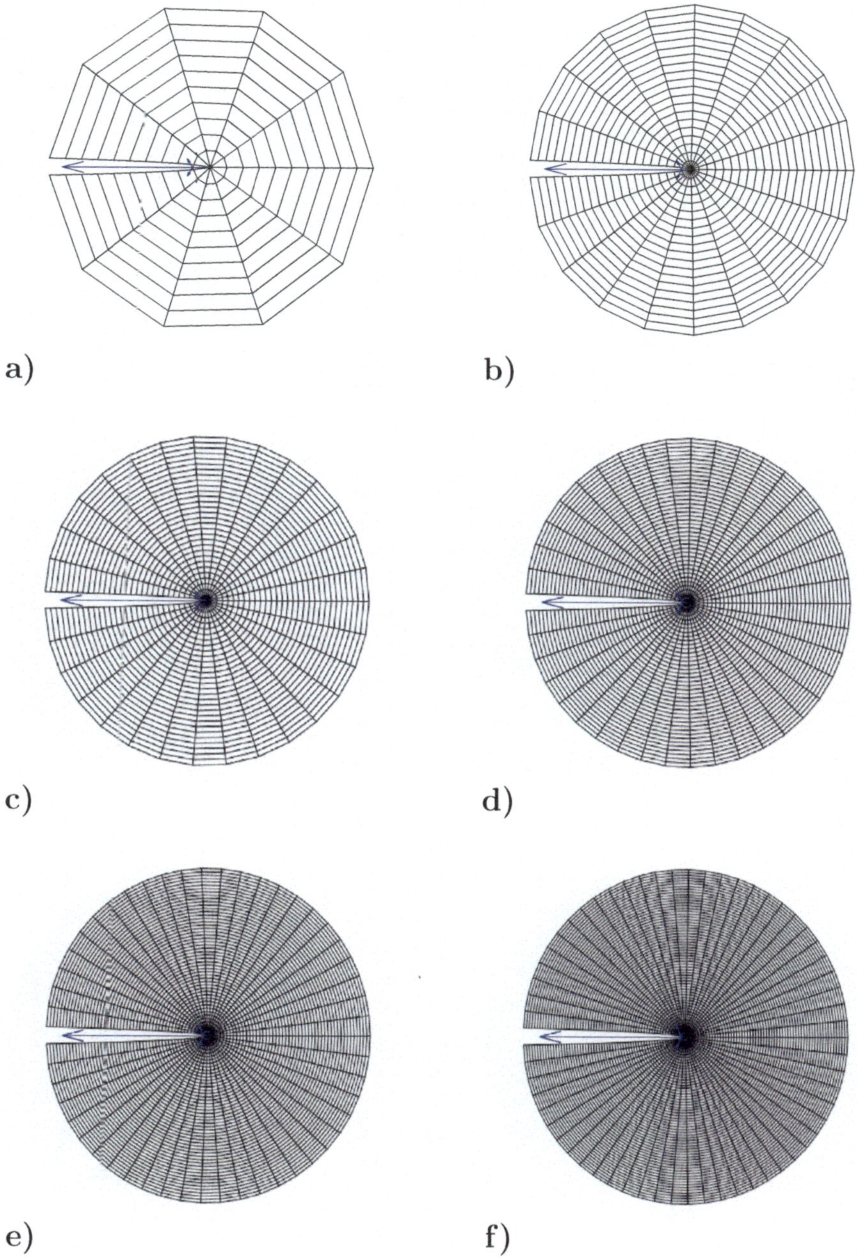

Fig. 13.6 Spatial Pacman configuration and material forces for prescribed Dirichlet BCs corresponding to the classical mode I solution with $J = 10^1$ ($\lambda = 1800$, $\mu = 3800$). Uniform mesh refinement with **a** 10×10, **b** 20×20, **c** 30×30, **d** 40×40, **e** 50×50, and **f** 60×60 (bi-linear) $Q1$ elements

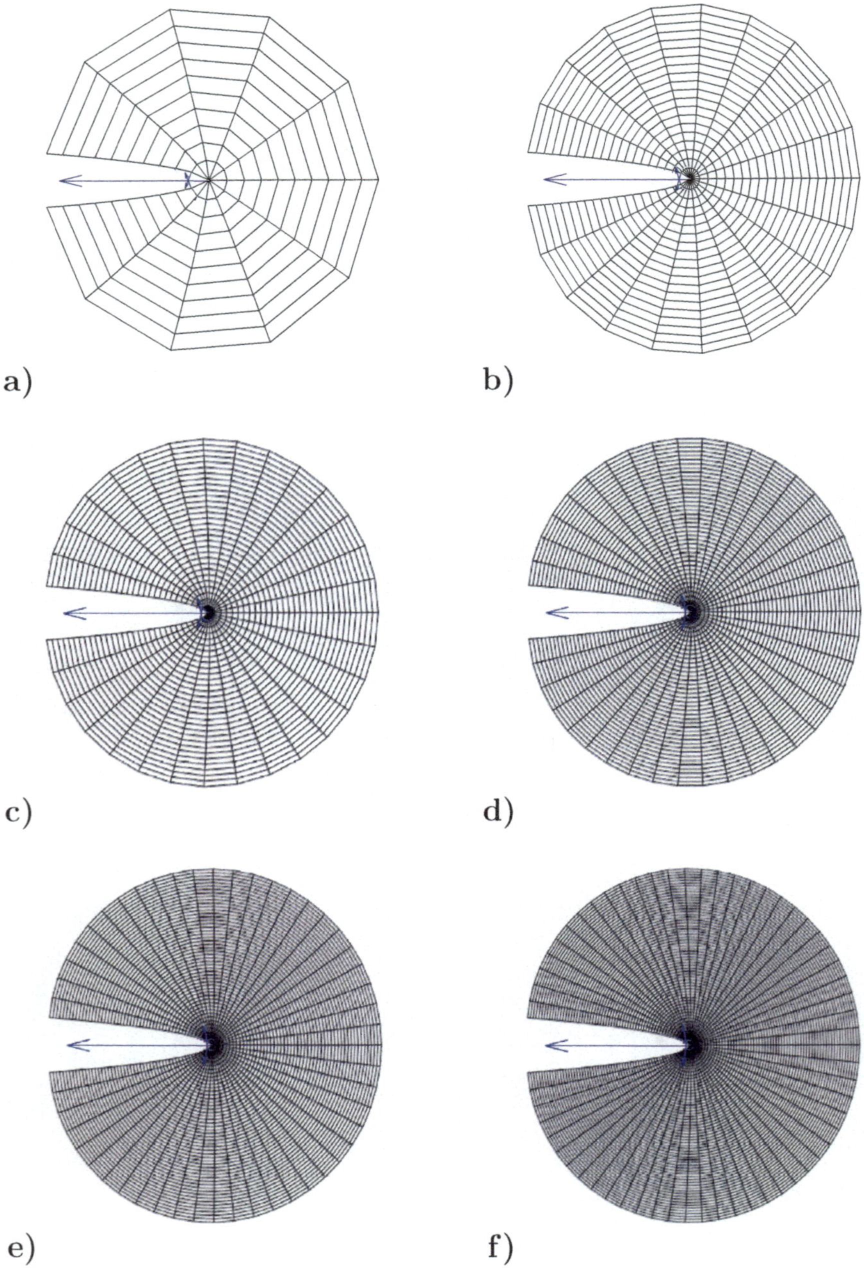

Fig. 13.7 Spatial Pacman configuration and material forces for prescribed Dirichlet BCs corresponding to the classical mode I solution with $J = 10^2$ ($\lambda = 1800$, $\mu = 3800$). Uniform mesh refinement with **a** 10×10, **b** 20×20, **c** 30×30, **d** 40×40, **e** 50×50, and **f** 60×60 (bi-linear) $Q1$ elements

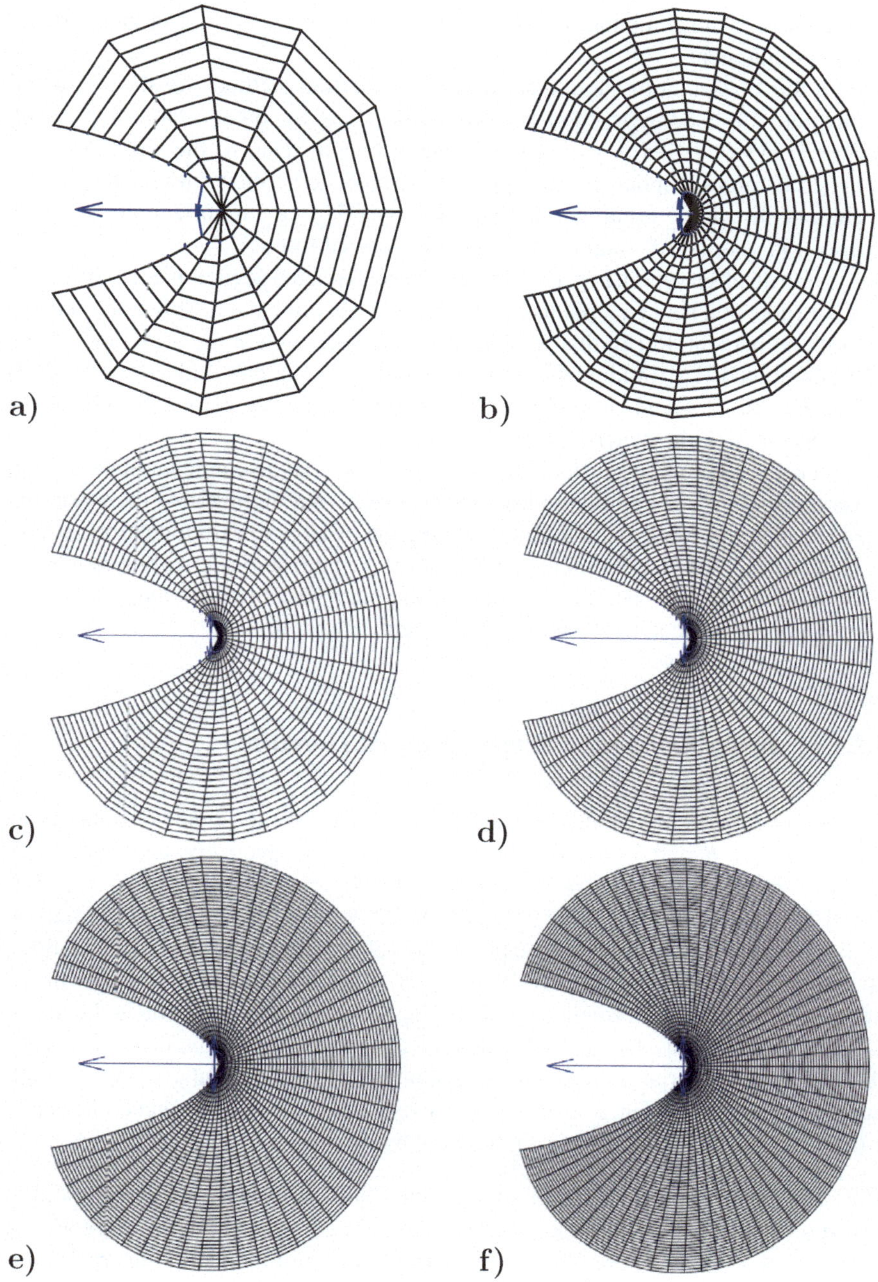

Fig. 13.8 Spatial Pacman configuration and material forces for prescribed Dirichlet BCs corresponding to the classical mode I solution with $J = 10^3$ ($\lambda = 1800$, $\mu = 3800$). Uniform mesh refinement with **a** 10×10, **b** 20×20, **c** 30×30, **d** 40×40, **e** 50×50, and **f** 60×60 (bi-linear) $Q1$ elements

singularities can computationally only be resolved in the limit of infinitely dense mesh refinement.

- As a third observation, note that common evaluation methods for the J-integral integrate the Eshelby traction in $\boldsymbol{e}_1$-direction along the outer contour (i.e. along $R = 1$ and $\Theta \in [+\pi, -\pi]$) of the Pacman domain. Due to the equilibrium of material forces implied by the quasi-static balance of momentum, this simply amounts to adding up all nodal material forces $\mathsf{F}^N_{\text{mat}} \cdot \boldsymbol{e}_1$ in horizontal, i.e. crack direction in order to compute the resultant material force F_{mat}.
- As a fourth observation, Fig. 13.3 depicts the percentaged deviation $\Delta := [|\mathsf{F}_{\text{mat}}| - J]/J$ (note that here F_{mat} points in negative $\boldsymbol{e}_1$-direction) as a function of J and the mesh refinement. Recall that we can only expect a match of the prescribed J and the computed F_{mat} (modulo the sign) in the infinitesimal limit. Indeed, for mesh **(f)** and $J = 10^{-1}$, a deviation of merely some 0.9% is depicted, thus demonstrating the accuracy of the material force method.
- As a fifth observation, for loading by larger J, we can also note an obvious convergence of F_{mat} upon refining from mesh **(a)**–**(f)**, thus highlighting the versatility of the material force method also in the regime of large to extreme deformations. However, for large deformations, the percentaged deviation Δ will of course not asymptotically approach to zero, since the applied Dirichlet boundary conditions only reflect the analytical solution in the infinitesimal limit.

13.5.2 Heterogeneous Pacman-Shaped Domain

For the heterogeneous Pacman-shaped domains, we set the Lamé constants in an outer ring $1 \leq R \leq 0.5$ to $\lambda = 1800$ and $\mu = 3800$, respectively, and scale these values by **(a)** $\kappa = 10^3$, **(b)** $\kappa = 10^2$, **(c)** $\kappa = 10^1$, **(d)** $\kappa = 10^{-1}$, **(e)** $\kappa = 10^{-2}$, and **(f)** $\kappa = 10^{-3}$ in the remaining inner core domain $0.5 \leq R \leq 0$. We use a uniform mesh with 30×30 (bi-linear) $Q1$ elements for the sake of demonstration. The resulting spatial configurations for $J = 10^{-1}$ to $J = 10^3$ are highlighted together with the nodal material forces (normalized to their respective summed-up value in the mesh) in Figs. 13.10, 13.11, 13.12, 13.13, and 13.14, respectively (again, for $J = 10^{-1}$ the spatial and material configurations are almost not distinguishable in the eye-ball-norm). As a visual guidance, in each plot, the respective sub-domain that is elastically 'stiffer' due to the choice of κ is highlighted by shading.

- As a first observation note that the case with a 'stiff' core asymptotically describes a rigid inclusion for $\kappa \to \infty$, whereas the case of a 'soft' core asymptotically describes a vacancy/pore for $\kappa \to 0$. In the former limiting case, the crack tip is situated at $R = 0.5$ rather than at $R = 0$, and in the latter limiting case, the crack ends freely at the interface to the vacancy/pore at $R = 0.5$.
- As a second observation, note that the resultant material force F_{mat} (that is equilibrated with the resultant material force at the boundary of the Pacman-shaped domain) does not coincide with the prescribed $J \neq |\mathsf{F}_{\text{mat}}|$, as clearly indicated

$J = 10^0$	a)	b)	c)	d)	e)	f)
Σ	95.38	59.67	−11.14	66.50	95.79	99.66
Δ	368.97	299.60	128.29	−51.94	−60.30	−61.18

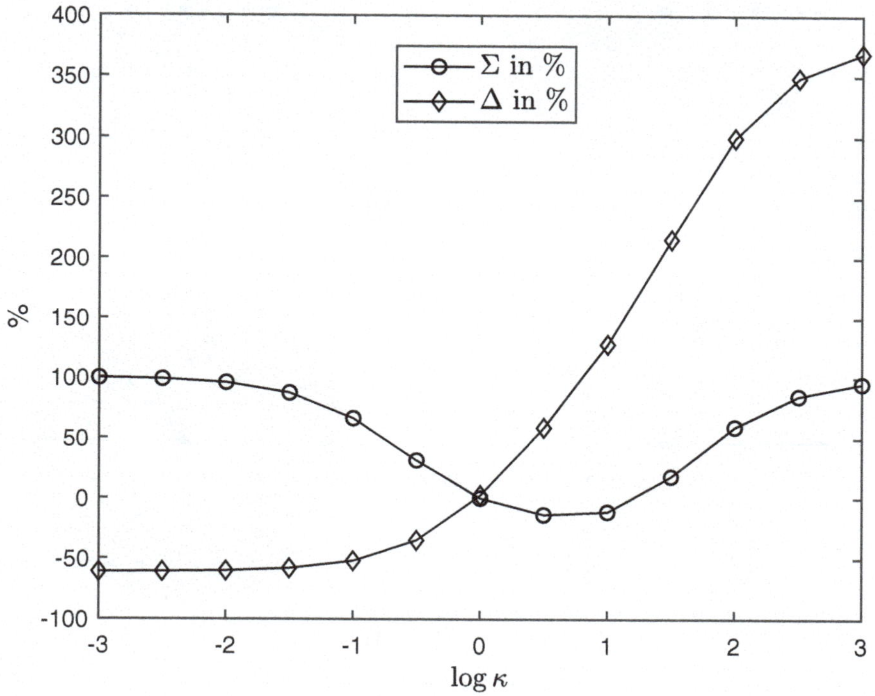

Fig. 13.9 Percentaged shielding $\Sigma := \mathsf{F}_{\text{ifc}}/\mathsf{F}_{\text{mat}}$ and deviation $\Delta := [|\mathsf{F}_{\text{mat}}| - J]/J$ for prescribed Dirichlet BCs corresponding to the classical mode I solution with $J = 10^0$ (30 × 30 elements). Outer domain $\kappa := [\lambda, \mu] = [1800, 3800]$, inner domain **a** $\kappa \times 10^3$, **b** $\kappa \times 10^2$, **c** $\kappa \times 10^1$, **d** $\kappa \times 10^{-1}$, **e** $\kappa \times 10^{-2}$, **f** $\kappa \times 10^{-3}$

by the percentaged deviation $\Delta := [|\mathsf{F}_{\text{mat}}| - J]/J$ in Fig. 13.9. Obviously, these discrepancies are foremost due to the heterogeneous material parameters with a substantial sub-domain being either 'stiffer' or 'softer' than the homogeneous reference case, but in lesser parts also due to the geometrical nonlinearities and the restricted mesh refinement.

- As a third observation, note that additional nodal material forces appear at the interface between the 'stiffer' and the 'softer' domain, thereby consistently pointing into the latter. In the current hyperelastic context, the interfacial nodal material forces (with resultant F_{ifc} in horizontal direction) indicate a potential increase of the total energy stored in the domain if the 'stiff' sub-domain increases at the expense of a decreasing 'soft' sub-domain. Positive dissipation thus restricts pos-

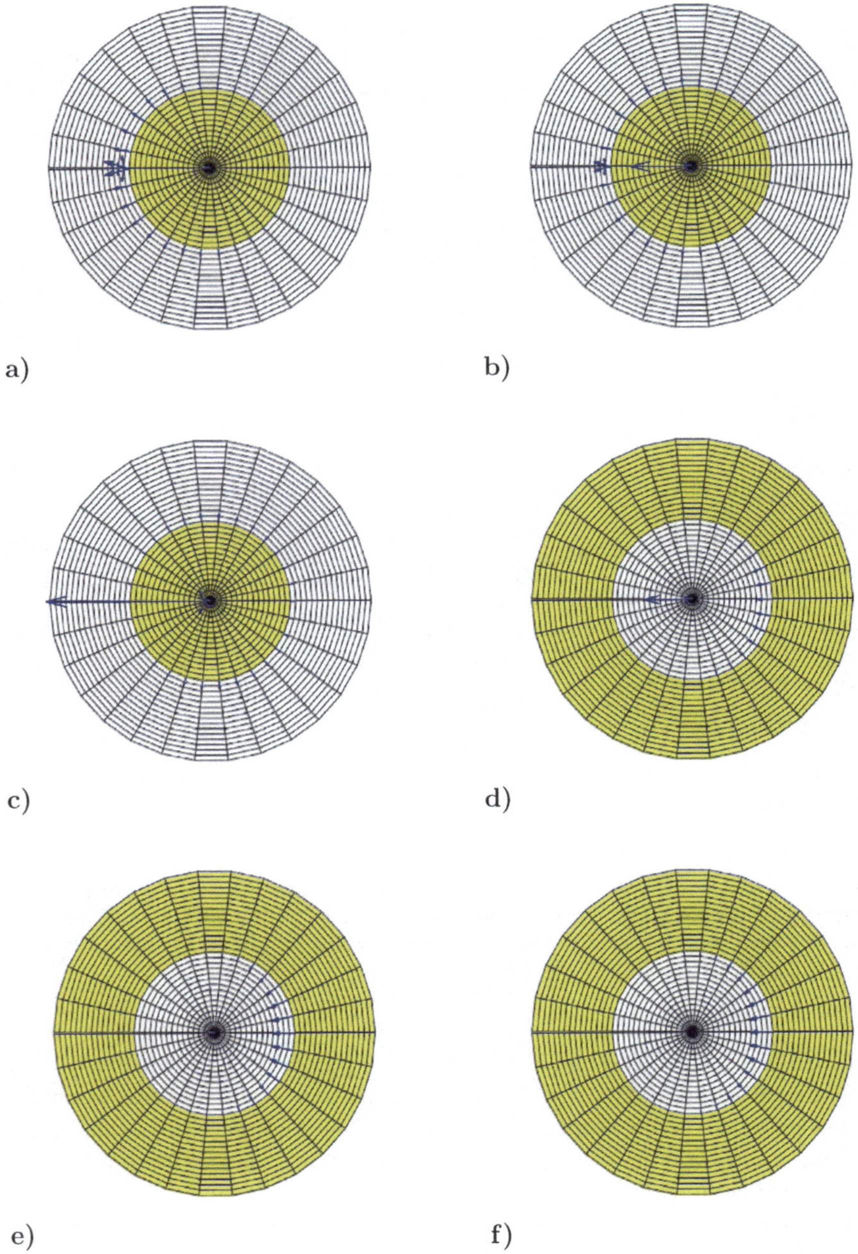

Fig. 13.10 Spatial Pacman configuration and material forces for prescribed Dirichlet BCs corresponding to the classical mode I solution with $J = 10^{-1}$ (30 × 30 elements). Outer domain $\kappa := [\lambda, \mu] = [1800, 3800]$, inner domain **a** $\kappa \times 10^3$, **b** $\kappa \times 10^2$, **c** $\kappa \times 10^1$, **d** $\kappa \times 10^{-1}$, **e** $\kappa \times 10^{-2}$, **f** $\kappa \times 10^{-3}$

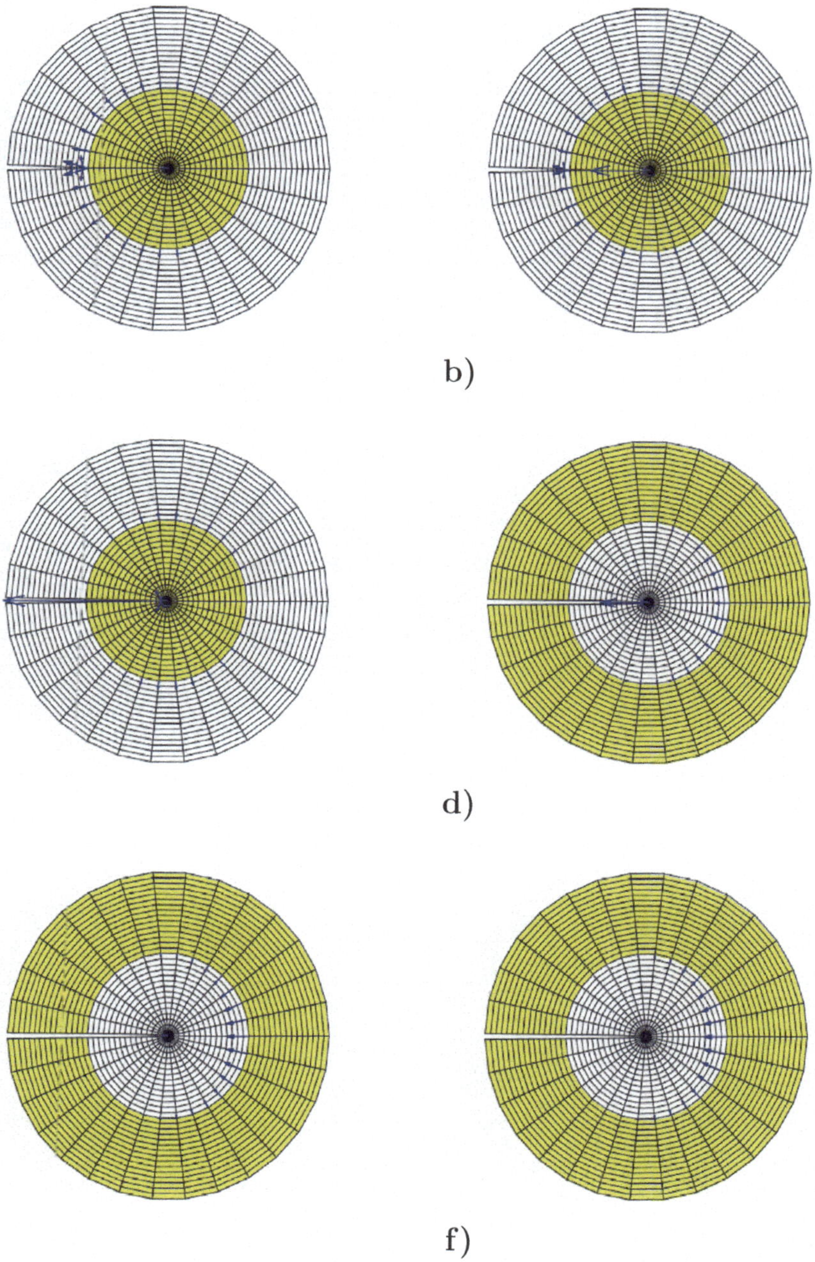

a) b) c) d) e) f)

Fig. 13.11 Spatial Pacman configuration and material forces for prescribed Dirichlet BCs corresponding to the classical mode I solution with $J = 10^0$ (30 × 30 elements). Outer domain $\kappa := [\lambda, \mu] = [1800, 3800]$, inner domain **a** $\kappa \times 10^3$, **b** $\kappa \times 10^2$, **c** $\kappa \times 10^1$, **d** $\kappa \times 10^{-1}$, **e** $\kappa \times 10^{-2}$, **f** $\kappa \times 10^{-3}$

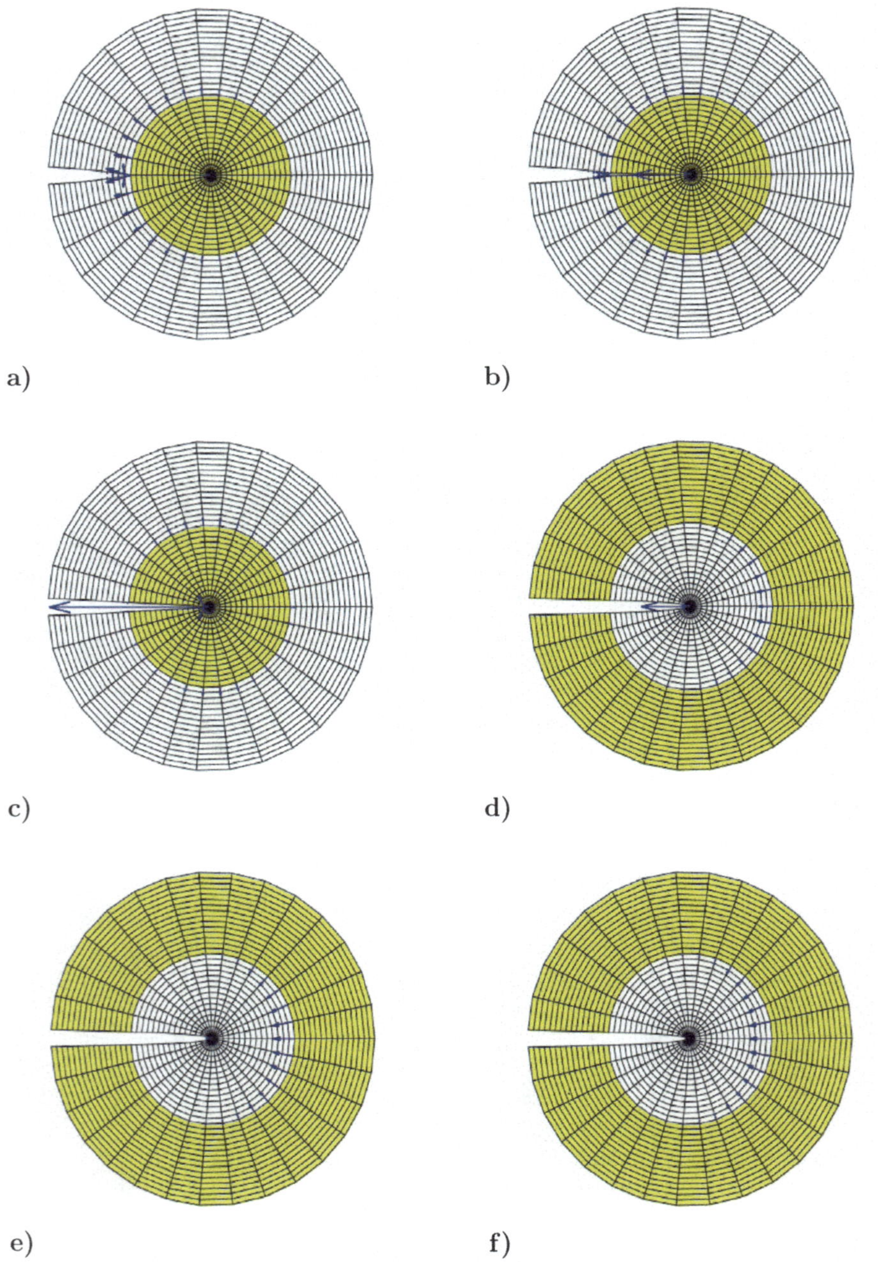

Fig. 13.12 Spatial Pacman configuration and material forces for prescribed Dirichlet BCs corresponding to the classical mode I solution with $J = 10^1$ (30×30 elements). Outer domain $\kappa := [\lambda, \mu] = [1800, 3800]$, inner domain **a** $\kappa \times 10^3$, **b** $\kappa \times 10^2$, **c** $\kappa \times 10^1$, **d** $\kappa \times 10^{-1}$, **e** $\kappa \times 10^{-2}$, **f** $\kappa \times 10^{-3}$

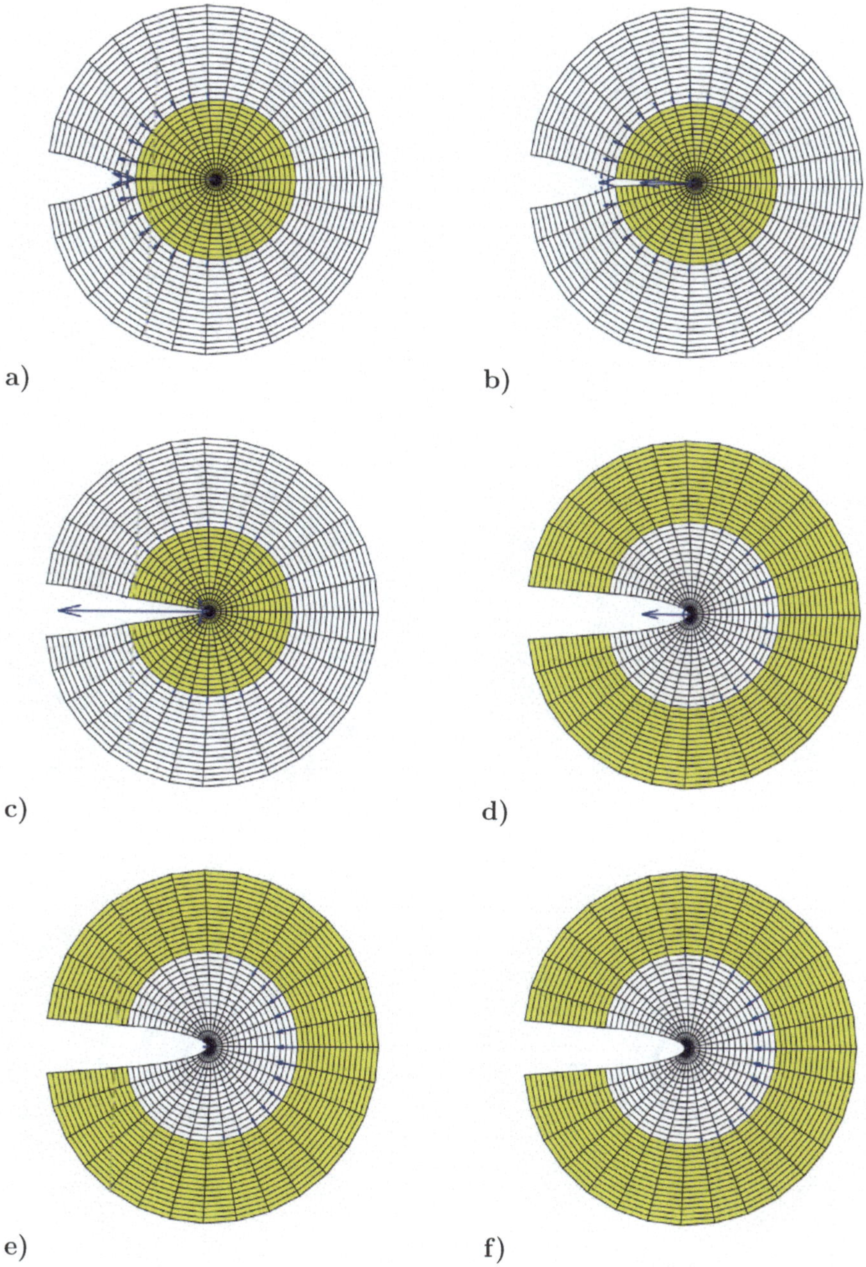

Fig. 13.13 Spatial Pacman configuration and material forces for prescribed Dirichlet BCs corresponding to the classical mode I solution with $J = 10^2$ (30×30 elements). Outer domain $\kappa := [\lambda, \mu] = [1800, 3800]$, inner domain **a** $\kappa \times 10^3$, **b** $\kappa \times 10^2$, **c** $\kappa \times 10^1$, **d** $\kappa \times 10^{-1}$, **e** $\kappa \times 10^{-2}$, **f** $\kappa \times 10^{-3}$

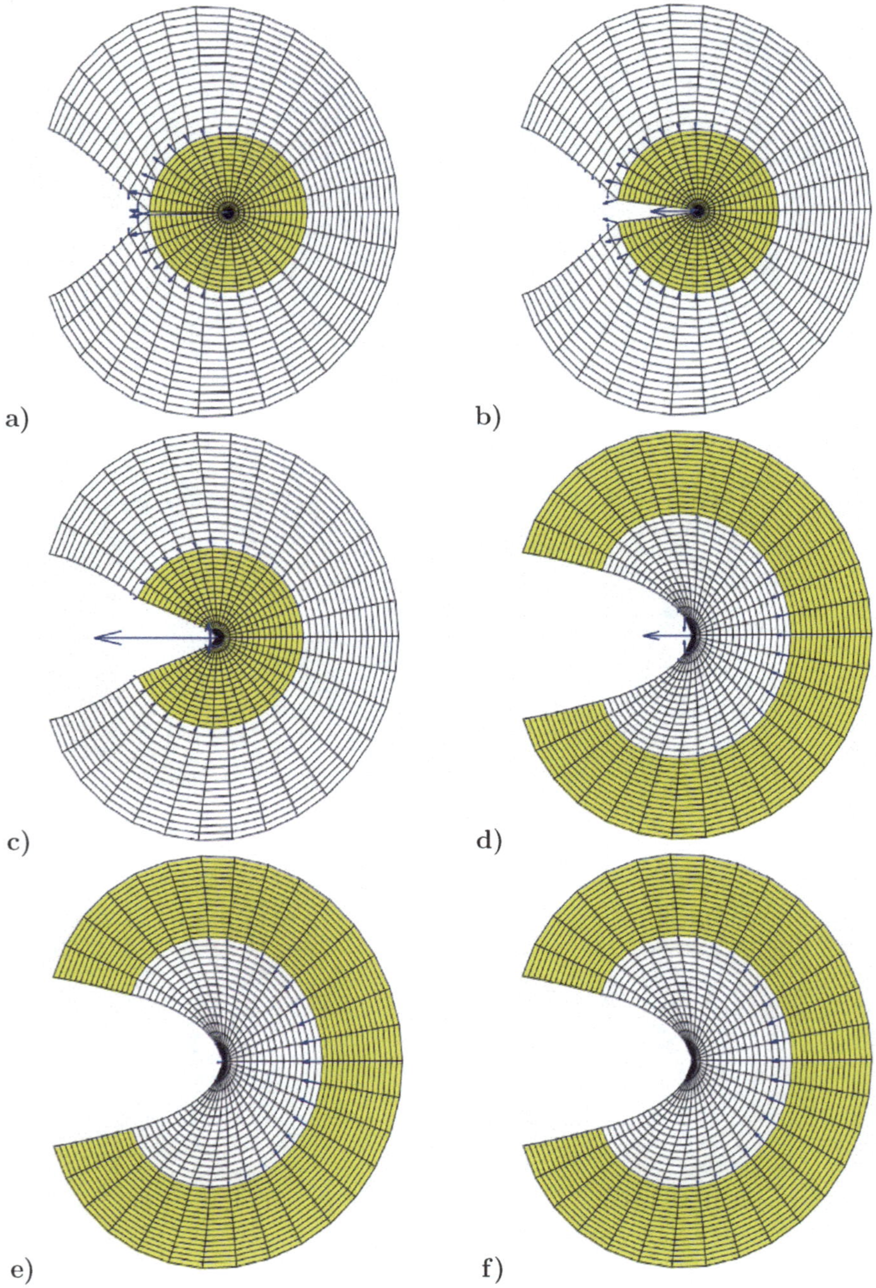

Fig. 13.14 Spatial Pacman configuration and material forces for prescribed Dirichlet BCs corresponding to the classical mode I solution with $J = 10^3$ (30×30 elements). Outer domain $\kappa := [\lambda, \mu] = [1800, 3800]$, inner domain **a** $\kappa \times 10^3$, **b** $\kappa \times 10^2$, **c** $\kappa \times 10^1$, **d** $\kappa \times 10^{-1}$, **e** $\kappa \times 10^{-2}$, **f** $\kappa \times 10^{-3}$

sible interface motion to the direction *opposite* to the here displayed interfacial nodal material forces, thus allowing a release of total energy stored in the domain.

- As a forth observation, note that the additional interfacial nodal material forces contribute to the equilibrium of material forces implied by the quasi-static balance of material momentum. As a consequence, the resultant material force $\mathsf{F}_{crk} := \mathsf{F}_{mat} - \mathsf{F}_{ifc}$ at the crack tip is *shielded* from the resultant material force F_{mat} (i.e. the resultant material force at the boundary of the Pacman-shaped domain) by the resultant material force F_{ifc} at the interface. The corresponding percentaged shielding factor $\Sigma := \mathsf{F}_{ifc}/\mathsf{F}_{mat}$ is highlighted in Fig. 13.9.
- As a fifth observation, note that the percentaged deviation Δ in Fig. 13.9 seems to saturate at some -61 and 380% for extreme cases of 'soft' and 'stiff' cores with $\kappa \to 0$ and $\kappa \to \infty$, respectively. At $\kappa = 10^0$ the percentaged deviation $\Delta = 2.41\%$ of the homogeneous case is retrieved. Likewise, the percentaged shielding factor Σ in Fig. 13.9 depends sensitively on κ. For $\kappa \to 0$ and $\kappa \to \infty$, respectively, Σ approaches 100%, indicating complete shielding of the crack tip due to the interfacial nodal material forces. At $\kappa = 10^0$, the percentaged shielding factor $\Sigma = 0\%$ of the homogeneous case is retrieved. In between, Σ varies nonlinearly with κ and approaches an even negative minimum of some -13.5% at approximately $\kappa = 10^{0.5}$, i.e. the perhaps unwanted case of crack tip loading exceeding the externally prescribed material force.

References

1. Ackermann D, Barth FJ, Steinmann P (1999) Theoretical and computational aspects of geometrically nonlinear problems in fracture mechanics. In: Proceedings (CD-ROM) of the European conference on computational mechanics ECCM'99 (ECCOMAS), August 31 to September 3, Munich, Germany 1999
2. Steinmann P (2000) Application of material forces to hyperelastostatic fracture mechanics. I. Continuum mechanical setting. Int J Solids Struct 37:7371–7391
3. Steinmann P, Ackermann D, Barth FJ (2001) Application of material forces to hyperelastostatic fracture mechanics. II. Computational setting. Int J Solids Struct 38:5509–5526

The manufacturer's authorised representative in the EU is Springer Nature Customer Service Centre GmbH, Europaplatz 3, 69115 Heidelberg, Germany. If you have any concerns regarding our products, please contact ProductSafety@springernature.com

Printed and bound by CPI Group (UK) Ltd, Croydon, CR0 4YY
01/10/2026
02225893-0013